MW00844802

Jessye Bemley

Markov Decision Processes

Markov Decision Processes

Discrete Stochastic Dynamic Programming

MARTIN L. PUTERMAN

University of British Columbia

WILEY-
INTERSCIENCE

A JOHN WILEY & SONS, INC., PUBLICATION

Copyright © 1994, 2005 by John Wiley & Sons, Inc. All rights reserved.

Published by John Wiley & Sons, Inc., Hoboken, New Jersey.
Published simultaneously in Canada.

No part of this publication may be reproduced, stored in a retrieval system or transmitted in any form or by any means, electronic, mechanical, photocopying, recording, scanning or otherwise, except as permitted under Sections 107 or 108 of the 1976 United States Copyright Act, without either the prior written permission of the Publisher, or authorization through payment of the appropriate per-copy fee to the Copyright Clearance Center, 222 Rosewood Drive, Danvers, MA 01923, (978) 750-8400, fax (978) 750-4470. Requests to the Publisher for permission should be addressed to the Permissions Department, John Wiley & Sons, Inc., 111 River Street, Hoboken, NJ 07030, (201) 748-6011, fax (201) 748-6008.

Limit of Liability/Disclaimer of Warranty: While the publisher and author have used their best efforts in preparing this book, they make no representation or warranties with respect to the accuracy or completeness of the contents of this book and specifically disclaim any implied warranties of merchantability or fitness for a particular purpose. No warranty may be created or extended by sales representatives or written sales materials. The advice and strategies contained herein may not be suitable for your situation. You should consult with a professional where appropriate. Neither the publisher nor author shall be liable for any loss of profit or any other commercial damages, including but not limited to special, incidental, consequential, or other damages.

For general information on our other products and services please contact our Customer Care Department within the U.S. at 877-762-2974, outside the U.S. at 317-572-3993 or fax 317-572-4002.

Wiley also publishes its books in a variety of electronic formats. Some content that appears in print, however, may not be available in electronic format.

Library of Congress Cataloging-in-Publication is available.

ISBN 0-471-72782-2

10 9 8 7 6

To my father-in-law
Dr. Fritz Katzenstein
1908–1993
who never lost his love for learning

Contents

Preface

The past decade has seen a notable resurgence in both applied and theoretical research on Markov decision processes. Branching out from operations research roots of the 1950's, Markov decision process models have gained recognition in such diverse fields as ecology, economics, and communications engineering. These new applications have been accompanied by many theoretical advances. In response to the increased activity and the potential for further advances, I felt that there was a need for an up-to-date, unified and rigorous treatment of theoretical, computational, and applied research on Markov decision process models. This book is my attempt to meet this need.

I have written this book with two primary objectives in mind: to provide a comprehensive reference for researchers, and to serve as a text in an advanced undergraduate or graduate level course in operations research, economics, or control engineering. Further, I hope it will serve as an accessible introduction to the subject for investigators in other disciplines. I expect that the material in this book will be of interest to management scientists, computer scientists, economists, applied mathematicians, control and communications engineers, statisticians, and mathematical ecologists. As a prerequisite, a reader should have some background in real analysis, linear algebra, probability, and linear programming; however, I have tried to keep the book self-contained by including relevant appendices. I hope that this book will inspire readers to delve deeper into this subject and to use these methods in research and application.

Markov decision processes, also referred to as stochastic dynamic programs or stochastic control problems, are models for sequential decision making when outcomes are uncertain. The Markov decision process model consists of decision epochs, states, actions, rewards, and transition probabilities. Choosing an action in a state generates a reward and determines the state at the next decision epoch through a transition probability function. Policies or strategies are prescriptions of which action to choose under any eventuality at every future decision epoch. Decision makers seek policies which are *optimal* in some sense. An analysis of this model includes

1. providing conditions under which there exist easily implementable optimal policies;
2. determining how to recognize these policies;
3. developing and enhancing algorithms for computing them; and
4. establishing convergence of these algorithms.

Surprisingly these analyses depend on the criterion used to compare policies. Because of this, I have organized the book chapters on the basis of optimality criterion.

The primary focus of the book is infinite-horizon discrete-time models with discrete state spaces; however several sections (denoted by *) discuss models with arbitrary state spaces or other advanced topics. In addition, Chap. 4 discusses finite-horizon models and Chap. 11 considers a special class of continuous-time discrete-state models referred to as semi-Markov decision processes.

This book covers several topics which have received little or no attention in other books on this subject. They include modified policy iteration, multichain models with average reward criterion, and sensitive optimality. Further I have tried to provide an in-depth discussion of algorithms and computational issues. The Bibliographic Remarks section of each chapter comments on relevant historical references in the extensive bibliography. I also have attempted to discuss recent research advances in areas such as countable-state space models with average reward criterion, constrained models, and models with risk sensitive optimality criteria. I include a table of symbols to help follow the extensive notation. As far as possible I have used a common framework for presenting results for each optimality criterion which

- explores the relationship between solutions to the optimality equation and the optimal value function;
- establishes the existence of solutions to the optimality equation;
- shows that it characterizes optimal (stationary) policies;
- investigates solving the optimality equation using value iteration, policy iteration, modified policy iteration, and linear programming;
- establishes convergence of these algorithms;
- discusses their implementation; and
- provides an approach for determining the structure of optimality policies.

With rigor in mind, I present results in a "theorem-proof" format. I then elaborate on them through verbal discussion and examples. The model in Sec. 3.1 is analyzed repeatedly throughout the book, and demonstrates many important concepts. I have tried to use simple models to provide counterexamples and illustrate computation; more significant applications are described in Chap. 1, the Bibliographic Remarks sections, and left as exercises in the Problem sections. I have carried out most of the calculations in this book on a PC using the spreadsheet Quattro Pro (Borland International, Scott's Valley, CA), the matrix language GAUSS (Aptech Systems, Inc., Kent, WA), and Bernard Lamond's package MDPS (Lamond and Drouin, 1992). Most of the numerical exercises can be solved without elaborate coding.

For use as a text, I have included numerous problems which contain applications, numerical examples, computational studies, counterexamples, theoretical exercises, and extensions. For a one-semester course, I suggest covering Chap. 1; Secs. 2.1 and 2.2; Chap. 3; Chap. 4; Chap. 5; Secs. 6.1, 6.2.1–6.2.4, 6.3.1–6.3.2, 6.4.1–6.4.2, 6.5.1–6.5.2, 6.6.1–6.6.7, and 6.7; Secs. 8.1, 8.2.1, 8.3, 8.4.1–8.4.3, 8.5.1–8.5.3, 8.6, and 8.8; and Chap. 11. The remaining material can provide the basis for topics courses, projects and independent study.

This book has its roots in conversations with Nico van Dijk in the early 1980's. During his visit to the University of British Columbia, he used my notes for a course on dynamic programming, and suggested that I expand them into a book. Shortly thereafter, Matt Sobel and Dan Heyman invited me to prepare a chapter on Markov decision processes for *The Handbook on Operations Research: Volume II, Stochastic Models*, which they were editing. This was the catalyst. My first version (180 pages single spaced) was closer to a book than a handbook article. It served as an outline for this book, but has undergone considerable revision and enhancement. I have learned a great deal about this subject since then, and have been encouraged by the breadth and depth of renewed research in this area. I have tried to incorporate much of this recent research.

Many individuals have provided valuable input and/or reviews of portions of this book. Of course, all errors remain my responsibility. I want to thank Hong Chen, Eugene Feinberg, and Bernard Lamond for their input, comments and corrections. I especially want to thank Laurence Baxter, Moshe Haviv, Floske Spieksma and Adam Shwartz for their invaluable comments on several chapters of this book. I am indebted to Floske for detecting several false theorems and unequal equalities. Adam used the first 6 chapters while in proof stage as a course text. My presentation benefited greatly from his insightful critique of this material. Linn Sennott deserves special thanks for her numerous reviews of Sects. 6.10 and 8.10, and I want to thank Pat Kennedy for reviewing my presentation of her research on Cooper's hawk mate desertion, and providing the beautiful slide which appears as Fig. 1.6.1. Bob Foley, Kamal Golabi, Tom McCormick, Evan Porteus, Maurice Queyranne, Matt Sobel, and Pete Veinott have also provided useful input. Several generations of UBC graduate students have read earlier versions of the text. Tim Lauck, Murray Carlson, Peter Roorda, and Kaan Katiriciougulu have all made significant contributions. Tim Lauck wrote preliminary drafts of Sects. 1.4, 1.6, and 8.7.3, provided several problems, and pointed out many inaccuracies and typos. I could not have completed this book without the support of my research assistant, Noel Paul, who prepared all figures and tables, most of the Bibliography, tracked down and copied many of the papers cited in the book, and obtained necessary permissions. I especially wish to thank the Natural Sciences and Engineering Research Council for supporting this project through Operating Grant A5527, The University of British Columbia Faculty of Commerce for ongoing support during the book's development and the Department of Statistics at The University of Newcastle (Australia) where I completed the final version of this book. My sincere thanks also go to Kimi Sugeno of John Wiley and Sons for her editorial assistance and to Kate Roach of John Wiley and Sons who cheerfully provided advice and encouragement.

Finally, I wish to express my appreciation to my wife, Dodie Katzenstein, and my children, Jenny and David, for putting up with my divided attention during this book's six year gestation period.

MARTIN L. PUTERMAN

Markov Decision Processes

CHAPTER 1

Introduction

Each day people make many decisions; decisions which have both immediate and long-term consequences. Decisions must not be made in isolation; today's decision impacts on tomorrow's and tomorrow's on the next day's. By not accounting for the relationship between present and future decisions, and present and future outcomes, we may not achieve good overall performance. For example, in a long race, deciding to sprint at the beginning may deplete energy reserves quickly and result in a poor finish.

This book presents and studies a model for sequential decision making under uncertainty, which takes into account both the outcomes of current decisions and future decision making opportunities. While this model may appear quite simple, it encompasses a wide range of applications and has generated a rich mathematical theory.

1.1 THE SEQUENTIAL DECISION MODEL

We describe the sequential decision making model which we symbolically represent in Figure 1.1.1. At a specified point in time, a decision maker, agent, or controller observes the state of a system. Based on this state, the decision maker chooses an action. The action choice produces two results: the decision maker receives an immediate reward (or incurs an immediate cost), and the system evolves to a new state at a subsequent point in time according to a probability distribution determined by the action choice. At this subsequent point in time, the decision maker faces a similar problem, but now the system may be in a different state and there may be a different set of actions to choose from.

The key ingredients of this sequential decision model are the following.

1. A set of decision epochs.
2. A set of system states.
3. A set of available actions
4. A set of state and action dependent immediate rewards or costs.
5. A set of state and action dependent transition probabilities.

1

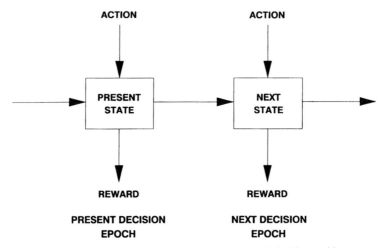

Figure 1.1.1 Symbolic representation of a sequential decision problem.

With the exception of some models which we refer to in the Afterword, we assume that all of these elements are known to the decision maker at the time of each decision.

Using this terminology, we describe the probabilistic sequential decision model as follows. At each decision epoch (or time), the system state provides the decision maker with all necessary information for choosing an action from the set of available actions in that state. As a result of choosing an action in a state, two things happen: the decision maker receives a reward, and the system evolves to a possibly different state at the next decision epoch. Both the rewards and transition probabilities depend on the state and the choice of action. As this process evolves through time, the decision maker receives a sequence of rewards.

At each decision epoch, the decision maker chooses an action in the state occupied by the system at that time. A *policy* provides the decision maker with a prescription for choosing this action in any possible future state. A *decision rule* specifies the action to be chosen at a particular time. It may depend on the present state alone or together with all previous states and actions. A policy is a sequence of decision rules. Implementing a policy generates a sequence of rewards. The sequential decision problem is to choose, prior to the first decision epoch, a policy to maximize a function of this reward sequence. We choose this function to reflect the decision maker's intertemporal tradeoffs. Possible choices for these functions include the expected total discounted reward or the long-run average reward.

This book focuses on a particular sequential decision model which we refer to as a *Markov decision process* model. In it, the set of available actions, the rewards, and the transition probabilities depend only on the current state and action and not on states occupied and actions chosen in the past. The model is sufficiently broad to allow modeling most realistic sequential decision-making problems.

We address the following questions in this book.

1. When does an optimal policy exist?
2. When does it have a particular form?
3. How do we determine or compute an optimal policy efficiently?

We will see that the choice of the optimality criterion and the form of the basic model elements has significant impact on the answers to these questions.

Often you see these models referred to as *dynamic programming models* or *dynamic programs*. We reserve the expression "dynamic programming" to describe an approach for solving sequential decision models based on inductive computation.

In the remainder of this chapter, we illustrate these concepts with significant and colorful applications from several disciplines. The Bibliographic Remarks section provides a brief historical review.

1.2 INVENTORY MANAGEMENT

Sequential decision models have been widely applied to inventory control problems and represent one of the earliest areas of application. The scope of these applications ranges from determining reorder points for a single product to controlling a complex multiproduct multicenter supply network. Some of the earliest and most noteworthy results in stochastic operations research concern the form of the optimal policy under various assumptions about the economic parameters. We describe an application of a model of this type.

Through local dealerships, Canadian Tire, Inc. operates a chain of automotive supply stores throughout Canada. The 21 stores in the Pacific region are operated by a single management group. Backup inventory for these 21 stores is maintained at a central warehouse in Burnaby, British Columbia. It stocks roughly 29,000 products. Periodically, inventory is delivered from the central warehouse to each of its stores to maintain target stock levels.

The timing of inventory replenishment varies with store size. At stores designated as "small," the inventory position of each product is reviewed once a week. For each product the inventory position (stock on hand) at the time of review determines the quantity, if any, to order. Orders arrive in about three days. Associated with an order for a particular product is a fixed charge associated with the time spent locating the item in the warehouse and shelving the item at the store. In addition to the fixed charge for filling the order, there is a daily carrying charge for keeping an item in inventory at a store. Management policy also dictates that at least 97.5% of demand be satisfied from stock on hand.

We now describe a sequential decision model for determining optimal reorder points and reorder levels for a single product at a single store. Decision epochs are the weekly review periods, and the system state is the product inventory at the store at the time of review. In a given state, actions correspond to the amount of stock to order from the warehouse for delivery at the store. Transition probabilities depend on the quantity ordered and the random customer demand for the product throughout the week. A decision rule specifies the quantity to be ordered as a function of the stock on hand at the time of review, and a policy consists of a sequence of such restocking functions. Management seeks a reordering policy which minimizes long-run average ordering and inventory carrying costs subject to the above constraint on the probability of being unable to satisfy customer demand.

Desirable properties for optimal policies in this setting are that they be simple to implement and not vary with time. Without the constraint on the probability of satisfying customer demand, the optimal policy may be shown to be of the following type: when the stock level falls below a certain threshold, order up to a target level;

otherwise do not order. With the inclusion of such a constraint, a policy of this form may not be optimal.

The importance of effective inventory control to effective cost management cannot be overemphasized. Sir Graham Day, chairman of Britain's Cadbury-Schweppes PLC notes (*The Globe and Mail*, October 20, 1992, p. C24):

> "I believe that the easiest money any business having any inventory can save lies with the minimization of that inventory."

The roots of sequential decision making lie in this discipline. The book by Arrow, Karlin, and Scarf (1958) provides a good overview of the foundations of mathematical inventory theory; Porteus (1991) provides a recent review.

1.3 BUS ENGINE REPLACEMENT

Markov decision process models have been applied to a wide range of equipment maintenance and replacement problems. In these settings, a decision maker periodically inspects the condition of the equipment, and based on its age or condition decides on the extent of maintenance, if any, to carry out. Choices may vary from routine maintenance to replacement. Costs are associated with maintenance and operating the equipment in its current status. The objective is to balance these two cost components to minimize a measure of long-term operating costs.

Howard (1960) provided a prototype for such models with his "automobile replacement problem." In it, an individual periodically decides whether or not to trade in an automobile and, if so, with what age automobile to replace it. Subsequently, many variants of this model have been studied and analyzed. In this section and the next, we describe two applications of such models.

Rust (1987) formulates and analyzes the following problem. Harold Zurcher, superintendent of maintenance at the Madison (Wisconsin) Metropolitan Bus Company, has the responsibility of keeping a fleet of buses in good working condition. One aspect of the job is deciding when to replace the bus engines.

Zurcher's replacement problem may be formulated as a Markov decision process model as follows. Replacement decisions are made monthly and the system state represents the accumulated engine mileage since the last replacement. Costs include an age-dependent monthly operating cost and a replacement cost. The monthly operating costs include a routine operating and maintenance cost component and an unexpected failure cost component. The failure cost accounts for the probability of breakdown for a bus of a given age and costs associated with towing, repair, and lost goodwill. If Zurcher decides to replace an engine, then the company incures a (large) replacement cost and, subsequently, the routine maintenance and operating cost associated with the replacement engine. Transition probabilities describe changes in accumulated mileage and the chance of an unplanned failure for a bus engine of a particular age. For each engine, Zurcher seeks an age-dependent replacement policy to minimize expected total discounted or long-run average costs.

The algorithms in Chaps. 4, 6, or 9 can be used to compute such an optimal policy for Harold Zurcher. However, the theory shows that, under reasonable assumptions, an optimal policy has a particularly simple and appealing form; at the first monthly inspection at which the mileage exceeds a certain level, referred to as a *control limit*, the engine must be replaced; otherwise it is not. Rust examines whether Zurcher adopts such a policy using data from the Madison Metropolitan Bus Company.

Operating, maintenance, and replacement costs vary with engine type. The table below summarizes Zurcher's data on replacement costs and average mileage at replacement for two main engine types.

Engine Type	Replacement Cost	Average Mileage at Replacement
1979 GMC T8H203	$9499	199,733
1975 GMC 5308A	$7513	257,336

This data shows that, although the replacement cost for a 1979 engine exceeded that of a 1975 engine by $2000, Zurcher decided to replace the 1979 engines 57,600 miles and 14 months earlier than the 1975 engines. This suggests that routine maintenance and operating costs differ for these two engine types and that they increase faster with mileage in the 1979 engines. Rust's analysis of the data suggests that these costs may be modeled by linear or "square-root" functions of age.

Further data suggests that Zurcher's decisions departed from a simple control limit policy. Between 1974 and 1985, 27 T8H203 engines and 33 5308A engines were replaced. The mileage at replacement varied from 124,800 to 273,400 for the T8H203 engine and between 121,200 and 387,300 for the 5308A engine. Thus we might infer that Zurcher is making his decisions suboptimally. Rust adopts a different viewpoint. He hypothesizes that Zurcher's decisions coincide with an optimal policy of a Markov decision process model; however, Zurcher takes into account many measurements and intangibles that are not known by the problem solver. In his extensive paper, Rust (1987) provides an approach for accounting for these factors, estimating model parameters, and testing this hypothesis. He concludes that, after taking these unobservables into account, Zurcher's behavior is consistent with minimizing long-run average operating cost.

1.4 HIGHWAY PAVEMENT MAINTENANCE

The Arizona Department of Transportation (ADOT) manages a 7,400 mile road network. Up to the mid 1970s its primary activity was construction of new roadways. As the Arizona roadway system neared completion, and because of changing federal guidelines, ADOT's emphasis shifted in the late 1970's to maintaining existing roads. Between 1975 and 1979, highway preservation expenditures doubled from $25 million to $52 million, and evidence suggested that such an increase would continue. By this time it was evident to ADOT management that a systematic centralized procedure for allocation of these funds was needed. In 1978, in conjunction with Woodward-Clyde Consultants of San Francisco, ADOT developed a pavement management system based on a Markov decision process model to improve allocation of its limited resources while ensuring that the quality of its roadways was preserved. In 1980, the first year of implementation, this system saved $14 million, nearly a third of Arizona's maintenance budget, with no decline in road quality. Cost savings over the next four years were predicted to be $101 million. Subsequently, this model was modified for use in Kansas, Finland, and Saudi Arabia. Related models have been developed for bridge and pipeline management. In this section, we describe the Arizona pavement management model. We base our presentation on Golabi, Kulkarni, and Way (1982), and additional information provided by Golabi in a personal communication.

The pavement management system relies on a dynamic long-term model to identify maintenance policies which minimize long-run average costs subject to constraints on road quality. To apply the model, the Arizona highway network was divided into 7,400 one-mile sections and nine subnetworks on the basis of road type, traffic density, and regional environment. For each category, a dynamic model was developed that specified the conditions of road segments, maintenance actions that could be used under each condition, and the expected yearly deterioration or improvement in pavement conditions resulting from each such action. In addition, costs associated with each maintenance action were determined. Developing categories for system states, actions, costs, and the state-to-state dynamics under different actions was a nontrivial task requiring data, models of road conditions, statistical analysis, and subject matter expertise.

We describe the management model for asphalt concrete highways; that for Portland cement concrete roadways had different states and actions. Decisions were made annually. The system state characterized the pavement condition of a one-mile segment by its roughness (three levels), its percentage of cracking (three levels), the change in cracking from the previous year (three levels), and an index which measured the time since the last maintenance operation and the nature of the operation (five levels). Consequently, a road segment could be described by one of 135 ($3 \times 3 \times 3 \times 5$) possible states, but, since some combinations were not possible, 120 states were used.

Actions corresponded to available pavement rehabilitation activities. These ranged from relatively inexpensive routine maintenance to costly actions such as thick resurfacing or recycling of the entire roadway. A list of possible actions and associated construction costs appear in Table 1.4.1 below. For each state, however, only about six of the actions were considered feasible.

Costs consisted of the action-dependent construction costs (Table 1.4.1) and annual routine maintenance costs (Table 1.4.2). Annual routine maintenance costs varied with the road condition and rehabilitation action. When only routine maintenance was carried out, these costs varied with the roughness and degree of cracking of the road segment; when a seal coat was applied, these costs varied only with roughness; and if any other rehabilitation action was taken, maintenance costs were independent of previous road condition. These costs were determined through a regression model based on existing data.

Transition probabilities specify the likelihood of yearly changes in road condition under the various maintenance actions. These were estimated using existing data, under the assumption that each dimension of the state description varied independently. Since in each state only a limited number of subsequent states could occur, most of the transition probabilities (97%) were zero.

The performance criteria was cost minimization subject to constraints on the proportion of roads in acceptable and unacceptable states. For example, ADOT policy requires that at least 80% of high traffic roadways must have a roughness level not exceeding 165 inches/mile, while at most 5% of these roads could have roughness exceeding 256 inches/mile. Similar constraints applied to levels of cracking.

Table 1.4.1   Rehabilitation Actions and Construction Costs

Action Index	Action Description[a]	Construction Cost $/yd^2$
1	Routine Maintenance	0
2	Seal Coat	0.55
3	ACFC	0.75
4	ACFC + AR	2.05
5	ACFC + HS	1.75
6	1.5 inch AC	1.575
7	1.5 inch AC + AR	2.875
8	1.5 inch AC + HS	2.575
9	2.5 inch AC	2.625
10	2.5 inch AC + AR	3.925
11	2.5 inch AC + HS	3.625
12	3.5 inch AC	3.675
13	3.5 inch AC + AR	4.975
14	3.5 inch AC + HS	4.675
15	4.5 inch AC	4.725
16	5.5 inch AC	5.775
17	Recycling (equivalent to 6 inch AC)	6.3

[a]Abbreviations used in table: ACFC-Asphalt concrete fine coat, AR-Asphalt Rubber, HS-Heater Scarifier, AC-Asphalt concrete

Table 1.4.2   Annual Routine Maintenance Costs

State After Rehabilitation Action		Rehabilitation Action[a]	Cost $/yd^2$
Roughness (in/mile)	Percentage of Cracking		
120 (±45)	5 (±5)	RM	0.066
120 (±45)	20 (±10)	RM	0.158
120 (±45)	45 (±15)	RM	0.310
120 (±45)	Any	SC	0.036
210 (±45)	5 (±5)	RM	0.087
210 (±45)	20 (±10)	RM	0.179
210 (±45)	45 (±15)	RM	0.332
210 (±45)	Any	SC	0.057
300 (±45)	5 (±5)	RM	0.102
300 (±45)	20 (±10)	RM	0.193
300 (±45)	45 (±15)	RM	0.346
300 (±45)	Any	SC	0.071
Any	Any	OT	0.036

[a]Action Abbreviations; RM-routine maintenance, SC-seal coat, OT-any other

This model is an example of a *constrained* average reward Markov decision process model and can be solved using the linear programming methodology in Chaps. 8 and 9. This model was designed not only to yield a single solution but also to interactively examine the consequences of regulatory policies and budget changes. Examples of solutions are too lengthy to be presented here, but one aspect of the solution is worth noting. Because of the addition of constraints, the optimal policy may be *randomized*. This means that in some states, it may be optimal to use a chance mechanism to determine the course of action. For example, if the road segment is cracked, 40% of the time it should be resurfaced with one inch of asphalt concrete (AC) and 60% of the time with two inches of AC. This caused no difficulty because the model was applied to individual one-mile road segments so that this randomized policy could be implemented by repairing 40% of them with one inch of AC and 60% with two inches of AC. Also, in a few instances, the model recommended applying a different maintenance action to a road segment than to its two adjacent segments. In such cases the solution was modified to simplify implementation yet maintain the same level of overall cost and satisfy road quality constraints.

In addition to producing significant cost reductions, the model showed that

"...corrective actions in the past were too conservative; it was common to resurface a road with five inches of asphalt concrete.... The policies recommended by the pavement management system...are less conservative; for example, a recommendation of three inches of overlay is rather rare and is reserved for the worst conditions. (Golabi, Kulkarni, and Way, 1982, p. 16)."

Observations such as this are consistent with the findings of many operations research studies. For example, preliminary results in the inventory control study described in Sect. 1.2 suggest that current in store inventory levels are 50% too high.

1.5 COMMUNICATIONS MODELS

A wide range of computer, manufacturing, and communications systems can be modeled by networks of interrelated queues (waiting lines) and servers. Efficient operation of these systems leads to a wide range of dynamic optimization problems. Control actions for these systems include rejecting arrivals, choosing routings, and varying service rates. These decisions are made frequently and must take into account the likelihood of future events to avoid congestion.

These models are widely applied and have had significant impact as noted by the following article in *The New York Times*, May 12, 1992, p. C2.

"More Dial Mom Than Expected"

Even greater numbers of people called their mothers on Mother's Day than AT&T had expected.

...A call-routing computer technique enabled the American Telephone and Telegraph Company to complete more calls than last year, when it logged 93.4 million calls.

On Sunday, there were about 1.5 million uncompleted calls, where customers got a recorded announcement advising them to call later, compared with 3.9 million last year.

A new computer technique called real-time network routing helped AT&T shepherd a larger number of calls through the labyrinth of telephone computers known as switches. By creative zigzagging around the country, AT&T could direct calls so that they were more likely to avoid congestion, especially in suburbs, which do not have the high-capacity telephone lines that big cities do.

We now describe a sequential decision process model for a particular communication system. Many packet communications systems are configured so that multiple terminals generating low rate, bursty traffic and must share a single channel to communicate with each other or with a central hub (Fig. 1.5.1).

This system architecture is typical of satellite broadcast networks where multiple earth stations communicate over the same radio frequency, and computer local area networks (LAN's) where many computers send job requests to a central file server over a single coaxial cable. Since a single channel may only carry one stream of traffic, the problem arises as to how to coordinate the traffic from the terminals to make the most efficient use of the channel.

The *Slotted ALOHA Protocol* is a popular and especially simple technique for providing such coordination. We describe the slotted ALOHA channel model and the mechanism by which it controls channel access. Stations communicate over a slotted ALOHA channel through equal-length packets of data. Time on the channel is divided into slots of the same length as the packets, and all terminals are synchronized so that packet transmissions always begin at the leading edge of a time slot and occupy exactly one slot. New packets are randomly generated at any idle terminal during a slot, and are transmitted in the following slot. If no other stations transmit a packet in that slot, the transmission is considered successful and the terminal returns to idle mode. If more than one terminal generates a packet, a collision occurs, the data become garbled, and the station goes into retransmission mode and must retransmit the packet in a future slot. If a collision occurs and all involved terminals always retransmit in the next slot, collisions will continue endlessly. To avoid this situation, the slotted ALOHA protocol specifies that stations in retransmission mode transmit in the next slot with a specified retransmission probability, thus achieving a random backoff between retransmission attempts. When a terminal successfully retransmits the packet, it returns to idle mode and waits for a new packet to be

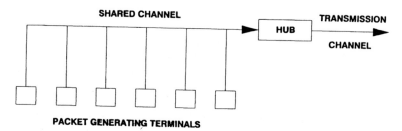

Figure 1.5.1 Multiple access channel configuration.

generated. We see then that, although the slotted ALOHA protocol does not avoid collisions on the channel, the use of a random retransmission backoff provides a scheme for effective contention resolution among terminals.

Since the message generating probability is fixed, the only means available to control channel access within this model is by regulating the retransmission probability. If it is held constant and the number of terminals in retransmission mode becomes large, the probability of a collision in the next slot will also become large. As the collisions become more frequent, newly arriving packets tend to become backlogged, increasing the number of terminals in retransmission mode. Thus, with a fixed retransmission probability, the system is prone to become highly congested, reducing the chance of a successful transmission to close to zero. This instability may be alleviated (for certain values of the packet generation probability) by taking into account the current number of terminals in retransmission mode when choosing a retransmission probability.

We now describe a Markov decision process model for this control problem. Decision epochs correspond to time slots, and the system state is the number of terminals in retransmission mode. Actions correspond to choosing a retransmission probability. The system generates a reward of one unit for each packet successfully transmitted, and transition probabilities combine the probabilities that new packets are generated in a time slot and a successful packet transmission occurs when the retransmission probability has been set at a particular level. The objective is to choose a retransmission probability-setting policy which maximizes the long-run average expected channel throughput (rate of successful packets per slot).

Feinberg, Kogan, and Smirnov (1985) show that the optimal retransmission probability is a monotonically decreasing function of the system state whenever the mean packet arrival rate (number of terminals times packet generation probability) is less than one. If this rate exceeds one, the system will become congested and other optimality criteria may be used. This control policy agrees with intuition in that the system will react to increasing congestion by decreasing retransmission probabilities and thus maintaining a reasonable probability of successful packet transmission.

In practical applications, the number of stations in retransmission mode and the packet generation probability are rarely known. They must be estimated on the basis of the history of channel observations (idle, successful, and collision slots). In this case, incorporation of both state and parameter estimation into the Markov decision process model is necessary to find the optimal retransmission policy. We provide references for models of this type in the Afterword.

1.6 MATE DESERTION IN COOPER'S HAWKS

Markov decision process models are becoming increasingly popular in behavioral ecology. They have been used in a wide range of contexts to gain insight into factors influencing animal behavior. Examples include models of social and hunting behavior of lions (Clark, 1987; Mangel and Clark, 1988), site selection and number of eggs laid by apple maggots and medflys (Mangel, 1987), daily vertical migration of sockeye salmon and zooplankton (Levy and Clark, 1988; Mangel and Clark 1988), changes in mobility of spiders in different habitats (Gallespie and Caraco, 1987), and singing versus foraging tradeoffs in birds (Houston and McNamara, 1986).

The theory of natural selection suggests that organisms predisposed to behavioral characteristics that allow them to adapt most efficiently to their environment have the greatest chance of reproduction and survival. Since any organism alive today has a substantial evolutionary history, we might infer that this organism has adopted optimal or near-optimal survival strategies which can be observed in day-to-day activity.

Models have been based on regarding the behavior of an organism as its reaction or response to its environment, conditional on its state of well being. Throughout its life, it makes behavioral choices which affect its chances of survival and successful reproduction. Investigators have used probabilistic sequential decision process models to determine state- and time-dependent strategies which maximizes a function of its survival and reproductive success probabilities and then compared model results to observed behavior. If there is "reasonable agreement," then the derived optimal policy may provide insight into the behavioral strategy of the organism.

We describe Kelly and Kennedy's (1993) use of this methodology in their study of mate desertion in Cooper's hawks (Acceipiter cooperii). Over a five-year period, they studied nesting behavior of several birds near Los Alamos National Laboratory in north-central New Mexico (Fig. 1.6.1.) They observed that more than 50% of the females deserted their nests before the young reached independence, and noted that the male of this species continued to feed the young regardless of whether or not a female was present. At issue was determining factors that influenced the female's decision to desert and the female's tradeoffs between her survival and that of her offspring.

In the study, the physical conditions of both the nestlings (young birds) and the female were monitored, assisted by the use of radiotelemetry. Females were trapped and tagged early in the breeding season, providing an opportunity to assess the initial health of the females. Birds with greater body mass had larger energy reserves and were considered healthier. Rather than disturb the nestlings, their health was determined by assuming that nestlings were initially healthy. A developmental model was

Figure 1.6.1 A female Cooper's hawk and her brood. (Photograph courtesy of Patricia L. Kennedy.)

used to account for parental hunting behavior, captures, and nestling growth rates.

Kelly and Kennedy developed a model for a single nesting season based on the assumption that behavioral choices of the female hawk maximized a weighted average of the probability of nestling survival and the probability of the female's survival to the next breeding season. The sequential decision model was used to determine an optimal behavioral strategy.

The nesting season was divided into four periods representing basic stages of development of the young.

1. Early nestling period.
2. Late nestling period.
3. Early fledgling dependence period.
4. Late fledgling dependence period.

The end of the late fledgling period marked the point at which the brood reaches independence.

The system state is a two-dimensional health index representing female and brood energy reserves. The states were constrained to lie between lower levels which represented the minimum physical condition for survival, and upper levels corresponding to limiting physical attributes of the birds.

Three basic behavioral strategies were observed for the female.

1. Stay at the nest to protect the young.
2. Hunt to supplement the food supplied by the male.
3. Desert the nest.

Decisions were assumed to have been made at the start of each of the above periods.

From one developmental stage to the next, the change in energy reserves of both the female and the young depends on the female's behavioral strategy and the amount of food captured, a random quantity. At the time of independence, the female's and brood's states of health depend on their initial energy reserves, the female's behavior, and the availability of food. The respective health indices at the end of the decision making period determine the probability of survival of the female and of the brood to the subsequent nesting period.

Using data estimated from the five-year study, results in the literature, and some intelligent guesswork, Kelly and Kennedy determined transition probabilities for the above model. They then solved the model to determine the optimal policy using inductive methods we describe in Chap. 4. Figure 1.6.2, which we adopt from their paper, shows the optimal policy under a specified degree of tradeoff between female and brood survival.

The four graphs show the optimal behavioral action as a function of the health of the female and the brood at each of the four decision periods. The vertical axis represents the female's health index and the horizontal axis represents the brood's health index. Low values indicate states of low-energy reserve.

Observe that, in all periods, if both the female's and the brood's health index exceed 4, the optimal strategy for the female is to stay at the nest and protect the young. At the other extreme, if the health index of both the female and the brood is at its lowest value, the optimal strategy for the female is to desert the nest.

PERIOD

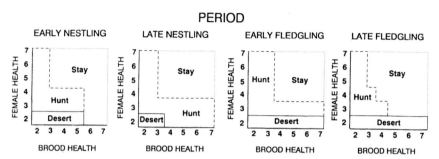

Figure 1.6.2 Symbolic representation of optimal desertion strategy under a specific tradeoff parameter choice. Quantities on axes denote health indices. (Adapted from Kelly and Kennedy, 1993.)

There are other patterns to this optimal strategy. For a fixed value of the brood health index, as the female's health index increases the optimal strategy changes from desert to hunt to stay. Similarly, if the female's health index is fixed, as the brood's energy reserves increase the strategy changes from desert to hunt to stay. Thus there is a form of monotonicity in the optimal strategy. One might conjecture that the behavioral strategy of the female will have this form under any parameter values. Observing such patterns can sometimes yield insight into theoretical results beyond a specific numerical scenario. In subsequent chapters, we will provide methods for identifying models in which optimal strategies have a particular form.

Kelly and Kennedy (p. 360–361) conclude

"The agreement of model predictions and observed strategies supported, but did not prove, the modelling hypotheses that:

1. a female's strategy during brood rearing maximizes the weighted average of the expected probability of survival of her current offspring and her future reproductive potential, and

2. the female's strategy choices were influenced by multiple factors including her state, the state of her brood, the risks to nestlings associated with each strategy, and the male's and female's foraging capabilities.

... dynamic state variable models are powerful tools for studying the complexities of animal behavior from an evolutionary standpoint because they lead to quantitative testable predictions about behavioral strategies."

1.7   SO WHO'S COUNTING

Games of chance and strategy provide natural settings for applying sequential decision models. Dubins and Savage (1965) in their monograph *How to Gamble if You Must* developed a model for gambling, not unlike the sequential decision model herein, and developed a rich mathematical theory for analyzing it. Their basic observation was that, even in an unfair game, some betting strategies might be better than others. Markov decision process models apply to such games of chance and also to a wide range of board and computer games. In this section we show how such a

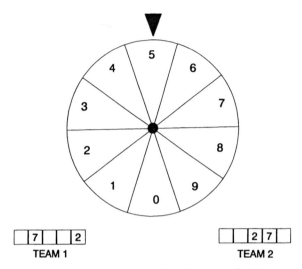

Figure 1.7.1 Spinner for "So Who's Counting."

model can be used to determine an optimal strategy in a challenging yet easy to describe television game show.

The mock game show *But Who's Counting* appeared on *Square One*, a mathematically oriented educational television program on the U.S. public broadcasting network. The game is played as follows. There are two teams of players. At each of five consecutive rounds of the game, a spinner (Fig. 1.7.1) produces a number between 0 and 9, each with equal probability. After each spin, the teams select an available digit of a five-digit number to place the number produced by the spinner. The team which generates the largest number wins the game.

Figure 1.7.1 illustrates the status of the game immediately following the third spin. At the two previous spins, the numbers 2 and 7 had appeared. At this point in time, the order in which they previously appeared is immaterial. Team 1 placed the 7 in the 1,000s place and the 2 in 1s place, while team 2 placed the 2 in the 100s place and the 7 in the 10s place. Each team must now decide where to place the 5. What would you do in this case if you were on team 1? On team 2?

Now ignore the competitive aspect of this game and suppose you were in the game alone with the objective of obtaining the highest five-digit number. Reflect on what strategy you would use. If you observed a 9, surely you would want to get the best out of it, and, thus, place it in the highest digit available. However, what would you do if you had a 5 or a 6?

We formulate the single-player game as a sequential decision problem. Decision epochs correspond to the instant immediately after the spinner identifies a number. We take as the system state the locations of the unoccupied digits and the number which has appeared on the spinner. Actions correspond to placing the number into one of the available digits and the reward equals the number times the place value for that digit. The objective is to choose a digit-placing strategy which maximizes the expected value of the five-digit number.

We can use the methods of Chap. 4 directly to derive an optimal policy. It has the property that the decision into which unoccupied digit to place the observed number

Table 1.7.1   Optimal Policy for "But Who's Counting."

Observed Number	Optimal Digit Locations				
	Spin 1	Spin 2	Spin 3	Spin 4	Spin 5
0	5	4	3	2	1
1	5	4	3	2	1
2	5	4	3	2	1
3	4	3	3	2	1
4	3	3	2	2	1
5	3	2	2	1	1
6	2	2	1	1	1
7	1	1	1	1	1
8	1	1	1	1	1
9	1	1	1	1	1

should be based on the number of unoccupied positions remaining, and not on their place values or the values of the previously placed digits. This observation enables us to summarize succinctly the optimal policy as in Table 1.7.1.

In this table, the entries represent the location of the unoccupied digit (counting from the left) into which to place the observed number. For example, consider the decision faced by a player on team 1 in Fig. 1.7.1 after observing a 5 on spin 3. Table 1.7.1 prescribes that the 5 should be placed in the 100's position. To proceed optimally, a player on team 2 should place the 5 in the 1000's position.

Furthermore, using methods in Chap. 4, we can show that using this policy yields an expected score of 78,734.12, compared to that of a random-digit choice policy which would result in an expected score of 49,999.5. Of course, in a particular game, this strategy may not always yield the greatest score, but in the long run, it will do best on average.

This problem is a special case of a *sequential assignment problem.* Ross (1983, p. 124) provides a clever approach for solving these problems in general. He establishes existence of and gives a method for computing a set of *critical levels* which in this context determine the optimal placement of the number. If the number is above the highest critical level, then it should be placed in the leftmost digit available. If it is between the highest and second highest, it should be placed in the second-leftmost unoccupied digit, and so on. His approach shows that the optimal policy would still be the same if we had any other increasing values for the contribution of digits to the total reward instead of 1, 10, 100, 1000, or 10,000.

It is not hard to think of variations of this problem. We may view it from a game theoretic point of view in which the objective is to derive a strategy which maximizes the probability of winning the game, or we may consider a single-person game in which the numbers have unequal probabilities.

HISTORICAL BACKGROUND

The books by Bellman (1957) and Howard (1960) popularized the study of sequential decision processes; however, this subject had earlier roots. Certainly some of the basic concepts date back to the calculus of variations problems of the 17th century. Cayley's paper (Cayley, 1875), which did not resurface until the 1960s, proposed an interesting problem which contains many of the key ingredients of a stochastic sequential decision problem. We describe and analyze this problem in detail in Chaps. 3 and 4.

The modern study of stochastic sequential decision problems began with Wald's work on sequential statistical problems during the Second World War. Wald embarked on this research in the early 1940's, but did not publish in until later because of wartime security requirements. His book (1947) presents the essence of this theory.

Pierre Massé, director of 17 French electric companies and minister in charge of French electrical planning, introduced many of the basic concepts in his extensive analysis of water resource management models (1946). Statistician Lucien Le Cam (1990), reflecting on his early days at Electricité de France, noted

"Massé had developed a lot of mathematics about programming for the future. What had become known in this country (the United States) as "dynamic programming," invented by Richard Bellman, was very much alive in Massé's work, long before Bellman had a go at it."

A description of Massé's reservoir management model appears in Gessford and Karlin (1958).

Arrow (1958, p. 13), in his colorful description of the economic roots of the dynamic stochastic inventory model, comments

"... it was Wald's work (rather than Massé's, which was unknown in this country at the time) which directly led to later work in multi-period inventory."

A precise time line with proper antecedants is difficult to construct. Heyman and Sobel (1984, p. 192) note

"The modern foundations were laid between 1949 and 1953 by people who spent at least part of that period as staff members at the RAND Corporation in Santa Monica, California. Dates of actual publication are not reliable guides to the order in which ideas were discovered during this period."

Investigators associated with this path breaking work include Arrow, Bellman, Blackwell, Dvoretsky, Girschik, Isaacs, Karlin, Kiefer, LaSalle, Robbins, Shapley, and Wolfowitz. Their work on games (Bellman and Blackwell, 1949; Bellman and LaSalle, 1949; Shapley, 1953), stochastic inventory models (Arrow, Harris, and Marschak,1951; Dvoretsky, Kiefer, and Wolfowitz, 1952), pursuit problems (Isaacs, 1955, 1965) and sequential statistical problems (Arrow, Blackwell, and Girshick, 1949; Robbins, 1952; Kiefer, 1953) laid the groundwork for subsequent developments.

Bellman in numerous papers identified common ingredients to these problems and through his work on functional equations, dynamic programming, and the principle of optimality, became the first major player. Bellman (1954) contains a concise presentation of many of his main ideas and a good bibliography of early work. His 1957 book contains numerous references to his own and other early research and is must reading for all investigators in the field. Karlin (1955) recognized and began studying the rich mathematical foundations of this subject.

CHAPTER 2

Model Formulation

This chapter introduces the basic components of a Markov decision process and discusses some mathematical and notational subtleties. Chapters 1 and 3 contain many examples of Markov decision processes. We encourage you to refer to those examples often to gain a clear understanding of the Markov decision process model. Section 2.2 illustrates these concepts and their interrelationship in the context of a one-period model.

A Markov decision process model consists of five elements: decision epochs, states, actions, transition probabilities, and rewards. We describe these in detail below.

2.1 PROBLEM DEFINITION AND NOTATION

A decision maker, agent, or controller (who we refer to as he with no sexist overtones intended) is faced with the problem, or some might say, the opportunity, of influencing the behavior of a probabilistic system as it evolves through time. He does this by making decisions or choosing actions. His goal is to choose a sequence of actions which causes the system to perform optimally with respect to some predetermined performance criterion. Since the system we model is ongoing, the state of the system prior to tomorrow's decision depends on today's decision. Consequently, decisions must not be made myopically, but must anticipate the opportunities and costs (or rewards) associated with future system states.

2.1.1 Decision Epochs and Periods

Decisions are made at points of time referred to as *decision epochs*. Let T denote the set of decision epochs. This subset of the non-negative real line may be classified in two ways: as either a discrete set or a continuum, and as either a finite or an infinite set. When discrete, decisions are made at all decision epochs. When a continuum, decisions may be made at

1. all decision epochs (continuously),
2. random points of time when certain events occur, such as arrivals to a queueing system, or
3. opportune times chosen by the decision maker.

17

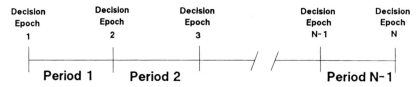

Figure 2.1.1 Decision Epochs and Periods.

When decisions are made continuously, the sequential decision problems are best analyzed using control theory methods based on dynamic system equations.

In discrete time problems, time is divided into *periods* or *stages*. We formulate models so that a decision epoch corresponds to the beginning of a period (see Fig. 2.1.1). The set of decision epochs is either finite, in which case $T \equiv \{1, 2, \ldots, N\}$ for some integer $N < \infty$, or infinite, in which case $T \equiv \{1, 2, \ldots\}$. We write $T = \{1, 2, \ldots, N\}$, $N \leq \infty$ to include both cases. When T is an interval, we denote it by either $T = [0, N]$ or $T = [0, \infty)$. Elements of T (decision epochs) will be denoted by t and usually referred to as "time t." When N is finite, the decision problem will be called a *finite horizon* problem; otherwise it will be called an *infinite horizon* problem. Most of this book will focus on infinite horizon models. We adopt the convention that, in finite horizon problems, decisions are *not* made at decision epoch N: we include it for evaluation of the final system state. Consequently, the last decision is made at decision epoch $N-1$. Frequently we refer to this as an $N-1$ period problem.

The primary focus of this book will be models with discrete T. A particular continuous time model (a semi-Markov decision process) will be discussed (Chapter 11).

2.1.2   State and Action Sets

At each decision epoch, the system occupies a *state*. We denote the set of possible system states by S. If, at some decision epoch, the decision maker observes the system in state $s \in S$, he may choose action a from the set of allowable actions in state s, A_s. Let $A = \bigcup_{s \in S} A_s$ (Fig. 2.1.2.) Note we assume that S and A_s do not vary with t. We expand on this point below.

The sets S and A_s may each be either

1. arbitrary finite sets,
2. arbitrary countably infinite sets,
3. compact subsets of finite dimensional Euclidean space, or
4. non-empty Borel subsets of complete, separable metric spaces.

In nondiscrete settings, many subtle mathematical issues arise which, while interesting, detract from the main ideas of Markov decision process theory. We expand on such issues in Section 2.3 and other sections of this book. These more technical sections will be indicated by asterisks. Otherwise, we assume that S and A_s are *discrete* (finite or countably infinite) unless explicitly noted.

Actions may be chosen either randomly or deterministically. Denote by $\mathscr{P}(A_s)$ the collection of probability distributions on (Borel) subsets of A_s and by $\mathscr{P}(A)$ the set of probability distributions on (Borel) subsets of A. (We may regard $q(\cdot) \in \mathscr{P}(A_s)$ as an

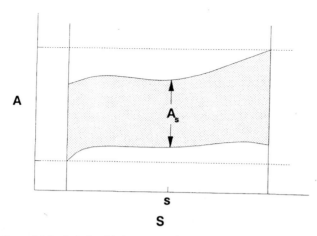

Figure 2.1.2 Relationship between action sets when S is a continuum.

element of $\mathscr{P}(A)$ with support A_s.) Choosing actions randomly means selecting a probability distribution $q(\cdot) \in \mathscr{P}(A_s)$, in which case action a is selected with probability $q(a)$. Degenerate probability distributions correspond to deterministic action choice.

This model may be generalized by allowing both or either of S and A_s to depend explicitly on t. Such generality is unnecessary for most applications and has little effect on theory. It can be included in the formulation above by setting $S = \bigcup_{t \in T} S_t$, where S_t denotes the set of possible states at time t and $A_s = \bigcup_{t \in T} A_{s,t}$, where $A_{s,t}$ denotes the set of allowable actions in state s at time t. Transition probabilities and rewards (defined below) must be modified accordingly.

We might restrict the model by requiring that $A_s = A$ for all $s \in S$. This simplifies notation and doesn't impact theory but limits the applicability of the model since applications as simple as those in Sections 3.1 and 3.3 are not easily included. (Problem 2.7)

Sometimes a distinction is made between a model in which decisions are made at every decision epoch, and one in which the system is uncontrolled and interventions or decisions are made only when necessary. We take the view that this distinction is unnecessary; the latter model is a special case of that model proposed here, in which, for each $s \in S$, A_s includes an action "do not intervene" (Problem 2.3).

2.1.3 Rewards and Transition Probabilities

As a result of choosing action $a \in A_s$ in state s at decision epoch t,

1. the decision maker receives a reward, $r_t(s, a)$ and
2. the system state at the next decision epoch is determined by the probability distribution $p_t(\cdot | s, a)$.

Let the real-valued function $r_t(s, a)$ defined for $s \in S$ and $a \in A_s$ denote the value at time t of the reward received in period t. When positive, $r_t(s, a)$ may be regarded as income, and when negative as cost. From the perspective of the models studied herein, it is immaterial how the reward is accrued during the period. We only require that its value or expected value be known before choosing an action, and that it not be effected by future actions. The reward might be

1. a lump sum received at a fixed or random time prior to the next decision epoch,
2. accrued continuously throughout the current period,
3. a random quantity that depends on the system state at the subsequent decision epoch, or
4. a combination of the above.

When the reward depends on the state of the system at the next decision epoch, we let $r_t(s, a, j)$ denote the value at time t of the reward received when the state of the system at decision epoch t is s, action $a \in A_s$ is selected, and the system occupies state j at decision epoch $t + 1$. Its expected value at decision epoch t may be evaluated by computing

$$r_t(s, a) = \sum_{j \in S} r_t(s, a, j) p_t(j|s, a). \tag{2.1.1}$$

In (2.1.1), the non-negative function $p_t(j|s, a)$ denotes the probability that the system is in state $j \in S$ at time $t + 1$, when the decision maker chooses action $a \in A_s$ in state s at time t. The function $p_t(j|s, a)$ is called a *transition probability function*. Note that many system transitions might occur in the time period between decision epoch t and decision epoch $t + 1$. We formulate the model so that transitions which occur between decision epochs do not influence the decision maker. Under most notions of optimality, all of the information necessary to make a decision at time t is summarized in $r_t(s, a)$ and $p_t(j|s, a)$; however, under some criteria we must use $r_t(s, a, j)$ instead of $r_t(s, a)$.

We usually assume that

$$\sum_{j \in S} p_t(j|s, a) = 1. \tag{2.1.2}$$

Not requiring equality in (2.1.2) increases the range of systems which we may model (cf. Rothblum and Whittle, 1982 and Chapter 11).

In finite horizon Markov decision processes, no decision is made at decision epoch N. Consequently, the reward at this time point is only a function of the state. We denote it by $r_N(s)$ and sometimes refer to it as a *salvage value* or *scrap value*.

We refer to the collection of objects

$$\{T, S, A_s, p_t(\cdot|s, a), r_t(s, a)\}$$

as a *Markov decision process*. The qualifier "Markov" is used because the transition probability and reward functions depend on the past only through the current state of the system and the action selected by the decision maker in that state. Under certain policies, the induced stochastic process need not be Markov. Some authors refer to the above collection of objects as a *Markov decision problem*; here that expression is reserved for a Markov decision process together with an optimality criterion.

2.1.4 Decision Rules

A *decision rule* prescribes a procedure for action selection in each state at a specified decision epoch. Decision rules range in generality from deterministic Markovian to randomized history dependent, depending on how they incorporate past information and how they select actions. Our primary focus will be on deterministic Markovian decision rules because they are easiest to implement and evaluate. Such decision rules are functions $d_t: S \rightarrow A_s$, which specify the action choice when the system occupies state s at decision epoch t. For each $s \in S$, $d_t(s) \in A_s$. This decision rule is said to be *Markovian* (memoryless) because it depends on previous system states and actions only through the current state of the system, and *deterministic* because it chooses an action with certainty.

We call a deterministic decision rule *history dependent* if it depends on the past history of the system as represented by the sequence of previous states and actions. That is, d_t is a function of the history $h_t = (s_1, a_1, \ldots, s_{t-1}, a_{t-1}, s_t)$ where s_u and a_u denote the state and action of the system at decision epoch u. *As convention, when the history is h_t, s_t denotes the state at decision epoch t.* The history h_t follows the recursion $h_t = (h_{t-1}, a_{t-1}, s_t)$. When using a history-dependent decision rule and observing h_t, the decision maker chooses actions from A_{s_t}. We let H_t denote the set of all histories h_t. Note that $H_1 = S$, $H_2 = S \times A \times S$ and $H_n = S \times A \times S \times \ldots \times S$, and that H_t satisfies the recursion $H_t = H_{t-1} \times A \times S$. For sets C and D, $C \times D$ denotes the *Cartesian product* of C and D. That is, $C \times D = \{(c,d): c \in C, d \in D\}$. A deterministic history-dependent decision rule d_t maps H_t into A, subject to the restriction that $d_t(h_t) \in A_{s_t}$.

A *randomized* decision rule d_t specifies a probability distribution $q_{d_t}(\cdot)$ on the set of actions. Randomized Markovian decision rules map the set of states into the set of probability distributions on the action space, that is $d_t: S \rightarrow \mathscr{P}(A)$, and randomized history-dependent decision rules according to $d_t: H_t \rightarrow \mathscr{P}(A)$. When Markovian, $q_{d_t(s_t)}(\cdot) \in \mathscr{P}(A_{s_t})$, and when history dependent, $q_{d_t(h_t)}(\cdot) \in \mathscr{P}(A_{s_t})$ for all $h_t \in H_t$. A deterministic decision rule may be regarded as a special case of a randomized decision rule in which the probability distribution on the set of actions is degenerate, that is, $q_{d_t(s_t)}(a) = 1$ or $q_{d_t(h_t)}(a) = 1$ for some $a \in A_s$.

We classify decision rules as history dependent and randomized (HR), history dependent and deterministic (HD), Markovian and randomized (MR), or Markovian and deterministic (MD) depending on their degree of dependence on past information and on their method of action selection. We denote the set of decision rules at time t by D_t^K, where K designates a class of decision rules ($K =$ HR, HD, MR, MD); D_t^K is called a *decision rule set*. A summary of this classification scheme appears in Table 2.1.1.

A fundamental question in Markov decision problem theory is

For a given optimality criterion, under what conditions is it optimal to use a deterministic Markovian decision rule at each stage?

The rewards and transition probabilities become functions on S or H_t after specifying decision rules. For $d_t \in D_t^{MD}$, the reward equals $r_t(s, d_t(s))$ and the transition probability equals $p_t(j|s, d_t(s))$ and for $d_t \in D_t^{HD}$, they equal $r_t(s, d_t(h_t))$ and $p_t(j|s\, d_t(h_t))$ whenever $h_t = (h_{t-1}, a_{t-1}, s)$. If d_t is a randomized Markov

Table 2.1.1   Classification of Decision Rules and Decision Rule Sets

History Dependence	Action Choice	
	Deterministic	Randomized
Markovian	$d_t(s_t) \in A_{s_t}$ D_t^{MD}	$q_{d_t(s_t)}(\cdot) \in \mathscr{P}(A_{s_t})$ D_t^{MR}
History Dependent	$d_t(h_t) \in A_{s_t}$ D_t^{HD}	$q_{d_t(h_t)}(\cdot) \in \mathscr{P}(A_{s_t})$ D_t^{HR}

decision rule the *expected* reward satisfies

$$r_t(s, d_t(s)) = \sum_{a \in A_s} r_t(s, a) q_{d_t(s)}(a) \tag{2.1.3}$$

and the transition probability satisfies

$$p_t(j|s, d_t(s)) = \sum_{a \in A_s} p_t(j|s, a) q_{d_t(s)}(a). \tag{2.1.4}$$

Analogous constructions apply to randomized history-dependent rules.

2.1.5   Policies

A *policy*, *contingency plan*, *plan*, or *strategy* specifies the decision rule to be used at all decision epoch. It provides the decision maker with a prescription for action selection under any possible future system state or history. A policy π is a sequence of decision rules, i.e., $\pi = (d_1, d_2, \ldots, d_{N-1})$ where $d_t \in D_t^K$ for $t = 1, 2, \ldots, N - 1$ for $N \leq \infty$, with K representing any of the above classes. Let Π^K denote the set of all policies of class K (K = HR, HD, MR, MD); that is, $\Pi^K = D_1^K \times D_2^K \times \cdots \times D_{N-1}^K$, $N \leq \infty$.

We call a policy *stationary* if $d_t = d$ for all $t \in T$. A stationary policy has the form $\pi = (d, d, \ldots)$; we denote it by d^∞. We let Π^{SD} be the set of stationary deterministic policies and Π^{SR} be the set of stationary randomized policies. We sometimes refer to stationary deterministic policies as *pure* policies. Stationary policies are fundamental to the theory of infinite horizon Markov decision processes.

The relationship between the various classes of policies is as follows: $\Pi^{SD} \subset \Pi^{SR} \subset \Pi^{MR} \subset \Pi^{HR}$, $\Pi^{SD} \subset \Pi^{MD} \subset \Pi^{MR} \subset \Pi^{HR}$, and $\Pi^{SD} \subset \Pi^{MD} \subset \Pi^{HD} \subset \Pi^{HR}$. Thus randomized, history-dependent policies are most general and stationary deterministic policies most specific.

*2.1.6   Induced Stochastic Processes, Conditional Probabilities, and Expectations

In this section we provide a formal model for the stochastic process generated by a Markov decision process. *To simplify presentation we assume discrete S and A.* Section 2.3.2 discusses extension of this construction to a more general model.

A probability model consists of three elements: a sample space Ω, a σ-algebra of (Borel) measurable subsets of Ω, $B(\Omega)$, and a probability measure P on $B(\Omega)$. An elaborate discussion of these concepts is beyond the scope of this book but note that when Ω is a *finite* set $B(\Omega)$ equals the set of all subsets of Ω and a probability measure is a probability mass function.

In a finite horizon Markov decision process, we choose

$$\Omega = S \times A \times S \times A \times \cdots \times A \times S = \{S \times A\}^{N-1} \times S,$$

and in an infinite horizon model, $\Omega = \{S \times A\}^{\infty}$. A typical element $\omega \in \Omega$ consists of a sequence of states and actions, that is

$$\omega = (s_1, a_1, s_2, a_2, \ldots, a_{N-1}, s_N),$$

and, in an infinite horizon model,

$$\omega = (s_1, a_1, s_2, a_2, \ldots).$$

We refer to ω as a *sample path*. Given a set C, let $\{C\}^k$ denote the Cartesian product of C with itself k times. In finite horizon models, with horizon length N, $B(\Omega) = B(\{S \times A\}^{N-1} \times S)$ and in infinite horizon models, $B(\Omega) = B(\{S \times A\}^{\infty})$.

We define the random variables X_t and Y_t which take values in S and A, respectively, by

$$X_t(\omega) = s_t \text{ and } Y_t(\omega) = a_t \tag{2.1.5}$$

for $t = 1, 2, \ldots, N$, $N \le \infty$. This means that when the observed sequence of states and actions is ω, the random variable X_t denotes the state at time t, and Y_t denotes the action at time t. Define the history process Z_t by

$$Z_1(\omega) = s_1 \quad \text{and} \quad Z_t(\omega) = (s_1, a_1, \ldots, s_t) \text{ for } 1 \le t \le N; N \le \infty.$$

Let the probability distribution $P_1(\cdot)$ denote the *initial distribution* of the system state. In most applications, we assume degenerate $P_1(\cdot)$, that is, $P_1(s_1) = 1$ for some $s_1 \in S$.

A randomized history-dependent policy $\pi = (d_1, d_2, \ldots, d_{N-1})$, $N \le \infty$, induces a probability P^{π} on $(\Omega, B(\Omega))$ through

$$P^{\pi}\{X_1 = s\} = P_1(s), \tag{2.1.6}$$

$$P^{\pi}\{Y_t = a | Z_t = h_t\} = q_{d_t(h_t)}(a), \tag{2.1.7}$$

$$P^{\pi}\{X_{t+1} = s | Z_t = (h_{t-1}, a_{t-1}, s_t), Y_t = a_t\} = p_t(s|s_t, a_t) \tag{2.1.8}$$

so that the probability of a sample path $\omega = (s_1, a_1, s_2, \ldots, s_N)$ is given by

$$P^{\pi}(s_1, a_1, s_2, \ldots, s_N) = P_1(s_1) q_{d_1(s_1)}(a_1) p_1(s_2|s_1, a_1) q_{d_2(h_2)}(a_2)$$

$$\cdots q_{d_{N-1}(h_{N-1})}(a_{N-1}) p_{N-1}(s_N|s_{N-1}, a_{N-1}). \tag{2.1.9}$$

For π in Π^{HD} or Π^{MD}, (2.1.9) simplifies to

$$P^{\pi}(s_1, a_1, s_2, \ldots, s_N) = P_1(s_1) p_1(s_2|s_1, a_1) \cdots p(s_N|s_{N-1}, a_{N-1}).$$

Note that the policy determines P^{π} explicitly through (2.1.7) and implicitly through (2.1.8). The Markovian nature of the model is reflected through (2.1.8).

For computation, we require conditional probabilities of the process from t onward conditional on the history at time t. Under the discreteness assumptions, we compute these probabilities as follows:

$$P^{\pi}(a_t, s_{t+1}, \ldots, s_N|s_1, a_1, \ldots, s_t) = \frac{P^{\pi}(s_1, a_1, \ldots, s_N)}{P^{\pi}(s_1, a_1, \ldots, s_t)} \qquad (2.1.10)$$

provided the quantity in the denominator is nonzero; otherwise the probability equals zero. We evaluate the denominator of (2.1.10) by summing $P^{\pi}(s_1, a_1, \ldots, s_N)$ over all sample paths which equal $s_1, a_1, \ldots, s_t$ in the first $2t - 1$ components. This reduces to

$$P^{\pi}(s_1, a_1, s_2, \ldots, s_t)$$
$$= P_1(s_1) q_{d_1(s_1)}(a_1) p_1(s_2|s_1, a_1) q_{d_2(h_2)}(a_2) \cdots q_{d_{t-1}(h_{t-1})}(a_{t-1}) p_{t-1}(s_t|s_{t-1}, a_{t-1})$$

so that upon substitution (2.1.10) simplifies to

$$P^{\pi}(a_t, s_{t+1}, \ldots, s_N|s_1, a_1, \ldots, s_t)$$
$$= q_{d_t(h_t)}(a_t) p(s_{t+1}|s_t, a_t) \cdots q_{d_{N-1}(h_{N-1})}(a_{N-1}) p(s_N|s_{N-1}, a_{N-1}). \qquad (2.1.11)$$

By similar arguments,

$$P^{\pi}(s_{t+1}, a_{t+1}, \ldots, s_N|s_1, \ldots, s_t, a_t)$$
$$= p(s_{t+1}|s_t, a_t) q_{d_{t+1}(h_{t+1})}(a_{t+1}) \cdots q_{d_{N-1}(h_{N-1})}(a_{N-1}) p(s_N|s_{N-1}, a_{N-1}) \qquad (2.1.12)$$

Observe that $P^{\pi}\{X_{t+1} = s|Z_t = (h_{t-1}, a_{t-1}, s_t), Y_t = a_t\}$ is a function of Z_t only through $X_t = s_t$; however, because of the dependence on the history through (2.1.7), the process need not be a Markov chain. For Markovian π (i.e., for $\pi \in \Pi^{MD}$ or $\pi \in \Pi^{MR}$), d_t depends on the history only through the current state of the process, so that (2.1.7) becomes

$$P^{\pi}\{Y_t = a|Z_t = (h_{t-1}, a_{t-1}, s_t)\} = P\{Y_t = a|X_t = s_t\} = q_{d_t(s_t)}(a).$$

Consequently,

$$P^{\pi}(a_t, s_{t+1}, \ldots, s_N|s_1, a_1, \ldots, s_t) = P^{\pi}(a_t, s_{t+1}, \ldots, s_N|s_t),$$

so that the induced stochastic processes $\{X_t; t \in T\}$ is a discrete time Markov chain.

When π is Markovian, we refer to the bivariate stochastic process $\{(X_t, r_t(X_t, Y_t));$ $t \in T\}$ as a *Markov reward process* (a Markov chain together with a real-valued function defined on its state space). This process represents the sequence of system states and stream of rewards received by the decision maker when using policy π.

Let W denote a (real-valued) random variable defined on $\{\Omega, B(\Omega), P^\pi\}$. When n is finite we define the *expected value* of W with respect to policy π by

$$E^\pi\{W\} = \sum_{\omega \in \Omega} W(\omega)P^\pi\{\omega\} = \sum_{w \in R^1} wP^\pi\{\omega: W(\omega) = w\}, \quad (2.1.13)$$

where, as above, $\omega = (s_1, a_1, s_2, a_2, \ldots, a_{N-1}, s_N)$. When N is infinite integrals replace sums in (2.1.13). Usually we will evaluate the expectation of quantities of the form

$$W(s_1, a_1, \ldots, s_N) = \sum_{t=1}^{N-1} r_t(s_t, a_t) + r_N(s_N). \quad (2.1.14)$$

Expression (2.1.13) does not provide an efficient procedure for computing such expectations and ignores the dynamic aspects of the problem. When $\{X_t; t \in T\}$ is a Markov chain, such an expectation can be calculated by standard matrix methods (Appendix A); however, induction methods are more suitable for history-dependent π.

Suppose we observe $h_t = (s_1, a_1, \ldots, s_t)$, and W is a function of $s_t, a_t, \ldots, s_N$. Then

$$E_{h_t}^\pi\{W(X_t, Y_t, \ldots, X_N)\} = \sum W(s_t, a_t, \ldots, s_N)P^\pi(a_t, s_{t+1}, \ldots, s_N | s_1, a_1, \ldots, s_t), \quad (2.1.15)$$

where the summation ranges over $(a_t, s_{t+1}, \ldots, s_N) \in A \times S \times \cdots \times S$ and the conditional probability $P^\pi\{\cdot | s_1, a_1, \ldots, s_t\}$ is evaluated according to (2.1.11). For Markovian π,

$$E_{s_t}^\pi\{W(X_t, Y_t, \ldots, X_N)\} \equiv \sum W(s_t, a_t, \ldots, s_N)P^\pi(a_t, s_{t+1}, \ldots, s_N | s_t). \quad (2.1.16)$$

Note that $E^\pi\{W\} = \sum_{s \in S} P_1(s)E_s^\pi\{W\}$ for any $\pi \in \Pi^{HR}$.

2.2 A ONE-PERIOD MARKOV DECISION PROBLEM

We now describe a one-period Markov decision problem (MDP) in detail to illustrate notation, indicate the flavor of analyses to follow, and provide a focus for discussing some technical issues in Sec. 2.3. We assume *finite* S and A_s for each $s \in S$.

In a one-period model $N = 2$ so that $T = \{1, 2\}$. We assume that whenever the system occupies state s' at the end of period 1, the decision maker receives a terminal reward $v(s')$, where v is a specified real-valued function on S. Suppose the decision maker finds the system in state s at the start of stage 1 and his objective is to choose an action $a \in A_s$ to maximize the sum of the immediate reward, $r_1(s, a)$, and the expected terminal reward. If the decision maker chooses a deterministic policy

$\pi = (d_1)$, which selects action $a' \in A_s$ at decision epoch 1, the total expected reward equals

$$r_1(s, a') + E_s^{\pi}\{v(X_2)\} = r_1(s, a') + \sum_{j \in S} p_1(j|s, a')v(j). \qquad (2.2.1)$$

To achieve his objective, the decision maker will choose an $a' \in A_s$ to make the expression in (2.2.1) as large as possible. To find such an action, evaluate (2.2.1) for each $a' \in A_s$ and select any action that achieves this maximum. Denote such an action by a^* and the maximum value of the expected total reward by

$$\max_{a' \in A_s} \left\{ r_1(s, a') + \sum_{j \in S} p_1(j|s, a')v(j) \right\}. \qquad (2.2.2)$$

Any maximizing action satisfies

$$r_1(s, a^*) + \sum_{s \in S} p_1(j|s, a^*)v(j) = \max_{a' \in A_s} \left\{ r_1(s, a') + \sum_{j \in S} p_1(j|s, a')v(j) \right\}.$$
$$(2.2.3)$$

The following additional notation simplifies describing maximizing actions. Let X be an arbitrary set, and $g(x)$ a real-valued function on X. Define

$$\arg\max_{x \in X} g(x) \equiv \{ x' \in X : g(x') \ge g(x) \text{ for all } x \in X \}.$$

Using this notation we see that any a_s^* which satisfies (2.2.3) may be represented by

$$a_s^* \in \arg\max_{a' \in A_s} \left\{ r_1(s, a') + \sum_{j \in S} p_1(j|s, a')v(j) \right\}. \qquad (2.2.4)$$

Because A_s is finite, at least one maximizing action exists. If it is unique, it is the optimal decision in state s. If not, choosing any maximizing action yields the maximum expected reward.

We stress that the operation "max" results in a value and "arg max" in a set. The following simple example illustrates this point.

Example 2.2.1. Suppose $X = \{a, b, c, d\}$, $g(a) = 5$, $g(b) = 7$, $g(c) = 3$, and $g(d) = 7$. Then

$$\max_{x \in X} \{g(x)\} = 7 \quad \text{and} \quad \arg\max_{x \in X} \{g(x)\} = \{b, d\}.$$

Note that when the maximum of a function is not attained, we instead seek a supremum. We write this as "sup." In this case the "arg max" is the empty set.

Example 2.2.2. Suppose $X = \{1, 2, \ldots\}$ and $g(x) = 1 - 1/x$. Then $\max_{x \in X}\{g(x)\}$ does not exist, arg $\max_{x \in X}\{g(x)\} = \varnothing$ but $\sup_{x \in X}\{g(x)\} = 1$. Further, for any $\varepsilon > 0$, there exists a subset of X, say X_ε, for which $g(x') > 1 - \varepsilon$ for all $x' \in X_\varepsilon$.

It is natural to ask whether the decision maker can obtain a larger reward by using a random mechanism to select actions in state s. If action $a \in A_s$ is selected with probability $q(a)$, the expected total reward equals

$$\sum_{a \in A_s} q(a)\left[r_1(s, a) + \sum_{j \in S} p_1(j|s, a)v(j)\right],$$

where $\sum_{a \in A_s} q(a) = 1$ and $q(a) \geq 0$ for $a \in A_s$.

Since

$$\max_{q \in \mathscr{P}(A_s)} \left\{\sum_{a \in A_s} q(a)\left[r_1(s, a) + \sum_{j \in S} p_1(j|s, a)v(j)\right]\right\}$$

$$= \max_{a' \in A_s} \left\{r_1(s, a') + \sum_{j \in S} p_1(j|s, a')v(j)\right\},$$

we cannot obtain a larger expected reward in state s by randomization. Note, however, that any decision rule which randomizes over actions satisfying (2.2.4) obtains the maximum expected reward.

If s is not known prior to determining the optimal action, for example, if it is determined by an initial distribution P_1, then the decision maker must choose an action for each possible $s \in S$, that is, he must specify a decision rule. To find an optimal decision rule, he maximizes (2.2.2) independently for each $s \in S$. Therefore any decision rule $d^*(s) = a_s^*$ where a_s^* satisfies (2.2.4), would yield the largest total expected reward. In this case, the deterministic policy $\pi^* = (d^*)$ maximizes the total expected reward within the class of all Markov randomized policies. Since $h_1 = s_1$ in this one-period model, we need not distinguish between Markov and history-dependent policies.

Thus we have provided a method for finding a $\pi^* = \Pi^{MD}$ for which

$$E^{\pi^*}\{r_1(X_1, Y_1) + v(X_2)\} \geq E^{\pi}\{r_1(X_1, Y_1) + v(X_2)\}$$

for all $\pi \in \Pi^{HR}$. Furthermore, the same choice of π^* satisfies

$$E_s^{\pi^*}\{r_1(X_1, Y_1) + v(X_2)\} \geq E_s^{\pi}\{r_1(X_1, Y_1) + v(X_2)\}$$

for all $\pi \in \Pi^{HR}$ and each $s \in S$.

*2.3 TECHNICAL CONSIDERATIONS

This section discusses some technical issues and provides a more general formulation of a Markov decision process model.

2.3.1   The Role of Model Assumptions

The model in Sec. 2.2 provides a focus for discussing the role of assumptions about S and A_s. Suppose first that S is a Borel measurable subset of Euclidean space. Let $\pi = (d_1) \in \Pi^{MD}$ and suppose $d_1(s) = a$. In this case (2.2.1) becomes

$$r_1(s, a) + E_s^\pi\{v(X_2)\} = r_1(s, a) + \int_S v(u)p_1(u|s, a)\, du \qquad (2.3.1)$$

if the density $p_1(u|s, a)$ exists. In greater generality the integral in (2.3.1) may be represented by the Lebesgue-Stieltjes integral $\int_S v(u)p_1(du|s, a)$. For the above expressions to be meaningful requires that $v(\cdot)p_1(\cdot|s, a)$ be Lebesgue integrable, or $v(\cdot)$ be Lebesgue-Stieltjes integrable with respect to $p_1(du|s, a)$ for each $s \in S$ and $a \in A_s$.

We now consider (2.3.1) as a function on S. Upon substitution of d_1, it becomes

$$r_1(s, d_1(s)) + \int_S v(u)p_1(u|s, d_1(s))\, du.$$

For analyzing multi-period models we require this to be a measurable and integrable function on S. This necessitates imposing assumptions on $r_1(s, \cdot)$, $p_1(u|s, \cdot)$, and $d_1(\cdot)$. At a minimum we require $d_1(s)$ to be a measurable function from S to A, which means that we must restrict the set of admissible decision rules to include only measurable functions. To identify measurable functions requires a topology on A.

In Sec. 2.2, we constructed an optimal decision rule d^* by setting $d^*(s)$ equal to a_s^* for each $s \in S$ where a_s^* satisfied (2.2.4). That is, we solved a separate maximization problem for each $s \in S$. For discrete S, $d^* \in D_1$, since D_1 consisted of the set of all functions from S to A, but, as discussed above, we require that d^* be measurable. To ensure this we need a *selection theorem* which ensures that $\times_{s \in S} A_s^*$ contains a measureable function from S to A where

$$A_s^* \equiv \operatorname*{arg\,max}_{a \in A_s} \left\{ r_1(s, a) + \int_S v(u)p_1(u|s, a)\, du \right\}.$$

Such a result does not hold without further assumptions on r_1, p_1, and A_s, so even in a one-period model an optimal policy need not exist (Sect. 6.3.5).

Assumptions about the form of the action sets affect the maximization in (2.2.3). When A_s is finite, a maximizing action always exists. If A_s is a compact subset of Euclidean space and the expression inside the brackets in (2.2.3) is upper semicontinuous (Appendix B), then the maximum is attained, but when A_s is countable the maximum need not exist. In this latter case, we replace "max" by "sup" (supremum) and seek actions or decision rules that yield values within some prespecified small $\varepsilon > 0$ of the supremum.

2.3.2   The Borel Model

A formulation in greater generality follows. We refer to $\{X, B(X)\}$ as a *Borel space* or *Polish space* if X is a Borel subset of a complete separable metric space and $\{B(X)\}$ its family of Borel subsets. For Borel spaces $\{X, B(X)\}$ and $\{Y, B(Y)\}$, we refer to q

as a *conditional probability on Y given X* or a *transition kernel* if for $x \in X$, $q(\cdot|x)$ is a probability measure on $B(Y)$ and for $G \in B(Y)$, $q(G|\cdot)$ is a Borel measurable function from X into $[0, 1]$.

In this generality a Markov decision process consists of

1. A Borel space $\{S, B(S)\}$.
2. A Borel space $(A, B(A))$ and a collection of sets A_s, for which $A_s \in B(A)$ for each $s \in S$. Let $B(A_s)$ denote the induced collection of Borel subsets of A_s. Furthermore, we require existence of a measurable function δ mapping S into A with $\delta(s) \in A_s$ for each $s \in S$.
3. A family $\mathscr{P}(A_s)$ of probability measures on $B(A_s)$ for each $s \in S$.
4. Real-valued reward functions $r_t(s, a)$ which for each $t \in T$ satisfy
   a. $r_t(\cdot, \cdot)$ is measurable with respect to $B(S \times A_s)$,
   b. $r_t(s, \cdot)$ is integrable with respect to all $q \in \mathscr{P}(A_s)$ for all $s \in S$.
5. Conditional probabilities $p_t(\cdot|s, a)$ which for each $t \in T$ satisfy
   a. $p_t(G|\cdot, \cdot)$ is measurable with respect to $B(S \times A_s)$ for $G \in B(S)$,
   b. $p_t(G|s, \cdot)$ is integrable with respect to each $q \in \mathscr{P}(A_s)$ for each $s \in S$ and $G \in B(S)$.

We now define sets of decision rules. As before, let $H_t = S \times A \times \cdots \times S$ denote the set of histories up to time t and now let $B(H_t)$ denote the derived family of Borel subsets on H_t. Then

$$D_t^{HD} = \{\delta: H_t \to A: \delta \text{ is measurable and } \delta(h_{t-1}, a_{t-1}, s) \in A_s \text{ for all } s \in S\}$$
$$D_t^{MD} = \{\delta: S \to A: \delta \text{ is measurable and } \delta(s) \in A_s \text{ for all } s \in S\},$$
$$D_t^{HR} = \{\text{conditional probabilities } q \text{ on } A \text{ given } H_t: q(A_s|h_{t-1}, a_{t-1}, s) = 1 \text{ for each } s \in S\},$$
$$D_t^{MR} = \{\text{conditional probabilities } q \text{ on } A \text{ given } S: q(A_s|s) = 1 \text{ for each } s \in S\}.$$

The policy sets Π^K, $K = $ HR, MR, HD, MD, are the Cartesian products of the corresponding decision sets.

Given $\delta \in D_t^{HR}$ for $h_t = (h_{t-1}, a_t, s_t) = H_t$ we generalize (2.1.3) to

$$r_t(s_t, \delta(h_t)) = \int_{A_{s_t}} r_t(s_t, a) q_{\delta(h_t)}(da|h_t). \tag{2.3.2}$$

Since r_t is measurable in S and $q_{\delta(h_t)}$ is a conditional probability, $r_t(\cdot, \delta(\cdot))$ is a measurable function on H_t. By a similar argument, the above assumptions imply that $p_t(G|\cdot, \delta(\cdot))$ is a measurable function on H_t for every Borel subset G of S. Likewise for $\delta \in D_t^K$, $r_t(\cdot, \delta(\cdot))$ and $p_t(G|\cdot, \delta(\cdot))$ are measurable functions on S for $K = $ MR, MD, HD.

The construction of induced stochastic processes in Section 2.1.6 applies with the following modifications

1. $B(S)$ and $B(A)$ are the respective Borel sets of S and A.
2. Ω is endowed with the product σ-algebra.
3. The coordinate mappings X_t and Y_t in (2.1.5) are measurable mappings from Ω to S and A, respectively.
4. The expressions in (2.1.7) and (2.1.8) are conditional probabilities.
5. For $\pi \in \Pi^{HR}$ the probability measure P^π satisfies

$$P^\pi(ds_1 da_1 ds_2 \cdots ds_n)$$

$$= P_1(ds_1)q_{d_1(s_1)}(da_1)p(ds_2|s_1, a_2) \cdots p(ds_N|s_{N-1}, a_{N-1}),$$

where its existence and uniqueness is guaranteed by a result of Ionescu-Tulcea (Hinderer, 1970; p. 149, Bertsekas and Shreve, 1978, pp. 140–144).

We regard the theory of Markov decision problems as sufficiently rich and complex without having to confront these additional mathematical subtleties. Several excellent books including Hinderer (1970), Bertsekas and Shreve (1978), Dynkin and Yushkevich (1979), and Hernandez-Lerma (1989) develop results at this level of generality. Our main focus will be models in which S and A_s are finite or countable. Results which hold in greater generality will be noted.

BIBLIOGRAPHIC REMARKS

Bellman (1954) coined the expression "Markov decision process." His early papers, many of which are summarized in his classic book (1957), provide the basis for the Markov decision process model. In this book, he introduces the concepts of states, actions, and transitions, and develops functional equations for finding optimal policies. Karlin (1955) elaborates on these concepts and provides a more abstract mathematical basis for these models. Shapley (1953) introduces and analyzes a two-person stochastic game which is very close in spirit to a Markov decision process. Howard's monograph (1960) played a major role in establishing the importance of the Markov decision process model.

Rigorous foundations for the Markov decision model were established by Dubins and Savage (1965) and Blackwell (1965). Blackwell's classic paper (1965) provides the model we analyze in this book. It defines state and action sets and rewards and transition probabilities in considerable generality, distinguishes randomized, deterministic (degenerate), Markov, and history-dependent policies and emphasizes the importance of stationary policies.

How to Gamble if You Must by Dubins and Savage (1965), which first appeared in mimeograph form in 1960, analyzes a gambling model which possesses a similar mathematical structure to the Markov decision model. Their development raises and addresses many of the mathematical subtleties that are still the subject of research efforts. Their terminology is quite different, so, when referring to their book, note the equivalence between fortunes and states, gambling houses and action sets, and gambles and transition probabilities. The gambling model does not explicitly contain a reward function, and the system state represents the gambler's current wealth.

Hinderer (1970) addresses foundational questions for both countable and arbitrary state spaces models. His model is more general than ours in that rewards and transition probabilities may depend on the entire history of the process. The books by Bertsekas and Shreve (1978) and Dynkin and Yushkevich (1979) provide a detailed analysis of many of the deeper measurability questions. These two books also provide considerable insight into the underlying probabilistic structure of the Markov decision process model.

Fleming and Rishel (1975) and Bertsekas (1987) provide nice treatments of dynamic optimization problems formulated in terms of stochastic control models. Fleming and Rishel's focus is on continuous time problems with continuous state spaces while Bertsekas presents results for discrete time problems with continuous state spaces. We discuss the relationship between these models and the Markov decision process model in Sec. 3.5.

Royden (1963) is a suitable reference for the topological foundations of this chapter and Breiman (1968) and Lamperti (1966) for the more advanced probabilistic concepts.

PROBLEMS

2.1. Derive $r_t(s, a)$ for a discrete-time Markov decision process in which a reward $g_t(s, a)$ arrives in a lump sum at a random point τ between two decision epochs according to a probability distribution $q_t(s, a)$.

2.2. Consider the discrete-time Markov decision process model of this chapter but suppose instead that decisions are made at every second period and are used for two consecutive periods. That is, if the decision maker chooses decision rule d_t at time t, he will use the identical decision rule at time $t + 1$. The next decision epoch is $t + 2$.

Reformulate this as a Markov decision process model. Clearly identify the reward and transition functions. Be sure to account for the possibility of a randomized decision rule.

2.3. Suppose a discrete-time Markov chain evolves according to a fixed transition law $p_t(s'|s)$ and it generates rewards $r_t(s)$ if the system occupies state s at time t. At any stage, the decision maker may either let the system evolve uninterrupted or instead intervene and choose an action from a state-dependent set A_s which determines a one-period transition law $p_t(s'|s, a)$ and reward $r_t(s, a)$.

  a. Formulate this as a Markov decision process. Clearly identify the action sets, rewards, and transition probabilities.

  b. Explicitly provide expression (2.2.2) for this model.

2.4. Consider a finite horizon Markov decision problem in which the only reward is received at termination. Define the reward functions for such a problem.

2.5. Consider a one-period model with $S = \{s_1, s_2\}$ $A_{s_1} = \{a_{1,1}, a_{1,2}\}$ and $A_{s_2} = \{a_{2,1}, a_{2,2}\}$; $r_1(s_1, a_{1,1}) = 5$, $r_1(s_1, a_{1,2}) = 10$, $r_1(s_2, a_{2,1}) = -1$, and $r_1(s_2, a_{2,2}) = 2$; and $p_1(s_1|s_1, a_{1,1}) = p_1(s_2|s_1, a_{1,1}) = 0.5$, $p_1(s_1|s_1, a_{1,2}) = 0$, $p_1(s_2|s_1, a_{1,2}) = 1$, $p_1(s_1|s_2, a_{2,1}) = 0.8$, $p_1(s_2|s_2, a_{2,1}) = 0.2$, $p_1(s_1|s_2, a_{2,2}) = 0.1$, and $p_1(s_2|s_2, a_{2,2}) = 0.9$.

a. Find the deterministic policy that maximizes the total one-period expected reward provided that the terminal reward $v(s)$ is identically 0.

b. Find the deterministic policy that maximizes the total one-period expected reward provided that the terminal reward $v(s_1) = d$ and $v(s_2) = e$. Investigate the sensitivity of this policy to the values of d and e.

c. Compute $r_1(s, d_1(s))$ and $p_1(j|s, d_1(s))$ for the randomized decision rule d_1 which in state s_1 chooses action $a_{1,1}$ with probability q and action $a_{1,2}$ with probability $1 - q$, and in state s_2 uses action $a_{2,1}$ with certainty.

d. Find the randomized policy that maximizes the total one-period expected reward provided that the terminal reward $v(s_1) = 5$ and $v(s_2) = -5$.

2.6. Consider the model in Problem 2.5. Suppose the initial state is determined by the distribution $P_1(s_1) = P_1(s_2) = 0.5$. Let $\pi = (d_1)$ be the randomized policy which in s_1 uses action $a_{1,1}$ with probability 0.6 and in s_2 uses action $a_{2,1}$ with probability 0.3. Let $v(s_1) = 0$ and $v(s_2) = 1$.

a. Compute P^π as defined by (2.1.9) for all possible sample paths.

b. Compute the expected total reward using this policy by using the representation for W in (2.1.14) and then computing the expectation using each formula in (2.1.13). Compare these to values computed using (2.2.1).

2.7. Consider the model in Problem 2.5. Reformulate it so that the set of actions is *not* state dependent. To do this, let $A = A_{s_1} \cup A_{s_2}$ and redefine the rewards and transition probabilitites. Note that this necessitates adding many superfluous components.

2.8. Prove for a one-period Markov decision problem that there always exists a deterministic policy with a reward at least as great as that of any randomized policy.

2.9. Suppose P^π has been given for each sample path of a deterministic Markov policy π in a two-period finite state and action Markov decision model. Find conditions under which you can recover the transition probabilities of the model and show how to compute them.

CHAPTER 3

Examples

In Chap. 1, we described some significant applications of Markov decision process models. In this chapter we provide further applications and, through them, illustrate notation and concepts from Chap. 2. With the exception of Secs. 3.1 and 3.2, in which the models are quite specific and used to illustrate many points throughout this book, we have chosen to present Markov decision process formulations for classes of models that have wide applicability. We delay discussing optimality criteria until Chaps. 4 and 5 because in most cases, especially those which have an economic basis, they do not influence the identification of states and actions. An exception is the "secretary problem" of Sec. 3.4.3. To place models in an optimization context, assume for now that the decision maker seeks a policy which maximizes the expected total reward.

3.1 A TWO-STATE MARKOV DECISION PROCESS

The following simple abstract example illustrates the basic components of a Markov decision process. We refer to it often in later chapters.

We assume stationary rewards and transition probabilities; that is the rewards and transition probabilities are the same at every decision epoch. At each decision epoch the system occupies either state s_1 or s_2. In state s_1, the decision maker chooses either action $a_{1,1}$ or action $a_{1,2}$; in state s_2, only action $a_{2,1}$ is available. Choosing action $a_{1,1}$ in s_1 provides the decision maker with an immediate reward of five units, and at the next decision epoch the system is in state s_1 with probability 0.5 and state s_2 with probability 0.5. If instead he chooses action $a_{1,2}$ in state s_1, he receives an immediate reward of ten units and at the next decision epoch the system moves to state s_2 with probability 1. In state s_2, the decision maker must choose action $a_{2,1}$. As a consequence of this choice, the decision maker incurs a cost of one unit and the system occupies state s_2 at the next decision epoch with certainty.

Figure 3.1.1, which illustrates this specific model, provides a convenient symbolic representation for any *stationary* model. In it, circles represent states and directed arcs represent possible transitions corresponding to actions. The first expression in brackets below the arc is the reward achieved when that action is selected and a transition between states connected by that arc occurs. The second quantity in brackets is the probability that the transition occurs. For example, if action $a_{1,1}$ is selected, the process travels the arc between s_1 and s_2 with probability 0.5 and, when

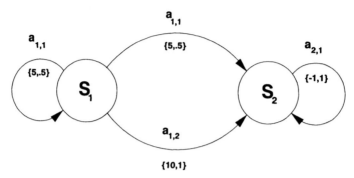

Figure 3.1.1 Symbolic representation of the two-state Markov decision process in this section.

it does so, the decision maker receives a reward of five units. Note that this representation allows the reward to depend on the subsequent state as well as the present state.

A formal description of this problem in terms of the notation of Chap. 2 follows.
Decision Epochs:

$$T = \{1, 2, \ldots, N\}, \quad N \le \infty.$$

States:

$$S = \{s_1, s_2\}.$$

Actions:

$$A_{s_1} = \{a_{1,1}, a_{1,2}\}, \quad A_{s_2} = \{a_{2,1}\}.$$

Rewards:

$$r_t(s_1, a_{1,1}) = 5 \qquad r_t(s_1, a_{1,2}) = 10 \qquad r_t(s_2, a_{2,1}) = -1$$
$$r_N(s_1) = 0 \qquad r_N(s_2) = 0 \qquad \text{if } N < \infty$$

Transition Probabilities:

$$p_t(s_1|s_1, a_{1,1}) = 0.5, \qquad p_t(s_2|s_1, a_{1,1}) = 0.5,$$
$$p_t(s_1|s_1, a_{1,2}) = 0, \qquad p_t(s_2|s_1, a_{1,2}) = 1,$$
$$p_t(s_1|s_2, a_{2,1}) = 0, \qquad p_t(s_2|s_2, a_{2,1}) = 1.$$

Suppose instead that the quantities below the arcs corresponding to action $a_{1,1}$ were $\{3, 0.5\}$ and $\{7, 0.5\}$. Then using the notation of (2.1.1)

$$r_t(s_1, a_{1,1}, s_1) = 3 \quad \text{and} \quad r_t(s_1, a_{1,1}, s_2) = 7.$$

In this case, the reward received when action $a_{1,1}$ is chosen in state s_1 depends on the state at the next decision epoch; if it is s_1, the reward is 3, and if it is s_2, the reward is

7. As before, the expected reward $r_t(s, a)$ is 5. This is because from (2.1.1)

$$r_t(s_1, a_{1,1}) = r_t(s_1, a_{1,1}, s_1)p_t(s_1|s_1, a_{1,1}) + r_t(s_1, a_{1,1}, s_2)p_t(s_2|s_1, a_{1,1})$$
$$= 3 \times 0.5 + 7 \times 0.5 = 5.$$

We now provide examples of the policies distinguished in Sec. 2.1.5. We assume that $N = 3$, that is, decisions are made at decision epochs 1 and 2. These policies may be represented by $\pi^K = (d_1^K, d_2^K)$ with $K = $ MD, MR, HD, or HR.

A deterministic Markov policy π^{MD}:

Decision epoch 1:

$$d_1^{MD}(s_1) = a_{1,1}, \qquad d_1^{MD}(s_2) = a_{2,1}.$$

Decision epoch 2:

$$d_2^{MD}(s_1) = a_{1,2}, \qquad d_2^{MD}(s_2) = a_{2,1}.$$

A randomized Markov policy π^{MR},

Decision epoch 1:

$$q_{d_1^{MR}(s_1)}(a_{1,1}) = 0.7, \qquad q_{d_1^{MR}(s_1)}(a_{1,2}) = 0.3$$
$$q_{d_1^{MR}(s_2)}(a_{2,1}) = 1.$$

Decision epoch 2:

$$q_{d_2^{MR}(s_1)}(a_{1,1}) = 0.4, \qquad q_{d_2^{MR}(s_1)}(a_{1,2}) = 0.6$$
$$q_{d_2^{MR}(s_2)}(a_{2,1}) = 1.$$

Because of the transition structure of the above model, the set of history-dependent policies equals the set of Markov policies. To see this, note that if the system is in state s_1 at any decision epoch, then the history must have been $(s_1, a_{1,1}, s_1, a_{1,1}, \ldots, s_1, a_{1,1})$ so that knowing the history would be redundant to the decision maker. In state s_2, the situation is different. Many possible histories could result in the process being in state s_2, but in s_2, there is only one available action $a_{2,1}$, so that the decision maker must choose it regardless of the history.

We modify the model to illustrate history-dependent policies (Figure 3.1.2). In s_1, we add an action $a_{1,3}$ which causes the system to remain in state s_1 with certainty, that is, $p_t(s_1|s_1, a_{1,3}) = 1$. We assume also that $r_t(s_1, a_{1,3}) = 0$. Note that $A_{s_1} = \{a_{1,1}, a_{1,2}, a_{1,3}\}$. We now provide an example of a history-dependent policy for this modified model.

A deterministic history-dependent-policy π^{HD}.

Decision epoch 1:

$$d_1^{HD}(s_1) = a_{1,1}, \qquad d_1^{HD}(s_2) = a_{2,1}.$$

Decision epoch 2:

Decision epoch 1 (s, a)	$d_2^{HD}(s, a, s_1)$	$d_2^{HD}(s, a, s_2)$
$(s_1, a_{1,1})$	$a_{1,3}$	$a_{2,1}$
$(s_1, a_{1,2})$	infeasible	$a_{2,1}$
$(s_1, a_{1,3})$	$a_{1,1}$	infeasible
$(s_2, a_{2,1})$	infeasible	$a_{2,1}$

A randomized history-dependent policy π^{HR}:

Decision epoch 1:

$$q_{d_1^{HR}(s_1)}(a_{1,1}) = 0.6, \qquad q_{d_1^{HR}(s_1)}(a_{1,2}) = 0.3, \qquad q_{d_1^{HR}(s_1)}(a_{1,3}) = 0.1,$$

$$q_{d_1^{HR}(s_2)}(a_{2,1}) = 1.$$

Decision epoch 2:
When the system is in state s_1 at decision epoch 2, probabilities are chosen according to the following probability distribution:

Decision epoch 1 (s, a)	$q_{d_2^{HR}(s,a,s_1)}(a)$		
	$a = a_{1,1}$	$a = a_{1,2}$	$a = a_{1,3}$
$(s_1, a_{1,1})$	0.4	0.3	0.3
$(s_1, a_{1,2})$	infeasible	infeasible	infeasible
$(s_1, a_{1,3})$	0.8	0.1	0.1
$(s_2, a_{2,1})$	infeasible	infeasible	infeasible

and, for every (s, a),

$$q_{d_2^{HR}(s, a, s_2)}(a_{2,1}) = 1.$$

Some comments regarding history-dependent decision rules follow. Since there is only one action in $s_2, a_{2,1}$, both deterministic and randomized policies choose it. In the history-dependent case, specifying actions for $h_2 = (s_1, a_{1,2}, s_1)$ or $h_2 = (s_2, a_{2,1}, s_1)$ is unnecessary since neither of these histories can occur. However, if action $a_{1,1}$ or $a_{1,3}$ were chosen at decision epoch 1, then several distinct histories are possible. For example, if we adopt π^{HR} for the model in Fig. 3.1.2, and the history at decision epoch 2 is $(s_1, a_{1,1}, s_1)$, then the randomized decision rule d_2^{HR} chooses action a_1, with probability 0.4, action $a_{1,2}$ with probability 0.3, and action $a_{1,3}$ with probability 0.3.

We return to the model in Fig. 3.1.1. Suppose the decision maker uses the deterministic Markov policy π^{MD}. Then the system evolves as a nonstationary two-period Markov chain with transition probability matrices

$$P_{d_1^{MD}} = \begin{bmatrix} 0.5 & 0.5 \\ 0 & 1 \end{bmatrix}, \qquad P_{d_2^{MD}} = \begin{bmatrix} 0 & 1 \\ 0 & 1 \end{bmatrix}.$$

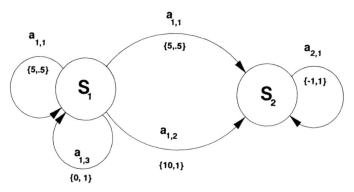

Figure 3.1.2 Symbolic representation of the modified model.

The corresponding rewards are

$$r_1\big(s_1, d_1^{\mathrm{MD}}(s_1)\big) = 5, \qquad r_1\big(s_2, d_1^{\mathrm{MD}}(s_2)\big) = -1,$$

and

$$r_2\big(s_1, d_2^{\mathrm{MD}}(s_1)\big) = 10, \qquad r_2\big(s_2, d_2^{\mathrm{MD}}(s_2)\big) = -1.$$

We leave it as an exercise (Problem 3.1) to derive the analogous expressions for the randomized Markov policy π^{MR}.

3.2 SINGLE-PRODUCT STOCHASTIC INVENTORY CONTROL

In Sec. 1.3, we described how Markov decision processes may be used to model inventory control systems. This section presents a simplified version of that model. We also include a numerical example which we refer to in later chapters. Inventory models were among the first problems solved using Markov decision problem methods, and their study has motivated many theoretical developments. These models are widely applicable; they can also be used to manage fisheries, forests, and cash balances. We expand on the model herein in Sec. 3.5.2.

3.2.1 Model Formulation

Each month, the manager of a warehouse determines current inventory (stock on hand) of a single product. Based on this information, he decides whether or not to order additional stock from a supplier. In doing so, he is faced with a tradeoff between the costs associated with keeping inventory and the lost sales or penalties associated with being unable to satisfy customer demand for the product. The manager's objective is to maximize some measure of profit (sales revenue less inventory holding and ordering costs) over the decision-making horizon. Demand for the product is random with a known probability distribution.

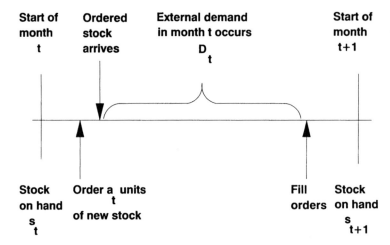

Figure 3.2.1 Timing of events in the inventory model of Sec. 3.2.1.

We formulate a model under the following set of simplifying assumptions. Generalizations which make the model more realistic are explored through problems at the end of the chapter.

1. The decision to order additional stock is made at the beginning of each month and delivery occurs instantaneously.
2. Demand for the product arrives throughout the month but all orders are filled on the last day of the month.
3. If demand exceeds inventory, the customer goes elsewhere to purchase the product; that is, there is no backlogging of unfilled orders so that excess demand is lost.
4. The revenues, costs, and the demand distribution do not vary from month to month.
5. The product is sold only in whole units.
6. The warehouse has capacity of M units.

Figure 3.2.1 illustrates the timing of arrival of orders and demands implied by the above assumptions.

Let s_t denote the inventory on hand at the beginning of month t, a_t the number of units ordered by the inventory manager in month t and D_t the random demand in month t. We assume that the demand has a known time-homogeneous probability distribution $p_j = P\{D_t = j\}$, $j = 0, 1, 2, \ldots$. The inventory at decision epoch $t + 1$, s_{t+1}, is related to the inventory at decision epoch t, s_t, through the system equation

$$s_{t+1} = \max\{s_t + a_t - D_t, 0\} \equiv [s_t + a_t - D_t]^+. \qquad (3.2.1)$$

Because backlogging is not permitted, the inventory level cannot be negative. Thus whenever $s_t + a_t - D_t < 0$, the inventory level at the subsequent decision epoch is 0.

We now describe the economic parameters of this model. We express them as values at the start of the month so that we are implicitly considering the time value of money when defining these quantities. We refer to them as *present values* to emphasize this point. The present value of the cost of ordering u units in any month is $O(u)$. We assume it is composed of a fixed cost $K > 0$ for placing orders and a variable cost $c(u)$ that increases with quantity ordered. Hence

$$O(u) = \begin{cases} K + c(u) & \text{if } u > 0 \\ 0 & \text{if } u = 0 \end{cases}. \tag{3.2.2}$$

The present value of the cost of maintaining an inventory of u units for a month (between receipt of the order and releasing inventory to meet demand) is represented by the nondecreasing function $h(u)$. In finite-horizon problems, the remaining inventory at the last decision epoch has value $g(u)$. Finally, if the demand is j units and sufficient inventory is available to meet demand, the manager receives revenue with present value $f(j)$. Assume $f(0) = 0$.

In this model the reward depends on the state of the system at the subsequent decision epoch; that is,

$$r_t(s_t, a_t, s_{t+1}) = -O(a_t) - h(s_t + a_t) + f(s_t + a_t - s_{t+1}).$$

For subsequent calculations, it is more convenient to work with $r_t(s_t, a_t)$, which we evaluate using (2.1.1). To this end we compute $F_t(u)$, the expected present value (at the start of month t) of the revenue received in month t when the inventory prior to receipt of customer orders is u units. It is derived as follows. If inventory u exceeds demand j, the present value of the revenue is $f(j)$. This occurs with probability p_j. If demand exceeds inventory, the present value of the revenue equals $f(u)$. This occurs with probability $q_u = \sum_{j=u}^{\infty} P_j$. Thus

$$F(u) = \sum_{j=0}^{u-1} f(j)p_j + f(u)q_u.$$

A Markov decision process formulation follows.

Decision epochs:

$$T = \{1, 2, \ldots, N\}, \qquad N \le \infty.$$

States (the amount of inventory on hand at the start of a month):

$$S = \{0, 1, 2, \ldots, M\}.$$

Actions (the amount of additional stock to order in month t):

$$A_s = \{0, 1, 2, \ldots, M - s\}.$$

Expected rewards (expected revenue less ordering and holding costs):

$$r_t(s, a) = F(s + a) - O(a) - h(s + a), \qquad t = 1, 2, \ldots, N - 1 \tag{3.2.3}$$

(the value of terminal inventory)

$$r_N(s) = g(s), \quad t = N.$$

Transition probabilities:

$$p_t(j|s, a) = \begin{cases} 0 & \text{if } M \geq j > s + a \\ p_{s+a-j} & \text{if } M \geq s + a \geq j > 0 \\ q_{s+a} & \text{if } M \geq s + a \text{ and } j = 0, \end{cases}$$

where q_{s+a} is defined above.

A brief explanation of the derivation of the transition probabilities follows. If the inventory on hand at the beginning of period t is s units and an order is placed for a units, the inventory prior to external demand is $s + a$ units (Assumption 1). An inventory level of $j > 0$ at the start of period $t + 1$ requires a demand of $s + a - j$ units in period t. This occurs with probability p_{s+a-j}. Because backlogging is not permitted (Assumption 3), if the demand in period t exceeds $s + a$ units, then the inventory at the start of period $t + 1$ is 0 units. This occurs with probability q_{s+a}. The probability that the inventory level exceeds $s + a$ units is 0, since demand is non-negative. Assumption 6 constrains the inventory always to be less than or equal to M.

Assumption 2 implies that the inventory throughout a month is $s + a$, so that the total monthly holding cost is $h(s + a)$. If instead, the demand D_t arrives at the beginning of a month, the expected holding cost $E\{h([s + a - D_t]^+)\}$ replaces the holding cost in (3.2.3).

Deterministic Markov decision rules assign the quantity of inventory to be ordered each month to each possible starting inventory position. A policy is a sequence of such ordering rules. An example of such a decision rule is:

Order sufficient stock to raise the inventory to Σ units whenever the inventory level at the beginning of a month is less than σ units. When the inventory level at the beginning of a month is σ units or greater, do not place an order.

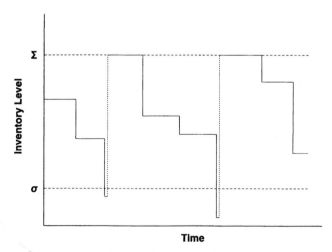

Figure 3.2.2 Realization of an inventory model under a (σ, Σ) policy.

This decision rule may be represented by

$$d_t(s) = \begin{cases} \Sigma - s & s < \sigma \\ 0 & s \geq \sigma. \end{cases}$$

We refer to a policy which uses decision rules of the above form as a (σ, Σ) policy. Figure 3.2.2 illustrates a realization of an inventory process under this policy. Such policies have practical appeal and are optimal under various assumptions on the model components. In practice, Σ is referred to as the *target stock* and $\Sigma - \sigma$ as the *minimum fill*.

3.2.2 A Numerical Example

We provide values for the parameters of the above model. We choose $K = 4$, $c(u) = 2u$, $g(u) = 0$, $h(u) = u$, $M = 3$, $N = 3$, $f(u) = 8u$, and

$$p_j = \begin{cases} \frac{1}{4} & \text{if } j = 0 \\ \frac{1}{2} & \text{if } j = 1 \\ \frac{1}{4} & \text{if } j = 2. \end{cases}$$

This model can be interpreted as follows. The inventory is constrained to be 3 or fewer units. All costs and revenues are linear. This means that for each unit ordered the per unit cost is 2, for each unit held in inventory for one month, the per unit cost is 1 and for each unit sold the per unit revenue is 8. The expected revenue when u units of stock are on hand prior to receipt of an order is given by

u	$F(u)$
0	0
1	$0 \times \frac{1}{4} + 8 \times \frac{3}{4} = 6$
2	$0 \times \frac{1}{4} + 8 \times \frac{1}{2} + 16 \times \frac{1}{4} = 8$
3	$0 \times \frac{1}{4} + 8 \times \frac{1}{2} + 16 \times \frac{1}{4} = 8$

Combining the expected revenue with the ordering and holding costs gives the expected profit in period t if the inventory level is s at the start of the period and an order for a units is placed. If $a = 0$, the ordering and holding cost equals s, and if a is positive, it equals $4 + s + 3a$. It is summarized in the table below where $\times$ denotes an infeasible action. Transition probabilities only depend on the total inventory on hand prior to receipt of orders. They are the same for any s and a which have the same value for $s + a$ and are functions of $s + a$ only. The information in the following tables defines this problem completely.

	$r_t(s, a)$					$p_t(j\|s, a)$			
a	0	1	2	3	j	0	1	2	3
s					$s + a$				
0	0	-1	-2	-5	0	1	0	0	0
1	5	0	-3	$\times$	1	$\frac{3}{4}$	$\frac{1}{4}$	0	0
2	6	-1	$\times$	$\times$	2	$\frac{1}{4}$	$\frac{1}{2}$	$\frac{1}{4}$	0
3	5	$\times$	$\times$	$\times$	3	0	$\frac{1}{4}$	$\frac{1}{2}$	$\frac{1}{4}$

3.3 DETERMINISTIC DYNAMIC PROGRAMS

Deterministic dynamic programs (DDPs) constitute an important and widely studied class of Markov decision processes. Applications include finding the shortest route in a network, critical path analysis, sequential allocation, and inventory control with known demands.

3.3.1 Problem Formulation

In a deterministic dynamic program, choice of an action determines the subsequent state with *certainty*. Standard formulations take this into account by using a transfer function instead of a transition probability to specify the next state. By a *transfer function*, we mean a mapping, $\tau_t(s, a)$, from $S \times A_s$ to S, which specifies the system state at time $t + 1$ when the decision maker chooses action $a \in A_s$ in state s at time t. To formulate a DDP as a Markov decision process, define the transition probability function as

$$p_t(j|s, a) = \begin{cases} 1 & \text{if } \tau_t(s, a) = j \\ 0 & \text{if } \tau_t(s, a) \neq j. \end{cases} \tag{3.3.1}$$

The rewards are given by $r_t(s, a)$.

3.3.2 Shortest Route and Critical Path Models

When the total reward is used to compare policies, every DDP with finite S, A, and T is equivalent to a shortest or longest route problem. We introduce such a problem and discuss this equivalence.

A *finite directed graph* consists of a set of nodes and directed arcs. By a *path* we mean a sequence of arcs that connects one node to another node. We call such a graph *acyclic* whenever there are no paths which begin and end at the same node. An example of an acyclic directed graph appears in Fig. 3.3.1. In it, nodes are labeled by integers $\{1, \ldots, 8\}$ and directed arcs are represented by lines with arrows. Node 1 is called the *origin* and node 8 the *destination*. The value above the arc give the "distance," "cost," or "time" associated with traversing the arc. A "shortest route problem" corresponds to finding a path or sequence of arcs from the origin to the

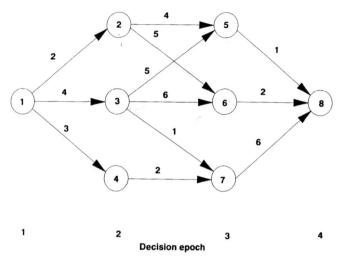

Figure 3.3.1 Example of a routing problem.

destination (node 1 to node 8) of minimal total value, while a "longest route problem" seeks a path of greatest total value.

Applications of shortest route problems include optimization of shipping and communication networks. Longest route problems are the basis for *critical path analysis*. In critical path analysis, each arc represents a task and its length the time to complete the task. The task cannot begin until completion of all tasks which immediately precede it (those with arcs ending at the node from which the designated arc originates). The length of the longest path from the first node to the last node gives the minimum time to complete all tasks and hence the entire project. It is the "critical path" because if any task on it is delayed, the time to complete the project increases. In addition to the critical path, earliest and latest start dates for each task may be easily obtained from a solution to this problem.

Formulation of a shortest route problem as a deterministic dynamic program involves identifying nodes with states, arcs with actions, transfer functions with transition probabilitities and values with rewards. However direct identification of decision epochs or stages is not always possible and not necessary for solution. In the example in Fig. 3.3.1, node 1 can be identified with decision epoch 1, nodes 2, 3, and 4 with decision epoch 2, nodes 5, 6, and 7 with decision epoch 3, and node 8 with decision epoch 4. However, if the example was modified to contain an arc connecting nodes 3 and 4 (Fig. 3.3.2), an obvious identification of decision epochs would not be available. The lower portion of Fig. 3.3.2 shows that adding nodes to the graph and modifying arc lengths allows identification of decision epochs.

Figure 3.3.3 provides a graphical representation for an arbitrary DDP. At each decision epoch, there is a node corresponding to each state. Arcs originate at each state, one for each action, and end at the next stage at the node determined by the corresponding transfer function. Arc lengths give rewards. We add a "dummy" origin "0" and destination "D." Arc lengths from nodes at stage N to the destination give terminal rewards $r_N(s)$. If a solution is sought for a particular initial state s' and the problem is one of maximization, then we give the arc from the origin to s' a large

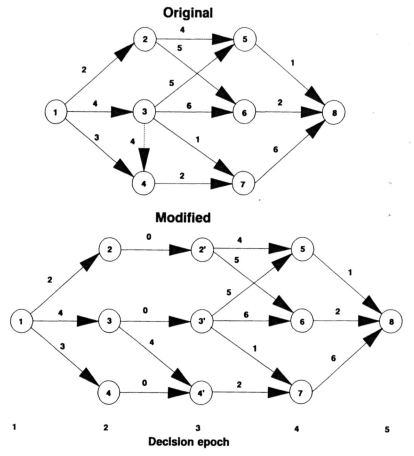

Figure 3.3.2 Identification of decision epochs in a modification of the routine problem of Figure 3.3.1.

positive length, say L, and set lengths of all other arcs starting from the origin equal to the zero as in Figure 3.3.3.

3.3.3 Sequential Allocation Models

A decision maker has M units of resource to utilize or consume over N periods. Denote by x_t the quantity of resource used or consumed in period t and by $f(x_1, \ldots, x_N)$ the utility of this allocation pattern to the decision maker. Regard negative $f(x_1, \ldots, x_N)$ as disutility or cost. The problem of choosing an optimal consumption pattern is given by the following mathematical program.

$$\text{maximize} \quad f(x_1, \ldots, x_N),$$

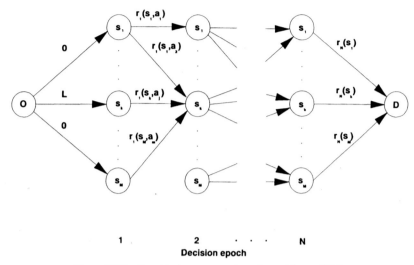

Figure 3.3.3   Graphical representation of an arbitrary DDP.

subject to

$$x_1 + \cdots + x_N = M$$

and

$$x_t \geq 0, \quad t = 1, \ldots, N.$$

Direct formulation as a DDP or a Markov decision process is easiest for *separable* objective functions $f(x_1, \ldots, x_N)$, that is those which satisfy

$$f(x_1, \ldots, x_N) = \sum_{t=1}^{N} g_t(x_t).$$

To define this as a Markov decision process model, let the state at time t represent the total quantity of resources available for consumption in periods $t, t+1, \ldots, N$ and the action represent the quantity chosen to be consumed in period t. If the state at time t is s, and a units are consumed in period t, then $s-a$ units are available for consumptions in periods $t+1, t+2, \ldots, N$. No decisions are made in period N because specifying the allocations in periods $1, \ldots, N-1$ fixes the allocation in period N. Note that, in this model, the sets of states and actions are not discrete; they are compact intervals of the real line.

The Markov decision process formulation for this problem follows.
Decision epochs:

$$T = \{1, \ldots, N\}, \qquad N < \infty.$$

States:

$$S = [0, M], \quad M < \infty.$$

Actions:

$$A_s = [0, s], \quad 0 \le s \le M.$$

Rewards:

$$r_t(s, a) = g_t(a), \quad s \in S, a \in A_s, t = 1, \dots, N - 1,$$

and

$$r_N(s) = g_N(s), \quad s \in S.$$

Transition probabilities:

$$p_t(j|s, a) = \begin{cases} 1 & j = s - a \\ 0 & \text{otherwise,} \end{cases} \quad s \in S, a \in A_s, t = 1, \dots, N - 1.$$

3.3.4 Constrained Maximum Likelihood Estimation

We now formulate the problem of finding maximum likelihood estimators of the parameters of a multinomial distribution as a sequential allocation model. The multinomial probability function may be represented as

$$p(y_1, y_2, \dots, y_N) = \frac{K!}{y_1! y_2! \cdots y_N!} q_1^{y_1} q_2^{y_2} \cdots q_N^{y_N},$$

where y_i is the number of observations of type i, q_i is the probability of an observation of type i, K is total number of observations, and N is the number of observation types. Note that $\sum_{i=1}^{N} y_i = K$. Since q_i is a probability, it is non-negative and satisfies

$$\sum_{i=1}^{N} q_i = 1. \tag{3.3.2}$$

Maximum likelihood estimators are found by regarding the observations $(y_1, \dots, y_N)$ as fixed and choosing values of the parameters $(q_1, \dots, q_N)$ to maximize the multinomial log likelihood

$$L(q_1, \dots, q_N) = \log(K!) - \sum_{i=1}^{N} \log(y_i!) + \sum_{i=1}^{N} y_i \log(q_i). \tag{3.3.3}$$

Since the first two terms in (3.3.3) do not effect the choice of the maximizing values of q_i, the problem of finding a maximum likelihood estimator becomes the following.

Maximize

$$\sum_{i=1}^{N} y_i \log(q_i) \tag{3.3.4}$$

subject to (3.3.2) and $q_i \geq 0$ for $i = 1, \ldots, N$.

This is a sequential allocation problem with $S = [0, 1]$ and $g_t(x_t) = y_t \log(x_t)$.

3.4 OPTIMAL STOPPING

This section formulates and provides examples of optimal stopping problems. These represent a colorful collection of problems that often have elegant structured solutions.

3.4.1 Problem Formulation

In an optimal stopping problem, the system evolves as an uncontrolled, possibly nonstationary, Markov chain with state space S'. At each decision epoch, the decision maker has two actions available in each state: to stop or to continue. If he decides to stop in state s at time t, he receives reward $g_t(s)$, and if he decides to continue, he incurs a cost $f_t(s)$ and the system evolves until the next decision epoch according to the transition probability matrix of the Markov chain. The problem may be finite horizon, in which case a function $h(s)$ represents the reward if the system reaches decision epoch N without stopping. The objective is to choose a policy to maximize the expected total reward. When stopped, it remains in the stopped state, yielding zero reward.

Applications of these models include selling an asset, choosing a candidate for a job (the secretary problem), exercising and valuing a financial option, sequential hypothesis testing, and gambling. A MDP formulation of the optimal stopping problem follows.

Decision epochs:

$$T = \{1, 2, \ldots, N\}, \qquad N \leq \infty.$$

States:

$$S = S' \cup \{\Delta\}.$$

Actions:

$$A_s = \begin{cases} \{C, Q\} & s \in S' \\ \{C\} & s = \Delta. \end{cases}$$

Rewards:

$$r_t(s, a) = \begin{cases} -f_t(s) & s \in S', \quad a = C \\ g_t(s) & s \in S', \quad a = Q \quad t < N, \\ 0 & s = \Delta, \end{cases}$$

$$r_N(s) = h(s).$$

Transition probabilities:

$$p_t(j|s, a) = \begin{cases} p_t(j|s) & s \in S', \quad j \in S', \quad a = C \\ 1 & s \in S', \quad j = \Delta, \quad a = Q \quad \text{or } s = j = \Delta, a = C \quad t < N \\ 0 & \text{otherwise.} \end{cases}$$

The state space consists of S', the state space for an uncontrolled Markov chain with transition probabilities $p_t(j|s)$, in addition to a distinguished absorbing state Δ. In S', two actions are available in each state: continue, C, or quit, Q. Continuation causes the process to move to state j in S' with probability $p_t(j|s)$. Termination moves the system to state Δ, where it remains forever and receives reward zero. Usually, for some states $g_t(s)$ is large, so that the decision maker's objective is to terminate the process in these states.

3.4.2 Selling an Asset

An "investor" owns an expensive property or asset which he expects to appreciate in value over time. At the end of each week, the investor decides whether to accept the best offer he received during the preceding week and sell the property, or else decline this offer and solicit new offers in the next week. We assume he must sell the property within one year, although we also consider infinite-horizon variants of this problem in Chap. 7.

We formulate this as a finite-horizon optimal stopping problem as follows. Decision epochs correspond to the times at which decisions are made. Number them 1 to 53 corresponding to the weeks of the year. The state is the best offer available at the end of a week (supposedly the market value of the property). The state space S' represents the set of all possible offers over the decision-making horizon. It can be either discrete or a continuum but it is usually bounded. The best available offer at decision epoch $t + 1$ is related to the best available offer at decision epoch t through the nonstationary transition probability matrix or function $p_t(j|s)$. Nonstationarity reflects changing market conditions. If the investor accepts an offer s, his reward is $s - K(s)$, where $K(s)$ includes fixed and variable costs associated with selling a property of value s. The continuation cost $f_t(s)$ includes advertising expenses, interest charges, and taxes associated with carrying the property for one week. If he holds the property for the entire year without sale, he must sell it at the end of the year and receive a terminal reward of $s - K(s)$.

Good policies for this problem use decision rules of the form

$$d_t(s) = \begin{cases} Q & s \geq B_t \\ C & s < B_t, \end{cases} \quad s \in S'.$$

We refer to this as a *control limit* policy; the number B_t is the control limit. Solving the problem reduces to determining the boundary B_t between the continuation and stopping regions. Of course, the solution depends on the forecasts of $p_t(j|s)$ for $t = 1, \ldots, 52$.

3.4.3   The Secretary Problem

This classic dynamic programming problem was first proposed by Cayley (1875) in the context of finding an optimal policy for playing a lottery. We provide a more colorful description. An employer seeks to hire an individual to fill a vacancy for a secretarial position. There are N candidates or applicants for this job, with N fixed and known by the employer. Candidates are interviewed sequentially. Upon completion of each interview, the employer decides whether or not to offer the job to the *current* candidate. If he does not offer the job to this candidate, that individual seeks employment elsewhere and is no longer eligible to receive an offer.

Formulation of this problem depends on the decision maker's objective. Here we assume that the employer wishes to maximize the probability of giving an offer to the best candidate.

A more rigorous statement of the problem follows. A collection of N objects is ranked from 1 to N, with that ranked 1 being the most desirable. The true rankings are unknown to the decision maker. He observes the objects one at a time in random order. He can either select the current object and terminate the search, or discard it and choose the next object. His objective is to maximize the probability of choosing the object ranked number 1. We assume that the decision maker's relative rankings are consistent with the absolute rankings. That is, if object A has a lower numerical ranking than object B, the decision maker will prefer object A to object B.

Decisions are made after observing each object, so that the horizon N equals the number of candidates. The state space $S' = \{0,1\}$; 1 denotes that the current object is the best object (rank closest to 1) seen so far, and 0 that a previous object was better.

In either state, the action Q means select the current object (give an offer to the current candidate) and C means do not select the current object and continue the search (interview the next candidate). Rewards are received only when stopping; that is, choosing action Q. They correspond to the probability of choosing the best candidate. In the notation above, the continuation cost $f_t(s) = 0$, $s = 0, 1$; the rewards at stopping, $g_t(0) = 0$ and $g_t(1) = t/N$; and the terminal rewards, $h(0) = 0$ and $h(1) = 1$. A derivation of g_t and h follows.

Suppose after observing t objects that the decision maker ranks the current object as the best of those seen so far. Then $g_t(1)$, the probability that object is the best among all objects, is determined as follows.

$$P \{\text{Best object is in first } t\}$$

$$= \frac{\text{Number of subsets of } \{1, 2, \ldots, N\} \text{ of size } t \text{ containing } 1}{\text{Number of subsets of } \{1, 2, \ldots, N\} \text{ of size } t}$$

$$= \frac{\dbinom{N-1}{t-1}}{\dbinom{N}{t}} = \frac{t}{N}$$

where $\dbinom{n}{r}$ denotes a binomial coefficient. If all objects have been seen, the last object must be chosen. If that object is best, that is $s = 1$, then the probability of choosing the best object is 1 so that $h(1) = 1$; otherwise, $h(0) = 0$.

Transition probabilities for the uncontrolled system are independent of the system state, that is, the status of the current object. Therefore $p_t(j|s) = p_t(j)$ for $s = 0, 1$. The probability that the subsequent candidate is the best among the first $t + 1$, $p_t(1|s) = 1/(t + 1)$, and the probability that the subsequent candidate is not the best among the first $t + 1$, $p_t(0|s) = t/(t + 1)$ for $s = 0, 1$.

Variants of this model include random numbers of candidates, maximization of the probability of choosing one of the k best candidates, minimization of the expected rank of the candidate who receives the offer, or a candidate selection rule in which, after interviewing the current candidate, the employer may make an offer to the current candidate and some designated subset of previous candidates whose availability depends on a specified probability distribution.

3.4.4 Call Options

In financial markets, a *call option* gives the owner the right to purchase an asset such as shares of common stock, at a fixed price W (the strike price) at any time prior to a specified expiry date. For example we might purchase a call option to buy 100 shares of Sun Microsystems at $40 per share at or before June 30. Suppose, at some day prior to June 30, the stock price reached $45 per share. Then we might exercise the option and buy 100 shares of Sun Microsystems stock for $4000 and immediately sell them for $4500 to make a profit of $500 less transaction costs (commissions). If the stock price lies below $40 per share, we would not exercise the option.

Options are traded in financial markets and values vary on a day-to-day basis in relationship to the price of the underlying stock. A fundamental problem in investment theory is how to determine the value of an option. Clearly its value should depend on using it as efficiently as possible. A sequential decision process formulation provides insight into how to do this.

We formulate a simplified form of the option holder's decision problem as an optimal stopping problem by ignoring the impact of dividends and the real life possibilities of selling the options on any day. At the start of each trading day prior to the expiry date, the option holder decides whether or not to exercise the option. Decision epochs correspond to the times when the option holder decides whether or not to exercise the option, and the state represents the market price of the underlying asset (stock). The choice of actions is to exercise the option (stop) or not exercise the option (continue). After the expiry date, the option has no value. By exercising an option for 100 shares with strike price W when the security has value s, the option holder receives $100(s - W) - K$, where K represents the transaction costs. In the above notation, $h(s) = g_t(s) = 100(s - W) - K$ for all t. Since not exercising the option has no financial implications, $f_t(s) = 0$. The transition probability function $p_t(j|s)$ gives the probability that the next day's stock price equals j given that the stock price on day t equals s. Clearly it is not affected by the action of the option holder.

There are many intricate uses of call options ranging from the conservative to the highly speculative. We refer the interested reader to references noted in the Bibliographic Remarks section.

3.5 CONTROLLED DISCRETE-TIME DYNAMIC SYSTEMS

In many settings, especially when the states represent real-valued quantities such as wealth or inventory levels, it is natural to describe the evolution of the system in terms of sample paths and system equations instead of in terms of transition probabilities. We refer to this as the *control theory* approach. In this section we introduce the control theory model, show how it may be reformulated as a Markov decision process, and provide illustrations from operations research, economics, and engineering.

3.5.1 Model Formulation

Let the state space S, control set U, and disturbance set W be subsets of R^k, R^m, and R^n, respectively. We allow, but do not require, that $k = m = n$, or even that they all equal 1. The state of the system at time $t + 1$, s_{t+1}, is related to the state of the system at time t, s_t, the control used at time t, u_t, and the disturbance at time t, w_t, through the system equation or "law of motion"

$$s_{t+1} = f_t(s_t, u_t, w_t) \qquad (3.5.1)$$

where f_t is a R^k-valued function mapping $S \times U \times W$ into S. (Note an inconsistency in notation in this section; in contrast to Sec. 3.1 where we distinguished states as s_1 and s_2, here s_t denotes the state at decision epoch t.) The unobservable disturbance w_t is the realization of a random variable W_t that assumes values in W, is independent of w_τ for all $\tau < t$, and has a distribution that does not depend on s_t or u_t. We assume that the distribution of W_t is expressed in terms of a density or probability function $q_t(\cdot)$.

When the system is in state s_t and the decision maker, who in this context is frequently referred to as "the controller," chooses control u_t from the set of admissible controls in state s_t, $U_{s_t} \subset U$, the system receives a reward of $g_t(s_t, u_t)$ [or incurs a cost $-g_t(s_t, u_t)$]. The externally generated disturbance effects this evolution by interacting with the state, s_t and control u_t, to determine the state s_{t+1} through (3.5.1). When the horizon is finite and the system terminates in state s_N at time N, it receives a terminal reward $g(s_N)$. The time evolution of this system is depicted in Fig. 3.5.1.

An equivalent Markov decision process formulation follows.

Decision epochs:

$$T = \{1, 2, \ldots, N\}, \qquad N \leq \infty.$$

States:

$$S \subset R^k.$$

Actions:

$$A_s = U_s.$$

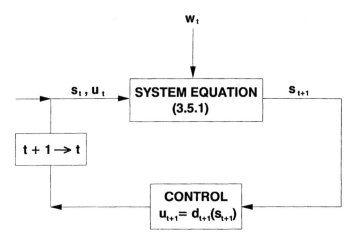

Figure 3.5.1 Time evolution of the discrete-time control theory model.

Rewards:

$$r_t(s, a) = g_t(s, a), \quad t < N, \quad r_N(s) = g_N(s), \quad t = N \text{ when } N < \infty.$$

Transition probabilities:

$$p_t(j|s, a) = P\big[W_t \in \{w \in W : j = f_t(s, a, w)\}\big] \qquad (3.5.2)$$

$$= \sum_{\{w \in W : j = f_t(s, a, w)\}} q_t(w). \qquad (3.5.3)$$

We see from the above that the only substantial difference in these two approaches to modeling stochastic sequential control problems is the way in which transition probabilities are incorporated. In the Markov decision process formulation, the transition probabilities are fundamental, while in the stochastic control formulation they must be derived. Equations (3.5.2) and (3.5.3) appear quite formidable, but in applications their evaluation is usually straightforward.

When S and W are continuous, $p_t(j|s, a)$ in (3.5.2) and $q_t(w)$ in (3.5.3) are densities, and integration replaces summation. In either case, the state space is usually discretized prior to numerical solution.

In control theory language, decision rules are referred to as *feedback* controls because they use the current state information to feedback into the system equation f_t and influence future states of the system as in Fig. 3.5.1. An *open loop* control is a decision rule d_t in which $d_t(s) = a$ for all $s \in S$. Since it uses the same action regardless of the state, it does not require observation of the system for implementation. In practice, open loop controls are easier to implement and evaluate but are usually suboptimal.

The system equation (3.5.1) may be generalized to include controls, states, and disturbances from some or all previous periods. In such generality it may be represented by

$$s_{t+1} = f_t(s_t, s_{t-1}, \ldots, s_1, u_t, u_{t-1}, \ldots, u_1, w_t, w_{t-1}, \ldots, w_1). \qquad (3.5.4)$$

This increases the flexibility of the model by allowing correlated disturbances and delays in the effect of control actions. Equation (3.5.4) can usually be reduced to (3.5.1) by reparameterizing the model in terms of vectors which account for previous systems states, controls, or disturbances.

3.5.2 The Inventory Control Model Revisited

We reformulate the single-product inventory model of Sec. 3.2 as a controlled discrete-time dynamic system and discuss some generalizations.

The state space $S = \{0, 1, \ldots, M\}$, the control set $U = \{0, 1, \ldots, M\}$, and the disturbance set $W = \{0, 1, \ldots\}$ are all subsets of R^1. In state s, the set of admissible controls is $U_s = \{0, 1, \ldots, M - s\} \subset U$. The system equation (3.5.1) is given by (3.2.1); that is,

$$s_{t+1} = [s_t + u_t - w_t]^+$$

where u_t is identified with a_t and w_t with D_t in (3.2.1). The rewards are as derived in Sec. 3.2.

We now illustrate the calculation of transition probabilities based on (3.5.2). Since

$$\{w \in W : j = [s + a - w]^+\} = \begin{cases} \varnothing & j > s + a \\ s + a - j & 0 < j \le s + a \\ \{s + a, s + a + 1, \ldots\} & j = 0 \end{cases}$$

where $\varnothing$ represents the empty set, the transition probabilities satisfy

$$p_t(j|s, a) = \begin{cases} 0 & j > s + a \\ p_{s+a-j} & 0 < j \le s + a \\ q_{s+a} & j = 0 \end{cases}$$

for $a \in \{0, 1, \ldots, M - s\}$ where $q_u = \sum_{j=u}^{\infty} p_j$.

We now describe some modifications of this model.

Backlogging

Suppose that unfilled orders may be backlogged, that is, orders which cannot be filled from current inventory are filled when sufficient stock arrives. We modify the above model as follows.

a. The state space $S = \{\ldots, -1, 0, 1, \ldots, M\}$, where negative states correspond to backlogged orders.

b. The holding cost $h(x)$ is also defined for negative values of x. When $x < 0$, it may be regarded as a penalty cost associated with unsatisfied demand. A typical form for $h(x)$ is

$$h(x) = \begin{cases} -\phi x & x < 0 \\ \eta x & x \ge 0 \end{cases}$$

where ϕ and η are positive constants.

c. The system equation is

$$s_{t+1} = s_t + u_t - w_t$$

Lag in Delivery of Orders (Lead Times)

Assume an L-month lag ($L > 0$) between the time of placing an order to replenish inventory and its delivery to the supply facility. That is, an order placed at the beginning of month t is not available to meet demand until the start of month $L + t$. We assume that demand that cannot be satisfied from stock on hand is never filled.

From system equation (3.5.4), we obtain

$$s_{t+1} = \left[s_t + u_{t-L} - w_t \right]^+,$$

where u_{t-L} is the quantity ordered in month $t - L$ for delivery to meet demand in month t.

Since the stock level at time t may exceed M, it is more natural to formluate this model with $S = \{0,1,\dots\}$ and $U_s = \{0,1,\dots\}$.

Correlated Demands

In practice, demands in successive periods may be correlated. The simplest representative scheme is one in which we assume a correlation between demand in month $t + 1$ and month t of ρ_t, $-1 \le \rho_t \le 1$. When ρ_t is negative, high demands are usually followed by low demands, and when ρ_t is positive, high demands usually follow high demands. We refer to such a correlation scheme as *first-order autocorrelation*.

This situation may be modeled directly using (3.5.4); however, we formulate it in terms of (3.5.1) by using a two-dimensional state vector in which the first component represents the inventory level and the second component the demand level. Let s_t^i represent the inventory level at the start of period t, and s_t^d represent the demand in period t. Assuming no backlogging and no restriction on warehouse capacity, $S = \{0, 1, \dots\} \times \{0, 1, \dots\} \subset R^2$. In this model, w_t denotes an externally generated disturbance that is uncorrelated with w_{t-1} and s_t^d. This model may be represented by the system of dynamic equations

$$s_{t+1}^i = \left[s_t^i + u_t - s_t^d \right]^+, \qquad (3.5.5)$$

$$s_{t+1}^d = \rho_t s_t^d + w_t. \qquad (3.5.6)$$

Equation (3.5.5) is the familiar inventory balance equation while equation (3.5.6) provides a representation for a first-order autocorrelated system commonly used in statistical modeling.

When $\rho_t = \rho$ for all t, we show that (3.5.6) yields a sequence of demands with approximate correlation ρ. For a pair of random variables X and Y, denote the covariance of X and Y by $\mathrm{cov}(X,Y)$ and the variance of X by $\mathrm{var}(X)$. Recall that the correlation between X and Y, denoted $\mathrm{corr}(X,Y) \equiv \mathrm{cov}(X,Y)/$ $[\mathrm{var}(X)\mathrm{var}(Y)]^{1/2}$. Using (3.5.6),

$$
\begin{aligned}
\mathrm{cov}\left(s_{t+1}^d, s_t^d\right) &= E\left\{\left(s_{t+1}^d - E\{s_{t+1}^d\}\right)\left(s_t^d - E\{s_t^d\}\right)\right\} \\
&= E\left\{\left(\rho s_t^d + w_t - E\{\rho s_t^d + w_t\}\right)\left(s_t^d - E\{s_t^d\}\right)\right\} \\
&= \rho\,\mathrm{var}\left(s_t^d\right) + \mathrm{cov}\left(s_t^d, w_t\right) = \rho\,\mathrm{var}\left(s_t^d\right).
\end{aligned}
$$

Since for large t, $\mathrm{var}(s_t^d) \approx \mathrm{var}(s_{t+1}^d)$, the conclusion follows.

When applying such a model, we often assume that ρ is unknown. Since ρ cannot be observed, the controller must estimate ρ prior to deciding on the order level. The model may also be generalized with ρ_t described by a dynamic equation.

Multiproduct Inventory, No Backlogging, and Unlimited Warehouse Capacity

Assume that the inventory manager controls the level of k products. They are managed together because of correlated demands and a fixed cost component associated with placing an order for *any* product. Let $S = S^1 \times S^2 \times \cdots \times S^k \subset R^k$, where $S^i = \{0, 1, \ldots\}$ is the set of possible stock levels of the ith product and $U = U^1 \times U^2 \times \cdots \times U^k \subset R^k$, where $U^i = \{0, 1, \ldots\}$ is the set of possible order quantities of product i, $i = 1, 2, \ldots, k$.

In this model, the dynamic equation holds for each product individually; that is

$$
s_{t+1}^i = \left[s_t^i + u_t^i - w_t^i\right]^+, \quad i = 1, 2, \ldots, k,
$$

where w_t^i is the realized demand for product i in month t. Transition probabilities are derived from the joint demand distribution $P\{W_t^1 = w_t^1, \ldots, W_t^k = w_t^k\}$. Under the assumed timing of placement and arrival of orders in Sec. 3.2, the holding cost can be represented by $h(s_t^1 + u_t^1, \ldots, s_t^k + u_t^k)$, and the ordering cost by

$$
0(u^1, \ldots, u^k) = \begin{cases} 0, & u^i = 0, \quad i = 1, \ldots, k \\ K + c(u^1, \ldots u^k), & u^i > 0 \quad \text{for some } i, \end{cases}
$$

where c is non-decreasing in each component of U.

3.5.3 Economic Growth Models

We formulate a dynamic model for optimal consumption and investment in a single-good planned economy. The good, which we refer to as capital, may be either invested or consumed. When invested, it produces capital according to a production function F_t, and when consumed it yields "utility" Ψ_t. The economic planner chooses a sequence of investments to maximize total utility of consumption. When most consumption occurs early in the planning period, little capital is available to finance subsequent growth. Conversely, little early consumption permits a big "splurge" at the end of the planning horizon.

More formally, let s_t denote the capital available for consumption in period t; $s_t \in S = [0, \infty)$. By restricting s_t to be non-negative, we are assuming the economy to be debt free. After observing the level of capital at decision epoch t, the planner chooses a level of consumption u_t from $U_{s_t} = [0, s_t]$, and invests the remaining capital $s_t - u_t$. The consumption generates immediate utility $\Psi_t(u_t)$ and the investment produces capital at the next decision epoch according to the dynamic equation

$$s_{t+1} = w_t F_t(s_t - u_t), \tag{3.5.7}$$

where the function F_t represents existing technology and anticipated inflation, and the quantity w_t denotes a random disturbance which accounts for unanticipated inflation and random shocks to the system. We regard w_t as the realization of a non-negative random variable with a mean of one.

We model the dynamics in (3.5.7) using a multiplicative disturbance instead of an additive disturbance. This implicitly assumes that the disturbance produces a constant percentage change over the range of possible capital levels; a value of 1.1 for w_t corresponds to a 10% appreciation due to unforeseen events, and a w_t of 0.9 corresponds to a 10% decrease. This percentage change effects all levels of capital similarly.

Taking logarithms of the expression in (3.5.7) yields the transformed system equation

$$\log(s_{t+1}) = \log[F_t(s_t - u_t)] + \log(w_t).$$

In this equation, the error enters additively on the logarithmic scale. For empirical investigations, we often assume that $\log(w_t)$ has a normal distribution with mean 0 and variance σ^2; consequently, w_t has a lognormal distribution.

Generalization of this model include multiple-goods economies and dynamic competitive equilibrium models.

3.5.4 Linear Quadratic Control

Linear systems with quadratic costs play a central role in control engineering. Such problems arise when a controller attempts to guide an observable system along a predetermined trajectory or maintain a system near some desirable operating level. Departures from this trajectory are penalized proportionally to the square of the distance. We refer to this as the *linear regulator* or *linear quadratic control* problem.

The state of the system $s_t \in S = R^k$ follows the linear dynamic equation

$$s_{t+1} = A_t s_t + B_t u_t + C_t w_t, \tag{3.5.8}$$

where the control $u_t \in U_{s_t} = R^m$, the disturbance w_t is the realization of an R^n-valued random variable with mean 0, and covariance matrix Σ_t, A_t is a $k \times k$ matrix, B_t is a $k \times m$ matrix, and C_t is a $k \times n$ matrix.

Normalizing the model so that the desirable trajectory is the origin (the zero vector in R^k), the cost components consist of a quadratic penalty for deviations from the origin and a quadratic penalty for using control u_t. In guidance applications, we may think of the u_t as fuel utilization at decision epoch t. We represent the combined cost in any period by

$$r_t(s_t, u_t) = s_t^T Q_t s_t + u_t^T R_t u_t, \tag{3.5.9}$$

where Q_t is a $k \times k$ positive definite matrix, R_t is an $m \times m$ positive definite matrix, and the superscript T denotes transposition. In a finite-horizon problem, the ultimate deviation from the objective is penalized at cost $s_N^T Q_N s_N$. The objective is to choose a sequence of controls to minimize total expected cost.

3.6   BANDIT MODELS

The expression *bandit model* refers to a sequential decision model in which, at each decision epoch, the decision maker observes the state of each of K Markov reward processes and, based on the states, the transition probabilities, and rewards of each, selects a process to use in the current period. The selected process changes state according to its transition probabilities, and the state of all other processes remain fixed.

The effects of choosing process i when it is in state $s^i \in S^i$, $i = 1, 2, \ldots, K$ are that

1. it changes state according to a transition law $p_t^i(j^i | s^i)$,
2. the decision maker receives a reward $r_t^i(s^i)$, and
3. all other processes remain in their current state.

At each decision epoch, the state of process i changes while all others remain fixed, so that a process which is attractive to choose at the current decision epoch may be in an unattractive state at the subsequent decision epoch. The decision maker's objective is to choose a process selection sequence to maximize some function of the expected total reward over the planning horizon.

This model receives its colorful name because it applies (in a rather complicated fashion) to the decision problem facing a casino gambler when deciding whether or not to play a particular slot machine or "one-armed bandit" when the probability of winning is unknown. When $K > 1$, we refer to the model as a multiarmed bandit; the gambler must choose between K different slot machines with unknown win probabilities.

After formulating this model as a Markov decision process, we describe some applications.

3.6.1   Markov Decision Problem Formulation

We formulate this model as a Markov decision problem.
Decision epochs:

$$T = \{1, 2, \ldots, N\}, \quad N \leq \infty.$$

States:

$$S = S^1 \times S^2 \times \cdots \times S^K.$$

Actions (process to be selected):

$$A_s = \{1, 2, \ldots, K\}, \quad s \in S.$$

Rewards:

$$r_t((s^1, s^2, \ldots, s^K), a) = r_t^i(s^i) \quad \text{if } a = i.$$

Transition probabilities:

$$p_t((u^1, \ldots, u^k) | (s^1, \ldots, s^k), i) = \begin{cases} p(j^i | s^i) & u^i = j^i \text{ and } u^m = s^m \quad \text{for } m \neq i \\ 0 & u^m \neq s^m \quad \text{for some } m \neq i \end{cases}$$

Actions correspond to the process chosen in the current period and rewards and transition probabilities reflect this choice. Observe that when action i is selected, only component i of the state vector changes. A Markovian decision rule assigns a specific process to each possible configuration of process states.

The Markov decision process formulation requires a very large state space, making direct computation prohibitive. The beauty of this model is that, under most optimality criteria, the optimal policy can be computed by decomposing it into K one-dimensional problems and analyzing each separately.

3.6.2   Applications

Applications of the bandit model include project selection, gambling, allocation of treatments in a clinical trial, random search, and job scheduling. We describe the details of the first three applications below.

Project Selection
In any period a decision maker (such as the author of this book) can choose to work on any of K available projects (such as writing a book, working on a research paper, responding to e-mail, or exercising). The state s^i represents the degree of completion, or status, of project i, and $r^i(s^i)$ the expected reward for working on project i for one period. For example, when $S^i = [0, 1]$, $s^i \in S^i$ represents the fraction of the project that has been completed. Hopefully, working on a project for an additional period will draw us closer to completion, but, as we all know, there is a chance that we might not be any further along and in fact we may be further behind. If state C corresponds to completion of the project, and a reward of R^i is received only upon completion of the project, then $r^i(s^i) = R^i p^i(C | s^i)$. Since most of us can only do one thing at a time, by working on project i, its state changes but the status of all other projects remain unaltered.

Gambling—The One-Armed Bandit
This model is considerably more complicated than any model encountered so far because it requires the state space to be a set of functions. A gambler in a smoke-filled casino may either pay c units and pull the lever on a slot machine that pays one unit with probability q and zero units with probability $1 - q$, or decide not to play. Unfortunately, the gambler does not know q; instead she summarizes her

beliefs regarding its values through a probability density $f(q)$, on $[0, 1]$. By playing the game several times, the gambler acquires information about the distribution of q and revises her assessed probability density accordingly.

We model this as a bandit model as follows. Let process 1 correspond to the "do not play" option; $S^1 = \{1\}$, $r^1(1) = 0$, and $p^1(1|1) = 1$. Let process 2 correspond to the "play" option. We let the state of the system be the gambler's current assessment of the density of q; that is,

$$S^2 = \{f: f \text{ is a density with support on } [0, 1]\}.$$

The quantity f is referred to as a *prior distribution* in a Bayesian setting. Since the probability that the gambler wins one unit is q,

$$r^2(f) = \int_0^1 qf(q)\, dq - c = E_f[Q] - c,$$

where Q denotes the random variable with density f, and E_f an expectation with respect to f.

Transitions in this model take place between densities according to Bayes' rule. If the current estimate of the density is f and the gambler decides to play and wins, then the revised density f' (posterior probability) will put more weight on larger values of p than f. The transition probabilities between densities satisfy

$$p^2(f'|f) = \begin{cases} E_f[Q], & f' = \dfrac{qf(q)}{E_f[Q]} \\[2ex] 1 - E_f[Q], & f' = \dfrac{(1-q)f(q)}{1 - E_f[Q]}. \end{cases}$$

To see this, observe that the probability of the gambler winning in the current round is $E_f[Q]$, and, if she wins, she revises her probability assessment of the value of Q according to Bayes' theorem as follows. Writing $Q \approx q$ instead of $Q \in [q, q + dq]$, we have

$$\begin{aligned} P(Q \approx q|\text{win}) &= \frac{P(\text{win}|Q \approx q)P(Q \approx q)}{\int_0^1 P(\text{win}|Q \approx q)P(Q \approx q)\, dq} \\[2ex] &= \frac{qf(q)}{\int_0^1 qf(q)\, dq}. \end{aligned} \tag{3.6.1}$$

This formulation is quite different than those encountered previously in that the system state corresponds to a non-negative real-valued function on $[0, 1]$. This makes the problem computationally infeasible. In practice such generality is unnecessary; instead we may adopt either of the following two approaches:

1. Choose a parametric family of densities that is closed under the calculation in (3.6.1). Sometimes this is referred to as a *conjugate family* of distributions. For

example, if the initial state $s_1^2 = f_1$ is a beta density, that is

$$f_1(q) = \frac{\Gamma(\alpha + \beta)}{\Gamma(\alpha)\Gamma(\beta)} q^{\alpha - 1}(1 - q)^{\beta - 1}, \quad 0 < p < 1,$$

where $\alpha > 0$ and $\beta > 0$ and $\Gamma(\cdot)$ denotes the gamma function ($\Gamma(\alpha) = (\alpha - 1)!$ if α is a non-negative integer), then s_n^2 will be a beta density for all n and hence can be described completely by the number of trials, the number of wins, and the initial parameters. Suppose that after n plays of the game there have been k wins, then s_{n+1}^2 will be a beta density with parameters $\alpha + k$ and $\beta + n - k$. Therefore, the state space can be represented instead as either $S^2 = (0, \infty) \times (0, \infty)$ corresponding to the parameters of the beta distribution or $S^2 = \{0, 1, \ldots\} \times \{0, 1, \ldots\}$ corresponding to the number of wins and losses.

2. In general, represent the state space as $S^2 = \{0, 1, \ldots\} \times \{0, 1, \ldots\}$ corresponding to the number of wins and losses and recompute f in state $(k, n - k)$ according to

$$f(q) = \frac{q^k(1 - q)^{n-k} f_1(q)}{\int_0^1 q^k(1 - q)^{n-k} f_1(q) \, dq}$$

prior to each play of the game.

Sequential Clinical Trials

Controlled clinical trials are the basis of most applied medical research. They date back to James Lind's studies of the use of citrus fruits for prevention of scurvy among British seamen. In a clinical trial, the investigator compares a new treatment with unknown properties to an existing treatment with known properties, or to an inactive treatment referred to as a placebo.

We provide a bandit model for a controlled clinical trial in which successive subjects are allocated to one of two treatments sequentially. We assume that there are two outcomes of the treatment, one of which can be designated as a "success" and the other as "failure," and, furthermore, that the treatment is fast acting so that its outcome is known prior to allocating the next subject to a treatment. This assumption may be valid in a trial comparing headache remedies, but not in oncology (cancer) trials, in which success is not determined until a subject has been in remission for several years.

Let process 1 represent the new treatment and process 2 the existing or placebo treatment, and let q^i represent the respective success probabilities. We assume that q^1 is unknown with prior distribution f^1, and that q^2 is unknown with prior distribution f^2. The prior distribution on the success probability of existing therapy may be derived from published studies or investigator experience. We assume a distribution for q^2 instead of a fixed probability because responses vary considerably from subject to subject and study to study. Further, it might be expeditious to initially assume that $f^1 = f^2$.

Let c^i denote the cost of treating a subject with treatment i, $i = 1, 2$, and R the reward associated with successful treatment. Note that instead of a reward for a success, we might include the cost to society (and the individual) of a treatment failure. Also c^1 and c^2 might differ considerably.

We represent the state of each process as the density of success probabilities. The model for the one-armed bandit in Sec. 3.6.1 now applies to each treatment separately, as do the modifications discussed at the end of that section. Consequently this is a two-armed bandit problem.

This approach to designing clinical trials differs from actual practice since we do not address the problem from the hypothesis-testing perspective. Instead we view the trial as continuing indefinitely, with the objective being to determine an adaptive treatment allocation sequence to maximize total expected reward or minimize expected cost. Once the evidence is convincing that one treatment dominates the other, then the allocation rule will assign subjects to that treatment. Of course, in practice we must be sure to have ample numbers of subjects receiving each treatment to ensure that results are convincing to the medical community, not biased by erroneous prior information and generalizable to a larger population.

3.6.3 Modifications

The bandit model above may be modified in several ways.

Restless Bandits

This model relaxes Assumption 3 at the beginning of this section, by allowing the other processes to change states between decision epochs. The effect of this modification is that the transition probability structure of Sec. 3.6.1 must be made more general. This modification is appropriate for the project selection models, but not realistic for models in which the state space represents a probability density. In the restless bandit model, the decomposition approach may not lead to optimal policies but can be used to obtain policies that are close to optimal.

This formulation might provide a conceptual model for a sequential clinical trial in which there is a time lag between treatment and its outcome.

Arm-Acquiring Bandits

In this model, the number of processes does not remain fixed over the decision-making horizon. Instead new arms (projects) become available in each decision epoch according to a Poisson process. Therefore K_t, the number of projects available at time t, increases with t. Of course this model may be modified so that projects also disappear.

The Tax Problem

In this model, Assumption 2 is modified. Instead of receiving a reward for choosing process k, a tax is charged for all projects that remain idle. That is, if project j is not selected at decision epoch t, the system incurs a cost $c^j(s^j)$ for that period. Therefore the cost of choosing project k when the state of the projects is $[s^1, s^2, \ldots, s^K]$ is $\sum_{k \neq j} c^j(s^j)$.

Continuous-Time Bandits

Instead of choosing processes periodically, the decision maker might instead be able to switch between processes at any time (preemptively) or at fixed random times corresponding to state changes in a process (nonpreemptively). In the latter case, the individual processes may be better modeled by semi-Markov processes and methods (Chap. 11).

3.7 DISCRETE-TIME QUEUEING SYSTEMS

Queueing systems model a wide range of phenomenon including multi-user computer systems, communication networks, manufacturing systems, supermarket checkout lines, or even a waiting room at a physician's office. A controlled queueing system consists of three components: queues, servers, and controllers. In a single-server uncontrolled queueing system, jobs arrive, enter the queue, wait for service, receive service, and depart from the system. In the absence of a controller, system behavior can be quite erratic with periods of long queues followed by periods in which the server remains idle. Controllers may improve system utilization by reducing expected throughput time or queue length.

In this section we consider discrete-time queueing systems, even though continuous-time formulations are usually more natural. Chapter 11 provides some continuous-time queueing control models.

We assume the system is observed every $\eta > 0$ units of time and that the controller makes decisions at times $0, \eta, 2\eta, \ldots$. These are the decision epochs for the model. We sometimes refer to the period between two decision epochs as a *time slot* and refer to time as *slotted*.

In this section, we formulate models for two types of controlled single-server queueing systems; one in which the controller decides on the number of arriving jobs to admit to the eligible queue (admission control), and one in which the controller adjusts the service rate (service rate control). For both models, we assume independent probability distributions for the number of arrivals and the number of service completions during each time slot. Section 1.5 and the problems at the end of this chapter provide other examples of queueing control models.

3.7.1 Admission Control

Jobs arrive for service and are placed in a "potential job queue." At each decision epoch, the controller observes the number of jobs in the system ("eligible queue" plus server), and on this basis decides how many jobs, if any, to admit from the potential job queue into the eligible queue. These jobs eventually receive service; those not admitted never enter the system. (Fig. 3.7.1).

We now provide a Markov decision process model for this system. Let X_t represent the number of jobs in the system immediately prior to decision epoch t, and Z_t the number of arrivals in period t. Arriving jobs during period $t - 1$ enter the "potential job queue." At decision epoch t, the controller admits u_t jobs from the potential job queue into the system. Let Y_t denote the number of "possible service completions" during period t. We include the adjective "possible" because if Y_t exceeds $X_t + u_t$, then only $X_t + u_t$ jobs are serviced. Table 3.7.1 represents the content of the components of the queueing system throughout period t. In it $t+$ denotes a point in time immediately after the control has been implemented but prior to any service completions.

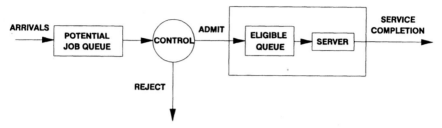

Figure 3.7.1 A queueing system with admission control.

We represent the system state by the pair (X_t, V_t), where V_t denotes the content of the potential job at decision epoch t. Following the notation of Sec. 3.5, the system equation (3.5.1) for these two components may be written as

$$X_{t+1} = [X_t + u_t - Y_t]^+,$$

$$V_{t+1} = Z_t.$$

Note that we can only admit as many jobs as are present, so that $0 \leq u_t \leq V_t$. Because of this, we must retain V_t in the state description.

The random variable Y_t assumes non-negative integer values and follows a time invariant probability distribution $f(n) = P(Y_t = n)$, $t = 0, 1, \ldots$, and Z_t assumes non-negative values and follows a time invariant probability distribution $g(n) = P(Z_t = n)$, $t = 0, 1, \ldots$. We assume no partial service completions during any period and that all serviced jobs depart the system at the end of the period.

The stationary reward structure consists of two components: a constant reward of R units for every completed service, and an expected holding cost of $h(x)$ per period when there are x jobs in the system.

A Markov decision process formulation follows.

Decision epochs:

$$T = \{0, \eta, 2\eta, \ldots, N\eta\}, \quad N \leq \infty.$$

States (S_1 is the number in the system, and S_2 the number in the potential job queue):

$$S = \{0, 1, \ldots\} \times \{0, 1, \ldots\} = S_1 \times S_2.$$

Table 3.7.1

Time	Potential Job Queue	System
t	Z_{t-1}	X_t
$t+$	0	$X_t + u_t$
$t+1$	Z_t	$[X_t + u_t - Y_t]^+$

Actions (number of jobs to admit):

$$A_{s_1, s_2} = \{0, 1, \ldots, s_2\}.$$

Rewards:

$$r_t(s_1, s_2, a) = R \cdot E\{\min(Y_t, s_1 + a)\} - h(s_1 + a). \tag{3.7.1}$$

Transition probabilities:

$$p_t(s_1', s_2' | s_1, s_2, a) = \begin{cases} f(s_1 + a - s_1')g(s_2') & a + s_1 > s_1' > 0 \\[2mm] \left[\displaystyle\sum_{i=s_1+a}^{\infty} f(i) \right] g(s_2') & s_1' = 0, \ a + s_1 > 0 \\[2mm] g(s_2') & s_1' = a + s_1 = 0 \\[2mm] 0 & s_1' > a + s_1 \geq 0. \end{cases}$$

Some comments on the model formulation follow. Note that the set of states need not be finite and that the reward is bounded above but may be unbounded below. It is natural to formulate the model as one in which the reward depends on the system state at the next decision epoch, as in (2.1.1), since it depends on the number of service completions in period t. That is the basis for the expression (3.7.1). The quantity $\min(Y_t, s_1 + a)$ gives the number of jobs completed in period t. To see this, note that if the number of possible services Y_t exceeds the number of eligible jobs $s_1 + a$, only $s_1 + a$ receive service. The expectation of this quantity satisfies

$$E\{\min(Y_t, s_1 + a)\} = \sum_{i=1}^{s_1+a-1} if(i) + (s_1 - a) \sum_{i=s_1+a}^{\infty} f(i).$$

Note that this is a concave, nondecreasing function of a. To derive the transition probabilities, we use the same argument and note that the number of arrivals is assumed to be independent of the number of service completions.

Example 3.7.1. We now consider a special case of this model with deterministic service rate of one job per period; that is, $Y_t \equiv 1$, so that $f(1) = 1$ and $f(j) = 0$ for $j \neq 1$. For this model (3.7.1) becomes

$$r_t(s_1, s_2, a) = \begin{cases} R - h(s_1 + a) & s_1 + a > 0 \\ 0 & s_1 + a = 0, \end{cases}$$

and the transition probabilities reduce to

$$p_t(s_1', s_2' | s_1, s_2, a) = \begin{cases} g(s_2') & s_1' = s_1 + a - 1 \geq 0 \text{ or } s_1' = s_1 + a = 0 \\ 0 & \text{otherwise.} \end{cases}$$

3.7.2 Service Rate Control

In this model, all arriving jobs enter the eligible queue and are served on a first come first serve (FCFS) basis. At each decision epoch, the controller observes the number

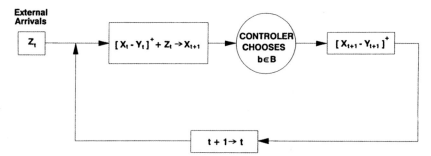

Figure 3.7.2   Dynamics in a service rate controlled queueing system.

of jobs in the system and selects the service rate from a set of probability distributions indexed by elements of a set B. In most applications, we model the service distribution by a parametric family such as the geometric or Poisson, so that B might represent the set of possible parameter values. For each $b \in B$, we let $f_b(n)$ denote the corresponding probability function. As in Sec. 3.7.1, let Y_t represent the number of possible service completions in time slot t.

The reward structure for this model augments that of the model in Sec. 3.7.1 by including a fixed cost K for changing the service rate and a per period cost $d(b)$ for using service distribution $f_b(\cdot)$ for one period. Suppose the controller uses $f_b(\cdot)$ in period t and uses $f_{b'}(\cdot)$ in period $t + 1$; then the extra cost component $c(b, b')$ equals

$$c(b, b') = \begin{cases} K + d(b') & b' \neq b \\ d(b) & b' = b. \end{cases}$$

The Markov decision process formulation below assumes that any arrivals during a period do not receive service in that period. Consequently the system equation becomes

$$X_{t+1} = [X_t - Y_t]^+ + Z_t, \tag{3.7.2}$$

where the distribution of Y_t depends on u_t, the index of the distribution chosen in period t (Fig. 3.7.2). Otherwise the notation follows that of Sec. 3.7.1.
Decision epochs:

$$T = \{0, \eta, \ldots, N\eta\}, \quad N \leq \infty.$$

States (number of jobs in the system and the current service rate):

$$S = \{0, 1, \ldots\} \times B.$$

Actions:

$$A_{s,a} = B.$$

Rewards:

$$r_t((s, a), a') = R \cdot E\{\min(Y_t, s)\} - h(s) - c(a, a').$$

Transition probabilities:

$$p_t\big(s', a' | (s, a), a'\big) = \sum_{n=0}^{s-1} f_{a'}(n) g(s' - s + n) + \left[\sum_{n=s}^{\infty} f_{a'}(n) \right] g(s').$$

We derive the transition probability by appealing to (3.7.2) and using the law of total probabilities. The first term in the expression for the transition probability is the sum over n, the number of service completions, of the product of the probability of n service completions and $s' - s + n$ arrivals, while the second term gives the probability of at least s service completions and s' arrivals.

BIBLIOGRAPHIC REMARKS

Bellman's path breaking book (1957) contains many interesting and novel applications, especially in the problem sections. They include optimal search, sequential allocation, experimental design, optimal replacement, scheduling, and inventory control. Many of these appeared in his earlier publications. Howard (1960) provides some colorful examples including strategy choice in baseball and taxicab routing. His automobile replacement model has often been used as a test problem for computational algorithms.

Inventory models have their origins in work of Arrow, Harris, and Marschak (1951) and Dvoretsky, Kiefer, and Wolfowitz (1952). These papers were a driving force in popularizing dynamic programming methods. The former paper introduces the concept of a penalty cost as follows:

"The organization—whether commercial or noncommercial—has a general idea of the value it would attach to the damage that would be caused by the nonavailability of an item. It knows the cost and the poorer performance of emergency substitutes. The penalty for depleted stocks may be very high: 'A horse, a horse, my kingdom for a horse,' cried defeated Richard III."

Arrow, Karlin, and Scarf (1958) provide many interesting models and applications. Conditions for the optimality of (s, S) or (σ, Σ) policies were first provided by Scarf (1960). There exists a voluminous literature on the inventory model and its applications. Noteworthy surveys include Veinott (1966) and Porteus (1991).

Early papers on shortest-path models include Minty (1957), Dantzig (1957), and Dijkstra (1959). Ford and Fulkerson (1962) made important contributions to theory and methods for these models. A review article by Pollack and Wiebenson (1960) and papers by Dreyfus (1969) and Denardo and Fox (1979) provide extensive bibliographies on such models. Denardo's text (1982) serves as an excellent reference for many aspects of deterministic dynamic programming models. The critical path model has its origins of the PERT charts of the early days of operations research. Kelly (1961) and Bigelow (1962) summarize and review some of the earlier references on this model and its application. This technique has become more popular of late with many readily available PC software packages. Sequential allocation models are among the first applications of dynamic programming (Bellman, 1957). Apparently the application to constrained maximum likelihood estimation is new.

Optimal stopping problems have roots in the sequential hypothesis testing models of Wald (1947) and subsequent papers by Wald and Wolfowitz (1948), Arrow, Blackwell, and Girshick (1949). Karlin (1962) proposed and solved the asset selling problem. The secretary problem, or candidate problem, originates in Cayley (1875), where he formulates it in the following way:

"A lottery is arranged as follows:—There are n tickets representing a, b, c pounds respectively. A person draws once; looks at his ticket; and if he pleases, draws again (out of the remaining $n - 1$ tickets); looks at his ticket, and if he pleases draws again (out of the remaining $n - 2$ tickets); and so on, drawing in all not more than k times; and he receives the value of the last drawn ticket. Supposing he regulates his drawings in the manner most advantageous to him according to the theory of probabilities, what is the value of his expectation?"

The model presented in this chapter relates to those studied by Gilbert and Mosteller (1966) and Chow, Moriguti, Robbins, and Samuels (1964). This appealing problem has generated a voluminous literature which explores many variants; recent references include Hlynka and Sheahan (1988) and Reeves and Flack (1988). The book by Chow, Robbins, and Siegmund (1971) provides a comprehensive study of the optimal stopping problem.

Option pricing models have been studied extensively in the financial literature. The classical references are the papers by Black and Scholes (1973) and Merton (1973). Cox and Rubenstein (1985) is a comprehensive book on this topic, with an extensive bibliography.

The discrete-time dynamic equation approach has its origins in continuous-time control theory and the calculus of variations. Bellman (1961) provides a lively presentation discussing some of its origins and many applications. The advent of computers and Bellman's dynamic programming approach seem to have gotten the ball rolling on the discrete-time dynamic model. Its earliest applications were to inventory models as described above; Arrow, Harris, and Marschak (1951) provide a dynamic equation for the inventory state. Applications in economics are the subject of Stokey and Lucas (1990). Simon (1956) and Theil (1957) provide early results on the linear quadratic control model with the latter paper among the first to use matrix methods. A special issue of *IEEE Transactions of Automatic Control*, edited by Athans (1971) focuses on the linear quadratic control problem providing an overview of theory together with applications in chemical process control and satellite navigation. The excellent book by Bertsekas (1987) approaches dynamic programming from the system equation perspective.

Bandit problems originate in the work of Thompson (1933 and 1935). They seem to have evolved fairly independently in the statistical and engineering/operations research literature with different emphases. The former approach is reflected in the gambling model, the latter in the project selection model. Gittins and Jones (1974) demonstrate the optimality of dynamic allocation indices and hence the separability of the model. Whittle (1980c) refines and extends these results using Markov decision process methods. Bandit problems are the subject of the book by Berry and Fristedt (1985). It contains a thorough annotated bibliography. Colton (1963) provides an early reference to the application of bandit models in clinical trials.

Queueing control represents an active area of current research and one of the main areas of application of Markov decision process methodology. Early research in queueing control primarily concerned single-server models; examples include Yadin and Naor (1967), Heyman (1968), Naor (1969), Sobel (1969), and Bell (1969). Sobel (1974) and Crabill, Gross, and Magazine (1977) survey this research. Stidham (1985) surveys admission control models paying particular attention to Markov decision process approaches. Most of the recent work on queueing control has been concerned with controlling networks of queues and allocating jobs of different classes to one of several servers. The book by Walrand (1988) and his handbook article (1991) provide good overviews of formulations and results for these models.

White (1985b) and (1988a) surveys a wide range of applications of Markov decision processes. He provides short summaries of each application and comments on the extent of implementation. Some areas of application included in the survey are fisheries, forestry, water resources, airline booking, and vehicle replacement. In the 1985 paper, White notes that

"I identified only five pieces of work that have been directly implemented..."

However, he goes on to conclude that

"...it appears that in the last few years more efforts are being made to model phenomena using Markov decision processes."

PROBLEMS

3.1. Derive transition probabilities and expected one-period rewards for the randomized Markov policy of Sec. 3.1.

3.2. Consider a generalization of the inventory model of Sec. 3.2 in which unfilled orders may be backlogged indefinitely with a cost of $b(u)$ if u units are backlogged for one period. Assume revenue is received at the end of the period in which orders are placed and that backlogging costs are charged only if a unit is backlogged for an entire month, in which case the backlogging cost is incurred at the beginning of that month.

   a. Identify the state space and derive transition probabilities and expected rewards.

   b. Calculate $r_t(s, a)$ and $p_t(j|s, a)$ for the above model with $b(u) = 3u$ and all other model parameters identical to Sec. 3.2.2.

   c. Formulate the model under the assumption that payment for an order is received at the end of the month in which the order is filled.

3.3. Consider the network in Fig. 3.3.1, but suppose in addition that it contains an arc connecting node 1 to node 5 with length 10. Give a modified network in which nodes are grouped by stages. Be sure to indicate all arc lengths in the modified network.

3.4. Formulate the following stochastic version of the shortest-route problem in Fig. 3.3.1 as a Markov decision problem. In this model, when the decision maker chooses to traverse a particular arc, there is a positive probability of actually traversing an alternative arc. The probabilities are as follows. If there are two adjacent arcs, such as at node 1, and the decision maker chooses to traverse the arc between nodes 1 and 2, then the probability of traversing that arc is 0.6 and the probability of traversing each of the adjacent arcs is 0.2. When there is only one adjacent arc, then the probability of traversing the chosen arc is 0.7. If there are no adjacent arcs, the chosen arc is traversed with certainty.

3.5. Provide a graphical representation for a deterministic version of the stochastic inventory model of Sec. 3.2, in which the demand is known with certainty prior to placing an order. Assume parameter values of Sec. 3.2.2, and that the demand in period 1 is two units and demand in period 2 is one unit.

3.6. In most situations where critical path analyses are applied, the length of time to complete a task is not known with certainty. Formulate such a model as a longest-route model and discuss appropriate and practical optimality criteria for such a model.

3.7. Provide a graphical representation for the sequential allocation model of Sec. 3.3 in which we restrict each x_n to be an integer.

3.8. At each round of a game of chance, a player may either pay a fee of C units and play the game or else quit and depart with his current fortune. Upon paying the fee he receives a payoff which is determined by a probability distribution $F(\cdot)$. The player's objective is to leave the game with the largest possible fortune.

   a. Formulate this gambling problem as an optimal stopping problem.

   b. Explicitly give transition probabilities and rewards when the cost of playing the game is one unit and the player receives two units with probability p and zero units with probability $1 - p$ if he plays.

3.9. When going out for dinner, I always try to park as close as possible to the restaurant. I do not like to pay to use a parking lot, so I always seek on-street parking. Assume my chosen restaurant is situated on a very long street running east to west which allows parking on one side only. I approach the restaurant from the east starting at a distance of Q units away. Because traffic is heavy I can only check one potential parking spot at a time.

   Formulate the parking problem as an optimal stopping problem. Assume the street is divided into one-car-length sections, and the probability that a parking spot at distance s from the restaurant is unoccupied is p_s independent of all others.

3.10. Formulate a variant of the secretary problem in which the objective is to maximize the probability that the decision maker chooses

   a. one of the two best candidates,

   b. one of the k best candidates.

3.11. Formulate the secretary problem when the employer's objective is to minimize the expected rank of the candidate who receives an offer. Assume the best candidate has rank 1 and the least desirable candidate has rank N.

3.12. Provide a dynamic equation formulation for a two-product inventory model in which there is a constraint on total warehouse capacity. Derive the appropriate transition probabilities.

3.13. Formulate the inventory model in which there is a random lag between placing and receiving orders; that is, the lead time is a random variable with a known probability distribution.

3.14. Provide a dynamic equation for an inventory model in which a fixed random fraction q of the stock deteriorates and is unavailable for use in the subsequent period.

3.15. (Love, 1985; Controlling a two city rental car agency) A rental car agency serves two cities with one office in each. The agency has M cars in total. At the start of each day, the manager must decide how many cars to move from one office to the other to balance stock. Let $f_i(q)$; $i = 1$ and 2; denote the probability that the daily demand for cars to be picked up at city i and returned to city i equals q. Let $g_i(q)$, $i = 1$ and 2; denote the probability that the daily demand for "one-way" rentals from city i to the other equals q. Assume that all rentals are for one day only, that it takes one day to move cars from city to city to balance stock, and, if demand exceeds availability, customers go elsewhere. The economic parameters include a cost of K per car for moving a car from city to city to balance stock, a revenue of R_1 per car for rentals returned to the rental location, and R_2 per car for one-way rentals.

a. Provide dynamic equations for this system and formulate it as a Markov decision process.

b. Suppose, in addition to relocating cars to balance stock, the manager may reserve a certain number of cars for one-way rental only. Formulate this modified model as a Markov decision process.

3.16. (Stengos and Thomas, 1980; The blast furnace problem) Two identical machines are used in a manufacturing process. From time to time, these machines require maintenance which takes three weeks. Maintenance on one machine costs c_1 per week while maintenance on both machines costs c_2 per week. Assume $c_2 > 2c_1$. The probability that a machine breaks down if it has been i periods since its last maintenance is p_i, with p_i nondecreasing in i. Maintenance begins at the start of the week immediately following breakdown, but preventive maintenance may be started at the beginning of any week. The decision maker must choose when to carry out preventive maintenance, if ever, and, if so, how many machines to repair. Assume the two machines fail independently.

a. What is the significance of the condition $c_2 > 2c_1$?

b. Formulate this a Markov decision problem.

c. Provide an educated guess about the form of the optimal policy when the decision maker's objective is to minimize expected operating cost.

3.17. (Rosenthal, White, and Young, 1978; A dynamic location model) A repairman who services Q facilities moves between location s and location j in any period according to the stationary transition probability $p(j|s)$. An equipment trailer which carries spare parts and tools may be located at any one of M sites. If the trailer is at site m and the repairman is at facility j, the cost of obtaining material from the trailer is $c(m, j)$. The cost of moving the trailer from site m to site j is $d(m, j)$. The decision maker's objective is to dynamically relocate the trailer so as to minimize expected costs. Assume that the decision maker observes the location of the repairman and trailer, relocates the trailer, and then the repairman moves and services a facility.

Formulate this as a Markov decision process clearly identifying states, actions, transition probabilities, and rewards. What other timing scenarios might be realistic?

3.18. (Stokey and Lucas, 1989; Employment seeking) At the start of each week, a worker receives a wage offer of w units per week. He may either work at that wage for the entire week or instead seek alternative employment. If he decides to work in the current week, then at the start of the next week, with probability p, he will have the same wage offer available while, with probability $1 - p$, he will be unemployed and unable to seek employment during that week. If he seeks alternative employment, he receives no income in the current week and obtains a wage offer of w' for the subsequent week according to a transition probability $p_t(w'|w)$. Assume his utility when receiving wage w in week t is $\Psi_t(w)$.

  a. Formulate the worker's decision problem as a Markov decision process in which his objective is to maximize his total expected utility over a finite horizon.

  b. Formulate the worker's decision problem as a bandit problem.

3.19. (Morrison, 1976 and 1986; A hockey coach's dilemma) A hockey team is losing with a few minutes remaining in a game. The coach may either replace his goalie by an extra player or leave his goalie in. By removing the goalie he increases the probability that his team scores the *only* goal in the next h seconds from p_G to p_{NG}; however, the probability of the opposing team scoring the only goal in that h-second interval increases from q_G to q_{NG}, where q_{NG} is much larger than q_G. Of course, $p_G + q_G < 1$ and $p_{NG} + q_{NG} < 1$. Assume that events in successive h-second intervals are independent.

Formulate the problem facing the coach as a discrete-time Markov decision process in which the objective is to maximize the probability of achieving at least a tie. To do this, let the state represent the point differential between the two teams and let the terminal reward be 1 if the point differential is non-negative and 0 if it is negative. (Note that a continuous-time formulation is more natural for this problem.)

3.20. Provide system equations and a Markov decision process formulation for a modification of the admission control problem of Sec. 3.7.1, in which each job in the "potential job" queue which is not admitted for service returns in the next period, with probability p independent of the number of new arrivals and the number of jobs in the "potential job queue."

3.21. Provide system equations and a Markov decision process formulation for the following modification of the admission control problem of Sec. 3.7.1, in which there are two parallel servers each accompanied by its own eligible job queue. In this model, the controller may allocate jobs from the potential job queue to either eligible job queue or decide to not serve a job in which case it leaves the system. See Fig. P3.21. Assume that the "possible service distribution" for server i, $i = 1$ and 2, equals $f_i(n)$, and that the holding cost equals $h_i(x)$ per period when there are x jobs in total at server i and its eligible queue. Provide a guess regarding the most efficient operating policy for this system.

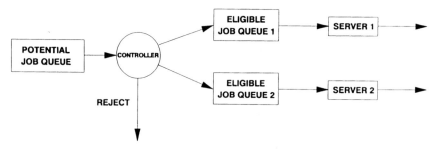

Figure P3.21   Admission control to a parallel queueing system.

3.22. Provide system equations and a Markov decision process formulation for the following modification of the admission control problem of Sec. 3.7.1, in which there are two servers in series each accompanied by its own eligible job queue. In this model, the controller may allocate some or all jobs from the potential job queue to the first queue only or decide to refuse to service the job, in which case it leaves the system. Jobs which complete service at the first queue immediately join the eligible job queue preceding the second server. See Fig. P3.22. Assume that the "possible service distribution" for server i, $i = 1$ and 2, equals $f_i(n)$, and that the holding cost equals $h_i(x)$ per period when there are x jobs in total at server i and its eligible queue. Provide a guess regarding the form of the most efficient operating policy for this system.

Figure P3.22   Admission control to a series queueing system.

3.23. Consider a variant of the queueing admission control model in Sec. 3.7.1 in which there are two classes of jobs and a single-server. Suppose $f(n_1, n_2)$ denotes the probability that n_1 class-1 jobs and n_2 class-2 jobs arrive in a time slot, and $g(n_1, n_2)$ denotes the probability of n_1 potential class-1 and n_2

potential class-2 service completions in a time slot. Let R_i denote the revenue from completing a class-i job and $h_i(u)$ denote the cost of holding u class-i jobs for one period. At the start of each time slot, the controller determines the number of jobs of each class to admit to service.

a. Provide dynamic equations for this model.

b. Formulate it as a Markov decision process.

c. Speculate on the form of an optimal policy.

3.24. A put option gives the holder the right to sell shares of a stock at a specified price at any time up to its expiry date. Formulate the problem of when to exercise a put option as a optimal stopping problem.

3.25. (Mangel and Clark, 1988) Suppose that in each period an animal "chooses" in which of K "patches" of land to forage. Each patch has a different predation risk, different energy reserve requirements, and also yields a random energy return. The energy return probability distribution varies from patch to patch. Suppose that the animal has a minimal energy threshold that will keep it alive, and that it has a maximum energy capacity. Formulate the patch selection problem as a Markov decision problem in which the animal's objective is to maximize its probability of survival for T periods.

3.26. (Mangel and Clark, 1988; The lazy adaptable lion) An adult female lion requires about 6 kg of meat per day and has a gut capacity of about 30 kg; which means that, with a full gut, it can go six days without eating. Zebras have an average edible biomass of 164 kg, large enough to meet the food demands of many lions. If a female selects to hunt in a group, the probability of a successful kill per chase has been observed to increase with group size to a point, and that each chase consumes 0.5 kg of energy. Suppose that the probabilities of a successful hunt are given by $p(1) = 0.15$, $p(2) = 0.33$, $p(3) = 0.37$, $p(4) = 0.40$, $p(5) = 0.42$, and $p(\geq 6) = 0.43$, where $p(n)$ represents the kill probability per chase for lions hunting in a group of size n.

Formulate this as a Markov decision process, in which the state represents the lion's energy reserves, the action is whether to hunt, and, if so, in what size group. Assume one hunt per day and that lion's objective is to maximize its probability of survival over T days. Assume that if the hunt is successful, lions in the hunting group share the meat equally.

3.27. (Feinberg, Kogan, and Smirnov, 1985). Consider the slotted ALOHA model of Sec. 1.5, in which the controller is faced with the problem of dynamically choosing the retransmission probability. Formulate it as a Markov decision process under the assumption that the number of packets generated in a time slot follow a Poisson distribution with parameter μ.

3.28. Consider the game "But Who's Counting" described in Sec. 1.7. Formulate the problem of maximizing the expected five-digit number as a Markov decision process. Hint: Let the state represent the occupied digits, the numbers in each location and the number observed on the spinner.

CHAPTER 4

Finite-Horizon Markov Decision Processes

This chapter concerns finite-horizon, discrete-time Markov decision problems. In it, we introduce and discuss many basic concepts including optimality criteria, inductive computation, The Principle of Optimality, optimality equations, and structured policies. Section 4.5 contains the key algorithm and Section 4.6 illustrates its use through numerical examples and applications. We suggest reviewing Chap. 2 before delving into this chapter. *(We assume throughout that $N < \infty$.)*

4.1 OPTIMALITY CRITERIA

As a result of choosing and implementing a policy, the decision maker receives rewards in periods $1, \ldots, N$. Since they are not known prior to implementing the policy, the decision maker must view the reward sequence as random. The decision maker's objective is to choose a policy so that the corresponding random reward sequence is as "large" as possible. This necessitates a method of comparing random reward sequences. We discuss approaches to comparing random sequences below and provide some justification for comparing policies on the basis of their expected total reward.

4.1.1 Some Preliminaries

Let $\pi = (d_1, d_2, \ldots, d_{N-1})$ denote a randomized history-dependent policy. That is, for each $t = 1, \ldots, N - 1$, $d_t \colon H_t \to \mathscr{P}(A)$. (Recall that $H_t = S \times A \times S \times \cdots \times A \times S$ is the set of all histories of states and actions prior to decision epoch t). As described in Sec. 2.1.6, each $\pi \in \Pi^{\mathrm{HR}}$ generates a probability distribution $P^\pi(\cdot)$ on (Borel subsets of) H_N, the set of histories (sample paths) up to time N. For each realization of the history $h_N = (s_1, a_1, s_2, \ldots, s_N)$ there corresponds a sequence of rewards $\{r_1(s_1, a_1), r_2(s_2, a_2), \ldots, r_{N-1}(s_{N-1}, a_{N-1}), r_N(s_N)\}$. Let $R_t \equiv r_t(X_t, Y_t)$ denote the random reward received in period $t < N$, $R_N \equiv r_N(X_N)$ denote the terminal reward, $R \equiv (R_1, R_2, \ldots, R_N)$ denote a random sequence of rewards, and $\mathfrak{R}$ the set of all possible reward sequences. A policy π induces a probability distribution $P_{\mathfrak{R}}^\pi(\cdot)$ on on (Borel subsets of) $\mathfrak{R}$:

$$P_{\mathfrak{R}}^{\pi}(\rho_1, \rho_2, \ldots, \rho_N) \equiv P^{\pi}\big[\{(s_1, a_1, \ldots, s_N): (r_1(s_1, a_1), \ldots, r_N(s_N))$$
$$= (\rho_1, \ldots, \rho_N)\}\big].$$

In this definition, $P^{\pi}[\cdot]$ denotes the probability distribution on the set of all realizations of the Markov decision process under π.

For a deterministic history-dependent π, we may write this reward sequence as

$$\{r_1(X_1, d_1(h_1)), \ldots, r_{N-1}(X_{N-1}, d_{N-1}(h_{N-1})), r_N(X_N)\}$$

and, for $\pi \in \Pi^{MD}$, as

$$\{r_1(X_1, d_1(X_1)), \ldots, r_{N-1}(X_{N-1}, d_{N-1}(X_{N-1})), r_N(X_N)\}.$$

We compare policies on the basis of the decision maker's preference for different realizations of R and the probability that each occurs. We now discuss approaches for such comparisons based on stochastic orderings, expected utility, and other criteria.

Let W be an arbitrary set and let $\succ$ denote a *partial ordering* on that set. By a partial ordering we mean a transitive, reflexive, and antisymmetric relationship between elements in the set. That is, for u, v, and w in W, $u \succ v$ and $v \succ w$ implies that $u \succ w$ (transitivity) and $w \succ w$ (reflexivity), and $v \succ w$ and $w \succ v$ implies $w = v$ (antisymmetry). Examples of partial orderings include component-wise ordering on R^n and $\subset$ on a collection of sets. Two elements u and v are *comparable* if either $u \succ v$ or $v \succ u$. A partial ordering is a *total ordering* or *linear ordering* on W if every pair of elements in W is comparable. Under the usual component-wise ordering the real line is totally ordered while n-dimensional Euclidean space is not.

Since the models we study are stochastic, we require partial orderings for random vectors or, equivalently, for probability distributions of random vectors. We first define stochastic orderings for random variables. We say that the random variable U is *stochastically greater than* the random variable V if

$$P[V > t] \leq P[U > t]$$

for all $t \in R^1$. If P^1 and P^2 are two probability distributions on the same real probability space, then P^1 is stochastically greater than P^2 if

$$P^2[(t, \infty)] \leq P^1[(t, \infty)]$$

for all $t < \infty$.

Many generalizations to random vectors have been proposed; we adopt the following. We say that the random vector $U = (U_1, \ldots, U_n)$ is *stochastically greater* than the random vector $V = (V_1, \ldots, V_n)$ if

$$E[f(V_1, \ldots, V_n)] \leq E[f(U_1, \ldots, U_n)] \tag{4.1.1}$$

for all $f: R^n \to R^1$ for which the expectation exists and which preserve the partial ordering on R^n (that is, if $v_i \leq u_i$ for $i = 1, \ldots, N$; $f(v_1, \ldots, v_N) \leq f(u_1, \ldots, u_N)$).

From a decision-making perspective, the following result which relates the vector and single component stochastic orderings provides a notion of ordering which is more consistent with sequential decision making. It is one of few results on stochastic orderings which takes the order of the vector components into account. We state it without proof.

Proposition 4.1.1. Suppose $U = (U_1, \ldots, U_n)$ and $V = (V_1, \ldots, V_n)$ are random vectors such that $P[V_1 > t] \leq P[U_1 > t]$ for all $t \in R^1$, and for $j = 2, \ldots, n$, $v_i \leq u_i$ for $i = 1, \ldots, j - 1$ implies

$$P\big[V_j > t | V_1 = v_1, \ldots, V_{j-1} = v_{j-1}\big] \leq P\big[U_j > t | U_1 = u_1, \ldots, U_{j-1} = u_{j-1}\big].$$

Then U is stochastically greater than V. Further, for any nondecreasing function g: $R^1 \rightarrow R^1$,

$$E\big[g(V_j) | V_1 = v_1, \ldots, V_{j-1} = v_{j-1}\big] \leq E\big[g(U_j) | U_1 = u_1, \ldots, U_{j-1} = u_{j-1}\big].$$

The following example illustrates a serious drawback of using stochastic partial orders to compare policies.

Example 4.1.1. Let $N = 2$ and suppose that there are two policies π and ν with distributions of rewards given by

$$P^\pi[R = (0,0)] = P^\pi[R = (0,1)] = P^\pi[R = (1,0)] = P^\pi[R = (1,1)] = \tfrac{1}{4}$$

and

$$P^\nu[R = (0,0)] = P^\nu[R = (1,1)] = \tfrac{1}{2}.$$

We compare these policies on the basis of $E^\pi[f(R_1, R_2)]$ and $E^\nu[f(R_1, R_2)]$. For $f(u,v) = u + v$,

$$E^\pi[f(R_1, R_2)] = E^\nu[f(R_1, R_2)] = 1.$$

For $f(u, v) = [u]^+[v]^+$,

$$\tfrac{1}{4} = E^\pi[f(R_1, R_2)] < E^\nu[f(R_1, R_2)] = \tfrac{1}{2}.$$

And, for $f(u, v) = \sup(u, v)$,

$$\tfrac{3}{4} = E^\pi[f(R_1, R_2)] > E^\nu[f(R_1, R_2)] = \tfrac{1}{2}.$$

Thus, even in this simple model, the policies π and ν are not comparable under the vector stochastic partial order defined through (4.1.1).

The difficulty with using a stochastic ordering to compare policies is that for the decision maker to prefer π to ν, then

$$E^\pi[f(R_1,\ldots,R_N)] \geq E^\nu[f(R_1,\ldots,R_N)] \tag{4.1.2}$$

must hold for a large class of functions f, many of which do not reflect the decision maker's attitude toward risk. By requiring that (4.1.2) hold only for a specified function, utility theory provides an attractive basis for policy comparison. By a *utility* $\Psi(\cdot)$, we mean a real-valued function on a set W which represents the decision maker's preference for elements in W. If he does not prefer v over w, then $\Psi(v) \leq \Psi(w)$. Conversely, if $\Psi(v) \leq \Psi(w)$, then the decision maker does not prefer v over w. He is indifferent to v and w if $\Psi(v) = \Psi(w)$. Under $\Psi(\cdot)$, all elements of W are comparable.

When the elements of W represent outcomes of a random process, *expected utility* provides a total ordering on equivalence classes of outcomes. For a discrete random vector Y, its expected utility is given by

$$E(\Psi(Y)) = \sum_{y \in W} \Psi(y)P\{Y = y\},$$

and for a continuous random vector by

$$E(\Psi(Y)) = \int_W \Psi(y)\, dF(y),$$

where $F(\cdot)$ represents the probability distribution of the W-valued random vector Y.

For finite-horizon Markov decision processes with discrete state spaces, the expected utility of policy π may be represented by

$$E^\pi[\Psi(R)] = \sum_{(\rho_1,\ldots,\rho_N) \in \Re} \Psi(\rho_1,\ldots,\rho_N)P_\Re^\pi\{(\rho_1,\ldots,\rho_N)\}.$$

Under the expected utility criterion, the decision maker prefers policy π to policy ν if

$$E^\pi[\Psi(R_1,\ldots,R_N)] > E^\nu[\Psi(R_1,\ldots,R_N)].$$

The policies are equivalent if $E^\pi[\Psi(R_1,\ldots,R_N)] = E^\nu[\Psi(R_1,\ldots,R_N)]$.

For arbitrary utility Ψ, the above computation is difficult to carry out because it requires explicit determination of the probability distribution of the vector R, or the joint distribution of the histories. However, by exploiting properties of conditional expectations, inductive computation appears possible.

This book will focus on models using *linear additive utility*, that is, if $(\rho_1,\ldots,\rho_N)$ represents a realization of the reward process, then

$$\Psi(\rho_1,\rho_2,\ldots,\rho_N) = \sum_{i=1}^{N} \rho_i.$$

Such utilities represent the preferences of a risk neutral decision maker who is indifferent to the timing of rewards. We may account for time preferences by including a discount factor in this expression. We discuss this alternative in Sec. 4.1.2. If $r_t(s, a)$ denotes the utility (instead of the reward) associated with choosing action a in state s, then $\Psi(r_1, r_2, \ldots, r_N) = \sum_{i=1}^{N} r_i$ denotes the decision maker's total utility. A formulation with r_t as utility may be appropriate when the state of the system represents a measure of wealth such as in the economic examples of Sec. 3.5.3. Multiplicative utilities of the form $\Psi(\rho_1, \rho_2, \ldots, \rho_N) = \prod_{i=1}^{N} u(\rho_i)$ can also be analyzed using Markov decision process methods.

As an alternative to expected utility criteria, the decision maker might use a criteria which explicitly takes into account both the total expected value and the variability of the sequence of rewards. As a consequence of using such a criteria, he might forego the policy with the largest expected total reward because its variance is too large. Policies with large reward variances are undesirable because, when using them, the chance of receiving a small total reward might be prohibitively high. Using such an approach in Example 4.1.1, a decision maker might prefer policy π with variance of the total reward of $\frac{1}{2}$ instead of policy ν which has variance 1, even though both have total expected reward of 1.

4.1.2 The Expected Total Reward Criterion

Let $v_N^\pi(s)$ represent the expected total reward over the decision making horizon if policy π is used and the system is in state s at the first decision epoch. For $\pi \in \Pi^{HR}$, it is defined by

$$v_N^\pi(s) \equiv E_s^\pi \left\{ \sum_{t=1}^{N-1} r_t(X_t, Y_t) + r_N(X_N) \right\}. \qquad (4.1.3)$$

For $\pi \in \Pi^{HD}$, the expected total reward can be expressed as

$$v_N^\pi(s) = E_s^\pi \left\{ \sum_{t=1}^{N-1} r_t(X_t, d_t(h_t)) + r_N(X_N) \right\} \qquad (4.1.4)$$

since at each decision epoch $Y_t = d_t(h_t)$.

Under the assumption that $|r_t(s, a)| \le M < \infty$ for $(s, a) \in S \times A$ and $t \le N$, and that S and A are discrete, $v_N^\pi(s)$ exists and is bounded for each $\pi \in \Pi^{HR}$ and each finite N. In greater generality, we also require measurability assumptions on r_t and π.

To account for the time value of rewards, we often introduce a discount factor. By a *discount factor*, we mean a scalar $\lambda, 0 \le \lambda < 1$ which measures the value at time n of a one unit reward received at time $n + 1$. A one-unit reward received t periods in the future has present value λ^t. For $\pi \in \Pi^{HR}$, the *expected total discounted reward*

$$v_{N, \lambda}^\pi(s) = E_s^\pi \left\{ \sum_{t=1}^{N-1} \lambda^{t-1} r_t(X_t, Y_t) + \lambda^{N-1} r_N(X_N) \right\}. \qquad (4.1.5)$$

Note that the exponent of λ is $t - 1$ and not t because r_t represents the value at decision epoch t of the reward received throughout period t, i.e., r_1 is the value at the first decision epoch of the reward received in period 1. Substitution of $\lambda = 1$ into (4.1.5) yields the expected total reward (4.1.3).

Taking the discount factor into account does not effect any theoretical results or algorithms in the finite-horizon case but might effect the decision maker's preference for policies. The discount factor plays a key analytic role in the infinite-horizon models we discuss in Chaps. 6–10.

4.1.3 Optimal Policies

Markov decision process theory and algorithms for finite-horizon models primarily concern determining a policy $\pi^* \in \Pi^{HR}$ with the largest expected total reward, and characterizing this value. That is, we seek a policy π^* for which

$$v_N^{\pi^*}(s) \geq v_N^{\pi}(s), \quad s \in S \tag{4.1.6}$$

for all $\pi \in \Pi^{HR}$. We refer to such a policy as an *optimal policy*. In some models such a policy need not exist, so instead we seek an *ε-optimal policy*, that is, for an $\varepsilon > 0$, a policy π_ε^* with the property that

$$v_N^{\pi_\varepsilon^*}(s) + \varepsilon > v_N^{\pi}(s), \quad s \in S \tag{4.1.7}$$

for all $\pi \in \Pi^{HR}$.

We seek to characterize the *value* of the Markov decision problem, v_N^*, defined by

$$v_N^*(s) \equiv \sup_{\pi \in \Pi^{HR}} v_N^{\pi}(s), \quad s \in S, \tag{4.1.8}$$

and when the supremum in (4.1.8) is attained, by

$$v_N^*(s) = \max_{\pi \in \Pi^{HR}} v_N^{\pi}(s), \quad s \in S. \tag{4.1.9}$$

The expected total reward of an optimal policy π^* satisfies

$$v_N^{\pi^*}(s) = v_N^*(s), \quad s \in S$$

and the value of an ε-optimal policy π_ε^* satisfies

$$v_N^{\pi_\varepsilon^*}(s) + \varepsilon > v_N^*(s), \quad s \in S.$$

As a consequence of the definition of the supremum, such a policy exists for any $\varepsilon > 0$.

We have implicitly assumed above that the decision maker wishes to choose an optimal policy for all possible initial-system states. In practice, all that might be required is an optimal policy for some specified initial state. Alternatively, he might seek a policy prior to knowing the initial state. In this case he seeks a $\pi \in \Pi^{HR}$ which maximizes $\sum_{s \in S} v_N^{\pi}(s) P_1\{X_1 = s\}$. Clearly he may find such a policy by maximizing $v_N^{\pi}(s)$ for each s for which $P_1\{X_1 = s\} > 0$.

We use the expression *Markov decision problem* (MDP) for a Markov decision process together with an optimality criteria. The problems defined above are discrete-time, finite-horizon Markov decision problems with expected total reward criteria.

4.2   FINITE-HORIZON POLICY EVALUATION

Markov decision problem theory and computation is based on using backward induction (dynamic programming) to recursively evaluate expected rewards. In this section, we introduce the fundamental recursion of dynamic programming by providing an efficient method to evaluate the expected total reward of a fixed policy.

Let $\pi = (d_1, d_2, \ldots, d_{N-1})$ be a randomized history-dependent policy. Let u_t^π: $H_t \to R^1$ denote the total expected reward obtained by using policy π at decision epochs $t, t + 1, \ldots, N - 1$. If the history at decision epoch t is $h_t \in H_t$, then define u_t^π for $t < N$ by

$$u_t^\pi(h_t) = E_{h_t}^\pi \left\{ \sum_{n=t}^{N-1} r_n(X_n, Y_n) + r_N(X_N) \right\}, \qquad (4.2.1)$$

where the conditional expectation is as defined in (2.1.15), and let $u_N^\pi(h_N) = r_N(s)$ when $h_N = (h_{N-1}, a_{N-1}, s)$. When $h_1 = s$, $u_1^\pi(s) = v_N^\pi(s)$. The difference between u_t^π and v_N^π is that v_N^π includes rewards over the *entire* future, while u_t^π only incorporates rewards from decision epoch t onward.

We now show how to compute v_N^π by inductively evaluating u_t^π. We call it the *finite-horizon policy evaluation algorithm*. To simplify the notation, assume that a deterministic $\pi \in \Pi^{HD}$ has been specified and that S is discrete. In practice, we will not need to evaluate randomized policies, because subsequent results establish that deterministic policies are optimal under the expected total reward criteria.

The Finite Horizon-Policy Evaluation Algorithm (for fixed $\pi \in \Pi^{HD}$)

1. Set $t = N$ and $u_N^\pi(h_N) = r_N(s_N)$ for all $h_N = (h_{N-1}, a_{N-1}, s_N) \in H_N$.
2. If $t = 1$, stop, otherwise go to step 3.
3. Substitute $t - 1$ for t and compute $u_t^\pi(h_t)$ for each $h_t = (h_{t-1}, a_{t-1}, s_t) \in H_t$ by

$$u_t^\pi(h_t) = r_t(s_t, d_t(h_t)) + \sum_{j \in S} p_t(j|s_t, d_t(h_t)) u_{t+1}^\pi(h_t, d_t(h_t), j), \qquad (4.2.2)$$

noting that $(h_t, d_t(h_t), j) \in H_{t+1}$.
4. Return to 2.

Some motivation for (4.2.2) follows. The expected value of policy π over periods $t, t + 1, \ldots, N$ when the history at epoch t is h_t equals the immediate reward received by selecting action $d_t(h_t)$ plus the expected reward over the remaining periods. The second term contains the product of the probability of being in state j at decision epoch $t + 1$ if action $d_t(h_t)$ is used, and the expected total reward obtained using policy π over periods $t + 1, \ldots, N$ when the history at epoch $t + 1$ is $h_{t+1} = (h_t, d_t(h_t), j)$. Summing over all possible j gives the desired expectation expressed in terms of u_{t+1}^π; that is,

$$u_t^\pi(h_t) = r_t(s_t, d_t(h_t)) + E_{h_t}^\pi\{u_{t+1}^\pi(h_t, d_t(h_t), X_{t+1})\}. \qquad (4.2.3)$$

This inductive scheme reduces the problem of computing expected total rewards over N periods to a sequence of $N - 1$ similar one-period calculations having immediate rewards r_t and terminal rewards u_{t+1}^π as in Section 2.2. This reduction is the essence of dynamic programming:

Multistage problems may be solved by analyzing a sequence of simpler inductively defined single-stage problems.

By doing this, we avoid the onerous task of explicitly determining the joint probability distribution of histories under π. The procedure for finding optimal policies in the next section generalizes these calculations.

We now formally show that this algorithm does in fact yield u_t^π. The proof is by induction where the index of induction is t. Contrary to usual mathematical convention, we apply the induction backwards. This simplifies the subscripts in the recursions.

Theorem 4.2.1. Let $\pi \in \Pi^{\text{HD}}$ and suppose u_t^π, $t \le N$, has been generated by the policy evaluation algorithm. Then, for all $t \le N$, (4.2.1) holds, and $v_N^\pi(s) = u_1^\pi(s)$ for all $s \in S$.

Proof. The result is obviously true when $t = N$. Suppose now that (4.2.1) holds for $t + 1, t + 2, \ldots, N$. Then by using representation (4.2.3) and the induction hypothesis

$$u_t^\pi(h_t) = r_t(s_t, d_t(h_t)) + E_{h_t}^\pi\left[E_{h_{t+1}}^\pi\left\{ \sum_{n=t+1}^{N-1} r_n(X_n, Y_n) + r_N(X_N) \right\} \right] \qquad (4.2.4)$$

$$= r_t(s_t, d_t(h_t)) + E_{h_t}^\pi\left\{ \sum_{n=t+1}^{N-1} r_n(X_n, Y_n) + r_N(X_N) \right\}. \qquad (4.2.5)$$

Since s_t and h_t are known at decision epoch t, $X_t = s_t$, so that the first term in (4.2.5) can be included inside the expectation yielding the desired result.   □

Observe that in (4.2.3), when taking an expectation conditional on the history up to time $t + 1$, X_{t+1} is known, but when taking conditional expectation with respect to the history up to time t, X_{t+1} is random. Thus, since the algorithm evaluates $u_{t+1}^\pi(h_{t+1})$ for all h_{t+1}, before evaluating u_t^π the expectation over X_{t+1} can be computed.

Fundamental to justifying the validity of this algorithm is the assumption of linear additive utility. A recursive scheme is also possible for models with multiplicative exponential utility.

In presenting and analyzing this algorithm, we have assumed discrete state space and arbitrary action sets. In theory, this algorithm is also valid for general state and sets provided that integration replaces summation in (4.2.2) and the functions in the expectations are measurable and integrable. From a computational perspective, this degree of generality is unnecessary since numerical calculations are implementable only for problems with finite-state and action sets. On the other hand, results on the structure of optimal policies require such generality.

A key feature of this algorithm is that the induction is backwards. This is required because of the probabilistic nature of the MDP. If the model were deterministic, policies could be evaluated by either forward or backward induction because expectations need not be computed. To evaluate a conditional expectation from decision epoch t onward, we require the conditional expectation from decision epoch $t + 1$ onward for each possible history at decision epoch $t + 1$. Proceeding in a forward fashion would require enumeration of all histories, and deriving the probability distribution of each under policy π.

We have stated the algorithm for nonrandomized history-dependent policies. To generalize it to randomized policies would require an additional summation in (4.2.2) to account for the probability distribution of the action at decision epoch t under decision rule d_t as follows:

$$u_t^\pi(h_t) = \sum_{a \in A_{s_t}} q_{d_t(h_t)}(a) \left\{ r_t(s_t, a) + \sum_{j \in S} p_t(j|s_t, a) u_{t+1}(h_t, a, j) \right\}. \quad (4.2.6)$$

Theorem 4.2.1 may be extended to $\pi \in \Pi^{HR}$ as follows.

Theorem 4.2.2. Let $\pi \in \Pi^{HR}$ and suppose u_t^π, $t \leq N$ has been generated by the policy evaluation algorithm with (4.2.6) replacing (4.2.2). Then, for all $t \leq N$, (4.2.1) holds and $v_N^\pi(s) = u_1^\pi(s)$ for all $s \in S$.

The policy evaluation algorithm for deterministic Markovian policies replaces (4.2.2) by

$$u_t^\pi(s_t) = r_t(s_t, d_t(s_t)) + \sum_{j \in S} p_t(j|s_t, d_t(s_t)) u_{t+1}^\pi(j). \quad (4.2.7)$$

Because histories need not be retained, this simplifies computation through reduced storage and fewer arithmetic operations. An algorithm based on (4.2.7) evaluates the right-hand side above *only once* for each history that includes s_t in its realization at time t. In contrast, the right-hand side of (4.2.2) requires evaluation for each history up to time t because a different decision rule might be used at time t for each realization.

When there are K states and L actions available in each state, at decision epoch t there are $K^{t+1}L^t$ histories. Since each pass through step 3 of the evaluation algorithm requires K multiplications, evaluation of a nonrandomized history-dependent decision rule requires $K^2 \sum_{i=0}^{N-1} (KL)^i$ multiplications. For Markov policies, the number of multiplications decreases to $(N - 1)K^2$.

4.3 OPTIMALITY EQUATIONS AND THE PRINCIPLE OF OPTIMALITY

In this section we introduce optimality equations (sometimes referred to as Bellman equations or functional equations) and investigate their properties. We show that solutions of these equations correspond to optimal value functions and that they also provide a basis for determining optimal policies. We assume either finite or countable S to avoid technical subtleties.

Let

$$u_t^*(h_t) = \sup_{\pi \in \Pi^{HR}} u_t^\pi(h_t). \tag{4.3.1}$$

It denotes the supremum over all policies of the expected total reward from decision epoch t onward when the history up to time t is h_t. For $t > 1$, we need not consider all policies when taking the above supremum. Since we know h_t, we only consider portions of policies from decision epoch t onward; that is, we require only the supremum over $(d_t, d_{t+1}, \ldots, d_{N-1}) \in D_t^{HR} \times D_{t+1}^{HR} \times \cdots \times D_{N-1}^{HR}$. When minimizing costs instead of maximizing rewards, we sometimes refer to u_t^* as a *cost-to-go* function.

The *optimality equations* are given by

$$u_t(h_t) = \sup_{a \in A_{s_t}} \left\{ r_t(s_t, a) + \sum_{j \in S} p_t(j|s_t, a) u_{t+1}(h_t, a, j) \right\} \tag{4.3.2}$$

for $t = 1, \ldots, N - 1$ and $h_t = (h_{t-1}, a_{t-1}, s_t) \in H_t$. For $t = N$, we add the boundary condition

$$u_N(h_N) = r_N(s_N) \tag{4.3.3}$$

for $h_N = (h_{N-1}, a_{N-1}, s_N) \in H_N$.

These equations reduce to the policy evaluation equations (4.2.1) when we replace the supremum over all actions in state s_t by the action corresponding to a specified policy, or equivalently, when A_s is a singleton for each $s \in S$.

The operation "sup" in (4.3.2) is implemented by evaluating the quantity in brackets for each $a \in A_{s_t}$ and then choosing the supremum over all of these values. When A_{s_t} is a continuum, the supremum might be found analytically. If the supremum in (4.3.2) is attained, for example, when each A_{s_t} is finite, it can be replaced by "max" so (4.3.2) becomes

$$u_t(h_t) = \max_{a \in A_{s_t}} \left\{ r_t(s_t, a) + \sum_{j \in S} p_t(j|s_t, a) u_{t+1}(h_t, a, j) \right\}. \tag{4.3.4}$$

A solution to the system of equations (4.3.2) or (4.3.4) and boundary condition (4.3.3) is a sequence of functions $u_t: H_t \to R$, $t = 1, \ldots, N$, with the property that u_N satisfies (4.3.3), u_{N-1} satisfies the $(N - 1)$th equation with u_N substituted into the right-hand side of the $(N - 1)$th equation, and so forth.

The optimality equations are fundamental tools in Markov decision theory, having the following important and useful properties.

a. Solutions to the optimality equations are the optimal returns from period t onward for each t.

b. They provide a method for determining whether a policy is optimal. If the expected total reward of policy π from period t onward satisfies this system of equations for $t = 1, \ldots, N$ it is optimal.

c. They are the basis for an efficient procedure for computing optimal return functions and policies.

d. They may be used to determine structural properties of optimal policies and return functions.

Before stating and proving the main result in this chapter, we introduce the following important yet simple lemma.

Lemma 4.3.1. Let w be a real-valued function on an arbitrary discrete set W and let $q(\cdot)$ be a probability distribution on W. Then

$$\sup_{u \in W} w(u) \geq \sum_{u \in W} q(u)w(u)$$

Proof. Let $w^* = \sup_{u \in W} w(u)$. Then,

$$w^* = \sum_{u \in W} q(u)w^* \geq \sum_{u \in W} q(u)w(u).$$

$\square$

Note that the lemma remains valid with W a Borel subset of a measurable space, $w(u)$ an integrable function on W, and the summation replaced by integration.

The following theorem summarizes the optimality properties of solutions of the optimality equation. Its inductive proof illustrates several dynamic programming principles and consists of two parts. First we establish that solutions provide upper bounds on u_t^* and then we establish existence of a policy π' for which $u_t^{\pi'}$ is arbitrarily close to u_t.

Theorem 4.3.2. Suppose u_t is a solution of (4.3.2) for $t = 1, \ldots, N - 1$, and u_N satisfies (4.3.3). Then

a. $u_t(h_t) = u_t^*(h_t)$ for all $h_t \in H_t$, $t = 1, \ldots, N$, and
b. $u_1(s_1) = v_N^*(s_1)$ for all $s_1 \in S$.

Proof. The proof is in two parts. First we establish by induction that $u_n(h_n) \geq u_n^*(h_n)$ for all $h_n \in H_n$ and $n = 1, 2, \ldots, N$.

Since no decision is made in period N, $u_N(h_N) = r_N(s_N) = u_N^\pi(h_N)$ for all $h_N \in H_N$ and $\pi \in \Pi^{HR}$. Therefore $u_N(h_N) = u_N^*(h_N)$ for all $h_N \in H_N$. Now assume that $u_t(h_t) \geq u_t^*(h_t)$ for all $h_t \in H_t$ for $t = n + 1, \ldots, N$. Let $\pi' = (d_1', d_2', \ldots, d_{N-1}')$ be an arbitrary policy in Π^{HR}. For $t = n$, the optimality equation is

$$u_n(h_n) = \sup_{a \in A_{s_n}} \left\{ r_n(s_n, a) + \sum_{j \in S} p_n(j|s_n, a) u_{n+1}(h_n, a, j) \right\}.$$

By the induction hypothesis

$$u_n(h_n) \geq \sup_{a \in A_{s_n}} \left\{ r_n(s_n, a) + \sum_{j \in S} p_n(j|s_n, a) u_{n+1}^*(h_n, a, j) \right\} \tag{4.3.5}$$

$$\geq \sup_{a \in A_{s_n}} \left\{ r_n(s_n, a) + \sum_{j \in S} p_n(j|s_n, a) u_{n+1}^{\pi'}(h_n, a, j) \right\} \tag{4.3.6}$$

$$\geq \sum_a q_{d_n'(h_n)}(a) \left\{ r_n(s_n, a) + \sum_{j \in S} p_n(j|s_n, a) u_{n+1}^{\pi'}(h_n, a, j) \right\} \tag{4.3.7}$$

$$= u_n^{\pi'}(h_n).$$

The inequality in (4.3.5) follows from the induction hypothesis and the non-negativity of p_n, and that in (4.3.6) from the definition of u_{n+1}^*. That in (4.3.7) follows from Lemma 4.3.1 with $W = A_{s_n}$ and w equal to the expression in brackets. The last equality follows from (4.2.6) and Theorem 4.2.2. Since π' is arbitrary,

$$u_n(h_n) \geq u_n^{\pi}(h_n) \quad \text{for all } \pi \in \Pi^{HR}.$$

Thus $u_n(h_n) \geq u_n^*(h_n)$ and the induction hypothesis holds.

Now we establish that for any $\varepsilon > 0$, there exists a $\pi' \in \Pi^{HD}$ for which

$$u_n^{\pi'}(h_n) + (N - n)\varepsilon \geq u_n(h_n) \tag{4.3.8}$$

for all $h_n \in H_n$ and $n = 1, 2, \ldots, N$. To do this, construct a policy $\pi' = (d_1, d_2, \ldots, d_{N-1})$ by choosing $d_n(h_n)$ to satisfy

$$r_n(s_n, d_n(h_n)) + \sum_{j \in S} p_n(j|s_n, d_n(h_n)) u_{n+1}(s_n, d_n(h_n), j) + \varepsilon \geq u_n(h_n).$$

$$\tag{4.3.9}$$

We establish (4.3.8) by induction. Since $u_N^{\pi'}(h_N) = u_N(h_N)$, the induction hypothesis holds for $t = N$. Assume that $u_t^{\pi'}(h_t) + (N - t)\varepsilon \geq u_t(h_t)$ for $t = n + 1, \ldots, N$.

Then it follows from Theorem 4.2.1 and (4.3.9) that

$$u_n^{\pi'}(h_n) = r_n(s_n, d_n(h_n)) + \sum_{j \in S} p_n(j|s_n, d_n(h_n)) u_{n+1}^{\pi'}(s_n, d_n(h_n), j)$$

$$\geq r_n(s_n, d_n(h_n)) + \sum_{j \in S} p_n(j|s_n, d_n(h_n)) u_{n+1}(s_n, d_n(h_n), j)$$

$$-(N - n - 1)\varepsilon$$

$$\geq u_n(h_n) - (N - n)\varepsilon.$$

Thus the induction hypothesis is satisfied and (4.3.8) holds for $n = 1, 2, \ldots, N$. Therefore for any $\varepsilon > 0$, there exists a $\pi' \in \Pi^{HR}$ for which

$$u_n^*(h_n) + (N - n)\varepsilon \geq u_n^{\pi'}(h_n) + (N - n)\varepsilon \geq u_n(h_n) \geq u_n^*(h_n)$$

so that (a) follows. Part (b) follows from the definitions of the quantities. □

Part (a) of Theorem 4.3.2 means that solutions of the optimality equation are the optimal value functions from period t onward, and result (b) means that the solution to the equation with $n = 1$ is the value function for the MDP, that is, it is the optimal value from decision epoch 1 onward.

The following result shows how to use the optimality equations to find optimal policies, and to verify that a policy is optimal. Theorem 4.3.3 uses optimality equations (4.3.4) in which maxima are attained. Theorem 4.3.4 considers the case of suprema.

Theorem 4.3.3. Suppose u_t^*, $t = 1, \ldots, N$ are solutions of the optimality equations (4.3.4) subject to boundary condition (4.3.3), and that policy $\pi^* = (d_1^*, d_2^*, \ldots, d_{N-1}^*) \in \Pi^{HD}$ satisfies

$$r_t(s_t, d_t^*(h_t)) + \sum_{j \in S} p_t(j|s_t, d_t^*(h_t)) u_{t+1}^*(h_t, d_t^*(h_t), j)$$

$$= \max_{a \in A_{s_t}} \left\{ r_t(s_t, a) + \sum_{j \in S} p_t(j|s_t, a) u_{t+1}^*(h_t, a, j) \right\} \quad (4.3.10)$$

for $t = 1, \ldots, N - 1$.
Then

a. For each $t = 1, 2, \ldots, N$,

$$u_t^{\pi^*}(h_t) = u_t^*(h_t), \quad h_t \in H_t. \quad (4.3.11)$$

b. π^* is an optimal policy, and

$$v_N^{\pi^*}(s) = v_N^*(s), \quad s \in S. \quad (4.3.12)$$

Proof. We establish part (a); part (b) follows from Theorem 4.2.1 and Theorem 4.3.2b. The proof is by induction. Clearly

$$u_N^{\pi^*}(h_n) = u_N^*(h_n), \quad h_n \in H_n.$$

Assume the result holds for $t = n + 1, \ldots, N$. Then, for $h_n = (h_{n-1}, d_{n-1}^*(h_{n-1})), s_n)$,

$$u_n^*(h_n) = \max_{a \in A_{s_n}} \left\{ r_n(s_n, a) + \sum_{j \in S} p_n(j|s_n, a) u_{n+1}^*(h_n, a, j) \right\}$$

$$= r_n(s_n, d_n^*(h_n)) + \sum_{j \in S} p_n(j|s_n, d_n^*(h_n)) u_{n+1}^{\pi^*}(h_n, d_n^*(h_n), j)$$

$$= u_n^{\pi^*}(h_n).$$

The second equality is a consequence of (4.3.10), and the induction hypothesis and the last equality follows from Theorem 4.2.1. Thus the induction hypothesis is satisfied and the result follows. □

We frequently write equation (4.3.10) as

$$d_t^*(h_t) \in \arg\max_{a \in A_{s_t}} \left\{ r_t(s_t, a) + \sum_{j \in S_t} p_t(j|s_t, a) u_{t+1}^*(h_t, a, j) \right\},$$

where "arg max" has been defined in Sec. 2.2.

The theorem implies that an optimal policy is found by first solving the optimality equations, and then for each history choosing a decision rule which selects any action which attains the maximum on the right-hand side of (4.3.10) for $t = 1, 2, \ldots, N$. When using this equation in computation, for each history the right-hand side is evaluated for all $a \in A_{s_t}$ and the set of maximizing actions is recorded. When there is more than one maximizing action in this set, there is more than one optimal policy.

Note that we have restricted attention to history-dependent deterministic policies in Theorem 4.3.3. This is because if there existed a history-dependent randomized policy which satisfied the obvious generalization of (4.3.10), as a result of Lemma 4.3.1, we could find a deterministic policy which satisfied (4.3.10). We expand on this point in the next section.

This theorem provides a formal statement of "The Principle of Optimality," a fundamental result of dynamic programming. An early verbal statement appeared in Bellman (1957, p. 83).

"An optimal policy has the property that whatever the initial state and initial decision are, the remaining decisions must constitute an optimal policy with regard to the state resulting from the first decision."

Denardo (1982, p. 15) provides a related statement.

"There exists a policy that is optimal for every state (at every stage)."

Any deterministic policy π^* satisfying (4.3.10) has these properties and such a policy must *exist* because maxima are attained. Alternatively, Theorem 4.3.3 provides *sufficient* conditions for verifying the optimality of a policy.

Note that "The Principle of Optimality" may not be valid for other optimality criteria. In Sec. 4.6.2, we provide an example in which it does not hold, albeit under a nonstandard optimality criteria.

In case the supremum in (4.3.2) is not attained, the decision maker must be content with ε-optimal policies. To account for this we modify Theorem 4.3.3 as follows. Arguments in the second part of the proof of Theorem 4.3.2 can be used to establish it.

Theorem 4.3.4. Let $\varepsilon > 0$ be arbitrary and suppose u_t^*, $t = 1, \ldots, N$ are solutions of the optimality equations (4.3.2) and (4.3.3). Let $\pi^\varepsilon = (d_1^\varepsilon, d_2^\varepsilon, \ldots, d_{N-1}^\varepsilon) \in \Pi^{HD}$ satisfy

$$
r_t(s_t, d_t^\varepsilon(h_t)) + \sum_{j \in S} p_t(j|s_t, d_t^\varepsilon(h_t))u_{t+1}^*(h_t, d_t^\varepsilon(h_t), j) + \frac{\varepsilon}{N-1}
$$

$$
\geq \sup_{a \in A_{s_t}} \left\{ r_t(s_t, a) + \sum_{j \in S} p_t(j|s_t, a)u_{t+1}^*(h_t, a, j) \right\} \tag{4.3.13}
$$

for $t = 1, 2, \ldots, N - 1$. Then,

a. For each $t = 1, 2, \ldots, N - 1$,

$$
u_t^{\pi^\varepsilon}(h_t) + (N - t)\frac{\varepsilon}{N-1} \geq u_t^*(h_t), \quad h_t \in H_t. \tag{4.3.14}
$$

b. π^ε is an ε-optimal policy, that is

$$
v_N^{\pi^\varepsilon}(s) + \varepsilon \geq v_N^*(s), \quad s \in S. \tag{4.3.15}
$$

4.4 OPTIMALITY OF DETERMINISTIC MARKOV POLICIES

This section provides conditions under which there exists an optimal policy which is deterministic and Markovian, and illustrates how backward induction can be used to determine the structure of an optimal policy.

From the perspective of application, we find it comforting that by restricting attention to nonrandomized Markov policies, which are simple to implement and evaluate, we may achieve as large an expected total reward as if we used randomized history-dependent policies. We show that when the immediate rewards and transition probabilities depend on the past only through the current state of the system (as assumed throughout this book), the optimal value functions depend on the history only through the current state of the system. This enables us to impose assumptions on the action sets, rewards, and transition probabilities which ensure existence of optimal policies which depend only on the system state.

Inspection of the proof of Theorem 4.3.2 reveals that it constructs an ε-optimal deterministic history-dependent policy. Theorem 4.3.3 and 4.3.4 identify optimal and ε-optimal policies. We summarize these results as follows.

Theorem 4.4.1.

a. For any $\varepsilon > 0$, there exists an ε-optimal policy which is deterministic history dependent. Any policy in Π^{HD} which satisfies (4.3.13) is ε-optimal.

b. Let u_t^* be a solution of (4.3.2) and (4.3.3) and suppose that for each t and $s_t \in S$, there exists an $a' \in A_{s_t}$ for which

$$r_t(s_t, a') + \sum_{j \in S} p_t(j|s_t, a')u_{t+1}^*(h_t, a', j)$$

$$= \sup_{a \in A_{s_t}} \left\{ r_t(s_t, a) + \sum_{j \in S} p_t(j|s_t, a)u_{t+1}^*(h_t, a, j) \right\} \quad (4.4.1)$$

for all $h_t = (s_{t-1}, a_{t-1}, s_t) \in H_t$. Then there exists a deterministic history-dependent policy which is optimal.

Proof. Part (a) follows from the second part of the proof of Theorem 4.3.2. The policy π^ε in Theorem 4.3.4 is optimal and deterministic. When there exists an $a' \in A_{s_t}$ for which (4.4.1) holds, the policy $\pi^* \in \Pi^{HD}$ of Theorem 4.3.3 is optimal. $\square$

We next show by induction that there exists an optimal policy which is Markovian and deterministic.

Theorem 4.4.2.   Let u_t^*, $t = 1, \ldots, N$ be solutions of (4.3.2) and (4.3.3). Then

a. For each $t = 1, \ldots, N$, $u_t^*(h_t)$ depends on h_t only through s_t.

b. For any $\varepsilon > 0$, there exists an ε-optimal policy which is deterministic and Markov.

c. If there exists an $a' \in A_{s_t}$ such that (4.4.1) holds for each $s_t \in S$ and $t = 1, 2, \ldots, N - 1$, there exists an optimal policy which is deterministic and Markov.

Proof. We show that (a) holds by induction. Since $u_N^*(h_N) = u_N^*(h_{N-1}, a_{N-1}, s) = r_N(s)$ for all $h_{N-1} \in H_{N-1}$ and $a_{N-1} \in A_{s_{N-1}}$, $u_N^*(h_N) = u_N^*(s_N)$. Assume now that (a) is valid for $n = t + 1, \ldots, N$. Then

$$u_t^*(h_t) = \sup_{a \in A_{s_t}} \left\{ r_t(s_t, a) + \sum_{j \in S} p_t(j|s_t, a)u_{t+1}^*(h_t, a, j) \right\},$$

which by the induction hypothesis gives

$$u_t^*(h_t) = \sup_{a \in A_{s_t}} \left\{ r_t(s_t, a) + \sum_{j \in S} p_t(j|s_t, a)u_{t+1}^*(j) \right\}. \quad (4.4.2)$$

Since the quantity in brackets depend on h_t only through s_t, (a) holds for all t.

Choose $\varepsilon > 0$, and let $\pi^\varepsilon = (d_1^\varepsilon, d_2^\varepsilon, \ldots, d_{N-1}^\varepsilon)$ be any policy in Π^{MD} satisfying

$$r_t(s_t, d_t^\varepsilon(s_t)) + \sum_{j \in S} p_t(j|s_t, d_t^\varepsilon(s_t))u_{t+1}^*(j) + \frac{\varepsilon}{N-1}$$

$$\geq \sup_{a \in A_{s_t}} \left\{ r_t(s_t, a) + \sum_{j \in S} p_t(j|s_t, a)u_{t+1}^*(j) \right\}. \tag{4.4.3}$$

Then, by part (a), π^ε satisfies the hypotheses of Theorem 4.3.4b so it is ε-optimal.

Part (c) follows by noting that under the hypotheses of Theorem 4.4.1b, there exists a $\pi^* = (d_1^*, d_2^*, \ldots, d_{N-1}^*) \in \Pi^{MD}$, which satisfies

$$r_t(s_t, d_t^*(s_t)) + \sum_{j \in S} p_t(j|s_t, d_t^*(s_t))u_{t+1}^*(j)$$

$$= \max_{a \in A_{s_t}} \left\{ r_t(s_t, a) + \sum_{j \in S} p_t(j|s_t, a)u_{t+1}^*(j) \right\}. \tag{4.4.4}$$

Therefore by part (a) and Theorem 4.3.3b, π^* is optimal.   □

Thus we have established that

$$v_N^*(s) = \sup_{\pi \in \Pi^{HR}} v_N^\pi(s) = \sup_{\pi \in \Pi^{MD}} v_N^\pi(s), \quad s \in S.$$

We now provide conditions under which the supremum in (4.4.1) is attained, so that we may easily determine when there exists a deterministic Markovian policy which is optimal. Following Appendix B, we say that a set X is *compact* if it is a compact subset of a complete separable metric space. In many applications we consider only compact subsets of R^n. The proof of part (c) below is quite technical and relies on properties of semicontinuous functions described in Appendix B. Note that more subtle arguments are required for general S.

Proposition 4.4.3. Assume S is finite or countable, and that

a. A_s is finite for each $s \in S$, or
b. A_s is compact, $r_t(s, a)$ is continuous in a for each $s \in S$, there exists an $M < \infty$ for which $|r_t(s, a)| \leq M$ for all $a \in A_s$, $s \in S$, and $p_t(j|s, a)$ is continuous in a for each $j \in S$ and $s \in S$ and $t = 1, 2, \ldots, N$, or
c. A_s is a compact, $r_t(s, a)$ is upper semicontinuous (u.s.c.) in a for each $s \in S$, there exists an $M < \infty$ for which $|r_t(s, a)| \leq M$ for all $a \in A_s$, $s \in S$, and for each $j \in S$ and $s \in S$, $p_t(j|s, a)$ is lower semi-continuous (l.s.c.) in a and $t = 1, 2, \ldots, N$.

Then there exists a deterministic Markovian policy which is optimal.

Proof. We show that there exists an a' which satisfies (4.4.1) under hypothesis (a), (b), or (c), in which case the result follows from Theorem 4.4.2. Note that as a consequence of Theorem 4.4.2c, we require that, for each $s \in S$, there exists an

$a' \in A_s$, for which

$$r_t(s, a') + \sum_{j \in S} p_t(j|s_t, a')u^*_{t+1}(j)$$

$$= \sup_{a \in A_s} \left\{ r_t(s, a) + \sum_{j \in S} p_t(j|s, a)u^*_{t+1}(j) \right\}. \qquad (4.4.5)$$

Clearly such an a' exists when A_s is finite so the result follows under a.

Suppose that the hypotheses in part (c) hold. By assumption $|r_t(s, a)| \le M$ for all $s \in S$ and $a \in A_s$, $|u^*_t(s)| \le NM$ for all $s \in S$ and $t = 1, 2, \ldots, N$. Therefore, for each t, $u^*_t(s) - NM \le 0$. Now apply Proposition B.3, with s fixed and X identified with S_Y, A_s identified with $q(x, y)$, $p_t(j|s, a)$ identified with $f(w, x)$ and $u^*_t(s) - NM$ identified with $f(x)$, to obtain that

$$\sum_{j \in S} p_t(j|s, a)\left[u^*_{t+1}(j) - NM\right]$$

is u.s.c., from which we conclude that $\sum_{j \in S} p_t(j|s, a)u^*_{t+1}(j)$ is u.s.c. By Proposition B.1.a, $r_t(s, a) + \sum_{j \in S} p_t(j|s, a)u^*_{t+1}(j)$ is u.s.c. in a for each $s \in S$. Therefore by Theorem B.2, the supremum over a in (4.4.5) is attained, from which the result follows.

Conclusion (b) follows from (c) since continuous functions are both upper and lower semicontinuous. □

To illustrate this result, consider the following example which we analyze in further detail in Sec. 6.4.

Example 4.4.1. Let $N = 3$; $S = \{s_1, s_2\}$; $A_{s_1} = [0, 2]$, and $A_{s_2} = \{a_{2,1}\}$; $r_t(s_1, a) = -a^2$, $r_t(s_2, a_{2,1}) = -\frac{1}{2}$, $p_t(s_1|s_1, a) = \frac{1}{2}a$, $p_t(s_2|s_1, a) = 1 - \frac{1}{2}a$, and $p_t(s_2|s_2, a_{2,1}) = 1$ for $t = 1, 2$, $r_3(s_1) = -1$, and $r_3(s_2) = -\frac{1}{2}$. (See Fig. 4.4.1)

Since A_{s_1} is compact and $r_t(s_1, \cdot)$ and $p_t(j|s_1, \cdot)$ are continuous functions on A_{s_1} there exists a deterministic Markov policy which is optimal.

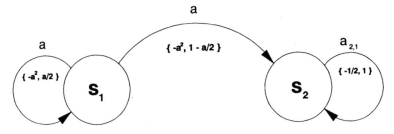

Figure 4.4.1 Graphical representation of Example 4.4.1.

4.5 BACKWARD INDUCTION

Backward induction provides an efficient method for solving finite-horizon discrete-time MDPs. For stochastic problems, enumeration and evaluation of *all* policies is the only alternative, but forward induction and reaching methods provide alternative solution methods for deterministic systems. The terms "backward induction" and "dynamic programming" are synonymous, although the expression "dynamic programming" often refers to all results and methods for sequential decision processes. This section presents the backward induction algorithm and shows how to use it to find optimal policies and value functions. The algorithm generalizes the policy evaluation algorithm of Sec. 4.2.

We present the algorithm for a model in which maxima are obtained in (4.3.2), so that we are assured of obtaining an optimal (instead of an ε-optimal) Markovian deterministic policy. The algorithm solves optimality equations (4.3.4) subject to boundary condition (4.3.3). Generalization to models based on (4.3.2) is left as an exercise.

The Backward Induction Algorithm

1. Set $t = N$ and

$$u_N^*(s_N) = r_N(s_N) \quad \text{for all } s_N \in S,$$

2. Substitute $t - 1$ for t and compute $u_t^*(s_t)$ for each $s_t \in S$ by

$$u_t^*(s_t) = \max_{a \in A_{s_t}} \left\{ r_t(s_t, a) + \sum_{j \in S} p_t(j|s_t, a) u_{t+1}^*(j) \right\}. \qquad (4.5.1)$$

Set

$$A_{s_t, t}^* = \operatorname*{arg\,max}_{a \in A_{s_t}} \left\{ r_t(s_t, a) + \sum_{j \in S} p_t(j|s_t, a) u_{t+1}^*(j) \right\}. \qquad (4.5.2)$$

3. If $t = 1$, stop. Otherwise return to step 2.

As a consequence of Theorem 4.4.2, which among other results shows that u_t^* depends on $h_t = (s_{t-1}, a_{t-1}, s_t)$ only through s_t, we need not evaluate u_t^* for all $h_t \in H_t$. This significantly reduces computational effort. Combining results from Theorems 4.3.2 and 4.4.2 yields the following properties of the iterates of the backward induction algorithm.

Theorem 4.5.1. Suppose u_t^*, $t = 1, \ldots, N$ and $A_{s_t, t}^*$ $t = 1, \ldots, N - 1$ satisfy (4.5.1) and (4.5.2); then,

a. for $t = 1, \ldots, N$ and $h_t = (h_{t-1}, a_{t-1}, s_t)$

$$u_t^*(s_t) = \sup_{\pi \in \Pi^{HR}} u_t^\pi(h_t), \quad s_t \in S.$$

b. Let $d_t^*(s_t) \in A_{s_t, t}^*$ for all $s_t \in S$, $t = 1, \ldots, N - 1$, and let $\pi^* = (d_1^*, \ldots, d_{N-1}^*)$. Then $\pi^* \in \Pi^{MD}$ is optimal and satisfies

$$v_N^{\pi^*}(s) = \sup_{\pi \in \Pi^{HR}} v_N^\pi(s), \quad s \in S$$

and

$$u_t^{\pi^*}(s_t) = u_t^*(s_t), \quad s_t \in S$$

for $t = 1, \ldots, N$.

This theorem represents a formal statement of the following properties of the backward induction algorithm.

a. For $t = 1, 2, \ldots, N - 1$, it finds sets $A_{s_t, t}^*$ which contain all actions in A_{s_t} which attain the maximum in (4.5.1).

b. It evaluates any policy which selects an action in $A_{s_t, t}^*$ for each $s_t \in S$ for all $t = 1, 2, \ldots, N - 1$.

c. It computes the expected total reward for the entire decision-making horizon, and from each period to the end of the horizon for any optimal policy.

Let $D_t^* \equiv \times_{s \in S} A_{s, t}^*$. Then any $\pi^* \in \Pi^* \equiv D_1^* \times \ldots \times D_{N-1}^*$ is an optimal policy. If more than one such π^* exists, each yields the same expected total reward. This occurs if for some s_t, $A_{s_t, t}^*$ contains more than one action. To obtain a particular optimal policy, it is only necessary to retain a single action from $A_{s_t, t}^*$ for each $t \leq N - 1$ and $s_t \in S$.

Although this chapter emphasizes models with finite S and A, the algorithm and results of Theorem 4.5.1 are valid in greater generality. It applies to models with countable, compact, or Polish state and action spaces. In nondiscrete models, regularity conditions are required to ensure that u_t is measurable, integrals exist, and maxima are attained. Often we discretize the state and action spaces prior to computation; however, backward induction may be used to find optimal policies when the maxima and maximizing actions can be determined analytically. More importantly, a considerable portion of stochastic optimization literature uses backward induction to characterize the form of optimal policies under structural assumptions on rewards and transition probabilities. When such results can be obtained, specialized algorithms may be developed to determine the best policy of that type. We expand on this point in Sec. 4.7.

When there are K states with L actions in each, the backward induction algorithm requires $(N - 1)LK^2$ multiplications to evaluate and determine an optimal policy. Since there are $(L^K)^{(N-1)}$ deterministic Markovian policies, and direct evaluation of each requires $(N - 1)K^2$ multiplications, this represents considerable reduction in computation.

A further advantage of using backward induction is that, at pass t through the algorithm, only r_t, p_t and u_{t+1}^* need be in high-speed memory. The data from iterations $t + 1$, $t + 2, \ldots, N$ are not required since they are summarized through u_{t+1}^*, and the data from decision epochs $1, 2, \ldots, t - 1$ are not needed until subsequent iterations. Of course, $A_{s, t}^*$ must be stored for all t to recover all optimal policies.

4.6 EXAMPLES

In this section, we use the backward induction algorithm to find optimal policies for several finite-horizon Markov decision problems. In the first two problems, we use enumeration to perform maximization. The deterministic model of Sec. 4.6.3 illustrates the use of analytic methods. In Sec. 4.6.4, we solve the secretary problem. Recall that u_t^*, $t = 1, 2, \ldots, N$ denote solutions of the optimality equations.

4.6.1 The Stochastic Inventory Model

We find the optimal policy for the model of Sec. 3.2. Because the model parameters are stationary, we drop the time index on the transition probability and the reward. Numerical values for these quantities appear in Sec. 3.2.2. Define $u_t^*(s, a)$ by

$$u_t^*(s, a) = r(s, a) + \sum_{j \in S} p(j|s, a) u_{t+1}^*(j).$$

We implement the backward induction algorithm as follows.

1. Set $t = 4$ and $u_4^*(s) = r_4(s) = 0$, $s = 0, 1, 2, 3$.
2. Since $t \neq 1$, continue. Set $t = 3$ and

$$u_3^*(s) = \max_{a \in A_s} \left\{ r(s, a) + \sum_{j \in S} p(j|s, a) u_4^*(j) \right\}, \quad s = 0, 1, 2, 3$$

$$= \max_{a \in A_s} \{ r(s, a) \}.$$

Inspecting the values of $r(s, a)$ shows that in each state the maximizing action is 0, that is, do not order. We obtain

s	$u_3^*(s)$	$A_{s,3}^*$
0	0	0
1	5	0
2	6	0
3	5	0

3. Since $t \neq 1$, continue. Set $t = 2$ and

$$u_2^*(s) = \max_{a \in A_s} \{ u_2^*(s, a) \},$$

where, for example,

$$u_2^*(0, 2) = r(0, 2) + p(0|0, 2)u_3^*(0) + p(1|0, 2)u_3^*(1) + p(2|0, 2)u_3^*(2) + p(3|0, 2)u_3^*(3)$$

$$= -2 + \left(\tfrac{1}{4} \right) \times 0 + \left(\tfrac{1}{2} \right) \times 5 + \left(\tfrac{1}{4} \right) \times 6 + 0 \times 5 = 2$$

The quantities $u_2^*(s, a)$, $u_2^*(s)$ and $A_{s,2}^*$ are summarized in the following table with $\times$'s denoting infeasible actions.

s	$a = 0$	$u_2^*(s, a)$ $a = 1$	$a = 2$	$a = 3$	$u_2^*(s)$	$A_{s,2}^*$
0	0	$\frac{1}{4}$	2	$\frac{1}{2}$	2	2
1	$\frac{25}{4}$	4	$\frac{5}{2}$	$\times$	$\frac{25}{4}$	0
2	10	$\frac{9}{2}$	$\times$	$\times$	10	0
3	$\frac{21}{2}$	$\times$	$\times$	$\times$	$\frac{21}{2}$	0

4. Since $t \neq 1$, continue. Set $t = 1$ and

$$u_1^*(s) = \max_{a \in A_s} \{u_1^*(s, a)\}.$$

The quantities $u_1^*(s, a)$, $u_1^*(s)$, and $A_{s,1}^*$ are summarized in the following table.

s	$a = 0$	$u_1^*(s, a)$ $a = 1$	$a = 2$	$a = 3$	$u_1^*(s)$	$A_{s,1}^*$
0	2	$\frac{33}{16}$	$\frac{66}{16}$	$\frac{67}{16}$	$\frac{67}{16}$	3
1	$\frac{129}{16}$	$\frac{98}{16}$	$\frac{99}{16}$	$\times$	$\frac{129}{16}$	0
2	$\frac{194}{16}$	$\frac{131}{16}$	$\times$	$\times$	$\frac{194}{16}$	0
3	$\frac{227}{16}$	$\times$	$\times$	$\times$	$\frac{227}{16}$	0

5. Since $t = 1$, stop.

This algorithm yields the optimal expected total reward function $v_4^*(s)$ and the optimal policy $\pi^* = (d_1^*(s), d_2^*(s), d_3^*(s))$, which is tabulated below. Note in this example that the optimal policy is *unique*.

s	$d_1^*(s)$	$d_2^*(s)$	$d_3^*(s)$	$v_4^*(s)$
0	3	2	0	$\frac{67}{16}$
1	0	0	0	$\frac{129}{16}$
2	0	0	0	$\frac{194}{16}$
3	0	0	0	$\frac{227}{16}$

The quantity $v_4^*(s)$ gives the expected total reward obtained using the optimal policy when the inventory at the start of the month is s units. The optimal policy has a particularly simple form; if at the start of month 1 the inventory is 0 units, order three units, otherwise do not order; if at the start of month 2 the inventory is two units, order two units, otherwise do not order, and do not order in month 3 for any inventory level. This policy is a nonstationary (σ, Σ) policy as defined in Sec. 3.2. The reordering point $\sigma_t = 0$ at each decision epoch and the target stock Σ_t equals 3 at decision epoch 1; 2 at decision epoch 2; and 0 at decision epoch 3. For example,

$$d_1^*(s) = \begin{cases} 0 & s > 0 \\ 3 & s = 0. \end{cases}$$

Policies of this form are optimal in inventory models in which the ordering cost has the form

$$O(u) = \begin{cases} 0 & u = 0 \\ K + cu & u > 0, \end{cases}$$

where $K > 0$ and $c > 0$, the shortage and holding cost $h(u)$ is convex, and backlogging of unfilled orders is permitted. A proof that (σ, Σ) policies are optimal under these assumptions may be based on backward induction; however, the technical details are quite complicated. We refer the interested reader to references in the bibliographic summary for more on results of this type.

4.6.2 Routing Problems

In this section, we apply backward induction to find the longest route in the model in Fig. 3.3.1. Here states correspond to nodes and action sets correspond to the collection of arcs originating at each node. Below, the action (i, j) corresponds to choosing the arc between nodes i and j.

1. Set $t = 4$ and $u_4^*(8) = 0$.
2. Since $t \neq 1$, continue. Set $t = 3$ and

$$u_3^*(5) = 1 + u_4^*(8) = 1, \quad u_3^*(6) = 2 + u_4^*(8) = 2,$$
$$u_3^*(7) = 6 + u_4^*(8) = 6.$$

Because there is only one action in each state,

$$A_{5,3}^* = (5,8), \quad A_{6,3}^* = (6,8), \quad A_{7,3}^* = (7,8).$$

3. Since $t \neq 1$, continue. Set $t = 2$ and

$$u_2^*(2) = \max\{4 + u_3^*(5), 5 + u_3^*(6)\} = \max\{5, 7\} = 7,$$
$$u_2^*(3) = \max\{5 + u_3^*(5), 6 + u_3^*(6), 1 + u_3^*(7)\} = \max\{6, 8, 7\} = 8,$$
$$u_2^*(4) = 2 + u_3^*(7) = 8.$$

The optimal action sets are

$$A_{2,2}^* = (2,6), \ A_{3,2}^* = (3,6), \ A_{4,2}^* = (4,7).$$

4. Since $t \neq 1$, continue. Set $t = 1$ and

$$u_1^*(1) = \max\{2 + u_2^*(2), 4 + u_2^*(3), 3 + u_2^*(4)\} = \max\{9, 12, 11\} = 12.$$

The optimal action set is

$$A_{1,1}^* = (1,3).$$

5. Since $t = 1$, stop.

Since the optimal action set at each state is a singleton, there is a unique optimal policy $\pi^* = (d_1^*, d_2^*, d_3^*)$ where

$$d_1^*(1) = (1,3),$$

$$d_2^*(2) = (2,6), \quad d_2^*(3) = (3,6), \quad d_2^*(4) = (4,7),$$

$$d_3^*(5) = (5,8), \quad d_3^*(6) = (6,8), \quad d_3^*(7) = (7,8).$$

This policy selects which arc to traverse from each node in order to travel the longest route from that node to node 8. Starting at node 1, the optimal path is $1 \rightarrow 3 \rightarrow 6 \rightarrow 8$. While starting at node 2, the optimal path is $2 \rightarrow 6 \rightarrow 8$.

From the critical path analysis perspective, if the distances represent weeks, this result means that the project requires 12 weeks to complete and the critical path consists of the tasks represented by arcs $(1, 3)$, $(3, 6)$, and $(6, 8)$. If any of these are delayed, the entire project will be delayed.

This example illustrates both statements of the Principle of Optimality in Sec. 4.3. Starting at node 1 and using the optimal policy moves the system from node 3 to node 6 to node 8. However, if the system started at node 3, the optimal policy would be to go to node 6 and then node 8. This is identical to the optimal policy for the whole problem as stated in the Principle of Optimality. Because the policy π^* above gives the optimal path from each node to node 8, the second statement of the Principle of Optimality is also satisfied.

We now provide an example in which the Principle of Optimality does not hold. Consider the network in Figure 4.6.1. Define the "value" of a path from a given node

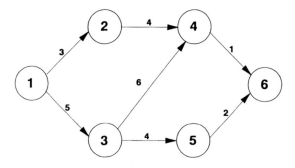

Figure 4.6.1 Network example for Sec. 4.6.2.

to node 6, to be the length of the *second-longest arc* on that path when there are two
or more arcs between that node and node 6; if there is a single arc, then its length is
the arc length. Our objective is to find the path from node 1 to node 6 with greatest
value. Clearly this is the routing $1 \rightarrow 3 \rightarrow 4 \rightarrow 6$. Its value is 5 corresponding to arc
$(1, 3)$. Applying the optimal decision in node 1 moves the system to node 3. If the
Principle of Optimality were valid, the optimal path from that node to node 6 would
be $3 \rightarrow 4 \rightarrow 6$. Clearly it is not. The optimal path is $3 \rightarrow 5 \rightarrow 6$ with value 2
corresponding to arc $(5, 6)$ while the value of $3 \rightarrow 4 \rightarrow 6$ is 1 corresponding to $(4, 6)$.
Under this optimality criterion, backward induction does not yield an optimal policy.
This is because to find an optimal policy using backward induction requires a
separable policy evaluation function such as expected total reward (see also Prob.
4.2).

4.6.3   The Sequential Allocation Model

In this section we adapt the backward induction algorithm to find an optimal solution
to the sequential allocation model of Sec. 3.3.3 and use it to solve the problem with
$f(x_1, \ldots, x_N) = -x_1^2 - x_2^2 - \cdots - x_N^2$. Since maximization of a negative function is
equivalent to minimization of a positive function, this problem has the following
nonlinear (quadratic) programming representation:

$$\text{Minimize } x_1^2 + x_2^2 + \cdots + x_N^2$$

subject to

$$x_1 + x_2 + \cdots + x_N = M \tag{4.6.1}$$

and $x_t \geq 0$, $t = 1, \ldots, N$. We solve this problem using backward induction with
"min" replacing "max" throughout.

   We first describe the algorithm for a separable sequential allocation problem. Let
$u_t(s)$ be the minimal cost if s units are available for use in periods t through N.
Clearly $u_N(s) = g_N(s)$. For $t = 1, \ldots, N - 1$,

$$u_t(s) = \operatorname*{minimum}_{(x_t, \ldots, x_N) \in B_t(s)} g_t(x_t) + \cdots + g_N(x_N),$$

where $B_t(s) = \{x_n : x_n \geq 0, n = t, \ldots, N \text{ and } x_t + \cdots + x_N = s\}$. To solve the problem,
we seek $u_1(M)$ and the corresponding allocations. These are found by first using
the backward induction algorithm to find $u_t(s)$ for each t. Since the problem is
deterministic, we require actions only at states that can be achieved by implementing
the optimal policy. We find these by applying the optimal decision rule in period 1 at
the specified initial system state, determining the resulting state at decision epoch 2,
applying the optimal decision epoch 2 decision rule at this state to determine the state
at decision epoch 3, etc.

   The optimality equations for this problem are

$$u_t(s) = \min_{a \in [0, s]} \{ g_t(a) + u_{t+1}(s - a) \}. \tag{4.6.2}$$

$0 \leq s \leq M$ and $t = 1, \ldots, N - 1$ and $u_N(s) = g_N(s)$. We derive them by noting that
the optimal value over periods t through N equals the minimal value that can be
found by allocating a units in period t and then optimally allocating the remaining
$s - a$ over periods $t + 1$ to N.

We solve problem (4.6.1) by applying backward induction for a few iterations to suggest the form of the optimal policy and the corresponding cost-to-go functions and then using mathematical induction to prove that the optimal policy has this form. Notation follows Sec. 3.3.3 with $g_t(s) = s^2$ for $t = 1, \ldots, N$. Implementation of the algorithm proceeds as follows.

1. Set $t = N$ and $u_N^*(s) = g_N(s) = s^2$, for $s \in [0, M]$.
2. Since $t \neq 1$, set $t = N - 1$ and

$$u_{N-1}^*(s) = \min_{a \in [0, s]} \{a^2 + u_N^*(s - a)\} = \min_{a \in [0, s]} \{a^2 + (s - a)^2\}.$$

Differentiating the expression in brackets with respect to a, setting the derivative equal to zero and solving for a shows that

$$\frac{s}{2} = \arg\min_{a \in [0, s]} \{a^2 + (s - a)^2\},$$

so that $A_{s, N-1}^* = s/2$ and $u_{N-1}^*(s) = s^2/2$.
3. Since $t \neq 1$, set $t = N - 2$, and

$$u_{N-2}^*(s) = \min_{a \in [0, s]} \{a^2 + \tfrac{1}{2}(s - a)^2\}.$$

We find that $A_{s, N-2}^* = s/3$ and $u_{N-2}^*(s) = s^2/3$.

The pattern is now obvious. We hypothesize that $A_{s, N-n+1}^* = s/n$ and that $u_{N-n+1}^*(s) = s^2/n$ and apply induction. As a consequence of the induction hypothesis, the optimality equation becomes

$$u_{N-n}^*(s) = \min_{a \in [0, s]} \left\{ a^2 + \frac{1}{n}(s - a)^2 \right\},$$

so that $A_{s, N-n}^* = s/(n + 1)$ and $u_{N-n}^*(s) = s^2/(n + 1)$. Thus $A_{s, N-n}^* = s/(n + 1)$ and $u_{N-n}^*(s) = s^2/(n + 1)$ for all n.

The optimal value of the objective function is $u_1^*(M) = M^2/N$ and the optimal decision in period 1 is the unique element of $A_{M, 1}^*$, M/N. When action s/N is used in period 1, the state of the system at decision epoch 2 is $M - (M/N) = [(N - 1)/N]M$, so the optimal decision in period 2 is the unique element of $A_{[(N-1)/N]M, 2}^*$ which equals $([(N - 1)/N]M)/N - 1)$ or M/N. Applying a formal induction argument shows that the optimal allocation in each period is M/N so that the optimal solution to the problem is $x_n = M/N$ for $t = 1, \ldots, N$.

The same procedure applied to (3.3.4) subject to (3.3.2) and $q_i \geq 0$, $i = 1, \ldots, N$ yields the following maximum likelihood estimates for the multinomial distribution:

$$\hat{q}_i = \frac{y_i}{y_1 + \cdots + y_N}, \quad i = 1, \ldots, N. \tag{4.6.3}$$

Demonstrating this is left as an exercise.

4.6.4 The Secretary Problem

We now solve the secretary problem of Sec. 3.4.3. Recall that N candidates are available. Let $u_t^*(1)$ denote the maximum probability of choosing the best candidate, when the current candidate has the highest relative rank among the first t interviewed, and $u_t^*(0)$ denote the maximum probability of choosing the best overall candidate if the current candidate does not have highest relative rank among the first t interviewed. Recall that Δ denotes the stopped state.

Then u_t^* satisfies the following recursions:

$$u_N^*(1) = h(1) = 1, \quad u_N^*(0) = h(0) = 0, \quad u_N^*(\Delta) = 0,$$

and, for $t < N$,

$$u_t^*(1) = \max\{g_t(1) + u_{t+1}^*(\Delta), \; -f_t(0) + p_t(1|1)u_{t+1}^*(1) + p_t(0|1)u_{t+1}^*(0)\}$$

$$= \max\left\{\frac{t}{N}, \; \frac{1}{t+1}u_{t+1}^*(1) + \frac{t}{t+1}u_{t+1}^*(0)\right\}, \tag{4.6.4}$$

$$u_t^*(0) = \max\{g_t(0) + u_{t+1}^*(\Delta), \; -f_t(0) + p_t(1|0)u_{t+1}^*(1) + p_t(0|0)u_{t+1}^*(0)\}$$

$$= \max\left\{0, \; \frac{1}{t+1}u_{t+1}^*(1) + \frac{t}{t+1}u_{t+1}^*(0)\right\}, \tag{4.6.5}$$

and

$$u_t^*(\Delta) = u_{t+1}^*(\Delta) = 0.$$

Noting that $u_t^* \geq 0$ allows simplification of (4.6.5) and (4.6.4) to:

$$u_t^*(0) = \frac{1}{t+1}u_{t+1}^*(1) + \frac{t}{t+1}u_{t+1}^*(0), \tag{4.6.6}$$

$$u_t^*(1) = \max\left\{\frac{t}{N}, u_t^*(0)\right\}. \tag{4.6.7}$$

Solution of (4.6.6) and (4.6.7) yields an optimal policy for this problem as follows. In state 1 when $t/N > u_t^*(0)$ the optimal action is to stop, when $t/N < u_t^*(0)$ the optimal action is to continue and when $t/N = u_t^*(0)$ either action is optimal. In state 0 the optimal action is to continue.

Applying induction formally or appealing to intuition suggests that the optimal policy has the form "Observe the first τ candidates; after that, select the first candidate who is better than all the previous ones." This translates to the policy $\pi^* = (d_1^*, d_2^*, \ldots, d_{N-1}^*)$, with $d_t^*(0) = C$, $d_t^*(1) = C$ if $t \leq \tau$ and $d_t^*(1) = Q$ if $t > \tau$.

We now show that an optimal policy has this form by demonstrating that if $u_\tau^*(1) > \tau/N$ or $u_\tau^*(1) = \tau/N = u_\tau^*(0)$ for some τ, in which case it is optimal to continue, then $u_t^*(1) > t/N$ for $t < \tau$ so that it is again optimal to continue. Assume either $u_\tau^*(1) > \tau/N$ so that from (4.6.7), $u_\tau^*(1) = u_\tau^*(0)$, or $u_\tau^*(1) = \tau/N = u_\tau^*(0)$ so that

$$u_{\tau-1}^*(0) = \frac{1}{\tau}u_\tau^*(1) + \frac{\tau-1}{\tau}u_\tau^*(0) = u_\tau^*(0) \geq \frac{\tau}{N}.$$

Thus

$$u_{\tau-1}^*(1) = \max\left\{\frac{\tau-1}{N}, u_{\tau-1}^*(0)\right\} \geq \frac{\tau}{N} > \frac{\tau-1}{N}.$$

Repeating this argument with $\tau - 1$ replacing τ and so forth shows that the result holds for all $t < \tau$. This implies that an optimal policy cannot have the form $d_t^*(1) = C$ for $t \leq t'$, $d_t^*(1) = Q$ for $t' < t \leq t''$, and $d_t^*(1) = C$ for $t > t''$.

By definition $\tau < N$. We show that, if $N > 2$, $\tau \geq 1$. Supposing it is not, then, for all t, $u_t^*(1) = t/N$, so that, by (4.6.6),

$$u_t^*(0) = \frac{1}{t+1}\frac{t+1}{N} + \frac{t}{t+1}u_{t+1}^*(0) = \frac{1}{N} + \frac{t}{t+1}u_{t+1}^*(0). \qquad (4.6.8)$$

Noting that $u_N^*(0) = 0$, and solving Eq. (4.6.8) by backward induction yields

$$u_t^*(0) = \frac{t}{N}\left[\frac{1}{t} + \frac{1}{t+1} + \cdots + \frac{1}{N-1}\right], \quad 1 \leq t < N, \qquad (4.6.9)$$

for $N > 2$. This implies that $u_1^*(0) > 1/N = u_1^*(1) \geq u_1^*(0)$; the second inequality is a consequence of (4.6.7). This is a contradiction, so we may conclude that $\tau \geq 1$ when $N > 2$.

Thus, when $N > 2$,

$$u_1^*(0) = u_1^*(1) = \cdots = u_\tau^*(0) = u_\tau^*(1), \qquad (4.6.10)$$

$u_t^*(1) = t/N$ for $t > \tau$, and

$$u_t^*(0) = \frac{t}{N}\left[\frac{1}{t} + \frac{1}{t+1} + \cdots + \frac{1}{N-1}\right], \quad t > \tau.$$

Thus it is optimal to continue to observe candidates as long as $u_t^*(0) > t/N$, so that

$$\tau = \max_{t \geq 1}\left\{\left[\frac{1}{t} + \frac{1}{t+1} + \cdots + \frac{1}{N-1}\right] > 1\right\}. \qquad (4.6.11)$$

When $N \leq 2$, choose $\tau = 0$, but of course, any policy is optimal in this case.

For example, suppose $N = 4$, since $1 + \frac{1}{2} + \frac{1}{3} > 1$ and $\frac{1}{2} + \frac{1}{3} < 1$, $\tau = 1$. This means that, with four candidates, the decision maker should observe the first candidate and then choose the next candidate whose relative rank is 1. Figure 4.6.2 shows how τ/N and $u_1^*(1)$, the probability of choosing the best candidate, vary with N. It suggests that, as N becomes large, τ/N and $u_1^*(1)$ approach limits. Let $\tau(N)$ denote the optimal value of τ when there are N candidates to consider. Further calculations

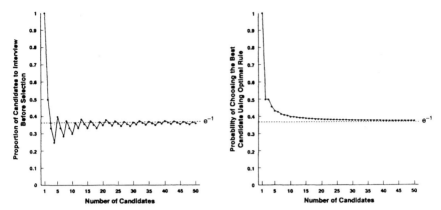

Figure 4.6.2   Graphical representation of the solution of the secretary problem.

reveal that $\tau(1000) = 368$, $\tau(10,000) = 3679$, and $\tau(100,000) = 36788$. From (4.6.10), it follows that, for N sufficiently large,

$$1 \approx \left[\frac{1}{\tau(N)} + \frac{1}{\tau(N) + 1} + \cdots + \frac{1}{N - 1} \right] \approx \int_{\tau(N)}^{N} \frac{1}{x} \, dx, \qquad (4.6.12)$$

so that, for large N,

$$\log\left(\frac{N}{\tau(N)} \right) \approx 1.$$

This implies that

$$\lim_{N \to \infty} \frac{\tau(N)}{N} = e^{-1},$$

and, as a consequence of (4.6.10), (4.6.11) and the first equality in (4.6.12),

$$u_1^*(0) = u_1^*(1) = u_{\tau(N)}^*(0) = \frac{\tau(N)}{N} \to e^{-1}$$

Thus with a large number of candidates the decision maker should observe Ne^{-1} or 36.8% of them and subsequently choose the first candidate with highest relative rank. The probability of choosing the best candidate using this policy is $1/e$ or 0.36788, which agrees to five places with the value when $N = 100,000$.

4.7   OPTIMALITY OF MONOTONE POLICIES

In Sec. 4.4, we provided conditions under which there exists a deterministic Markov optimal policy. In this section we investigate more detailed results concerning the structure of an optimal policy. We provide conditions which ensure that optimal policies are monotone (nondecreasing or nonincreasing) in the system state, interpret them in the context of a product price determination model, and show that they are satisfied in an equipment replacement model. For such a concept to be meaningful, we require that the state have a physical interpretation and some natural ordering. Throughout this book the expression "monotone policy" refers to a monotone deterministic Markov policy.

4.7.1 Structured Policies

One of the principal uses of Markov decision process methods is to establish the existence of optimal policies with special structure. The importance of results regarding the optimality of structured policies lies in

a. their appeal to decision makers,
b. their ease in implementation, and
c. their enabling efficient computation.

The last point is particularly noteworthy because, when the optimal policy has simple form, specialized algorithms can be developed to search only among policies that have the same form as the optimal policy. This avoids the need for the less efficient general backward induction algorithm.

Examples of policies with simple structure are (σ, Σ) policies in inventory models, and control limit or critical number policies in queueing control models. A *control limit policy* is a deterministic Markov policy composed of decision rules of the form

$$d_t(s) = \begin{cases} a_1 & s < s^* \\ a_2 & s \geq s^*, \end{cases}$$

where a_1 and a_2 are distinct actions and s^* is a control limit. We interpret such a policy as follows; when the state of the system is less than s^*, it is optimal to use action a_1, and when the system state is s^* or greater, it is optimal to use action a_2. In a queueing control model, s might represent the number of customers in the system, a_1 the minimum service rate, and a_2 the maximum service rate; while, in a replacement problem, s might represent the age or condition of the component, a_1 the action not to replace, and a_2 the action to replace. If we establish that such policies are optimal, the problem of finding an optimal policy reduces to that of determining s^*.

4.7.2 Superadditive Functions

We begin with some technical material. Let X and Y be partially ordered sets and $g(x, y)$ a real-valued function on $X \times Y$. We say that g is *superadditive* if for $x^+ \geq x^-$ in X and $y^+ \geq y^-$ in Y,

$$g(x^+, y^+) + g(x^-, y^-) \geq g(x^+, y^-) + g(x^-, y^+). \tag{4.7.1}$$

If the reverse inequality above holds, $g(x, y)$ is said to be *subadditive*. Thus $g(x, y)$ is

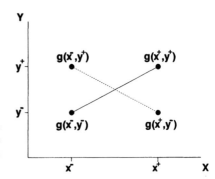

Figure 4.7.1 In a superadditive function, the sum of the quantities joined by the solid line exceeds that of the sum of the quantities joined by the dashed line.

subadditive if $-g(x, y)$ is superadditive. We sometimes refer to functions satisfying (4.7.1) as *supermodular* and expression (4.7.1) as the *quadrangle inequality* (Fig. 4.7.1).

Equivalently, we say that $g(x, y)$ is superadditive if it has *monotone increasing differences*; that is,

$$g(x^+, y^+) - g(x^+, y^-) \geq g(x^-, y^+) - g(x^-, y^-). \qquad (4.7.2)$$

Examples of superadditive functions include the following.

1. $g(x, y) = h(x) + e(y)$, where h and e are arbitrary functions on X and Y respectively.
2. $g(x, y) = h(x + y)$, where $X = Y = R^1$ and h is convex increasing.
3. $g(x, y) = xy$, where $X = Y = R^1$.

In particular, $(x + y)^2$ and $-(x - y)^2$ are superadditive while $(x - y)^2$ and $-(x + y)^2$ are subadditive. A consequence of (4.7.2) is that, when $X = Y = R^1$, and $g(x, y)$ is twice differentiable, then g is superadditive whenever

$$\frac{\partial^2 g(x, y)}{\partial x \, \partial y} \geq 0.$$

The following lemma provides the key property of superadditive functions from our perspective.

Lemma 4.7.1. Suppose g is a superadditive function on $X \times Y$ and for each $x \in X$, $\max_{y \in Y} g(x, y)$ exists. Then

$$f(x) = \max\left\{ y' \in \arg\max_{y \in Y} g(x, y) \right\} \qquad (4.7.3)$$

is monotone nondecreasing in x.

Proof. Let $x^+ \geq x^-$ and choose $y \leq f(x^-)$. Then, by the definition of f,

$$g(x^-, f(x^-)) - g(x^-, y) \geq 0,$$

and by (4.7.1),

$$g(x^-, y) + g(x^+, f(x^-)) \geq g(x^-, f(x^-)) + g(x^+, y).$$

Rewriting the second inequality as

$$g(x^+, f(x^-)) \geq g(x^+, y) + [g(x^-, f(x^-)) - g(x^-, y)]$$

and combining it with the first inequality above yields

$$g(x^+, f(x^-)) \geq g(x^+, y)$$

for all $y \leq f(x^-)$. Consequently, $f(x^+) \geq f(x^-)$.   □

When $\arg\max_{y \in Y} g(x, y)$ contains a unique element, (4.7.3) may be written as $f(x) = \arg\max_{y \in Y} g(x, y)$. Problems at the end of the chapter address analogous conditions for minimization problems.

4.7.3 Optimality of Monotone Policies

We now provide conditions on the states, actions, rewards, and transition probabilities under which monotone policies are optimal. Assume that S represents the set of non-negative integers and $A_s = A'$ for all $s \in S$. Define

$$q_t(k|s, a) = \sum_{j=k}^{\infty} p_t(j|s, a)$$

for $t = 1, \ldots, N - 1$. It represents the probability that the state at decision epoch $t + 1$ exceeds $k - 1$ when choosing action a in state s at decision epoch t.

We demonstrate optimality of monotone policies by

1. Inductively showing that the optimal value functions from t onward, $u_t^*(s)$, are nondecreasing or nonincreasing in S, and
2. then showing that

$$w_t(s, a) = r_t(s, a) + \sum_{j=0}^{\infty} p_t(j|s, a) u_t^*(j) \tag{4.7.4}$$

is superadditive or subadditive.

We begin with the following important technical lemma.

Lemma 4.7.2. Let $\{x_j\}, \{x_j'\}$ be real-valued non-negative sequences satisfying

$$\sum_{j=k}^{\infty} x_j \geq \sum_{j=k}^{\infty} x_j' \tag{4.7.5}$$

for all k, with equality holding in (4.7.5) for $k = 0$.
    Suppose $v_{j+1} \geq v_j$ for $j = 0, 1, \ldots$, then

$$\sum_{j=0}^{\infty} v_j x_j \geq \sum_{j=0}^{\infty} v_j x_j', \tag{4.7.6}$$

where limits in (4.7.6) exist but may be infinite.

Proof. Let k be arbitrary and $v_{-1} = 0$. Then

$$\sum_{j=0}^{\infty} v_j x_j = \sum_{j=0}^{\infty} x_j \sum_{i=0}^{j} (v_i - v_{i-1}) = \sum_{j=0}^{\infty} (v_j - v_{j-1}) \sum_{i=j}^{\infty} x_i \tag{4.7.7}$$

$$= \sum_{j=1}^{\infty} (v_j - v_{j-1}) \sum_{i=j}^{\infty} x_i + v_0 \sum_{i=0}^{\infty} x_i \geq \sum_{j=1}^{\infty} (v_j - v_{j-1}) \sum_{i=j}^{\infty} x_i' + v_0 \sum_{i=0}^{\infty} x_i'$$

$$= \sum_{j=0}^{\infty} v_j x_j'.$$

The last equality follows by reversing the steps in (4.7.7). As a consequence of representation (4.7.7), both limits in (4.7.6) exist. $\square$

Note that Lemma 4.7.2 allows us to derive an important property of stochastically ordered random variables. Suppose Y and Y' denote random variables with $x_j = P\{Y = j\}$ and $x_j' = P\{Y' = j\}$. When (4.7.5) holds, Y is stochastically greater than Y' (Sec. 4.1). As a consequence of (4.7.6), it follows that, for any nondecreasing function $f(j)$, $E\{f(Y)\} \geq E\{f(Y')\}$.
    We now provide conditions under which u_t^* is monotone. We assume that A', r_t, and p_t are such that

$$\max_{a \in A'} \left\{ r_t(s, a) + \sum_{j=0}^{\infty} p_t(j|s, a) u(j) \right\} \tag{4.7.8}$$

is attained for all monotone u. This occurs under the conditions in Proposition 4.4.3. Note Proposition 4.7.3 is also valid when "sup" replaces "max" in (4.7.8).

Proposition 4.7.3. Suppose the maximum in (4.7.8) is attained and that

1. $r_t(s, a)$ is nondecreasing (nonincreasing) in s for all $a \in A'$ and $t = 1, \ldots,$ $N - 1$,
2. $q_t(k|s, a)$ is nondecreasing in s for all $k \in S$, $a \in A'$, and $t = 1, \ldots, N - 1$, and
3. $r_N(s)$ is nondecreasing (nonincreasing) in s.

Then $u_t^*(s)$ is nondecreasing (nonincreasing) in s for $t = 1, \ldots, N$.

Proof. We prove the result in the nondecreasing case by using backward induction to show that $u_t^*(s)$ is nondecreasing for all t. Since $u_N^*(s) = r_N(s)$, the result holds for $t = N$ by 3.

Assume now that $u_n^*(s)$ is nondecreasing for $n = t + 1, \ldots, N$. By assumption, there exists an $a_s^* \in A'$ which attains the maximum in

$$u_t^*(s) = \max_{a \in A'} \left\{ r_t(s, a) + \sum_{j=0}^{\infty} p_t(j|s, a) u_{t+1}^*(j) \right\},$$

so that

$$u_t^*(s) = r_t(s, a_s^*) + \sum_{j=0}^{\infty} p_t(j|s, a_s^*) u_{t+1}^*(j).$$

Let $s' \geq s$. By (1) and (2), the induction hypothesis, and Lemma 4.7.2 applied with $x_j' = p_t(j|s, a_s^*)$, $x_j = p_t(j|s', a_s^*)$ and $v_j = u_{t+1}^*(j)$,

$$u_t^*(s) \leq r_t(s', a_s^*) + \sum_{j=0}^{\infty} p_t(j|s', a_s^*) u_{t+1}^*(j)$$

$$\leq \max_{a \in A'} \left\{ r_t(s', a) + \sum_{j=0}^{\infty} p_t(j|s', a) u_{t+1}^*(j) \right\} = u_t^*(s').$$

Thus $u_t^*(s)$ is nondecreasing, the induction hypothesis is satisfied, and the result follows.   □

The following theorem provides conditions under which there exist monotone optimal policies. Note there might exist other optimal policies which are not monotone.

Theorem 4.7.4.   Suppose for $t = 1, \ldots, N - 1$ that

1. $r_t(s, a)$ is nondecreasing in s for all $a \in A'$,
2. $q_t(k|s, a)$ is nondecreasing in s for all $k \in S$ and $a \in A'$,
3. $r_t(s, a)$ is a superadditive (subadditive) function on $S \times A'$,
4. $q_t(k|s, a)$ is a superadditive (subadditive) function on $S \times A'$ for all $k \in S$, and
5. $r_N(s)$ is nondecreasing in s.

Then there exist optimal decision rules $d_t^*(s)$ which are nondecreasing (nonincreasing) in s for $t = 1, \ldots, N - 1$.

Proof. We prove the result in the superadditive case. We show that $w_t(s, a)$ defined in (4.7.4) is superadditive whenever q and r are. By condition (4), and the definition of superadditivity, for $s^- \leq s^+$ and all $k \in S$,

$$\sum_{j=k}^{\infty} [p_t(j|s^-, a^-) + p_t(j|s^+, a^+)] \geq \sum_{j=k}^{\infty} [p_t(j|s^-, a^+) + p_t(j|s^+, a^-)].$$

By Proposition 4.7.3, $u_t^*(s)$ is nondecreasing in s for all t, so applying Lemma 4.7.2 yields

$$\sum_{j=0}^{\infty} [p_t(j|s^-,a^-) + p_t(j|s^+,a^+)]u_t(j) \geq \sum_{j=0}^{\infty} [p_t(j|s^-,a^+) + p_t(j|s^+,a^-)]u_t(j).$$

Thus for each t, $\sum_{j=0}^{\infty} p_t(j|s,a)ut(j)$ is superadditive.

From condition (3), $r_t(s,a)$ is superadditive and, since the sum of superadditive functions is superadditive, $w_t(s,a)$ is superadditive. The result follows from Lemma 4.7.1.   □

Theorem 4.7.5 below provides alternative conditions which lead to monotone optimal policies. It follows from the last three lines of the proof of Theorem 4.7.4. Note that, although condition (4) includes the function u which is not part of the basic problem data, the condition may be verified directly in many applications.

Theorem 4.7.5.   Suppose for $t = 1, \ldots, N - 1$ that

1. $r_t(s,a)$ is nonincreasing in s for all $a \in A'$,
2. $q_t(k|s,a)$ is nondecreasing in s for all $k \in S$ and $a \in A'$,
3. $r_t(s,a)$ is a superadditive function on $S \times A'$,
4. $\sum_{j=0}^{\infty} p_t(j|s,a)u(j)$ is a superadditive function on $S \times A'$ for nonincreasing u, and
5. $r_N(s)$ is nonincreasing in s.

Then there exist optimal decision rules $d_t^*(s)$ which are nondecreasing in s for $t = 1, \ldots, N$.

For applications to queueing and inventory models, the assumption that $A_s = A'$ for all $s \in S$ is often overly restrictive. Slight modifications of the theorem hypotheses above allow extension to A_s which satisfy

1. $A_s \subset A'$ for $s \in S$,
2. $A_s \subset A_{s'}$ for $s' \geq s$, and
3. for each $s, a \in A_s$ and $a' \leq a$ implies $a' \in A_s$.

4.7.4   A Price Determination Model

We apply Theorem 4.7.4 to a product price determination model to interpret conditions under which the optimal price level is monotone nondecreasing in sales. A statement of the problem follows.

A marketing manager wishes to determine optimal price levels based only on current sales. Let the system state $s \in S = \{0, 1, \ldots\}$ represent the monthly sales of a particular product. At the start of each month, the manager notes sales in the previous month. Based on this information, he selects the product price a, for the current month from the set $A' = \{a: a_L \leq a \leq a_U\}$, where a_L and a_U represent the minimum and maximum price levels. Price effects both current revenue and sales.

The quantity $r_t(s, a)$ represents expected revenue in month t if the previous month's sales were s and the price in month t was set at a. The probability that sales in month t are j units when the price is a and sales in month $t - 1$ were s equals $p_t(j|s, a)$. The product lifetime is $N - 1$ months so that $r_N(s) = 0$ for all s. The manager's objective is to choose an adaptive price setting rule that maximizes expected total revenue over the product lifetime.

We discuss the hypotheses of Theorem 4.7.4 in the context of this model. Assume that $r_t(s, a)$ and $p_t(j|s, a)$ are continuous in a for fixed s and that A' is compact so that the maximum in (4.7.8) is attained.

1. That $r_t(s, a)$ is nondecreasing in s for fixed a means that, for a fixed price a, the expected revenue in the current month will be greater when previous months sales are greater.
2. That $q_t(k|s, a)$ is nondecreasing in s for fixed a and k means that the probability that sales exceed k in the current month is greater when the previous month's sales are greater. In other words, sales one month ahead are stochastically increasing with respect to current sales.
3. From (4.7.2), the superadditivity of $r_t(s, a)$ implies

$$r_t(s^+, a^+) - r_t(s^+, a^-) \geq r_t(s^-, a^+) - r_t(s^-, a^-)$$

   for $s^+ \geq s^-$ and $a^+ \geq a^-$. This condition is satisfied whenever the incremental effect on revenue of decreasing price is greater when the previous month's sales is greater. If $r_t(s, a)$ is differentiable in a, superadditivity is equivalent to the partial derivative of $r_t(s, a)$ with respect to a being nondecreasing in s.
4. Superadditivity of $q_t(k|s, a)$ implies that

$$q_t(k|s^+, a^+) - q_t(k|s^+, a^-) \geq q_t(k|s^-, a^+) - q_t(k|s^-, a^-)$$

   This condition is satisfied when the incremental effect of a price decrease on the probability that sales exceed a fixed level is greater if current sales are greater.

When all of these conditions are satisfied, the optimal pricing level is a nondecreasing function of sales. Thus the higher the current sales, the higher the optimal price in the subsequent period.

4.7.5 An Equipment Replacement Model

In this section, we apply Theorem 4.7.5 to the following equipment replacement model to establish that a control limit policy is optimal.

The condition of a piece of equipment used in a manufacturing process deteriorates over time. Let the state $s \in S = \{0, 1, \ldots\}$, represent the condition of the equipment at each decision epoch. The higher the value of s, the worse the condition of the equipment. At each decision epoch, the decision maker can choose one of two actions from the set $A' = \{0, 1\}$. Action 0 corresponds to operating the equipment as is for an additional period, while action 1 corresponds to scrapping the equipment and replacing it immediately with a new and identical piece of equipment. We assume that, in each period, the equipment deteriorates by i states with probability $p(i)$

independent of the state at the beginning of the period. The transition probabilities for this model satisfy

$$p_t(j|s,0) = \begin{cases} 0, & j < s \\ p(j-s), & j \geq s, \end{cases}$$

and $p_t(j|s,1) = p(j)$, $j \geq 0$.

The reward consists of three parts: a fixed income of R units per period; a state-dependent operating cost $h(s)$, with $h(s)$ nondecreasing in s; and a replacement cost of K units. Consequently,

$$r_t(s,a) = \begin{cases} R - h(s), & a = 0 \\ R - K - h(0), & a = 1. \end{cases}$$

The scrap value of the equipment at the end of the planning horizon is $r_N(s)$ which we assume to be nonincreasing in s.

We show that this model satisfies the hypotheses of Theorem 4.7.5.

1. By the assumption on h, $r_t(s, a)$ is non-increasing in s for $a = 0, 1$.

2. For $a = 1$, $q_t(k|s, a)$ is independent of s so it is nondecreasing. For $a = 0$, define

$$\Delta q_t(k, s) \equiv q_t(k|s + 1, 0) - q_t(k|s, 0)$$

so that

$$\Delta q_t(k, s) = \begin{cases} \sum_{j=k}^{\infty} [p(j - s - 1) - p(j - s)] = p(k - s - 1), & k > s \\ 0, & k \leq s. \end{cases}$$

Thus $\Delta q_t(k, s) \geq 0$, and it follows that q_t is nondecreasing.

We apply the following easily proved lemma to verify conditions (3) and (4) of Theorem 4.7.5.

Lemma 4.7.6. Let $g(s, a)$ be a real-valued function on $S \times A$, with $A = \{0, 1\}$ and $S = \{0, 1, \ldots\}$. If $g(s, a)$ satisfies

$$[g(s + 1, 1) - g(s + 1, 0)] - [g(s, 1) - g(s, 0)] \geq 0 \qquad (4.7.9)$$

for all s, it is superadditive.

3. Since

$$[r_t(s + 1, 1) - r_t(s + 1, 0)] - [r_t(s, 1) - r_t(s, 0)] = h(s + 1) - h(s) \geq 0,$$

where the last inequality follows from the assumed monotonicity of h, Lemma 4.7.6 establishes that $r_t(s, a)$ is superadditive.

4. We show that $\sum_{j=0}^{\infty} p_t(j|s, a) u(j)$ is superadditive for nonincreasing u. The left-hand side of (4.7.9) applied to this quantity equals

$$\sum_{j=0}^{\infty} p(j)u(j) - \sum_{j=s+1}^{\infty} p(j-s-1)u(j) - \sum_{j=0}^{\infty} p(j)u(j) + \sum_{j=s}^{\infty} p(j-s)u(j)$$

$$= \sum_{j=0}^{\infty} p(j)[u(j+s) - u(j+s+1)] \geq 0.$$

The inequality follows because u is nonincreasing and p is non-negative. Thus, by Lemma 4.7.6, $\sum_{j=0}^{\infty} p_t(j|s, a) u(j)$ is superadditive.

Consequently, this model satisfies the hypotheses of Theorem 4.7.5, so that for each t there is an optimal decision rule which is nondecreasing. Since there are only two actions, a control limit policy is optimal. That is, for each t there exists an s_t^* with the property that if s exceeds s_t^*, the optimal decision is to replace the equipment; and if $s \leq s_t^*$ then it is optimal to operate the equipment as is for another period.

4.7.6 Monotone Backward Induction

We now develop an efficient backward induction algorithm for finding optimal monotone decision rules. We assume that for each t there is a monotone optimal decision rule, that $S = \{0, 1, \ldots, M\}$ with M finite and $A_s = A'$ for all $s \in S$.

The Monotone Backward Induction Algorithm

1. Set $t = N$ and

$$u_N^*(s) = r_N(s) \quad \text{for all } s \in S.$$

2. Substitute $t - 1$ for t, set $s = 1$ and $A_1' = A'$.
   2a. Set

$$u_t^*(s) = \max_{a \in A_s'} \left\{ r_t(s, a) + \sum_{j \in S} p_t(j|s, a) u_{t+1}^*(j) \right\}.$$

   2b. Set

$$A_{s,t}^* = \arg\max_{a \in A_s'} \left\{ r_t(s, a) + \sum_{s \in S} p_t(j|s, a) u_{t+1}^*(j) \right\}.$$

   2c. If $s = M$, go to 3, otherwise set

$$A_{s+1}' = \{ a \in A' : a \geq \max[a' \in A_{s,t}^*] \}$$

   2d. Substitute $s + 1$ for s, and return to 2a.
3. If $t = 1$, stop, otherwise return to (2).

Any decision rule which chooses actions from $A_{s,t}^*$ in state s at decision epoch t is monotone and optimal.

This algorithm differs from the backward induction algorithm of Sec. 4.5 in that maximization is carried out over the sets A_s' which become smaller with increasing s. In the worse case, $A_s' = A'$ for all s and computational effort equals that of backward induction; however, when an optimal decision rule is strictly increasing, the sets A_s' will decrease in size with increasing s and hence reduce the number of actions which need to be evaluated in step (2). If at some s', $A_{s'}'$ contains a single element, say a^*, then no further maximization is necessary since that action will be optimal at all $s \geq s'$. In this case we need not carry out any further maximizations at iteration t, and instead set

$$u_t^*(s) = r_t(s, a^*) + \sum_{j \in S} p_t(j|s, a^*) u_{t+1}^*(j)$$

for all $s \geq s'$. For this reason, such an algorithm is particularly appealing in a two action model such as those in Secs. 3.4 and 4.7.5. Note also that in countable state models, this algorithm provides the only way of explicitly finding an optimal control limit policy.

BIBLIOGRAPHIC REMARKS

Heyman and Sobel (1984) provide background on utility theory relevant to sequential decision making. Our discussion on stochastic partial orderings in Sec. 4.1 follows Chap. 17 in Marshall and Olkin (1979). Example 4.1 is adopted from Example A.2 therein. Howard and Matheson, (1972), Jaquette (1976), and Rothblum (1984) study models with multiplicative utilities. Kreps (1977) adopts a different approach which defines utilities on histories instead of reward sequences. His model may be transformed into one with rewards by expanding the state space. White (1988b) and Sobel (1988) provide surveys of Markov decision processes with nonstandard optimality criteria such as expected utility, preference orderings, mean and variance comparisons, and constraints. These topics are currently receiving considerable attention in Markov decision process theory research.

Section 4.2 introduces inductive computation. The algorithm therein is a special case of the general dynamic programming algorithm of Sec. 4.5 in which each A_s contains a single action.

Bellman's book (1957) presented the optimality equations and the Principle of Optimality together with references to his earlier papers (dating back to 1952) which introduced and illustrated many of the key ides of dynamic programming. Karlin (1955), recognizing the need for a formal proof of the Principle of Optimality, introduced a formal mathematical structure for the analysis of dynamic programming models and demonstrated its validity within this context. Hinderer (1970) discussed the relationship between the Principle of Optimality and the optimality equations in considerable generality, distinguishing and identifying many mathematical subleties. The results in Sec. 4.4 are in the spirit of Chap. 2 of Derman (1970).

Proofs of the optimality of (σ, Σ) or (s, S) policies appear in Heyman and Sobel (1984) and Bertsekas (1987). As mentioned earlier, the first proof of this result was provided by Scarf (1960). Notable extensions include works of Veinott (1966), Porteus (1971), Schal (1976), and Zheng (1990). Kalin (1980) extends these concepts to multidimensional problems. Johnson (1968), Federgruen and Zipkin (1984), Tijms (1986) and Zheng and Federgruen (1991) provide algorithms for computing σ and Σ.

Our solution of the secretary problem in Sec. 4.6.4 follows Bather (1980). The result seems to have some empirical validity. We note the following article which appeared in *The Globe and Mail* on September 12, 1990, page A22.

"The Last Shall be First

The last person interviewed for a job gets it 55.8 per cent of the time according to Runzheimer Canada, Inc. Early applicants are hired only 17.6 per cent of the time; the management-consulting firm suggests that job-seekers who find they are among the first to be grilled 'tactfully ask to be rescheduled for a later date.' Mondays are also poor days to be interviewed and any day just before quitting time is also bad."

The elegant paper of Serfozo (1976) and the development in Albright and Winston (1979) provide the basis for Sec. 4.7. Our proof of Lemma 4.7.2 is adapted from Marshall and Olkin (1979), which contains considerable details on superadditive functions. They attribute the concept of superadditivity to Fan (1967). Derman (1970) gives an alternative proof of Lemma 4.7.2. Topkis (1978) provides a general theory of submodular functions on lattices which extends our results. His work appears to be responsible for introducing these ideas in the operations research literature. Further discussions of monotonicity appear in Ross (1983) and Heyman and Sobel (1984). Stidham (1985) discusses the use of superadditivity in queueing control models.

PROBLEMS

4.1. Provide a policy evaluation algorithm analogous to that in Sec. 4.2, which computes the expected total reward for a randomized history-dependent policy. Use it to evaluate the randomized history-dependent problem defined in Sec. 3.1.

4.2. (Howard and Matheson, 1972) Consider a Markov decision problem in which the decision maker wishes to maximize the expected utility of the random reward over a finite horizon where the utility is multiplicative and has the form

$$\Psi(r_1, \ldots, r_N) = \psi(r_1)\psi(r_2) \cdots \psi(r_N).$$

a. Provide a backward algorithm for solving this problem when the reward is a function of the current state, the current action, and the subsequent state. How would the algorithm simplify when the reward depends only on the current state and action?

b. Solve a two-period ($N = 3$) version of the example in Sec. 3.1 with $\psi(x) = e^{\gamma x}$, $r_3(s) = 0$, and $\gamma = 0.2$.

c. Show by example that there exist random variables X and Y for which $E[X] > E[Y]$ but $E[e^{\gamma X}] < E[e^{\gamma Y}]$. Conclude that different policies may be optimal under the expected total reward and the expected utility criteria.

4.3. Prove Theorem 4.3.4.

4.4. For the separable sequential allocation problem of Sec. 4.6.3 show that if g_t is concave and has a maximum on $[0, M]$ then the optimal allocation pattern is to consume $M/(N - 1)$ units each period.

4.5. For the separable sequential allocation problem, show that, if g_t is convex, then an optimal consumption pattern is to consume M units in one period and 0 units in all other periods.

4.6. Solve Problem 3.4 using backward induction. Explicitly state the optimal policy.

4.7. Show that the maximum likelihood estimates for parameters of a multinomial probability distribution are given by (4.6.3).

4.8. For the secretary problem, show by induction that, for all $t < \tau$, $u_t^*(1) > t/n$.

4.9. Solve the secretary problem, in which the objective is to maximize the probability of choosing one of the two best candidates, i.e., a candidate with rank 1 or 2.

4.10. Solve the secretary problem when the objective is to maximize the expected rank of the candidate selected.

4.11. Verify directly that each of the functions in Sec. 4.7.2 is superadditive.

4.12. Show that a twice differentiable function $g(x, y)$ on $R^1 \times R^1$ is superadditive (subadditive) whenever its second mixed partial derivatives are non-negative (nonpositive).

4.13. Prove the following generalization of Lemma 4.7.6. Let $g(s, a)$ be a function on $S \times A$, where $S = A = \{0, 1, \ldots\}$, and suppose

$$g(s + 1, a + 1) + g(s, a) \geq g(s, a + 1) + g(s + 1, a)$$

for all $a \in A$ and $s \in S$. Then g is superadditive.

4.14. Show that the sum of superadditive functions is superadditive. Is the product of superadditive functions superadditive?

4.15. Generalize Theorem 4.7.4 to a model in which S is the non-negative real line.

4.16. Prove a result analogous to Lemma 4.7.1 for g subadditive. That is, show that, whenever g is subadditive, $f(x)$ defined by (4.7.3) is nonincreasing.

4.17. Show that, if $g(x, y)$ is superadditive (subadditive) on $X \times Y$, then

$$f(x) = \min\left\{y': y' = \underset{\in Y}{\arg\min}\, g(x, y)\right\}$$

is nonincreasing, (nondecreasing) in x.

4.18. In a deterministic problem show that the graph corresponding to an optimal policy is always a maximal spanning tree.

4.19. Prove Proposition 4.7.3 in the case where all quantities are nonincreasing.

4.20. Solve a two-period ($N = 3$) version of the replacement model in Sec. 4.7.5 using the monotone backward induction algorithm. Take $p(j)$ to be geometric with $\pi = 0.4$, $R = 0$, $K = 5$, $h(s) = 2s$, and $r_3(s) = \max\{5 - s, 0\}$. Recall that a geometric random variable has a probability mass function $p(j) = (1 - \pi)\pi^j$ for $j = 0, 1, 2, \ldots$, and equals 0 otherwise. Compare the computational effort to that for the general backward induction algorithm.

4.21. (A simple bandit model) Suppose there are two projects available for selection in each of three periods. Project 1 yields a reward of one unit and always occupies state s and the other, project 2, occupies either state t or state u. When project 2 is selected, and it occupies state u, it yields a reward of 2 and moves to state t at the next decision epoch with probability 0.5. When selected in state t, it yields a reward of 0 and moves to state u at the next decision epoch with probability 1. Assume a terminal reward of 0, and that project 2 does not change state when it is not selected.

Using backward induction determine a strategy that maximizes the expected total reward.

4.22. Consider the hockey model described in Problem 3.19. Suppose your team is losing by one goal with two minutes remaining in the game, and decisions to "pull the goalie" may be made every 20 seconds. Assume that $p_G = q_G = 0.02$.
  a. Assume that $p_{NG} = 0.05$ and $q_{NG} = 0.10$; find the strategy that maximizes the probability of winning the game; of obtaining at least a tie. Find the indicated probabilities.
  b. Repeat part (a) when $q_{NG} = 0.20$.
  c. Determine the optimal strategy as a function of the model parameters.

4.23. Consider the employment seeking model in Problem 3.18.
  a. Impose assumptions on $p_t(w|w')$ and $\Psi_t(w)$ so that the optimal policy is monotone in the wage w.
  b. Explicitly solve the problem with $N = 3$, when $\Psi_t(w) = w$, and $p_t(w|w')$ is uniform on $[0, W]$.

4.24. Prove the following modification of Theorem 4.7.4 in which we drop the assumption that $A_s = A'$ for all $s \in S$.

Suppose that

   i. $A_s \subset A'$ for $s \in S$,

  ii. $A_s \subset A_{s'}$ for $s' \geq s$, and

 iii. for each s, $a \in A_s$ and $a' \in a$ implies $a' \in A_s$,

and, for $t = 1, \ldots, N - 1$,

 1. $r_t(s, a)$ is nondecreasing in s for all $a \in A_s$,

 2. $q_t(k|s, a)$ is nondecreasing in s for all $k \in S$ and $a \in A_s$,

 3. $r_t(s, a)$ is a superadditive (subadditive) function,

 4. $q_t(k|s, a)$ is a superadditive (subadditive) function for all $k \in S$, and

 5. $r_N(s)$ is nondecreasing in s.

Then there exist optimal decision rules $d_t^*(s)$ which are nondecreasing (nonincreasing) in s for $t = 1, \ldots, N$.

4.25. Use the following approach to establish that u_t^*, $t = 1, 2, \ldots, N$, satisfies the optimality equations (4.3.2) subject to (4.3.3).

a. Show inductively that

$$u_t^*(h_t) \geq \sup_{a \in A_s} \left\{ r_t(s_t, a) + \sum_{j \in S} p_t(j|s_t, a) u_{t+1}^*(h_t, a, j) \right\}$$

by observing that, since

$$u_{t+1}^*(h_{t+1}) = \sup_{\pi \in \Pi^{HR}} u_{t+1}^\pi(h_{t+1})$$

for every $\varepsilon > 0$, there exists a $\pi' \in \Pi^{HR}$ such that $u_{t+1}^{\pi'}(h_{t+1}) \geq u_{t+1}^*(h_{t+1}) - \varepsilon$. Show that this implies that, for all $a \in A_s$,

$$u_t^*(h_t) \geq r_t(s_t, a) + \sum_{j \in S} p_t(j|s_t, a) u_{t+1}^{\pi'}(h_t, a, j) - \varepsilon,$$

from which the result follows.

b. Show inductively that

$$u_t^*(h_t) \leq \sup_{a \in A_s} \left\{ r_t(s_t, a) + \sum_{j \in S} p_t(j|s_t, a) u_{t+1}^*(h_t, a, j) \right\}$$

by using the result that, for every $\varepsilon > 0$, there exists a $\pi' = (d_1, d_2, \ldots, d_{N-1}) \in \Pi^{HR}$, for which $u_t^\pi(h_t) \geq u_t^*(h_t) - \varepsilon$, to establish that

$$u_t^*(h_t) \leq \sum_{a \in A_s} q_{d_t(h_t)}(a) \left[r_t(s_t, a) + \sum_{j \in S} p_t(j|s_t, a) u_{t+1}^{\pi'}(h_t, a, j) \right] + \varepsilon.$$

c. Combine (a) and (b) to conclude the result.

4.26. Numerically solve the following version of the service rate control model of Sec. 3.7.2. Objectives are to determine if there is any obvious structure for the optimal policy and to investigate its sensitivity to model parameters.

Assume that there is a finite system capacity of eight units; that is, if arriving jobs cause the system content to exceed eight units, excess jobs do not enter the system and are lost. Let $R = 5$, $h(s) = 2s$, $B = \{0, 1, 2\}$, $K = 3$, $d(0) = 0$, $d(1) = 2$, and $d(2) = 5$. Take $r_N(s, b) = 0$ for all s and b and $N = 10$. Assume that jobs arrive followong a Poisson distribution with a rate of 1.5 jobs per period so that

$$g(s) = \frac{e^{-1.5}(1.5)^s}{s!}, \quad s = 0, 1, 2, \ldots,$$

and that the service distribution $f_b(s)$ satisfies $f_0(0) = 0.8$, $f_0(1) = 0.2$; $f_1(0) = 0.5$, $f_1(1) = 0.5$; and $f_2(0) = 0.2$, $f_2(1) = 0.8$.

Of course this problem requires developing a computer program to carry out calculations.

4.27. Find an optimal policy for the game "But Who's Counting" described in Sec. 1.7, using the description of the state space suggested in Problem 3.28. Show that the optimal policy depends only on the position of the unoccupied digits; that is, it does not depend on the numbers previously placed in the occupied digits.

4.28. In the model in Problem 3.25, suppose there are three patchs in which the animal can forage. In patch 1, the risk of predation is 0, the probability of finding food is 0.0, and its energy value is 0. In patch 2, the risk of predation is 0.004 in each period, the probability of finding food is 0.4, and the energy gain is 3 if food is found. In patch 3, the predation risk is 0.020, the probability of finding food is 0.6, and its energy gain is 5. Foraging in any patch uses one unit of energy reserves. Energy reserves below 4 indicate death and the animal's maximum energy capacity is ten units. Solve this problem for 20 foraging periods to find a patch selection strategy which maximizes the animal's probability of survival over this period.

4.29. Consider the model in Problem 3.26 regarding hunting group size selection in lions.

  a. Find a strategy (determining whether or not to hunt, and selection of hunting group size) that maximizes the probability of survival of the lion over 30 hunting days, assuming one hunt per day seeking a prey of zebras.

  b. Investigate the sensitivity of this policy to the number of hunts per day and the edible biomass of available prey.

  c. Suppose a lion's stomach content is known at the beginning of the 30-day period. How could the model results be compared to field observations?

4.30. Suppose the current price of some stock is $30 per share, and its daily price increases by $0.10 with probability 0.6, remains the same with probability 0.1 and decreases by $0.10 with probability 0.3. Find the value of a call option to purchase 100 shares of this stock at $31 per share any time in the next 30 days by finding an optimal policy for exercising this option. Assume a transaction cost of $50. Investigate the effect of the time value of money and the stock price change distribution on the optimal policy and option value.

4.31. Suppose the current price of some stock is $30 per share and its daily price increases by $0.10 with probability 0.4, remains the same with probability 0.1, and decreases by $0.10 with probability 0.5. Find the value of a put option to sell 100 shares of this stock at $29 per share any time in the next 30 days by finding an optimal policy for exercising this option. Assume a transaction cost of $50. Investigate the effect of the time value of money and the stock price change distribution on the optimal policy and option value.

4.32. Find an optimal policy for a variant of the inventory model of Secs. 3.2.2 and 4.6.2 which allows backlogging of unfilled orders, and in which there is a penalty cost for backlogged orders of 0.5 per unit per period.

4.33. (Canadian Tire inventory control, Section 1.2) Daily demand for paint brushes at a particular store follows the demand distribution:

demand	0	1	2	3	4
probability	0.7	0.15	0.1	0.04	0.01

The stock level is reviewed in the evening every four days and when warranted an order is placed at the central warehouse to augment stock. Orders arrive two days later (a two day lead time) and are available to meet demand on the morning of the third day following the review. For example, an order placed Sunday evening can be used to meet demand on Wednesday. Demand not satisfed from stock on hand is never filled. Management imposes a penalty to account for this.

a. Find a reodering policy that minimizes expected total ordering, holding and shortage costs under the assumption that the fixed cost for placing an order is $0.20, the daily per unit holding cost is $0.01 and the per unit penalty cost for unfilled orders is $0.50. Assume a planning horizon of 84 days and assign a value of $0.70 for each unit on hand at the end of this period.

b. Investigate the sensitivity of the optimal policy to the penalty cost.

c. Investigate the sensitivity of the expected total cost to the review period. In particular what is the effect of reviewing the inventory level every 7 days instead. (Hint: Write a system equation which describes the system state every review period as a function of the system state at the previous review period and the daily demand at each day between reviews. Note further that the two day demand can be found by computing the convolution of the above demand distribution with itself. Of course, solving this problem involves writing a computer program.)

CHAPTER 5

Infinite-Horizon Models: Foundations

This chapter provides an introduction to the infinite-horizon models of Chaps. 6–10 by

1. defining, illustrating, and discussing optimality criteria for infinite-horizon MDP's,

2. establishing the relationship between stochastic processes derived from history-dependent and Markov policies, and

3. providing notation for subsequent chapters.

Throughout the next four chapters, we assume stationary (time-homogeneous) problem data, that is, the rewards $r(s, a)$, the transition probabilities $p(j|s, a)$, and the decision sets D^K (K = HR, HD, MR, and MD) do not vary from decision epoch to decision epoch. Further, we assume discrete (finite or countable) S, except where noted.

Infinite-horizon models require evaluation of infinite sequences of rewards at all states in S. Consequently, we need some notions of convergence of functions on S. In most cases when we write convergence we mean *pointwise convergence*, that is, limits are defined separately for each s in S. In models with expected total discounted reward, we analyze convergence of algorithms and series in terms of convergence in supremum norm. Appendix C provides background on normed spaces, operators, and matrices.

We say that the limit of a series *exists* whenever the series has a unique limit point, even though it might be $+\infty$ or $-\infty$. This distinguishes a divergent series, that is, one with limits of either $+\infty$ or $-\infty$, from a non-convergent series which has multiple limit points.

Stationary policies assume a particularly important role in infinite-horizon models. To avoid statements such as "the stationary policy which uses decision rule d every period," we use the notation $d^\infty = (d, d, \dots)$ to denote this policy.

5.1   THE VALUE OF A POLICY

In a stationary infinite-horizon Markov decision process, each policy $\pi = (d_1, d_2, \ldots)$, induces a bivariate discrete-time reward process; $\{(X_t, r(X_t, Y_t)); \ t = 1, 2, \ldots\}$. The first component X_t represents the state of the system at time t and the second component represents the reward received when using action Y_t in state X_t. The decision rule d_t determines the action Y_t as follows.

For deterministic d_t,

$$Y_t = d_t(X_t) \quad \text{for } d_t \in D^{MD} \quad \text{and} \quad Y_t = d_t(Z_t) \quad \text{for } d_t \in D^{HD},$$

where the random variable Z_t denotes the history up to time t. For randomized d_t,

$$P\{Y_t = a\} = q_{d_t(X_t)}(a) \quad \text{for } d_t \in D^{MR}$$

and

$$P\{Y_t = a\} = q_{d_t(Z_t)}(a) \quad \text{for } d_t \in D^{HR}.$$

For Markovian π, $\{(X_t, r(X_t, Y_t)); \ t = 1, 2, \ldots\}$ is a Markov reward process.

The following functions assign values to the reward streams generated by a *fixed* policy when the system starts in a fixed state.

a. The *expected total reward* of policy $\pi \in \Pi^{HR}$, v^π is defined to be

$$v^\pi(s) \equiv \lim_{N \to \infty} E_s^\pi \left\{ \sum_{t=1}^N r(X_t, Y_t) \right\} = \lim_{N \to \infty} v_{N+1}^\pi(s), \tag{5.1.1}$$

where as in (4.1.3), $v_{N+1}^\pi(s)$ denotes the expected total reward in a model with N decision epochs and terminal reward 0. Note that the limit in (5.1.1) may be $+\infty$ or $-\infty$; as in a one-state Markov decision process with a single action and a nonzero reward. In some models, for instance that in Example 5.1.2 below, the limit in (5.1.1) need not exist. When the limit exists and when interchanging the limit and expectation is valid, for example under the conditions in Section 5.2, we write

$$v^\pi(s) = E_s^\pi \left\{ \sum_{t=1}^{\infty} r(X_t, Y_t) \right\}. \tag{5.1.2}$$

Consequently this performance measure is not always appropriate. In the next section, we distinguish infinite-horizon models which have finite expected total reward. Example 5.1.1 below provides further insight into this issue.

b. The *expected total discounted reward* of policy $\pi \in \Pi^{HR}$ is defined to be

$$v_\lambda^\pi(s) \equiv \lim_{N \to \infty} E_s^\pi \left\{ \sum_{t=1}^N \lambda^{t-1} r(X_t, Y_t) \right\} \qquad (5.1.3)$$

for $0 \le \lambda < 1$. The limit in (5.1.3) exists when

$$\sup_{s \in S} \sup_{a \in A_s} |r(s,a)| = M < \infty, \qquad (5.1.4)$$

in which case, $|v_\lambda^\pi(s)| \le (1 - \lambda)^{-1} M$ for all $s \in S$ and $\pi \in \Pi^{HR}$. When the limit exists and interchaning the limit and expectation are valid, for example when (5.1.4) holds, we write

$$v_\lambda^\pi(s) = E_s^\pi \left\{ \sum_{t=1}^\infty \lambda^{t-1} r(X_t, Y_t) \right\}. \qquad (5.1.5)$$

Note that $v^\pi(s) = \lim_{\lambda \uparrow 1} v_\lambda^\pi(s)$ whenever (5.1.2) holds.

In (5.1.3), the present value of the reward received in the first period is not multiplied by λ because we assume that rewards are received immediately after choice of the decision rule.

c. The *average reward* or *gain* of policy $\pi \in \Pi^{HR}$ is defined by

$$g^\pi(s) \equiv \lim_{N \to \infty} \frac{1}{N} E_s^\pi \left\{ \sum_{t=1}^N r(X_t, Y_t) \right\} = \lim_{N \to \infty} \frac{1}{N} v_{N+1}^\pi(s). \qquad (5.1.6)$$

When the limit in (5.1.6) does not exist, we define the *lim inf average reward*, g_- and the *lim sup average reward* g_+ by

$$g_-^\pi(s) \equiv \liminf_{N \to \infty} \frac{1}{N} v_{N+1}^\pi(s), \qquad g_+^\pi(s) \equiv \limsup_{N \to \infty} \frac{1}{N} v_{N+1}^\pi(s).$$

The reward functions provide upper and lower bounds on the average reward per period attainable by policy π. Note that g^π exists if and only if $g_-^\pi = g_+^\pi$.

When $\lim_{N \to \infty} E_s^\pi[r(X_N, Y_N)]$ exists, for example, when S is infinite, π is Markovian and stationary and the corresponding Markov chain is aperiodic, it equals $g^\pi(s)$ as defined in (5.1.6). (See Problem 5.2.)

We calculate and discuss the above quantities in the following examples.

Example 5.1.1. Consider the model in Sec. 3.1 (Fig. 3.1.1). The only deterministic Markovian decision rules are d and e where $d(s_1) = a_{1,1}$, $e(s_1) = a_{1,2}$ and $d(s_2) = e(s_2) = a_{2,1}$. For $N \ge 1$, $v_N^{e^\infty}(s_1) = 10 - (N - 1)$, $v_N^{d^\infty}(s_1) \le 10 - 0.5(N - 1)$ and $v_N^{d^\infty}(s_2) = v_N^{e^\infty}(s_2) = -N$ provided $r_N = 0$. Therefore $\lim_{N \to \infty} v_N^{e^\infty}(s) = \lim_{N \to \infty} v_N^{d^\infty}(s) = -\infty$ for $s = s_1, s_2$. However, if instead, $r(s_2, a_{2,1}) = 0$, finite limits exist with $v^{d^\infty}(s_1) = v^{e^\infty}(s_1) = 10$ and $v^{d^\infty}(s_2) = v^{e^\infty}(s_2) = 0$.

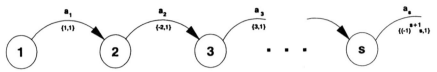

Figure 5.1.1 Symbolic representation of model in Example 5.1.2.

With the original data, the expected total discounted rewards, for $0 \le \lambda < 1$ may be shown to be equal to

$$v_\lambda^{e^\infty}(s_1) = 10 - \frac{\lambda}{1 - \lambda} = \frac{10 - 11\lambda}{1 - \lambda}, \qquad v_\lambda^{e^\infty}(s_2) = -\frac{1}{1 - \lambda}$$

and

$$v_\lambda^{d^\infty}(s_1) = \frac{5 - 5.5\lambda}{(1 - 0.5\lambda)(1 - \lambda)} = \frac{10 - 11\lambda}{(2 - \lambda)(1 - \lambda)}, \qquad v_\lambda^{d^\infty}(s_2) = -\frac{1}{1 - \lambda}.$$

Observe that both $v_\lambda^{d^\infty}(s)$ and $v_\lambda^{e^\infty}(s)$ diverge as λ increases to 1. The expected average reward or gain of both policies is -1 since, under each policy, the process is absorbed in state s_2 where the reward per period is -1, i.e., $g^{d^\infty}(s) = g^{e^\infty}(s) = -1$ for $s = s_1, s_2$.

The following countable state example with an unbounded reward function illustrates some other possible limiting behavior for these quantities.

Example 5.1.2. Let $S = \{1, 2, \dots\}$, and suppose for each $s \in S$ there is a single action a_s with $p(s + 1|s, a_s) = 1$, $p(j|s, a_s) = 0$ for $j \ne s + 1$, and $r(s, a_s) = (-1)^{s+1}s$. (See Fig. 5.1.1)

Starting in any state, only one policy is available; denote it by π. Assuming that $r_N(s) = 0$, $v_1^\pi(1) = 0$, $v_2^\pi(1) = 1$, and, in general, for $N > 1$,

$$v_N^\pi(1) = \begin{cases} k & N = 2k \\ -k & N = 2k + 1, \end{cases} \qquad k = 1, 2, \dots.$$

Thus $\lim_{N \to \infty} v_N^\pi(1)$ does not exist because

$$\limsup_{N \to \infty} v_N^\pi(1) = +\infty \qquad \text{and} \qquad \liminf_{N \to \infty} v_N^\pi(1) = -\infty, \qquad (5.1.7)$$

and the gain $g^\pi(1)$ does not exist because

$$g_+^\pi(1) = \limsup_{N \to \infty} N^{-1} v_{N+1}^\pi(1) = \tfrac{1}{2}$$

and

$$g_-^\pi(1) = \liminf_{N \to \infty} N^{-1} v_{N+1}^\pi(1) = -\tfrac{1}{2}.$$

Even though $r(s, a_s)$ is unbounded, we can use some tricks for summing infinite series to show that the expected total discounted reward is finite and satisfies

$$v_\lambda^\pi(1) = 1/(1 + \lambda)^2. \tag{5.1.8}$$

5.2 THE EXPECTED TOTAL REWARD CRITERION

The most obvious objective in an infinite-horizon model is to find a policy π with the largest value of

$$v^\pi(s) = \lim_{N \to \infty} v_N^\pi(s). \tag{5.2.1}$$

In many applications, $v_N^\pi(s)$ diverges for some or all policies, or, as in Example 5.1.2, $\lim_{N \to \infty} v_N^\pi(s)$ need not exist. We consider three approaches to analysis in such situations.

a. Restrict attention to models in which $\lim_{N \to \infty} v_N^\pi(s)$ exists for all policies.
b. Restrict attention to models in which the limit exists and is finite for at least one policy π^*, with the property that

$$v^{\pi^*}(s) \geq \limsup_{N \to \infty} v_N^\pi(s)$$

for all other π.
c. Use optimality criteria which are sensitive to the rate at which $v_N^\pi(s)$ diverges.

In this section, we concentrate on approach (a) by providing conditions on rewards and transition probabilities which ensure that the limit in (5.2.1) exists for all policies and is finite for at least one. Sections 5.4.2 and 5.4.3 expand on approach (c).

In Example 5.1.2, the limit in (5.2.1) did not exist because of the presence of arbitrarily large negative and positive rewards. Noting this, we impose a condition which ensures the existence of this limit. Define

$$v_+^\pi(s) \equiv E_s^\pi \left\{ \sum_{t=1}^\infty r^+(X_t, Y_t) \right\} \tag{5.2.2}$$

and

$$v_-^\pi(s) \equiv E_s^\pi \left\{ \sum_{t=1}^\infty r^-(X_t, Y_t) \right\} \tag{5.2.3}$$

where $r^-(s, a) \equiv \max\{-r(s, a), 0\}$ and $r^+(s, a) \equiv \max\{r(s, a), 0\}$. Both of these quantities are well defined because the summands are non-negative. The most general model we consider assumes:

> *For all $\pi \in \Pi^{HR}$, for each $s \in S$, either $v_+^\pi(s)$ or $v_-^\pi(s)$ is finite.*

Under this assumption, $v^\pi(s) = \lim_{N \to \infty} v_N^\pi(s)$ exists and satisfies

$$v^\pi(s) = v_+^\pi(s) - v_-^\pi(s). \tag{5.2.4}$$

On the basis of these quantities we distinguish some particular models we will further analyze using the expected total reward optimality criterion.

Positive bounded models
For each $s \in S$, there exists an $a \in A_s$, for which $r(s, a) \geq 0$ and $v_+^\pi(s)$ is finite for all $\pi \in \Pi^{HR}$.

Negative models
For each $s \in S$ and $a \in A_s$, $r(s, a) \leq 0$ and, for *some* $\pi \in \Pi^{HR}$, $v^\pi(s) > -\infty$ for all $s \in S$.

Convergent models
For each $s \in S$, both $v_+^\pi(s)$ and $v_-^\pi(s)$ are finite for all $\pi \in \Pi^{HR}$.

The categories above exclude models such as that in Example 5.1.2 in which both $v_+^\pi(s)$ and $v_-^\pi(s)$ are infinite. Note that $\lim_{N \to \infty} v_N^\pi(s)$ exists for each $s \in S$ for negative, positive bounded, and convergent models.

In a positive bounded model, at each state there exists an action with a non-negative reward so that there exists a stationary policy with a non-negative expected total reward. Consequently, in a positive bounded model, the optimal value function is non-negative. Further, in a finite-state positive bounded model, the boundedness assumption assures that under every policy the system is absorbed in a set of states with non-positive reward as was the case in Example 5.1.1 in which $r(s_2, a_{2,1}) = 0$.

In a negative model, $v_+^\pi(s) = 0$ for each $\pi \in \Pi^{HR}$. We also require that $v_-^\pi(s) < \infty$ for at least one $\pi \in \Pi^{HR}$, so that the expected total reward criterion distinguishes between policies. Convergent models require that both quantities in (5.2.4) are finite, or, equivalently, that

$$E_s^\pi \left\{ \sum_{t=1}^\infty |r(X_t, Y_t)| \right\} < \infty$$

for each $\pi \in \Pi^{HR}$ and $s \in S$.

When we seek to maximize expected total reward, negative models are *not* equivalent to positive models with signs reversed, because in positive models the decision maker seeks policies with a large expected total reward, and in negative models policies with an expected total reward close to zero.

Examples of positive models include

a. optimal stopping problems (Sec. 3.4);

b. problems in which the objective is to maximize the probability of reaching a certain desirable state or set of states, such as in a gambling model where the state represents the gamblers current wealth; and

c. problems in which the objective is to maximize the time to reach a undesirable set, such as the end of a computer game.

Negative models arise in the context of minimization of expected total costs when costs are non-negative. Changing signs converts all costs to negative rewards, and minimization to maximization. The condition that at least one policy has $v^\pi(s) > -\infty$ means that there exists a policy with finite total expected cost. In a finite-state model, under such a policy the system is absorbed in a set of states with zero reward. Other examples of negative models include

a. minimizing the probability of reaching an undesirable state,
b. minimizing the expected time (or cost) to reach a desirable target state, and
c. optimal stopping with minimum expected total cost criterion.

In a positive bounded model, the decision maker seeks policies which prolong termination, so that he may continue to obtain non-negative rewards as long as possible, while in a negative model the decision maker seeks termination so that he may avoid incurring costs. This suggests that methods of analysis may differ between these two cases. We discuss these models in detail in Chap. 7 by deriving results which hold in general, and then analyzing positive bounded and negative models separately. Section 10.4 uses a different approach to analyze these models.

5.3   THE EXPECTED TOTAL DISCOUNTED REWARD CRITERION

Discounted models play a central role in this book because they apply directly to economic problems, because they have a very complete theory, and because numerous computational algorithms are available for their solution. Discounting arises naturally in applications in which we account for the time value of the rewards. The discount factor λ measures the present value of one unit of currency received one period in the future, so that $v_\lambda^\pi(s)$ represents the expected total present value of the income stream obtained using policy π. From a mathematical perspective, discounting together with the finite reward assumption (5.1.4) ensures convergence of series (5.1.3).

Discounting may be viewed in another way. The decision maker values policies according to the expected total reward criterion; however, the horizon length ν is random and independent of the actions of the decision maker. Such randomness in the horizon length might be due to bankruptcy of the firm in an economic model, failure of a system in a production model, or death of an animal in an ecological model.

Let $v_\nu^\pi(s)$ denote the expected total reward obtained by using policy π when the horizon length ν is random. We define it by

$$v_\nu^\pi(s) \equiv E_s^\pi \left[E_\nu \left\{ \sum_{t=1}^\nu r(X_t, Y_t) \right\} \right], \tag{5.3.1}$$

where E_ν denotes expectation with respect to the probability distribution of ν.

We now direct attention to a model in which the horizon length ν follows a geometric distribution with parameter λ, $0 \leq \lambda < 1$ independent of the policy. That is,

$$P\{\nu = n\} = (1 - \lambda)\lambda^{n-1}, \qquad n = 1, 2, \ldots .$$

Such probability distributions arise when modeling the time to the first "failure" in a sequence of independent, identically distributed Bernoulli trails with "success" probability λ. Generalizations include the negative binomial distribution which provides a model for the time to the kth failure in such a sequence.

The following result relates these two criteria in the case of geometrically distributed ν.

Proposition 5.3.1. Suppose that (5.1.4) holds and ν has a geometric distribution with parameter λ. Then $v_\nu^\pi(s) = v_\lambda^\pi(s)$ for all $s \in S$.

Proof. Rewriting (5.3.1) gives

$$v_\nu^\pi(s) = E_s^\pi \left\{ \sum_{n=1}^\infty \sum_{t=1}^n r(X_t, Y_t)(1 - \lambda)\lambda^{n-1} \right\}.$$

As a consequence of (5.1.4), and the assumption that $\lambda < 1$, for each realization of $\{(X_t, Y_t); t = 1, 2, \ldots\}$, the series converges so that we may reverse the order of summation within the brackets to obtain

$$= E_s^\pi \left\{ \sum_{t=1}^\infty \sum_{n=t}^\infty r(X_t, Y_t)(1 - \lambda)\lambda^{n-1} \right\}.$$

Using the identity

$$\sum_{n=1}^\infty \lambda^{n-1} = \frac{1}{1 - \lambda}$$

yields

$$v_\nu^\pi(s) = E_s^\pi \left\{ \sum_{t=1}^\infty \lambda^{t-1} r(X_t, Y_t) \right\} = v_\lambda^\pi(s). \qquad \square$$

In terms of a Markov decision problem, the calculation in the above proof corresponds to augmenting the state space S with an absorbing state Δ and a corresponding set of actions $A_\Delta = \{a_{\Delta,1}\}$ and transforming $p(j|s,a)$ and $r(s, a, j)$ to $\bar{p}(j|s, a)$ and $\bar{r}(s, a, j)$ as follows:

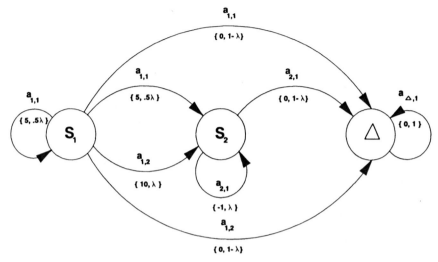

Figure 5.3.1 Modified version of Example 5.1.1 showing effect of random termination.

$$\bar{p}(j|s, a) = \begin{cases} \lambda p(j|s, a), & j \neq \Delta, s \neq \Delta \\ 1 - \lambda, & j = \Delta, s \neq \Delta \\ 1, & j = \Delta, s = \Delta, a = a_{\Delta, 1}, \end{cases}$$

$$\bar{r}(s, a, j) = \begin{cases} r(s, a, j), & j \neq \Delta \\ 0, & j = \Delta, \text{ or } s = \Delta, a = a_{\Delta, 1}. \end{cases}$$

Note we require the rewards to depend on both the current and subsequent states.

Example 5.3.1. Figure 5.3.1 illustrates the effect of this transformation on the data in Sec. 3.1. We use the notation of Example 5.1.1. Observe that $v_\lambda^{e^*}(s_2)$ satisfies the recursion

$$v_\lambda^{e^*}(s_2) = -1 + \lambda v_\lambda^{e^*}(s_2)$$

so that $v_\lambda^{e^*}(s_2) = -1/(1 - \lambda)$ and $v_\lambda^{e^*}(s_1) = 10 + \lambda v_\lambda^{e^*}(s_2) = 10 - \lambda/(1 - \lambda)$, as was shown directly in Example 5.1.1.

Several extensions of the random horizon model and its equivalent MDP formulation include models in which

1. the random variable ν has a probability distribution that is not geometric (Problem 5.9);

2. the "success" probability λ depends on the state of the system, for example, in models of equipment deterioration; and

3. the "success" probability λ depends on both the state and action.

5.4 OPTIMALITY CRITERIA

We begin this section by presenting optimality criteria based on the policy value functions introduced in Sec. 5.1; quite simply, a policy with the largest value function is optimal. A difficulty with such optimality criteria is that there may be several optimal policies for each criterion, yet each will have a distinct reward process. Because of this, we provide more refined optimality criteria in Sec. 5.4.2 and 5.4.3.

5.4.1 Criteria Based on Policy Value Functions

We say that a policy π^* is *total reward optimal* whenever

$$v^{\pi^*}(s) \geq v^{\pi}(s) \qquad \text{for each } s \in S \text{ and all } \pi \in \Pi^{HR}.$$

This criterion applies to the positive bounded, negative, and convergent models distinguished in Sec. 5.2. Define the *value* of the MDP by

$$v^*(s) \equiv \sup_{\pi \in \Pi^{HR}} v^{\pi}(s). \tag{5.4.1}$$

An optimal policy $\pi^* \in \Pi^K$ ($K = $ HR, HD, MR, or MD) exists whenever

$$v^{\pi^*}(s) = v^*(s) \qquad \text{for all } s \in S.$$

We say that a policy π^* is *discount optimal* for fixed λ, $0 \leq \lambda < 1$, whenever

$$v_{\lambda}^{\pi^*}(s) \geq v_{\lambda}^{\pi}(s) \qquad \text{for each } s \in S \text{ and all } \pi \in \Pi^{HR}.$$

In discounted models, the value of the MDP, $v_{\lambda}^*(s)$, is defined by

$$v_{\lambda}^*(s) \equiv \sup_{\pi \in \Pi^{HR}} v_{\lambda}^{\pi}(s). \tag{5.4.2}$$

A discount-optimal policy $\pi^* \in \Pi^K$ ($K = $ HR, HD, MR, or MD) exists whenever

$$v_{\lambda}^{\pi^*}(s) = v_{\lambda}^*(s) \qquad \text{for all } s \in S.$$

When $g^{\pi}(s)$ exists for all $s \in S$ and $\pi \in \Pi^{HR}$, we say that a policy π^* is *average optimal* or *gain optimal* whenever

$$g^{\pi^*}(s) \geq g^{\pi}(s) \qquad \text{for each } s \in S \text{ and all } \pi \in \Pi^{HR}.$$

When using this criterion, the value or optimal gain g^* is defined by

$$g^*(s) = \sup_{\pi \in \Pi^{HR}} g^\pi(s). \qquad (5.4.3)$$

An average optimal policy $\pi^* \in \Pi^K$ (K = HR, HD, MR, or MD) exists whenever

$$g^{\pi^*}(s) = g^*(s) \qquad \text{for all } s \in S.$$

When the limits in (5.1.6), which define $g^\pi(s)$, do not exist for some policies, for example when induced Markov chains are periodic, we consider related concepts of optimality. In this case we call a policy π^* *average optimal* if the infimum of its limit points is at least as great as *any* limit point of any other policy. That is, for each $s \in S$,

$$g_-^{\pi^*}(s) = \liminf_{N \to \infty} \frac{1}{N} v_{N+1}^{\pi^*}(s) \geq \limsup_{N \to \infty} \frac{1}{N} v_{N+1}^\pi(s) = g_+^\pi(s)$$

for all $\pi \in \Pi^{HR}$. Weaker optimality criteria may be based on comparing the same limit points for all policies. We say that a policy π^* is *lim sup average optimal* if the supremum of the limit points of its average reward sequences is at least as great as any limiting average reward of any other policy. That is, for each $s \in S$,

$$g_+^{\pi^*}(s) \geq g_+^\pi(s)$$

for all $\pi \in \Pi^{HR}$. We say that a policy π^* is *lim inf average optimal* if the infimum of limit points of its average reward sequences is at least as great as *the smallest* limiting average reward of any other policy. That is, for each $s \in S$,

$$g_-^{\pi^*}(s) \geq g_-^\pi(s)$$

for all $\pi \in \Pi^{HR}$. Examples in Chap. 8 distinguish these criteria.

We will primarily be concerned with models in which $\lim_{N \to \infty} N^{-1} v_{N+1}^\pi(s)$ exists. However, even in this case, the average optimality criterion has limitations in distinguishing policies, as illustrated by the following simple example.

Example 5.4.1. Take $S = \{s_1, s_2, s_3\}$ and choose the action sets, rewards, and transition probabilities as in Figure 5.4.1: $A_{s_1} = \{a_{1,1}\}$, $r(s_1, a_{1,1}) = 0$, and $p(s_2|s_1, a_{1,1}) = 1$; $A_{s_2} = \{a_{2,1}, a_{2,2}\}$, $r(s_2, a_{2,1}) = 1$, $r(s_2, a_{2,2}) = 0$, $p(s_1|s_2, a_{2,1}) = 1$, and $p(s_3|s_2, a_{2,2}) = 1$; and $A_{s_3} = \{a_{3,1}\}$, $r(s_3, a_{3,1}) = 1$, and $p(s_2|s_3, a_{3,1}) = 1$.

Action choice in s_2 determines stationary policies. Denote the stationary policy which uses action $a_{2,1}$ by d^∞, and that which uses $a_{2,2}$ by e^∞. Clearly $g^{d^\infty}(s) = g^{e^\infty}(s) = 0.5$ for all $s \in S$, and the average reward for every other policy also equals 0.5. Consequently, all policies are average optimal.

This example shows that the average reward criteria does not distinguish between policies which might have different appeal to a decision maker. Starting in state s_2, policy d^∞ with reward stream $(1, 0, 1, 0, \dots)$ would clearly be preferred to e^∞ with reward stream $(0, 1, 0, 1, \dots)$ because it provides a reward of one unit in the first

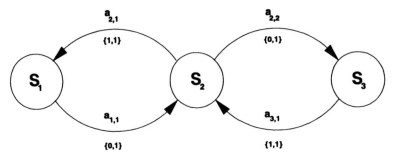

Figure 5.4.1 Symbolic representation of Example 5.4.1.

period which can be used or invested to yield a reward exceeding one in the second period (assuming positive interest rates). We call a criterion such as the average reward *unselective* because under it, there may be several optimal policies, each with a different reward stream.

In the next two subsections, we discuss several more selective criteria based on either

a. the comparative finite-horizon expected total reward as the number of periods becomes large, or

b. the comparative expected discounted reward as the discount factor λ increases to 1.

*5.4.2 Overtaking Optimality Criteria

We base the criteria in this section on the comparative asymptotic behavior of v_N^π, without requiring that $\lim_{N \to \infty} v_N^\pi$ exists. We say that a policy π^* is *overtaking optimal*, if, for each $s \in S$,

$$\liminf_{N \to \infty} [v_N^{\pi^*}(s) - v_+^\pi(s)] \geq 0 \qquad \text{for all } \pi \in \Pi^{HR}. \tag{5.4.4}$$

Example 5.4.1 (ctd.). We evaluate the limit in (5.4.4) for policies d^∞ and e^∞ of Example 5.4.1. Observe that the $\{v_N^{d^*}(s_2): N \geq 1\} = \{1, 1, 2, 2, 3, 3, \ldots\}$ and $\{v_N^{e^*}(s_2): N \geq 1\} = \{0, 1, 1, 2, 2, 3, \ldots\}$ so that $\{v_N^{d^*}(s_2) - v_N^{e^*}(s_2): N \geq 1\} = \{1, 0, 1, 0, 1, 0, \ldots\}$. Thus

$$\liminf_{N \to \infty} \left[v_N^{d^*}(s_2) - v_N^{e^*}(s_2)\right] = 0 \quad \text{and} \quad \liminf_{N \to \infty} \left[v_N^{e^*}(s_2) - v_N^{d^*}(s_2)\right] = -1,$$

so that d^∞ dominates e^∞ in the overtaking optimal sense. In fact, we can show that d^∞ is overtaking optimal.

The following example shows that an overtaking optimal policy need not exist.

Example 5.4.2. (Denardo and Miller, 1968) Take $S = \{s_1, s_2, s_3\}$ and choose the action sets, rewards, and transition probabilities as follows (see Fig. 5.4.2): $A_{s_1} = \{a_{1,1}, a_{1,2}\}$, $r(s_1, a_{1,1}) = 1$, $r(s_1, a_{1,2}) = 0$, $p(s_2|s_1, a_{1,1}) = 1$, and $p(s_3|s_1, a_{1,2}) = 1$; $A_{s_2} = \{a_{2,1}\}$, $r(s_2, a_{2,1}) = 0$, and $p(s_3|s_2, a_{2,1}) = 1$; and $A_{s_3} = \{a_{3,1}\}$, $r(s_3, a_{3,1}) = 2$, and $p(s_2|s_3, a_{3,1}) = 1$.

Let d^∞ be the stationary policy which uses action $a_{1,1}$ in state s_1 and e^∞ be the stationary policy which uses $a_{1,2}$ in state s_1. Clearly $g^{d^\infty}(s) = g^{e^\infty}(s) = 1$, so that both policies are average optimal. Since $\{v_N^{d^\infty}(s_1) - v_N^{e^\infty}(s_1): N \geq 1\} = \{1, -1, 1, -1, 1, -1, \ldots\}$,

$$\liminf_{N \to \infty} \left[v_N^{d^\infty}(s_1) - v_N^{e^\infty}(s_1) \right] = \liminf_{N \to \infty} \left[v_N^{e^\infty}(s_1) - v_N^{d^\infty}(s_1) \right] = -1.$$

Hence neither d^∞ nor e^∞ dominates the other in terms of overtaking optimality. In fact, for this model there exists *no* overtaking optimal policy (even after admitting randomized policies as in Problem 5.8).

This example shows that this criterion is *overselective*, that is, there exists no optimal policy with respect to this criterion. We next consider a less selective criterion. Call a policy π^* *average-overtaking optimal* if, for each $s \in S$,

$$\liminf_{N \to \infty} \frac{1}{N} \sum_{t=1}^{N} \left[v_t^{\pi^*}(s) - v_t^{\pi}(s) \right] \geq 0 \qquad \text{for all } \pi \in \Pi^{HR}. \tag{5.4.5}$$

Example 5.4.2 (ctd.). We evaluate the limit in (5.4.5) for policies d^∞ and e^∞ from Example 5.4.2. Observe that the $\{\sum_{t=1}^{N} v_t^{d^\infty}(s_1): N \geq 1\} = \{1, 2, 5, 8, 13, 18, \ldots\}$ and $\{\sum_{t=1}^{N} v_t^{e^\infty}(s_1): N \geq 1\} = (0, 2, 4, 8, 12, 18, \ldots\}$ so that $\{\sum_{t=1}^{N} [v_t^{d^\infty}(s_1) - v_t^{e^\infty}(s_1)]: N \geq 1\} =$

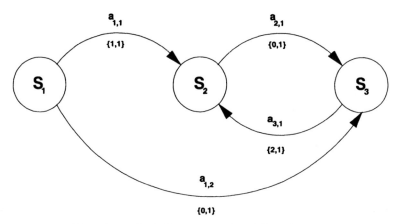

Figure 5.4.2 Symbolic representation of Example 5.4.2.

$\{1, 0, 1, 0, 1, 0, \ldots\}$. Thus

$$\liminf_{N \to \infty} \frac{1}{N} \sum_{t=1}^{N} \left[v_t^{d^\times}(s_1) - v_t^{e^\times}(s_1) \right] = \liminf_{N \to \infty} \frac{1}{N} \sum_{t=1}^{N} \left[v_t^{e^\times}(s_1) - v_t^{d^\times}(s_1) \right] = 0,$$

so that d^∞ and e^∞ are equivalent under the average-overtaking optimality criterion. In fact both d^∞ and e^∞ are average-overtaking optimal.

Note further that

$$\liminf_{N \to \infty} \sum_{t=1}^{N} \left[v_t^{d^\times}(s_1) - v_t^{e^\times}(s_1) \right] = 0 \tag{5.4.6}$$

and

$$\liminf_{N \to \infty} \sum_{t=1}^{N} \left[v_t^{e^\times}(s_1) - v_t^{d^\times}(s_1) \right] = -1,$$

so that clearly d^∞ is superior to e^∞ under the criterion implicit in (5.4.6).

We say that a policy π^* is *cumulative-overtaking optimal* whenever

$$\liminf_{N \to \infty} \sum_{t=1}^{N} \{ v_t^{\pi^*}(s) - v_t^{\pi}(s) \} \geq 0 \qquad \text{for all } \pi \in \Pi^{HR}. \tag{5.4.7}$$

In Example 5.4.2, d^∞ is cumulative overtaking optimal.

It is easy to see that, if π^* is overtaking optimal, it is average-overtaking optimal; the above example illustrates that the converse need not hold. Further, whenever a policy is cumulative-overtaking optimal it is also average-overtaking optimal, so that the cumulative-overtaking optimality criterion is more selective than the average-over-taking optimality criterion.

The average-overtaking optimal criterion may be generalized further. Let $v_{N,1}^{\pi}$ denote the expected total return for policy π up to period N, and define $v_{N,n}^{\pi}$ recursively for $n \geq 1$ by

$$v_{N,n}^{\pi}(s) = \sum_{j=1}^{N} v_{j,n-1}^{\pi}(s).$$

We say that a policy $\pi^* \in \Pi$ is *n-average optimal* for $n = -1, 0, 1, \ldots$ if, for all $s \in S$,

$$\liminf_{N \to \infty} N^{-1} \left[v_{N,n+2}^{\pi^*}(s) - v_{N,n+2}^{\pi}(s) \right] \geq 0 \qquad \text{for all } \pi \in \Pi^{HR}.$$

for all $\pi \in \Pi$. Observe that (-1) − average optimality and gain optimality are equivalent, as are 0-average optimality and average-overtaking optimality. Overtaking optimality and cumulative-overtaking optimality are distinct criteria.

5.4.3 Sensitive Discount Optimality Criteria

The family of sensitive-discount optimality criteria are based on the comparative limiting behavior of the expected total discounted reward as the discount rate increases to 1. We say that a policy π^* is *n-discount optimal* for $n = -1, 0, 1, \ldots$, if, for each $s \in S$,

$$\liminf_{\lambda \uparrow 1} (1 - \lambda)^{-n} [v_\lambda^{\pi^*}(s) - v_\lambda^\pi(s)] \geq 0 \qquad \text{for all } \pi \in \Pi^{HR} \qquad (5.4.8)$$

Further, a policy is ∞-*discount optimal* if it is n-discount optimal for all $n \geq -1$ and that a policy π^* is *1-optimal* or *Blackwell optimal* if for each $s \in S$ there exists a $\lambda^*(s)$ such that

$$v_\lambda^{\pi^*}(s) - v_\lambda^\pi(s) \geq 0 \qquad \text{for all } \pi \in \Pi^{HR} \text{ for } \lambda^*(s) \leq \lambda < 1. \qquad (5.4.9)$$

Blackwell (1962) proposed this criterion for finite S, in which case $\lambda^* \equiv \sup_{s \in S} \lambda^*(s)$ is attained. In countable-state problems, this supremum might equal 1; we call policies *strongly Blackwell optimal* when $\lambda^* < 1$.

We will subsequently show that

a. average or gain optimality is equivalent to (-1)-discount optimality, and

b. Blackwell optimality is equivalent to ∞-discount optimality.

We will also justify referring to 0-discount optimality as *bias optimality*.

Blackwell optimality is the most selective of the n-discount optimality criteria as it implies n-discount optimality for all finite n. Hence, it implies gain and bias optimality. In general, n-discount optimality implies m-discount optimality for all $m < n$, so that bias optimality ($n = 0$) is more selective than gain optimality ($n = -1$).

The following example illustrates some of these ideas.

Example 5.4.3. We again consider the data from the example in Sec. 3.1 and note the expected discounted rewards computed in Example 5.1.1. Since

$$v_\lambda^{d^*}(s_1) - v_\lambda^{e^*}(s_1) = \frac{11\lambda - 10}{2 - \lambda}, \qquad (5.4.10)$$

it follows that, for $s = s_1, s_2$,

$$\liminf_{\lambda \uparrow 1} (1 - \lambda) [v_\lambda^{d^*}(s) - v_\lambda^{e^*}(s)] = 0.$$

Thus d^∞ and e^∞ are equivalent in the (-1)-discount optimality or average optimality sense (as already noted in Sec. 5.1) and in fact are both (-1)-discount optimal. But, since

$$\liminf_{\lambda \uparrow 1} [v_\lambda^{d^*}(s_1) - v_\lambda^{e^*}(s_1)] = 1,$$

d^∞ dominates e^∞ with respect to the 0-discount (bias) optimality criterion and can be shown to be 0-discount optimal. Note further that, for $n \geq 1$,

$$\liminf_{\lambda \uparrow 1} (1 - \lambda)^{-n} [v_\lambda^{d^*}(s_1) - v_\lambda^{e^*}(s_1)] = +\infty,$$

so that d^∞ is n-discount optimal for all $n \geq 0$. From the discussion preceding this example, this means that d^∞ is Blackwell optimal. We can also see this from (5.4.10), which shows that $v_\lambda^{d^*}(s_1) \geq v_\lambda^{e^*}(s_1)$ for $\lambda > \frac{10}{11}$.

5.5 MARKOV POLICIES

In this section, we show that given any history-dependent policy and starting state, there exists a randomized Markov policy with the same expected total reward, expected total discounted reward, and average reward. We derive this result by constructing for each history-dependent policy a randomized Markov policy with the same joint probability distribution of states and actions. The important consequence of this result is that for *each fixed initial state* we may restrict attention to Markov policies when computing and implementing optimal policies. In subsequent chapters, we will show that, for most Markov decision problems, we can further restrict attention to deterministic Markov policies when seeking optimal policies. State-action frequencies, which are the essence of Theorem 5.5.1, also play an important role in linear programming formulations of Markov decision processes and models with constraints. To simplify the proof of the main result, we assume finite or countable S, but the results hold in greater generality (Strauch, 1966).

Theorem 5.5.1.   Let $\pi = (d_1, d_2, \dots) \in \Pi^{HR}$. Then, for each $s \in S$, there exists a policy $\pi' = (d_1', d_2', \dots) \in \Pi^{MR}$, satisfying

$$P^{\pi'}\{X_t = j, Y_t = a | X_1 = s\} = P^{\pi}\{X_t = j, Y_t = a | X_1 = s\} \qquad (5.5.1)$$

for $t = 1, 2, \dots$.

Proof.   Fix $s \in S$. For each $j \in S$ and $a \in A_j$, define the randomized Markov decision rule d_t' by

$$q_{d_t'(j)}(a) \equiv P^{\pi}\{Y_t = a | X_t = j, X_1 = s\} \qquad (5.5.2)$$

for $t = 1, 2, \dots$. Let $\pi' = (d_1', d_2', \dots)$, so that, as a consequence of (5.5.2),

$$P^{\pi'}\{Y_t = a | X_t = j\} = P^{\pi'}\{Y_t = a | X_t = j, X_1 = s\} = P^{\pi}\{Y_t = a | X_t = j, X_1 = s\}. \qquad (5.5.3)$$

We show by induction that (5.5.1) holds with d_t' defined through (5.5.2). Clearly it holds with $t = 1$. Assume (5.5.1) holds for $t = 2, 3, \dots, n-1$. Then

$$P^{\pi}\{X_n = j | X_1 = s\} = \sum_{k \in S} \sum_{a \in A_k} P^{\pi}\{X_{n-1} = k, Y_{n-1} = a | X_1 = s\} p(j|k, a)$$

$$= \sum_{k \in S} \sum_{a \in A_k} P^{\pi'}\{X_{n-1} = k, Y_{n-1} = a | X_1 = s\} p(j|k, a) \qquad (5.5.4)$$

$$= P^{\pi}\{X_n = j | X_1 = s\},$$

where the equality in (5.5.4) follows from the induction hypothesis. Therefore

$$P^{\pi'}\{X_n = j, Y_n = a | X_1 = s\} = P^{\pi'}\{Y_n = a | X_n = j\}P^{\pi'}\{X_n = j | X_1 = s\}$$

$$= P^{\pi}\{Y_n = a | X_n = j, X_1 = s\}P^{\pi}\{X_n = j | X_1 = s\}$$

$$= P^{\pi}\{Y_n = j, Y_n = a | X_1 = s\}, \qquad (5.5.5)$$

where the equality in the line above (5.5.5) follows from (5.5.4) and (5.5.3). Thus the induction hypothesis is satisfied and (5.5.1) holds for $t = 1, 2, \ldots$. □

Note that, in the above theorem, the derived randomized Markov policy π' depends on the initial state X_1. When the state at decision epoch 1 is chosen according to a probability distribution, the Markov policy will not depend on the initial state but might depend on the probability distribution. We state this result without proof.

Corollary 5.5.2. For each distribution of X_1, and any history-dependent policy π, there exists a randomized Markov policy π' for which

$$P^{\pi'}\{X_t = j, Y_t = a\} = P^{\pi}\{X_t = j, Y_t = a\}.$$

The key step in the proof of Theorem 5.5.1 is the construction of an equivalent Markov policy though (5.5.2). We illustrate this construction with an example.

Example 5.5.1. We illustrate the construction in the proof of Theorem 5.5.1 for the model in Fig. 3.1.2. Set $N = 3$ and define the randomized history-dependent policy $\pi = (d_1, d_2)$ by

$$q_{d_1(s_1)}(a_{1,1}) = 0.6, \qquad q_{d_1(s_1)}(a_{1,2}) = 0.3, \qquad q_{d_1(s_1)}(a_{1,3}) = 0.1,$$

$$q_{d_1(s_2)}(a_{2,1}) = 1,$$

and,

	$q_{d_2(s, a, s_1)}(a)$		
History	$a = a_{1,1}$	$a = a_{1,2}$	$a = a_{1,3}$
$(s_1, a_{1,1})$	0.4	0.3	0.3
$(s_1, a_{1,3})$	0.8	0.1	0.1

and, for every (s, a),

$$q_{d_2(s, a, s_2)}(a_{2,1}) = 1.$$

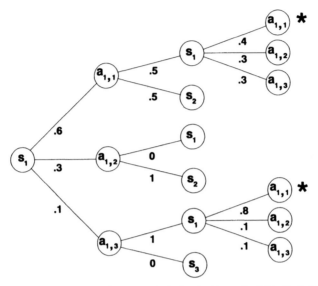

Figure 5.5.1 Tree diagram illustrating the calculation in Example 5.5.1. Probabilities on the right most arcs correspond to the randomized history-dependent policy π (* indicates quantities used in the example).

We now construct $\pi' = (d'_1, d'_2)$. Of course $d'_1 = d_1$. Applying (5.5.2) (see Fig. 5.5.1) yields

$$
q_{d'_2(s_1)}(a_{1,1}) = P^\pi\{Y_2 = a_{1,1} | X_2 = s_1\} = P^\pi\{Y_2 = a_{1,1} | X_2 = s_1, X_1 = s_1\}
$$

$$
= \frac{P^\pi\{Y_2 = a_{1,1}, X_2 = s_1, X_1 = s_1\}}{P^\pi\{X_2 = s_1, X_1 = s_1\}}
$$

$$
= \frac{(0.6)(0.5)(0.4) + (0.1)(1.0)(0.8)}{(0.6)(0.5) + (0.1)(1.0)} = 0.5
$$

We leave it as an exercise to compute $q_{d'_2(s_1)}(a)$ for $a = a_{1,2}, a_{1,3}$.

One might conjecture that, if in the hypotheses of Theorem 5.5.1, $\pi \in \Pi^{HD}$, then π' could be chosen from Π^{MD}. This is not the case because even though the action chosen by π is deterministic, the distribution of state and actions at any decision epoch is random, so that d'_t defined by (5.5.2) must be random (Problem 5.11). We return to this point in Sec. 6.2.

Noting that

$$
v_N^\pi(s) = \sum_{t=1}^{N-1} \sum_{j \in S} \sum_{a \in A_j} r(j,a) P^\pi\{X_t = j, Y_t = a | X_1 = s\}
$$

$$
+ \sum_{j \in S} \sum_{a \in A_j} r_N(j) P^\pi\{X_N = j, Y_N = a | X_1 = s\}
$$

and

$$v_\lambda^\pi(s) = \sum_{t=1}^{\infty} \sum_{j \in S} \sum_{a \in A_j} \lambda^{t-1} r(j, a) P^\pi \{ X_t = j, Y_t = a | X_1 = s \},$$

the following important theorem follows directly from Theorem 5.1.1.

Theorem 5.5.3. Suppose $\pi \in \Pi^{HR}$, then for each $s \in S$ there exists a $\pi' \in \Pi^{MR}$ (which possibly varies with s) for which

a. $v_N^{\pi'}(s) = v_N^\pi(s)$ for $1 \le N < \infty$;
b. $v_\lambda^{\pi'}(s) = v_\lambda^\pi(s)$ for $0 \le \lambda < 1$;
c. $g_+^{\pi'}(s) = g_+^\pi(s)$, $g_-^{\pi'}(s) = g_-^\pi(s)$, and $g^\pi(s) = g^{\pi'}(s)$, whenever $g_+^{\pi'}(s) = g_-^{\pi'}(s)$ and $g_+^\pi(s) = g_-^\pi(s)$, and
d. if $r_N(s) = 0$ and $v^\pi(s) \equiv \lim_{N \to \infty} v_N^\pi(s)$ exists, $v^{\pi'}(s) = v^\pi(s)$.

5.6 VECTOR NOTATION FOR MARKOV DECISION PROCESSES

This section provides notation we will use in the remainder of the book. We refer the reader to Appendix C and the references therein for background on normed linear space concepts.

Let V denote the set of bounded real valued functions on S with componentwise partial order and norm $\|v\| \equiv \sup_{s \in S} |v(s)|$. Further, let V_M denote the subspace of V of Borel measurable functions. For finite or countable S (endowed with a discrete topology), all real-valued functions are measurable so that V and V_M coincide, but, when S is a continuum, V_M is a proper subset of V. Let $e \in V$ denote the function with all components equal to 1; that is, $e(s) = 1$ for all $s \in S$.

For discrete S, we often refer to elements of V as vectors and linear operators on V as matrices. When using the above norm on V, the corresponding matrix norm is given by

$$\|H\| \equiv \sup_{s \in S} \sum_{j \in S} |H(j|s)|,$$

where $H(j|s)$ denotes the (s, j)th component of H.

For discrete S, let $|S|$ denote the number of elements in S. For $d \in D^{MD}$, define $r_d(s)$ and $p_d(j|s)$ by

$$r_d(s) \equiv r(s, d(s)) \quad \text{and} \quad p_d(j|s) \equiv p(j|s, d(s)), \qquad (5.6.1)$$

and for $d \in D^{MR}$ by

$$r_d(s) \equiv \sum_{a \in A_s} q_{d(s)}(a) r(s, a) \quad \text{and} \quad p_d(j|s) \equiv \sum_{a \in A_s} q_{d(s)}(a) p(j|s, a) \quad (5.6.2)$$

In either case, let r_d denote the $|S|$-vector, with sth component $r_d(s)$, and P_d the $|S| \times |S|$ matrix with (s, j)th entry given by $p_d(j|s)$. We refer to r_d as the *reward vector* and P_d as the *transition probability matrix* corresponding to Markovian decision rule d. For $0 \leq \lambda \leq 1$, $r_d + \lambda P_d v$ equals the expected total one-period discounted (undiscounted when $\lambda = 1$) reward obtained using decision rule d with terminal reward v.

Lemma 5.6.1. Suppose S is discrete, $|r(s,a)| \leq M$ for all $a \in A_s$ and $s \in S$, and $0 \leq \lambda \leq 1$. Then, for all $v \in V$ and $d \in D^{MR}$, $r_d + \lambda P_d v \in V$.

Proof. As a consequence of the assumption on $r(s,a)$, $\|r_d\| \leq M$ for all $d \in D^{MR}$, so that $r_d \in V$. When P_d is a probability matrix, $\|P_d\| = 1$, so that $\|P_d v\| \leq \|P_d\| \|v\| = \|v\|$. Consequently, $P_d v \in V$ for all $v \in V$, so that $r_d + \lambda P_d v \in V$. $\qquad\square$

Under policy $\pi = (d_1, d_2, \dots) \in \Pi^{MR}$, the (s,j)th component of the t-step transition probability matrix P_π^t satisfies

$$P_\pi^t(j|s) = \left[P_{d_t} P_{d_{t-1}} \cdots P_{d_1} \right](j|s) = P^\pi(X_{t+1} = j | X_1 = s). \qquad (5.6.3)$$

Expectations with respect to the Markov chain corresponding to this policy are computed according to

$$E_s^\pi\{v(X_t)\} = P_\pi^{t-1} v(s) = \sum_{j \in S} P_\pi^{t-1}(j|s) v(j) \qquad (5.6.4)$$

for $v \in V$ and $1 \leq t < \infty$.

As a consequence of this representation for the expectation, and the definition of v_λ^π, for $0 \leq \lambda \leq 1$,

$$v_\lambda^\pi = \sum_{t=1}^\infty \lambda^{t-1} P_\pi^{t-1} r_{d_t}, \qquad (5.6.5)$$

provided the limit exists.

BIBLIOGRAPHIC REMARKS

Bellman (1957) and Howard (1960) were the first researchers to explicitly consider the infinite-horizon Markov decision process model. Bellman considered a model without reward and provided an iterative procedure for obtaining a policy with maximum growth rate. Howard introduced the model studied here and provided an algorithm for finding optimal policies under the discounted and average reward criteria. Howard (1978) later related that his development of this model was motivated by a consulting project with the catalog division of Sears, Roebuck and Company, which he participated in while a graduate student at MIT. The problem he modeled was that of determining an optimal catalog mailing policy based on customer's purchasing behavior in the current period. He noted that

"The net result was a predicted few percent increase in profitability of the catalog operation, which, however, amounted to several million dollars per year." (p. 203)

Blackwell was an early and significant contributor to Markov decision problem theory. He studied the expected total-reward criterion (1961), (-1)-, 0-, and ∞-discount optimality (1962), discounted models in considerable generality (1965), and positive models (1967). Strauch (1966), provided a thorough analysis of negative models. Hordijk (1974) and van Hee, Hordijk, and van der Wal (1976) analyzed convergent models. Derman (1970) established the relationship between discounting and random termination according to a geometric distribution as presented in Proposition 5.3.1; Haviv and Puterman (1992) extended this result to negative binomial stopping times.

The overtaking optimality criteria appears to originate in the economic accumulation models of von Weizsacker (1965) and Gale (1967). Veinott (1966a) introduced the average overtaking optimality criterion in the Markov decision process framework. Denardo and Miller (1968) contributed to this theory by introducing and studying the overtaking optimality criterion. Other contributors to this theory include Sladky (1974), Hordijk and Sladky (1977), Denardo and Rothblum (1979), and Leizarowitz (1987).

Rothblum and Veinott (1975) and Rothblum (1984) generalize and unify the overtaking optimality criteria through the introduction of the family of (n, k)-optimality criteria. The equivalence between their criteria and those referred to in Sec. 5.3 are that: $(0, 0)$-optimality is equivalent to overtaking optimality, $(0, 1)$-optimality is equivalent to average-overtaking optimality, and $(n, 1)$-optimality is equivalent to n-average optimality. Their concept of $(1, 0)$-optimality might be characterized as cumulative-overtaking optimality.

Using a partial Laurent series expansion of the expected discounted reward, Blackwell (1962) introduced two optimality criteria that were latter shown by Veinott (1969) and (1974) to be specific sensitive optimality criteria; Blackwell's notion of a *nearly optimal* policy is equivalent to Veinott's notion of a 0-discount optimal policy, and Blackwell's concept of an *optimal* policy is equivalent to Veinott's concept of an ∞-discount optimal policy. The family of sensitive discount optimality criteria provide a rich and unified framework for analysis of undiscounted Markov decision problems. Dekker (1985), Lasserre (1988) and Dekker and Hordijk (1988, 1991 and 1992) studied sensitive discount optimality criteria in countable-state problems.

Derman and Strauch (1966) provided Theorem 5.5.1, which demonstrated that, in countable-state models for any history-dependent policy and any initial state, there exists a randomized Markov policies with equivalent-state action frequencies. Strauch (1966) proved this result in greater generality. Results on the existence of optimal deterministic policies and stationary policies are easiest to derive separately for each optimality criterion, and appear in appropriate chapters.

PROBLEMS

5.1. Verify the calculations of the expected total reward and the expected total discounted reward of policies d^{∞} and e^{∞} in Example 5.1.1.

5.2. The purpose of this example is to explore the need for averaging when computing the gain. Suppose $S = \{1, 2\}$, $A_1 = \{a\}$, $A_2 = \{b\}$, $p(2|1, a) = p(1|2, b) = 1$, and $r(1, a) = r(2, b) = 1$. Let δ be the unique decision rule.

   a. Show that $\lim_{N \to \infty} E_s^{\delta^{\infty}}\{r(X_N, Y_N))\}$ exists for $s \in S$, and equals $g^{\delta}(s)$ as defined in (5.1.6).

   b. Suppose instead that $r(1, a) = 2$ and $r(2, b) = 1$, then show that the limit in part (a) does not exist, but that defined in (5.1.6) does.

5.3. Modify the rewards in Example 5.1.2 to show that $\lim\inf_{N\to\infty} v_N^\pi$ and $\lim\sup_{N\to\infty} v_N^\pi$ may be finite and unequal.

5.4. Show that a nonstationary model with $S = \{s\}$, $A_s = \{a\}$, $r_t(s, a) = (-1)^{t+1}t$, and $p_t(s|s, a) = 1$ for $t = 1, 2, \ldots$ gives the same results as Example 5.1.2.

5.5. Show for the data in Example 5.1.2 that

$$\lim_{N\to\infty} \frac{1}{2}N^{-2} \sum_{n=1}^{N} v_n^\pi(1) \qquad (*)$$

exists. Interpret this quantity in the context of the problem and discuss the advantages and disadvantages of using g^π defined by $(*)$ as an optimality criterion. Veinott (1969) refers to the limit in $(*)$ as a Cesaro limit of order 2.

5.6. Show that when the limit in (5.1.6) exists, it has the same value as the limit in $(*)$ in Problem 5.5.

5.7. Consider the following example in which $S = \{s_1, s_2, s_3, s_4\}$, $A_{s_1} = \{a_{1,1}\}$, $r(s_1, a_{1,1}) = 0$, and $p(s_1|s_1, a_{1,1}) = 1$; $A_{s_2} = \{a_{2,1}, a_{2,2}\}$, $r(s_2, a_{2,1}) = 2$, $r(s_2, a_{2,2}) = 1$, $p(s_1|s_2, a_{2,1}) = 1$, and $p(s_3|s_2, a_{2,2}) = 1$; $A_{s_3} = \{a_{3,1}\}$, $r(s_3, a_{3,1}) = -1$, and $p(s_4|s_3, a_{3,1}) = 1$; and $A_{s_4} = \{a_{4,1}\}$, $r(s_4, a_{4,1}) = 1$, and $p(s_3|s_4, a_{4,1}) = 1$. Let δ denote the decision rule which chooses action $a_{2,1}$ in state s_2 and η denote the policy which chooses action $a_{2,2}$ in s_2.

a. Show that this model is not a positive, negative, or convergent model, and that the expected total-reward criterion cannot be used to compare policies.

b. Show that δ^∞ is overtaking optimal.

5.8. The purpose of this problem is extend the conclusions of Example 5.4.2 to the family of randomized policies. Consider the data of Example 5.4.2 augmented by actions $a_{1,q}$ in s_1 with $r(s_1, a_{1,q}) = 1$, $p(s_2|s_1, a_{1,q}) = q$, and $p(s_3|s_1, a_{1,q}) = 1 - q, 0 < q < 1$.

a. Observe that using action $a_{1,q}$ in state s_1 is equivalent to randomizing between action $a_{1,2}$ and $a_{1,1}$, and show that there exists no overtaking optimal policy within Π^{HR}.

b. Show that all $\pi \in \Pi^{HR}$ are average-overtaking optimal.

5.9. (Haviv and Puterman, 1992) Show that when ν has a negative binomial distribution with parameters 2 and λ, that is

$$P\{\nu = n\} = (n - 1)(1 - \lambda)^2\lambda^{n-2}, \qquad n = 2, 3, \ldots;$$

then

$$v_\nu^\pi(s) = v_\lambda^\pi(s) + (1 - \lambda)\frac{\partial v_\lambda^\pi(s)}{\partial \lambda},$$

where $v_\nu^\pi(s)$ is defined in (5.3.1).

5.10. Extend results in the previous problem to a negative binomial stopping time with parameters k and λ.

5.11 a. Show, using arguments similar to those in Example 5.5.1, that for the deterministic history-dependent policy of Sec. 3.1 there exists a deterministic Markov policy π', such that $P^{\pi'}\{X_2 = s_1, Y_2 = a_{1,3}|X_1 = s_1\}$ equals the same quantity for the given deterministic history-dependent policy.

  b. Consider the deterministic history-dependent policy $\kappa = (d_1^\kappa, d_2^\kappa, \ldots)$ which alternately chooses actions $a_{1,1}$ and $a_{1,3}$ in s_1 at successive decision epochs; that is, d_1^κ and d_2^κ are as given in Sec. 3.1, and

$$d_3^\kappa(s_1, a_{1,1}, s_1, a_{1,3}, s_1) = a_{1,1}, \qquad d_4^\kappa(s_1, a_{1,1}, s_1, a_{1,3}, s_1, a_{1,1}, s_1) = a_{1,3}$$

and so forth. Show that for $n \geq 3$ there exists no deterministic Markov policy with equivalent-state action frequencies to κ but that there exists a randomized Markov policy π' for which

$$P^{\pi'}\{X_n = s_1, Y_n = a_{1,3}|X_1 = s_1\} = P^\kappa\{X_n = s_1, Y_n = a_{1,3}|X_1 = s_1\}$$

for $n \geq 3$.

5.12. Prove Corollary 5.5.2.

5.13. Establish equality (5.1.8).

5.14. Consider the following deterministic two-state model: $S = \{s_1, s_2\}$, $A_{s_1} = \{a_{1,1}, a_{1,2}\}$, $A_{s_2} = \{a_{2,1}, a_{2,2}\}$, $r(s_1, a_{1,1}) = 4$, $r(s_1, a_{1,2}) = 3$, $r(s_2, a_{2,1}) = 5$, and $r(s_2, a_{2,2}) = 7$; and $p(s_1|s_1, a_{1,1}) = 1$, $p(s_2|s_1, a_{1,2}) = 1$, $p(s_2|s_2, a_{2,1}) = 1$, and $p(s_1|s_2, a_{2,2}) = 1$.

  a. Compute the expected total discounted reward λ and the gain for each stationary policy.

  b. Find all stationary policies which are

   1. gain optimal

   2. (-1)-discount optimal

   3. 0-discount optimal

   4. 1-discount optimal

   5. Blackwell optimal.

  c. Find a λ_0 such that the Blackwell optimal policy is discount optimal for all $\lambda \geq \lambda_0$.

5.15. Compute $q_{d_2'(s_1)}(a_{1,2})$ and $q_{d_2'(s_1)}(a_{1,3})$ for the model in Example 5.5.1.

CHAPTER 6

Discounted Markov Decision Problems

In this chapter, we study infinite-horizon Markov decision processes with the expected total discounted reward optimality criterion. These models are the best understood of all infinite-horizon Markov decision problems. Results for these models provide a standard for the theory of models with other optimality criteria. Results for discounted models are noteworthy in that they hold regardless of the chain structure of Markov chains generated by stationary policies.

We focus on the optimality equation and use the theory of equations on normed linear spaces to establish the existence of its solution and the convergence of algorithms for solving it. Section 6.2 derives properties of solutions of the optimality equation and provides conditions that ensure existence of optimal policies. Throughout this chapter, we pay particular attention to algorithms for finding optimal policies, including

a. value iteration (Sec. 6.3),

b. policy iteration (Sec. 6.4),

c. modified policy iteration (Sec. 6.5), and

d. linear programming (Sec. 6.9).

Section 6.6 provides some important concepts that we use to enhance performance of value iteration, policy iteration, and modified policy iteration. Section 6.7.3 provides our recommended algorithm, and Sec. 6.7.4 reports computational experience with these algorithms. In Sec. 6.10 we address specific questions that arise when analyzing countable-state models, including unbounded reward functions and finite-state approximations. Section 6.11 illustrates one of the main uses of Markov decision process theory, determining the structure of optimal policies. We discuss models with general state spaces in Secs. 6.2.5 and 6.11.3. Throughout, we use the model in Fig. 3.1.1 from Sec. 3.1 to illustrate concepts. *For an overview of material in this comprehensive chapter, we recommend Sections 6.1, 6.2.1–6.2.4, 6.3.2, 6.4.2, 6.5.1, 6.6.2–6.6.3, 6.7.2, 6.7.3, 6.9, and 6.11.1.*

We impose the following assumptions in this chapter:

Assumption 6.0.1. *Stationary rewards and transition probabilities*; $r(s, a)$ and $p(j|s, a)$ do not vary from decision epoch to decision epoch.

Assumption 6.0.2. *Bounded rewards*; $|r(s, a)| \leq M < \infty$ for all $a \in A_s$ and $s \in S$ (except in Secs. 6.10 and 6.11).

Assumption 6.0.3. *Discounting*; future rewards are discounted according to a discount factor λ, with $0 \leq \lambda < 1$.

Assumption 6.0.4. *Discrete state spaces* (except in Secs. 6.2.5 and 6.11.3); S is finite or countable.

Occasionally we restate them for emphasis.

We assume familiarity with most linear space concepts in Appendix C and Sec. 5.6. Recall that V denotes the partially ordered normed linear space of bounded real-valued functions on S with supremum norm and componentwise partial order (Sec. C.1), and V_M denotes its subset of bounded measurable functions. When S is finite or countable, which is the main focus of this chapter, measurability is not an issue, and $V_M = V$. In Sec. 6.2.5, we discuss results for general state spaces, so therein we distinguish V and V_M. Except where noted, we use the expression *convergence* to refer to convergence in norm.

6.1 POLICY EVALUATION

In this section we develop recursions for the expected total discounted reward of a Markov policy π. These recursions provide the basis for algorithms and proofs in this chapter.

As a consequence of Theorem 5.5.3b, we need not consider history-dependent policies, since, for each $s \in S$ given any $\pi \in \Pi^{HR}$, there exists a $\pi' \in \Pi^{MR}$ with identical total discounted reward; so that

$$v_\lambda^*(s) \equiv \sup_{\pi \in \Pi^{HR}} v_\lambda^\pi(s) = \sup_{\pi \in \Pi^{MR}} v_\lambda^\pi(s).$$

Let $\pi = (d_1, d_2, \ldots) \in \Pi^{MR}$. As in (5.1.5), the expected total discounted reward of this policy is defined by

$$v_\lambda^\pi(s) = E_s^\pi \left\{ \sum_{t=1}^\infty \lambda^{t-1} r(X_t, Y_t) \right\}. \tag{6.1.1}$$

Following (5.6.5), and letting $P_\pi^0 \equiv I$, (6.1.1) may be expressed in vector notation as

$$v_\lambda^\pi = \sum_{t=1}^\infty \lambda^{t-1} P_\pi^{t-1} r_{d_t}$$

$$= r_{d_1} + \lambda P_{d_1} r_{d_2} + \lambda^2 P_{d_1} P_{d_2} r_{d_3} + \cdots$$

$$= r_{d_1} + \lambda P_{d_1} \left(r_{d_2} + \lambda P_{d_2} r_{d_3} + \lambda^2 P_{d_2} P_{d_3} r_{d_4} + \cdots \right). \tag{6.1.2}$$

So that

$$v_\lambda^\pi = r_{d_1} + \lambda P_{d_1} v_\lambda^{\pi'}, \tag{6.1.3}$$

where $\pi' = (d_2, d_3, \ldots)$. Equation (6.1.3) shows that the discounted reward corresponding to policy π equals the discounted reward in a one-period problem in which the decision maker uses decision rule d_1 in the first period and receives the expected total discounted reward of policy π' as a terminal reward. This interpretation is perhaps more obvious when it is expressed in component notation as

$$v_\lambda^\pi(s) = r_{d_1}(s) + \sum_{j \in S} \lambda p_{d_1}(j|s) v_\lambda^{\pi'}(j). \tag{6.1.4}$$

Equations (6.1.3) and (6.1.4) are valid for any $\pi \in \Pi^{MR}$; however, when π is stationary so that $\pi' = \pi$, they simplify further. Let $d^\infty \equiv (d, d, \ldots)$ denote the stationary policy which uses decision rule $d \in D^{MR}$ at each decision epoch. For this policy, (6.1.4) becomes

$$v_\lambda^{d^\infty}(s) = r_d(s) + \sum_{j \in S} \lambda p_d(j|s) v_\lambda^{d^\infty}(j), \tag{6.1.5}$$

and (6.1.3) becomes

$$v_\lambda^{d^\infty} = r_d + \lambda P_d v_\lambda^{d^\infty}. \tag{6.1.6}$$

Thus, $v_\lambda^{d^\infty}$ satisfies the system of equations

$$v = r_d + \lambda P_d v.$$

In fact, when $0 \le \lambda < 1$, we show below that it is the unique solution.

For each $v \in V$, define the linear transformation L_d by

$$L_d v \equiv r_d + \lambda P_d v. \tag{6.1.7}$$

As a consequence of Lemma 5.6.1, $L_d: V \to V$. In this notation, equation (6.1.6) becomes

$$v_\lambda^{d^\infty} = L_d v_\lambda^{d^\infty}. \tag{6.1.8}$$

This means that $v_\lambda^{d^\infty}$ is a *fixed point* of L_d in V. Fixed-point theorems provide important tools for analyzing discounted models, and are the subject of Sec. 6.2.3.

The above discussion together with Corollary C.4 in Appendix C leads to the following important result.

Theorem 6.1.1. Suppose $0 \leq \lambda < 1$. Then for any stationary policy d^* with $d \in D^{MR}$, $v_\lambda^{d^*}$ is the unique solution in V of

$$v = r_d + \lambda P_d v. \tag{6.1.9}$$

Further, $v_\lambda^{d^*}$ may be written as

$$v_\lambda^{d^*} = (I - \lambda P_d)^{-1} r_d. \tag{6.1.10}$$

Proof. Rewriting (6.1.9) yields

$$(I - \lambda P_d)v = r_d.$$

Since $\|P_d\| = 1$ and $\lambda = \|\lambda P_d\| \geq \sigma(\lambda P_d)$, Corollary C.4 establishes for $0 \leq \lambda < 1$ that $(I - \lambda P_d)^{-1}$ exists, so that from (5.6.5)

$$v = (I - \lambda P_d)^{-1} r_d = \sum_{t=1}^{\infty} \lambda^{t-1} P_d^{t-1} r_d = v_\lambda^{d^*}. \qquad \square$$

Example 6.1.1. We illustrate Theorem 6.1.1 for the model in Sec. 3.1. We evaluate the stationary policy δ^∞ which chooses action $a_{1,1}$ in s_1 and $a_{2,1}$ in s_2. The policy evaluation equations (6.1.9) expressed in component notation are

$$v(s_1) = 5 + 0.5\lambda v(s_1) + 0.5\lambda v(s_2),$$
$$v(s_2) = -1 + \lambda v(s_2)$$

so that

$$v_\lambda^{\delta^*}(s_1) = \frac{5 - 5.5\lambda}{(1 - 0.5\lambda)(1 - \lambda)}, \qquad v_\lambda^{\delta^*}(s_2) = -\frac{1}{1 - \lambda}. \tag{6.1.11}$$

Letting γ^∞ denote the stationary policy which chooses action $a_{1,2}$ in s_1 we find that

$$v_\lambda^{\gamma^*}(s_1) = \frac{10 - 11\lambda}{1 - \lambda}, \qquad v_\lambda^{\gamma^*}(s_2) = -\frac{1}{1 - \lambda}. \tag{6.1.12}$$

The matrix $(I - \lambda P_d)^{-1}$ plays a crucial role in the theory of discounted Markov decision problems. We will often use its following positivity and order-preserving properties. When s is discrete, we use a superscript "T" to denote transposition; that is, u^T denotes the transpose of u. Note we do not distinguish the scalar and vector 0 so that $u \geq 0$ means $u(s) \geq 0$ for all $s \in S$.

Lemma 6.1.2. Suppose $0 \leq \lambda < 1$ and $u \in V$ and $v \in V$. Then, for any $d \in D^{MR}$,

a. if $u \geq 0$, then $(I - \lambda P_d)^{-1} u \geq 0$ and $(I - \lambda P_d)^{-1} u \geq u$;
b. if $u \geq v$, then $(I - \lambda P_d)^{-1} u \geq (I - \lambda P_d)^{-1} v$; and
c. if $u \geq 0$, then $u^T (I - \lambda P_d)^{-1} \geq 0$ and $u^T (I - \lambda P_d)^{-1} \geq u^T$.

Proof. Since $\sigma(\lambda P_d) < 1$, Corollary C.4 and the non-negativity of all elements of P_d imply that

$$(I - \lambda P_d)^{-1} u = u + \lambda P_d u + \lambda^2 P_d u + \cdots \geq u \geq 0.$$

Part (b) follows from replacing u by $u - v$ in (a), and part (c) follows from (a) by transposition. $\square$

When $(I - \lambda P_d)^{-1} u \geq 0$ for $u \geq 0$, we often write $(I - \lambda P_d)^{-1} \geq 0$. In this case we refer to $(I - \lambda P_d)^{-1}$ as a *positive* operator on V.

6.2 OPTIMALITY EQUATIONS

The optimality equation and its solution play a central role in the theory of discounted Markov decision problems. In this section we show the following.

1. The optimality equation has a unique solution in V.
2. The value of the discounted MDP satisfies the optimality equation.
3. The optimality equation characterizes stationary optimal policies.
4. Optimal policies exist under reasonable conditions on the states, actions, rewards, and transition probabilities.

We introduce one of the key tools for analysis of discounted models, the Banach Fixed-Point Theorem, in Sec. 6.2.3.

6.2.1 Motivation and Definitions

Under Assumptions 6.0.1–6.0.4, the finite-horizon optimality equations (4.3.2) may be expressed as

$$v_n(s) = \sup_{a \in A_s} \left\{ r(s, a) + \sum_{j \in S} \lambda p(j|s, a) v_{n+1}(j) \right\}. \tag{6.2.1}$$

Passing to the limit in (6.2.1) suggests that equations of the following form will characterize values and optimal policies in infinite-horizon models:

$$v(s) = \sup_{a \in A_s} \left\{ r(s, a) + \sum_{j \in S} \lambda p(j|s, a) v(j) \right\}. \tag{6.2.2}$$

We refer to this system of equations (6.2.2) as *the optimality equations* or *Bellman equations*.

We use the following vector notation in this chapter. For $v \in V$, define the (nonlinear) operator $\mathscr{L}$ on V by

$$\mathscr{L}v \equiv \sup_{d \in D^{MD}} \{r_d + \lambda P_d v\}, \qquad (6.2.3)$$

where we compute the supremum in (6.2.3) with respect to the componentwise partial ordering on V. This means that we find it by evaluating the supremum over A_s for each $s \in S$ separately as on the right-hand side of (6.2.2). For discrete S and r_d bounded it follows from Lemma 5.6.1 that $\mathscr{L}v \in V$ for all $v \in V$. For unbounded $r(s, a)$, we use a different vector space V_w to ensure that $\mathscr{L}$ maps V_w into V_w (Sec. 6.10), while for nondiscrete S we require additional assumptions for the reward and transition probabilities to ensure that $\mathscr{L}v \in V_M$ for $v \in V_M$ (see Problem 6.7a and Sec. 6.2.5).

When the supremum on the right-hand side of (6.2.3) is attained for all $v \in V$, we define L by

$$Lv \equiv \max_{d \in D^{MD}} \{r_d + \lambda P_d v\}. \qquad (6.2.4)$$

Note that we take the supremum in (6.2.3) and the maximum in (6.2.4) over the set of deterministic Markovian decision rules instead of over the set of randomized Markovian decision rules. Using Lemma 4.3.1, we now show that we obtain the same value of the supremum as if we were to allow randomization.

The following result holds for $0 \le \lambda \le 1$ and will be used in the next chapter with $\lambda = 1$.

Proposition 6.2.1.   For all $v \in V$ and $0 \le \lambda \le 1$,

$$\sup_{d \in D^{MD}} \{r_d + \lambda P_d v\} = \sup_{d \in D^{MR}} \{r_d + \lambda P_d v\}. \qquad (6.2.5)$$

Proof.   Since $D^{MR} \supset D^{MD}$, the right-hand side of (6.2.5) must be as least as great as the left-hand side. To establish the reverse inequality, choose $v \in V$, $\delta \in D^{MR}$ and apply Lemma 4.3.1 at each $s \in S$ with $W = A_s$, $q(\cdot) = q_\delta(\cdot)$, and

$$w(\cdot) = r(s, \cdot) + \sum_{j \in S} \lambda p(j|s, \cdot) v(j)$$

to show that

$$\sup_{a \in A_s} \left\{ r(s, a) + \sum_{j \in S} \lambda p(j|s, a) v(j) \right\} \ge \sum_{a \in A_s} q_\delta(a) \left[r(s, a) + \sum_{j \in S} \lambda p(j|s, a) v(j) \right].$$

Therefore, for any $\delta \in D^{MR}$,

$$\sup_{d \in D^{MD}} \{r_d + \lambda P_d v\} \geq r_\delta + \lambda P_\delta v,$$

from which it follows that the left-hand side of (6.2.5) is as least as great as the right-hand side.   □

We now represent the optimality equations in vector notation. *To simplify notation we use D instead of D^{MD} to denote the set of deterministic Markovian decision rules.* The optimality equation (6.2.2) may be expressed as

$$v = \sup_{d \in D} \{r_d + \lambda P_d v\} = \mathcal{L} v. \tag{6.2.6}$$

When the supremum in (6.2.2) or (6.2.6) is attained for all $v \in V$, for example when A_s is finite, we replace "sup" by "max." In component notation the optimality equations then become

$$v(s) = \max_{a \in A_s} \left\{ r(s, a) + \sum_{j \in S} \lambda p(j|s, a) v(j) \right\}, \tag{6.2.7}$$

and in vector notation,

$$v = \max_{d \in D} \{r_d + \lambda P_d v\} = L v. \tag{6.2.8}$$

Noting that solutions of the optimality equation are *fixed points* of $\mathcal{L}$ or L on V provides a unifying perspective from which to derive existence results.

6.2.2 Properties of Solutions of the Optimality Equations

The following theorem shows that sub-solutions and super-solutions of the optimality equations (6.2.6) or (6.2.8) provide lower and upper bounds on v_λ^*. Consequently, when we have a solution, both bounds hold so that the solution must equal v_λ^*. This result provides a key algorithmic and theoretical tool and applies for general S. Note that part (c) of this theorem does not establish the existence of a solution to $\mathcal{L} v = v$; it means that if this equation has a solution, then it equals v_λ^*. The proof of this theorem relies on the observation that when $0 \leq \lambda < 1$, the tail of the expected discounted reward series becomes arbitrarily small.

Theorem 6.2.2. Suppose there exists a $v \in V$ for which

a. $v \geq \mathcal{L} v$, then $v \geq v_\lambda^*$;

b. $v \leq \mathcal{L} v$, then $v \leq v_\lambda^*$;

c. $v = \mathcal{L} v$, then v is the only element of V with this property and $v = v_\lambda^*$.

Proof. First we prove part (a). Choose $\pi = (d_1, d_2, \ldots) \in \Pi^{MR}$. From Proposition 6.2.1,

$$v \geq \sup_{d \in D^{MD}} \{r_d + \lambda P_d v\} = \sup_{d \in D^{MR}} \{r_d + \lambda P_d v\},$$

so that

$$v \geq r_{d_1} + \lambda P_{d_1} v \geq r_{d_1} + \lambda P_{d_1} \left(r_{d_2} + \lambda P_{d_2} v \right) = r_{d_1} + \lambda P_{d_1} r_{d_2} + \lambda^2 P_{d_1} P_{d_2} v$$

By induction, it follows that, for $n \geq 1$,

$$v \geq r_{d_1} + \lambda P_{d_1} r_{d_2} + \cdots + \lambda^{n-1} P_{d_1} \cdots P_{d_{n-1}} r_{d_n} + \lambda^n P_\pi^n v.$$

Thus,

$$v - v_\lambda^\pi \geq \lambda^n P_\pi^n v - \sum_{k=n}^{\infty} \lambda^k P_\pi^k r_{d_{k+1}}. \tag{6.2.9}$$

Choose $\varepsilon > 0$. Since $\|\lambda^n P_\pi^n v\| \leq \lambda^h \|v\|$, and $0 \leq \lambda < 1$, for n sufficiently large

$$-(\varepsilon/2)e \leq \lambda^n P_\pi^n v \leq (\varepsilon/2)e$$

where e denotes a vector of 1's. As a result of Assumption 6.0.2,

$$-\frac{\lambda^n M e}{1 - \lambda} \leq \sum_{k=n}^{\infty} \lambda^k P_\pi^k r_{d_{n+1}}$$

so by choosing n sufficiently large, the second expression on the right-hand side of (6.2.9) can be bounded above and below by $(\varepsilon/2)e$. Since the left-hand side of (6.2.9) does not depend on n, it follows that

$$v(s) \geq v_\lambda^\pi(s) - \varepsilon$$

for all $s \in S$ and $\varepsilon > 0$. Since ε was arbitrary, Theorem 5.5.3 implies that for each $s \in S$

$$v(s) \geq \sup_{\pi \in \Pi^{MR}} v_\lambda^\pi(s) = \sup_{\pi \in \Pi^{HR}} v_\lambda^\pi(s) = v_\lambda^*(s)$$

establishing part (a). We establish the result in part (b), as follows. Since $v \leq \mathscr{L} v$, for arbitrary $\varepsilon > 0$ there exists a $d \in D^{MD}$ such that $v \leq r_d + \lambda P_d v + \varepsilon e$. From Lemma 6.1.2b,

$$v \leq (I - \lambda P_d)^{-1}(r_d + \varepsilon e) = v_\lambda^{d^\infty} + (1 - \lambda)^{-1} \varepsilon e.$$

Therefore $v \leq \sup_{\pi \in \Pi^{HR}} v_\lambda^\pi + (1 - \lambda)^{-1} \varepsilon e$. The result follows since ε was arbitrary. Part (c) follows by combining (a) and (b). □

Note that the above result holds with L replacing $\mathscr{L}$; only the proof is simpler.

6.2.3   Solutions of the Optimality Equation

In this section, we apply the Banach fixed-point theorem to establish existence of a solution to the optimality equation. We include its constructive proof because it provides the basis for the value iteration algorithm. This section assumes familiarity with material in Appendix C. If you are uncomfortable with that theory, you might relate it to notions of functions on R^n.

Let U be a Banach space. Recall that this means that U is a complete normed linear space. We say that an operator $T: U \to U$ is a *contraction mapping* if there exists a λ, $0 \le \lambda < 1$ such that

$$\|Tv - Tu\| \le \lambda \|v - u\|$$

for all u and v in U.

Theorem 6.2.3. (Banach Fixed-Point Theorem) Suppose U is a Banach space and $T: U \to U$ is a contraction mapping. Then

a. there exists a unique v^* in U such that $Tv^* = v^*$; and
b. for arbitrary v^0 in U, the sequence $\{v^n\}$ defined by

$$v^{n+1} = Tv^n = T^{n+1}v^0 \tag{6.2.10}$$

converges to v^*.

Proof. Let $\{v^n\}$ be defined by (6.2.10). Then, for any $m \ge 1$,

$$\|v^{n+m} - v^n\| \le \sum_{k=0}^{m-1} \|v^{n+k+1} - v^{n+k}\| = \sum_{k=0}^{m-1} \|T^{n+k}v^1 - T^{n+k}v^0\|$$

$$\le \sum_{k=0}^{m-1} \lambda^{n+k}\|v^1 - v^0\| = \frac{\lambda^n(1 - \lambda^m)}{(1 - \lambda)}\|v^1 - v^0\|. \tag{6.2.11}$$

Since $0 \le \lambda < 1$, it follows from (6.2.11) that $\{v^n\}$ is a Cauchy sequence; that is, for n sufficiently large, $\|v^{n+m} - v^n\|$ can be made arbitrarily small. From the completeness of U, it follows that $\{v^n\}$ has a limit $v^* \in U$.

We now show that v^* is a fixed point of T. Using properties of norms and contraction mappings, it follows that

$$0 \le \|Tv^* - v^*\| \le \|Tv^* - v^n\| + \|v^n - v^*\|$$
$$= \|Tv^* - Tv^{n-1}\| + \|v^n - v^*\| \le \lambda\|v^* - v^{n-1}\| + \|v^n - v^*\|.$$

Since $\lim_{n \to \infty} \|v^n - v^*\| = 0$, both quantities on the right-hand side of the above inequality can be made arbitrarily small by choosing n large enough. Consequently, $\|Tv^* - v^*\| = 0$, from which we conclude that $Tv^* = v^*$.

We leave the proof of uniqueness of v^* as an exercise. □

To apply this theorem to the discounted model, we now show that L is a contraction mapping on V, the space of bounded functions on S with supremum norm. A proof that $\mathscr{L}$ is a contraction mapping follows similar lines and is left as an exercise. Note that, under additional technical conditions, this result applies to more general state spaces, for example $S = R^n$ (c.f., Problem 6.7).

Proposition 6.2.4. Suppose that $0 \leq \lambda < 1$; then L and $\mathscr{L}$ are contraction mappings on V.

Proof. Since S is discrete, L maps V into V. Let u and v be in V, fix $s \in S$, assume that $Lv(s) \geq Lu(s)$, and let

$$a_s^* \in \arg\max_{a \in A_s} \left\{ r(s, a) + \sum_{j \in S} \lambda p(j|s, a) v(j) \right\}.$$

Then

$$0 \leq Lv(s) - Lu(s) \leq r(s, a_s^*) + \sum_{j \in S} \lambda p(j|s, a_s^*) v(j)$$

$$- r(s, a_s^*) - \sum_{j \in S} \lambda p(j|s, a_s^*) u(j)$$

$$= \lambda \sum_{j \in S} p(j|s, a_s^*) [v(j) - u(j)] \leq \lambda \sum_{j \in S} p(j|s, a_s^*) \|v - u\| = \lambda \|v - u\|.$$

Repeating this argument in the case that $Lu(s) \geq Lv(s)$ implies that

$$|Lv(s) - Lu(s)| \leq \lambda \|v - u\|$$

for all $s \in S$. Taking the supremum over s in the above expression gives the result.   □

We now state and prove the main result of this section and a fundamental result in the theory of discounted Markov decision problems. Note that, under additional technical considerations, we may extend this result to more general state spaces.

Theorem 6.2.5. Suppose $0 \leq \lambda < 1$, S is finite or countable, and $r(s, a)$ is bounded.

a. Then there exists a $v^* \in V$ satisfying $Lv^* = v^*$ ($\mathscr{L}v^* = v^*$). Further, v^* is the only element of V with this property and equals v_λ^*.
b. For each $d \in D^{\text{MR}}$, there exists a unique $v \in V$ satisfying $L_d v = v$. Further, v is the unique solution and equals $v_\lambda^{d^\infty}$.

Proof. Since V is a complete normed linear space, Proposition 6.2.4 establishes that L and $\mathscr{L}$ are contractions so that the hypotheses of Theorem 6.2.3 are satisfied. Therefore there exists a unique solution $v^* \in V$ to $Lv = v$ or $\mathscr{L}v = v$. Theorem 6.2.2(c) establishes that $v^* = v_\lambda^*$.

Part (b) follows from part (a) by choosing $D = \{d\}$.   □

Observe that the proof of part (b) provides an alternative proof to Theorem 6.1.1. This simple argument reduces many results we obtain for optimality equations to results for linear equations.

6.2.4 Existence of Optimal Policies

From the perspective of both computation and implementation, we would like to restrict attention to stationary policies when seeking optimal policies. In this section, we show that the existence of a decision rule which attains the supremum in (6.2.3) for $v = v_\lambda^*$ implies the existence of a stationary optimal policy. We show how to identify such a policy and provide conditions on the rewards and transition probabilities which ensure attainment of this supremum.

We begin this section with a result which provides an obvious method for identifying optimal policies in infinite-horizon models. To implement it, pick a policy π, evaluate it, and check whether its value satisfies the optimality equation.

Theorem 6.2.6. A policy $\pi^* \in \Pi^{HR}$ is optimal if and only if $v_\lambda^{\pi^*}$ is a solution of the optimality equation.

Proof. Suppose π^* is optimal; then $v_\lambda^{\pi^*} = v_\lambda^*$. From Theorem 6.2.5(a) it follows that $v_\lambda^{\pi^*}$ satisfies $\mathscr{L}v = v$. Suppose $\mathscr{L}v_\lambda^{\pi^*} = v_\lambda^{\pi^*}$; then Theorem 6.2.2(c) implies that $v_\lambda^{\pi^*} = v_\lambda^*$, so that π^* is optimal. $\square$

For $v \in V$, call a decision rule $d_v \in D^{MD}$ *v-improving* if

$$d_v \in \arg\max_{d \in D}\{r_d + \lambda P_d v\}. \tag{6.2.12}$$

Equivalently,

$$r_{d_v} + \lambda P_{d_v} v = \max_{d \in D}\{r_d + \lambda P_d v\} \quad \text{or} \quad L_{d_v} v = Lv. \tag{6.2.13}$$

In component notation, d_v is v improving if for all $s \in S$;

$$r(s, d_v(s)) + \sum_{j \in S} \lambda p(j|s, d_v(s))v(j) = \max_{a \in A_s}\left\{r(s,a) + \sum_{j \in S} \lambda p(j|s,a)v(j)\right\}.$$

Note that the expression "v-improving" is a slight misnomer for two reasons. First the expected discounted reward of d_v^∞ need not be greater than or equal to v; $v_\lambda^{(d_v)^\infty} \geq v$ when $r_{d_v} + \lambda P_{d_v} v \geq v$. Second, even if $r_{d_v} + \lambda P_{d_v} v \geq v$, $v_\lambda^{(d_v)^\infty}$ exceeds v in some component only if

$$r_{d_v}(s') + \lambda P_{d_v} v(s') > v(s')$$

for at least one $s' \in S$ (Problem 6.4).

Decision rules $d \in D^{MD}$, which are v_λ^* improving, are of particular note. Many authors refer to them as *conserving*. That is, a decision rule d^* is conserving if

$$L_{d^*} v_\lambda^* \equiv r_{d^*} + \lambda P_{d^*} v_\lambda^* = v_\lambda^*, \tag{6.2.14}$$

or, alternatively, if

$$d^* \in \underset{d \in D}{\arg\max} \{r_d + \lambda P_d v_\lambda^*\}. \tag{6.2.15}$$

The following theorem, which shows that such decision rules are optimal, is one of the most important results of this chapter. It provides one of the main tools of Markov decision process theory: a method for identifying stationary optimal policies.

Theorem 6.2.7.   Let S be discrete, and suppose that the supremum is attained in (6.2.3) for all $v \in V$. Then

a. there exists a conserving decision rule $d^* \in D^{MD}$;
b. if d^* is conserving, the deterministic stationary policy $(d^*)^\infty$ is optimal; and
c. $v_\lambda^* = \sup_{d \in D^{MD}} v_\lambda^{d^\infty}$.

Proof.   Part (a) follows from noting that $v_\lambda^* \in V$ and that the supremum in (6.2.3) is attained. By Theorem 6.2.2(c), v_λ^* is the unique solution of $Lv = v$. Therefore, from (6.2.14),

$$v_\lambda^* = Lv_\lambda^* = r_{d*} + \lambda P_{d*} v_\lambda^* = L_{d*} v_\lambda^*,$$

so, from Theorem 6.1.1,

$$v_\lambda^{(d^*)^\infty} = v_\lambda^*.$$

Result (c) is an immediate consequence of (b).   □

Note that part (c) implies that the supremum of the expected total discounted reward over the set of deterministic stationary policies equals that over the set of all policies; that is, for all $s \in S$,

$$\sup_{d \in D^{MD}} v_\lambda^{d^\infty}(s) = \sup_{\pi \in \Pi^{HR}} v_\lambda^\pi(s).$$

Note further that the *same* stationary policy is optimal for all $s \in S$.
We restate Theorem 6.2.7 in component notation.

Corollary 6.2.8.   Suppose for each $v \in V$ and $s \in S$, there exists an $a_s^v \in A_s$, such that

$$r(s, a_s^v) + \sum_{j \in S} \lambda p(j|s, a_s^v) v(j) = \sup_{a \in A_s} \left\{ r(s, a) + \sum_{j \in S} \lambda p(j|s, a) v(j) \right\}.$$

Then there exists a deterministic stationary optimal policy $(d^*)^\infty$. Further, if $d^*(s) = a_s^*$ where

$$a_s^* \in \arg\max_{a \in A_s} \left\{ r(s, a) + \sum_{j \in S} \lambda p(j|s, a) v_\lambda^*(j) \right\},$$

then $(d^*)^\infty$ is optimal.

Thus in the discounted case, with S discrete, attainment of the supremum in (6.2.3) implies the existence of conserving decision rules and consequently of optimal stationary deterministic policies. We now provide other conditions which ensure the existence of stationary optimal policies. Note that these results hold for arbitrary S.

Theorem 6.2.9. Suppose there exists

a. a conserving decision rule, or
b. an optimal policy.

Then there exists a deterministic stationary policy which is optimal.

Proof. If (a) holds, the proof of Theorem 6.2.7(b) applies directly. We now establish (b). Suppose that there exists an optimal policy $\pi^* \in \Pi^{HR}$. Represent π^* by $\pi^* = (d', \pi')$, with $d' \in D^{MR}$. Then,

$$v_\lambda^{\pi^*} = r_{d'} + \lambda P_{d'} v_\lambda^{\pi'} \leq r_{d'} + \lambda P_{d'} v_\lambda^{\pi^*} \leq \sup_{d \in D} \left\{ r_d + \lambda P_d v_\lambda^{\pi^*} \right\} = \mathscr{L} v_\lambda^{\pi^*} = v_\lambda^{\pi^*},$$

where the last equality follows from Theorem 6.2.6. Hence d' is conserving and the result follows from (a). □

We now provide sufficient conditions for attainment of the supremum in (6.2.3) and, consequently, for the existence of a conserving decision rule and a stationary optimal policy. Note that a similar result, Proposition 4.4.3, was established for finite-horizon models with bounded rewards. The proof in that case applies.

Theorem 6.2.10. Assume S is discrete, and either

a. A_s is finite for each $s \in S$, or
b. A_s is compact, $r(s, a)$ is continuous in a for each $s \in S$, and, for each $j \in S$ and $s \in S$, $p(j|s, a)$ is continuous in a, or
c. A_s is compact, $r(s, a)$ is upper semicontinuous (u.s.c.) in a for each $s \in S$, and for each $j \in S$ and $s \in S$, $p(j|s, a)$ is lower semicontinuous (l.s.c.) in (a).

Then there exists an optimal deterministic stationary policy.

We now illustrate some of the concepts in this section by expanding on the calculations in Example 6.1.1.

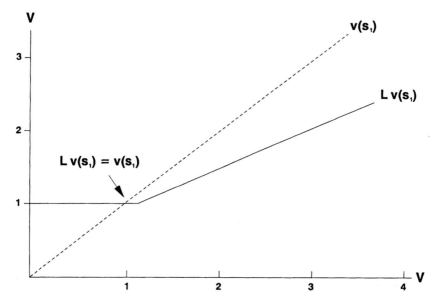

Figure 6.2.1 Graphical representation of the solution of (6.2.16) with $\lambda = 0.9$.

Example 6.2.1. We again consider the model in Sec. 3.1. Since A_s is finite for $s = s_1, s_2$, there exists an optimal stationary policy. We find it by solving the optimality equations and applying the above results.

The optimality equations are

$$v(s_1) = \max\{5 + 0.5\lambda v(s_1) + 0.5\lambda v(s_2), 10 + \lambda v(s_2)\},$$

$$v(s_2) = -1 + \lambda v(s_2).$$

Solving the second equation directly yields $v(s_2) = -1/(1 - \lambda)$, so upon substitution, the first equation becomes

$$v(s_1) = \max\left\{5 - 0.5\frac{\lambda}{1 - \lambda} + 0.5\lambda v(s_1), 10 - \frac{\lambda}{1 - \lambda}\right\} \equiv Lv(s_1). \quad (6.2.16)$$

Solving (6.2.16) yields $v_\lambda^*(s_1) = 10$, $v_\lambda^*(s_2) = -1$ for $\lambda = 0$, $v_\lambda^*(s_1) = 9$, $v_\lambda^*(s_2) = -2$ for $\lambda = 0.5$, and $v_\lambda^*(s_1) = 1$, $v_\lambda^*(s_2) = -10$ for $\lambda = 0.9$. (Figure 6.2.1)

Setting $\lambda = 0.9$, and substituting v_λ^* into the right-hand side of (6.2.16), shows that

$$\arg\max_{a \in A_s}\left\{r(s, a) + \sum_{j \in S} p(j|s, a)v_\lambda^*(j)\right\} = \arg\max\{0.95, 1\} = a_{1,2}.$$

Thus the decision rule γ defined by $\gamma(s_1) = a_{1,2}$ and $\gamma(s_2) = a_{2,1}$ is conserving (v_λ^* improving), and it follows from Theorem 6.2.7(b) that the stationary policy γ^∞ is optimal.

We now characterize the optimal policy as a function of λ. To do this, we use results in Example 6.1.1. Recall that δ^∞ denotes the stationary policy which uses action $a_{1,1}$ in state s_1. Combining (6.1.11) and (6.1.12) we find that $v_\lambda^{\gamma^\infty}(s_1) = v_\lambda^{\delta^\infty}(s_1)$ when

$$\frac{10 - 11\lambda}{1 - \lambda} = \frac{5 - 5.5\lambda}{(1 - 0.5\lambda)(1 - \lambda)}$$

or

$$\lambda^2 - \tfrac{21}{11}\lambda + \tfrac{10}{11} = 0.$$

This equation has roots $\lambda = 1$ and $\lambda = \tfrac{10}{11}$, so it follows that we are indifferent between the two policies when $\lambda = \tfrac{10}{11}$, and prefer δ^∞ to γ^∞ when λ exceeds $\tfrac{10}{11}$. This makes sense, as the discount rate increases to 1, the consequences of future actions become more important, and action $a_{1,2}$ which moves the system to the undesireable state s_2, becomes less attractive.

The following simple example shows that, when the supremum in (6.2.3) is not attained, there need not exist optimal polices.

Example 6.2.2. Let S consist of a single state s with action set $A_s = \{1, 2, \ldots\}$ and $r(s, a) = 1 - a^{-1}$. Then clearly $v_\lambda^*(s) = (1 - \lambda)^{-1}$, but there exists no policy π with this value. Note, however, that for any $\varepsilon > 0$, we can find a policy with value that is within ε of $(1 - \lambda)^{-1}$.

When optimal policies do not exist, we seek ε-optimal policies. Call a policy π_ε^* ε-optimal for $\varepsilon > 0$ if, for all $s \in S$,

$$v_\lambda^{\pi_\varepsilon^*}(s) \geq v_\lambda^*(s) - \varepsilon, \tag{6.2.17}$$

or, in vector notation,

$$v_\lambda^{\pi_\varepsilon^*} \geq v_\lambda^* - \varepsilon e$$

Theorem 6.2.11. Suppose S is finite or countable, then for all $\varepsilon > 0$ there exists an ε-optimal deterministic stationary policy.

Proof. From Theorem 6.2.5, $\mathscr{L}v_\lambda^* = v_\lambda^*$. Choose $\varepsilon > 0$ and select $d_\varepsilon \in D^{MD}$ to satisfy

$$r_{d_\varepsilon} + \lambda P_{d_\varepsilon} v_\lambda^* \geq \sup_{d \in D^{MD}} \{r_d + \lambda P_d v_\lambda^*\} - (1 - \lambda)\varepsilon e = v_\lambda^* - (1 - \lambda)\varepsilon e.$$

Since $v_\lambda^{(d_\varepsilon)^\infty} = (I - \lambda P_{d_\varepsilon})^{-1} r_{d_\varepsilon}$, rearranging terms and multiplying through by $(I - \lambda P_{d_\varepsilon})^{-1}$ shows that

$$v_\lambda^{(d_\varepsilon)^\infty} \geq v_\lambda^* - \varepsilon e,$$

so that $(d_\varepsilon)^\infty$ is ε-optimal. $\square$

*6.2.5 General State and Action Spaces

We now discuss the applicability of the above theory to a general model. We consider a stationary version of the model formulated in Sec. 2.3. The following example from Blackwell (1965) shows that, even when the supremum is attained in (6.2.3),

   a. there need not exist measurable ε-optimal policies;
   b. for v in $V_M, \mathscr{L}v$ need not be in V_M; and
   c. the value of the MDP, v_λ^*, need not be measurable.

Example 6.2.3. Let $S = [0, 1]$ and $A_s = [0, 1]$ for all $s \in S$. Let B be a Borel subset of $[0, 1] \times [0, 1]$ with projection W on S which is non-Borel. Dynkin and Yushkevich (1979; pp. 251–253) show how to construct such a function. Suppose $r(s, a) = I_{\{B\}}(s, a)$, where $I_{\{B\}}(s, a)$ denotes the indicator function of B; that is, it equals 1 if $(s, a) \in B$, and 0 otherwise, and $p(s|s, a) = 1$ for all $a \in A_s$, so that the system remains at s under any action choice.

Let $\pi = (d_1, d_2, \dots) \in \Pi^{HR}$. Note that $d_1(\cdot)$ is a function only of s. As discussed in Sec. 2.3.3, we restrict attention to measurable decision rules. Because the history at decision epoch 1 equals s_1, d_1 is measurable and $r_{d_1}(s) = r(s, d_1(s))$ is measurable. Therefore $Q = \{s: r_{d_1}(s) = 1\}$ is a Borel subset of S and by definition it is contained in W. Hence there exists an $s_0 \in W - Q$ for which

$$v_\lambda^\pi(s_0) < \lambda + \lambda^2 + \lambda^3 + \cdots = \lambda(1 - \lambda)^{-1}$$

and a $\pi' \in \Pi^{HR}$ for which $v_\lambda^{\pi'}(s_0) = (1 - \lambda)^{-1}$. Therefore, for all $\varepsilon < 1$,

$$v_\lambda^\pi(s_0) + \varepsilon < v_\lambda^*(s_0).$$

Since this argument may be repeated for any $\pi \in \Pi^{HR}$ (s_0 will vary), there exists no ε-optimal policy for $0 < \varepsilon < 1$.

Since $r(s, a)$ assumes at most two values and $p(s|s, a) = 1$ for each $s \in S$, $\mathscr{L}v = Lv$ for all $v \in V_M$, so that the optimality equation for this model is

$$v(s) = \max_{a \in [0, 1]} \{I_{\{B\}}(s, a) + \lambda v(s)\} = Lv(s).$$

Note that $L0(s) = I_{\{W\}}(s)$, so that L does not map V_M into V_M, and Theorem 6.2.3 cannot be used to establish existence of a solution to the optimality equation. Nevertheless, $Lv = v$ has the unique solution $v_\lambda^*(s) = (1 - \lambda)^{-1}$ for $s \in W$, and $v_\lambda^*(s) = 0$ for $s \in S - W$ which is not Borel measurable.

Note, however, that v_λ^* is *universally measurable*; that is it is measurable with respect to the smallest σ-algebra which contains completions of Borel sets with respect to all probability measures on S.

Several approaches may be used to remedy the difficulties encountered in this example.

   a. Impose regularity conditions on $r(s, a)$, $p(j|s, a)$, and A_s, so that selection theorems may be used to ensure that L or $\mathscr{L}$ maps V_M into V_M.

b. Use outer integrals to avoid measurability issues.

c. Extend notions of measurability.

We provide references to these approaches in the Bibliographic Remarks section of this chapter.

We conclude this section by stating a theorem (without proof) which summarizes some important existence results that hold for the discounted model under the assumptions in Sec. 2.3.

Let P be a probability measure on S and $\varepsilon > 0$. We say that a policy $\pi \in \Pi^{HR}$ is (P, ε)-optimal if

$$P\{s: v_\lambda^\pi(s) > v_\lambda^*(s) - \varepsilon\} = 1. \tag{6.2.18}$$

By choosing P to be degenerate at s, we see that this is more general then ε-optimality.

Theorem 6.2.12. Let S be a Polish space, P a probability measure on the Borel subsets of S, and $\varepsilon > 0$. Then

a. there exists a (P, ε)-optimal (Borel measurable) stationary policy;

b. if each A_s is countable, there exists an ε-optimal stationary policy;

c. if each A_s is finite, there exists an optimal stationary policy; and

d. if each A_s is a compact metric space, $r(s, a)$ is a bounded u.s.c. function on A_s for each $s \in S$, and $p(B|s, a)$ is continuous in a for each Borel subset B of S and $s \in S$, there exists an optimal stationary policy.

In Sec. 6.11.3 we establish part (d) of the above theorem. Note that the conditions in (d) exclude the above example.

6.3   VALUE ITERATION AND ITS VARIANTS

Value iteration is the most widely used and best understood algorithm for solving discounted Markov decision problems. You might already be familiar with it under other names including successive approximations, over-relaxation, backward induction, pre-Jacobi iteration, or even dynamic programming. The appeal of this algorithm may perhaps be attributed to its conceptual simplicity, its ease in coding and implementation, and its similarity to approaches used in other areas of applied mathematics. In addition to providing a simple numerical tool for solving these models, it can be used to obtain results regarding the structure of optimal policies.

This section begins by presenting the value iteration algorithm in its most basic form. We then introduce several variants which enhance convergence.

We advise you to delay implementation of this algorithm until delving more deeply into this chapter. Subsequent sections provide several enhancements to this algorithm and more efficient algorithms for solving discounted Markov decision problems.

In this section, we assume that the maximum in (6.2.2) or equivalently in (6.2.3) is attained for all $v \in V$, which occurs, for example, when A_s is finite for each $s \in S$. We do not require this assumption to establish convergence of the algorithm; however, we adopt it to simplify exposition. It includes almost any model we would solve numerically. Consequently, we will be concerned with solving the system of equations

$$v(s) = \max_{a \in A_s} \left\{ r(s,a) + \lambda \sum_{j \in S} p(j|s,a)v(j) \right\}.$$

We begin this section with a discussion of some general concepts regarding convergence rates of algorithms.

6.3.1 Rates of Convergence

Let $\{y_n\} \subset V$ be a sequence which converges to y^*, that is, $\lim_{n \to \infty} \|y_n - y^*\| = 0$. We say that $\{y_n\}$ converges *at order* (at least) α, $\alpha > 0$ if there exists a constant $K > 0$ for which

$$\|y_{n+1} - y^*\| \le K \|y_n - y^*\|^{\alpha} \qquad (6.3.1)$$

for $n = 1, 2, \ldots$. *Linear convergence* corresponds to α at least 1; *quadratic convergence* to α at least 2. We say that $\{y^n\}$ converges *superlinearly* if

$$\limsup_{n \to \infty} \frac{\|y_{n+1} - y^*\|}{\|y_n - y^*\|} = 0.$$

Note that a superlinearly convergent sequence is known to converge faster than a linearly convergent sequence.

If $\{y_n\}$ converges at order α, we define the *rate of convergence* $\hat{K}$ to be the smallest K such that (6.3.1) holds for all n. Of course, the smaller the rate of convergence the faster y_n converges to y^*. Most often, we use the rate of convergence to compare linearly convergent sequences. The requirement that (6.3.1) holds for *all* n can cause this measure to be insensitive to the true convergence properties of a sequence. To acquire a better overall view of the speed of convergence, define the *asymptotic average rate of convergence* (AARC) as

$$\limsup_{n \to \infty} \left[\frac{\|y_n - y^*\|}{\|y_0 - y^*\|} \right]^{1/n}.$$

Of course, we assume $y_0 \neq y^*$ in the above definition. Note that this definition may be extended to series which converge at order greater than 1 by raising the quantity in the denominator to the power α^n.

Given a non-negative real-valued function $f(n)$, defined on the integers, we say that the sequence $\{y_n\}$ is $O(f(n))$ whenever

$$\limsup_{n \to \infty} \frac{\| y_n - y^* \|}{f(n)}$$

is finite. In this case we write $y_n = y^* + O(f(n))$. When $f(n) = \beta^n$, with $0 < \beta < 1$ we say that convergence is *geometric at rate β*.

Suppose a sequence converges linearly with AARC equal to Φ. This means that for any $\varepsilon > 0$, there exists an N such that, for $n \geq N$,

$$\| y_n - y^* \| \leq (\Phi + \varepsilon)^n \| y_0 - y^* \|,$$

so that convergence is $O((\Phi + \varepsilon)^n)$.

We now extend these concepts to algorithms. In this book, we represent iterative algorithms by mappings $T: V \to V$. Given an initial value y_0, the algorithm generates iterates $Ty_0, T^2 y_0, \dots$. In general, $y_n = T^n y_0$. We distinguish *local* and *global* convergence rates and orders. We say that an algorithm converges *locally* with a specified order, rate of convergence, or asymptotic average rate of convergence if the sequence $\{T^n y_0\}$ for fixed y_0 converges with that order, rate, or AARC. We say that the algorithm converges *globally* at a specified order, rate, or AARC if it converges locally for *all* y_0. Thus, global algorithmic convergence rates and orders measure the worst-case performance of the algorithm values, while local convergence rates and orders measure performance for a particular starting value.

We interpret the asymptotic average rate of convergence of an algorithm as follows. Suppose we wish to know the number of iterations n, required to reduce the error $\| y_n - y^* \|$ by a fraction ϕ of the initial error $\| y_0 - y^* \|$. If the asymptotic average rate of convergence equals ρ, we find n by solving $\phi^{1/n} = \rho$ or $n \approx \log(\phi)/\log(\rho)$ iterations. Since ϕ and ρ are less than 1, the closer ρ is to 1, the more iterations required to obtain a given reduction in the error. For example, if $\rho = 0.9$, it requires 22 iterations to reduce error by a factor of 10 ($\phi = 0.1$).

When comparing two iterative methods, with asymptotic average rates of convergence ρ_1 and ρ_2, the number of iterations required to reduce the initial error by a fixed amount using method 1, n_1, is related to that using the second method, n_2, by

$$\frac{n_1}{n_2} = \frac{\log(\rho_2)}{\log(\rho_1)}.$$

6.3.2 Value Iteration

The following value iteration algorithm finds a stationary ε-optimal policy, $(d_\varepsilon)^\infty$, and an approximation to its value.

Value Iteration Algorithm

1. Select $v^0 \in V$, specify $\varepsilon > 0$, and set $n = 0$.

2. For each $s \in S$, compute $v^{n+1}(s)$ by

$$v^{n+1}(s) = \max_{a \in A_s} \left\{ r(s, a) + \sum_{j \in S} \lambda p(j|s, a) v^n(j) \right\}. \tag{6.3.2}$$

3. If

$$\|v^{n+1} - v^n\| < \varepsilon(1 - \lambda)/2\lambda, \tag{6.3.3}$$

go to step 4. Otherwise increment n by 1 and return to step 2.

4. For each $s \in S$, choose

$$d_\varepsilon(s) \in \arg\max_{a \in A_s} \left\{ r(s, a) + \sum_{j \in S} \lambda p(j|s, a) v^{n+1}(j) \right\} \tag{6.3.4}$$

and stop.

The key step in the algorithm, (6.3.2), may be expressed in vector notation as

$$v^{n+1} = Lv^n, \tag{6.3.5}$$

and (6.3.4) may be expressed as

$$d_\varepsilon \in \arg\max_{d \in D} \left\{ r_d + \lambda P_d v^{n+1} \right\}$$

The following theorem provides the main results regarding convergence of the above value iteration algorithm.

Theorem 6.3.1. Let $v^0 \in V$, $\varepsilon > 0$, and let $\{v^n\}$ satisfy (6.3.5) for $n \geq 1$. Then

a. v^n converges in norm to v_λ^*,
b. finite N for which (6.3.3) holds for all $n \geq N$,
c. the stationary policy $(d_\varepsilon)^\infty$ defined in (6.3.4) is ε-optimal, and
d. $\|v^{n+1} - v_\lambda^*\| < \varepsilon/2$ whenever (6.3.3) holds.

Proof. Parts (a) and (b) follow immediately from Theorem 6.2.3. Suppose now that (6.3.3) holds for some n and d^ε satisfies (6.3.4). Then

$$\left\| v_\lambda^{(d_\varepsilon)^\infty} - v_\lambda^* \right\| \leq \left\| v_\lambda^{(d_\varepsilon)^\infty} - v^{n+1} \right\| + \left\| v^{n+1} - v_\lambda^* \right\|. \tag{6.3.6}$$

Since $v_\lambda^{(d_\varepsilon)^\infty}$ is a fixed point of L_{d_ε}, and, as a consequence of (6.3.4), $L_{d_\varepsilon} v^{n+1} = Lv^{n+1}$, the first expression on the right-hand side of (6.3.6) satisfies

$$
\begin{aligned}
\left\| v_\lambda^{(d_\varepsilon)^*} - v^{n+1} \right\| &= \left\| L_{d_\varepsilon} v_\lambda^{(d_\varepsilon)^*} - v^{n+1} \right\| \\
&\leq \left\| L_{d_\varepsilon} v_\lambda^{(d_\varepsilon)^*} - L v^{n+1} \right\| + \left\| L v^{n+1} - v^{n+1} \right\| \\
&= \left\| L_{d_\varepsilon} v_\lambda^{(d_\varepsilon)^*} - L_{d_\varepsilon} v^{n+1} \right\| + \left\| L v^{n+1} - L v^n \right\| \\
&\leq \lambda \left\| v_\lambda^{(d_\varepsilon)^*} - v^{n+1} \right\| + \lambda \left\| v^{n+1} - v^n \right\|,
\end{aligned}
$$

where the last inequality follows because L and L_{d_ε} are contraction mappings on V. Rearranging terms yields

$$
\left\| v_\lambda^{(d_\varepsilon)^*} - v^{n+1} \right\| \leq \frac{\lambda}{1 - \lambda} \left\| v^{n+1} - v^n \right\|.
$$

Applying a similar argument to that used to derive (6.2.11) to the second expression on the right-hand side of (6.3.6) results in the inequality

$$
\left\| v^{n+1} - v_\lambda^* \right\| \leq \frac{\lambda}{1 - \lambda} \left\| v^{n+1} - v^n \right\|.
$$

Thus when (6.3.3) holds, each of the expressions on the right-hand side of (6.3.6) are bounded by $\varepsilon/2$, so that

$$
\left\| v_\lambda^{(d_\varepsilon)^*} - v_\lambda^* \right\| \leq \varepsilon.
$$

This establishes parts (c) and (d). $\square$

Thus the value iteration algorithm finds a stationary policy that is ε-optimal within a finite number of iterations. Of course, $(d_\varepsilon)^\infty$ might be optimal, but the algorithm as stated above provides no means of determining this. By combining this algorithm with methods for identifying suboptimal actions (Sec. 6.7), we can often ensure that the algorithm terminates with an optimal policy. In practice, choosing ε small enough ensures that the algorithm stops with a policy that is very close to optimal.

Note that ε-optimality might also be due to nonattainment of the supremum in (6.2.3). In that case, stopping criterion (6.3.3) must be modified to ensure ε-optimality (Problem 6.8).

The conclusions of Theorem 6.3.3 are not restricted to models with discrete state spaces and finite action sets. Convergence of value iteration is ensured whenever L is a contraction mapping on V_M. This means that value iteration will converge in norm even when S is compact or Borel. Unfortunately, numerical evaluation of the maximization in (6.3.2) is only practical when S is finite. For more general state spaces, the maximization can only be carried out by using special structure of the rewards, transition probabilities, and value functions to determine the structure of maximizing decision rules, such as was done in Sec. 4.7 for finite-horizon MDP's. General alternatives include discretization and/or truncation.

We now provide a condition under which the iterates of value iteration are monotone. This simple result will be useful for comparing convergence rates of algorithms.

Proposition 6.3.2.

a. Let $u \in V$ and $v \in V$ with $v \geq u$. Then $Lv \geq Lu$.
b. Suppose for some N that $Lv^N \leq (\geq)v^N$, then $v^{N+m+1} \leq (\geq)v^{N+m}$ for all $m \geq 0$.

Proof. First we show that L is monotone. Let $\delta \in \arg\max_{d \in D}\{r_d + \lambda P_d u\}$. Then, since $P_\delta u \leq P_\delta v$,

$$Lu = r_\delta + \lambda P_\delta u \leq r_\delta + \lambda P_\delta v \leq \max_{d \in D}\{r_d + \lambda P_d v\} = Lv.$$

Hence (a) follows. As a result of part (a), $L^m v \geq L^m u$ for all $m \geq 1$. Thus

$$v^{N+m+1} = L^m Lv^N \geq L^m v^N = v^{N+m}. \qquad \square$$

Consequently, value iteration converges monotonically to v_λ^* whenever $Lv^0 \leq (\geq)v^0$. This occurs when $r(s, a) \geq 0$ or $r(s, a) \leq 0$ and $v^0 = 0$. In the latter case, Theorem 6.2.2 implies that v^0 is a lower (upper) bound on v_λ^*.

The following theorem summarizes convergence rate properties of value iteration. Parts (c) and (d) provide error bounds; (a)–(d) follow because L is a contraction mapping.

Theorem 6.3.3. Let $v^0 \in V$ and let $\{v^n\}$ denote the iterates of value iteration. Then the following global convergence rate properties hold for the value iteration algorithm:

a. convergence is linear at rate λ,
b. its asymptotic average rate of convergence equals λ,
c. it converges $O(\lambda^n)$,
d. for all n,

$$\|v^n - v_\lambda^*\| \leq \frac{\lambda^n}{1 - \lambda}\|v^1 - v^0\|, \tag{6.3.7}$$

e. for any $d_n \in \arg\max_{d \in D}\{r_d + \lambda P_d v^n\}$,

$$\|v_\lambda^{(d_n)^\infty} - v_\lambda^*\| \leq \frac{2\lambda^n}{1 - \lambda}\|v^1 - v^0\|. \tag{6.3.8}$$

Proof. For any v^0 in V, the iterates of value iteration satisfy

$$\|v^{n+1} - v_\lambda^*\| = \|Lv^n - Lv_\lambda^*\| \leq \lambda\|v^n - v_\lambda^*\|. \tag{6.3.9}$$

Choosing $v^0 = v_\lambda^* + ke$, where k is a nonzero scalar, gives

$$v^1 - v_\lambda^* = \lambda(v^0 - v_\lambda^*).$$

Thus for this sequence, (6.3.9) holds with equality so the rate of convergence of value iteration equals λ.

Iterating (6.3.9), dividing both sides by $\|v^0 - v_\lambda^*\|$, and taking the nth root shows that

$$\limsup_{n \to \infty} \left[\frac{\|v^n - v_\lambda^*\|}{\|v^0 - v_\lambda^*\|} \right]^{1/n} \leq \lambda.$$

That equality holds follows by again choosing $v^0 = v_\lambda^* + ke$.

To obtain (c), iterate (6.3.9) and divide both sides by λ^n to show that

$$\limsup_{n \to \infty} \frac{\|v^n - v_\lambda^*\|}{\lambda^n} \leq \|v^0 - v_\lambda^*\|.$$

Choosing v^0 as above shows that equality holds.

Parts (d) and (e) follow from a similar argument used to prove Theorem 6.3.1(c).   □

Error bounds in (d) and (e) provide estimates of the number of additional iterations required to obtain an ε-optimal policy or a good approximation to v_λ^*. For example, if $\|v^1 - v^0\| = 1$, and $\lambda = 0.95$, (6.3.8) implies that it would require 162 iterations to obtain a 0.01-optimal policy.

Example 6.3.1. We use value iteration to solve the model in Fig. 3.1.1. We set $\varepsilon = 0.01$, $\lambda = 0.95$, and choose $v^0(s_1) = v^0(s_2) = 0$. The recursions (6.3.3) become

$$v^{n+1}(s_1) = \max\{5 + 0.5\lambda v^n(s_1) + 0.5\lambda v^n(s_2), 10 + \lambda v^n(s_2)\},$$
$$v^{n+1}(s_2) = -1 + \lambda v^n(s_2).$$

Table 6.3.1 shows that $Lv' \leq v'$ and convergence is monotone henceforth

Observe that value iteration converges very slowly. It requires 162 iterations to satisfy stopping criterion (6.3.3)

$$\|v^{n+1} - v^n\| < (0.01)(0.05)/1.90 = 0.00026.$$

Noting that $v^{162}(s_1) = -8.566$ and $v^{162}(s_2) = -19.955$, it follows that

$$\max\{5 + 0.5\lambda v^{162}(s_1) + 0.5\lambda v^{162}(s_2), 10 + \lambda v^{162}(s_2)\} = \max\{-8.566, -8.995\},$$

so that by (6.3.4), $d_\varepsilon(s_1) = a_{1,1}$. Since $d_\varepsilon(s_2) = a_{2,1}$, the stationary policy $(d_\varepsilon)^\infty$ is ε-optimal for $\varepsilon = 0.01$. Calculations in Sec. 6.2, show that this policy is optimal.

*6.3.3 Increasing the Efficiency of Value Iteration with Splitting Methods

Theorem 6.3.3 and the calculations in Example 6.3.1 show that for λ close to 1, convergence of value iteration may be quite slow. In this section, we provide several variants of value iteration which converge linearly but at a faster rate. These

Table 6.3.1 Sequence of Iterates for Value Iteration in Example 6.3.1

n	$v^n(s_1)$	$v^n(s_2)$	$\|v^n - v^{n-1}\|$
0	0	0	
1	10.00000	-1	10.0
2	9.27500	-1.95	0.95
3	8.47937	-2.8525	0.9025
4	7.67276	-3.70988	0.857375
5	6.88237	-4.52438	0.814506
6	6.12004	-5.29816	0.773781
7	5.39039	-6.03325	0.735092
8	4.69464	-6.73159	0.698337
9	4.03244	-7.39501	0.66342
10	3.40278	-8.02526	0.630249
20	-1.40171	-12.8303	0.377354
30	-4.27865	-15.7072	0.225936
40	-6.00119	-17.4298	0.135276
50	-7.03253	-18.4611	0.080995
60	-7.65003	-19.0786	0.048495
70	-8.01975	-19.4483	0.029035
80	-8.24112	-19.6697	0.017385
90	-8.37366	-19.8022	0.010409
100	-8.45302	-19.8816	0.006232
120	-8.52898	-19.9576	0.002234
130	-8.54601	-19.9746	0.001338
140	-8.55621	-19.9848	0.000801
150	-8.56232	-19.9909	0.000480
160	-8.56597	-19.9945	0.000287
161	-8.56625	-19.9948	0.000273
162	-8.56651	-19.9951	0.000259
163	-8.56675	-19.9953	0.000246

algorithms generalize algorithms for solving linear systems based on splitting $I - \lambda P_d$ (for fixed $d \in D$) as

$$I - \lambda P_d = Q_d - R_d. \tag{6.3.10}$$

For a matrix A we write $A \geq 0$ if all components of A are non-negative; for matrices A and B of the same size we write $A \geq B$ if $A - B \geq 0$. When $Q_d^{-1} \geq 0$ and $R_d \geq 0$, we call the decomposition (Q_d, R_d) a *regular splitting*. The simplest regular splitting of $I - \lambda P_d$ is $Q_d = I$ and $R_d = \lambda P_d$, which corresponds to value iteration.

We focus on the Gauss-Seidel value iteration algorithm to illustrate this approach, but results will be more general and include other variants of value iteration. Gauss-Seidel value iteration substitutes updated values of components of v^{n+1} into recursion (6.3.2) as soon as they become available. In the following statement of this algorithm we label the states by $s_1, s_2, \ldots, s_N$ and evaluate them in order of their subscripts. We regard a summation over an empty set of indices as 0.

Gauss-Seidel Value Iteration Algorithm.

1. Specify $v^0(s)$ for all $s \in S, \varepsilon > 0$, and set $n = 0$.
2. Set $j = 1$ and go to 2(a).
  2a. Compute $v^{n+1}(s_j)$ by

$$v^{n+1}(s_j) = \max_{a \in A_{s_j}} \left\{ r(s_j, a) + \lambda \left[\sum_{i<j} p(s_i|s_j, a)v^{n+1}(s_i) + \sum_{i \geq j} p(s_i|s_j, a)v^n(s_i) \right] \right\},$$

$$(6.3.11)$$

  2b. If $j = N$, go to 3. Else increment j by 1 and go to 2(a).
3. If

$$\|v^{n+1} - v^n\| < \varepsilon(1 - \lambda)/2\lambda, \qquad (6.3.12)$$

go to step 4. Otherwise increment n by 1 and return to step 2.
4. For each $s \in S$, choose

$$d^\varepsilon(s) \in \arg\max_{a \in A_s} \left\{ r(s, a) + \sum_{j \in S} \lambda p(j|s, a)v^{n+1}(j) \right\} \qquad (6.3.13)$$

and stop.

We now represent a pass through step 2 as a regular splitting. To do this, let d denote a decision rule corresponding to the maximizing actions obtained when evaluating (6.3.11) for $s_1, s_2, \ldots, s_N$, and write $P_d = P_d^L + P_d^U$, where

$$P_d^L = \begin{bmatrix} 0 & 0 & \cdot & \cdot & \cdot & 0 \\ p_{21} & 0 & & & & 0 \\ p_{31} & p_{32} & 0 & & & \\ \cdot & & & \cdot & & \\ & & & & \cdot & \\ p_{N1} & & & & p_{N,N-1} & 0 \end{bmatrix}, \quad P_d^U = \begin{bmatrix} p_{11} & p_{12} & \cdot & \cdot & \cdot & p_{1N} \\ 0 & p_{22} & \cdot & \cdot & \cdot & p_{2N} \\ 0 & 0 & p_{33} & & & p_{3N} \\ \cdot & & & \cdot & & \\ \cdot & & & & \cdot & \\ 0 & \cdot & \cdot & \cdot & 0 & p_{NN} \end{bmatrix}.$$

It follows that step 2 of the Gauss-Seidel value iteration algorithm may be written as

$$v^{n+1} = (I - \lambda P_d^L)^{-1}(\lambda P_d^U)v^n + (I - \lambda P_d^L)^{-1}r_d.$$

Letting $Q_d = (I - \lambda P_d^L)$ and $R_d = \lambda P_d^U$, it is easy to see that (Q_d, R_d) is a regular splitting of $I - \lambda P_d$. Thus Gauss-Seidel value iteration may be expressed as

$$v^{n+1} = Q_d^{-1}R_d v^n + Q_d^{-1}r_d. \qquad (6.3.14)$$

Note that although we represent Gauss-Seidel in terms of the inverse of Q_d, there is no need to explicitly evaluate this inverse, it simply represents the calculations in step 2.

We now provide a general convergence theorem for methods based on regular splittings.

Theorem 6.3.4. Suppose that (Q_d, R_d) is a regular splitting of $I - \lambda P_d$ for all $d \in D$ and that

$$\alpha \equiv \sup_{d \in D} \|Q_d^{-1} R_d\| < 1. \tag{6.3.15}$$

Then:

a. for all $v^0 \in V$, the iterative scheme

$$v^{n+1} = \max_{d \in D} \{Q_d^{-1} r_d + Q_d^{-1} R_d v^n\} \equiv Tv^n \tag{6.3.16}$$

converges to v_λ^*.
b. The quantity v_λ^* is the unique fixed point of T.
c. The sequence $\{v^n\}$ defined by (6.3.16) converges globally with order one at a rate less than or equal to α, its global asymptotic average rate of convergence is less than or equal to α, and it converges globally $O(\beta^n)$ where $\beta \le \alpha$.

Proof. We first show that (6.3.15) implies that T is a contraction mapping. Assume for some $s \in S$ and $u, v \in V$ that $Tv(s) - Tu(s) \ge 0$. Let

$$d_v \in \arg\max_{d \in D} \{Q_d^{-1} r_d + Q_d^{-1} R_d v\}.$$

Then

$$0 \le Q_{d_v}^{-1} r_{d_v}(s) + Q_{d_v}^{-1} R_{d_v} v(s) - Q_{d_v}^{-1} r_{d_v}(s) - Q_{d_v}^{-1} R_{d_v} u(s) \le \|Q_{d_v}^{-1} R_{d_v}\| \|v - u\|.$$

Applying the same argument in the case $0 \le Tu(s) - Tv(s)$, and again appealing to (6.3.15), establishes that

$$\|Tv - Tu\| \le \alpha \|v - u\|, \tag{6.3.17}$$

so T is a contraction mapping on V. Hence by Theorem 6.2.3, $\{v^n\}$ converges to the unique fixed point of T which we denote by v^*.

We now show that $v^* = v_\lambda^*$. Since v^* is a fixed point of T, for all $d \in D$

$$v^* \ge Q_d^{-1} r_d + Q_d^{-1} R_d v^*.$$

Consequently,

$$(I - Q_d^{-1} R_d) v^* \ge Q_d^{-1} r_d.$$

From (6.3.15), $\sigma(Q_d^{-1} R_d) < 1$, so from Appendix C, Corollary C.4, $(I - Q_d^{-1} R_d)^{-1}$

exists and satisfies

$$\left(I - Q_d^{-1}R_d\right)^{-1} = \sum_{n=0}^{\infty} \left(Q_d^{-1}R_d\right)^n.$$

From the definitions of Q_d and R_d, $(I - Q_d^{-1}R_d)^{-1} \geq 0$. Therefore,

$$v^* \geq \left(I - Q_d^{-1}R_d\right)^{-1}Q_d^{-1}r_d = (Q_d - R_d)^{-1}r_d = (I - \lambda P_d)^{-1}r_d = v_\lambda^{d^*}.$$

Since, from Theorem 6.2.7, there exists an optimal stationary policy, $v^* \geq v_\lambda^*$, but by assumption there exists a $d^* \in D$ such that $v^* = v_\lambda^{(d^*)^*}$, so that $v^* = v_\lambda^*$.

Part (c) follows by a similar argument to that in the proof of Theorem 6.3.3.   □

The following proposition gives conditions which allow us to conclude that Gauss-Seidel value iteration and other algorithms based on regular splittings converge. It also provides an approach to comparing rates of convergence of algorithms based on splitting methods.

Proposition 6.3.5. Let P be a transition probability matrix and (Q_1, R_1) and (Q_2, R_2) be regular splittings of $I - \lambda P$, where $0 \leq \lambda < 1$. Then, if $R_2 \leq R_1 \leq \lambda P$,

$$\|Q_2^{-1}R_2\| \leq \|Q_1^{-1}R_1\|. \tag{6.3.18}$$

Proof. For $i = 1$ or 2, $I - Q_i = \lambda P - R_i$ so that $\lambda P \geq R_i \geq 0$ implies $\lambda P \geq I - Q_i \geq 0$. Consequently, $1 > \lambda = \|\lambda P\| \geq \|I - Q_i\|$. Therefore Corollary C.3 implies that $Q_i^{-1} = I + (I - Q_i) + (I - Q_i)^2 + \cdots$. Because $R_2 \leq R_1$, $I - Q_2 \geq I - Q_1$. Combining these two observations implies that $Q_2^{-1} \geq Q_1^{-1}$. Since $(I - \lambda P)e \geq 0$,

$$Q_2^{-1}(I - \lambda P)e \geq Q_1^{-1}(I - \lambda P)e \geq 0$$

Consequently,

$$Q_2^{-1}(Q_2 - R_2)e \geq Q_1^{-1}(Q_1 - R_1)e \geq 0,$$

which implies that $0 \leq Q_2^{-1}R_2e \leq Q_1^{-1}R_1e$. Therefore $\|Q_2^{-1}R_2e\| \leq \|Q_1^{-1}R_1e\|$. Since $Q_i^{-1}R_i \geq 0$, for $i = 1, 2$,

$$\|Q_i^{-1}R_ie\| = \|Q_i^{-1}R_i\|,$$

from which the result follows. (Note that, in the above equality, the quantity on the left-hand side is a vector norm and that on the right-hand side is a matrix norm).   □

The following stronger result may be used to provide a more clear distinction between rates of convergence of algorithms based on splittings. Problem 6.14 provides an outline of its proof, which uses the Perron-Froebenius Theorem for positive matrices. Recall that $\sigma(\cdot)$ denotes the spectral radius of a matrix.

Proposition 6.3.6. Suppose the hypotheses of Proposition 6.3.5 hold, and $R_1 - R_2 \neq 0$. Then

$$\sigma(Q_2^{-1}R_2) < \sigma(Q_1^{-1}R_1). \tag{6.3.19}$$

We now apply this theory to Gauss-Seidel value iteration.

Theorem 6.3.7. For arbitrary $v^0 \in V$, the iterates of Gauss-Seidel value iteration $\{v_{GS}^n\}$ converge to v_λ^*. Further, the convergence is global with order one, at rate less than or equal λ, its global asymptotic average rate of convergence is less than or equal to λ, and it converges globally $O(\beta^n)$ with $\beta \leq \lambda$.

Proof. Applying Proposition 6.3.5, with $R_1 = \lambda P_d$ and $R_2 = \lambda P_d^U$, implies that, for all $d \in D$,

$$\left\| (I - \lambda P_d^L)^{-1} \lambda P_d^U \right\| \leq \|\lambda P_d\| = \lambda < 1.$$

Consequently, by Theorem 6.3.4, $\{v_{GS}^n\}$ converges to v_λ^* at the hypothesized rates. □

The following result shows that when evaluating a fixed stationary policy d^∞, the global asymptotic average rate of convergence of Gauss-Seidel is bounded by $\sigma(Q_d^{-1}R_d)$, which under mild conditions is strictly less than λ. In practice, we usually will observe such an enhanced convergence rate for the MDP model because the Gauss-Seidel value iteration algorithm will eventually identify an optimal policy, which will then be evaluated at all subsequent iterations.

Proposition 6.3.8. Let $d \in D$, let $v^0 \in V$, and define the sequence $\{v^n\}$ by

$$v^{n+1} = Q_d^{-1}R_d v^n + Q_d^{-1}r_d \equiv T_d^G v^n, \tag{6.3.20}$$

where $Q_d = (I - \lambda P_d^L)$ and $R_d = \lambda P_d^U$. Then v^n converges to $v_\lambda^{d^\infty}$ with global asymptotic average rate of convergence bounded by $\sigma(Q_d^{-1}R_d)$. Further, if P_d is not upper triangular, then the asymptotic average rate of convergence is strictly less than λ.

Proof. Assume $v^0 \neq v_\lambda^{d^\infty}$. Since $v_\lambda^{d^\infty}$ is a fixed point of T_d^G,

$$\left[\frac{\|v^n - v_\lambda^{d^\infty}\|}{\|v^0 - v_\lambda^{d^\infty}\|} \right]^{1/n} = \left[\frac{\left\| (T_d^G)^n v^0 - (T_d^G)^n v_\lambda^{d^\infty} \right\|}{\|v^0 - v_\lambda^{d^\infty}\|} \right]^{1/n}.$$

Because

$$\left\| (T_d^G)^n v^0 - (T_d^G)^n v_\lambda^{d^\infty} \right\| \leq \left\| (Q_d^{-1}R_d)^n \right\| \|v^0 - v_\lambda^{d^\infty}\|,$$

it follows that

$$\limsup_{n \to \infty} \left[\frac{\|v^n - v_\lambda^{d^\infty}\|}{\|v^0 - v_\lambda^{d^\infty}\|} \right]^{1/n} \le \limsup_{n \to \infty} \left[\left\| \left(Q_d^{-1} R_d \right)^n \right\|^{1/n} \right] = \sigma \left(Q_d^{-1} R_d \right).$$

When P_d is not upper triangular, $P_d - P_d^U \neq 0$, so that, as a consequence of Proposition 6.3.6, $\sigma(Q_d^{-1} R_d) < \lambda$. □

We now provide a bound on the local AARC for the Gauss-Seidel value iteration algorithm.

Theorem 6.3.9. For all v^0 for which $Lv^0 \ge v^0$, the local AARC for Gauss-Seidel is at most

$$\alpha \equiv \sup_{d \in D} \sigma \left(Q_d^{-1} R_d \right).$$

Further, suppose that, for all $d \in D$, P_d is not upper triangular; then $\alpha < \lambda$.

Proof. Let $T^G v \equiv \max_{d \in D} T_d^G v$, where T_d^G is defined in (6.3.20). Since $T^G v^0 \ge Lv^0 \ge v^0$, by Theorem 6.2.2(b) and the same proof as that used to establish Proposition 6.3.2, $v_\lambda^* \ge v^{n+1} \ge v^n$ for all n. Let

$$d^* \in \arg\max_{d \in D} \left\{ Q_d^{-1} r_d + Q_d^{-1} R_d v_\lambda^* \right\}.$$

Then

$$v^{n+1} \le Q_{d^*}^{-1} r_d + Q_{d^*}^{-1} R_{d^*} v^n,$$

so that

$$0 \le v_\lambda^* - v^n \le \left(Q_{d^*}^{-1} R_{d^*} \right) \left(v_\lambda^* - v^{n-1} \right). \tag{6.3.21}$$

Applying (6.3.21) iteratively establishes that

$$0 \le v_\lambda^* - v^n \le \left(Q_{d^*}^{-1} R_{d^*} \right)^n \left(v_\lambda^* - v^0 \right).$$

Taking norms in this expression yields

$$\| v_\lambda^* - v^n \| \le \left\| \left(Q_{d^*}^{-1} R_{d^*} \right)^n \right\| \| v_\lambda^* - v^0 \|.$$

Therefore,

$$\limsup_{n \to \infty} \left[\frac{\|v^n - v_\lambda^*\|}{\|v^0 - v_\lambda^*\|} \right]^{1/n} \le \limsup_{n \to \infty} \left[\left\| \left(Q_{d^*}^{-1} R_{d^*} \right)^n \right\|^{1/n} \right] = \sigma \left(Q_{d^*}^{-1} R_{d^*} \right).$$

Since by assumption for all $d \in D$, P_d is not upper triangular, $P_d - P_d^U \neq 0$, so that, as a consequence of Proposition 6.3.6, $\sigma(Q_{d^*}^{-1} R_{d^*}) \le \alpha < \lambda$. □

Since in practice it is easy to initiate Gauss-Seidel from a v^0 which satisfies $Lv^0 \geq v^0$, Gauss-Seidel will converge $O((\alpha + \varepsilon)^n)$ with $\alpha + \varepsilon < \lambda$. Equivalently, for each $\varepsilon > 0$, there exists an N for which

$$\|v^{n+1} - v_\lambda^*\| \leq (\alpha + \varepsilon)^n \|v^1 - v_\lambda^*\|$$

for all $n \geq N$.

We now illustrate these results with an example.

Example 6.3.2.   Suppose D consists of a single decision rule with transition matrix P and reward r given by

$$P = \begin{bmatrix} 0.2 & 0.4 & 0.4 \\ 0.3 & 0.3 & 0.4 \\ 0.5 & 0.5 & 0 \end{bmatrix}, \qquad r = \begin{bmatrix} 1 \\ 2 \\ 3 \end{bmatrix}.$$

For $\lambda = 0.9$, $v_\lambda^* = (I - \lambda P)^{-1}r = (18.82, 19.73, 20.35)^T$. Noting that

$$P^L = \begin{bmatrix} 0 & 0 & 0 \\ 0.3 & 0 & 0 \\ 0.5 & 0.5 & 0 \end{bmatrix}, \qquad P^U = \begin{bmatrix} 0.2 & 0.4 & 0.4 \\ 0 & 0.3 & 0.4 \\ 0 & 0 & 0 \end{bmatrix},$$

it follows that $Q^{-1}R = (I - \lambda P^L)^{-1}(\lambda P^U)$ is given by

$$Q^{-1}R = \begin{bmatrix} 0.1800 & 0.3600 & 0.3600 \\ 0.0486 & 0.3672 & 0.4572 \\ 0.1028 & 0.3272 & 0.3677 \end{bmatrix}.$$

Observe that the first row of $Q^{-1}R$ equals that of λP. This is because ordinary value iteration and Gauss-Seidel value iteration are identical for the first component. Consequently, $\|Q^{-1}R\| = \|\lambda P\| = \lambda$, although other row sums of $Q^{-1}R$ are less than λ. This difference is reflected in the spectral radius, which equals 0.9 for value iteration and 0.84 for Gauss-Seidel. This means that the asymptotic average rate of convergence for Gauss-Seidel is less than that of value iteration. Based on the discussion following the definition of the asymptotic average rate of convergence, it would require 27 iterations using Gauss-Seidel and 44 iterations using value iteration to reduce the initial error by a factor of 100. Thus Gauss-Seidel value iteration would require 39% fewer iterations to achieve the same precision as value iteration. Numerical calculations bear this out. Solving for v_λ^* to an accuracy of 0.1, starting from $v^0 = (0\ 0\ 0)^T$, requires 51 iterations using ordinary value iteration and 31 iterations using Gauss-Seidel value iteration; a reduction of 39% as predicted by the asymptotic average rates of convergence. Figure 6.3.1 illustrates and compares the iterates of these two algorithms.

The performance of Gauss-Seidel value iteration may be enhanced by reordering rows prior to computing (6.3.11) in step 2(a) or reversing the direction of the evaluation of states at each pass through the algorithm.

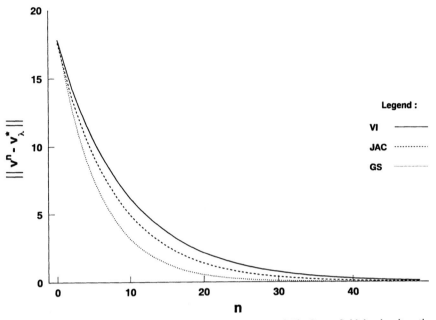

Figure 6.3.1 Comparison of convergence of value iteration (VI), Gauss-Seidel value iteration (GS), and Jacobi value iteration (JAC) in Example 6.3.2.

We conclude this section by illustrating two other iterative methods. *Jacobi value iteration* corresponds to splitting $I - \lambda P_d$ according to

$$
Q_d = \begin{bmatrix}
1 - \lambda p_{11} & 0 & \cdot & & \cdot & \cdot & 0 \\
0 & 1 - \lambda p_{22} & 0 & & & & 0 \\
0 & 0 & 1 - \lambda p_{33} & 0 & & & \cdot \\
\cdot & & & \cdot & \cdot & & \\
\cdot & & & & \cdot & \cdot & 0 \\
0 & & & & & 0 & 1 - \lambda p_{NN}
\end{bmatrix},
$$

$$
R_d = \lambda \begin{bmatrix}
0 & p_{12} & \cdot & \cdot & & \cdot & p_{1N} \\
p_{21} & 0 & p_{23} & & & & p_{2N} \\
p_{31} & p_{32} & 0 & p_{34} & & & p_{3N} \\
\cdot & & & \cdot & & & \cdot \\
\cdot & & & & \cdot & & \cdot \\
p_{N1} & \cdot & \cdot & \cdot & p_{N,N-1} & & 0
\end{bmatrix}.
$$

Since Q_d is diagonal, its inverse is particularly easy to compute. When $p_{ii} > 0$ for all i, $\|Q_d^{-1} R_d\| < \lambda$, so that the rate of convergence of Jacobi value iteration is less than that of value iteration.

Jacobi value iteration is based on writing $(I - \lambda P_d)v$ in component notation as

$$(1 - \lambda p_{ii})v(s_i)$$
$$= \lambda p_{i1}v(s_1) + \cdots + \lambda p_{i,i-1}v(s_{i-1}) + \lambda p_{i,i+1}v(s_{i+1}) + \cdots + \lambda p_{iN}v(s_N) \tag{6.3.22}$$

and then dividing both sides of (6.3.22) by $(1 - \lambda p_{ii})$ to obtain an expression for $v(s_i)$ in terms of the other components of v.

Example 6.3.2 (ctd.). Applying Jacobi value iteration, we find that $\|Q^{-1}R\| = 0.9$ (since $p_{33} = 0$), $\sigma(Q^{-1}R) = 0.88$, and that we require 42 iterations to attain an accuracy of 0.1. This represents a smaller improvement over value iteration than Gauss-Seidel, in agreement with the greater average asymptotic rate of convergence. Refer to Figure 6.3.1 for a comparison of this algorithm with value iteration and Gauss-Seidel value iteration.

Other variants based on splitting include

1. combined Jacobi and Gauss-Seidel value iteration;
2. block Jacobi value iteration, which is based on using a representation similar to (6.3.22) involving blocks of k components of v, and solving for these components in terms of the remaining components by inverting $k \times k$ matrix;
3. block Gauss-Seidel value iteration which substitutes updated information after several component values become available; and
4. block Gauss-Seidel and block Jacobi combined.

The last three methods appear well suited for parallel computation.

A further class of algorithms may be based on *over-relaxation* or *under-relaxation*. Let (Q_d, R_d) be regular splittings of $I - \lambda P_d$ and for $0 < \omega < 2$ define

$$v^{n+1} = v^n + \omega \left[\max_{d \in D} \{Q_d^{-1} r_d + Q_d^{-1} R_d v^n\} - v^n \right]. \tag{6.3.23}$$

Choosing $\omega = 1$ reduces this scheme to a splitting method; over-relaxation ($\omega > 1$) corresponds to incrementing v^n by a larger quantity than in the related splitting method, while under-relaxation ($\omega < 1$) corresponds to a smaller increment. Over-relaxation appears to be better suited then under-relaxation when iterates are monotone.

Analysis of this algorithm uses similar methods to those above and is based on rewriting (6.3.23) as

$$v^{n+1} = \max_{d \in D} \{\omega Q_d^{-1} r_d + [\omega Q_d^{-1} R_d + (1 - \omega) I] v^n\}.$$

An issue of practical concern is how to choose ω to assure and accelerate convergence. Kushner and Kleinman (1971) and Reetz (1973) propose and analyze a combined Gauss-Seidel successive over-relaxation algorithm. Reetz shows that choosing $\omega = \min_{a \in A_s, s \in S} [1 - \lambda p(s|s, a)]^{-1}$ assures convergence at an asymptotic average rate no more than λ. Numerical work by Kushner and Kleinman suggests choosing ω in the vicinity of 1.4, while a numerical study by Porteus and Totten (1978) suggests that values of ω in the vicinity of 1.2 result in more rapid convergence than ordinary value iteration when evaluating a fixed decision rule. Herzberg and Yechiali (1991) investigate this issue further and allow ω to vary with n.

The above discussion suggests the following interesting related optimization problem: find ω^* to minimize $\sigma(\omega Q_d^{-1} R_d + (1 - \omega)I)$.

6.4 POLICY ITERATION

Policy iteration, or approximation in policy space, is another method for solving infinite-horizon Markov decision problems. While value iteration may be regarded as an application of a general approach for finding fixed points, policy iteration appears to relate directly to the particular structure of Markov decision problems. It applies to stationary infinite-horizon problems. It is not an efficient approach for solving finite-horizon problems (Problem 6.16).

Section 6.4.2 establishes convergence of the policy iteration algorithm for models with finite-state and action sets. Results in Secs. 6.4.3 and 6.4.4 apply to any model in which, for all $v \in V$, there exists a $d_v \in D$ such that

$$d_v \in \arg\max_{d \in D} \{r_d + \lambda P_d v\}$$

or, equivalently for each $v \in V$ and $s \in S$, there exists an action $a_v^s \in A_s$ satisfying

$$a_v^s \in \arg\max_{a \in A_s} \left\{ r(s, a) + \lambda \sum_{j \in S} p(j|s, a)v(j) \right\}.$$

This generalization includes finite-state models with compact action sets and continuous reward and transition functions, countable-state models with compact action sets, bounded continuous reward and transition functions, and more general models under appropriate assumptions.

6.4.1 The Algorithm

The policy iteration algorithm for discounted Markov decision problems follows.

The Policy Iteration Algorithm

1. Set $n = 0$, and select an arbitrary decision rule $d_0 \in D$.
2. (Policy evaluation) Obtain v^n by solving

$$\left(I - \lambda P_{d_n} \right) v = r_{d_n}. \tag{6.4.1}$$

3. (Policy improvement) Choose d_{n+1} to satisfy

$$d_{n+1} \in \arg\max_{d \in D} \{r_d + \lambda P_d v^n\}, \qquad (6.4.2)$$

setting $d_{n+1} = d_n$ if possible.

4. If $d_{n+1} = d_n$, stop and set $d^* = d_n$. Otherwise increment n by 1 and return to step 2.

This algorithm yields a sequence of deterministic Markovian decision rules $\{d_n\}$ and value functions $\{v^n\}$. The sequence is finite whenever the stopping criterion $d_n = d_{n+1}$ holds, and infinite otherwise. In Sec. 6.4.2, we show that finite termination occurs with certainty in finite-state and action problems; more detailed analysis is necessary when there are an infinite number of policies.

We refer to step 2 as policy evaluation because, by solving (6.4.1), we obtain the expected discounted reward of policy $(d_n)^\infty$. This may be implemented through Gaussian elimination or any other linear equation solution method. Step 3 selects a v^n-improving decision rule. Since this decision rule is not necessarily unique, the specification that $d_{n+1} = d_n$ if possible, is included to avoid cycling.

We stress that implementation of the maximization in (6.4.2) is *componentwise*; if not, step 3 would require enumeration of all decision rules. This means that, for each $s \in S$, we choose $d_{n+1}(s)$ so that

$$d_{n+1}(s) \in \arg\max_{a \in A_s} \left\{ r(s, a) + \lambda \sum_{j \in S} p(j|s, a) v^n(j) \right\}. \qquad (6.4.3)$$

Thus, when invoking step 3, we determine the set of all v_{d_n}-improving decision rules before selecting a particular decision rule. An alternative specification of the algorithm would be to retain the entire set of v^n-improving decision rules and terminate when it repeats. This modification is unnecessary since, at termination, $v^n = v_\lambda^*$, so that all optimal decision rules are available by finding $\arg\max_{d \in D} \{r_d + \lambda P_d v_\lambda^*\}$.

In step 3, we might instead choose d_{n+1} to be any decision rule for which

$$r_{d_{n+1}} + \lambda P_{d_{n+1}} v^n \geq r_{d_n} + \lambda P_{d_n} v^n, \qquad (6.4.4)$$

with strict inequality in at least one component. In finite-state and action problems, such an algorithm will find an optimal policy, but it could terminate with a suboptimal policy in problems with compact action sets.

The following result establishes the monotonicity of the sequence $\{v^n\}$. It is a key property of this algorithm and holds in complete generality.

Proposition 6.4.1. Let v^n and v^{n+1} be successive values generated by the policy iteration algorithm. Then $v^{n+1} \geq v^n$.

Proof. Let d_{n+1} satisfy (6.4.2). Then it follows from (6.4.1) that

$$r_{d_{n+1}} + \lambda P_{d_{n+1}} v^n \geq r_{d_n} + \lambda P_{d_n} v^n = v^n.$$

Rearranging terms yields

$$r_{d_{n+1}} \geq \left(I - \lambda P_{d_{n+1}}\right) v^n.$$

Multiplying both sides by $(I - \lambda P_{d_{n+1}})^{-1}$ and applying Lemma 6.1.2(b) gives

$$v^{n+1} = \left(I - \lambda P_{d_{n+1}}\right)^{-1} r_{d_{n+1}} \geq v^n. \qquad \square$$

6.4.2 Finite State and Action Models

This section analyzes the convergence of the policy iteration algorithm for finite-state and action models. The following theorem demonstrates the finite convergence of policy iteration algorithm terminates in a finite number of iterations. At termination v_n is a solution of the optimality equation and the policy $(d*)^\infty$ is discount optimal.

Theorem 6.4.2. Suppose S is finite and, for each $s \in S$, A_s is finite. Then the policy iteration algorithm terminates in a finite number of iterations, with a solution of the optimality equation and a discount optimal policy $(d*)^\infty$.

Proof. From Proposition 6.4.1, the values v^n of successive stationary policies generated by the policy iteration algorithm are nondecreasing. Therefore, since there are only finitely many deterministic stationary policies, the algorithm must terminate under the stopping criterion in step 4, in a finite number of iterations. At termination, $d_{n+1} = d_n$, so that

$$v^n = r_{d_{n+1}} + \lambda P_{d_{n+1}} v^n = \max_{d \in D} \left\{ r_d + \lambda P_d v^n \right\}.$$

Thus v^n solves the optimality equation and, since $d_n = d*$, $v^n = v_\lambda^{(d*)^\infty}$. That $(d*)^\infty$ is discount optimal follows from Theorem 6.2.6. $\square$

The proof of Theorem 6.4.2 may be restated as follows. If an improvement on the current policy is possible, then the algorithm will find it, and a further iteration is necessary. If not, the algorithm stops, the current policy is optimal, and its value function satisfies the optimality equation. This theorem and its proof also provide an alternative derivation of Theorem 6.2.5 for finite S. That is, it demonstrates the existence of a solution to the optimality equation without using the Banach fixed-point theorem.

We illustrate the use of policy iteration with our standard example.

Example 6.4.1. Consider the model in Sec. 3.1. Choose $\lambda = 0.95$ $d_0(s_1) = a_{1,2}$ and $d_0(s_2) = a_{2,1}$. Then by solving (6.4.1), which in this model equals

$$v(s_1) - 0.95 v(s_2) = 10,$$

$$0.05 v(s_2) = -1,$$

we find that $v^0(s_1) = -9$ and $v^0(s_2) = -20$. We then apply step 3 and evaluate (6.4.3) in state s_1 as follows:

$$\max\{5 + 0.475v^0(s_1) + 0.475v^0(s_2),\, 10 + 0.95v^0(s_2)\} = \max\{-8.775, -9\},$$

so that $d_1(s_1) = a_{1,1}$ and $d_1(s_2) = a_{2,1}$.

Since $d_1 \neq d_0$, the stopping criterion does not hold. Therefore we increment n by 1 and return to step 2. Again, solving (6.4.1), which in this case equals

$$0.525v(s_1) - 0.475v(s_2) = 5,$$

$$0.05v(s_2) = -1$$

yields $v^1(s_1) = -8.571$ and $v^1(s_2) = -20$. Applying step 3 shows that $d_2 = d_1$. Thus we set $d^* = d_1$, and conclude that $(d^*)^\infty$ is the optimal policy.

Recalling Example 6.3.1, in which this model was solved using value iteration, we note the significant reduction in effort obtained through the use of policy iteration.

6.4.3 Nonfinite Models

When the set of deterministic stationary policies is not finite, for example when S is finite and A_s is compact, or S is countable, the proof of Theorem 6.4.2 is no longer valid since there is no guarantee that the stopping criteria in step 4 will ever be satisfied. In such cases, we use a different approach to demonstrate convergence. Drawing a parallel to the analysis of value iteration, other issues of concern are

a. What is the consequence of starting computation at step 3 (instead of at step 1) with an arbitrary $v^0 \in V$?
b. What are the order and the rate of convergence of the algorithm?

These questions are best answered by providing a recursive representation for the iterates of policy iteration similar to that used to analyze value iteration. We now do this. Define the operator $B: V \to V$ by

$$Bv \equiv \max_{d \in D}\{r_d + (\lambda P_d - I)v\}. \tag{6.4.5}$$

Hence $Bv = Lv - v$. The optimality equation (6.2.8) can thus be reexpressed as

$$Bv = 0. \tag{6.4.6}$$

Using this notation, solving the optimality equation can be regarded as finding a zero of B instead of a fixed point of L. We will show that policy iteration is equivalent to using a vector space version of Newton's method for finding a zero of B.

For $v \in V$, let D_v denote the set of v-improving decision rules. That is, $d_v \in D_v$ if

$$d_v \in \arg\max_{d \in D}\{r_d + (\lambda P_d - I)v\}. \tag{6.4.7}$$

Note that the identity I in (6.4.7) has no effect on the selection of a maximizing decision rule.

The following proposition provides the basis for our "geometric" analysis of this algorithm.

Proposition 6.4.3.   For $u, v \in V$ and any $d_v \in D_v$,

$$Bu \geq Bv + \left(\lambda P_{d_v} - I \right)(u - v). \tag{6.4.8}$$

Proof.   From the definition of B,

$$Bu \geq r_{d_v} + \left(\lambda P_{d_v} - I \right)u \tag{6.4.9}$$

and

$$Bv = r_{d_v} + \left(\lambda P_{d_v} - I \right)v. \tag{6.4.10}$$

Subtracting (6.4.10) from (6.4.9) gives the result.   □

We refer to (6.4.8) as the *support inequality* and regard it as a vector space generalization of the gradient inequality which defines convex functions in R^n. Thus, in a generalized sense, the operator B is "convex" and $\lambda P_{d_v} - I$ is the "support" of B at v. We find it convenient to refer to D_v as the set of *supporting decision rules* at v.

Figure 6.4.1 illustrates this result. In it, there are four decision rules denoted by 1, 2, 3, and 4. We see that, at each $v \in V$,

$$Bv = \max_{i=1,2,3,4} \left\{ r_i + (\lambda P_i - I)v \right\}$$

and, further, that the support inequality (6.4.8) holds. Observe also that $Bv_\lambda^* = 0$.

The following recursion provides the basis for the analysis of policy iteration.

Proposition 6.4.4.   Suppose the sequence $\{v^n\}$ is obtained from the policy iteration algorithm. Then, for any $d_{v^n} \in D_{v^n}$,

$$v^{n+1} = v^n - \left(\lambda P_{d_{v^n}} - I \right)^{-1} Bv^n. \tag{6.4.11}$$

Proof.   From the definition of D_{v^n} and v^{n+1},

$$v^{n+1} = \left(I - \lambda P_{d_{v^n}} \right)^{-1} r_{d_{v^n}} - v^n + v^n$$

$$= \left(I - \lambda P_{d_{v^n}} \right)^{-1} \left[r_{d_{v^n}} + (\lambda P_{d_{v^n}} - I)v^n \right] + v^n. \tag{6.4.12}$$

The result follows by noting that the expression in [ ]'s in (6.4.12) is Bv^n and that $(I - \lambda P_{d_{v^n}})^{-1} = -(\lambda P_{d_{v^n}} - I)^{-1}$.   □

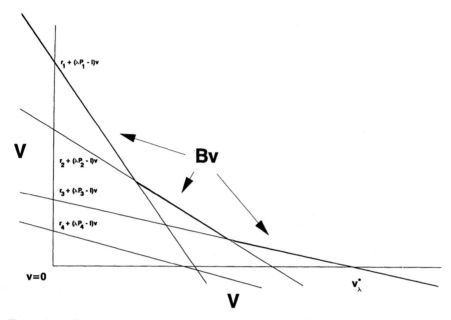

Figure 6.4.1 Geometric representation of Bv. Each linear operator $r_i + \lambda P_i v$ maps V to V. The horizontal axis represents their domain and the vertical axis their range. Bv is the maximum at each $v \in V$.

This result provides a closed form representation for the values generated by policy iteration. By the definition of D_{v^n}, (6.4.11) holds for any supporting decision rule d_{v^n} so that it is valid for any decision rule d_{n+1} which may be obtained in step 3 of the policy iteration algorithm. Whenever the algorithm repeats a decision rule, it uses the same support for two successive iterations. This means that the supporting decision rule must correspond to an optimal policy as illustrated by decision rule 3 in Fig. 6.4.1.

Noting the analogy between the support in V and the derivative in R^1, (6.4.11) can be regarded as a vector space version of Newton's method recursion

$$x_{n+1} = x_n - \left[f'(x_n) \right]^{-1} f(x_n)$$

for finding a zero of a real-valued function $f(x)$. If $f(x)$ is convex decreasing and has a zero, then starting Newton's method at a point at which the function is positive ensures that the iterates converge monotonically to the zero. This observation is the basis for Theorem 6.4.6 below. Its proof compares the iterates of policy iteration to those of value iteration, and shows that if policy iteration and value iteration begin at the same point, then the iterates of policy iteration are always bounded below by those of value iteration and above by v_λ^*. A more general approach, provided by Puterman and Brumelle (1979), requires only that there exists a y such that $By \leq 0$.

Define $V_B \equiv \{v \in V; \ Bv \geq 0\}$. Since $Bv = Lv - v$, Theorem 6.2.2(b) implies that if $v \in V_B$, then v is a lower bound for v_λ^*. The following lemma compares the iterates of policy iteration generated by the operator Z to those of value iteration which are generated by the operator L. It shows that successive iterates of policy iteration are monotone and lie in V_B.

Lemma 6.4.5. Let $v \in V_B$, $d_v \in D_v$, and suppose $v \geq u$. Then

a. $Zv \equiv v + (I - \lambda P_{d_v})^{-1} Bv \geq Lu$,   (6.4.13)

b. $Zv \in V_B$, and

c. $Zv \geq v$.

Proof. Since, for $v \in V_B$, $Bv \geq 0$, it follows from Lemma 6.1.2a. that

$$Zv = v + (I - \lambda P_{d_v})^{-1} Bv \geq v + Bv = Lv \geq Lu. \qquad (6.4.14)$$

To establish (b), apply Proposition 6.4.3 to obtain

$$B(Zv) \geq Bv + (\lambda P_{d_v} - I)(Zv - v) = Bv - Bv = 0.$$

The third result follows from the definition of Zv, the assumption that $Bv \geq 0$, and Lemma 6.1.2a. □

Theorem 6.4.6. The sequence of values $\{v^n\}$ generated by policy iteration converges monotonically and in norm to v_λ^*.

Proof. Let $u^k = L^k v^0$. We show inductively that $u^k \leq v^k \leq v_\lambda^*$ and $v^k \in V_B$. Let $k = 0$. Then

$$Bv^0 \geq r_{d_0} + (\lambda P_{d_0} - I) v^0 = 0,$$

so that $v^0 \in V_B$ and, by Theorem 6.2.2(b), $v^0 \leq v_\lambda^*$. Since by definition $u^0 = v^0$, the induction hypothesis holds for $k = 0$.

Now assume that result holds for $k \leq n$. In particular, $u^k \leq v^k \leq v_\lambda^*$ and $Bv^k \geq 0$. Since $v^{n+1} = Zv^n$, from Lemma 6.4.5 it follows that $v^{n+1} \in V_B$, $v^n \leq v_\lambda^*$, and $v^{n+1} \geq Lu^n = u^{n+1}$. Hence the induction hypothesis holds.

From Theorem 6.3.1, u^n converges to v_λ^* in norm. Thus, since $u^n \leq v^n \leq v_\lambda^*$ for all n, v^n converges to v_λ^* in norm. Monotonicity of $\{v^n\}$ was established in Proposition 6.4.1. □

The following corollary addresses one of the questions posed at the beginning of this section.

Corollary 6.4.7. Suppose the policy iteration algorithm is initiated in step 3 with an arbitrary $v^0 \in V$. Then the conclusions of Theorem 6.4.6 hold.

Proof. Apply policy iteration beginning with any $d_0 \in D_{v^0}$. The result follows from Theorem 6.4.6. □

The above results establish the convergence of policy iteration for arbitrary state and action spaces under the assumption that there is a maximizing decision rule at each $v \in V$ or V_M. This includes models with A_s compact, $p(j|s, a)$ and $r(s, a)$ continuous in a for each $s \in S$, and S either finite or compact.

Implementation of policy iteration when the set of decision rules is not finite requires a stopping rule to obtain finite convergence. That, in step 3 of the value iteration algorithm ensures finite convergence to an ε-optimal policy. When S is not finite, step 3 cannot be implemented unless a special structure can be exploited to determine the form of the maximizing decision rule. In that case, policy iteration algorithms can be devised which search only among a specific class of policies.

6.4.4 Convergence Rates

We now address the second question posed at the beginning of this section: at what rate do the values generated by policy iteration converge to the optimal value function? Our proof of Theorem 6.4.6 shows that its order is at least linear. The following theorem provides conditions under which it is quadratic.

Theorem 6.4.8. Suppose $\{v^n\}$ is generated by policy iteration, that $d_{v^n} \in D_{v^n}$ for each n, and there exists a K, $0 < K < \infty$ for which

$$\|P_{v^n} - P_{d_{v_\lambda^*}}\| \leq K\|v^n - v_\lambda^*\| \tag{6.4.15}$$

for $n = 1, 2, \ldots$. Then

$$\|v^{n+1} - v_\lambda^*\| \leq \frac{K\lambda}{1 - \lambda}\|v^n - v_\lambda^*\|^2. \tag{6.4.16}$$

Proof. Let

$$U_n \equiv \left(\lambda P_{d_{v^n}} - I\right) \quad \text{and} \quad U_* \equiv \left(\lambda P_{d_{v_\lambda^*}} - I\right).$$

From Proposition 6.4.3,

$$Bv^n \geq Bv_\lambda^* + U_*(v^n - v_\lambda^*) = U_*(v^n - v_\lambda^*),$$

so that, by the positivity of $-U_n^{-1}$ (Lemma 6.1.2),

$$U_n^{-1}Bv^n \leq U_n^{-1}U_*(v^n - v_\lambda^*). \tag{6.4.17}$$

From Theorem 6.4.6, Lemma 6.1.2 and (6.4.17),

$$0 \leq v_\lambda^* - v^{n+1} = v_\lambda^* - v^n + U_n^{-1}Bv^n \leq U_n^{-1}U_n(v_\lambda^* - v^n) - U_n^{-1}U_*(v_\lambda^* - v^n).$$

Rearranging terms and taking norms implies that

$$\|v_\lambda^* - v^{n+1}\| \leq \|U_n^{-1}\| \|U_n - U_*\| \|v_\lambda^* - v^n\|.$$

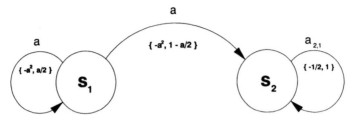

Figure 6.4.2   Graphical representation of Example 6.4.2.

Since $\|U_n^{-1}\| \le (1 - \lambda)^{-1}$ and

$$\|U_n - U_*\| = \lambda\|P_{d_r n} - P_{d_r *}\|,$$

the result follows from (6.4.15).   □

This theorem implies that when (6.4.15) holds, policy iteration converges at least at order 2 (quadratically) to the optimal value function. This accounts for the fast convergence of policy iteration in practice.

The following example illustrates the use of policy iteration in a problem with a compact set of actions.

Example 6.4.2. Let $S = \{s_1, s_2\}$, $A_{s_1} = [0, 2]$, $A_{s_2} = \{a_{2,1}\}$, $r(s_1, a) = -a^2$, $r(s_2, a_{2,1}) = -0.5$, $p(s_1|s_1, a) = 0.5a$, $p(s_2|s_1, a) = 1 - 0.5a$, and $p(s_2|s_2, a_{2,1}) = 1$. Figure 6.4.2 illustrates this model.

The optimality equation $Bv = 0$ is given by

$$0 = \max_{a \in [0, 2]} \left\{-a^2 + \lambda(0.5a)v(s_1) + \lambda(1 - 0.5a)v(s_2) - v(s_1)\right\} \quad (6.4.18)$$

and

$$0 = -0.5 + \lambda v(s_2) - v(s_2)$$

We apply policy iteration to find an optimal policy when $\lambda = 0.9$. Because there is only a single action in state s_2, we can simplify subsequent calculations by solving the second equation above and substituting its solution into (6.4.18). Taking this approach, we find that $v_\lambda^*(s_2) = -5$. Substituting into (6.4.18) reduces the equation to

$$0 = \max_{a \in [0, 2]} \left\{-a^2 + 0.45av(s_1) - 4.5 + 2.25a - v(s_1)\right\}. \quad (6.4.19)$$

For fixed a,

$$v(s_1) = (1 - 0.45a)^{-1}(-a^2 + 2.25a - 4.5), \quad (6.4.20)$$

Table 6.4.1 Computational Results for Example 6.4.2

		Policy Iteration	
n	$d_n(s_1)$	$v^n(s_1)$	$v_\lambda^*(s_1) - v^n(s_1)$
0	0.000000000000000	−4.500000000000000	1.33×10^{-02}
1	0.112500000000000	−4.486668861092825	9.49×10^{-06}
2	0.115499506254114	−4.486659370799152	4.81×10^{-12}
3	0.115501641570191	−4.486659370794342	0^a
4	0.115501641571273	−4.486659370794342	0

		Value Iteration	
n	$d_n(s_1)$	$v^n(s_1)$	$v_\lambda^*(s_1) - v^n(s_1)$
0	^{b}	−4.500000000000000	1.33×10^{-02}
1	0.112500000000000	−4.487343750000001	6.84×10^{-04}
2	0.115347656250000	−4.486694918197633	3.55×10^{-05}
3	0.115493643405533	−4.486661218332917	1.85×10^{-06}
4	0.115501225875094	−4.486659466821352	9.60×10^{-08}
5	0.115501619965196	−4.486659375785417	4.99×10^{-09}
6	0.115501640448281	−4.486659371053757	2.59×10^{-10}
7	0.115501641512905	−4.486659370807826	1.35×10^{-11}
8	0.115501641568239	−4.486659370795043	7.01×10^{-13}
9	0.115501641571116	−4.486659370794379	3.64×10^{-14}
10	0.115501641571265	−4.486659370794344	1.78×10^{-15}
11	0.115501641571273	−4.486659370794342	0
12	0.115501641571273	−4.486659370794342	0

^{a}Zero represents less than 10^{-16}.
^{b}Value iteration was initiated with $v^0(s_1) = -4.5$, the value obtained from the first iteration of policy iteration algorithm.

and for fixed $v(s_1)$ the action which attains the maximum in (6.4.19) satisfies

$$a_r = \begin{cases} 0 & v(s_1) < -5 \\ 0.225 \cdot v(s_1) + 1.125 & -5 \le v(s_1) \le 3.889 \\ 2 & v(s_1) > 3.889. \end{cases} \quad (6.4.21)$$

We implement policy iteration by choosing an action, substituting it into (6.4.20) to obtain its value, and then substituting the resulting value into (6.4.21) to obtain the maximizing action at the subsequent iteration. Initiating this procedure at $d_0(s_1) = 0$ yields the results in Table 6.4.1. Observe that policy iteration obtains the maximizing action to 16-decimal place accuracy in three iterations. The last column of this table illustrates that the rate of convergence of the values is indeed quadratic since the number of decimal places of accuracy doubles at each iteration.

We also provide computational results for value iteration beginning from $v^0 = (-4.5, -5)^T$. Observe that the convergence in that case is much slower and requires

11 iterations to obtain the maximizing decision rule to the same degree of precision as policy iteration.

This example also enables us to illustrate verification of condition (6.4.15) in Theorem 6.4.8. Restricting attention to $v \in V$ of the form $v = (v(s_1), -5)^T$, we see that $d_v = (a_v, a_{2,1})^T$, where a_v is defined by (6.4.21). For $-5 \le v(s_1) \le 3.889$,

$$P_{d_v} = \begin{bmatrix} 0.1125v(s_1) + 0.5625 & 1 - 0.1125v(s_1) - 0.5625 \\ 0 & 1 \end{bmatrix},$$

so that for u and v satisfying $-5 \le u(s_1) \le 3.889$ and $-5 \le v(s_1) \le 3.889$, and $u(s_1) = v(s_1) = -5$,

$$\| P_{d_v} - P_{d_u} \| = 0.225 \| v - u \|.$$

Since $-4.5 \le v_\lambda^*(s_1) \le 0$, $v_\lambda^*(s_2) = -5$, and the iterates of policy iteration are monotone, (6.4.21) must hold, and the convergence will be quadratic with rate 2.025.

Note that condition (6.4.15) is awkward to verify since v_λ^* is not known. One approach to applying this result is to show that if (6.4.15) is violated at v', then v' cannot be a solution of the optimality equation. Alternative weaker results appear in the following corollary.

Corollary 6.4.9. Suppose $\{v^n\}$ is generated by policy iteration, that $d_{v^n} \in D_{v^n}$ for each n and there exits a K, $0 < K < \infty$ for which either

$$\| P_{d_v} - P_{d_u} \| \le K \| v - u \| \qquad \text{for all } v, v \in V \tag{6.4.22}$$

or

$$\| P_{d_v} - P_{d_{v_\lambda^*}} \| \le K \| v - v_\lambda^* \| \qquad \text{for all } u, v \in V \tag{6.4.23}$$

Then (6.4.16) holds for $n = 0, 1, \ldots$.

In terms of the parameters of the model, sufficient conditions for (6.4.22) to hold are that, for each $s \in S$,

a. A_s is compact and convex,
b. $p(j|s, a)$ is affine in a, and
c. $r(s, a)$ is strictly concave and twice continuously differentiable in a.

Observe that these conditions are satisfied in Example 6.4.2.

When A_s is finite, (6.4.15) need not hold because P_{d_v} will not be unique at several $v \in V$. If a rule such as that in step 3 of the policy iteration algorithm is used to break ties, the algorithm provides a unique support at each v. Thus convergence will be quadratic although K might be large. Other conditions which imply (6.4.15) may be derived from selection theorems in Fleming and Rishel (1975).

The following corollary provides a convergence rate result which holds under weaker hypotheses than in the above theorem.

Corollary 6.4.10. Suppose $\{v^n\}$ is generated by policy iteration, $d_{v^n} \in D_{v^n}$ for $n \geq 0$ and

$$\lim_{n \to \infty} \left\| P_{d_{v^n}} - P_{d_{v_\lambda^*}} \right\| = 0. \tag{6.4.24}$$

Then

$$\lim_{n \to \infty} \frac{\left\| v^{n+1} - v_\lambda^* \right\|}{\left\| v^n - v_\lambda^* \right\|} = 0. \tag{6.4.25}$$

Using the nomenclature of Sec. 6.3.1, Corollary 6.4.10 implies that whenever (6.4.24) holds, the sequence generated by policy iteration converges superlinearly to v_λ^*. This means that the convergence is asymptotically faster than any first-order algorithm. In terms of model parameters, if the conditions for quadratic convergence above are relaxed to require only that $r(s, a)$ be strictly concave in a, then convergence of policy iteration will be superlinear.

Puterman and Brumelle (1979) obtained the following error bound for the iterates of policy iteration. We refer the reader to that reference for a proof.

Theorem 6.4.11. Suppose $\{v^n\}$ is generated by policy iteration and there exists a K, $0 < K < \infty$ such that (6.4.15) holds. If $h \equiv L\|v^1 - v^0\| < 1$, where $L = K\lambda/(1 - \lambda)$, then, for $n = 1, 2, \ldots$,

$$\|v^n - v_\lambda^*\| \leq [L \log(2)]^{-1} \int_{a_n}^{\infty} u^{-1} e^{-u} \, du,$$

where $a_n \equiv -2^{n-1} \log(h)$.

6.5 MODIFIED POLICY ITERATION

We usually implement the evaluation step of the policy iteration algorithm by using Gaussian elimination to solve the linear system

$$(I - \lambda P_{d_n})v = r_{d_n}. \tag{6.5.1}$$

In a model with M states, this requires on the order of M^3 multiplications and divisions. For large M, obtaining an exact solution of (6.5.1) may be computationally prohibitive. Moreover, the geometric representation in Figure 6.5.1 suggests that it is not necessary to determine v^n prcisely to identify an improved policy.

We now explain why. Suppose we begin policy iteration at step 3 with v^0 as indicated in Figure 6.5.1. Then decision rule a will be v^0-improving, so that it would be set equal to d_1 in the improvement step. At the next iteration, solving (6.5.1) would yield v^1. Note that b is the v^1-improving decision rule but also that it is v-improving for any v such that $w \leq v \leq v^1$. Therefore we do not need to explicitly determine v^1 to identify the new decision rule; we only require an element of V which exceeds w.

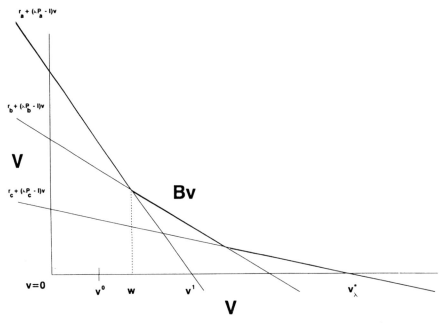

Figure 6.5.1   Geometric motivation for modified policy iteration (see text).

This can be obtained with a few iterations of successive approximations with the fixed decision rule $d_1 = a$.

6.5.1   The Modified Policy Iteration Algorithm

Let $\{m_n\}$ denote a sequence of non-negative integers.

The Modified Policy Iteration Algorithm (MPI).

1. Select $v^0 \in V_B$, specify $\varepsilon > 0$, and set $n = 0$.
2. (Policy improvement) Choose d_{n+1} to satisfy

$$d_{n+1} \in \underset{d \in D}{\arg\max}\{r_d + \lambda P_d v^n\}, \tag{6.5.2}$$

setting $d_{n+1} = d_n$ if possible (when $n > 0$).

3. (Partial policy evaluation).
   a. Set $k = 0$ and

$$u_n^0 \equiv \max_{d \in D}\{r_d + \lambda P_d v^n\}. \tag{6.5.3}$$

   b. If $\|u_n^0 - v^n\| < \varepsilon(1 - \lambda)/2\lambda$, go to step 4. Otherwise go to (c).

c. If $k = m_n$, go to (e). Otherwise, compute u_n^{k+1} by

$$u_n^{k+1} = r_{d_{n+1}} + \lambda P_{d_{n+1}} u_n^k = L_{d_{n+1}} u_n^k. \tag{6.5.4}$$

d. Increment k by 1 and return to (c).

e. Set $v^{n+1} = u_n^{m_n}$, increment n by 1, and go to step 2.

4. Set $d_\varepsilon = d_{n+1}$ and stop.

This algorithm combines features of both policy iteration and value iteration. Like value iteration, it is an iterative algorithm and begins with a value v^0 for which $Bv^0 \geq 0$. The stopping criterion used in step 3(b) is identical to that of value iteration; when it is satisfied, the resulting policy is ε-optimal. The computation of u_n^0 in step 3(a) requires no additional work because it already has been evaluated in step 2 when determining the arg max in (6.5.2).

Like policy iteration, the algorithm contains an improvement step, step 2, and an evaluation step, step 3; however, the evaluation is not done exactly. Instead it is carried out iteratively in step 3(c), which is repeated m_n times at iteration n. As a consequence of the second equality in (6.5.4), step 3 may be represented by

$$v^{n+1} = \left(L_{d_{n+1}} \right)^{m_n+1} v^n. \tag{6.5.5}$$

The *order sequence* $\{m_n\}$ may be

a. fixed for all iterations ($m_n = m$);

b. chosen according to some prespecified pattern (the geometric representation in Fig. 6.5.1 suggests m_n should be increasing in n); or

c. selected adaptively, for example, by requiring $\|u^{m_n+1} - u^{m_n}\| < \varepsilon_n$ where ε_n is fixed or variable.

We will show that the algorithm converges for any order sequence, and relate the rate of convergence to the order sequence. We now illustrate this algorithm through our usual example.

Example 6.5.1. We apply modified policy iteration to the model in Fig. 3.1.1 with $\lambda = 0.95$. We choose $m_n = 5$ for all n, $v^0 = (0, 0)^T$, and $\varepsilon = 0.01$. We apply the same recursion as was used for solving this problem with value iteration in Example 6.3.1. We find that $d_1(s_1) = a_{1,2}$, $d_2(s_1) = a_{2,1}$, and

$$u_0^0(s_1) = \max\{5, 10\} = 10,$$
$$u_0^0(s_2) = -1.$$

We implement step 3(c) by computing

$$u_0^1(s_1) = r_{d_1}(s_1) + \lambda P_{d_1} u_0^0(s_1) = 10 + 0.95(-1) = 9.05,$$
$$u_0^1(s_2) = r_{d_1}(s_2) + \lambda P_{d_1} u_0^0(s_2) = -1 + 0.95(-1) = -1.95.$$

We repeat this step four more times to obtain $u_0^5 = (4.967, -6.033)^T$. Since $k = 5$, we go to step 3(e), set $v^1 = u_0^5$ and return to step 2 to again implement the policy improvement step. We now find that $d_2(s_1) = a_{1,1}$, $d_2(s_2) = a_{2,1}$, and $u_1^0 = (4.494, -6.731)^T$. We now proceed to step 3(c).

We find that it requires 28 passes through the algorithm to satisfy the stopping criterion in step 3(b). This means that we perform a total of $6 \times 28 = 168$ successive substitution steps, but only evaluate the maximum in (6.5.3) 28 times. Thus we obtain similar precision as value iteration with considerably fewer evaluations of

$$\max_{a \in A_s} \left\{ r(s, a) + \lambda \sum_{j \in S} p(j|s, a)v^n(j) \right\}.$$

In this problem, the computational savings are minimal; however, in problems with large action sets, the improvements are often significant. We expand on this point and discuss enhancements to this algorithm in Sec. 6.7.4.

6.5.2 Convergence of Modified Policy Iteration

Our analysis relies on using Corollary C.4 in Appendix C to rewrite representation (6.4.11) for policy iteration as

$$v^{n+1} = v^n + \left(I - \lambda P_{d_{n+1}} \right)^{-1} Bv^n = v^n + \sum_{k=0}^{\infty} \left(\lambda P_{d_{n+1}} \right)^k Bv^n, \quad (6.5.6)$$

in which d_{n+1} is any v^n-improving decision rule. Truncating the series at m_n suggests the following recursive scheme:

$$v^{n+1} = v^n + \sum_{k=0}^{m_n} \left(\lambda P_{d_{n+1}} \right)^k Bv^n, \quad (6.5.7)$$

which we now show corresponds to modified policy iteration.

Proposition 6.5.1. Suppose the sequence $\{v^n\}$ is generated by the modified policy iteration algorithm. Then $\{v^n\}$ satisfies (6.5.7).

Proof. Expand Bv^n in (6.5.7) to obtain

$$v^{n+1} = v^n + \sum_{k=0}^{m_n} \left(\lambda P_{d_{n+1}} \right)^k \left[r_{d_{n+1}} + \lambda P_{d_{n+1}} v^n - v^n \right]$$

$$= r_{d_{n+1}} + \lambda P_{d_{n+1}} r_{d_{n+1}} + \cdots + \left(\lambda P_{d_{n+1}} \right)^{m_n} r_{d_{n+1}} + \left(\lambda P_{d_{n+1}} \right)^{m_n+1} v^n$$

$$= \left(L_{d_{n+1}} \right)^{m_n+1} v^n.$$

The result follows by noting (6.5.5). □

Equation (6.5.7) shows that modified policy iteration includes value iteration and policy iteration as extreme cases. Setting $m_n = 0$ for all n yields

$$v^{n+1} = v^n + Bv^n = Lv^n,$$

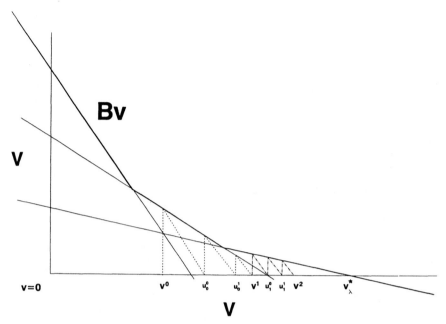

Figure 6.5.2 Graphical representation of two iterations of the modified policy iteration algorithm with $m_0 = m_1 = 2$.

so that modified policy iteration of fixed order 0 is value iteration. Letting m_n approach infinity shows that it also includes policy iteration as a special case.

The modified policy iteration algorithm corresponds to performing one value iteration step and then m_n successive approximation steps with fixed decision rule d_{n+1}. Figure 6.5.2 illustrates this for modified policy iteration with $m_0 = m_1 = 2$. In it, note that a "successive approximation step" corresponds to moving up to the supporting line for Bv and then returning to the axis on a "45-degree" line. This graphical representation is justified by noting that $u_0^0 = v^0 + Bv^0$, so that the "distance" between v^0 and u^0 on the horizontal axis is the same as that between v^0 and Bv^0 in the vertical direction. This pattern is repeated for two further steps, at which point Bv^1 is evaluated and iterations occur along a new supporting line.

The second equality in the proof of Proposition 6.5.1 shows that v^{n+1} represents the expected total discounted reward obtained by using the stationary policy d_{n+1} in a problem with finite horizon m_n and terminal reward v^n. Since v^n has a similar interpretation, v^{n+1} represents the expected total discounted reward of a policy which uses d_{n+1} for m_n periods, d_n for m_{n-1} periods, and so forth, in an $(m_0 + m_1 + \cdots + m_n)$ period problem with terminal reward v^0.

We now demonstrate the convergence of modified policy iteration under the assumption that $v^0 \in V_B$. Direct substitution shows that setting

$$v^0(s) = (1 - \lambda)^{-1} \min_{s' \in S} \min_{a \in A_s'} r(s', a) \tag{6.5.8}$$

for all $s \in S$ ensures that $v^0 \in V_B$.

Introduce the operator $U^m: V \to V$ defined by

$$U^m v \equiv \max_{d \in D} \left\{ \sum_{k=0}^{m} (\lambda P_d)^k r_d + (\lambda P_d)^{m+1} v \right\} \qquad (6.5.9)$$

and let $W^m: V \to V$ be given by

$$W^m v \equiv v + \sum_{k=0}^{m} \left(\lambda P_{d_v} \right)^k B v, \qquad (6.5.10)$$

where d_v denotes any v-improving decision rule. Note that W^m provides an operator representation for (6.5.7) and that $U^0 = L$. Clearly an algorithm based on (6.5.9) is computationally unattractive; we introduce it as a tool for showing convergence of modified policy iteration. We derive properties of these operators in the following lemmas.

Lemma 6.5.2. Let U^m be defined by (6.5.9) and let $w^0 \in V$. Then

a. U^m is a contraction mapping with constant λ^{m+1};
b. the sequence $w^{n+1} = U^m w^n$, $n = 0, 1, \ldots$ converges in norm to v_λ^*;
c. v_λ^* is the unique fixed point of U^m; and
d. $\|w^{n+1} - v_\lambda^*\| \le \lambda^{m+1} \|w^n - v_\lambda^*\|$.

Proof. Part (a) follows from arguments identical to those used in the proof of Proposition 6.2.4. Applying Theorem 6.2.3 shows that w^n converges to the fixed point w^* of U^m. We now show that $w^* = v_\lambda^*$. Let d^* be a v_λ^*-improving decision rule. Then

$$v_\lambda^* = L^m v_\lambda^* = \sum_{k=0}^{m} (\lambda P_{d*})^k r_{d*} + (\lambda P_{d*})^{m+1} v_\lambda^* \le U^m v_\lambda^*.$$

Iterating this expression shows that $v_\lambda^* \le (U^m)^n v_\lambda^*$ for all n, so that $v_\lambda^* \le w^*$. Since $w^* = U^m w^* \le L^m w^*$, letting $m \to \infty$ shows that $w^* \le v_\lambda^*$. Therefore $w^* = v_\lambda^*$. Part (d) is immediate. $\qquad \square$

Lemma 6.5.3. For $u \in V$ and $v \in V$ satisfying $u \ge v$, $U^m u \ge W^m v$. Furthermore, if $u \in V_B$, then $W^m u \ge U^0 v = Lv$.

Proof. Let $d_v \in D$ be v-improving. Then

$$U^m u - W^m v \ge \sum_{k=0}^{m} \left(\lambda P_{d_v} \right)^k r_{d_v} + \left(\lambda P_{d_v} \right)^{m+1} u - \sum_{k=0}^{m} \left(\lambda P_{d_v} \right)^k r_{d_v} - \left(\lambda P_{d_v} \right)^{m+1} v$$

$$\ge \left(\lambda P_{d_v} \right)^{m+1} (u - v) \ge 0.$$

Now suppose $u \in V_B$ and d_u is u-improving; then

$$W^m u = u + \sum_{k=0}^{m} \left(\lambda P_{d_u} \right)^k B u \ge u + B u$$

$$= Lu \ge r_{d_v} + \lambda P_{d_v} u \ge r_{d_v} + \lambda P_{d_v} v = Lv. \qquad \square$$

Lemma 6.5.4. Suppose $u \in V_B$, then, for any m, $W^m u \in V_B$.

Proof. Let $w \equiv W^m u$. Then from Proposition 6.4.3,

$$Bw \geq Bu + \left(\lambda P_{d_u} - I\right)(w - u) = Bu + \left(\lambda P_{d_u} - I\right) \sum_{k=0}^{m} \left(\lambda P_{d_u}\right)^k Bu$$

$$= \left(\lambda P_{d_u}\right)^{m+1} Bu \geq 0 \qquad \square$$

We now show that modified policy iteration converges by comparing its iterates to those of value iteration and U^m.

Theorem 6.5.5. Suppose $v_0 \in V_B$. Then, for any order sequence $\{m_n\}$

i. the iterates of modified policy iteration $\{v^n\}$ converge monotonically and in norm to v_λ^*, and

ii. the algorithm terminates in a finite number of iterations with an ε-optimal policy.

Proof. The proof is by induction. Define the sequences $\{y^n\}$ and $\{w^n\}$ by $y^0 = w^0 = v^0$, $y^{n+1} = Ly^n$, and $w^{n+1} = U^{m_n} w^n$. We show by induction that $v^n \in V_B$, $v^{n+1} \geq v^n$, and $w^n \geq v^n \geq y^n$.

By assumption, the induction hypothesis is satisfied for $n = 0$. Assume now it holds for $k = 1, 2, \ldots, n$. Then applying Proposition 6.5.1 shows that $v^{n+1} = W^{m_n} v^n$, so Lemma 6.5.4 implies that $v^{n+1} \in V_B$. Thus

$$v^{n+1} - v^n + \sum_{m=0}^{m_n} \left(\lambda P_{d_n}\right)^m B v^n \geq v^n.$$

Since we assume that $w^n \geq v^n \geq y^n$ and $v^n \in V_B$, Lemma 6.5.4 implies that $w^{n+1} \geq v^{n+1} \geq y^{n+1}$. Thus the induction hypothesis is satisfied and these results hold for all n. Since both w^n and y^n converge to v_λ^*, the result follows. $\square$

As was the case for value iteration and policy iteration, the result is valid for arbitrary state spaces providing there exists a

$$d_v \in \arg\max_{d \in D} \{r_d + \lambda P_d v\}$$

for each $v \in V_B$. This result can also be easily modified to the case when this maximum is not attained.

Theorem 6.5.5 shows that the iterates of modified policy iteration always exceed those of modified policy iteration of fixed order 0 (value iteration). One might conjecture that the iterates of modified policy iteration of fixed order $m + k$ ($k \geq 0$) always dominate those of modified policy iteration of fixed order m when started at the same initial value. The following intricate example shows that this conjecture is false.

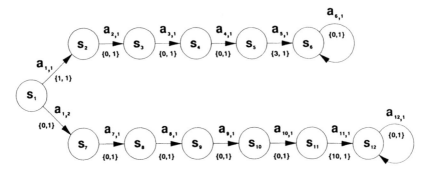

Figure 6.5.3   Symbolic representation of Example 6.5.1.

Example 6.5.1.   (van der Wal and van Nunen, 1977) Let $S = \{s_1, \ldots, s_{12}\}$, $A_{s_1} = \{a_{1,1}, a_{1,2}\}$, and $A_{s_i} = \{a_{i,1}\}$ for $i \neq 1$. All transitions are deterministic and are as illustrated in Fig. 6.5.3. Rewards are $r(s_1, a_{1,1}) = 0$, $r(s_1, a_{1,2}) = 1$, $r(s_5, a_{5,1}) = 3$, $r(s_{11}, a_{11,1}) = 10$, and $r(s_i, a_{i,1}) = 0$ otherwise. Let $1 > \lambda > (\frac{1}{3})^{1/4}$.

We compare the iterates x^n of modified policy iteration with $m_n = 2$ to y^n obtained from modified policy iteration with $m_n = 3$ for all n, starting at $x^0 = y^0 = 0$. We find that

$$x^1 = (1, 0, 3\lambda^2, 3\lambda, 3, 0, 0, 0, 10\lambda^2, 10\lambda, 10, 0)^T,$$

$$x^2 = (1 + 10\lambda^4, 3\lambda^3, 3\lambda^2, 3\lambda, 3, 0, 10\lambda^4, 10\lambda^3, 10\lambda^2, 10\lambda, 10, 0)^T,$$

$$y^1 = (1, 3\lambda^3, 3\lambda^2, 3\lambda, 3, 0, 0, 10\lambda^3, 10\lambda^2, 10\lambda, 10, 0)^T,$$

$$y^2 = (3\lambda^4, 3\lambda^3, 3\lambda^2, 3\lambda, 3, 0, 10\lambda^4, 10\lambda^3, 10\lambda^2, 10\lambda, 10, 0)^T.$$

Therefore $y^1 > x^1$ but $x^2 > y^2$. Note that the lower bound on λ ensures that

$$a_{1,1} = \arg\max\{r(s_1, a_{1,1}) + \lambda y^1(s_2), r(s_1, a_{1,2}) + \lambda y^1(s_7)\},$$

otherwise $x^2 = y^2$. Note also that $x^2 = v_\lambda^*$.

6.5.3   Convergence Rates

We now show that modified policy iteration converges at least linearly and provide a bound on its rate of convergence.

Theorem 6.5.6.   Suppose $Bv^0 \geq 0$ and $\{v^n\}$ is generated by modified policy iteration, that d_n is a v^n-improving decision rule, and d^* is a v_λ^*-improving decision rule. Then

$$\|v^{n+1} - v_\lambda^*\| \leq \left(\frac{\lambda(1 - \lambda^{m_n})}{(1 - \lambda)} \|P_{d_n} - P_{d^*}\| + \lambda^{m_n + 1} \right) \|v^n - v_\lambda^*\|. \quad (6.5.11)$$

Proof. As a consequence of Theorem 6.5.5 and Proposition 6.4.3 applied at v_λ^*,

$$0 \le v_\lambda^* - v^{n+1} = v_\lambda^* - v^n - \sum_{k=0}^{m_n} \left(\lambda P_{d_n} \right)^k B v^n$$

$$\le v_\lambda^* - v^n + \sum_{k=0}^{m_n} \left(\lambda P_{d_n} \right)^k (I - \lambda P_{d^*})(v^n - v_\lambda^*)$$

$$= \lambda (P_{d_n} - P_{d^*}) \sum_{k=0}^{m_n-1} \left(\lambda P_{d_n} \right)^k (v^n - v_\lambda^*) - \lambda^{m_n+1} P_{d_n}^{m_n} P_{d^*}(v^n - v_\lambda^*).$$

Taking norms yields (6.5.11).   □

The following corollary is immediate.

Corollary 6.5.7. Suppose the hypotheses of Theorem 6.5.6 are satisfied and

$$\lim_{n \to \infty} \| P_{d_n} - P_{d^*} \| = 0; \tag{6.5.12}$$

then, for any $\varepsilon > 0$, there exists an N for which

$$\| v^{n+1} - v_\lambda^* \| \le (\lambda^{m_n+1} + \varepsilon) \| v^n - v_\lambda^* \| \tag{6.5.13}$$

for all $n \ge N$.

Expression (6.5.13) demonstrates the appeal of this algorithm. Whenever the transition matrices of successive decision rules approach that of an optimal decision rule, the convergence rate of the algorithm is bounded by that of $m_n + 1$ steps of value iteration. Computationally this represents a significant improvement over value iteration because modified policy iteration avoids maximization over the set of decision rules at each pass through the algorithm. *Therefore in practice, value iteration should never be used. Implementation of modified policy iteration requires little additional programming effort yet attains superior convergence.*

Conditions ensuring the validity of (6.5.12) were discussed in the previous section. It is always satisfied in finite-state and action problems in which the rule "choose $d_{n+1} = d_n$ if possible" is used to uniquely specify a v^n-improving policy in step 2.

As alluded to above, optimal choice of the sequence $\{m_n\}$ remains an open question. The following corollary shows that if m_n becomes large as n increases, modified policy iteration attains superlinear convergence, as was the case for policy iteration under (6.5.12) (cf. Corollary 6.4.10).

Corollary 6.5.8. Suppose the hypotheses of Theorem 6.5.6 hold, (6.5.12) is satisfied, and $m_n \to \infty$. Then

$$\limsup_{n \to \infty} \frac{\| v^{n+1} - v_\lambda^* \|}{\| v^n - v_\lambda^* \|} = 0.$$

6.5.4 Variants of Modified Policy Iteration

By regarding modified policy iteration as value iteration in which maximization over the sets of actions is carried out intermittently, we see that it is natural to incorporate the variants of value iteration discussed in Sec. 6.3.3. In this section, we show how the approach of Sec. 6.5.1 can be adapted to analyze a modified policy iteration algorithm which includes Gauss-Seidel, Jacobi, and/or overrelaxation.

A consequence of Theorem 6.3.4 is that for any regular splitting (Q_d, R_d) of $I - \lambda P_d$ with $\|Q_d^{-1} R_d\| < 1$, v_λ^* satisfies

$$v_\lambda^* = \max_{d \in D} \{Q_d^{-1} r_d + Q_d^{-1} R_d v_\lambda^*\} \equiv T v_\lambda^*. \qquad (6.5.14)$$

Using (6.5.7) as motivation, we may solve (6.5.14) by the iterative scheme

$$v^{n+1} = v^n + \sum_{k=0}^{m_n} \left(Q_{d_n}^{-1} R_{d_n}\right)^k (T v^n - v^n), \qquad (6.5.15)$$

in which

$$d_n \in \arg\max_{d \in D} \{Q_d^{-1} r_d + Q_d^{-1} R_d v^n\}.$$

The following theorem summarizes properties of algorithms based on (6.5.15). Because (Q_d, R_d) is a regular splitting of $I - \lambda P_d$, $Q_d^{-1} R_d \geq 0$, so that the montonicity properties used in proving Theorem 6.5.5 apply here. We leave the proof of this result as an exercise.

Theorem 6.5.9. Suppose $(T - I)v^0 \geq 0$. Then for any order sequence $\{m_n\}$, the sequence defined by (6.5.15) converges monotonically and in norm to v_λ^*. Further, if $\max_{d \in D} \sigma(Q_d^{-1} R_d) < \lambda$, then its rate of convergence is bounded by that of modified policy iteration.

Since the Gauss-Seidel variant of value iteration corresponds to a regular splitting of $I - \lambda P_d$, a Gauss-Seidel version of modified policy iteration will converge. We conclude this section by providing such an algorithm. Label the states $s_1, s_2, \ldots, s_N$ and specify a sequence $\{m_n\}$.

Gauss-Seidel Modified Policy Iteration Algorithm

1. Select a $v^0(s)$ for which $L v^0(s) \geq v^0(s)$ for all $s \in S$, set $n = 0$, and choose $\varepsilon > 0$.
2. (Policy improvement) Set $j = 1$ and go to 2(a).
   a. Set

$$u_n^0(s_j) = \max_{a \in A_{s_j}} \left\{ r(s_j, a) + \lambda \left[\sum_{i < j} p(s_i | s_j, a) u_n^0(s_i) + \sum_{i \geq j} p(s_i | s_j, a) v^n(s_i) \right] \right\}$$

and choose

$$d_{n+1}(s_j) \in \underset{a \in A_{s_j}}{\arg\max} \left\{ r(s_j, a) + \lambda \left[\sum_{i<j} p(s_i|s_j, a) u_n^0(s_i) + \sum_{i\geq j} p(s_i|s_j, a) v^n(s_i) \right] \right\},$$

setting $d_{n+1}(s_j) = d_n(s_j)$ if possible (when $n > 0$).

   b. If $j = N$ go to step 3. Otherwise, increment j by 1 and go to step 2(a).

3. (Partial evaluation)

   a. If $\|u_n^0 - v^n\| < \varepsilon(1 - \lambda)/2\lambda$, go to step 4. Otherwise set $k = 0$ and go to (b).

   b. If $k = m_n$ go to (f). Otherwise set $j = 1$ and go to (c).

   c. Compute $v_n^{k+1}(s_j)$ by

$$u_n^{k+1}(s_j) = r(s_j, d_{n+1}(s_j)) + \lambda \left[\sum_{i<j} p(s_i|s_j, d_{n+1}(s_j)) u_n^{k+1}(s_i) \right.$$
$$\left. + \sum_{j\geq i} p(s_i|s_j, d_{n+1}(s_j)) u_n^k(s_i) \right]$$

   d. If $j = N$, go to (e). Otherwise increment j by 1 and return to (c).

   e. Increment k by 1 and return to (b).

   f. Set $v^{n+1}(s) = u_n^{m_n}(s)$ for all $s \in S$, increment n by 1 and go to step 2.

4. Set $d_\varepsilon = d_{n+1}$.

This algorithm terminates in step 4 with an ε-optimal policy $(d^\varepsilon)^\infty$. The calculation in step 2 is the same as that of the Gauss-Seidel value iteration algorithm, while that in step 3 corresponds to applying Gauss-Seidel to evaluate the fixed policy d_{n+1}. Note that the second calculation in step 2 requires no additional work, since u^0 is determined at the same time as d_{n+1}. Theorem 6.5.9 guarantees that the algorithm reaches step 4 in a finite number of iterations.

6.6 SPANS, BOUNDS, STOPPING CRITERIA, AND RELATIVE VALUE ITERATION

In this section we discuss the span seminorm and bounds for v_λ^*. We use them together to improve algorithmic efficiency and provide insight into the sequence of decision rules generated by these algorithms. Bounds provide the basis for stopping criteria, action elimination procedures, and turnpike theory. Also, we provide a respect value iteration algorithm and show that it is equivalent to value iteration with respect to a stopping criterion based on the span seminorm.

We assume *finite* S so that, for each $v \in V$, $\min_{s \in S} v(s)$ and $\max_{s \in S} v(s)$ exist. Some results may be generalized by replacing "min" and "max" by "inf" and "sup;" however, proofs become more tedious and results of less practical significance.

6.6.1 The Span Seminorm

Our approach in preceding sections used the supremum norm to analyze and verify convergence of algorithms. Most results relied on the property of norms that, for

$v \in V$ and $d \in D$,

$$\|P_d v\| \le \|P_d\| \|v\| = \|v\|. \tag{6.6.1}$$

Choosing $v = e$ shows that the bound in (6.6.1) is tight; that is, $\|P_d e\| = \|e\| = 1$.

By using the span seminorm, we may obtain a sharper bound than (6.6.1). For $v \in V$, define

$$\Lambda(v) \equiv \min_{s \in S} v(s) \quad \text{and} \quad \Upsilon(v) \equiv \max_{s \in S} v(s) \tag{6.6.2}$$

and define the *span* of v devoted $sp(v)$ by

$$sp(v) \equiv \max_{s \in S} v(s) - \min_{s \in S} v(s) = \Upsilon(v) - \Lambda(s). \tag{6.6.3}$$

The span has the following properties.

1. $sp(v) \ge 0$ for $v \in V$.
2. $sp(u + v) \le sp(u) + sp(v)$ for u and $v \in V$,
3. $sp(kv) = |k| sp(v)$ for $v \in V$ and $k \in R^1$,
4. $sp(v + ke) = sp(v)$ for all $k \in R^1$,
5. $sp(v) = sp(-v)$, and
6. $sp(v) \le 2\|v\|$.

Properties 1–3 imply that $sp(v)$ is a *seminorm* on V. However, it is not a norm because of property 4; that is, $sp(v) = 0$ does not imply $v = 0$. If $sp(v) = 0$, then $v = ke$ for some scalar k. Property 6 will be used to relate norm convergence to span convergence.

The following proposition relates $sp(P_d v)$ to $sp(v)$ and is fundamental to what follows. For scalar a, let $[a]^+ \equiv \max(a, 0)$.

Proposition 6.6.1. Let $v \in V$ and $d \in D$. Then

$$sp(P_d v) \le \gamma_d \, sp(v), \tag{6.6.4}$$

where

$$\gamma_d = 1 - \min_{s, u \in S \times S} \sum_{j \in S} \min\{P_d(j|s), P_d(j|u)\} \tag{6.6.5}$$

$$= \tfrac{1}{2} \max_{s, u \in S \times S} \sum_{j \in S} |P_d(j|s) - P_d(j|u)| = \max_{s, u \in S \times S} \sum_{j \in S} [P_d(j|s) - P_d(j|u)]^+.$$

Further, $0 \le \gamma_d \le 1$ for all $d \in D$, and there exists a $v \in V$ such that $sp(P_d v) = \gamma_d \, sp(v)$.

Proof. We drop the subscript d and let $P(j|i)$ denote the components of an arbitrary transition probability matrix. Let $b(i, k; j) \equiv \min\{P(j|i), P(j|k)\}$. Then for any $v \in V$

$$\sum_{j \in S} P(j|i)v(j) - \sum_{j \in S} P(j|k)v(j)$$

$$= \sum_{j \in S} [P(j|i) - b(i,k;j)]v(j) - \sum_{j \in S} [P(j|k) - b(i,k;j)]v(j)$$

$$\leq \sum_{j \in S} [P(j|i) - b(i,k;j)]\Upsilon(v) - \sum_{j \in S} [P(j|k) - b(i,k;j)]\Lambda(v)$$

$$= \left[1 - \sum_{j \in S} b(i,k;j)\right] sp(v)$$

Therefore,

$$sp(Pv) \leq \max_{i,k \in S \times S} \left[1 - \sum_{j \in S} b(i,k;j)\right] sp(v),$$

from which (6.6.4) immediately follows. Application of the easily derived scalar identities

$$|x - y| = (x + y) - 2\min(x, y)$$

and

$$[x - y]^+ = x - \min(x, y)$$

yields the two alternative representations for γ_d.

We now show that there exists a $v' \in V$ such that (6.6.4) holds with equality. If $\gamma_d = 0$, P has equal rows, so that $sp(Pv) = 0 = 0 \cdot sp(v)$ for all $v \in V$. Suppose $\gamma_d > 0$. Define i^* and j^* in S by

$$\sum_{j \in S} [P(j|i^*) - P(j|k^*)]^+ = \max_{i,k \in S \times S} \sum_{j \in S} [P(j|i) - P(j|k)]^+ = \gamma_d$$

and define $v'(j)$ by

$$v'(j) = \begin{cases} 1 & \text{if } P(j|i^*) > P(j|k^*) \\ 0 & \text{if } P(j|k^*) \geq P(j|i^*). \end{cases}$$

Then, noting that $sp(v') = 1$,

$$sp(Pv) \geq \sum_{j \in S} P(j|i^*)v'(j) - \sum_{j \in S} P(j|k^*)v'(j) = \sum_{j \in S} [P(j|i^*) - P(j|k^*)]^+$$

$$= \gamma_d sp(v')$$

Combining this with (6.6.4) shows that $sp(Pv') = \gamma_d sp(v')$.   □

The norm and span each measure different properties of a vector. The norm measures how close it is to zero while the span measures how close it is to being constant. This is particularly relevant in Markov chain theory when investigating convergence of probabilities to limiting distributions.

Proposition 6.6.1 illustrates the "averaging" property of a transition matrix. By multiplying a vector by a transition matrix, the resulting vector has components which are more nearly equal.

The quantity γ_d in Proposition 6.6.1 has been studied extensively (cf. Senata, 1981) and is referred to as the *Hajnal measure* or *delta coefficient* of P_d. It provides an upper bound on the subradius (modulus of the second largest eigenvalue) of P_d, $\sigma_s(P_d)$. As a consequence of (6.6.5), γ_d equals 0 if all rows of P_d are equal, and equals 1 if at least two rows of P_d are orthogonal. For example, when P_d is an identity matrix, $\gamma_d = 1$. From a different perspective we note that $\gamma_d < 1$ if, for each pair of states, there exists at least one state which they both reach with positive probability in one step under P_d.

Note that

$$\gamma_d \leq 1 - \sum_{j \in S} \min_{s \in S} P_d(j|s) \equiv \gamma_d',$$

which is easier to evaluate.

We now evaluate these quantities for the matrix in Example 6.3.2.

Example 6.6.1. Recall that P_d was defined by

$$P_d = \begin{bmatrix} 0.2 & 0.4 & 0.4 \\ 0.3 & 0.3 & 0.4 \\ 0.5 & 0.5 & 0.0 \end{bmatrix}.$$

We find γ_d by computing $b(i, k; j) \equiv \min\{P_d(j|i), P_d(j|k)\}$ and choosing the minimum of $\sum_{j \in S} b(i, k; j)$ over i and k.

$$b(1,2;1) = 0.2, b(1,2;2) = 0.3, b(1,2;3) = 0.4, \sum_{j \in S} b(1,2;j) = 0.9,$$

$$b(1,3;1) = 0.2, b(1,3;2) = 0.4, b(1,3;3) = 0, \sum_{j \in S} b(1,3;j) = 0.6,$$

$$b(2,3;1) = 0.3, b(2,3;2) = 0.3, b(2,3;3) = 0, \sum_{j \in S} b(2,3;j) = 0.6.$$

Thus $\gamma_d = 1 - 0.6 = 0.4$. Since the subradius of P_d, $\sigma_s(P_d) = 0.4$, $\gamma_d = \sigma_s(P_d)$. Choosing $v = (1, 2, 3)^T$, we see that $P_d v = (2.2, 2.1, 1.5)^T$, so that $sp(Pv) = 0.7$ since $sp(v) = 2$, (6.6.4) holds. Note that $\gamma_d' = 0.5$.

The following theorem plays the role of the Banach fixed-point theorem (Theorem 6.2.3) when analyzing convergence of value-iteration-type algorithms with respect to the span seminorm. Its proof is similar to that of Theorem 6.2.3 and is left as an exercise.

Theorem 6.6.2. Let $T: V \to V$ and suppose there exists an α, $0 \leq \alpha < 1$ for which

$$sp(Tv - Tu) \leq \alpha sp(v - u) \tag{6.6.6}$$

for all u and v in V.

a. Then there exists a $v^* \in V$ for which $sp(Tv^* - v^*) = 0$.

b. For any $v^0 \in V$, define the sequence $\{v^n\}$ by $v^n = T^n v^0$. Then

$$\lim_{n \to \infty} sp(v^n - v^*) = 0. \tag{6.6.7}$$

c. For all n,

$$sp(v^{n+1} - v^*) \le \alpha^n sp(v^0 - v^*). \tag{6.6.8}$$

For an operator $T: V \to V$, we say that v^* is a *span fixed point* of T if $sp(Tv^* - v^*) = 0$. As a consequence of property 4 above, this means that

$$Tv^* = v^* + ke$$

for any scalar k.

We say that T is a *span contraction* whenever it satisfies (6.6.6) with $\alpha < 1$. This theorem guarantees the existence of a span fixed point and convergence of value iteration to it whenever T is a span contraction. For discounted models, we show below that L and related operators based on splittings are span contractions with α often less than λ. Consequently, convergence to a span fixed point is often faster than to a (norm) fixed point. Note, however, that in discounted models, these two fixed points coincide, so that by using the span to monitor convergence we will identify a limit in fewer iterations. Also, this result extends to value iteration for undiscounted models in which $\lambda = 1$ but α is often less than 1.

6.6.2 Bounds on the Value of a Discounted Markov Decision Processes

This section presents upper and lower bounds for the optimal value function in terms of v and Bv. They are of considerable importance computationally because they can be used to

a. provide stopping criteria for nonfinite iterative algorithms,
b. provide improved approximations to v_λ^* when iterates satisfy stopping criteria, and
c. identify suboptimal actions throughout the iterative process.

The following is our main result. In it, take the summation to be zero when the upper index is less than the lower index, that is, when $m = -1$.

Theorem 6.6.3.   For $v \in V$, $m \ge -1$, and any v-improving decision rule d_v,

$$G_m(v) \equiv v + \sum_{k=0}^{m} \left(\lambda P_{d_v} \right)^k Bv + \lambda^{m+1}(1 - \lambda)^{-1} \Lambda(Bv)e \le v_\lambda^{(d_v)^\infty}$$

$$\le v_\lambda^* \le v + \sum_{k=0}^{m} \left(\lambda P_{d_v} \right)^k Bv + \lambda^{m+1}(1 - \lambda)^{-1} \Gamma(Bv)e \equiv G^m(v). \tag{6.6.9}$$

Further, $G_m(v)$ is nondecreasing in m and $G^m(v)$ is nonincreasing in m.

Proof. We derive the lower bound for v_λ^*. To obtain the upper bounds, reverse the role of v and v_λ^* in the following.

By Proposition 6.4.3 and Theorem 6.2.5,

$$0 = Bv_\lambda^* \geq Bv + \left(\lambda P_{d_v} - I\right)\left(v_\lambda^* - v\right).$$

Since $\left(I - \lambda P_{d_v}\right)^{-1} \geq 0$,

$$0 \geq v - v_\lambda^* + \left(I - \lambda P_{d_v}\right)^{-1} Bv.$$

Rearranging terms, expanding $(I - \lambda P_{d_v})^{-1}$ as in Corollary C.4, and taking into account the definition of $\Lambda(Bv)$, gives

$$v_\lambda^* \geq v + \sum_{k=0}^{m} \left(\lambda P_{d_v}\right)^k Bv + \sum_{k=m+1}^{\infty} \left(\lambda P_{d_v}\right)^k [\Lambda(Bv)]e$$

$$= v + \sum_{k=0}^{m} \left(\lambda P_{d_v}\right)^k Bv + \frac{\lambda^{m+1}}{1-\lambda}[\Lambda(Bv)]e, \qquad (6.6.10)$$

which gives the desired result. It is easy to see from (6.6.10) that $G_n(v)$ is nondecreasing in m. Letting m approach $+\infty$ yields the tightest lower bound. $\square$

Choosing m equal to -1 and to 0 gives the following important corollary.

Corollary 6.6.4.   For any $v \in V$ and v-improving d_v,

$$v + (1 - \lambda)^{-1}\Lambda(Bv)e \leq v + Bv + \lambda(1 - \lambda)^{-1}\Lambda(Bv)e \leq v_\lambda^{(d_v)^\infty}$$

$$\leq v_\lambda^* \leq v + Bv + \lambda(1 - \lambda)^{-1}\Upsilon(Bv)e$$

$$\leq v + (1 - \lambda)^{-1}\Upsilon(Bv)e.$$

The bounds in Theorem 6.6.3 may be applied during any iterative procedure by replacing v in (6.6.9) by v^n, the current estimate of v_λ^*. Of course, any bound above may be used, but in practice we choose the tightest bound which is available without *further computation*. Those in Corollary 6.6.4 apply to value iteration, those in Theorem 6.6.2 with $m \geq 1$ apply to modified policy iteration, and the lower bound $v_\lambda^{(d_v)^\infty} \leq v_\lambda^*$ together with either upper bound in Corollary 6.6.4, apply to policy iteration. Note when using value iteration that $v^n + Bv^n = Lv^n = v^{n+1}$.

We illustrate these bounds with the following example.

Example 6.6.2.   Choose $v = 0$. Then

$$Bv(s) = \max_{a \in A_s} \{r(s, a)\} \equiv r^*(s)$$

so that the bounds in Corollary 6.6.4 become

$$\frac{1}{1-\lambda}\Lambda(r^*)e \le r^* + \frac{\lambda}{1-\lambda}\Lambda(r^*)e \le v_\lambda^{d_0^\infty} \le v_\lambda^* \le r^* + \frac{\lambda}{1-\lambda}\Upsilon(r^*)e$$

$$\le \frac{1}{1-\lambda}\Upsilon(r^*)e.$$

Applying these to the model in Fig. 3.1.1 with $\lambda = 0.9$ yields

$$\begin{bmatrix} -10 \\ -10 \end{bmatrix} \le \begin{bmatrix} 1 \\ -10 \end{bmatrix} \le v_\lambda^{d_0^\infty} \le v_\lambda^* \le \begin{bmatrix} 100 \\ 89 \end{bmatrix} \le \begin{bmatrix} 100 \\ 100 \end{bmatrix}.$$

Note that the inner lower bound equals v_λ^* and the inside bounds are tighter than the outer bounds. Finding the bounds in Theorem 6.6.3 for $m > 1$ requires additional calculation.

6.6.3 Stopping Criteria

The stopping criterion in the algorithms of Sec. 6.2–6.4 used the conservative-norm-based bounds derived in the proof of Theorem 6.3.3. Example 6.3.1 suggests that using such bounds results in a large number of unnecessary iterations to confirm the optimality of a policy identified early in the iterative process. We now use the results in Sec. 6.6.1 and 6.6.2 to obtain sharper criteria for terminating these algorithms.

The following proposition provides an improved stopping criterion.

Proposition 6.6.5. Suppose for $v \in V$ and $\varepsilon > 0$ that

$$sp(Lv - v) = sp(Bv) < \frac{(1-\lambda)}{\lambda}\varepsilon; \qquad (6.6.11)$$

then,

$$\left\| Lv + \lambda(1-\lambda)^{-1}\Lambda(Bv)e - v_\lambda^* \right\| < \varepsilon \qquad (6.6.12)$$

and

$$\left\| v_\lambda^{d_v^\infty} - v_\lambda^* \right\| < \varepsilon \qquad (6.6.13)$$

for any v-improving decision rule d_v.

Proof. To obtain (6.6.12), apply Corollary 6.6.4 to obtain

$$0 \le v_\lambda^* - v - Bv - \lambda(1-\lambda)^{-1}\Lambda(Bv)e \le \lambda(1-\lambda)^{-1}sp(Bv)e.$$

The result follows by taking norms and recalling that $Lv = Bv + v$. Inequality (6.6.13) follows from Corollary 6.6.4 by noting that, when $w \le x \le y \le z$, $0 \le y - x \le z - w$. □

We use (6.6.11) as a stopping criterion in iterative algorithms. It replaces the stopping rule in step 3 of the value iteration algorithm or step 3(b) of the modified policy iteration algorithm. Note that, in both of these algorithms, $Bv^n = Lv^n - v^n$ is available in the nth improvement step, prior to testing whether the stopping criterion holds. Determining $sp(Bv^n)$ requires little additional effort. Although policy iteration is a finite algorithm, the above stopping criteria can be included after step 3, if an ε-optimal policy is sought.

Proposition 6.6.5 may be interpreted as follows. When (6.6.11) holds for small ε, Bv^n is nearly constant, so that v^{n+1} will differ from v^n by nearly a constant amount. Since $\arg\max_{d \in D}\{r_d + \lambda P_d v^n\} = \arg\max_{d \in D}\{r_d + \lambda P_d(v^n + ke)\}$ for any scalar k, we would not expect any appreciable differences in decision rules chosen at subsequent iterations. Consequently any v_n-improving decision rule will be close to optimal, as indicated by (6.6.13).

We refer to the quantity $v + Bv + (1 - \lambda)^{-1}\lambda\Lambda(Bv)e$ in (6.6.12) as a *lower bound extrapolation*. It gives an improved approximation to v_λ^* upon termination of an iterative algorithm with (6.6.11). One might conjecture that convergence of the algorithms would be enhanced if such extrapolations were used at each iteration. Unfortunately this is not the case because the set of improving decision rules is invariant under addition of a scalar to a value function. Porteus (1980a) provides other extrapolations.

Another advantage of using (6.6.11) as a stopping criterion is that $sp(Bv^n)$ converges to zero at a faster rate than $\|Bv^n\|$. We demonstrate this by applying the following result regarding the contraction properties of L with respect to the span seminorm.

Theorem 6.6.6. Define γ by

$$\gamma \equiv \max_{s \in S, a \in A_s, s' \in S, a' \in A_{s'}} \left[1 - \sum_{j \in S} \min[p(j|s, a), p(j|s', a')] \right]. \quad (6.6.14)$$

Then, for any u and v in V,

$$sp(Lv - Lu) \leq \lambda\gamma sp(v - u). \quad (6.6.15)$$

Proof. Let $s^* \equiv \arg\max_{s \in s}\{Lv(s) - Lu(s)\}$ and $s_* = \arg\min_{s \in s}\{Lv(s) - Lu(s)\}$. Then

$$Lv(s^*) - Lu(s^*) \leq L_{d_v}v(s^*) - L_{d_u}u(s^*) = \lambda P_{d_v}(v - u)(s^*)$$

and

$$Lv(s_*) - Lu(s_*) \geq L_{d_u}v(s_*) - L_{d_u}u(s_*) = \lambda P_{d_u}(v - u)(s_*).$$

Therefore,

$$sp(Lv - Lu) \leq \lambda P_{d_v}(v - u)(s^*) - \lambda P_{d_u}(v - u)(s_*)$$

$$\leq \max_{s \in S} \lambda P_{d_v}(v - u)(s) - \min_{s \in S} \lambda P_{d_u}(v - u)(s)$$

$$\leq sp(\lambda[P_{d_v}/P_{d_u}](v - u))$$

where, for $|S| \times |S|$ matrices H_1 and H_2, H_1/H_2 denotes the $2|S| \times |S|$ "stacked" matrix in which the rows of H_2 follow the rows of H_1. Applying Proposition 6.6.1 establishes

$$sp(Lv - Lu) \leq \lambda \gamma_{d_v, d_u} sp(v - u),$$

where γ_{d_v, d_u} denotes the delta coefficient of the matrix P_{d_v}/P_{d_u}. The result follows by noting that γ_{d_v, d_u} is at most γ.   □

Choosing $u = Lv$ in Theorem 6.6.6 leads to the following important result.

Corollary 6.6.7.   For all $v \in V$, $sp(B^2 v) \leq \lambda \gamma sp(Bv)$.

Given a sequence of iterates $\{v^n\}$ of one of the above algorithms, and recalling that $Bv^n = Lv^n - v^n$, this corollary assures that $sp(Bv^n)$ will eventually satisfy (6.6.11). More importantly, this condition will be satisfied sooner than the analogous norm-based condition. For example, for value iteration,

$$\|v^{n+2} - v^{n+1}\| = \|Bv^{n+1}\| = \|B^2 v^n\| \leq \lambda \|Bv^n\| = \lambda \|v^{n+1} - v^n\|$$

and

$$sp(v^{n+2} - v^{n+1}) = sp(B^2 v^n) \leq \lambda \gamma sp(Bv^n) = \lambda \gamma sp(v^{n+1} - v^n).$$

When $\gamma < 1$, the span-based bound will converge to zero faster than the norm-based bound. Even greater reductions will occur with other algorithms. The proof of Theorem 6.6.6 shows that the true rate of convergence depends on the coefficients of the sequence of decision rules chosen by the algorithm, so, in practice, convergence of $sp(Bv^n)$ may be at a faster rate than $\lambda \gamma$.

Note that γ may be computed by forming a matrix with rows $p(\cdot | s, a)$ and s and a running over all state-action pairs. The quantity γ is the delta coefficient of this matrix, and we compute it following the approach of Example 6.6.1. The following bound on γ:

$$\gamma \leq 1 - \sum_{j \in S} \min_{s \in S, a \in A_s} p(j|s, a) \equiv \gamma' \qquad (6.6.16)$$

is easier to compute.

We may apply the span seminorm stopping criterion in conjunction with any of the algorithms in Secs. 6.3–6.5. Analysis of algorithms based on regular splittings appears more complex because implicit in the definition of the delta coefficient is the assumption that the "transition matrices" have equal row sum. However, empirical evidence also suggests enhanced performance in these cases.

6.6.4   Value Iteration and Relative Value Iteration

We investigate the implications of these concepts for value iteration. Combining Theorem 6.6.6 with Theorem 6.6.2 yields the following result.

Corollary 6.6.8.   Let $v^0 \in V$, let δ be defined through (6.6.14), and suppose $\{v^n\}$ has been generated using value iteration. Then

a. $\lim_{n \to \infty} sp(v^n - v_\lambda^*) = 0$;

b. for all n,

$$sp\left(v^{n+1} - v_\lambda^*\right) \leq (\lambda\gamma)^n sp\left(v^0 - v_\lambda^*\right); \tag{6.6.17}$$

c. and

$$sp\left(v^{n+1} - v^n\right) \leq (\lambda\gamma)^n sp\left(v^1 - v^0\right). \tag{6.6.18}$$

Equation (6.6.18) provides an estimate of the number of iterations required to identify an ε-optimal policy through (6.6.11). For example, if $sp(v^1 - v^0) = 1$, $\lambda = 0.9$, $\gamma = 0.7$ and $\varepsilon = 0.1$, we require ten iterations to identify a 0.1-optimal policy using value iteration.

Example 6.6.3.   We illustrate the above ideas in the context of the model in Fig. 3.1.1 with $\lambda = 0.95$. Recall that we solved it using value iteration with a norm-based stopping criterion in Example 6.3.1. Table 6.3.1 showed that, using that stopping criterion, it required 162 iterations to identify a 0.01-optimal policy. We now apply value iteration with the span-based stopping rule (6.6.11). This means that, with $\varepsilon = 0.01$, we stop as soon as

$$sp(v^{n+1} - v^n) = sp(Bv^n) < (1 - \lambda)\varepsilon/\lambda = 0.000\,526$$

which occurs at iteration 12.

Hence, using an identical algorithm but changing the stopping criterion results in reducing computational effort by 93%. The reason for this is that, in this model, $\gamma = \gamma' = 0.5$, so it follows from (6.6.18) that

$$sp(v^{n+1} - v^n) \leq (0.475)^n sp(v^1 - v^0),$$

while Theorem 6.3.3 shows that

$$\|v^{n+1} - v^n\| \leq (0.95)^n \|v^1 - v^0\|.$$

Table 6.6.1 compares the behavior of the span and the norm in this example.

At iteration 12, $Lv^{12}(s_1) = 2.804$ and $Lv^{12}(s_2) = -8.624$. Applying the lower bound extrapolation (note that $\Lambda(Bv^{12}) = -0.598$) yields the approximation $Lv^{12}(s_1) + \lambda/(1 - \lambda)\Lambda(Bv^{12}) = -8.566$ and $Lv^{12}(s_2) + \lambda/(1 - \lambda)\Lambda(Bv^{12}) = -19.995$. Since $v_\lambda^*(s_1) = -8.571$ and $v_\lambda^*(s_2) = -20$, we see that (6.6.12) holds with $\varepsilon = 0.01$.

Note that changing the stopping criterion in the above example resulted in dramatic reduction in computational effort; however, this might not be the case in many applications, particularly those with *sparse* transition matrices.

Table 6.6.1 Comparison of $\|Bv^n\|$ and $sp(Bv^n)$ when value iteration is applied to the model in Sec. 3.1.

n	$\|v^n - v^{n-1}\|$	$sp(v^n - v^{n-1})$
1	10	11
2	0.950 000	0.225 000
3	0.902 500	0.106 875
4	0.857 375	0.050 765
5	0.814 506	0.024 113
6	0.773 781	0.011 453
7	0.735 092	0.005 440
8	0.698 337	0.002 584
9	0.663 420	0.001 227
10	0.630 249	0.000 583
20	0.377 354	3.409E-07
30	0.225 936	1.993E-10
40	0.135 276	1.163E-13
50	0.080 995	< E-14
60	0.048 495	< E-15

Several investigators have proposed using *relative value iteration* algorithms to accelerate convergence. The following relative value iteration finds an ε-optimal stationary policy.

Relative Value Iteration Algorithm

1. Select $u^0 \in V$, choose $s^* \in S$, specify $\varepsilon > 0$, set $w^0 = u^0 - u^0(s^*)e$, and set $n = 0$.

2. Set $u^{n+1} = Lw^n$ and $w^{n+1} = u^{n+1} - u^{n+1}(s^*)e$.

3. If $sp(u^{n+1} - u^n) < (1 - \lambda)\varepsilon/\lambda$, go to step 4. Otherwise increment n by 1 and return to step 2.

4. Choose $d_\varepsilon \in \arg\max_{d \in D}\{r_d + \lambda P_d u^n\}$.

This algorithm differs from value iteration in that it normalizes u^n by subtracting the constant $u^n(s^*)$ after each iteration. This results in enhanced norm convergence, but does not effect span convergence because the span is invariant to subtraction of a constant. Thus when using stopping criterion (6.6.11), the computational effort of these two algorithms is *identical*.

The motivation for this algorithm comes from models with maximal average reward criterion (Chap. 8) in which the iterates of value iteration usually *diverge*, but, because of the normalization in step 2, the iterates of relative value iteration may *converge*.

6.7 ACTION ELIMINATION PROCEDURES

In this section, we develop methods for identifying and eliminating nonoptimal actions in value iteration, policy iteration, and modified policy iteration algorithms. The advantages of using action elimination procedures include the following:

a. Reduction in size of the action sets to be searched at each iteration of the algorithm because

$$r(s, a) + \sum_{j \in S} \lambda p(j|s, a) v^n(j)$$

need not be evaluated for actions which have already been identified as nonoptimal.

b. Identification of optimal policies in models with *unique* optimal policies. Using action elimination provides the only way of ensuring that value iteration and modified policy iteration terminate with an optimal, instead of an ε-optimal, policy. When all but one action is eliminated in each state, the stationary policy which uses the remaining decision rule is necessarily optimal. This will occur within finitely many iterations in finite-state and action models.

6.7.1 Identification of Nonoptimal Actions

The following result enables us to identify actions that cannot be part of an optimal stationary policy. It serves as the basis for action elimination procedures and turnpike theorems. For $v \in V$, $s \in S$, and $a \in A_s$, define the real-valued function $B(s, a)v$ by

$$B(s, a)v \equiv r(s, a) + \sum_{j \in S} \lambda p(j|s, a) v(j) - v(s). \qquad (6.7.1)$$

Proposition 6.7.1. If

$$B(s, a') v_\lambda^* < 0, \qquad (6.7.2)$$

then

$$a' \notin \underset{a \in A_s}{\arg\max} \left\{ r(s, a) + \sum_{j \in S} \lambda p(j|s, a) v_\lambda^*(j) \right\} \qquad (6.7.3)$$

and any stationary policy which uses action a' in state s cannot be optimal.

Proof. From Theorem 6.2.5, $Bv_\lambda^*(s) = 0$ for all $s \in S$. Since $Bv_\lambda^*(s) = \max_{a \in A_s} B(s, a)v_\lambda^*$, $B(s, a)v_\lambda^* \leq 0$ for all $a \in A_s$. By Corollary 6.2.8,

$$a' \in \underset{a \in A_s}{\arg\max} \left\{ r(s, a) + \sum_{j \in S} \lambda p(j|s, a') v_\lambda^*(j) \right\}$$

implies that $B(s, a') v_\lambda^* = 0$, so that, if (6.7.3) holds, (6.7.2) follows. □

We restate this result in terms of decision rules as follows.

Corollary 6.7.2. Suppose for $d' \in D$ and some $s' \in S$ that

$$B(s, d'(s'))v^*_\lambda < 0.$$

Then, for all $s \in S$, $v^{(d')^\infty}_\lambda(s) \le v^*_\lambda(s)$, with strict inequality for some s.

Figure 6.7.1 illustrates this result. The vertical axis represents a single component of V and the horizontal axis all of V. There are three decision rules d_1, d_2, and d_3. The stationary policy d^∞_2 is optimal. Noting that $B(s, d(s))v$ is the sth component of $r_d + (\lambda P_d - I)v$, we see that $B(s, d_1(s))v^*_\lambda < 0$ and $B(s, d_3(s))v^*_\lambda < 0$, so that the actions chosen by these decision rules in state s cannot be optimal, and that d^∞_1 and d^∞_3 are not optimal.

Since v^*_λ is unknown, the result in Proposition 6.7.1 cannot be used in practice to identify nonoptimal actions. Instead, we substitute upper and lower bounds for v^*_λ, such as those in Theorem 6.6.3 or Corollary 6.6.4, to obtain an implementable elimination rule as follows.

Proposition 6.7.3. Suppose there exists v^L and v^U in V such that $v^L \le v^*_\lambda \le v^U$. Then, if

$$r(s, a') + \sum_{j \in S} \lambda p(j|s, a')v^U(j) < v^L(s), \tag{6.7.4}$$

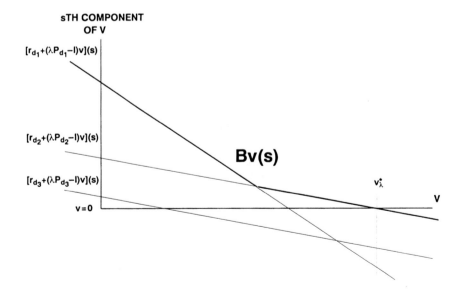

Figure 6.7.1 Graphical representation of Proposition 6.7.1.

any stationary policy which uses action a' in state s is nonoptimal.

Proof. From (6.7.2), and the definitions of v^L and v^U,

$$r(s, a') + \sum_{j \in S} \lambda p(j|s, a')v_\lambda^*(j) \le r(s, a') + \sum_{j \in S} \lambda p(j|s, a')v^U(j) < v^L(s) \le v_\lambda^*(s)$$

The result now follows from Proposition 6.7.1. $\square$

Theorem 6.7.10 below provides an analogous result, which allows identification of actions that will not be selected at the subsequent improvement step in any iterative algorithm. It provides the basis for temporary action elimination procedures.

6.7.2 Action Elimination Procedures

We now discuss inclusion of action elimination procedures in an algorithm. Figure 6.7.2 provides a schematic representation of an iterative algorithm showing where action elimination lies in the algorithmic flow. Note that, by applying evaluation prior to improvement, we are able to incorporate action elimination with minimal redundant computation. In value iteration, the policy evaluation step is superfluous since evaluation is accomplished at the same time as obtaining an improved policy.

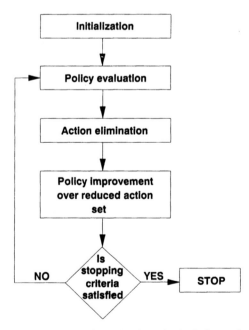

Figure 6.7.2   Flowchart of modified policy iteration including action elimination.

When we apply action elimination, we obtain Bv^n, $\Lambda(Bv^n)$, and $\Upsilon(Bv^n)$ from the previous improvement step, and v^{n+1} from the preceding evaluation step. Recall that

$$v^{n+1} = \begin{cases} Lv^n & \text{for value iteration} \\ v^n + \sum_{k=0}^{m_n} (\lambda P_{d_{v^n}})^k Bv^n & \text{for modified policy iteration} \\ (I - \lambda P_{d_{v^n}})^{-1} r_{d_{v^n}} & \text{for policy iteration.} \end{cases}$$

We choose bounds v^L and v^U, which allow evaluation of the quantities in (6.7.4) without further computation. For all three algorithms, we use the weakest upper bound $v^U = v^n + (1 + \lambda)^{-1}\Upsilon(Bv^n)e$ and take $v^L = G_0(v^n)$ for value iteration, $G_{m_n}(v^n)$ for modified policy iteration and $G_\infty(v^n)$, which equals the expected discounted reward of policy $d_{v^n}^\infty$, for policy iteration where $G_m(v^n)$ is defined in (6.6.9). The following result provides the basis for eliminating actions in value iteration, modified policy iteration, and policy iteration.

Theorem 6.7.4. Action a' is identified as nonoptimal in state s at iteration n of modified policy iteration if

$$r(s, a') + \sum_{j \in S} \lambda p(j|s, a')v^n(j) + \frac{\lambda}{1 - \lambda}\Upsilon(Bv^n) < G_{m_n}(v^n)(s). \quad (6.7.5)$$

Of course, in any of these algorithms we can instead use the weaker lower bound which may be written as $G_0(v^n)(s) = Lv^n + \lambda(1 - \lambda)^{-1}\Lambda(Bv^n)$. Using this bound and rearranging terms yields the following result which applies to all of these algorithms.

Corollary 6.7.5. Action a' is nonoptimal in state s at iteration n if

$$\frac{\lambda}{1 - \lambda}sp(Bv^n) < Lv^n(s) - r(s, a) - \sum_{j \in S} \lambda p(j|s, a)v^n(j). \quad (6.7.6)$$

Hubner (1977) showed that (6.7.5) may be refined further by replacing $\lambda/(1 - \lambda)$ by $\lambda\gamma/(1 - \lambda\gamma)$ or $\lambda\gamma_{s,a}/(1 - \lambda\gamma)$, where γ is defined in (6.6.14) and

$$\gamma_{s,a} \equiv \max_{a' \in A_s} \left\{ 1 - \sum_{j \in S} \min[p(j|s, a), p(j|s, a')] \right\}.$$

In some applications, particularly when γ is much smaller than 1, these quantities will be smaller than $\lambda/(1 - \lambda)$ and may identify more actions as suboptimal; if $\lambda = 0.9$ and $\gamma = 0.5$, $\lambda/(1 - \lambda) = 9$ while $\lambda\gamma/(1 - \lambda\gamma) = 0.82$. For modified policy iteration or policy iteration, we may use *both* (6.7.5) and the refinement of (6.7.6) to eliminate actions.

We would hope that as n increases, more actions will be eliminated through (6.7.5) or (6.7.6). This will occur in (6.7.5) if the quantity on the left-hand side increases with n and that on the right-hand side decreases with n, that is, if the bounds become tighter. We now show this to be the case. We first show that, at successive iterates of modified policy iteration, the bounds in Theorem 6.6.3 are monotone. Note that the proof uses both representations for iterates of modified policy iteration, namely

$$v^{n+1} = v^n + \sum_{k=0}^{m_n} \left(\lambda P_{d_{v^n}} \right)^k B v^n = L_{d_{v^n}}^{m_n+1} v^n,$$

where d_{v^n} is a v^n-improving decision rule. In the following result, we use p and q to represent the orders of modified policy iteration at successive iterations.

Proposition 6.7.6. Let $v \in V$, p and q be arbitrary non-negative integers, and define

$$v' = v + \sum_{k=0}^{p} \left(\lambda P_{d_v} \right)^k B v,$$

where d_v is any v-improving decision rule. Then

$$G_p(v) \le G_q(v') \le v_\lambda^* \le G^q(v') \le G^p(v).$$

Proof. We establish the lower bound inequalities only.
From Theorem 6.6.3, $G_{p-1}(v) \le G_p(v)$, if

$$v + \sum_{k=0}^{p-1} \left(\lambda P_{d_v} \right)^k B v + \frac{\lambda^p}{1-\lambda} \Lambda(Bv)e \le v + \sum_{k=0}^{p} \left(\lambda P_{d_v} \right)^k B v + \frac{\lambda^{p+1}}{1-\lambda} \Lambda(Bv)e.$$

By the monotonicity of the operator L, the fact that $L(v + ke) = Lv + \lambda ke$, the observation that $L_d v \le Lv$ for all $d \in D$, and the above inequality

$$L_{d_v} \left[v + \sum_{k=0}^{p-1} \left(\lambda P_{d_v} \right)^k B v + \frac{\lambda^p}{1-\lambda} \Lambda(Bv)e \right]$$

$$\le L \left[v + \sum_{k=0}^{p} \left(\lambda P_{d_v} \right)^k B v + \frac{\lambda^{p+1}}{1-\lambda} \Lambda(Bv)e \right]$$

$$\le L \left[v' + \frac{\lambda^{p+1}}{1-\lambda} \Lambda(Bv)e \right] = L_{d_v} v' + \frac{\lambda^{p+2}}{1-\lambda} \Lambda(Bv)e.$$

Since $L_{d_v} v = Lv = v + Bv$,

$$L_{d_v} \left[v + \sum_{k=0}^{p-1} \left(\lambda P_{d_v} \right)^k B v + \frac{\lambda^p}{1-\lambda} \Lambda(Bv)e \right]$$

$$= L_{d_v} v + \sum_{k=1}^{p} \left(\lambda P_{d_v} \right)^k B v + \frac{\lambda^{p+1}}{1-\lambda} \Lambda(Bv)e$$

$$= v + \sum_{k=0}^{p} \left(\lambda P_{d_v} \right)^k B v + \frac{\lambda^{p+1}}{1-\lambda} \Lambda(Bv)e = v' + \frac{\lambda^{p+1}}{1-\lambda} \Lambda(Bv)e,$$

it follows that

$$v' + \frac{\lambda^{p+1}}{1-\lambda} \Lambda(Bv)e \leq L_{d_v}v' + \frac{\lambda^{p+2}}{1-\lambda} \Lambda(Bv)e. \tag{6.7.7}$$

By the monotonicity of L_{d_v},

$$L_{d_v}v' + \frac{\lambda^{p+2}}{1-\lambda} \Lambda(Bv)e \leq L_{d_v}^2 v' + \frac{\lambda^{p+3}}{1-\lambda} \Lambda(Bv)e. \tag{6.7.8}$$

Iterating (6.7.8) and applying (6.7.7) yields

$$v' + \frac{\lambda^{p+1}}{1-\lambda} \Lambda(Bv)e \leq L_{d_v}^{q+1} v' + \frac{\lambda^{q+p+2}}{1-\lambda} \Lambda(Bv)e. \tag{6.7.9}$$

We now show that

$$\lambda^{p+1}\Lambda(Bv)e \leq \Lambda(Bv')e. \tag{6.7.10}$$

To do this, note that

$$Bv' = Lv' - v' = L\left[L_{d_v}^{p+1}v\right] - L_{d_v}^{p+1}v$$
$$\geq L_{d_v}^{p+2}v - L_{d_v}^{p+1}v = L_{d_v}^p\left[L_{d_v}v - v\right] = L_{d_v}^{p+1}[Bv].$$

Therefore

$$Bv' \geq L_{d_v}^{p+1}[\Lambda(Bv)e] = \lambda^{p+1}\Lambda(Bv)e,$$

which implies (6.7.10). Combining this with (6.7.9) gives

$$G_p(v) = v' + \frac{\lambda^{p+1}}{1-\lambda} \Lambda(Bv)e \leq L_{d_v}^{q+1} v' + \frac{\lambda^{q+1}}{1-\lambda} \Lambda(Bv')e = G_q(v'),$$

which is the desired result. That these quantities are bounded above by v_λ^* follows from Theorem 6.6.3. □

We obtain the following corollary, which applies to value iteration, by choosing $p = q = 0$ in the above proposition.

Corollary 6.7.7. Let $v \in V$ and $v' = Lv$. Then

$$Lv + \frac{\lambda}{1-\lambda} \Lambda(Bv)e \leq Lv' + \frac{\lambda}{1-\lambda} \Lambda(Bv')e \leq v_\lambda^* \leq Lv' + \frac{\lambda}{1-\lambda} \Upsilon(Bv')e$$

$$\leq Lv + \frac{\lambda}{1-\lambda} \Upsilon(Bv)e.$$

We also require the following easily derived result.

Lemma 6.7.8. Let $v \in V$ and $v' = Lv$. Then

$$v_\lambda^* \le v' + \frac{1}{1-\lambda} \Upsilon(Bv')e \le v + \frac{1}{1-\lambda} \Upsilon(Bv)e.$$

The next result shows that, if an action is identified as nonoptimal at some iteration of the modified policy iteration algorithm, it will be identified as nonoptimal by inequality (6.7.5) in Theorem 6.7.4 at all successive iterations.

Proposition 6.7.9. Suppose action a is identified as nonoptimal in state s at iteration n of modified policy iteration, then it will be identified as nonoptimal at iteration ν for all $\nu > n$.

Proof. Let $\{v^n\}$ be generated by modified policy iteration and $\{m_n\}$ denote the corresponding sequence of orders. Using the relationship between successive iterates in Lemma 6.7.8 and substituting them into (6.7.4) yields

$$r(s, a) + \sum_{j \in S} \lambda p(j|s, a)v^n(j) + \frac{\lambda}{1-\lambda} \Upsilon(Bv^n)$$

$$\le r(s, a) + \sum_{j \in S} \lambda p(j|s, a)v^\nu(j) + \frac{\lambda}{1-\lambda} \Upsilon(Bv^\nu).$$

From Proposition 6.7.6,

$$G_{m_n}(v^n)(s) \le G_{m_\nu}(v^\nu)(s) \le v_\lambda^*(s).$$

Thus the difference between the quantities on the right- and left-hand sides of (6.7.5) at iteration ν is at least as large as that at iteration n. From this observation we conclude the result.   □

The above results are concerned with identifying nonoptimal actions. The following result, which we state without proof, is the basis for an algorithm which eliminates actions for *one iteration only*. That is, it may be used to identify actions which will not be components of v^{n+1}-improving decision rules, but may be optimal. Such actions need not be evaluated in the next improvement step, but may require evaluation at subsequent iterations. Algorithms based on this inequality have excellent computational properties but tedious implementation. We refer the interested reader to Puterman and Shin (1982) for details. Section 8.5.7 analyzes a method of this type in the context of average reward models.

Theorem 6.7.10. Let $\{v^n\}$ be generated by modified policy iteration, and let d_{n+1} be any v^{n+1}-improving decision rule. Then $d_{n+1}(s)$ will not equal a' if, for some $\nu \le n$,

$$r(s, a') + \sum_{j \in S} \lambda p(j|s, a')v^\nu(j) + \lambda \sum_{k=\nu}^{n} \Upsilon(v^{k+1} - v^k) - v^{n+1}(s)$$
$$< \lambda^{m_n+1}\Lambda\left(P_{d_{,n}}^{m_n} Bv^n\right).$$

As above, this result applies to policy iteration and value iteration. In the latter case, it may be sharpened by using the span seminorm and the delta coefficient.

6.7.3 Modified Policy Iteration with Action Elimination and an Improved Stopping Criterion

In this section we describe an algorithm which takes into account the enhancements of the preceding sections. We recommend the use of this algorithm (or a modification which includes a temporary action elimination procedure based on Theorem 6.7.10) for solving discounted MDPs which do not possess special structure. Its flow follows Fig. 6.7.2. We assume that the states in S have been ordered as $s_1, s_2, \ldots, s_N$. The quantities M and δ control the order of modified policy iteration.

The Modified Policy Iteration Algorithm with Action Elimination

1. Initialization
   Select $\varepsilon > 0$, $\delta > 0$, an integer $M \geq 0$, and set $n = 0$. For each $s \in S$, set $A_s^0 = A_s$, and let

$$d_0(s) \in \arg\max_{a \in A_s} r(s, a).$$

   Set

$$v^0(s) = \frac{1}{1-\lambda} \Lambda(r_{d_0}), \qquad u_0^0(s) = r_{d_0}(s) + \frac{\lambda}{1-\lambda} \Lambda(r_{d_0})$$

$$\Lambda(Bv^0) = 0 \quad \text{and} \quad \Upsilon(Bv^0) = sp(r_{d_0}).$$

2. Policy Evaluation
   a. If $M = 0$, set $k = 0$ and go to step 2(e); otherwise, set $j = 1$, $k = 1$, and go to step 2(b).
   b. Set

$$u_n^k(s_j) = r_{d_n}(s_j) + \lambda \sum_{i<j} p_{d_n}(s_i|s_j) u_n^k(s_i) + \lambda \sum_{i \geq j} p_{d_n}(s_i|s_j) u_n^{k-1}(s_i).$$

   c. If $j = N$ go to step 2(d). Otherwise, increment j by 1 and return to step 2(b).
   d. If $k = M$, or

$$sp(u_n^k - u_n^{k-1}) < \delta,$$

   go to step 2(e). Otherwise, increment k by 1, set $j = 1$, and return to step 2(b).
   e. Set $m_n = k$, $v^{n+1}(s) = u_n^k(s)$ for all $s \in S$, and go to step 3.

3. Action Elimination

a. For each $s \in S$, let E_s^{n+1} be the set of a $a \in A_s^n$ for which

$$r(s, a) + \lambda \sum_{j \in S} p(j|s, a) v^n(j)$$

$$< v^{n+1}(s) + \frac{\lambda^{m_n + 1}}{1 - \lambda} \Lambda(Bv^n) - \frac{\lambda}{1 - \lambda} \Upsilon(Bv^n). \quad (6.7.11)$$

Set $A_s^{n+1} = A_s^n - E_s^{n+1}$.

b. If, for each $s \in S$, A_s^{n+1} contains one element, stop and set $d^*(s)$ equal to that element; otherwise continue.

4. Policy Improvement

For each $s \in S$ and all $a \in A_s^{n+1}$ compute and store

$$r(s, a) + \lambda \sum_{j \in S} p(j|s, a) v^{n+1}(j) \equiv L(s, a) v^{n+1}. \quad (6.7.12)$$

Set

$$u_{n+1}^0(s) = \max_{a \in A_s^{n+1}} \left\{ r(s, a) + \lambda \sum_{j \in S} p(j|s, a) v^{n+1}(j) \right\},$$

let

$$d_{n+1}(s) \in \arg\max_{a \in A_s^{n+1}} \left\{ r(s, a) + \lambda \sum_{j \in S} p(j|s, a) v^{n+1}(j) \right\},$$

and set

$$\Lambda(Bv^{n+1}) = \min_{s \in S} \left\{ u_{n+1}^0(s) - v^{n+1}(s) \right\}$$

and

$$\Upsilon(Bv^{n+1}) = \max_{s \in S} \left\{ u_{n+1}^0(s) - v^{n+1}(s) \right\}.$$

5. Stopping Criterion

If

$$sp(Bv^{n+1}) < \frac{1 - \lambda}{\lambda} \varepsilon,$$

stop, set $d_\varepsilon = d_{n+1}$ and $v^\varepsilon = u_{n+1}^0 + \lambda(1 - \lambda)^{-1}\Lambda(Bv^{n+1})e$. Otherwise, increment n by 1 and return to step 2.

Several comments regarding this implementation follow.

Initialization: In this step, ε specifies the precision of the algorithm, and δ and M determine the order of policy evaluation, m_n. The choice of v^0 ensures that $Bv^0 \geq 0$, which implies that the iterates $\{v^n\}$, will be monotone nondecreasing. Direct calculation verifies that $\Lambda(Bv^0) = 0$ and $\Upsilon(Bv^0) = sp(r_{d_0})$. Note that, if $sp(r_{d_0}) < \lambda^{-1}(1 - \lambda)\varepsilon$, the algorithm can be terminated with the ε-optimal policy d_0^∞. When programming this algorithm, the first pass through the action elimination step may be implemented separately and without additional computation since

$$r(s, a) + \sum_{j \in s} \lambda p(j|s, a)v^0(j) = r(s, a) + \frac{\lambda}{1 - \lambda}\Lambda(r_{d_0})$$

Evaluation: The computation of v^{n+1} in the evaluation step is similar to that in the Gauss-Seidel Modified Policy Iteration Algorithm of Sec. 6.5; however, the choice of the order has been more clearly specified. Choosing both δ and M adds flexibility to the specification of the algorithm. If $\delta = 0$, then $m_n = M$ for all n, while, if δ is positive and not too small, then m_n may be less than M at several iterations. Note also that the sensitive span stopping criterion, which converges at rate $\lambda\gamma_{d_n}$, has been incorporated. Choosing $M = 0$ reduces this algorithm to value iteration with action elimination. Note also that δ can be replaced by a sequence $\{\delta_n\}$.

Action Elimination: The action elimination step has been organized and located in the algorithmic flow so that all quantities in (6.7.11) are available without additional computation; $\Lambda(Bv^n)$, $\Upsilon(Bv^n)$, and $L(s, a)v^n$ were computed in the previous improvement step, while v^{n+1} was evaluated in step 2. Note that E_s^{n+1} is the set of eliminated actions in state s. If, at any iteration, A_s^{n+1} is a singleton for all $s \in S$, then the algorithm can be terminated, and the resulting stationary policy $(d^*)^\infty$ is optimal. A further pass through the evaluation step would provide a good approximation to its value.

Improvement: Our presentation of step 4 was motivated by clarity of exposition. Implementation requires only one pass through the set of states and one evaluation of $L(s, a)v^n$ for each state action pair. We start with $s_1 \in S$, and evaluate (6.7.12) for the "first" $a \in A_s^{n+1}$, say a_1; we then set

$$u_{n+1}^0(s_1) = r(s_1, a_1) + \sum_{j \in S} p(j|s_1, a_1)v^{n+1}(j)$$

and $d_{n+1}(s_1) = a_1$. We loop through $A_{s_1}^{n+1}$ evaluating (6.7.12) for each a', updating $u_{n+1}^0(s_1)$ and $d_{n+1}(s_1)$ according to $u = u_{n+1}^0(s_1)$, $d = d_{n+1}(s_1)$,

$$u_{n+1}^0(s_1) = \max\left\{r(s_1, a') + \sum_{j \in S} \lambda p(j|s_1, a')v^{n+1}(j), u\right\}$$

and

$$d_{n+1}(s_1) = \begin{cases} d & \text{if } L(s_1, a')v^{n+1} \leq u \\ a' & \text{if } L(s_1, a')v^{n+1} > u, \end{cases}$$

until we evaluate all actions in $A_{s_1}^{n+1}$. We then set $\Lambda(Bv^{n+1}) = \Upsilon(Bv^{n+1}) = u_{n+1}^0(s_1) - v^{n+1}(s_1)$.

We perform similar calculations for s_2, and then update $\Lambda(Bv^{n+1})$ and $\Upsilon(Bv^{n+1})$ according to $1 = \Lambda(Bv^{n+1})$, $u = \Upsilon(Bv^{n+1})$,

$$\Lambda(Bv^{n+1}) = \min\{u_{n+1}^0(s_2) - v^{n+1}(s_2), 1\}$$

and

$$\Upsilon(Bv^{n+1}) = \max\{u_{n+1}^0(s_2) - v^{n+1}(s_2), u\}.$$

This step requires $N^2\Sigma_{s \in S}|A_s^{n+1}|$ multiplications and storage of $N\Sigma_{s \in S}|A_s^{n+1}|$ entries.

Since bounds and stopping criteria have been expressed and analyzed in terms of Bv^n, we have chosen not to use a Gauss-Seidel implementation of the improvement step as in Sec. 6.5.3 because Bv^n would not have been available. Reetz (1976) and Ohno (1981) consider action elimination for value iteration and modified policy iteration with Gauss-Seidel improvement steps.

Stopping: The algorithm terminates either in step 3 with an optimal policy, or in step 5 with an ε-optimal policy. In the former case, all but one action has been identified as nonoptimal, so that the policy which uses it must be optimal. In the latter case, the sensitive span-based bound quickly detects an ε-optimal policy and $\|v^\varepsilon - v_\lambda^*\| < \varepsilon$.

This algorithm includes value iteration ($M = 0$) and may be easily adapted to policy iteration. For policy iteration, replace step 2 by a step which solves $(I - \lambda P_{d_n})v^{n+1} = r_{d_n}$. This can be achieved efficiently in the above algorithm by specifying M to be large and δ small. By incorporating step 5 in the context of policy iteration, we terminate with an ε-optimal policy.

6.7.4   Numerical Performance of Modified Policy Iteration with Action Elimination

We now summarize some published computational experience with algorithms similar to that in the preceding section. We note that results were reported in terms of CPU times which depend greatly on programming efficiency, machine speed, and architecture. Puterman and Shin (1982) applied the above algorithm with $\varepsilon = \delta = 0.1$ and $M = \infty$ to Howard's (1960) automobile replacement problem, with $|S| = 40$ states and $|A_s| = 41$ for all $s \in S$, and to a sparse randomly generated test problem with $|S| = 40$ states and $|A_s| = 100$ for all s. They solved each problem with four values of λ. Results in terms of average CPU time in seconds (over the four values of λ) are summarized in Figure 6.7.3.

For the Howard problem, CPU times for policy iteration were 16% lower than for modified policy iteration. Little further reduction was obtained through elimination of nonoptimal actions. One-step action elimination reduced overall CPU times by 41% for policy iteration and 34% for modified policy iteration. For the randomly generated problem, CPU times for modified policy iteration and policy iteration were similar. For each algorithm, permanent action elimination reduced computational times by 23% and temporary action elimination by 60%.

Ohno and Ichiki (1987) applied a variant of the above algorithm to find optimal service rates for a tandem queueing model. (Note that this is a continuous-time model but, as a consequence of results in Sec. 11.5, the methods described here apply

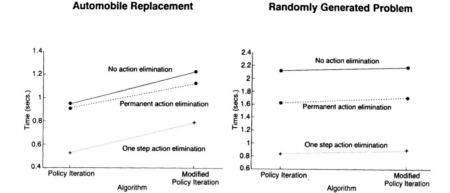

Figure 6.7.3 CPU times for the above algorithm and its modifications. (*Source*: Adapted from Puterman and Shin (1982) Table III.)

directly.) Their implementation used a decreasing sequence of δ_n's in step 2(d) defined by $\delta_0 = \delta$, and $\delta_{n+1} = \delta_n \lambda^{m_n+1}$. They solved a two-stage model with 624 states and 5,034 actions in total, and a three-stage system with 3,779 states and 43,813 actions. For the two-stage model, they compared value iteration, Gauss-Seidel value iteration with and without permanent action elimination, and Gauss-Seidel modified policy iteration with various specifications of M and δ. They sought a 0.01-optimal policy. Their results are summarized in Fig. 6.7.4; for comparison, note that Gauss-Seidel value iteration without action elimination required 2.52 seconds to identify a policy with a value that was within 0.065 of the optimal value. In each of the cases reported in Fig. 6.7.4, the algorithm found an optimal policy.

These results show that action elimination did not enhance the performance of Gauss-Seidel value iteration, but, through adaptive choice of m_n ($M = 4$ and $\delta = 1$), a 25% reduction in CPU time was obtained. Using this choice for M and δ, Ohno and

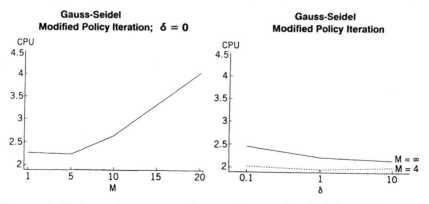

Figure 6.7.4 CPU times for two-stage tandem queueing control model. (*Source*: Adapted from Ohno and Ichiki (1987) Table I.)

Ichiki solved the three-stage model in 33.9 seconds. Since there were 3779 states, policy iteration would have required the computationally intensive task of solving a linear system with 3779 equations.

Taking these results together we may conclude that

1. permanent action elimination procedures can reduce computational times in problems with a large number of actions,
2. Gauss-Seidel modified policy iteration with an adaptive choice of m_n and permanent action elimination can lead to a reduction in computation time when compared to ordinary and Gauss-Seidel value iteration with and without action elimination.
3. temporary action elimination algorithms can significantly reduce overall computation times, and
4. large practical problems can be solved with these algorithms.

6.8   CONVERGENCE OF POLICIES, TURNPIKES, AND PLANNING HORIZONS

Up to this point, results in this chapter have focused on properties of sequences of values $\{v^n\}$ generated by iterative algorithms. Here we investigate behavior of the sequence of v^n-improving decision rules by combining results from Secs. 6.6.1, 6.6.2, and 6.7.1. These results provide a basis for approximating finite-horizon models by infinite-horizon models and implementing MDP's in changing environments.

Let $\{v^n\}$ represent a sequence of iterates of value iteration, and $\{D_n^*\}$ the corresponding sequence of sets of v^n-improving decision rules; that is,

$$D_n^* = \left\{ \delta \in D : r_\delta + \lambda P_\delta v^n = \max_{d \in D} \{ r_d + \lambda P_d v^n \} \right\}$$

Let D^* denote the set of v_λ^*-improving decision rules.

The following theorem relates D_n^* to D^*.

Theorem 6.8.1. Suppose S and A_s are finite. Then for any $v^0 \in V$, there exists an n^* such that, for all $n \geq n^*$, $D_n^* \subset D^*$. If $D^* = D$, $n^* = 0$, otherwise

$$n^* \leq \left\lfloor \frac{\log(\lambda^{-1}(1 - \lambda)c) - \log(sp(Bv^0))}{\log(\lambda\gamma)} \right\rfloor^+ + 1 \equiv n_1^*, \qquad (6.8.1)$$

where γ is defined in (6.6.14),

$$c = \inf_{d \in D/D^*} \|v_\lambda^* - L_d v_\lambda^*\| > 0, \qquad (6.8.2)$$

and $\lfloor \cdot \rfloor^+$ denotes the positive integer part of the quantity in brackets.

Proof. If $D^* = D$, $D_n^* \subset D^*$ for all n, so that $n^* = 0$. Suppose that D^* is a proper subset of D. From Theorem 6.2.7(b), $L_d v_\lambda^* - v_\lambda^* = 0$ implies d^∞ is optimal, so that if $\|L_d v_\lambda^* - v_\lambda^*\| > 0$, $d \in D/D^*$. Consequently, the assumed finiteness of D implies that $c > 0$.

Let $d_n \in D_n^*$. Applying the upper bound $G^0(v^n)$ and lower bound $v^n + (1 - \lambda)^{-1}\Lambda(Bv^m)e$ for v_λ^*, it follows that $\|v_\lambda^* - L_{d_n} v_\lambda^*\| < c$ whenever

$$ce > Lv^n + \lambda(1 - \lambda)^{-1}\Upsilon(Bv^n)e - L_{d_n}\left[v^n + (1 - \lambda)^{-1}\Lambda(Bv^n)e\right]$$

$$= \lambda(1 - \lambda)^{-1}sp(Bv^n)e.$$

From Corollary 6.6.8(c), $sp(Bv^n) \le (\lambda\gamma)^n sp(Bv^0)$, so that the above inequality is satisfied whenever

$$\frac{(1 - \lambda)c}{\lambda} > (\lambda\gamma)^n sp(Bv^0).$$

Since $\lambda\gamma < 1$, there exists an n^* such that, for all $n \ge n^*$, the above inequality holds. Taking logarithms shows that n^* satisfies (6.8.1). □

For λ near 1, $\log((1 - \lambda)\lambda^{-1}c)$ may be considerably less than 0. Since the denominator is also negative, this suggests that the bound in (6.8.1) may be quite large when λ is near 1. For this reason, and for application when $\lambda = 1$, we prefer a bound on n^* which does not contain $\log(1 - \lambda)$ in the numerator. We provide such a bound in Proposition 6.8.3 below. It is based on Lemma 6.8.2, the proof of which we leave as an exercise.

Lemma 6.8.2. Let γ be as defined in (6.6.14). Then, for any $u \in V$, $d \in D$ and $d' \in D$

$$-\gamma sp(u) \le P_d u - P_{d'} u \le \gamma sp(u) \tag{6.8.3}$$

and

$$\|P_d u - P_{d'} u\| \le \gamma sp(u). \tag{6.8.4}$$

Proposition 6.8.3. Suppose the hypotheses of Theorem 6.8.1 hold; then

$$n^* \le \left\lfloor \frac{\log(c) - \log(sp(v_\lambda^* - v^0))}{\log \lambda\gamma} \right\rfloor^+ + 1 \equiv n_2^*. \tag{6.8.5}$$

Proof. Let $\{v^n\}$ be a sequence of iterates of value iteration and $d \in D$. We show first that

$$Lv^n - L_d v^n \ge Lv_\lambda^* - L_d v_\lambda^* - \lambda\gamma sp(v_\lambda^* - v^n). \tag{6.8.6}$$

Since

$$Lv_\lambda^* - L_dv_\lambda^* = L[v^n + (v_\lambda^* - v^n)] - L_d[v^n + (v_\lambda^* - v^n)]$$
$$= Lv^n - L_dv^n + \lambda P_{d^*}(v_\lambda^* - v^n) - \lambda P_d(v_\lambda^* - v^n),$$

where d^* is any v_λ^*-improving decision rule, (6.8.6) follows from (6.8.3). From Corollary 6.6.8(b), it follows that

$$sp(v_\lambda^* - v^n) \le (\lambda\gamma)^n sp(v_\lambda^* - v^0).$$

Therefore, if $n^* \ge n_2^*$, for $n \ge n^*$ and $d \in D/D^*$, $Lv^n - L_dv^n > 0$, so that $d \notin D_n^*$.
□

The following important result is an immediate consequence of Theorem 6.8.1.

Corollary 6.8.4. Suppose D^* consists of a single decision rule. Then there exists an n^* such that, for all $n \ge n^*$, $D_n^* = D^*$.

When D^* contains more than one decision rule, the inclusion of D_n^* in D^* in Theorem 6.8.1 may indeed be proper. Further, even if $D = D^*$, D_n^* may be a proper subset of D^* for all n. The following example illustrates this erratic behavior.

Example 6.8.1. (Shapiro, 1968). Let $S = \{s_1, s_2, s_3, s_4, s_5\}$, $A_{s_1} = \{a_{1,1}, a_{1,2}\}$ and $A_{s_i} = \{a_i\}$, $i = 2, 3, 4, 5$. All transitions are deterministic and are indicated together with rewards in Fig. 6.8.1. Let d_1 denote the decision rule which uses action $a_{1,1}$ in s_1, and d_2 denote the decision rule which uses action $a_{1,2}$; each decision rule uses the same actions in the remaining states. Choose $\lambda = 0.75$. Solving $(I - \lambda P_{d_i})v = r_{d_i}$ for $i = 1, 2$ shows that $v_\lambda^{d_1^\infty}(s_1) = v_\lambda^{d_2^\infty}(s_1) = 44$. Thus $D^* = D$ and $n^* = 0$.
We now investigate the behavior of D_n^*. Let

$$\Delta^n \equiv r_{d_1}(s_1) + \lambda P_{d_1}v^n(s_1) - r_{d_2}(s_1) - \lambda P_{d_2}v^n(s_1)$$

When $\Delta^n > 0$, $D_n^* = \{d_1\}$; when $\Delta^n < 0$, $D_n^* = \{d_2\}$; and when $\Delta^n = 0$, $D_n^* = \{d_1, d_2\}$ = D. Starting with $v^0(s_i) = 0$, $i = 1, 2, \dots, 5$ we see that $\Delta^0 = 2$, $\Delta^1 = -1$, $\Delta^2 = 18/16$, and, in general,

$$\Delta^n = \begin{cases} 2\left[\frac{3}{4}\right]^n & n \text{ even} \\ -\left[\frac{3}{4}\right]^{n-1} & n \text{ odd.} \end{cases}$$

Therefore, for n even, $D_n^* = \{d_1\}$; for n odd, $D_n^* = \{d_2\}$; and, in the limit, $\Delta^* = 0$, so that $D^* = \{d_1, d_2\}$.

The results concerning the existence of n^* shed light on the use of infinite-horizon models as approximations to finite-horizon models. In this context, a natural question is "What is the relationship between the optimal policy for the infinite-horizon

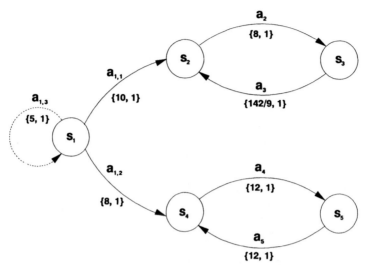

Figure 6.8.1 Graphical representation of Example 6.8.1. (Action indicated by - - - is referred to in Example 6.8.2.)

problem and that for the finite-horizon problem with the same rewards and transition probabilities?" Combining the following observation with Corollary 6.8.4 provides an answer.

The quantity v^n equals the expected total reward in an n-period problem with terminal reward (scrap value) v^0, obtained by using the policy $\{d_{n-1}, d_{n-2}, \ldots, d_0\}$, where $d_i \in D_i^*$, $i = 0, 1, \ldots, n - 1$. As a consequence of Theorem 4.3.3, any policy obtained in this fashion is optimal for the n-period problem. If there exists a *unique* optimal policy d^*, Corollary 6.8.4 ensures the existence of an n^* with the property that, for all $n \geq n^*$, $D_n^* = \{d^*\}$. Figure 6.8.2 depicts the relationship between D_n^* and n^*.

Reinterpreting this result in the context of the above question, it follows that in an n-period problem with $n > n^*$, the optimal policy uses the v_λ^*-improving decision rule d^* at the first $n - n^*$ decision epochs. The decision rules to be used in the remaining

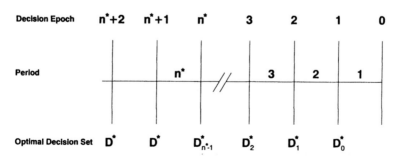

Figure 6.8.2 Illustration of planning horizon notation.

n^* periods must be determined by backward induction. The optimal infinite-horizon decision rule is often referred to as a *turnpike*. It is reached in a finite-horizon model after traveling n^* periods (in a reverse time sense) on the nonstationary "side roads." For example, if $n^* = 2$ and $n = 4$, it would be optimal to use d^* when there are four and three decision epochs remaining, and then to determine the optimal policy at the next-to-last and last decision epoch using backward induction.

Another interpretation of this result is that it is optimal to use d^* for the first decision in a finite-horizon model in which the number of periods is at least n^*. Thus it is not necessary to know the horizon specifically, only that it exceeds n^*. For this reason, we also refer to n^* as a *planning horizon*. When there exists a unique optimal policy to the infinite-horizon model, the quantities n_1^* and n_2^* in (6.8.1) and (6.8.5) provide bounds on the length of the planning horizon.

We can apply these results in yet another way. Specify some $n' > n^*$, and behave as if the problem has horizon length n'. In the first period, we know that d^* is optimal, so use it. At the next decision epoch assume again that n' periods remain, so again use d^*. Proceeding in this fashion, we provide alternative justification for using the stationary policy d^*. We refer to this approach as using a *rolling horizon policy*. Note that if D^* contains more than one decision rule, then we know only that we should use some element of D^* at each decision epoch. Of course, if at some point the rewards and/or transition probabilities were to change, we could resolve the infinite-horizon model and again use the above approach.

Example 6.8.2. Example 6.8.1 provides a model in which the optimal policy is not unique and there exists no planning horizon or turnpike. However, if the horizon is known to be even, d_1 would be the optimal decision to use in the first period, while, if the horizon is odd, d_2 would be optimal. Suppose now we add another action in state $s_1, a_{1,3}$ (see Fig. 6.8.1), which returns the system to state s_1 with probability 1 and yields a reward of five units. Denoting by d_3 the decision rule which uses $a_{1,3}$ whenever the system state is s_1, we see that d_3 is nonoptimal. In this extended model, $D_0^* = D^*$ is a proper subset of D, $n^* = 0$, and, for any horizon length, the optimal policy is in D^*. If we adopt a rolling horizon policy, we would always use a decision rule in D^*.

We further illustrate these concepts with the model in Sec. 3.1.

Example 6.8.3. Let d_1 denote the decision rule which uses action $a_{1,1}$ in s_1, and d_2 denote the decision rule which uses action $a_{1,2}$ in state s_1. Each uses action $a_{2,1}$ in state s_2. When $\lambda = 0.9$, $D^* = \{d_2\}$ and $v_\lambda^* = (1, -10)^T$, and, for $\lambda = 0.95$, $D^* = \{d_1\}$ and $v_\lambda^* = (-8.57, -20)^T$.

Choose $v^0 = 0$. For $\lambda = 0.9$, $D_n^* = \{d_2\}$ for all n; for $\lambda = 0.95$, $D_0^* = \{d_2\}$ and $D_n^* = \{d_1\}$ for $n \geq 1$. Therefore, when $\lambda = 0.9$, $n^* = 0$ and, for $\lambda = 0.95$, $n^* = 1$. In the case when $\lambda = 0.9$, the planning horizon is of length 0, so that d_2 is the optimal decision in all periods. When $\lambda = 0.95$, the planning horizon is of length 1, so we know that we should use d_1 in the first period of any problem with two or more decision epochs. Alternatively, when $n > 1$, it is optimal use d_1 at all but the final decision epoch.

We now evaluate the above bounds on n^*. Applying (6.6.14) shows that $\gamma = 0.5$. When $\lambda = 0.9$ and $v^0 = 0$, $Bv^0 = (10, -1)^T$ and $c = 0.05$, so that $n_1^* = 10$ and

$n_2^* = 7$, while, for $\lambda = 0.95$, $n_1^* = 9$ and $n_2^* = 5$. In each case, the bounds overestimate n^* considerably, but n_2^* is superior. Note, however, that these bounds are several orders of magnitude closer to n^* than norm-based bounds which have appeared in the literature. These observations are further supported by calculations in Hinderer and Hubner (1977).

6.9 LINEAR PROGRAMMING

This section explores the relationship between discounted Markov decision problems and linear programs (LP's). At present, linear programming has not proven to be an efficient method for solving large discounted MDPs; however, innovations in LP algorithms in the past decade might change this. We study the linear programming formulation because of its elegant theory, the ease in which it allows inclusion of constraints, and its facility for sensitivity analysis. Appendix D provides background on linear programming and the Bibliographic Remarks section provides references.

6.9.1 Model Formulation

Theorem 6.2.2a showed that, whenever $v \in V$ satisfies

$$v \geq r_d + \lambda P_d v \tag{6.9.1}$$

for all $d \in D$, then v is an upper bound for the value of the MDP, v_λ^*. This observation provides the basis for a primal linear programming formulation of the discounted Markov decision problem.

Choose $\alpha(j)$, $j \in S$ to be positive scalars which satisfy $\sum_{j \in S} \alpha(j) = 1$. Note that any positive constants would do, but the added condition that they sum to 1 allows interpretation as a probability distribution on S.

Primal Linear Program

$$\text{Minimize } \sum_{j \in S} \alpha(j)v(j)$$

subject to

$$v(s) - \sum_{j \in S} \lambda p(j|s, a)v(j) \geq r(s, a)$$

for $a \in A_s$ and $s \in S$, and $v(s)$ unconstrained for all $s \in S$.

We will find it more informative to analyze this model through its dual LP which follows.

Dual Linear Program

$$\text{Maximize} \sum_{s \in S} \sum_{a \in A_s} r(s, a) x(s, a)$$

subject to

$$\sum_{a \in A_j} x(j, a) - \sum_{s \in S} \sum_{a \in A_s} \lambda p(j|s, a) x(s, a) = \alpha(j) \qquad (6.9.2)$$

and $x(s, a) \geq 0$ for $a \in A_s$ and $s \in S$.

Observe that the primal LP has $|S|$ columns and $\sum_{s \in S} |A_s|$ rows, while the dual LP has $|S|$ rows and $\sum_{s \in S} |A_s|$ columns. Consequently, we prefer to solve the problem using its dual formulation. In the case of the three-stage tandem queueing example referred to in Sec. 6.7.4, the dual LP would have 3,779 rows and 43,813 columns, so that solution by standard codes would be time consuming.

6.9.2   Basic Solutions and Stationary Policies

Recall that $x(s, a)$ is said to be a *feasible solution* to the dual LP if it is non-negative and satisfies (6.9.2) for all $j \in S$. The following result relates feasible solutions of the dual LP to stationary randomized policies in the discounted MDP.

Theorem 6.9.1

a. For each $d \in D^{MR}$, $s \in S$ and $a \in A_s$ define $x_d(s, a)$ by

$$x_d(s, a) \equiv \sum_{j \in S} \alpha(j) \sum_{n=1}^{\infty} \lambda^{n-1} P^{d^{\infty}}(X_n = s, Y_n = a | X_1 = j). \quad (6.9.3)$$

Then $x_d(s, a)$ is a feasible solution to the dual problem.

b. Suppose $x(s, a)$ is a feasible solution to the dual problem, then, for each $s \in S$, $\sum_{a \in A_s} x(s, a) > 0$. Define the randomized stationary policy d_x^{∞} by

$$P\{d_x(s) = a\} = \frac{x(s, a)}{\sum_{a' \in A_s} x(s, a')}. \qquad (6.9.4)$$

Then $x_{d_x}(s, a)$ as defined in (6.9.3) is a feasible solution to the dual LP and $x_{d_x}(s, a) = x(s, a)$ for all $a \in A_s$ and $s \in S$.

Proof

a. Clearly, $x_d \geq 0$. Let $\delta(j|k)$ denote the (k, j)th element of I. Then

$$\sum_{s \in S} \sum_{a \in A_s} \lambda p(j|s, a) x_d(s, a)$$

$$= \sum_{s \in S} \sum_{a \in A_s} \lambda p(j|s, a) \sum_{k \in S} \alpha(k) \sum_{n=1}^{\infty} \lambda^{n-1} P^{d^\infty}(X_n = s, Y_n = a | X_1 = k)$$

$$= \sum_{k \in S} \alpha(k) \left[\sum_{n=1}^{\infty} \lambda^n P^{d^\infty}(X_{n+1} = j | X_1 = k) \right]$$

$$= \sum_{k \in S} \alpha(k) \left[\sum_{n=1}^{\infty} \lambda^{n-1} P^{d^\infty}(X_n = j | X_1 = k) - \delta(j|k) \right]$$

$$= \sum_{a \in A_j} x(j, a) - \alpha(j) \qquad (6.9.5)$$

establishing (6.9.2).

b. Let $x(s, a)$ be a feasible solution of the dual, and define $u(s)$ by

$$u(s) \equiv \sum_{a \in A_s} x(s, a).$$

That $u(s) > 0$ follows from (6.9.2), the non-negativity of $x(s, a)$ and the positivity of α. Rewriting (6.9.2) yields

$$\alpha(j) = u(j) - \sum_{s \in S} \sum_{a \in A_s} \lambda p(j|s, a) x(s, a)$$

$$= u(j) - \sum_{s \in S} \sum_{a \in A_s} \lambda p(j|s, a) x(s, a) \frac{u(s)}{\sum_{a' \in A_s} x(s, a')},$$

which, from (6.9.4),

$$= u(j) - \sum_{s \in S} \sum_{a \in A_s} \lambda p(j|s, a) P\{d_x(s) = a\} u(s)$$

$$= u(j) - \sum_{s \in S} \lambda p_{d_x}(j|s) u(s).$$

Expressing this in matrix notation, yields

$$\alpha^T = u^T [I - \lambda P_{d_x}],$$

so, from Corollary C.4,

$$u^T = \alpha^T [I - \lambda P_{d_x}]^{-1} = \alpha^T \left[\sum_{n=1}^{\infty} (\lambda P_{d_x})^{n-1} \right].$$

Thus, it follows from (6.9.3) that

$$u(s) = \sum_{k \in S} \alpha(k) \sum_{n=1}^{\infty} \lambda^{n-1} \sum_{a \in A_s} P^{d_x^{\infty}}(X_n = s, Y_n = a | X_1 = k) = \sum_{a \in A_s} x_{d_x}(s, a).$$

Hence,

$$\sum_{a \in A_s} x(s, a) = \sum_{a \in A_s} x_{d_x}(s, a). \tag{6.9.6}$$

From the definition of $x_{d_x}(s, a)$ and (6.9.6),

$$x_{d_x}(s, a) = \sum_{j \in S} \alpha(j) \sum_{n=1}^{\infty} \lambda^{n-1} P^{d_x^{\infty}}(X_n = s | X_1 = j) q_{d_x(s)}(a)$$

$$= \sum_{j \in S} \alpha(j) \sum_{n=1}^{\infty} \lambda^{n-1} P^{d_x^{\infty}}(X_n = s | X_1 = j) \frac{x(s, a)}{\sum_{a' \in A_s} x(s, a')}. \tag{6.9.7}$$

Since

$$\sum_{a \in A_s} x_{d_x}(s, a) = \sum_{j \in S} \alpha(j) \sum_{n=1}^{\infty} \lambda^{n-1} P^{d_x^{\infty}}(X_n = s | X_1 = j),$$

it follows from (6.9.6) and (6.9.7) that

$$x_{d_x}(s, a) = \sum_{a \in A_s} x_{d_x}(s, a) \frac{x(s, a)}{\sum_{a' \in A_s} x_{d_x}(s, a')} = x(s, a). \qquad \square$$

The quantity $x(s, a)$ defined in (6.9.3) represents the total discounted joint probability under initial-state distribution $\{\alpha(j)\}$ that the system occupies state s and chooses action a. When multiplied by $r(s, a)$ and summed over all state-action pairs, it yields the expected total discounted reward of policy $(d_x)^{\infty}$. That is,

$$\sum_{s \in S} \alpha(s) v_{\lambda}^{d_x^{\infty}}(s) = \sum_{s \in S} \sum_{a \in A_s} x(s, a) r(s, a). \tag{6.9.8}$$

Combining parts (a) and (b) of Theorem 6.9.1 implies that for any $d \in D^{MR}$

$$\sum_{s \in S} \alpha(s) v_{\lambda}^{d^{\infty}}(s) = \sum_{s \in S} \sum_{a \in A_s} x_d(s, a) r(s, a) \tag{6.9.9}$$

The first part of Theorem 6.9.1 identifies feasible solutions with randomized stationary policies. Part (b) shows how to generate a randomized policy from a feasible solution and, through the equivalence of x_{d_x} and x, establishes a one-to-one relationship between stationary randomized policies and solutions in the following sense.

Let X denote the set of $x(s, a)$ satisfying (6.9.2) and $x(s, a) \geq 0$ for $a \in A_s$ and $s \in S$; $\mathscr{X}$ denote the mapping from x into D^{MR} defined by (6.9.4) and $\mathscr{D}$ denote the mapping from D^{MR} into X defined by (6.9.3). Theorem 6.9.1 implies the following result.

Corollary 6.9.2. $\mathscr{X}$ and $\mathscr{D}$ are 1-1 and onto so that

$$\mathscr{X} = \mathscr{D}^{-1} \quad \text{and} \quad \mathscr{D} = \mathscr{X}^{-1}.$$

Recall that x is a *basic feasible solution* of an LP if it cannot be expressed as a nontrivial convex combination of any other feasible solutions of the LP. A key property of basic feasible solutions is that when a LP with m rows has a bounded optimal solution, then any basic feasible solution has at most m positive components. We now establish a one-to-one relationship between stationary deterministic policies and extreme point (basic feasible) solutions to the dual LP.

Proposition 6.9.3.

a. Let x be a basic feasible solution to the dual LP. Then $d_x \in D^{MD}$.
b. Suppose that $d \in D^{MD}$, then x_d is a basic feasible solution to the dual LP.

Proof.

a. In the proof of Theorem 6.9.1, we showed that, for each $s \in S$, $\sum_{a \in A_s} x(s, a) > 0$. Since the dual LP has $|S|$ rows, for each s, $x(s, a_s) > 0$ for only one $a_s \in A_s$. Thus, d_x as defined by (6.9.4) equals 1 or 0, so it is deterministic.
b. Suppose $x_d(s, a)$ is feasible but not basic. Then, there exist basic feasible solutions w and z, and β, $0 \leq \beta \leq 1$, such that $x(s, a) = \beta w(s, a) + (1 - \beta) z(s, a)$. Since $w(s, a)$ and $z(s, a)$ are basic and distinct, it follows from the first conclusion in part (b) of Theorem 6.9.1 that $\sum_{a \in A_s} w(s, a) > 0$ and $\sum_{a \in A_s} z(s, a) > 0$ for each $s \in S$. Since $w \neq z$, for some $s \in S$, there exists a and a' for which $w(s, a) > 0$ and $w(s, a') > 0$. Consequently x must have more than one nonzero entry for this $s \in S$ and d_x must be randomized. Since $d = d_x$, this contradicts the hypothesis that d is deterministic. $\square$

6.9.3 Optimal Solutions and Optimal Policies

We now use results from the previous section to establish the relationship between optimal policies and optimal solutions of the dual linear program.

Theorem 6.9.4. Suppose Assumption 6.0.2 holds. Then:

a. There exists a bounded optimal basic feasible solution x^* to the dual LP.
b. Suppose x^* is an optimal solution to the dual linear program, then $(d_{x^*})^\infty$ is an optimal policy.
c. Suppose x^* is an optimal basic solution to the dual linear program, then $(d_{x^*})^\infty$ is an deterministic optimal policy.

d. Suppose $(d^*)^\infty$ is an optimal policy for the discounted Markov decision problem. Then x_{d^*} is an optimal solution for the dual linear program.

e. Suppose $(d^*)^\infty$ is a deterministic optimal policy for the discounted Markov decision problem. Then x_{d^*} is an optimal basic solution for the dual linear program.

Proof. Let x be a feasible solution to the dual. From Corollary 6.9.2, there exists a decision rule $d \in D^{MR}$ for which $x = x_d$. From (6.9.9) and the assumed boundedness of $r(s, a)$,

$$\sum_{s \in S} \sum_{a \in A_s} x_d(s, a) r(s, a) = \sum_{s \in S} \alpha(s) v_\lambda^{d^\infty}(s) \le (1 - \lambda)^{-1} \max_{s \in S} \max_{a \in A_s} |r(s, a)|$$

so that the dual objective function is bounded. Therefore part (a) follows from Theorem D.1a.

We establish (b) as follows. Since the primal linear program is the dual of the dual, from Theorem D.1(c) and (d), the primal linear program has an optimal solution $v^*(s)$ for which

$$\sum_{s \in S} \alpha(s) v^*(s) = \sum_{s \in S} \sum_{a \in A_s} x^*(s, a) r(s, a) = \sum_{s \in S} \alpha(s) v_\lambda^{(d_{x^*})^\infty}(s)$$

where the second equality follows from (6.9.8). Since $v^*(s) \ge v_\lambda^*(s)$ (Theorem 6.2.2a) and $\alpha(s) > 0$, it follows from the above equality and the definition of $v_\lambda^*(s)$ that $(d_{x^*})^\infty$ is optimal. Part (c) follows from (b) and Proposition 6.9.4(a).

We now prove (d). Let $x(s, a)$ be a feasible solution to the dual. Then Corollary 6.9.2 implies there exists a $d \in D^{MR}$ for which $x_d(s, a) = x(s, a)$ for all $s \in S$ and $a \in A_s$. From (6.9.9), the optimality of $(d^*)^\infty$ and the positivity of α

$$\sum_{s \in S} \sum_{a \in A_s} x_{d^*}(s, a) r(s, a) = \sum_{s \in S} \alpha(s) v_\lambda^{(d^*)^\infty}(s) \ge \sum_{s \in S} \alpha(s) v_\lambda^{d^\infty}(s)$$

$$= \sum_{s \in S} \sum_{a \in A_s} x_d(s, a) r(s, a) = \sum_{s \in S} \sum_{a \in A_s} x(s, a) r(s, a)$$

which establishes the optimality of x_{d^*}. Part (e) follows from (d) and Proposition 6.9.3(b).   □

You might wonder about the effect of α on the optimal policy. We now show that the optimal policy does not depend on its value. Note that here we use the expression *basis* to refer to the columns of the constraint matrix which determine the basic feasible solution.

Proposition 6.9.5. For any positive vector α, the dual linear program has the same optimal basis. Hence, $(d_{x^*})^\infty$ does not depend on α.

Proof. From LP sensitivity analysis theory, altering α affects only the feasibility of the current basis, not its optimality. Hence the result will follow if we show that the basis corresponding to x^* remains feasible for any positive α. Since $x^*(s, a) = 0$ if

$a \neq d_{x^*}(s)$, (6.9.2) may be expressed in matrix form as

$$(x^*)^T (I - \lambda P_{d_{x^*}}) = \alpha^T.$$

From Lemma 6.1.2(c), it follows that, for any $\alpha > 0$, $\alpha^T (I - \lambda P_{d_{x^*}})^{-1} \geq 0$. Thus the same basis remains feasible for all $\alpha > 0$. Consequently, from Theorem 6.9.1(b), $(d_{x^*})^\infty$ is optimal for all $\alpha > 0$. $\square$

6.9.4 An Example

We illustrate the LP formulation and some of the above results by again analyzing the example in Fig. 3.1.1.

Example 6.9.1. For this model the primal LP may be written as follows. Minimize

$$\alpha(s_1)v(s_1) + \alpha(s_2)v(s_2)$$

subject to

$$(1 - 0.5\lambda)v(s_1) - 0.5\lambda v(s_2) \geq 5,$$

$$v(s_1) - \lambda v(s_2) \geq 10,$$

$$(1 - \lambda)v(s_2) \geq -1,$$

while the dual LP is given by the following. Maximize

$$5x(s_1, a_{1,1}) + 10x(s_1, a_{1,2}) - x(s_2, a_{2,1})$$

subject to

$$(1 - 0.5\lambda)x(s_1, a_{1,1}) + x(s_1, a_{1,2}) = \alpha(s_1)$$

$$- 0.5\lambda x(s_1, a_{1,1}) - \lambda x(s_1, a_{1,2}) + (1 - \lambda)x(s_2, a_{2,1}) = \alpha(s_2), \quad (6.9.10)$$

and $x(s_1, a_{1,1}) \geq 0$, $x(s_1, a_{1,2}) \geq 0$, and $x(s_2, a_{2,1}) \geq 0$.

Solving the dual with $\lambda = 0.95$ and $\alpha(s_1) = \alpha(s_2) = 0.5$, we find an objective function value of -14.2857, dual solution $x^*(s_1, a_{1,1}) = 0.9523$, $x^*(s_1, a_{1,2}) = 0$, $x^*(s_2, a_{1,2}) = 19.0476$, and primal solution $v^*(s_1) = -8.5714$ and $v^*(s_2) = -20$. From this we obtain the optimal policy $(d_{x^*})^\infty$, with $d_{x^*}(s_1) = a_{1,1}$ and $d_{x^*}(s_2) = a_{2,1}$. Note that these agree with previous results computed in other ways.

In view of (6.9.8), we can add constraints on the model. Letting $c(s, a)$ denotes a cost associated with choosing action a in state (s), we can introduce the constraint that the expected total discounted cost does not exceed C through the inequality

$$\sum_{s \in S} \sum_{a \in A_s} c(s, a)x(s, a) \leq C. \quad (6.9.11)$$

Note that, to ensure that the policy generated by (6.9.4) from a solution of the dual linear program with this additional constraint is optimal for the constrained model, we need to establish the existence of an optimal *stationary* policy for the constrained model. We defer this analysis to Sec. 8.9, where we study a constrained model under the average reward criterion.

Example 6.9.1 (ctd.). We solve a constrained version of the model in Fig. 3.1.1. Suppose that $c(s_1, a_{1,1}) = 3$, $c(s_1, a_{1,2}) = 2$, and $c(s_2, a_{2,1}) = 1$, and $C = 21$. Then we solve (6.9.10) together with the extra constraint

$$3x(s_1, a_{1,1}) + 2x(s_1, a_{1,2}) + x(s_2, a_{2,1}) \leq 21$$

to obtain $x^*(s_1, a_{1,1}) = 0.3390$, $x^*(s_1, a_{1,2}) = 0.3220$, $x^*(s_2, a_{2,1}) = 19.339$, and an objective function value of -14.4237. In light of (6.9.4), we find this corresponds to the randomized decision rule d^*, in which

$$q_{d^*(s_1)}(a_{1,1}) = 0.512 \quad \text{and} \quad q_{d^*(s_1)}(a_{1,2}) = 0.488.$$

As a consequence of Theorem 6.9.1, under the additional constraint the optimal solution of the dual LP corresponds to the randomized stationary policy $(d^*)^\infty$.

We now discuss the relationship between the simplex algorithm and the dynamic programming solution algorithms, and comment briefly on some computational issues. When the dual problem is solved by the simplex algorithm with block pivoting (one a for each s), it is equivalent to policy iteration. When policy iteration is implemented by changing only the action which gives the maximum improvement over all states, it is equivalent to solving the dual problem by the usual simplex method. Modified policy iteration is equivalent to a variant of the simplex method, in which a basic feasible solution is evaluated by iterative relaxation instead of direct solution.

Numerical results of Koehler (1976) show that modified policy iteration is considerably more attractive computationally than the simplex-method-based linear programming codes for solving discounted Markov decision problems. For example, to solve a model with 2,000 states and 10 actions in each, phase II of the linear programming algorithm implemented using MPSX requires 12 times longer than modified policy iteration of fixed order 3, (LP: 3.34 min, MPI: 0.28 min). A LP formulation of the three-stage tandem queueing model of Ohno and Ichiki described in Sec. 6.6.6 requires 3,779 rows and 43,813 columns. We would expect far greater computational times using LP than the 33.9 sec required by their implementation of modified policy iteration.

Some tradeoffs to consider when considering the use of linear programming to solve discounted MDPs include

1. the additional effort required to generate the linear programming tableau,
2. the inability of standard linear programming codes to incorporate the easily available initial basic feasible solution,
3. the inability of LP algorithms to incorporate information regarding structure of optimal policies,
4. the difficulty of identifying optimal policies through (6.9.4) in large problems in which many $x^*(s, a)$ are near zero,

5. the facility of sensitivity analysis using linear programming, and
6. the ability to include additional constraints in linear programming models.

6.10   COUNTABLE-STATE MODELS

Although most previous results in this chapter apply to countable-state space models, their practical significance is limited because

1. we have assumed bounded rewards, and
2. we have not provided methods for implementing algorithms when the state space is not finite.

In this section we address these issues and illustrate them through examples.

To simplify notation, we assume that $S = \{0, 1, \dots\}$. Results hold for arbitrary countable S and may be extended to models with S equal to $[0, \infty)$ or to an unbounded Borel subset of Euclidean space.

6.10.1   Unbounded Rewards

Countable-state spaces provide natural settings for applications of MDPs to queueing control, inventory management, or economic planning. In these applications, the state represents system content or wealth, so that, *a priori*, there is no obvious upper bound for the system state. Practice often dictates that rewards or costs are nondecreasing (for example, linear) with respect to the state. Hence our Assumption 6.0.2 that $|r(s, a)| \leq M < \infty$ limits application of previous results to these models.

We used the assumption of bounded one-period rewards to show that, for all $d \in D$, $L_d v = r_d + \lambda P_d v$ was bounded whenever v was bounded (that is, L_d: $V \to V$), and most importantly, to establish that L and $\mathcal{L}$ were contraction mappings on V (Proposition 6.2.4). Consequently, the contraction mapping fixed-point theorem (Theorem 6.2.3) established the existence of solutions to the optimality equation, and the convergence of value iteration. We now extend these results to models with unbounded rewards by modifying V so that it includes many unbounded functions, and using the concept of a J-stage contraction mapping. To this end we impose assumptions on rewards and transition probabilities. We state these assumptions in sufficient generality to derive key results and to include a wide range of applications, and then show that they are satisfied when more easily verifiable conditions hold.

Let w be an arbitrary positive real-valued function on S satisfying $\inf_{s \in S} w(s) > 0$, for example, $w(s) = \max(s, 1)$ or $w(s) = \log(s + \delta)$, with $\delta > 1$. Define the *weighted supremum norm* $\| \cdot \|_w$ for real-valued functions v on S by

$$\|v\|_w = \sup_{s \in S} w(s)^{-1} |v(s)|,$$

and let V_w be the space of real-valued functions v on S satisfying $\|v\|_w < \infty$. When $w(s) = 1$ for all $s \in S$, $V_w = V$. We note that convergence in V_w implies pointwise convergence, because if $\|v^n - v\|_w < \varepsilon$ for some $\varepsilon > 0$, then $|v_n(s) - v(s)| < \varepsilon w(s)$

for each $s \in S$. It is not hard to see that every Cauchy sequence of elements in V_w converges to an element of V_w, so that V_w is a Banach space.

Let H denote an $|S| \times |S|$ matrix with (s, j)th component $h(j|s)$. When $H: V_w \rightarrow V_w$, it follows from (C.2) in Appendix C that the norm of H, denoted $\|H\|_w$ satisfies

$$\|H\|_w = \sup_{s \in S} w(s)^{-1} \sum_{j \in S} |h(j|s)| w(j). \tag{6.10.1}$$

We analyze the countable-state model under the following assumption.

Assumption 6.10.1. There exists a constant $\mu < \infty$ such that

$$\sup_{a \in A_s} |r(s, a)| \leq \mu w(s). \tag{6.10.2}$$

Assumption 6.10.2

a. There exists a constant κ, $0 \leq \kappa < \infty$, for which

$$\sum_{j \in S} p(j|s, a) w(j) \leq \kappa w(s) \tag{6.10.3}$$

for all $a \in A_s$ and $s \in S$.

b. For each λ, $0 \leq \lambda < 1$, there exists an α, $0 \leq \alpha < 1$ and an integer J such that

$$\lambda^J \sum_{j \in S} P_\pi^J(j|s) w(j) \leq \alpha w(s) \tag{6.10.4}$$

for all $\pi = (d_1, \ldots, d_J)$ where $d_k \in D^{MD}$; $1 \leq k \leq J$.

We may express (6.10.2) in vector notation as

$$\|r_d\|_w \leq \mu;$$

(6.10.3) as

$$P_d w \leq \kappa w,$$

or, equivalently, as

$$\|P_d\|_w \leq \kappa \tag{6.10.5}$$

for all $d \in D^{MD}$; and (6.10.4) as

$$\lambda^J P_\pi^J w \leq \alpha w,$$

or, equivalently,

$$\|\lambda^J P_\pi^J\|_w \leq \alpha \qquad (6.10.6)$$

for all $\pi \in \Pi^{MD}$.

When (6.10.2) holds, $r(s, a)$ increases by at most rate w in s for any $a \in A_s$. Based on this observation, a suitable choice for w is

$$w(s) = \max\left\{ \sup_{a \in A_s} |r(s, a)|, 1 \right\}$$

in which case $\mu = 1$.

Assumption 6.10.2 may be interpreted in terms of expectations by noting that (6.10.3) and (6.10.4) may be rewritten as

$$E^\pi\{w(X_{n+1})|X_n = s, Y_n = a\} \leq \kappa w(s)$$

for all $s \in S$ and $a \in A_s$ and

$$E^\pi\{w(X_{n+J})|X_n = s\} \leq \alpha\lambda^{-J}w(s)$$

for $s \in S$ and $n = 1, 2, \ldots$. These conditions do not restrict allowable transitions. Any state may be visited, but the probability of reaching a distant state must be small under any policy.

The following simple example, which has reward and transition structure similar to many inventory and queueing models, illustrates these assumptions and provides some insight into when they hold.

Example 6.10.1.　Let $A_s = \{0, 1, 2, \ldots\}$, $r(s, a) = s$, and suppose that $p(j|s, a) = 1$ if $j = s + a$ and 0 otherwise. Taking $w(s) = \max(s, 1)$ shows that (6.10.2) holds with $\mu = 1$. Since

$$\sum_{j \in S} p(j|s, a)w(j) = w(s + a) = s + a,$$

neither (6.10.3) nor (6.10.4) hold. However, if $A_s = \{0, 1, \ldots, M\}$,

$$\sum_{j \in S} P_d(j|s)w(j) = s + a \leq s + M \leq (1 + M)w(s)$$

for all $d \in D^{MD}$, so that (6.10.3) holds with $\kappa = (1 + M)$. For any $\pi \in \Pi^{MD}$,

$$\lambda^J \sum_{j \in S} P_\pi^J(j|s)w(j) \leq \lambda^J(s + MJ) \leq \lambda^J(1 + MJ)w(s).$$

Consequently, for J sufficiently large, $\lambda^J(1 + MJ) < 1$, so that Assumption 6.10.2(b) holds.

Suppose now that $r(s, a) = s^2$, and $A_s = \{0, 1, \ldots, M\}$. Then (6.10.2) holds with $w(s) = \max(s^2, 1)$. Hence

$$\lambda^J \sum_{j \in S} P_\pi^J(j|s) w(j) \leq \lambda^J (s + JM)^2 \leq \lambda^J (1 + 2JM + J^2 M^2) w(s),$$

from which it follows that Assumption 6.10.2 holds.

We leave it as an exercise to show that Assumption 6.10.2 holds with $r(s) = \log(s + \delta)$ for $\delta > 1$ or $r(s) = \beta^{-1} s^\beta$ for $0 < \beta < 1$.

In a model with bounded rewards, Assumptions 6.10.1 and 6.10.2 hold with $w(s) = 1$. When $|r(s, a)| \leq \mu$ for all $a \in A_s$ and $s \in S$ in such a model,

$$-\mu(1 - \lambda)^{-1} \leq v_\lambda^\pi(s) \leq \mu(1 - \lambda)^{-1}$$

for all $s \in S$ and $\pi \in \Pi^{MR}$. We now obtain a similar bound under Assumptions 6.10.1 and 6.10.2. Note that the bound will also be valid for all $\pi \in \Pi^{HR}$.

Proposition 6.10.1. Suppose Assumptions 6.10.1 and 6.10.2 hold. Then, for each $\pi \in \Pi^{MR}$ and $s \in S$,

$$|v_\lambda^\pi(s)| \leq \frac{\mu}{1 - \alpha} \left[1 + \lambda\kappa + \cdots + (\lambda\kappa)^{J-1}\right] w(s) \qquad (6.10.7)$$

and

$$\|v_\lambda^\pi\|_w \leq \frac{\mu}{1 - \alpha} \left[1 + \lambda\kappa + \cdots + (\lambda\kappa)^{J-1}\right]. \qquad (6.10.8)$$

Proof. We derive the upper bounds in (6.10.7). For any $\pi = (d_1, d_2, \ldots) \in \Pi^{MD}$,

$$v_\lambda^\pi = r_{d_1} + \lambda P_{d_1} r_{d_2} + \lambda P_{d_1} P_{d_2} r_{d_3} + \cdots + \lambda^J P_{d_1} \cdots P_{d_J} r_{d_{J+1}} + \cdots$$
$$\leq \mu w + \lambda\kappa\mu w + \lambda^2 \kappa^2 \mu w + \cdots + \lambda^{J-1} \kappa^{J-1} \mu w + \alpha\mu w + \alpha\lambda\kappa\mu w$$
$$+ \cdots + \alpha\lambda^{J-1} \kappa^{J-1} \mu w + \alpha^2 \mu w + \alpha^2 \lambda\kappa\mu w + \cdots,$$

where the inequality follows by applying Assumption 6.10.2b to obtain the bound in terms of α, and then Assumptions 6.10.1 and 6.10.2a to obtain $P_\pi^k r_{d_{k+1}} \leq \mu \lambda^k \kappa^k w$ for $k = 0, 1, \ldots, J - 1$. The result follows immediately from the above inequality. $\square$

In most applications $\kappa > 1$ (cf. Example 6.10.1), so neither L nor $\mathcal{L}$ are contraction mappings on V_w. Consequently, the results of Sec. 6.2 do not apply directly. Instead we show that, under Assumptions 6.10.1 and 6.10.2, L and $\mathcal{L}$ are J-stage contractions on V_w.

We say that an operator T on a Banach space U is a *J-stage contraction* if there exists an integer J and a scalar λ', $0 \leq \lambda' < 1$, such that, for all u and v in U,

$$\|T^J u - T^J v\| \leq \lambda' \|u - v\|.$$

J-stage contraction mappings inherit many of the important properties of contraction mappings. Most notably we have the following theorem.

Theorem 6.10.2. Suppose U is a Banach space, $T: U \to U$ is a J-stage contraction for some $J \geq 1$, and there exists a $B, 0 \leq B < \infty$ for which

$$\|Tu - Tv\| \leq B\|u - v\| \tag{6.10.9}$$

for all u and v in U. Then

    a. there exists a unique v^* in U for which $Tv^* = v^*$; and

    b. for arbitrary v^0 in U, the sequence $\{v^n\}$ defined by $v^{n+1} = Tv^n$ converges to v^*.

Proof. Since T^J is a contraction, Theorem 6.2.3 implies that the subsequence $\{v^{kJ}\}$ converges to v^* with $v^* = T^J v^*$. Then, for any k and $1 \leq j < J$,

$$\|v^{kJ+j} - v^*\| \leq \|v^{kJ} - v^*\| + \|v^{kJ+j} - v^{kJ}\|. \tag{6.10.10}$$

Applying the triangle inequality repeatedly and noting (6.10.9) shows that the second expression on the right-hand side of (6.10.10) satisfies

$$\|v^{kJ+j} - v^{kJ}\| \leq \|T^J\|^k \left[1 + B + B^2 + \cdots + B^{J-1}\right]\|v^1 - v^0\|.$$

Choosing k sufficiently large implies that both expressions on the right-hand side of (6.10.10) can be made arbitrarily small, implying that $\{v^n\}$ converges to v^*. The same argument used in the proof of Theorem 6.2.3 establishes that v^* is a fixed point of T.   □

To apply this result, we show that L and $\mathscr{L}$ are J-stage contractions under Assumptions 6.10.1 and 6.10.2.

Proposition 6.10.3. Suppose Assumptions 6.10.1 and 6.10.2 hold. Then L and $\mathscr{L}$ are J-stage contractions on V_w.

Proof. We prove the result for L only. For each $v \in V_w$, there exists a $d \in D^{MD}$ for which $Lv = r_d + \lambda P_d v$. Consequently $\|Lv\|_w \leq \mu + \lambda\kappa\|v\|_w$, so $Lv \in V_w$.

Let u and v be in V_w and assume $L^J v(s) \geq L^J u(s)$. Upon letting $\pi = (d_1, d_2, \ldots, d_J)$, denote the sequence of decision rules which achieve the maximum in $Lv, L(Lv), \ldots, L(L^{J-1})v$; respectively, we have

$$0 \leq L^J v(s) - L^J u(s) \leq L_{d_J} \cdots L_{d_2} L_{d_1} v(s) - L_{d_J} \cdots L_{d_2} L_{d_1} u(s)$$

$$= \lambda^J P_{d_J} \cdots P_{d_2} P_{d_1}(v - u)(s) \leq \lambda^J P_\pi^J w(s)\|v - u\|_w \leq \alpha\|v - u\|_w.$$

Repeating this argument under the assumption that $L^J u(s) \geq L^J v(s)$ establishes the result.   □

We apply Theorem 6.10.2 to obtain the main result of this section: that under suitable conditions, our earlier results hold for models with unbounded rewards.

Theorem 6.10.4. Let S be a countable set and suppose that Assumptions 6.10.1 and 6.10.2 hold. Then the following holds true.

a. The optimality equation

$$v = \sup_{d \in D} \{r_d + \lambda P_d v\}$$

has a unique solution $v_\lambda^* \in V_w$.

b. For any $\varepsilon > 0$, there exists a $d_\varepsilon \in D$ such that $v_\lambda^{(d_\varepsilon)^\infty} + \varepsilon e \geq v_\lambda^*$.

c. If there exists a $d^* \in D$ satisfying

$$d^* \in \arg\max_{d \in D} \{r_d + \lambda P_d v_\lambda^*\},$$

then $(d^*)^\infty$ is an optimal policy.

d. For any $v^0 \in V_w$, $\lim_{n \to \infty} \|L^n v^0 - v_\lambda^*\| = 0$.

e. For any $v^0 \in V_w$ satisfying $Lv^0 \geq v^0$, the sequence of iterates of policy iteration and modified policy iteration converge to v_λ^*.

Proof. As a consequence of Assumption 6.10.2(a), (6.10.9) holds with $B = \lambda \kappa$. Therefore (a) and (d) follow from Theorem 6.10.2 and Proposition 6.10.3. Part (b) is a restatement of Theorem 6.2.7, and part (c) a restatement of Theorem 6.2.11. The proof of Theorem 6.5.5 establishes (e). □

Thus, in principle, we can solve countable-state MDPs with unbounded rewards using value iteration, policy iteration, or modified policy iteration. In practice, this clearly is not possible, since any algorithm would require an infinite number of operations at each iteration. Remedies include the following.

1. Truncating the state space and bounding the discrepancy between the value of the truncated model and the optimal value for the countable-stage problem.
2. Analytically determining the structure of v^n, and of v^n-improving decision rules, and then developing special algorithms that only evaluate decision rules with that structure.
3. Showing that optimal policies partition the set of states into finitely many (usually one) recurrent classes and a countable set of transient states, and then solving a problem on a finite set of states which is guaranteed to include the set of recurrent classes under the optimal policy.

Section 6.10.2 provides a framework for analyzing truncated models, while the second approach above is illustrated in Sec. 6.11. We leave investigation of the third approach to you.

We conclude this section by providing some more easily verifiable conditions which imply Assumption 6.10.2.

Proposition 6.10.5

a. Suppose there exists a constant $L > 0$, for which

$$\sum_{j \in S} p(j|s, a) w(j) \le w(s) + L \qquad (6.10.11)$$

for $a \in A_s$ and $s \in S$. Then Assumption 6.10.2 holds.

b. Suppose there exists an integer M and a constant $L > 0$, for which

$$\sum_{j \in S} p(j|s, a) y(j)^k \le [y(s) + L]^k \qquad (6.10.12)$$

for $a \in A_s$, $s \in S$, and $k = 1, 2, \ldots, M$, where $y(s) = w(s)^{1/M}$. Then Assumption 6.10.2 holds.

c. Suppose that Assumption 6.10.1 holds with a function $w(s)$ which is a restriction of a differentiable concave nondecreasing function on $[0, \infty)$ to $\{0, 1, 2, \ldots\}$ and

$$\sum_{j \in S} j p(j|s, a) \le K \qquad (6.10.13)$$

for all $a \in A_s$ and $s \in S$ and some $K > 0$. Then Assumption 6.10.2 holds.

Proof. Since (a) is a special case of (b) with $M = 1$, we prove (b). Since

$$\sum_{j \in S} p(j|s, a) w(j) \le \left[w(s)^{1/M} + L \right]^M \le \left(1 + \frac{L}{\inf_{s \in S} w(s)} \right)^M w(s).$$

Assumption 6.10.2a holds with $\kappa = (1 + L/\inf_{s \in S} w(s))^M$.

We now show by induction that (6.10.12) implies that

$$\sum_{j \in S} P_\pi^n(j|s) y^M(j) \le (y(s) + nL)^M \qquad (6.10.14)$$

for all $\pi \in \Pi^{MD}$, $s \in S$ and n. By hypothesis, (6.10.14) holds with $n = 1$. Assume it holds for $k = 1, 2, \ldots, n - 1$. Choose $\pi = (d_1, d_2, \ldots) \in \Pi^{MD}$. Then

$$\sum_{j \in S} P_\pi^n(j|s) y^M(j) \le \sum_{j \in S} P_{d_1}(j|s)(y(j) + (n - 1)L)^M$$

$$= \sum_{j \in S} P_{d_1}(j|s) \left\{ \sum_{i=0}^{M} \binom{M}{i} [y(j)]^i [(n - 1)L]^{M-i} \right\}$$

$$\le \sum_{i=0}^{M} \binom{M}{i} [y(s) + L]^i [(n - 1)L]^{M-i}$$

$$= (y(s) + nL)^M,$$

where $\binom{M}{i}$ denotes a binomial coefficient and the inequality follows from (6.10.12).

Thus (6.10.14) holds for all n. Therefore

$$\sum_{j \in S} \lambda^n P_\pi^n(j|s) y^M(j) \le \lambda^n (y(s) + nL)^M \le \lambda^n (1 + nL/y')^M [y(s)]^M,$$

where $y' = \inf_{s \in S} y(s)$. Choosing J so that $\alpha \equiv \lambda^J (1 + JL/y')^M < 1$ shows that Assumption 6.10.2(b) holds under hypothesis (b).

We establish part (c) as follows. From the gradient inequality and the hypothesis that $w(s)$ is concave increasing, which implies that the derivative $w'(s)$ is nonincreasing and $w'(s) \ge 0$, it follows that, for all $s \in S$ and $j \ge s$,

$$w(j) \le w(s) + w'(s)(j - s) \le w(s) + w'(0)(j - s) \le w(s) + w'(0)j.$$

Therefore

$$\sum_{j \in S} p(j|s, a) w(j) \le \sum_{j \in S} p(j|s, a)[w(s) + w'(0)j] \le w(s) + w'(0)K,$$

where K is defined in (6.10.13). Therefore the hypothesis of part (a) of the theorem is satisfied, from which the result follows.   □

The following example applies the above result in the context of an inventory control model.

Example 6.10.2.   We consider a stationary unbounded version of the inventory model of Sec. 3.2 in which the objective is to maximize expected total discounted income (revenue from product sold less ordering and holding costs). Notation and timing of events follows Sec. 3.2. For this model we let $S = \{0, 1, 2, \dots\}$, $A_s = \{0, 1, \dots, M'\}$, and

$$p(j|s, a) = \begin{cases} 0 & j > s + a \\ p_{s+a-j} & s + a \ge j > 0 \\ q_{s+a} & j = 0, \end{cases}$$

where p_k denotes the probability of a demand of k units in any period, and q_k the probability of demand of at least k units in any period. The reward satisfies

$$r(s, a) = F(s + a) - O(a) - h \cdot (s + a),$$

where

$$F(s + a) = \sum_{j=0}^{s+a-1} bjp_j + b \cdot (s + a)q_{s+a},$$

with $b > 0$ representing the per unit price, $O(a) = K + ca$ for $a > 0$ and $O(0) = 0$ the ordering cost, and $h > 0$ the cost of storing one unit of product for one period.

Since $A_s = \{0, 1, \ldots, M'\}$

$$|r(s, a)| \le b \cdot (s + M') + K + cM' + h \cdot (s + M') \equiv \delta + \beta s,$$

with $\delta = K + bM' + cM' + hM'$ and $\beta = b + h$. Since $\delta > 0$ we can choose $w(s) = \delta + \beta s$, so that Assumption 6.10.1 holds with $\mu = 1$. Since

$$\sum_{j=0}^{\infty} w(j)p(j|s, a) = \sum_{j=1}^{s+a} w(j)p_{s+a-j} + w(0)q_{s+a}$$

$$= \sum_{j=0}^{s+a-1} w(s + a - j)p_j + w(0)q_{s+a}$$

$$= \delta + \beta \sum_{j=0}^{s+a-1} (s + a - j)p_j \le \delta + \beta(s + a) \le w(s) + \beta M'$$

(6.10.11) holds with $L = \beta M'$, so Proposition 6.10.5(a) establishes Assumption 6.10.2.

The assumption that $M' < \infty$ need not restrict applicability of this model. To see this, let s be an arbitrary state and suppose A_s contains arbitrarily large components; then, for a_s sufficiently large, $O(a_s) + h \cdot (s + a_s) > F(s + a_s)$, so that $r(s, a_s) < 0$. Consequently, we would never order a quantity larger than a_s in state s. We can choose set $M' = \max_{s \in S} a_s < \infty$.

*6.10.2   FINITE-STATE APPROXIMATIONS TO COUNTABLE-STATE DISCOUNTED MODELS

This section investigates using finite-state approximations to countable-state MDPs. In it, we provide conditions under which the optimal value for the finite-state approximation converges to that of the countable-state problem. We also suggest a computational approach based on finite-state approximations.

In this section, Assumptions 6.10.1 and 6.10.2 hold. The following simple example illustrates concepts to be discussed here.

Example 6.10.3. Let S be the non-negative integers. $A_s = \{a_s\}$, $r(s, a_s) = \mu$, and $p(j|s, a_s) = 1$ if $j = s + 1$, and 0 otherwise. Then for any $s, v_\lambda^*(s) = \mu(1 - \lambda)^{-1}$. Suppose we wish to truncate the model at state N by replacing the state space by $\{0, 1, \ldots, N\}$ and action a_N by an action a_N' for which $r(N, a_N') = 0$ and $p(N|N, a_N') = 1$. Let $v_*^{N,0}$ denote the expected total discounted reward of the truncated model. (We justify this notation below.) Then $v_*^{N,0}(0) = \mu(1 - \lambda)^{-1}(1 - \lambda^N)$. Since

$$v_\lambda^*(0) - v_*^{N,0}(0) = \mu\lambda^N(1 - \lambda)^{-1},$$

by choosing N sufficiently large, we can assure that $v_\lambda^*(0) - v_*^{N,0}(0)$ is arbitrarily small. Note also that $v_*^{N,0}(0) < v_\lambda^*(0)$ for all N, and that for each state the value for the truncated model converges monotonically to v_λ^*.

We have chosen $r(s, a)$ bounded; we may perform a similar analysis for unbounded $r(s, a_s)$, such as $r(s, a_s) = s^2$.

Let $S_N = \{0, 1, \ldots, N\}$ denote a truncation of the state space to the first $N + 1$ states. Fix $u \in V_w$ (for example, $u = 0$), and define, for $v \in V_w$,

$$v^{N,u}(s) = \begin{cases} v(s) & s \leq N \\ u(s) & s > N. \end{cases}$$

For $d \in D$, let $L_d: V_w \to V_w$ be defined by $L_d v = r_d + \lambda P_d v$. Define the operator $L_d^{N,u}: V_w \to V_w$ by

$$L_d^{N,u}v(s) = \begin{cases} r_d(s) + \lambda \sum_{j \leq N} p_d(j|s)v(j) + \lambda \sum_{j > N} p_d(j|s)u(j) & s \leq N \\ u(s) & s > N \end{cases}$$

The above notation is the most transparent, and the results herein of most practical significance when $u = 0$. In that case,

$$L_d^{N,0}v(s) = r(s, d(s)) + \lambda \sum_{j \leq N} p(j|s, d(s))v(j) \qquad s \leq N$$

and $L_d^{N,0}v(s) = 0$ for $s > N$. Of course, for these results to be of practical significance, we would choose u so that $\sum_{j > N} p(j|s, a)u(j)$ may be easily evaluated.

For fixed N, $d \in D$, and $u \in V_w$, $L_d^{N,u}$ is an N-stage contraction on V_w so, as a consequence of Theorem 6.10.2, it has a unique fixed point in V_w. Denote this fixed point by $v_d^{N,u}$. Note that $v_d^{N,u}(s) = u(s)$ for $s > N$. Refer to $v_d^{N,u}$ as an N-state approximation to $v_\lambda^{d^\infty}$.

Let $D_N \equiv \times_{s \leq N} A_s$ denote the set of deterministic Markovian decision rules on S_N. Let $v_*^{N,u} = \sup_{d \in D_N} v_d^{N,u}$. Define the operator $L^{N,u}: V_w \to V_w$ by

$$L^{N,u}v \equiv \max_{d \in D_N} L_d^{N,u}v \qquad (6.10.15)$$

and $\mathscr{L}^{N,u}$ to be the same operator with "sup" replacing "max" when necessary. We can easily show that, under Assumptions 6.10.1 and 6.10.2, $\mathscr{L}^{N,u}$ is an N-stage contraction on V_w, so that it has a unique fixed point $v_*^{N,u}$.

We now provide conditions under which an N-state approximation converges to $v_\lambda^{d^\infty}$ for a fixed decision rule d. The following lemma relates N and $N - 1$ state approximations.

Lemma 6.10.6. Suppose for some $u \in V_w$ that $L_d u \geq (\leq)u$. Then, for all k and N, $(L_d^{N,u})^k u \geq (\leq) (L_d^{N-1,u})^k u$.

Proof. We prove the result for the inequality $\geq$ only. A proof in the other case is identical. The proof is by induction on k. We show inductively that $(L_d^{N,u})^k u \geq (L_d^{N-1,u})^k u \geq u$.

Consider the case $k = 1$. Note that $(L_d^{N,u})u$ and $(L_d^{N-1,u})u$ differ in state N only. Then, by hypothesis,

$$L_d^{N,u}u(N) = r_d(N) + \lambda \sum_{j=0}^{\infty} p_d(j|N)u(j) = L_d u(N) \geq u(N) = L_d^{N-1,u}u(N),$$

so the induction hypothesis is satisfied in this case.

Suppose now that $(L_d^{N,u})^k u \geq (L_d^{N-1,u})^k u \geq u$ for $k < K$. For $s > N$,

$$\left(L_d^{N,u}\right)^K u(s) = \left(L_d^{N-1,u}\right)^K u(s) = u(s),$$

while, for $s \leq N$,

$$\left(L_d^{N,u}\right)^K u(s) = r_d(s) + \lambda \sum_{j \leq N} p_d(j|s)\left(L_d^{N,u}\right)^{K-1} u(j) + \lambda \sum_{j > N} p_d(j|s)u(j)$$

$$\geq r_d(s) + \lambda \sum_{j \leq N-1} p_d(j|s)\left(L_d^{N-1,u}\right)^{K-1} u(j) + \lambda \sum_{j > N-1} p_d(j|s)u(j)$$

$$= \left(L_d^{N-1,u}\right)^K u(s),$$

so the first inequality in the induction hypothesis holds. Combining the induction hypothesis that $(L_d^{N-1,u})^{K-1} u(s) \geq u(s)$ with the representation for $(L_d^{N-1,u})^K u(s)$ above shows that $(L_d^{N-1,u})^K u \geq u$, which establishes the validity of the induction hypothesis. $\square$

Proposition 6.10.7. Suppose Assumptions 6.10.1 and 6.10.2 hold and, for some $u \in V_w$, $L_d u \geq u$ or $L_d u \leq u$. Then, for each $s \in S$, $v_d^{N,u}(s)$ converges monotonically to $v_\lambda^{d^\infty}(s)$.

Proof. Fix $s \in S$ and suppose $L_d u \geq u$. For each N, $(L_d^{N,u})^k u(s)$ converges monotonically to $v_d^{N,u}(s)$ in k. As a consequence of Lemma 6.10.6, $v_d^{N,u}(s)$ is monotonically increasing in N. Because $L_d u \geq u$, Theorem 6.2.2(b) implies that $u(s) \leq v_\lambda^{d^\infty}(s)$, so it follows that $v_d^{N,u}(s) \leq v_\lambda^{d^\infty}(s)$ and, consequently, $\{v_d^{N,u}(s)\}$ converges. Denote this limit by $v'(s)$.

We show that v' is a fixed point of L_d from which the result follows. Choose $\varepsilon > 0$. Then, for any n,

$$0 \leq L_d v'(s) - L_d v_d^{N,u}$$

$$= \lambda \sum_{j \leq n} p_d(j|s)\left[v'(j) - v_d^{N,u}(j)\right] + \lambda \sum_{j > n} p_d(j|s)\left[v'(j) - v_d^{N,u}(j)\right]. \quad (6.10.16)$$

The second summation in (6.10.16) may be bounded by

$$\lambda \sum_{j > n} p_d(j|s)w(j)\|v' - v_d^{N,u}\|_w \leq \lambda \sum_{j > n} p_d(j|s)w(j)\|u - v_\lambda^*\|_w.$$

By Assumption 6.10.2(a), for each $s \in S$, $\sum_{n=0}^{\infty} p_d(j|s)w(j) \leq \kappa w(s)$, so, for each $s \in S$, we can choose an n' so that $\sum_{j > n} p_d(j|s)w(j) < \varepsilon$ for all $n \geq n'$.

Choose $n \geq n'$. Then $v_d^{N,u}$ converges uniformly to v' on $\{0, 1, \ldots, n\}$. Consequently the first summation in (6.10.16) can be made less than ε by choosing N sufficiently large. Therefore $L_d v_d^{N,u}(s)$ converges monotonically to $L_d v'(s)$. By the definition of $v_d^{N,u}$, $L_d v_d^{N,u} = L_d^{N,u} v_d^{N,u} = v_d^{N,u}$. The convergence of $v_d^{N,u}$ to v', shows that v' is a fixed point of L_d. The conclusion follows by the uniqueness of fixed points for J-stage contractions, as established in Theorem 6.10.2. $\qquad\square$

When $L_d u \geq u$, the approximation $v_d^{N,u} \leq v_\lambda^{d^\infty}$ for all N. In this case we refer to $v_d^{N,u}$ as a *lower approximation* to $v_\lambda^{d^\infty}$. On the other hand, when $L_d u \leq u$, $v_d^{N,u} \geq v_\lambda^{d^\infty}$ for all N, so that we refer to $v_d^{N,u}$ as an *upper approximation* to $v_\lambda^{d^\infty}$.

Through Theorems 6.10.8 and 6.10.9, we provide the main results of this section: that the optimal value function for the N-state approximation converges pointwise and monotonically to the optimal value for the original problem. Note that to obtain convergence of the upper approximation in Theorem 6.10.9, we require a uniform bound on the tails of the transition probabilities which is not required when using the lower approximation.

Theorem 6.10.8. Suppose Assumptions 6.10.1 and 6.10.2 hold, and there exists a $u \in V_w$ such that $L_d u \geq u$ for all $d \in D$. Then, for each $s \in S$, $v_*^{N,u}(s)$ converges monotonically from below to $v_\lambda^*(s)$.

Proof. Choose $\varepsilon > 0$. Then, as a result of Theorem 6.10.4(b), there exists a $d_\varepsilon \in D$ such that, for each $s \in S$,

$$v_\lambda^{(d_\varepsilon)^\infty}(s) \leq v_\lambda^*(s) \leq v_\lambda^{(d_\varepsilon)^\infty}(s) + \varepsilon. \qquad (6.10.17)$$

Since, by hypothesis $L_{d_\varepsilon} u \geq u$, Proposition 6.10.7 implies that $v_{d_\varepsilon}^{N,u}(s)$ converges monotonically to $v_\lambda^{(d_\varepsilon)^\infty}(s)$, so that there exists an N' such that, for all $N \geq N'$,

$$v_{d_\varepsilon}^{N,u}(s) \leq v_\lambda^{(d_\varepsilon)^\infty}(s) \leq v_{d_\varepsilon}^{N,u}(s) + \varepsilon. \qquad (6.10.18)$$

For all N,

$$v_{d_\varepsilon}^{N,u}(s) \leq v_*^{N,u}(s) \leq v_\lambda^*(s). \qquad (6.10.19)$$

Combining (6.10.17) and (6.10.18) implies that, for $N \geq N'$,

$$v_{d_\varepsilon}^{N,u}(s) \leq v_\lambda^{(d_\varepsilon)^\infty}(s) \leq v_\lambda^*(s) \leq v_\lambda^{(d_\varepsilon)^\infty}(s) + \varepsilon \leq v_{d_\varepsilon}^{N,u}(s) + 2\varepsilon.$$

From (6.10.19), it follows that

$$v_*^{N,u}(s) \leq v_\lambda^*(s) \leq v_*^{N,u}(s) + 2\varepsilon,$$

which implies the result. $\qquad\square$

We now give conditions on model parameters under which Theorem 6.10.8 holds. The assumption that there exists a u such that $L_d u \geq u$ for all $d \in D$ is a considerably stronger assumption then that of the existence of a u for which $Lu \geq u$. The former condition implies that $u \leq v_\lambda^{d^\infty}$ for all $d \in D$, while the latter implies only that $u \leq v_\lambda^*$.

We may choose $u(s) = -M/(1 - \lambda)$ when either

1. $|r(s, a)| \leq M < \infty$ for all $s \in S$ and $a \in A_s$, in which case Assumptions 6.10.1 and 6.10.2 hold with $w(s) = 1$; or
2. $r(s, a) \geq -M > -\infty$, in which case Assumptions 6.10.1 and 6.10.2 hold with $w(s) = \max\{\sup_{a \in A_s} r(s, a), 1\}$.

The inventory model of Example 6.10.2 satisfies neither of these conditions. In it, the reward $r(s, a)$ is bounded from above but not from below, and it appears that there exists no $u \in V_w$ for which

$$r(s, a) + \lambda \sum_{j \in S} p(j|s, a)u(j) \geq u(s)$$

for all $s \in S$ and $a \in A_s$.

In such cases we can still obtain convergence of $v_*^{N, u}(s)$ to $v_\lambda^*(s)$, but instead we require an upper approximation. We choose u satisfying $L_d u \leq u$ for all $d \in D$ or, equivalently $Lu \leq u$. Theorem 6.2.2a ensures that such a u is an upper bound for v_λ^*. The following example from Fox (1971) shows that further conditions are required to guarantee convergence of these upper approximations.

Example 6.10.4. Let $S = \{0, 1, \ldots\}$, $A_s = \{0, 1, \ldots\}$ for all $s \in S$, $r(s, a) = -0.5$, $p(j|s, a) = 1$, if $j = s + a$ and 0 otherwise, and $\lambda = 0.5$. Then

$$L_d v(s) = -0.5 + 0.5 v(s + d(s)).$$

Clearly any $u(s) \leq -1$ satisfies $L_d u \geq u$ for all $d \in D$, in which case the conclusions of Theorem 6.10.8 are valid. Suppose instead we choose a u for which $L_d u \leq u$ for all $d \in D$, for example $u = 0$. Then $v_*^{N, 0}(s) = -0.5$ for $s \leq N$, but $v_\lambda^*(s) = -1$, so that $v_*^{N, 0}$ does not converge to v_λ^* and the conclusion of Theorem 6.10.8 does not hold with this choice for u. The additional condition in Theorem 6.10.9 excludes this possibility and assures convergence 0 such approximations.

Theorem 6.10.9. Suppose Assumptions 6.10.1 and 6.10.2 hold, and that there exists a $u \in V_w$ such that $L_d u \leq u$ for all $d \in D$. Then, if, for some $s \in S$,

$$\lim_{N \to \infty} \sup_{a \in A_s} \sum_{j > N} p(j|s, a)w(j) = \lim_{N \to \infty} \sup_{d \in D} \sum_{j > N} p_d(j|s)w(j) = 0, \quad (6.10.20)$$

$v_*^{N, u}(s)$ converges monotonically from above to $v_\lambda^*(s)$ as $N \to \infty$.

Proof. As a consequence of Lemma 6.10.6 and Theorem 6.10.5, $v_d^{N, u} \leq v_d^{N-1, u}$ for all $d \in D$. Consequently for all $d \in D$, $v_d^{N, u} \leq v_*^{N-1, u}$, so that $v_\lambda^* \leq v_*^{N, u} \leq v_*^{N-1, u}$. Consequently $v_*^{N, u}$ converges pointwise. Denote the limit by v'.

We now show that v' is a fixed point of $\mathscr{L}$, from which it follows that $v' = v_\lambda^*$. Noting that the supremum of a difference is greater than the difference of suprema, it follows that, for any $s \in S$ and $m \in S$,

$$|\mathscr{L}v_*^{N,u}(s) - \mathscr{L}v'(s)| \leq \sup_{d \in D}\left\{\lambda \sum_{j \leq m} p_d(j|s)|v_*^{N,u}(j) - v'(j)|\right\}$$

$$+ \sup_{d \in D}\left\{\lambda \sum_{j > m} p_d(j|s)|v_*^{N,u}(j) - v'(j)|\right\}.$$

The second expression may be bounded by

$$\sup_{d \in D}\left\{\lambda \sum_{j > m} p_d(j|s)w(j)\|u - v_\lambda^*\|_w\right\},$$

which as a consequence of (6.10.20) can be made arbitrarily small by choosing m sufficiently large. For such an m, the first expression above may be bounded by $\lambda\sum_{j \leq m}|v_*^{N,u}(j) - v'(j)|$, which can also be made arbitrarily small as a consequence of the uniform convergence of $v_*^{N,u}$ to v' on $\{0, 1, \ldots, m\}$. Therefore, $\mathscr{L}v_*^{N,u}(s)$ converges to $\mathscr{L}v'(s)$.

Since $v_*^{N,u} = \mathscr{L}^{N,u}v_*^{N,u} = \mathscr{L}v_*^{N,u}$, the convergence of $v_*^{N,u}$ to v together with the demonstrated convergence of $\mathscr{L}v_*^{N,u}$ to $\mathscr{L}v'$ implies v' is a fixed point of $\mathscr{L}$. Consequently, by Theorem 6.10.4a., $v' = v_\lambda^*$.   □

When $r(s, a)$ is bounded, we can choose $w(s) = 1$ for all $s \in S$ so that (6.10.20) reduces to

$$\lim_{N \to \infty} \sup_{a \in A_s} \sum_{j > N} p(j|s, a) = 0. \tag{6.10.21}$$

This condition does not hold in Example 6.10.4, in which the limit in (6.10.21) equals 1 because arbitrarily large transitions may occur. Condition (6.10.21) holds when A_s is compact and $p(j|s, a)$ is u.s.c. in A; (6.10.20) holds if instead $w(j)p(j|s, a)$ is u.s.c. in a.

When the hypotheses of both Theorems 6.10.8 and 6.10.9 are satisfied, for example when r is bounded, they provide upper and lower bounds on v_λ^* which can be used to determine the adequacy of an approximation. Of course, to use them requires solving two problems, one starting with an upper bound and the other with a lower bound. Note also that, as a consequence of the monotonicity of the approximations, these bounds become tighter with increasing N. We summarize these observations as follows:

Corollary 6.10.10. Suppose Assumptions 6.10.1 and 6.10.2 hold, that there exists ψ and ω in V_w such that $L_d\psi \geq \psi$ and $L_d\omega \leq \omega$ for all $d \in D$, and that (6.10.20) holds. Then, for all N,

$$v_*^{N,\psi} \leq v_*^{N+1,\psi} \leq v_\lambda^* \leq v_*^{N+1,\omega} \leq v_*^{N,\omega}.$$

The results above pertain to the relationship between the optimal solution of a finite-state approximation and the optimal solution of the countable-state model. Instead we might wish to use an iterative approach to obtain an adequate finite-state approximation. The following is one such approach. Investigation of its theoretical and numerical properties remains open.

We distinguish a set of states $S_\nu = \{0, 1, \ldots, \nu\}$ at which we require an accurate approximation to the value of the countable-state model. We choose a $u \in V_w$ such that either $L_d u \geq u$ for all $d \in D$ or that $L_d u \leq u$ for all $d \in D$ and (6.10.20) holds. Recall that D_n denotes the set of all deterministic Markov decision rules on S_n.

Approximate Modified Policy Iteration Algorithm

1. Initialization.
   Set $v^0 = u$, set $n = 0$, choose $\varepsilon > 0$, choose an integer ν and a sequence of nonnegative integers $\{m_n\}$.
2. Policy Improvement.
   Choose d_{n+1} to satisfy

$$d_{n+1} \in \arg\max_{d \in D_n} \{r_d + \lambda P_d v^n\}.$$

3. Partial Approximate Policy Evaluation.
   Increment n by 1 and set

$$v^n = \left(L_{d_n}^{n-1, u}\right)^{m_n + 1} v^{n-1}.$$

4. Stopping criterion.
   If

$$\max_{s \leq \nu}\{v^n(s) - v^{n-1}(s)\} - \min_{s \leq \nu}\{v^n(s) - v^{n-1}(s)\} < \varepsilon,$$

stop and choose

$$d_\nu^*(s) \in \arg\max_{a \in A_s}\left\{r(s, a) + \lambda \sum_{j \in S} p(j|s, a)v^n(j)\right\}$$

for $s \leq \max(\nu, n)$, and arbitrary for $s > \max(\nu, n)$. Otherwise return to step 2.

This algorithm differs from the usual modified policy iteration algorithm with respect to the choice and implementation of policy evaluation and the stopping criterion. At each iteration, it uses an approximation based on an additional state but, throughout, a stopping criterion based only on S_ν.

*6.10.3   Bounds for Approximations

This section provides a method for generating *a priori* bounds on the precision of an N-state approximation. Such bounds enable us to determine the degree of approximation (value of N) necessary to ensure that

$$\left| v_\lambda^*(s) - v_*^{N, u}(s) \right|$$

is sufficiently small for s in a distinguished subset S^* of the state space. Although we concentrate on approximating countable-state models by finite-state models here, the concepts in this section also apply to problems in which the size of the state space has been reduced through state aggregation. The approach is quite general and allows many different bounds.

We assume in this section that $u \in V_w$ has been chosen so that the hypotheses of either Theorems 6.10.8 or 6.10.9 and both Assumptions 6.10.1 and 6.10.2 of Sec. 6.10.1 hold.

Choose an integer ν and let $\{S^i\}$ for $i = 0, \ldots, \nu$ denote a sequence of nested subsets of S of the form $\{0, 1, \ldots, k\}$, with $S^0 = \varnothing$ and $S^{\nu+1} = S$. That is,

$$\varnothing = S^0 \subseteq S^1 \subseteq \cdots \subseteq S^\nu \subseteq S^{\nu+1} = S.$$

We choose this sequence so that $S^* \subseteq S^j$ for some j.

Recall that for two sets A and B with $A \subseteq B$, B/A denotes the complement of A in B. For $1 \leq i \leq \nu$ and $1 \leq k \leq \nu$, let

$$q(S^i, S/S^k) = \sup_{s \in S^i} \sup_{a \in A_s} \left\{ \sum_{j \in S/S^k} p(j|s, a) \right\}.$$

Now define an "aggregate" transition matrix Q with components

$$Q(i, \nu + 1) \equiv q(S^i, S/S^\nu), \quad 1 \leq i \leq \nu,$$
$$Q(i, j) \equiv q(S^i, S/S^{j-1}) - q(S^i, S/S^j), \quad 1 \leq i \leq \nu, 1 \leq j \leq \nu.$$

The quantity $Q(i, j)$ provides a bound on the probability of a transition from any state in S^i to the set S^j/S^{j-1}. Let Q denote the $\nu \times \nu$ matrix with elements $Q(i, j)$. Note Q does not include the components $Q(i, \nu + 1)$.

Let

$$\Delta(j) \equiv \sup_{s \in S^j} \left| v_\lambda^*(s) - v_*^{N, u}(s) \right|$$

and

$$b(j) \equiv \sup_{s \in S^j} \sup_{a \in A_s} \left\{ \lambda \sum_{j \in S/S^\nu} p(j|s, a) \left| v_\lambda^*(j) - u(j) \right| \right\}.$$

The following theorem provides a general approach for generating bounds on Δ.

Theorem 6.10.11. Let $\{S^i\}$; $i = 0, 1, \ldots, \nu + 1$, Δ, Q and b be defined as above, and suppose $0 \leq \lambda < 1$. Then

$$\Delta \leq (I - \lambda Q)^{-1} b. \tag{6.10.22}$$

Proof. Choose N so that $s^\nu = \{0, 1, \ldots, N\}$. Since v_λ^* and $v_*^{N, u}$ are fixed points of $\mathscr{L}$ and $\mathscr{L}^{N, u}$, it follows that, for any $s \in S^\nu$,

$$\left| v_\lambda^*(s) - v_*^{N, u}(s) \right| \leq \sup_{a \in A_s} \left\{ \lambda \sum_{j \in S} p(j|s, a) \left| v_\lambda^*(j) - v_*^{N, u}(j) \right| \right\}.$$

Consequently for $i = 1, 2, \ldots, \nu$,

$$\Delta(i) \le \sup_{s \in S^i} \sup_{a \in A_s} \left\{ \lambda \sum_{j \in S} p(j|s, a) |v_\lambda^*(j) - v_*^{N, u}(j)| \right\}$$

$$\le \sup_{s \in S^i} \sup_{a \in A_s} \left\{ \lambda \sum_{j \in S/S^\nu} p(j|s, a) |v_\lambda^*(j) - u(j)| \right\}$$

$$+ \sup_{s \in S^i} \sup_{a \in A_s} \left\{ \lambda \sum_{k=1}^{\nu} \sum_{j \in S^k/S^{k-1}} p(j|s, a) |v_\lambda^*(j) - v_*^{N, u}(j)| \right\}. \quad (6.10.23)$$

By definition, the first expression in (6.10.23) equals b(i). Since $\Delta(j)$ increases with j, the second expression in (6.10.23) may be bounded by

$$\lambda \sup_{s \in S^i} \sup_{a \in A_s} \left\{ \sum_{k=1}^{\nu} \sum_{j \in S^k/S^{k-1}} p(j|s, a) \Delta(k) \right\}. \quad (6.10.24)$$

For each $s \in S^i$, we apply Lemma 4.7.2 with $v_k = \Delta(k)$, $x_k = Q(i, k)$, and

$$x_k' = \sup_{a \in A_s} \left\{ \sum_{j \in S^k/S^{k-1}} p(j|s, a) \right\}$$

to establish that

$$\lambda \sup_{a \in A_s} \left\{ \sum_{k=1}^{\nu} \sum_{j \in S^k/S^{k-1}} p(j|s, a) \Delta(k) \right\} \le \lambda \sum_{k=1}^{\nu} Q(i, k) \Delta(k).$$

Consequently, (6.10.24) may be bounded by $\lambda \sum_{k=1}^{\nu} Q(i, k) \Delta(k)$. Combining these results yields

$$\Delta(i) - \lambda \sum_{k=1}^{\nu} Q(i, k) \Delta(k) \le b(i).$$

Since the row sums of Q are less than or equal to 1, $(I - \lambda Q)^{-1}$ exists and is nonnegative. The result follows by multiplying both sides of the above inequality by $(I - \lambda Q)^{-1}$. $\square$

Corollary 6.10.12. Inequality (6.10.22) holds with $b(j)$ replaced by

$$b'(j) = \sup_{s \in S^j} \{ \lambda \kappa w(s) \|v_\lambda^* - u\|_w \}.$$

Proof. From the definition of $b(j)$ above and Assumption 6.10.2a, it follows that

$$b(j) \le \sup_{s \in S^j} \sup_{a \in A_s} \left\{ \lambda \sum_{k \in S/S^\nu} p(k|s, a) w(k) \right\} \|v_\lambda^* - u\|_w \le b'(j). \quad (6.10.25)$$

The result follows from the nonnegativity of $(I - \lambda Q)^{-1}$. $\square$

Corollary 6.10.13. Suppose $|r(s, a)| \leq \mu$ for all $s \in S$ and $a \in A_s$. Then (6.10.22) holds with $b(j)$ replaced by

$$b'(j) = Q(j, \nu + 1)\frac{2\mu\lambda}{1 - \lambda}. \tag{6.10.26}$$

Proof. Since r is bounded, choose $w(s) = 1$ and $u = \pm(1 - \lambda)^{-1}\mu e$. Thus

$$b(j) = \lambda Q(j, \nu + 1) \sup_{s \in S/S^\nu} |v_\lambda^*(s) - u(s)| \leq Q(j, \nu + 1)\frac{2\mu\lambda}{1 - \lambda} = b'(j)$$

and the result follows from the nonnegativity of $(I - \lambda Q)^{-1}$. □

We illustrate Theorem 6.10.11 with a simple example. Sec. 6.10.4 provides a more substantial application.

Example 6.10.3 (ctd.). As before, let S denote the nonnegative integers, $A_s = \{a_s\}$, $r(s, a_s) = \mu > 0$, and $p(s + 1|s, a_s) = 1$. Clearly Theorem 6.10.8 applies when we select $u = 0$. Choose $\nu = N$, $S^k = \{0, 1, \ldots, k - 1\}$ and $S^* = \{0\}$. It is easy to see that, for any $k \leq \nu$, $Q(k, k + 1) = 1$ and $Q(k, j) = 0$ for $j \neq k + 1$. Since $r(s, a_s) = \mu$ and $u(s) = 0$, we can choose $b(\nu) = \mu\lambda(1 - \lambda)^{-1}$ and $b(j) = 0$ for $j < \nu$. Consequently,

$$\Delta \leq \begin{bmatrix} 1 & -\lambda & 0 & \cdot & \cdot & & 0 \\ 0 & 1 & -\lambda & 0 & & & 0 \\ \cdot & & \cdot & & & & \cdot \\ \cdot & & & \cdot & & & \cdot \\ \cdot & & & & \cdot & & \cdot \\ 0 & & & 0 & 1 & -\lambda \\ 0 & & & & 0 & 1 \end{bmatrix}^{-1} \begin{bmatrix} 0 \\ 0 \\ \vdots \\ 0 \\ \mu\lambda(1 - \lambda)^{-1} \end{bmatrix}.$$

implying that $\Delta(1) \leq \mu\lambda^N(1 - \lambda)^{-1}$, in agreement with results in Example 6.10.3. Hence we can choose N to ensure that $\Delta(1)$ is sufficiently small.

6.10.4 An Equipment Replacement Model

This section applies results from Sections 6.10.1–6.10.3 to a stationary version of the replacement model of Section 4.7.5.

In that model $S = \{0, 1, \ldots\}$ represents the condition of the equipment at each decision epoch. State 0 corresponds to new equipment and the greater the state index the poorer the condition of the equipment. At each decision epoch the available actions are to replace the equipment ($a = 1$) or continue to operate it as is ($a = 0$). Between decision epochs, the equipment deteriorates by i states with probability $p(i)$ so that the transition probabilities for this model satisfy

$$p(j|s, 0) = \begin{cases} 0 & j < s \\ p(j - s) & j \geq s \end{cases}$$

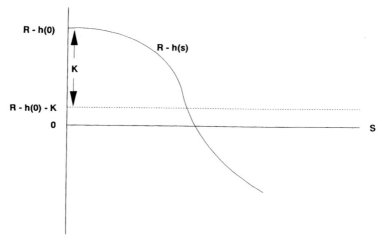

Figure 6.10.1   Graphical representation of reward structure for the replacement model.

and $p(j|s, 1) = p(j)$, $j \geq 0$. The reward is given by

$$r(s, a) = \begin{cases} R - h(s) & a = 0 \\ R - h(0) - K \equiv R' & a = 1 \end{cases}$$

where $R > 0$ denotes the fixed income per period, $K > 0$ the fixed replacement cost and $h(s)$ the expected cost of operating equipment in state s for one period. We assume $h(s)$ is non-decreasing in s, that is, it costs more to operate the equipment in poorer condition. Figure 6.10.1 will clarify subsequent discussion.

Optimality Equations. We begin by showing that Assumptions 6.10.1 and 6.10.2 hold for various choices of $h(s)$. Let $\beta \equiv R + K$. We verify Assumption 6.10.2 by applying Proposition 6.10.5. Let Y denote a random variable with probability distribution $\{p(j)\}$.

Suppose $h(s) = s$. Then Assumption 6.10.1 holds with $w(s) = s + \beta$ and $\mu = 1$. In this case

$$\sum_{j \in S} p(j|s, 0)w(j) = \sum_{j=s}^{\infty} p(j - s)j + \beta = \sum_{k=0}^{\infty} p(k)(k + s) + \beta = s + E[Y] + \beta$$

and

$$\sum_{j \in S} p(j|s, 1)w(j) = \sum_{j=0}^{\infty} p(j)j + \beta = E[Y] + \beta$$

so that Proposition 6.10.5a holds with $L = E[Y]$ whenever $E[Y] < \infty$.

If instead $h(s) = s^2$, Assumption 6.10.1 holds with $\mu = 1$ and $w(s) = (s + \beta)^2$. To verify Assumption 6.10.2 in this case, we apply Proposition 6.10.5(b) with $M = 2$. Since $y(s) = [(s + \beta)^2]^{1/2} = s + \beta$

$$\sum_{j \in S} p(j|s, 0) y(j)^2 = \sum_{k=0}^{\infty} p(k)(s + \beta + k)^2$$

$$= (s + \beta)^2 + E[Y](s + \beta) + E[Y^2] \le \left(s + \beta + E[Y^2]^{1/2}\right)^2$$

whenever $E[Y^2] < \infty$, and also

$$\sum_{j \in S} p(j|s, 0) y(j) = s + \beta + E[Y] \le s + \beta + E[Y^2]^{1/2}$$

Proposition 6.10.5b holds with $L = E[Y^2]^{1/2}$.

When h is the restriction of a concave differentiable nondecreasing function on $[0, \infty)$ to S such as $\log(s + 1)$, and $E[Y] < \infty$, Proposition 6.10.5c implies that Assumption 6.10.2 holds.

Therefore Theorem 6.10.4 implies that with any of the above choices for h, the optimality equation

$$v(s) = \max\left\{R - K - h(0) + \lambda \sum_{j=0}^{\infty} p(j) v(j), R - h(s) + \lambda \sum_{j=0}^{\infty} p(j) v(s + j)\right\}$$

has a unique solution v_λ^* in V_w which can be found (in theory) by value iteration or any other iterative algorithm.

Convergence of N-state Approximations. We now consider the N-state approximations of Section 6.10.2. Since $r(s, a) \le R - h(0)$, $v_\lambda^*(s) \le [R - h(0)]/[1 - \lambda]$. Choose

$$u(s) = [R - h(0)]/[1 - \lambda]$$

for all $s \in S$. We show that with this choice of u, $L_d u \le u$ for all $d \in D$. Pick $s \in S$. When $d(s) = 0$, noting that $h(s) \ge h(0)$ implies

$$L_d u(s) = R - h(s) + \lambda \frac{R - h(0)}{1 - \lambda} \le \frac{R - h(0)}{1 - \lambda} = u(s)$$

This result follows by a similar argument when $d(s) = 1$ because $K \ge 0$.

For any $s \in S$

$$\sum_{j > N} p(j|s, 1) \le \sum_{j > N} p(j|s, 0) = \sum_{j > N} p(j - s)$$

Therefore for any $\varepsilon > 0$, for N' sufficiently large

$$\sum_{j>N'} p(j) < \varepsilon$$

so by choosing $N > s + N'$

$$\max_{a=0,1} \sum_{j>N} p(j|s, a) < \varepsilon$$

which shows that the hypotheses of Theorem 6.10.9 are satisfied. Consequently for each $s \in S$, $v_*^{N,u}$ decreases monotonically in N and converges to v_λ^*.

Suppose now that in addition $h(s)$ is bounded above by γ, for example, $h(s) = \gamma(1 - e^{-s})$ for some $\gamma > 0$. Then for all $d \in D$, $L_d u' \geq u'$ when

$$u'(s) = -(\alpha + K)/(1 - \lambda)$$

Consequently Theorem 6.10.8 implies that $v_*^{N,u'}$ increases monotonically with N and converges to v_λ^*. Thus by solving approximations based on using $u'(s)$ and $u(s)$ whenever $s > N$ we obtain the *a posteriori* bounds

$$v_*^{N,u'} \leq v_\lambda^* \leq v_*^{N,u}$$

A strategy for using such bounds would be to choose an N, solve both N-stage approximations and stop if $v_*^{N,u}(s) - v_*^{N,u'}(s)$ is sufficiently small for all s in a designated subset of S. Otherwise increase N and resolve the problem beginning iterations at $v_*^{N,u'}$ and $v_*^{N,u}$. Note that we need not solve the N-state approximations exactly to obtain bounds on v_λ^*. As a consequence of Proposition 6.3.2, the iterates $(L^{N,u})^k u$ and $(L^{N,u'})^k u'$ are monotone in k. This yields the looser bound

$$\left(L^{N,u'}\right)^k u' \leq v_\lambda^* \leq \left(L^{N,u}\right)^k u$$

which may be applied at each iteration of the respective value iteration algorithms.

A Priori Bounds

We provide *a priori* bounds on the discrepancy between $v_*^{N,u}$ and v_λ^*. We fix n and seek a bound on

$$\Delta(1) = \sup_{s \leq n} \left| v_*^{N,u}(s) - v_\lambda^*(s) \right|$$

We first analyze this model under the following assumptions:

   i. there exists a $\delta < \infty$ for which $\delta \geq h(s) > K + h(0)$ for some $s \in S$, and
   ii. $\{p(j)\}$ has finite support, that is, there exists an $M > 0$ such that $\sum_{j=0}^{M} p(j) = 1$.

We include the lower bound on $h(s)$ in Assumption i. because if $R - h(s) > R - K - h(0)$ for all $s \in S$, the optimal decision would be to never replace the equipment. Assumption i. also implies that $|r(s, a)| \le \rho$ where

$$\rho = \max\{|R - \delta|, |R - h(0) - K|, |R - h(0)|\}.$$

To initiate finite state approximations choose $u(s)$ equal to either $\rho(1 - \lambda)^{-1}$ or $-\rho(1 - \lambda)^{-1}$ for all s.

Choose an integer $K' > 0$ and let $S^0 = \varnothing$, $S^1 = \{0, 1, \ldots, n\}$, $S^2 = \{0, 1, \ldots, n + M\}$ and $S^k = \{0, 1, \ldots, n + (k - 1)M\}$ for $k = 1, 2, \ldots, K'$. We leave it as an exercise to verify that the aggregate $K' \times K'$ transition matrix Q satisfies

$$
Q = \begin{bmatrix}
p(0) & 1 - p(0) & 0 & & 0 \\
0 & p(0) & 1 - p(0) & 0 & 0 \\
\vdots & & & & \vdots \\
\vdots & & & & \vdots \\
0 & & 0 & p(0) & 1 - p(0) \\
0 & & & & p(0)
\end{bmatrix}
\tag{6.10.27}
$$

and $b(j) = 0$ for $j < K'$ and

$$b(K') = \frac{2\lambda\rho}{1 - \lambda}[1 - p(0)].$$

Thus we may obtain a bound on $\Delta(1)$ by inverting $(I - \lambda Q)$ and applying Theorem 6.10.11.

Instead we use a probabilistic approach for bounding $\Delta(1)$ which avoids computing $(I - \lambda Q)^{-1}$. Let $\{Y_n: n = 0, 1, \ldots\}$ denote a Markov chain with state space $\{1, 2, \ldots\}$ and with a transition matrix which when restricted to $\{1, 2, \ldots, K'\}$ equals Q. Let the random variable Z denote the number of transitions required for this chain to reach state $K' + 1$. Then for $j \ge K'$

$$P\{Z = j | Y_0 = 1\} = \binom{K' + j}{j}[1 - p(0)]^{K'}[p(0)]^j$$

and $P\{Z = j | Y_0 = 1\} = 0$ for $j < K'$. In other words, Z conditional on $Y_0 = 1$ follows a negative binomial distribution so that

$$\Delta(1) \le \sum_{j=K'}^{\infty} \lambda^j P\{Z = j | Y_0 = 1\} b(K') = \lambda^{K'} b(K') \sum_{j=K'}^{\infty} \lambda^{j-K'} P\{Z = j | Y_0 = 1\}$$

Noting that the second expression above equals $\lambda^{K'} b(K')$ times the probability generating function of a negative binomial probability distribution, it follows that

$$\Delta(1) \le \left[\frac{1 - p(0)}{1 - \lambda p(0)}\right]^{K'} \frac{2\mu\lambda^{K'+1}(1 - p(0))}{1 - \lambda}
\tag{6.10.28}$$

We evaluate this bound for an arbitrarily chosen set of model parameter values. Suppose $\lambda = 0.9$, $p(0) = 0.2$ and $\rho = 1$, then the bound in (6.10.28) becomes $14.4(0.878)^{K'}$ so that to ensure $\Delta(1) \leq 0.1$ requires $K' = 39$. If $n = 9$ and $M = 5$ this would mean that we would have to solve at most a 205 state problem to ensure that $|v_*^{N,u}(s) - v_\lambda^*(s)| < 0.1$ for $s = 0, 1, \ldots, 9$. Calculations below suggest that this bound is very conservative.

Note that by choosing a finer decomposition of S we may obtain a tighter bound for $\Delta(1)$.

We now provide an priori bounds without the upper bound on $h(s)$ in Assumption i. Since $r(s, a)$ is not bounded from below, we choose $u(s) = [R - h(0)]/[1 - \lambda]$ for all $s \in S$. Let $N' = n + (K' - 1)M$. Corollary 6.10.12 shows that the above analysis remains valid when we choose

$$b(K') = \lambda \kappa w(K') \| v_\lambda^* - u \|_w$$

where κ is implicitly defined through Assumption 6.10.2a. When $h(s) = s$ we use expression (6.10.25) directly to obtain a bound. Since

$$\sum_{j=0}^{\infty} p(j|s, a)w(j) \leq s + E[Y] + \beta$$

where $\beta = R + K$ and

$$|v_\lambda^*(s)| \leq \frac{s + \beta}{1 - \lambda} + \frac{\lambda(\beta + E[Y])}{(1 - \lambda)^2} \tag{6.10.29}$$

it follows that

$$\| v_\lambda^* - u \|_w \leq \frac{1}{1 - \lambda} \left| 1 - R + h(0) + \frac{\lambda(\beta + E[Y])}{\beta(1 - \lambda)} \right|$$

and we can choose

$$b(K') = \frac{1}{1 - \lambda} \left| 1 - R + h(0) + \frac{\lambda(\beta + E[Y])}{\beta(1 - \lambda)} \right| (N' + E[Y] + \beta)$$

We now relax Assumption ii. that $\{p(j)\}$ has finite support. Instead of providing analytic results, we analyze a particular example numerically. We assume that

$$p(j) = \frac{e^{-\alpha}\alpha^j}{j!} \qquad j = 0, 1, \ldots$$

that is, the change of state follows a Poisson distribution with parameter α. We again impose Assumption i. which ensures that $|r(s, a)| \leq \rho$. In our calculations we set $\rho = 1$.

To facilitate generation of Q, we choose $S^i = \{0, 1, \ldots, i - 1\}$ for $i = 1, 2, \ldots, \nu$. We seek a value of $N = \nu - 1$ for which

$$\Delta(1) = \left| v_*^{N,u}(0) - v_\lambda^*(0) \right| < 0.1$$

and investigate the sensitivity of N to γ and λ. To do this, we apply Theorem 6.10.11 and evaluate $(I - \lambda Q)^{-1} b$ numerically. For this model

$$Q = \begin{bmatrix} p(0) & p(1) & p(2) & \cdots & & p(\nu - 1) \\ 0 & p(0) & p(1) & \cdots & & p(\nu - 2) \\ \vdots & & & \cdot & \cdot & \vdots \\ 0 & \cdot & \cdot & & p(0) & p(1) \\ 0 & \cdot & \cdot & & 0 & p(0) \end{bmatrix}$$

and

$$b(k) = \frac{\rho\lambda}{1 - \lambda}\left(1 - \sum_{j=0}^{\nu-k} p(j)\right)$$

for $k = 1, 2, \ldots, \nu$.

Figure 6.10.2 shows the relationship between N, λ, and α. Observe that N increases with both λ and α and, for example, with $\alpha = 0.5$ and $\lambda = 0.9$, $N = 27$. Additional calculations with these parameter choices show that to ensure an approximation that is accurate to 0.1 in states 0 through 10, that is, $\Delta(11) < 0.1$, requires $N = 36$.

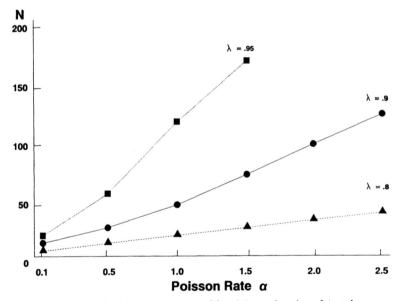

Figure 6.10.2   Value of N which ensures that $\Delta(1) < 0.1$ as a function of λ and α.

6.11   THE OPTIMALITY OF STRUCTURED POLICIES

This section illustrates one of the most significant applications of Markov decision methodology; establishing that an infinite-horizon discounted MDP has an optimal policy of a particular form. In it, we provide a general method for establishing such results and illustrate it with examples. We have already encountered some of the concepts of this section in the context of finite-horizon models. In Sec. 4.7 we used induction to establish optimality of monotone policies under various assumptions on rewards and transition probabilities. Also, results stated without proof in Sec. 6.2.5 relied on concepts to be discussed in this section.

The analysis in this section generalizes the inductive arguments used in Sec. 4.7 by combining them with the results of Sec. 6.3 on convergence of value iteration. In Sec. 4.7, we showed that whenever v^n has a structure which admits a v^n-improving decision rule of a particular form and Lv^N has the same structure as v^n, then, in any finite-horizon model, there exists a optimal policy of that form. To extend this, we focus on determining structures for $\{v^n\}$ which are preserved when passing to the limit. When v_λ^* retains the structure of v^n, there exists a

$$d^* \in \operatorname*{arg\,max}_{d \in D} \{r_d + \lambda P_d v_\lambda^*\} \qquad (6.11.1)$$

with the specified form. Hence we may choose an optimal stationary policy which has this form.

In Sec. 6.11.1 we provide a general framework for determining the structure of and computing optimal policies. We illustrate it with examples in Secs. 6.11.2 and 6.11.3.

6.11.1   A General Framework

Let V^σ and D^σ denote subsets of V (or V_w for some suitably chosen w) and D that contain elements which possess particular properties. We refer to V^σ as the set of *structured values* and D^σ as the set of *structured decision rules*. We choose the structure of these two sets to be compatible in the sense of hypotheses (a) and (b) of the following theorem. A *structured policy* π is a sequence of decision rules $(d_1, d_2, \ldots)$ with $d_n \in D^\sigma$ for $n = 1, 2, \ldots$. Let $\Pi^\sigma \subset \Pi^{MD}$ denote the set of structured policies.

We first consider models in which the maximum of $L_d v$ is attained over D and rewards are bounded.

Theorem 6.11.1.   Suppose for all $v \in V$ there exists a $d \in D$ such that $L_d v = Lv$, that $\|r_d\| \leq M < \infty$ for all $d \in D$, and that

  a. $v \in V^\sigma$ implies $Lv \in V^\sigma$;
  b. $v \in V^\sigma$ implies there exists a $d' \in D^\sigma \cap \operatorname{arg\,max}_{d \in D} L_d v$; and
  c. V^σ is a closed subset of V, that is, for any convergent sequence $\{v^n\} \subset V^\sigma$, $\lim_{n \to \infty} v^n \in V^\sigma$.

Then there exists an optimal stationary policy $(d^*)^\infty$ in Π^σ for which d^* satisfies (6.11.1).

Proof. Choose $v^0 \in V$ and set $v^n = Lv^{n-1}$ for all n. As a consequence of (a), $v^n \in V^\sigma$ for all n and by Theorem 6.3.1(a), v^n converges to v_λ^* in norm. It follows from (c) that $v_\lambda^* \in V^\sigma$. Theorem 6.2.7(b) implies that whenever $d^* \in \arg\max_{d \in D} L_d v_\lambda^*$, $(d^*)^\infty$ is an optimal policy. As a consequence of hypothesis (b), we can choose $d^* \in D^\sigma$. Hence $(d^*)^\infty \in \Pi^\sigma$.   □

Condition (a) ensures that v^n defined through induction (value iteration) remains structured for all n provided that v^0 is structured. Thus by (b) there exists a structured v^n-improving decision rule. Condition (c) ensures that the structure is preserved in the limit. Note that it is not necessary that the value iteration algorithm used in the proof begin at some v^0 in V^σ; all that is required is that, for some N, $v^N \in V^\sigma$. The hypotheses of the above theorem then ensure that $v^n \in V^\sigma$ for all $n \geq N$. For application, conditions (a) and (b) impose restrictions on A_s, $r(s, a)$, and $p(j|s, a)$, and usually require considerable effort to verify. Verification of (c) relies on results from analysis and is usually more straightforward. It requires that the structure of V^σ be preserved under uniform convergence on S.

Figure 6.11.1 below illustrates Theorem 6.11.1. In it we note that $L: V^\sigma \to V^\sigma$, that $L_\delta: V^\sigma \to V^\sigma$ for the structured decision rule $\delta \in D^\sigma$, but that for $v \in V^\sigma$, $L_\gamma v$ need not be in V^σ when γ is not in D^δ.

The following theorem relaxes the assumption that the maximum of $L_d v$ over D be attained. We state it without proof.

Theorem 6.11.2. Suppose that

a. $v \in V^\sigma$ implies $\mathscr{L}v \in V^\sigma$;

b. $v \in V^\sigma$ implies that for any $\varepsilon > 0$ there exists a $d \in D^\sigma$ such that

$$L_d v + \varepsilon e \geq \mathscr{L}v;$$

c. V^σ is a closed subset of V.

Then for any $\varepsilon > 0$ there exists an ε-optimal stationary policy in Π^σ.

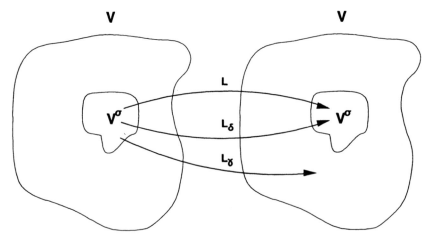

Figure 6.11.1   Illustration of Theorem 6.11.1.

We now extend Theorem 6.11.1 to unbounded rewards. A proof follows that of Theorem 6.11.1, only, instead of using Theorem 6.3.1 to establish convergence of value iteration, appeal to Theorem 6.10.4. Hypotheses (c) requires preservation of structure under pointwise convergence.

Theorem 6.11.3. Suppose that $S = \{0, 1, \ldots\}$, and Assumptions 6.10.1 and 6.10.2 hold, that for all $v \in V_w$ there exists a $d \in D$ such that $L_d v = Lv$, and

a. $v \in V^\sigma$ implies $Lv \in V^\sigma$,
b. $v \in V^\sigma$ implies there exists a $d' \in D^\sigma \cap \arg\max_{d \in D} L_d v$, and
c. V^σ is a closed subset of V_w.

Then there exists an optimal stationary policy $(d^*)^\infty$ in Π^σ for which d^* satisfies (6.11.1).

Theorem 6.11.1 provides the basis for establishing convergence of *structured* value iteration, policy iteration, and modified policy iteration algorithms. Such algorithms account for problem structure by seeking improved decision rules *only* within the set of structured decision rules. The proof of Theorem 6.11.1 showed that the iterates of a value iteration algorithm which maximizes only over the set of structured decision rules converges to v_λ^*. We state that as follows.

Corollary 6.11.4. Suppose that the hypotheses of Theorem 6.11.1 hold. Then for any $v^0 \in V^\sigma$, the sequence

$$v^{n+1} = \max_{d \in D^\sigma} \{r_d + \lambda P_d v^n\}$$

converges in norm to v_λ^*.

We now establish such a result for policy iteration.

Theorem 6.11.5. Suppose for all $v \in V$ there exists a $d \in D$ such that $L_d v = Lv$. Let $\{v^n\}$ denote the sequence of iterates of a policy iteration algorithm in which $d_0 \in D^\sigma$ and the decision rule chosen in the improvement step satisfies

$$d_{n+1} \in \arg\max_{d \in D^\sigma} \{r_d + \lambda P_d v^n\}. \tag{6.11.2}$$

Then if hypotheses (b) and (c) of Theorem 6.11.1 hold, $\|r_d\| \leq M < \infty$ for all $d \in D$, and, in addition, $L_d v \in V^\sigma$ for all $v \in V^\sigma$ and $d \in D^\sigma$, then v^n converges monotonically and in norm to v_λ^*.

Proof. We show that $d_n \in D^\sigma$ implies $v^n \in V^\sigma$. Applying Theorem 6.2.3 to $L_d v = v$ shows that, for any $v \in V$, the sequence $\{u^m\}$ defined by $u^{m+1} = L_{d_n} u^m$ converges to v^n. Pick $u^0 \in V^\sigma$, then, by hypothesis, $u^m \in V^\sigma$ for every m so that, by (c), $\lim_{m \to \infty} u^m = v^n \in V^\sigma$. By (b), there exists a $d'_{n+1} \in \{\arg\max_{d \in D} L_d v^n\} \cap D^\sigma$. Hence d'_{n+1} satisfies (6.11.2).

Therefore beginning policy iteration with $d_0 \in D^\sigma$, and using (6.11.2), we obtain the same sequence $\{v^n\}$ as if we used the original specification for the policy iteration algorithm. Hence, by Theorem 6.4.6, $\{v^n\}$ converges to v_λ^*. $\square$

A similar argument shows that a modified policy iteration algorithm which uses (6.11.2) in the improvement step converges to v_λ^*, provided $v_0 \in V^\sigma$ and $Bv^0 \geq 0$.

Because (6.11.2) can be used instead of the general improvement step, the set of actions to be searched over in a structured algorithm may be *smaller* than in the corresponding general algorithm. Consequently, computational effort will be reduced by using a structured algorithm. This contrasts with a structured linear programming approach in which imposing structure results in adding constraints to the primal formulation and consequently *increases* computational effort.

The operations research, economics, and engineering literature contain numerous applications of Theorems 6.11.1–6.11.3 to problems in queueing control, inventory control, and optimal control theory. We provide many references to such applications in the Bibliographic Remarks section at the end of this chapter. Applications in inventory theory and optimal control usually assume that the state space is one dimensional or multidimensional Euclidean space. The next section provides another illustration of the above concepts.

6.11.2   Optimal Monotone Policies

We apply Theorem 6.11.3 to establish optimality of monotone policies under the subadditivity and ordering assumptions of Sec. 4.7. The observation that the pointwise limit of nondecreasing functions is nondecreasing allows easy extension of the results of Sec. 4.7.2 to infinite-horizon models. *We choose S to be the set of non-negative integers, $A_s = A'$ for all $s \in S$, and impose Assumptions 6.10.1 and 6.10.2 (to ensure convergence of value iteration with unbounded rewards).* Definitions of superadditivity and subadditivity appear in Sec. 4.7.2.

The following theorems provide conditions which ensure optimality of monotone policies. The proof of Theorem 6.11.6 illustrates the choice of V^σ and D^σ. We omit the proof of Theorem 6.11.7 since it is identical to that of Theorem 6.11.6.

Theorem 6.11.6.   Suppose that Assumptions 6.10.1 and 6.10.2 hold and that

a. $r(s, a)$ is nondecreasing in s for all $a \in A'$,

b. $q(k|s, a) \equiv \sum_{j=k}^{\infty} p(j|s, a)$ is nondecreasing in s for all $k \in S$ and $a \in A'$,

c. $r(s, a)$ is a superadditive (subadditive) function on $S \times A'$, and

d. $q(k|s, a)$ is a superadditive (subadditive) function on $S \times A'$ for all $k \in S$.

Then there exists an optimal stationary policy $(d^*)^\infty$ for which $d^*(s)$ is nondecreasing (nonincreasing) in s.

Proof.   We establish this result by verifying the conditions of Theorem 6.11.3 for superadditive $q(k|s, a)$ and $r(s, a)$. Let $V^\sigma = \{v \in V_w : v$ is nondecreasing on $S\}$ and let $D^\sigma = \{d \in D : d$ is nondecreasing on $S\}$. The proof of Proposition 4.7.3 shows that $Lv \in V^\sigma$ whenever $v \in V^\sigma$. The proof of Theorem 4.7.4 establishes that for every $v \in V^\sigma$ there exists a $d' \in D^\sigma \cap \arg\max_{d \in D} L_d v$. Since the (pointwise) limit of any sequence v^n of nondecreasing functions on S is nondecreasing, condition (c) of Theorem 6.11.3 holds. Thus the hypotheses of Theorem 6.11.3 are satisfied so that the result follows.   □

Theorem 6.11.7. Suppose that Assumptions 6.10.1 and 6.10.2 hold and that

1. $r(s, a)$ is nonincreasing in s for all $a \in A'$,
2. $q(k|s, a)$ is nondecreasing in s for all $k \in S$ and $a \in A'$,
3. $r(s, a)$ is a superadditive function on $S \times A'$, and
4. $\sum_{j=0}^{\infty} p(j|s, a)u(j)$ is a superadditive function on $S \times A'$ for nonincreasing u.

Then there exists an optimal stationary policy $(d^*)^{\infty}$ in which $d^*(s)$ is nondecreasing in s.

For applications to queueing and inventory models, the assumption that $A_s = A'$ for all s may be overly restrictive. As in Sec. 4.7, slight modifications of these theorems allow extensions to A_s which satisfy

a. $A_s \subset A'$ for all $s \in S$,
b. $A_s \subset A_{s'}$ for $s' \geq s$, and
c. for each s, $a \in A_s$ and $a' \leq a$ implies $a' \in A_s$.

We now provide an implementation of the structured policy iteration algorithm of the preceding section, which finds a monotone optimal policy. We assume that the hypotheses of either Theorem 6.11.6 or 6.11.7 hold so that such a policy exists. Let $S = \{0, 1, \ldots, N\}$, with N finite and $A_s = A'$ for all $s \in S$. For countable S, we may regard this as solving an N-state approximation. Let D^{σ} denote the set of nondecreasing decision rules.

Monotone Policy Iteration Algorithm

1. Choose $d_0 \in D^{\sigma}$ and set $n = 0$.
2. Find v^n by solving

$$(I - \lambda P_{d_n})v = r_{d_n}.$$

3. Set $s = 0$, $A'_0 = A'$.
   a. Set

   $$A_s^* = \underset{a \in A'_s}{\arg\max} \left\{ r(s, a) + \lambda \sum_{j \in S} p(j|s, a)v^n(j) \right\}.$$

   b. If $s = N$, go to step 3(d), or else, set

   $$A'_{s+1} = \{a \in A'_s : a \geq \max\{a' \in A_s^*\}\}.$$

   c. Substitute $s + 1$ for s and return to step 3(a).
   d. Pick $d_{n+1} \in D^{\sigma} \cap \times_{s \in S} A_s^*$, setting $d_{n+1} = d_n$ if possible.

4. If $d_{n+1} = d_n$, stop and set d^* equal to d_n. Otherwise, replace n by $n + 1$ and return to step 2.

Theorem 6.11.5 ensures that this algorithm finds an optimal solution for finite A'. It differs from the general policy iteration algorithm of Sec. 6.4 in that improvement is carried out over the derived sets A'_s which decrease in size with increasing s. In the worse case, $A'_s = A'$ for all s and the computational effort equals that of the general policy iteration algorithm. When an optimal decision rule is strictly increasing, the sets A'_s will decrease in size with increasing s and hence reduce the number of actions which need to be evaluated in the improvement step 3. If, at some $u \in S$, A^*_u contains a single element, say a^*, then no further maximization is necessary since that action will be optimal at all $s \geq u$. In such a case $d_{n+1}(s) = a^*$ for all $s \geq u$.

Example 6.11.1. We establish the optimality of control limit policies for the stationary equipment replacement model of Sec. 6.10.4 by verifying the hypotheses of Theorem 6.11.7. We choose the operating cost function $h(s)$ so that Assumptions 6.10.1 and 6.10.2 hold. Section 4.7.5 verifies hypotheses 1–4 of Theorem 6.11.7; that is,

1. $r(s, a)$ is nonincreasing in s for $a = 0, 1$,
2. $q(k|s, a)$ is nondecreasing,
3. $r(s, a)$ is superadditive and
4. $\sum_{j=0}^{\infty} p(j|s, a)u(j)$ is superadditive for nonincreasing u.

Consequently, there exists a monotone optimal stationary policy. Since there are only two actions, a control limit policy is optimal, and it can be found using the monotone policy iteration algorithm.

*6.11.3 Continuous and Measurable Optimal Policies

In this section we apply Theorem 6.11.1 to establish the existence of continuous and measurable optimal policies under various assumptions on the states, the actions, the transition probabilities, and the rewards. Results are quite technical, and depend on selection theorems which appear here and in Appendix B. We assume that the models satisfy the technical assumptions of Sec. 2.3.

We first provide conditions under which there exist continuous optimal policies. We will use the following two technical lemmas to show that induction and improvement preserve structure. We refer to Lemma 6.11.9 as a "selection theorem" since it describes the property of a function which selects the "arg max" over Y at each $x \in X$. Such results are fundamental to existence theory in all nondiscrete MDP's.

Lemma 6.11.8. Suppose $g(x, y)$ is a continuous real-valued function on $X \times Y$, where X is a Borel subset and Y a compact subset of complete separable metric spaces. Define

$$f(x) \equiv \max_{y \in Y} g(x, y). \tag{6.11.3}$$

Then $f(x)$ is continuous at each $x \in X$.

Proof. Let $x^* \in X$ and choose a sequence $\{x_n\}$ converging to x^*. Define the sequence $\{y^n\}$, and y^* through $f(x_n) = g(x_n, y_n)$ and $f(x^*) = g(x^*, y^*)$. Then

$$f(x^*) = g(x^*, y^*) = \lim_{n \to \infty} g(x_n, y^*) \leq \lim_{n \to \infty} g(x_n, y_n) = \lim_{n \to \infty} f(x_n). \quad (6.11.4)$$

Now suppose x_n converges to x^* and let y_n be defined as above. Since Y is compact, we may choose a convergent subsequence $\{y'_n\}$. Let its limit be denoted by y'. Therefore

$$f(x^*) \geq g(x^*, y') = \lim_{n \to \infty} g(x'_n, y'_n) = \lim_{n \to \infty} f(x'_n) = \lim_{n \to \infty} f(x_n). \quad (6.11.5)$$

Combining (6.11.4) and (6.11.5) implies v is continuous at x^*. $\qquad\square$

Lemma 6.11.9. Suppose the hypotheses of Lemma 6.11.8, hold and, in addition, $\arg\max_{y \in Y} g(x, y)$ is unique for each $x \in X$. Then

$$h(x) = \arg\max_{y \in Y} g(x, y) \qquad\qquad (6.11.6)$$

is continuous for all $x \in X$.

Proof. Pick $x^* \in X$ and suppose x_n converges to x^* and $y_n \equiv \arg\max_{y \in Y} g(x_n, y)$ converges to y'. Then

$$g(x^*, y') = \lim_{n \to \infty} g(x_n, y_n) = \lim_{n \to \infty} f(x_n) = f(x^*),$$

where the last equality follows from Lemma 6.11.8. Since $\arg\max_{y \in Y} g(x, y)$ is unique, $y' = h(x^*)$, so that h is continuous at x^*. $\qquad\square$

We apply these lemmas to obtain conditions under which there exists a continuous optimal policy. To simplify statement of the following theorem, we assume that $A_s = A'$ for all $s \in S$ where A' is a compact convex subset of Euclidean space. The result holds when A' is a convex subset of a normed linear topological space. Extensions to variable A_s are possible.

Theorem 6.11.10. Suppose S is a Borel subset of Euclidean space:

a. $A_s = A'$ for all $s \in S$, where A' is a compact convex subset of Euclidean space;
b. $r(s, a)$ is a bounded real-valued continuous function on $S \times A'$;
c. $\int_S w(u) p(du|s, a)$ is continuous on $S \times A'$ for any bounded continuous function $w(\cdot)$ on S; and either
d. $r(s, a)$ is strictly concave in a for each $s \in S$ and $\int_S w(u) p(du|s, a)$ is linear in a for each $s \in S$ and bounded continuous function $w(u)$, or
d′. $r(s, a) \geq 0$ and strictly concave in a for each $s \in S$, and $\int_S w(u) p(du|s, a)$ is strictly concave in a for each $s \in S$ and nonnegative bounded continuous function $w(u)$.

Then there exists a continuous optimal stationary policy $(d^*)^\infty$.

Proof. We prove the result under condition (d). We leave the proof under (d') as an exercise. Choose V^σ to be the set of bounded continuous functions on S, and D^σ to be the set of continuous functions from S to A'.

For $v \in V^\sigma$, let

$$g_v(s, a) \equiv r(s, a) + \lambda \int_S v(u) p(du|s, a).$$

As a result of (a), (b) and (c), g_v is continuous on $S \times A'$, so that by Lemma 6.11.8 $\max_{a \in A'} g_v(s, a) = Lv(s)$ is continuous on s. Thus hypothesis (a) of Theorem 6.11.1 is satisfied.

Under (d), $g_v(s, a)$ is strictly concave in a for any $v \in V^\sigma$, so that $\arg\max_{a \in A'} g_v(s, a)$ is unique. Thus, by Lemma 6.11.9, there exists a continuous v-improving decision rule so that hypothesis (b) of Theorem 6.11.1 is satisfied.

Since convergence in V is uniform convergence, the limit of any sequence in V^σ is continuous, so that (c) of Theorem 6.11.1 holds. Hence there exists an optimal stationary policy $(d^*)^\infty$ with d^* in D^σ. □

Example 6.11.2. This example provides an illustration of a simple model which has a continuous optimal control. Let $S = [0, 1]$, $A_s = A' = [0, 1]$ for all $s \in S$, and $r(s, a) = -(s - a)^2$. When $a < 1$, $p(u|s, a) = 0$ if $u < a$ or $u > 1$ and $(1 - a)^{-1}$ if $a \leq u \leq 1$ for all $s \in S$ (i.e., $p(u|s, a)$ denotes the density of a uniform random variable on $[a, 1]$); when $a = 1$, $p(1|s, a) = 1$ for all $s \in S$.

The optimality equation may be written as

$$v(s) = \max_{0 \leq a \leq 1} \left\{ -(s - a)^2 + \frac{\lambda}{1 - a} \int_a^1 v(u) \, du \right\}.$$

It is easy to see that $v_\lambda^*(s) = 0$, and that the optimal policy is $(d^*)^\infty$ where $d^*(s) = s$. This function is continuous (and even linear) in s, as guaranteed by Theorem 6.11.10.

Choosing instead $r(s, a) = a - (0.5s - a)^2$ makes this problem considerably more difficult to solve, but Theorem 6.11.10 once again assures the existence of a continuous optimal policy.

We now state a weaker and more technically demanding result which ensures the existence of a measurable optimal policy. We appeal to Appendix B for appropriate generalizations of Lemmas 6.11.8 and 6.11.9, and note that several extensions of this result appear in the literature.

Theorem 6.11.11. Suppose S is a Borel subset of Euclidean space:

a. $A_s = A'$ for all $s \in S$, where A' is a compact subset of a Polish space;
b. $r(s, a)$ is a bounded real-valued upper semicontinuous function on $S \times A'$; and
c. for any bounded upper semicontinuous function $w(\cdot)$ on S, $\int_S w(u) p(du|s, a)$ is continuous in S for each $a \in A'$ and upper semicontinuous in a for each $s \in S$.

Then there exists an optimal stationary policy $(d^*)^\infty$ with the property that d^* is a measurable function from S to A'.

Proof. Choose V^σ to be the set of bounded u.s.c. functions on S, and D^σ to be the set of measurable functions from S to A'. As a result of Proposition B.1(a), and Proposition B.5(a) in Appendix B, $v \in V^\sigma$ implies $Lv \in V^\sigma$. Theorem B.5(b) guarantees that there exists a $d \in D^\sigma$ such that Theorem 6.11.1(b) holds. Since convergence of $\{v^n\}$ in V is uniform convergence, Proposition B.1(e) implies that condition (c) of Theorem 6.11.1 is satisfied. Hence the result follows from Theorem 6.11.1. □

BIBLIOGRAPHIC REMARKS

Discounted models appear to have been first considered and analyzed in generality by Howard (1960); however, application of this criterion to inventory models dates at least to Bellman, Glicksberg, and Gross (1955). Blackwell (1962 and 1965) provided the fundamental theoretical papers on discounted models. Example 6.2.3, which appeared in Blackwell (1965), motivated considerable theoretical research on measurability issues (cf. Bertseksas and Shreve, 1978 and Dynkin and Yushkevich, 1979).

The use of value iteration in stochastic sequential decision problems originates in the work of Shapley (1953), who applied it to demonstrate the existence of solutions to stochastic sequential games. Blackwell (1965) used the Banach fixed-point theorem to demonstrate the existence of solutions of the optimality equations, and Denardo (1967) extended these ideas to include and solve a wide variety of stochastic sequential decision problems.

Our presentation of the material on value iteration draws from a wide range of sources. Methods based on splitting have their origins in the numerical analysis literature. Excellent references include the books by Varga (1962), Ortega and Rheinboldt (1970), and Young (1971). Hastings (1968 and 1969) and Kushner and Kleinman (1968) independently suggested the use of Gauss-Seidel iteration to accelerate value iteration. Berman and Plemmons (1979, p. 207) note that

"...the Gauss-Seidel method was apparently unknown to Gauss and not recommended by Seidel,"

although

"probably the earliest mention of iterative methods for linear systems dates back to Gauss..."

Kushner and Kleinman (1971) proposed the use of relaxation methods in conjunction with Gauss-Seidel and Jacobi iteration. Reetz (1973) investigated the convergence properties of successive over-relaxation methods and provides conditions under which the convergence rate exceeds that of ordinary value iteration. See Kushner and Dupuis (1992), for a nice overview of this work and its application to approximating stochastic control problems.

Veinott (1969b) develops the concept of a *positively similar* MDP and shows that, given any MDP, there exists a positively similar MDP with maximum norm sufficiently close to the maximal spectral radius. Consequently value iteration algorithms for the positive similar MDP converge geometrically at a rate close to the maximal spectral radius (cf. Theorems 6.3.4 and 6.3.9). Porteus (1975) extends this result in several ways, and provides a unified approach for analyzing iterative methods; our approach in Sec. 6.3.3 is in that spirit.

Van Nunen (1976a and 1976b) derived some of the algorithms in Sec. 6.3.3 using an approach based on stopping times. Porteus and Totten (1978) and Porteus (1980 and 1981) investigate numerical properties of these algorithms for evaluating the discounted reward of fixed policies. Bertsekas and Castanon (1989) provide a promising variant of value iteration which incorporates aggregation methods.

Policy iteration is usually attributed to Howard (1960), and to Bellman (1955 and 1957) who referred to it as "approximation in policy space." Our development in the finite case and the proof of Theorem 6.4.2 are in the spirit of Howard's work. Our approach in the nonfinite case follows Puterman and Brumelle (1979), wherein results are developed in a considerably more abstract setting. Their work is based on Kantorovich's (1939) generalization of Newton's method to function spaces, and relies heavily on results in the numerical analysis literature, most notably Vandergraft (1967) and Ortega and Rheinboldt (1970). The relationship between Newton's method and policy iteration was noted by Kalaba (1959) and Pollatschek and Avi-Itzhak (1969) in other contexts.

The modified policy iteration scheme was suggested by Morton (1971) when he proposed using a variant of value iteration with

"...one full iteration alternating with 5 or 6 "cheap" fixed iterations in the early stages when straight modified [relative in our terminology] value iteration might converge slowly, one full iteration alternating with one or two in the middle stages, switching completely to the cheap iterations after the same policy began to repeat, terminated by a full iteration just to check the policy."

This idea was formalized by van Nunen (1976a) who referred to this algorithm as "value-oriented successive approximation." Puterman and Shin (1978) independently developed this algorithm. They referred to it as "modified policy iteration," reflecting the perspective from which they derived it. Section 6.5 is based on Puterman and Shin's approach. Rothblum (1979) provides an alternative proof of the convergence of modified policy iteration based on J-stage contraction mappings, and Dembo and Haviv (1984) provide some results regarding selection of the order of the algorithm. Ohno and Ichiki (1987) have investigated numerical properties of modified policy iteration and its variants in the context of optimally controlling a tandem queueing system. Example 6.5.1 was provided by van der Wal and van Nunen (1977). The approach and results in Sec. 6.5.3 appear to be original.

Bather (1973a) was the first to use the span seminorm for analyzing Markov decision processes. Our development in Sec. 6.6.1 combines results in Hubner (1977) and Senata (1981, p. 80–83). Theorem 6.6.6 also follows Hubner (1977). The book by Isaacson and Madsen (1976) is a good reference on ergodic theory for discrete Markov chains.

Relative value iteration was first proposed by White (1963) in the context of undiscounted models with average reward criterion. It was subsequently studied by Morton (1971) and Morton and Wecker (1977). In the latter paper, they provide conditions which ensure convergence of relative value iteration together with a wide range of results on ergodicity properties of products of transition matrices.

MacQueen (1967) introduced the concept of bounds on the optimal return. These were extended and improved in the discounted case by Porteus (1971 and 1975), Hubner (1977), Grinold (1973), and Puterman and Shin (1982). Porteus (1971 and 1980a) and Porteus and Totten (1978) suggested and explored the use of extrapolations. Theorem 6.6.3, which relates bounds at subsequent iterations of modified policy iteration, appears to be new.

The material in Sec. 6.7 has its origins in MacQueen (1967), who used Proposition 6.7.1 together with bounds on the optimal value function to identify and eliminate nonoptimal actions. Contributors to research in this area include Hastings and Mello (1973), Grinold (1973), who applied these results to policy iteration, Hubner (1979), who incorporated the delta coefficient in these procedures, and Puterman and Shin (1982), who applied them to modified policy iteration.

One-step-ahead elimination procedures were provided by Hastings and van Nunen (1977) and Hubner (1977) for value iteration, and by Puterman and Shin (1982) for modified policy iteration. White (1978) surveys the use of action elimination algorithms.

The modified policy iteration algorithm in Sec. 6.7.3 combines features of that of Ohno (1980), Puterman and Shin (1982), and Ohno and Ichiki (1987). The latter two references also investigate computational properties of these algorithms. Thomas, Hartley, and Lavercombe (1983) and Hartley, Lavercombe, and Thomas (1986) also look into computational properties of value iteration and policy iteration algorithms. In the first of these papers they observe

"We solved a 2000 state, average of seven actions per state problem in under two minutes, using algorithms in this paper, which suggests that it is the storage of data rather than the speed of solution, that will be the worry in the future."

A paper of Archibald, McKinnon, and Thomas (1993) investigate parallel implementation of value iteration type algorithms with promising results. In a personal communication to me in 1991, L. C. Thomas wrote

"We are able to solve randomly generated problems of up to 300,000 states with no real difficulty."

Shapiro (1968) was the first to consider planning horizon and turnpike results for Markov decision processes. These were subsequently improved upon by Hinderer and Hubner (1977). Our development follows Hinderer and Hubner and draws on background material from Hubner (1977). A characterization for D^* in terms of sets of decision rules that are within ε_n of v^n-improving decision rules has been provided by Federgruen and Schweitzer (1978). Hopp, Bean, and Smith (1987) provide planning horizon results for nonstationary models.

The linear programming formulation of the discounted infinite-horizon MDP was presented by d'Epenoux (1960). The development in Sec. 6.9 draws extensively on Derman (1970) and Kallenberg (1983). References on linear programming compatible with the presentation herein include Chvatal (1983) and Goldfarb and Todd (1989). Eaves (1978) and Koehler (1979) explore the relationship between the complementary pivot theory approach to LP and Markov decision processes.

Harrison (1972) was the first author to deal explicitly with discounted models with unbounded rewards. Subsequent contributions extending this approach include Lippman (1973 and 1975a), van Nunen (1976), Wessels (1977), and van Nunen and Wessels (1978). Cavasoz-Cadena (1986) unifies many of these ideas. Our presentation draws from all these sources. Lippman's and van Nunen and Wessel's papers provide examples which satisfy Assumptions 6.10.1 and 6.10.2. Miller (1974) develops related results in the context of an optimal consumption model.

Fox (1971) proposed solving countable-state models by finite truncation and demonstrated (pointwise) convergence of these approximations for models with bounded rewards. White (1980a) provided a proof of convergence under weaker conditions and *a priori* bounds on the accuracy of the approximation. White (1979, 1980b, and 1982) extended Fox's results to models with unbounded rewards and provided other approximation schemes. Whitt (1978, 1979a, and 1979b) provided a framework for analyzing a wider range of approximations to MDP's and considerably extended results of Fox and White. Our approach combines these results, paying particular attention to convergence conditions which include applications. The numerical properties of these schemes do not appear to have been investigated. Cavasoz-Cadena (1986 and 1987) and Hernandez-Lerma (1986) also analyze finite-state approximations. Seneta (1966 and 1981) in a different context investigated the effect of truncations of countable state Markov chains on ergodic properties.

Bellman, Glicksberg, and Gross (1955), Karlin (1960), Derman (1963), and Iglehart (1963) are some early references which address the optimality of structured policies in infinite-horizon discounted MDP's. The approach in Sec. 6.11.1 was motivated by Schal (1975) and Porteus (1982). Serfozo (1976) provides a basic framework for establishing the optimality of monotone policies for MDP's which has been extended by White (1985a) among others. We refer the interested reader to Chap. 7 of Heyman and Sobel (1984) and Chap. 6 of Bertsekas (1987) for numerous applications of and references to structural results for MDP's.

The material in Sec. 6.11.3 on the optimality of continuous and measurable controls has its roots in Blackwell (1965). The main existence theorems appear in Maitra (1968), Furukawa (1972), and Himmelberg, Parthasarathy and vanVleck (1976). Schal (1972) and Fleming and Rishel (1975, pp. 170–171) provide related results. The books of Hinderer (1970), Bertsekas and Shreve (1978), and Dynkin and Yushkevich (1979) are excellent references for results at this level of generality. Hogan (1973) provides a particularly nice overview on selection theorems in his survey on point-to-set maps and provides the motivation for our proof of Lemmas 6.11.8 and 6.11.9.

PROBLEMS

6.1. Each quarter the marketing manager of a retail store divides customers into two classes based on their purchase behavior in the previous quarter. Denote the classes as L for low and H for high. The manager wishes to determine to which classes of customers he should send quarterly catalogs. The cost of sending a catalog is \$15 per customer and the expected purchase depends on the customer's class and the manager's action. If a customer is in class L and receives a catalog, then the expected purchase in the current quarter is $20, and if a class L customer does not receive a catalog his expected purchase is $10. If a customer is in class H and receives a catalog, then his expected purchase is $50, and if a class H customer does not receive a catalog his expected purchase is $25.

The decision whether or not to send a catalog to a customer also affects the customer's classification in the subsequent quarter. If a customer is class L at the start of the present quarter, then the probability he is in class L at the subsequent quarter is 0.3 if he receives a catalog and 0.5 if he does not. If a customer is class H in the current period, then the probability that he remains in class H in the subsequent period is 0.8 if he receives a catalog and 0.4 if he does not. Assume a discount rate of 0.9 and an objective of maximizing expected total discounted reward.

a. Formulate this as an infinite-horizon discounted Markov decision problem.

b. Find an optimal policy using policy iteration starting with the stationary policy which has greatest one-step reward.

c. For $\varepsilon = 0.1$, find an ε-optimal policy using ordinary value iteration, Gauss-Seidel value iteration, and relative value iteration starting with $v_0^T = (0,0)$. In each, compare the number of iterations required under the supremum norm (6.3.3) and span seminorm (6.6.11) stopping criteria.

d. For $\varepsilon = 0.1$, find an ε-optimal policy using modified policy iteration and Gauss-Seidel modified policy iteration. Be sure to choose an appropriate starting value.

e. Formulate the problem as a linear program, giving its primal and dual. Solve both and interpret the solutions.

f. Find an optimal policy under the constraint that the total expected discounted cost of sending catalogs cannot exceed 99.

g. Apply the action elimination algorithm of Sec. 6.7.3. Record the total number of actions eliminated at each iteration.

h. What is the turnpike horizon for this problem? Compare it with the bounds in (6.8.1) and (6.8.5).

6.2. Consider the data in Problem 4.28, but in addition assume that the probability the animal dies in any period independent of the patch chosen is 0.001. Find a foraging policy that maximizes the probability of survival using policy iteration. Proposition 5.3.1 suggests how to convert this to a discounted model.

6.3. Show that $|r(s, a)| \le M$ implies $\|v_\lambda^*\| \le (1 - \lambda)^{-1}M$.

6.4. a. Show by example that there exists a v-improving decision rule for which
$v_\lambda^{(d_v)\infty} < v$.

b. Show by example that it is possible that d_v is v-improving, $r_{d_v} + \lambda P_{d_v} v \ge v$ and $v_\lambda^{(d_v)\infty} = v$.

   c. Prove that $v_\lambda^{(d_v)\infty}(s'') > v(s'')$ for some $s'' \in S$ if $r_{d_v}(s') + \lambda P_{d_v} v(s') \geq v(s')$ for some $s' \in S$.

6.5. Complete the proof of Theorem 6.2.3 by establishing that if T is a contraction mapping on a Banach space U, then its fixed point is unique.

6.6. Generalize the proof of Proposition 6.2.4 to show that $\mathscr{L}$ is a contraction mapping on V.

6.7. Suppose that $S = [0, 1]$, A_s is compact for each $s \in S$, and $r(s, a)$ and $p(j|s, a)$ are continuous on $S \times A$.

   a. Show that $\mathscr{L}$ has a fixed point in V_M by establishing that $\mathscr{L}: V_M \to V_M$.

   b. Generalize this result to $S = R^n$.

6.8. Suppose the supremum in (6.2.2) is not attained. Provide an alternative stopping criterion to that in (6.3.3) which ensures that we find an ε-optimal policy when using value iteration to solve $\mathscr{L}v = v$.

6.9. Show that step 2 of the Gauss-Seidel value iteration algorithm may be written as (6.3.16) and that $Q_d = I - \lambda P_d^L$ and $R_d = \lambda P_d$ is a regular splitting of $I - \lambda P_d$.

6.10. Show that if $Lv^0 \geq v^0$, then the iterates of Gauss-Seidel dominate those of value iteration and that they converge monotonically.

6.11. Let $\{v^n\}$ denote a sequence of iterates of Gauss-Seidel value iteration. Show that if $\mathrm{sp}(v^{n+1} - v^n) < \lambda^{-1}(1 - \lambda)\varepsilon$, then a stationary policy derived from a maximizing decision rule at iteration n, is ε-optimal. Note that the result requires generalizing Proposition 6.6.5 to the iterates of Gauss-Seidel.

6.12. For Jacobi value iteration, state and prove analogous theorems to Theorems 6.3.7 and 6.3.9.

6.13. (Combined point-Jacobi–Gauss-Seidel value iteration).

   a. Explicitly state an algorithm based on combining these two methods.

   b. Represent the iterative step in terms of a splitting of $I - \lambda P_d$.

   c. Verify that the splitting is regular.

   d. State and prove a result similar to Theorem 6.3.7 for this algorithm.

   e. Apply this combined algorithm to Example 6.3.1 to estimate its average asymptotic rate of convergence and the number of iterations required to achieve a solution accurate to 0.01.

6.14. This problem breaks down the proof of Proposition 6.3.6 into several steps. Use the following loosely stated form of the Perron-Frobenius Theorem where necessary.

   If A is a non-negative matrix, then its spectral radius is positive, the corresponding eigenvector is non-negative and, if $0 \leq B \leq A$ with $A - B \neq 0$, then $\sigma(A) > \sigma(B)$.

a. If $(I - A)^{-1}$ exists and is non-negative, then

$$\sigma\left(A(I - A)^{-1}\right) = \frac{\sigma(A)}{1 - \sigma(A)}.$$

b. Use (a) to show that

$$\sigma(A) = \frac{\sigma\left(A(I - A)^{-1}\right)}{1 + \sigma\left(A(I - A)^{-1}\right)}.$$

c. If (Q, R) is a regular splitting of an invertible matrix B, then

$$\sigma(Q^{-1}R) = \frac{\sigma(B^{-1}R)}{1 + \sigma(B^{-1}R)}.$$

d. Use (c) to prove Proposition 6.3.6.

6.15. (Veinott, 1969 and Porteus, 1975). In this problem, we drop the assumption that the row sums of P_d are identically 1 and assume only that they are less than or equal to 1.

Given a finite state and action infinite-horizon MDP with rewards r_d and transition probabilities P_d, we say that a MDP with rewards $\tilde{r}_d$ and transition probabilities $\tilde{P}_d$ is *positively similar* to the original MDP if there exists a diagonal matrix B with positive entries for which $\tilde{r}_d = Br_d$ and $\tilde{P}_d = BP_dB^{-1}$. In the following, use $\sim$'s to denote quantities for the positively similar MDP.

a. Show that $\tilde{v}_\lambda^* = Bv_\lambda^*$.

b. Show that if d^∞ is optimal for the original MDP, it is optimal for the positively similar MDP.

c. Show that if the iterates of value iteration for the original MDP converge geometrically at rate α, then the same applies to the iterates of value iteration for the positively similar MDP.

d. For any MDP, given $\varepsilon > 0$, there exists a positively similar MDP for which

$$\max_{d \in D} \|\tilde{P}_d\| < \max_{d \in D} \sigma(P_d) + \varepsilon.$$

Hint: Let u be the solution of $u = \max_{d \in D}\{e + \lambda P_d u\}$, choose B to be the matrix with diagonal entries $1/u(s)$, and note that, for a matrix H with nonnegative entries, $\|H\| = He$.

e. Suppose $\alpha \equiv \max_{d \in D} \sigma(P_d)$. Show that, for each $\varepsilon > 0$, value iteration converges $O((\alpha + \varepsilon)^n)$.

f. Suppose no P_d is upper triangular. Use Proposition 6.3.6 and the above results to show that Gauss-Seidel value iteration converges $O(\alpha^n)$ for $\alpha < \lambda$.

6.16. Devise a policy iteration algorithm for a finite-horizon MDP. Clearly indicate how policy evaluation and policy improvement would be implemented and discuss its computational requirements.

6.17. Provide conditions which guarantee convergence of policy iteration for the model in Sec. 6.2.5.

6.18. Prove Corollary 6.4.9.

6.19. (Puterman and Brumelle, 1979). Consider the following two-state discounted Markov decision process; $S = \{s_1, s_2\}$, $A_{s_1} = A_{s_2} = [0, 1]$, $r(s_1, a) = -a$, $r(s_2, a) = -1 + a/12$, $p(s_1|s_1, a) = a^2/2$, and $p(s_2|s_2, a) = a^2/4$.

a. Verify that the rate of convergence of policy iteration is quadratic.

b. Solve it in the case that $\lambda = 0.5$, using policy iteration starting with $d_0(s_1) = d_0(s_2) = 0$.

c. Compare results in (b) to those obtained using value iteration.

6.20. (Eaton and Zadeh, 1962). Let η and δ be two decision rules. Define a new decision rule ω by

$$\omega(s) = \begin{cases} \delta(s) & \text{if } v_\lambda^{\delta^\infty}(s) \geq v_\lambda^{\eta^\infty}(s) \\ \eta(s) & \text{if } v_\lambda^{\eta^\infty}(s) > v_\lambda^{\delta^\infty}(s). \end{cases}$$

Show that $v_\lambda^{\omega^\infty} \geq \max\{v_\lambda^{\delta^\infty}, v_\lambda^{\eta^\infty}\}$.

6.21. Verify the calculations in Example 6.4.2.

6.22. Develop a theory of the convergence of policy iteration without assuming that the maximum in (6.2.4) is attained. That is, replace "max" by "sup" in this expression.

6.23. Provide an alternative proof for the existence of a solution to the optimality equations by demonstrating the convergence of policy iteration under the condition that there exists a y^0 such that $By^0 \leq 0$.

6.24. Prove Theorem 6.5.9.

6.25. Show that v^0 defined by (6.5.8) satisfies the condition $(T - I)v^0$ in the hypothesis of Theorem 6.5.9.

6.26. Establish properties $1 - 6$ of the span seminorm in Section 6.6.1.

6.27. Derive the scalar identities

$$|x - y| = (x + y) - 2\min(x, y)$$

and

$$(x - y)^+ = x - \min(x, y)$$

and apply them to obtain the alternative representations for γ_d in Proposition 6.6.1.

6.28. Prove Theorem 6.6.2. Note that it requires replacing V by the space of equivalence classes of functions which differ by an additive constant.

6.29. Derive the upper bounds in Theorem 6.6.3.

6.30. (Thomas, 1981). Suppose v^{n-1} and v^n denote successive iterates of value iteration and $0 \le \alpha < 1$. Then

$$v^{n+1} + \frac{1}{1-\alpha}\left\{\alpha Bv^n + \frac{\lambda}{1-\lambda}\Lambda(Bv^n - \alpha Bv^{n-1})e\right\} \le v_\lambda^*$$

$$\le v^{n+1} - \frac{1}{1-\alpha}\left\{Bv^n + \frac{\lambda}{1-\lambda}\Lambda(\alpha Bv^n - Bv^{n-1})e\right\}.$$

Numerically compare these to the bounds in Corollary 6.6.4 using data from Problem 6.1.

6.31. Evaluate γ defined in (6.6.14) and γ' as defined by (6.6.16) for data from Problem 6.1. Investigate the relationship between the rate of convergence of value iteration with respect to the span seminorm and these quantities.

6.32. (Hubner, 1977). Show that action a' is nonoptimal in state s at iteration n of value iteration if

$$v^{n+1}(s) - r(s,a) - \sum_{j \in S} \lambda p(j|s,a)v^n(j) > \frac{\lambda\gamma}{1-\lambda\gamma}\mathrm{sp}(Bv^n).$$

Further show that, in the above inequality, the quantity $\lambda\gamma/(1 - \lambda\gamma)$ may be replaced by the potentially smaller quantity $\lambda\gamma_{s,a}/(1 - \lambda\gamma)$, where

$$\gamma_{s,a} = \max_{a' \in A_s}\left\{1 - \sum_{j \in S}\min[p(j|s,a), p(j|s,a)]\right\}.$$

Compare action elimination algorithms based on these procedures for the data in Problem 6.1.

6.33. (Hubner, 1977). For $u \in V$, $d \in D$, and $d' \in D$:
a. Show that

$$-\gamma \, \mathrm{sp}(u) \le P_d u - P_{d'}u \le \gamma \, \mathrm{sp}(u)$$

and that $\|P_d u - P_{d'}u\| \le \gamma \, \mathrm{sp}(u)$. Hint: follow the steps in the first part of the proof of Theorem 6.6.6.
b. Find a sharper bound on the difference in part (a) by replacing γ by a quantity which depends only on d and d' as in the previous problem.

6.34. Let $\{v^n\}$ be a sequence of values generated by the value iteration algorithm. Use the result in Proposition 6.7.6 to establish that the bounds $G_{-1}(v^n)$ and $G^{-1}(v^n)$, as defined in (6.6.9), converge to v_λ^*. Generalize this result to $G_m(v^n)$ and $G^m(v^n)$ for $m > -1$ and for iterates of other algorithms.

6.35. Prove Corollary 6.7.5.

6.36. Prove Lemma 6.7.8. Hint: the ideas used to prove Proposition 6.7.6 apply here in considerably simpler form.

6.37. Obtain the relationship between the upper bounds in Proposition 6.7.6.

6.38. For the algorithm in Sec. 6.7.3, verify that $Bv^0 \geq 0$, $\Lambda(Bv^0) = 0$, and $\Upsilon(Bv^0) = \mathrm{sp}(r_{d_0})$.

6.39. Verify all calculations in Example 6.8.1.

6.40. Show that the dual of the primal linear program equals that given in Section 6.9.1.

6.41. Show for the dual linear program, with the added constraint that $\sum_{a \in A_s} x(s', a) = 0$ for some $s' \in S$, that an optimal policy is to choose d^* arbitrary for s' and equal to a_s whenever $x^*(s, a_s) > 0$.

6.42. Suppose you have solved a discounted Markov decision process under maximization and have computed $v_\lambda^*(s)$ and an optimal policy d^∞ for which $v_\lambda^{d^\infty} = v_\lambda^*$.

  a. A new action a' becomes available in state s'. How can you determine whether d^∞ is still optimal without resolving the problem? If it is not, how can you find a new optimal policy and its value?

  b. Suppose action a^* is optimal in state s^*, that is $d(s^*) = a^*$, and you find that the return in state s^* under action a^* decreases by Δ. Provide an efficient way for determining whether d^∞ is still optimal and, if not, for finding a new optimal policy and its value.

6.43. Let S be countable and suppose that the matrix $H: V_w \to V_w$. Using the definition of a matrix norm from (C.2) in Appendix C, show that $\|H\|_w$ satisfies (6.10.1).

6.44. Show that, when Assumptions 6.10.1 and 6.10.2 hold, $L_d v \in V_w$ for all $v \in V_w$ and $d \in D$.

6.45. Prove that V_w is a Banach space by showing that, if $\{v_n\}$ is a Cauchy sequence in V_w, there exists a $v \in V_w$ for which $\lim_{n \to \infty} \|v_n - v\| = 0$.

6.46. Prove Proposition 6.10.3 for the operator $\mathscr{L}$.

6.47. Show under Assumptions 6.10.1 and 6.10.2 that $L_d^{N,u}$, $L^{N,u}$, and $\mathcal{L}^{N,u}$ are J-stage contractions on V_w, and that $v_*^{N,u}$ is a fixed point of $\mathcal{L}^{N,u}$.

6.48. (Rosenthal, White, and Young, 1978). Consider the following version of the dynamic location model presented in Problem 3.17. There are $Q = 4$ work sites, with site 1 denoting the home office and 2, 3, and 4 denoting remote sites. The cost of relocating the equipment trailer is $d(k, j) = 300$ for $k \neq j$; the cost $c(k, j)$ of using the equipment trailer is 100 if the work force is at site $k > 1$, and trailer is at site $j \neq k$ with $j > 1$; 50 if $j = k$ and $j > 1$ and 200 if the work force is at remote site $j > 1$, and the trailer is at the home office, site 1. If the work force is at site 1, no work is carried out, so the cost of using the trailer in this case can be regarded to be 0. Assume that the probability of moving between sites in one period $p(j|s)$ is given by the matrix

$$
P = \begin{bmatrix}
0.1 & 0.3 & 0.3 & 0.3 \\
0.0 & 0.5 & 0.5 & 0.0 \\
0.0 & 0.0 & 0.8 & 0.2 \\
0.4 & 0.0 & 0.0 & 0.6
\end{bmatrix}.
$$

Assuming the discount rate $\lambda = 0.95$, find a relocation policy which minimizes the expected discounted cost and describe the structure of the optimal policy.

6.49. Suppose w satisfies Assumption 6.10.1 and, in addition,

$$
\sum_{s \in S} p(j|s, a)w(j) \leq w(s) + L
$$

for some $L > 0$. Show that

$$
|v_\lambda^*(s)| \leq \frac{\mu}{1 - \lambda}\left(w(s) + \frac{\lambda L}{1 - \lambda}\right).
$$

Apply this result to the model in Sec. 6.10.4 to obtain (6.10.29).

6.50. Consider the following variant of the replacement model of Sec. 6.10.4 which includes the possibility of equipment failure. Let s denote the age of the equipment. For any s, there is a probability p_s that the equipment fails and a probability $1 - p_s$ that it ages by one period. In addition to the costs in Sec. 6.10.4, there is a fixed cost of a failure F. If the equipment fails, the system moves to state 0 during the period in which the failure occurs, it must be replaced instantaneously at cost K, and then starts the subsequent period in state 0. As before, the actions in any period are to replace or operate as is. Assume that the operating cost $h(s) = hs^2$.

a. Formulate this as a discounted Markov decision problem by defining $r(s, a)$ and $p(j|s, a)$.

b. Give the optimality equation and provide a bounding function w for which the optimality has a unique solution in V_w.

c. Prove that a finite-state approximation converges to v_λ^*.

   d. Find *a priori* bounds on the precision of the N-state approximation.

   e. Show that a control limit policy is optimal.

6.51. Verify that x_k and x'_k defined in the proof of Theorem 6.10.11 satisfy the hypotheses of Lemma 4.7.2.

6.52. Show that the matrix Q defined in Section 6.10.3 has row sums bounded by 1.

6.53. Consider the inventory model of Example 6.10.2 in which A_s is finite for each $s \in S$, the cost of storing s units for one period $h(s) = hs$, and the revenue from selling j units in a period equals bj.

   a. Establish that N-state approximations converge to the optimal value function by showing that condition (6.10.20) holds, and that there exists a $u \in V_W$ such that $L_d u \leq u$ for all $d \in D$.

   b. Using the result in Theorem 6.10.11, obtain a bound on N, the degree of approximation necessary to ensure that $|v_\lambda^*(0) - v_*^{N, u}(0)| < \varepsilon$ for some prespecified ε under the assumptions that $\{p(j)\}$ has finite support.

   c. Repeat (b), assuming that $\{p(j)\}$ has a Poisson distribution. Investigate the relationship between N and the model parameters.

6.54. Prove that iterates of the approximate modified policy iteration algorithm of Sec. 6.10.2 converge pointwise to v_λ^* under the assumptions of Theorems 6.10.8 or 6.10.9.

6.55. Using Corollary 6.10.12, obtain bounds on $|v_\lambda^*(0) - v_*^{4, u}(0)|$ based on choosing $S_1 = \{0, 1\}$, $S_2 = \{0, 1, 2, 3\}$, and $S_3 = \{0, 1, 2, 3, 4\}$ in the following model. Let $S = \{0, 1, \ldots\}$, $\lambda = 0.9$ and $A_s = \{a_s\}$ for $s = 0, 1, 2, 3$, $r(s, a_s) = s + 1$, $s = 0, 1, 2, 3$, and $A_4 = \{a_{4,1}, a_{4,2}\}$, with $r(4, a_{4,1}) = 5$ and $r(4, a_{4,2}) = 6$ and $0 \leq r(s, a) \leq 10$ for all other s and $a \in A_s$. Let $p(0|0, a_0) = 0.1$, $p(1|0, a_0) = 0.1$, $p(2|0, a_0) = 0.2$, $p(3|0, a_0) = 0.2$, and $p(4|0, a_0) = 0.3$; $p(0|1, a_1) = 0.2$, $p(1|1, a_1) = 0.1$, $p(2|1, a_1) = 0$, $p(3|1, a_1) = 0.2$, and $p(4|1, a_1) = 0.1$; $p(0|2, a_2) = 0.4$, $p(1|2, a_2) = 0.1$, $p(2|2, a_2) = 0$, $p(3|2, a_2) = 0$, and $p(4|2, a_2) = 0.5$; $p(0|3, a_3) = 0.1$, $p(1|3, a_3) = 0.1$, $p(2|3, a_3) = 0.4$, $p(3|3, a_3) = 0$ and $p(4|3, a_3) = 0.3$; $p(0|4, a_{4,1}) = 0.1$, $p(1|4, a_{4,1}) = 0.6$, $p(2|4, a_{4,1}) = 0$, $p(3|4, a_{4,1}) = 0.1$, and $p(4|4, a_{4,1}) = 0.1$; $p(0|4, a_{4,2}) = 0.2$, $p(1|4, a_{4,2}) = 0.2$, $p(2|4, a_{4,2}) = 0.2$, $p(3|4, a_{4,2}) = 0.1$, and $p(4|4, a_{4,1}) = 0.2$. Note that probabilities of transitions to states outside of S_4 are not specified. (Exact specification is not required to obtain bounds.)

6.56. Provide conditions on $O(\cdot)$, h, and $F(\cdot)$ in Example 6.10.2 which ensure that $M' < \infty$.

6.57. Prove Theorem 6.11.2.

6.58. Prove Theorem 6.11.3.

6.59. Suppose conditions on a finite-state discounted MDP are such that a monotone optimal policy exist. Provide primal and dual LP's which find it.

6.60. Prove Theorem 6.11.10 under hypothesis d'.

6.61. Perform two iterations of value iteration with $v^0 = 0$ for the alternative reward function in Example 6.11.2.

6.62. Consider an infinite horizon discounted version of Problem 4.33. Find an optimal policy when $\lambda = .95$ and note its structure.

6.63. (A simple bandit model). A decision maker observes a discrete-time system which moves between states $\{s_1, s_2, s_3, s_4\}$ according to the following transition probability matrix:

$$P = \begin{bmatrix} 0.3 & 0.4 & 0.2 & 0.1 \\ 0.2 & 0.3 & 0.5 & 0.0 \\ 0.1 & 0.0 & 0.8 & 0.1 \\ 0.4 & 0.0 & 0.0 & 0.6 \end{bmatrix}.$$

At each point of time, the decision maker may leave the system and receive a reward of $R = 20$ units, or alternatively remain in the system and receive a reward of $r(s_i)$ units if the system occupies state s_i. If the decision maker decides to remain in the system, its state at the next decision epoch is determined by P. Assume a discount rate of 0.9 and that $r(s_i) = i$.

a. Formulate this model as a Markov decision process.

b. Use policy iteration to find a stationary policy which maximizes the expected total discounted reward.

c. Find the smallest value of R so that it is optimal to leave the system in state 2.

d. Show for arbitrary P and $r(\cdot)$ that there exists a value R_{s_i} in each state such that it is optimal to leave the system in state s_i only if $R \geq R_{s_i}$.

6.64. (Feinberg and Shwartz, 1994). The problem illustrates some of the difficulties that arise when using a weighted discounted optimality criteria. Consider a model with $S = \{s_1, s_2\}$; $A_{s_1} = \{a_{1,1}, a_{1,2}\}$ and $A_{s_2} = \{a_{2,1}\}$; $r(s_1, a_{1,1}) = 1$, $r(s_1, a_{1,2}) = 0$, and $r(s_2, a_{2,1}) = 2$; and $p(s_1|s_1, a_{1,1}) = 1$, $p(s_2|s_1, a_{1,2}) = 1$, and $p(s_2|s_2, a_{2,1}) = 1$.

a. Show that, when $\lambda \leq 0.5$, the optimal stationary policy uses action $a_{1,1}$ in s_1, and, if $\lambda \geq 0.5$, the optimal stationary policy uses action $a_{1,2}$ in s_1.

b. Let λ_1 and λ_2 denote two discount rates and suppose we evaluate policies according to weighted discounted reward criteria

$$v_W^\pi(s) = v_{\lambda_1}^\pi(s) + v_{\lambda_2}^\pi(s).$$

Suppose $\lambda_1 = 0.2$ and $\lambda_2 = 0.6$. Show that the nonstationary policy which uses action $a_{1,1}$ for one period and $a_{2,1}$ for all subsequent periods has a larger weighted discounted reward than any stationary deterministic policy.

    c. Find a randomized stationary policy which has the largest weighted dis-
counted reward among the set of randomized stationary policies, and show
that its weighted reward exceeds that of any deterministic stationary policy.
Show also that the nonstationary policy in (b) has a larger weighted discounted reward than the best randomized stationary policy. Hint: Replace
this model by one with $A_{s_1} = [0, 1]$, $r(s_1, a) = a$, $p(s_1|s_1, a) = a$, and
$p(s_2|s_1, a) = 1 - a$.

    Note: Feinberg and Shwartz establish the existence of an ε-optimal Marko-
vian strategy for a general version of this model.

6.65. Consider an infinite-horizon discounted version of the discrete-time queueing
admission control model with a deterministic service rate that was presented in
Example 3.7.1. Assume that the per period holding cost $h(x)$ is linear in the
system content and that there is an infinite waiting room.

    a. Provide the optimality equations for this model and choose a bounding
function w so that they have a unique solution in V_w.

    b. Show that N-state approximations converge to v_λ^*.

    c. Find *a priori* bounds on the precision of the approximation under the
assumption that the arrival distribution $g(n)$ is geometric with parameter α.

    d. Show that a control limit policy is optimal.

6.66. Solve a discounted version of the service rate control model of Problem 4.26
with $\lambda = 0.95$. Investigate the sensitivity of the optimal policy to the discount
rate and compare the efficiency of numerical algorithms.

6.67. Individuals face the problem of allocating personal wealth between investment
and consumption to maximize their lifetime utility. Suppose that we represent
wealth by a value in $[0, \infty)$, and when the individual allocates x units of wealth
to consumption in a particular month, he receives $\log(x + 1)$ units of utility.
Wealth not allocated to consumption is invested for one period and appreciates
or depreciates at interest rate $\rho \in [-1, M]$ which varies according to a probability distribution $F(\rho)$ where $M < \infty$. This means that each invested unit of
wealth has value $(1 + \rho)$ in the next period.

    a. Formulate this as a continuous-state infinite-horizon discounted Markov
decision problem in which the objective is to maximize expected total
discounted utility. To derive transition probabilities, first give a dynamic
equation which models the change in system state.

    b. Verify that an appropriate generalization of Assumptions 6.10.1 and 6.10.2
hold for this model.

    c. Speculate on the form of an optimal policy for this model.

CHAPTER 7

The Expected Total-Reward Criterion

Alternatives to the expected total discounted reward criterion in infinite-horizon models include the expected total reward and the expected average reward criteria. This chapter focuses on the expected total-reward criterion. The analyses in this chapter will be more intricate than those in Chap. 6; however, through them, we gain considerable insight into the subtleties of the Markov decision process model and a greater appreciation of the role of discounting. Implicit in the models considered here are restrictions on the reward and/or transition functions, without which the expected total rewards may be unbounded or not even well defined. When these restrictions do not apply, the average and sensitive optimality criteria which will be the subjects of Chaps. 8–10 may be appropriate.

This chapter is organized as follows. Section 7.1 formulates the model and provides some results which hold under a fairly general assumption. Sections 7.2 and 7.3 analyze positive and negative models, and Sec. 7.4 provides a tabular comparison of results for these two models. Section 10.4 uses results of Chap. 10 to extend and simplify some aspects of the analysis of this chapter.

7.1 MODEL CLASSIFICATION AND GENERAL RESULTS

Our approach to analyzing discounted models in Chap. 6 will provide a framework for this chapter. In it we will try to establish the fundamental results of that chapter for models with expected total-reward optimality criterion. We provide counterexamples to show when they do not hold and add conditions, if necessary and when possible, to ensure validity.

We briefly summarize relevant results of Chap. 6.

a. For any $d \in D^{\text{MR}}$, $r_d + \lambda P_d v = v$ has a unique solution $v_\lambda^{d^\infty}$.
b. For any $v \in V$,

$$v^* \equiv \sup_{\pi \in \Pi^{\text{MR}}} v_\lambda^\pi = \sup_{\pi \in \Pi^{\text{MD}}} v_\lambda^\pi = \sup_{d \in D^{\text{MD}}} v_\lambda^{d^\infty}.$$

c. If $v \geq (\leq)\mathscr{L}v$, then $v \geq (\leq)v_\lambda^*$.

d. There exists a unique solution to $\mathscr{L}v = v$ in V which equals v_λ^*.

e. If d^* is conserving, then $(d^*)^\infty$ is an optimal policy.

f. If there exists an optimal policy, there exists a stationary optimal policy.

g. Value iteration converges to v_λ^* for any $v^0 \in V$.

h. Policy iteration converges to v_λ^*.

i. Iteratively improving bounds on v_λ^* are available.

Crucial for establishing these results was the existence of a discount factor $0 \leq \lambda < 1$ which ensured that

1. the operators $\mathscr{L}$ and L are contraction mappings on V;

2. $(I - \lambda P_d)^{-1}$ exists and is positive; and

3. for $v \in V$, and any $\pi \in \Pi^{MR}$,

$$\lim_{n \to \infty} \lambda^{n-1} P_\pi^n v(s) = \lim_{n \to \infty} \lambda^{n-1} E_s^\pi \{ v(X_n) \} = 0 \qquad (7.1.1)$$

for all $s \in S$.

Since these conditions do not hold for the models in this chapter, we use different methods of analysis. We assume, except where noted, finite or countable S and $\sum_{j \in S} p(j|s, a) = 1$ for $a \in A_s$. We write D for D^{MD} to simplify notation.

7.1.1 Existence of The Expected Total Reward

For any $\pi \in \Pi^{HR}$ and $N < \infty$, let

$$v_N^\pi(s) \equiv E_s^\pi \left\{ \sum_{t=1}^{N-1} r(X_t, Y_t) \right\}.$$

Under the expected total-reward criterion, we compare policies on the basis of

$$v^\pi(s) = E_s^\pi \left\{ \sum_{t=1}^{\infty} r(X_t, Y_t) \right\} = \lim_{N \to \infty} v_N^\pi(s). \qquad (7.1.2)$$

We may regard this as the expected total discounted reward with $\lambda = 1$. Define the *value* of the MDP under the expected total-reward criterion v^* by

$$v^*(s) \equiv \sup_{\pi \in \Pi^{HR}} v^\pi(s). \qquad (7.1.3)$$

Our objectives in this chapter include characterizing v^* and finding a policy with expected total reward equal to, or within ε of, v^*.

Without further assumptions, we have no guarantee that the limit in (7.1.2) is finite or even exists; for instance, in Example 5.1.2,

$$\limsup_{N \to \infty} v_N^\pi(s) > \liminf_{N \to \infty} v_N^\pi(s)$$

for all $s \in S$. We address this complication by restricting attention to models in which v^π is well defined.

As in Chap. 5, we let $r^+(s, a) = \max\{r(s, a), 0\}$ and $r^-(s, a) = \max\{-r(s, a), 0\}$ and define

$$v_+^\pi(s) \equiv \lim_{N \to \infty} E_s^\pi \left\{ \sum_{t=1}^{N-1} r^+(X_t, Y_t) \right\}$$

and

$$v_-^\pi(s) \equiv \lim_{N \to \infty} E_s^\pi \left\{ \sum_{t=1}^{N-1} r^-(X_t, Y_t) \right\}.$$

Since the summands are non-negative, both of the above limits exist. Observe that the limit defining $v^\pi(s)$ exists whenever $v_+^\pi(s)$ or $v_-^\pi(s)$ is finite, in which case

$$v^\pi(s) = v_+^\pi(s) - v_-^\pi(s).$$

Noting this, we impose the following finiteness assumption which assures that $v^\pi(s)$ is well defined.

Assumption 7.1.1.   For all $\pi \in \Pi^{HR}$ and $s \in S$, either $v_+^\pi(s)$ or $v_-^\pi(s)$ is finite.

In this chapter, we focus on two models in which this assumption holds; positive bounded models and negative models. In *positive bounded* models, we assume further that

a. $v_+^\pi(s) < \infty$ for all $s \in S$ and $\pi \in \Pi^{HR}$, and
b. for each $s \in S$, there exists *at least* one $a \in A_s$ with $r(s, a) \geq 0$.

By a *negative model*, we mean one in which, in addition to Assumption 7.1.1,

a. $v_+^\pi(s) = 0$ for all $s \in S$ and $\pi \in \Pi^{HR}$, and
b. there exists a $\pi \in \Pi^{HR}$ with $v^\pi(s) > -\infty$ for all $s \in S$.

The expression "positive model" was first used in the literature to distinguish models in which $r(s, a) \geq 0$, however, the results we present here hold under the less restrictive model description above.

We emphasize that these two classes of models are distinct. In positive models, an optimal policy is the one with expected total reward furthest from zero, while in negative models, it is the one with expected total reward closest to zero. Negative models usually concern minimization of expected total costs. Surprisingly, results are quite different for these two models, and, with the exception of some results concerning properties of the optimality equation and characterization of optimal policies we analyze each separately.

7.1.2 The Optimality Equation

In this section we show that the value of a MDP with expected total-reward criterion is a solution of the optimality equation. The following preliminary result is an immediate consequence of Theorem 5.5.3(d). It simplifies subsequent analyses by allowing us to restrict attention to Markovian policies.

Proposition 7.1.1. Suppose Assumption 7.1.1 holds. Then, for each $s \in S$,

$$v^*(s) \equiv \sup_{\pi \in \Pi^{HR}} v^\pi(s) = \sup_{\pi \in \Pi^{MR}} v^\pi(s).$$

Note that even when the suprema in Proposition 7.1.1 are attained, the result does not imply that there exists an optimal randomized Markovian policy. This is because for each s, a different $\pi \in \Pi^{MR}$ may attain the maximum.

As previously noted, Proposition 6.2.1 holds when $\lambda = 1$. We restate it here for convenience.

Lemma 7.1.2. For any $v \in V$,

$$\sup_{d \in D^{MR}} \{r_d + P_d v\} = \sup_{d \in D^{MD}} \{r_d + P_d v\}. \tag{7.1.4}$$

We now state and prove an important result for models with expected total reward criterion. Note that this result is valid for models with general state space, but we restrict attention here to finite- and countable-state models.

Theorem 7.1.3. Suppose Assumption 7.1.1 holds. Then the value of the MDP, v^*, satisfies the equation

$$v = \sup_{d \in D} \{r_d + P_d v\} \equiv \mathscr{L} v. \tag{7.1.5}$$

Proof. We show that v^* satisfies both $v^* \leq \mathscr{L} v^*$ and $v^* \geq \mathscr{L} v^*$, from which the result follows.

Choose $\varepsilon > 0$ and let $\pi' \in \Pi^{HR}$ denote a policy which satisfies $v^{\pi'} \geq v^* - \varepsilon e$. Then, by definition of v^*, for all $d \in D$,

$$v^* \geq r_d + P_d v^{\pi'}.$$

Therefore

$$v^* \geq \sup_{d \in D} \{r_d + P_d v^{\pi'}\} \geq \sup_{d \in D} \{r_d + P_d v^*\} - \varepsilon e,$$

from which it follows that $v^* \geq \mathscr{L} v^*$.

Next we prove the reverse inequality. Fix $s \in S$. From the definition of v^* and Proposition 7.1.1, for any $\varepsilon > 0$ there exists a $\pi = (d_1, d_2, \ldots) \in \Pi^{\mathrm{MR}}$ (possibly varying with s) which satisfies $v^{\pi}(s) \geq v^*(s) - \varepsilon$. Letting π' denote the policy $(d_2, d_3, \ldots)$, it follows that

$$v^*(s) - \varepsilon \leq r_{d_1}(s) + P_{d_1} v^{\pi'}(s) \leq r_{d_1}(s) + P_{d_1} v^*(s)$$

$$\leq \sup_{d \in D^{\mathrm{MR}}} \{r_d(s) + P_d v^*(s)\} = \sup_{d \in D^{\mathrm{MD}}} \{r_d(s) + P_d v^*(s)\},$$

where the last equality follows from Lemma 7.1.2. Since ε was arbitrary, $v^*(s) \leq \mathscr{L} v^*(s)$. Hence $v^* \leq \mathscr{L}^* v$ and the result follows. $\qquad \square$

Corollary 7.1.4 follows by applying Theorem 7.1.3 to a model with a single decision rule d, so that $v^* = v^{d^{\infty}}$.

Corollary 7.1.4. Let $d \in D^{\mathrm{MR}}$. Then $v^{d^{\infty}}$ is a solution of

$$L_d v \equiv r_d + P_d v = v. \tag{7.1.6}$$

We note that the above derivations also apply to discounted models; however, as discussed above, we approached those models differently. Unfortunately, in contrast to discounted models, for any scalar c,

$$\sup_{d \in D} \{r_d + P_d(v^* + ce)\} = v^* + ce, \tag{7.1.7}$$

so that the optimality equation *does not* uniquely characterize v^*.

We express the optimality equation in component notation as

$$v(s) = \sup_{a \in A_s} \left\{ r(s, a) + \sum_{j \in S} p(j|s, a)v(j) \right\}. \tag{7.1.8}$$

When the supremum over A_s is attained for each $s \in S$, we express the optimality equation in vector notation as

$$v = \max_{d \in D} \{r_d + P_d v\} \equiv L v.$$

Note that neither L nor $\mathscr{L}$ are contractions on V, but that each is *monotone* as we show below. We summarize some useful technical results in the following lemma.

Lemma 7.1.5

a. For any $u \leq v$ in V, $\mathcal{L}u \leq \mathcal{L}v$ and $Lu \leq Lv$.

b. For any scalar c and $v \in V$, $\mathcal{L}(v + ce) = \mathcal{L}v + ce$.

c. Let $\{v_n\}$ be a monotone increasing sequence of functions in V which converges pointwise to $v \in V$. Then, for each $d \in D$,

$$\lim_{n \to \infty} L_d v_n = L_d v.$$

Proof. For any $\varepsilon > 0$, there exists a $d \in D^{MD}$ such that

$$\mathcal{L}u \leq r_d + P_d u + \varepsilon e \leq r_d + P_d v + \varepsilon e \leq \mathcal{L}v + \varepsilon e.$$

Since ε was arbitrary, result (a) follows.

Since P_d is a probability matrix, part (b) follows. Part (c) follows from the monotone convergence theorem (cf. Royden, 1963, p. 72).   □

7.1.3 Identification of Optimal Policies

The results in this section are concerned with optimal policies, stationary optimal policies, and conserving decision rules. Call a decision rule $d \in D^{MD}$ *conserving* if

$$r_d + P_d v^* = v^*. \tag{7.1.9}$$

We refer to these rules as conserving because by using them for an additional period, we conserve the expected total reward at the optimal level v^*. Since, by Lemma 7.1.2,

$$\sup_{d \in D^{MR}} \{r_d + P_d v^*\} = \sup_{d \in D^{MD}} \{r_d + P_d v^*\} = \mathcal{L}v^* = v^*,$$

we need only consider conserving decision rules in D^{MD}.

The following result, which is an immediate consequence of Theorem 7.1.3, relates values of optimal policies to solutions of the optimality equation.

Theorem 7.1.6. Suppose $\pi^* \in \Pi^{HR}$ is optimal. Then v^{π^*} satisfies the optimality equation.

We now show that a necessary condition for a stationary policy d^∞ to be optimal is that it be conserving. In the discounted case, Theorem 6.2.7 established that the existence of a conserving decision rule was also a sufficient condition for the derived stationary decision rule to be optimal. Unfortunately this is not true for positive models, as examples in the next section illustrate. When conserving decision rules also satisfy condition (7.1.10) below, the derived stationary policies are optimal.

Theorem 7.1.7.

a. Suppose d^∞ is optimal, then d is conserving.

b. If $d \in D^{MD}$ is conserving, and for each $s \in S$

$$\limsup_{N \to \infty} P_d^N v^*(s) \le 0, \qquad (7.1.10)$$

then d^∞ is optimal.

Proof.

a. From Corollary 7.1.4, and the optimality of d^∞,

$$r_d + P_d v^* = r_d + P_d v^{d^\infty} = v^{d^\infty} = v^*.$$

b. Since d is conserving, for any N,

$$v^* = \sum_{n=0}^{N} P_d^{n-1} r_d + P_d^N v^*. \qquad (7.1.11)$$

Choose $s \in S$. As a consequence of Assumption 7.1.1, given $\varepsilon > 0$, for N sufficiently large, the first term on the right-hand side of (7.1.11) can be bounded by $v^{d^\infty}(s) + \varepsilon$; (7.1.10) implies that the second term can be made smaller than ε. Therefore $v^*(s) \le v^{d^\infty}(s) + 2e$, so that $v^{d^\infty}(s) = v^*(s)$. $\quad\square$

We refer to decision rules which satisfy (7.1.10) as *equalizing*. Since (7.1.10) may be written as

$$\limsup_{N \to \infty} E_s^{d^\infty}\{v^*(X_{N+1})\} \le 0,$$

a decision rule is equalizing whenever it drives the system to states in which there is little opportunity for positive future rewards. Theorem 7.1.7(b) may be restated as

Any stationary policy derived from a conserving and equalizing decision rule is optimal.

Note that we implicitly used this result for discounted models because (7.1.1) ensured that all decision rules were equalizing. In negative models, $v^* \le 0$, so that all decision rules are equalizing, and consequently conserving decision rules generate optimal stationary policies.

7.1.4 Existence of Optimal Policies

Under Assumption 7.1.1, we can establish the existence of a stationary optimal policy in finite-state and action models. The proof relies on existence results for discounted models and the following technical lemma which we state without proof.

Lemma 7.1.8. Suppose Assumption 7.1.1 holds, and $\{\lambda_n\}$ is a nondecreasing sequence of discount factors converging to 1. Then, for each $s \in S$ and $\pi \in \Pi^{HR}$,

$$v^\pi(s) = \lim_{n \to \infty} v^\pi_{\lambda_n}(s).$$

Theorem 7.1.9. Suppose that S and, for each $s \in S$, A_s are finite. Then there exists a stationary deterministic optimal policy.

Proof. By Theorem 6.2.7, for every λ, $0 \le \lambda < 1$ there exists a stationary optimal policy for the discounted model. Since D^{MD} is finite, for any monotone sequence $\{\lambda_n\}$ with λ_n converging to 1, there exists a subsequence $\{\lambda_{n_k}\}$ and a decision rule d^* such that

$$v^{(d^*)^\infty}_{\lambda_{n_k}} = v^*_{\lambda_{n_k}} \quad \text{for} \quad k = 1, 2, \dots.$$

From Lemma 7.1.8, it follows that, for any $\pi \in \Pi^{HR}$ and $s \in S$,

$$v^\pi(s) = \lim_{k \to \infty} v^\pi_{\lambda_{n_k}}(s) \le \lim_{k \to \infty} v^{(d^*)^\infty}_{\lambda_{n_k}}(s) = v^{(d^*)^\infty}(s).$$

Therefore $(d^*)^\infty$ is an optimal policy. $\quad\square$

In Sec. 7.3.2, we obtain a considerable generalization of this result for negative models.

7.2 POSITIVE BOUNDED MODELS

Blackwell (1967) characterized positive models as ones in which $r(s, a) \ge 0$ for *all* $a \in A_s$ and $s \in S$. This definition excludes some interesting applications; indeed, key results hold under the following weaker assumptions.

Assumption 7.2.1. For all $s \in S$ and $\pi \in \Pi^{HR}$, $v^\pi_+(s) < \infty$.

Assumption 7.2.2. For each $s \in S$, there exists *at least* one $a \in A_s$ with $r(s, a) \ge 0$.

As a result of Proposition 7.1.1,

$$v^*(s) = \sup_{\pi \in \Pi^{HR}} v^\pi(s) = \sup_{\pi \in \Pi^{MR}} v^\pi(s) \tag{7.2.1}$$

for each $s \in S$. Note that even if the supremum in (7.2.1) is attained, there need not exist an ε-optimal Markov randomized policy, since the policy which attains the supremum in (7.2.1) might vary with s. Note also that for finite S and A_s, Theorem 7.1.9 implies that the supremum is attained by a policy in Π^{SD}, and further, that $v^*(s)$ is finite.

7.2.1 The Optimality Equation

This section explores the role of the optimality equation for positive models. Letting $V^+ = \{v: S \rightarrow R: v \geq 0 \text{ and } v(s) < \infty \text{ for each } s \in S\}$, we establish some properties of positive models.

Proposition 7.2.1. In a positive bounded model,

a. $\mathscr{L}0 \geq 0$;   (7.2.2)

b. there exists a $d \in D^{MD}$ with $v^{d^\infty} \geq 0$;

c. $v^* \geq 0$;

d. $\mathscr{L}: V^+ \rightarrow V^+$; and

e. for any $d \in D^{MD}$, $r_d \leq 0$ on each recurrent state of P_d.

Proof. By Assumption 7.2.2, $r_{d'} \geq 0$ for some $d' \in D^{MD}$, so that $\mathscr{L}0 \geq r_{d'} \geq 0$. Since

$$v^{d^\infty} = \sum_{n=1}^{\infty} (P_d)^{n-1} r_d$$

for any $d \in D^{MD}$, $v^* \geq v^{(d')^\infty} \geq 0$, so that (b) and (c) hold. Part (d) follows from Assumption 7.2.1, part (a) and Lemma 7.1.5(a).

Suppose s is recurrent under d, and $r_d(s) > 0$. Since the expected number of visits to s under d^∞ is infinite,

$$E_s^{d^\infty} \left\{ \sum_{t=1}^{\infty} r^+(X_t, Y_t) \right\} = \infty,$$

contradicting Assumption 7.2.1.   □

Note that, for finite S, $r_d(s) \leq 0$ on any recurrent state of P_d is also a sufficient condition for Assumption 7.2.1 to hold; however, for countable S we require further conditions.

The following generalization of Theorem 6.2.2a for discounted models provides upper bounds for v^* and enables us to identify v^* as the *minimal* solution of the optimality equation.

Theorem 7.2.2. Suppose that there exists a $v \in V^+$, for which $v \geq \mathscr{L}v$. Then $v \geq v^*$.

Proof. Let $\pi = (d_1, d_2, \ldots) \in \Pi^{MR}$. Since $v \geq \mathscr{L}v$, by Lemma 7.1.2 and the definition of $\mathscr{L}$, $v \geq r_d + P_d v$ for all $d \in D^{MR}$. Hence

$$v \geq r_{d_1} + P_{d_1} v \geq r_{d_1} + P_{d_1} r_{d_2} + P_{d_1} P_{d_2} v$$

$$\geq \sum_{n=1}^{N} P_\pi^{n-1} r_{d_n} + P_\pi^N v = v_{N+1}^\pi + P_\pi^N v$$

Since $v \in V^+$, $P_{\pi}^N v \geq 0$, so that $v \geq v_{N+1}^{\pi}$ for all N. Thus $v \geq v^{\pi}$ for all $\pi \in \Pi^{MR}$. The result follows from Proposition 7.1.1.   □

Since Theorem 7.2.2 holds when $v = \mathcal{L}v$, Theorem 7.1.3 shows that v^* satisfies the optimality equation, and Proposition 7.2.1 shows that $v^* \geq 0$, we have the following important characterization of v^* and of $v^{d^{\infty}}$.

Theorem 7.2.3.

a. The quantity v^* is the minimal solution of $v = \mathcal{L}v$ in V^+.
b. For any $d \in D^{MR}$, $v^{d^{\infty}}$ is the minimal solution of $v = L_d v$ in V^+.

The following simple example illustrates the above concepts and shows that in contrast to the discounted case, the implication that $v \leq v^*$ whenever $v \leq \mathcal{L}v$ does not hold without additional conditions.

Example 7.2.1. Let $S = \{s_1, s_2\}$, $A_1 = \{a_{1,1}, a_{1,2}\}$, $A_2 = \{a_{2,1}\}$, $r(s_1, a_{1,1}) = 5$, $r(s_1, a_{1,2}) = 8$, $r(s_2, a_{2,1}) = 0$, $p(s_1|s_1, a_{1,1}) = p(s_2|s_1, a_{1,1}) = 0.5$, $p(s_2|s_1, a_{1,2}) = 1$ and $p(s_2|s_2, a_{2,1}) = 1$ (Fig. 7.2.1). Let δ and γ denote the two available deterministic decision rules, with δ choosing action $a_{1,1}$ in s_1 and γ choosing action $a_{1,2}$ in s_1. Policy δ^{∞} is optimal, so that $v^*(s_1) = 10$ and $v^*(s_2) = 0$.
The optimality equation $Lv = v$ is given by

$$v(s_1) = \max\{5 + 0.5v(s_1) + 0.5v(s_2), 8 + v(s_2)\} \qquad (7.2.3)$$

and $v(s_2) = v(s_2)$. Since (7.2.3) holds for any v satisfying $v(s_1) - v(s_2) = 10$, the optimality equation does not have a unique solution (Fig. 7.2.2). Theorem 7.2.3 distinguishes v^* as the minimal solution of $v(s_1) - v(s_2) = 10$, $v(s) \geq 0$.
Since any vector v with $v(s_1) - v(s_2) \leq 10$ satisfies $v \leq Lv$, in contrast to Chap. 6, this condition does not ensure $v \leq v^*$. For example, $v(s_1) = 12$ and $v(s_2) = 6$, which satisfies that equation is not a lower bound on v^*.

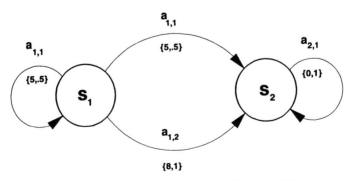

Figure 7.2.1 Symbolic representation of Example 7.2.1.

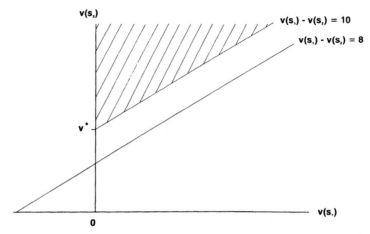

Figure 7.2.2   Graphical representation of (7.2.3). Shaded region indicates v satisfying $v \leq Lv$.

We now provide additional conditions which ensure that a v satisfying $v \leq \mathscr{L}v$ is a lower bound on v^*. This yields an alternative characterization of v^* as a solution of the optimality equation. Condition (7.2.4) is the appropriate restriction of (7.1.10) for positive models.

Theorem 7.2.4

a. Suppose there exists a $v \in V^+$ such that $v \leq \mathscr{L}v$, and

$$\limsup_{N \to \infty} E_s^{\pi}\{v(X_N)\} = \limsup_{N \to \infty} P_{\pi}^N v(s) = 0 \qquad \text{for all} \quad \pi \in \Pi^{\text{MD}}. \quad (7.2.4)$$

Then $v \leq v^*$.

b. Suppose $\mathscr{L}v = v$ and (7.2.4) holds. Then $v = v^*$.

Proof. We prove part (a); part (b) is an immediate consequence of it. Suppose $v \leq \mathscr{L}v$ and (7.2.4) holds. Given $\varepsilon > 0$, choose (ε_i) $i = 1, 2, \ldots$, so that $\varepsilon_i > 0$ and $\sum_{i=1}^{\infty} \varepsilon_i < \varepsilon$. Since $v \leq \mathscr{L}v$, there exists a $d_N \in D^{\text{MD}}$ for which $v \leq r_{d_N} + P_{d_N}v + \varepsilon_N e$. Applying $\mathscr{L}$ to this inequality, and noting Lemma 7.1.5, establishes that there exists a $\pi = (d_1, d_2, \ldots) \in \Pi^{\text{MD}}$ such that, for any N,

$$v \leq \sum_{n=1}^{N} P_{\pi}^{n-1} r_{d_n} + P_{\pi}^N v + \sum_{i=1}^{N} \varepsilon_i.$$

Thus

$$v \leq \lim_{N \to \infty} \sum_{n=1}^{N-1} P_\pi^N r_{d_n} + \lim_{N \to \infty} \sup P_\pi^N v + \varepsilon e \leq v^\pi + \varepsilon e \leq v^* + \varepsilon e.$$

The result follows because ε was chosen arbitrarily. □

We now return to Example 7.2.1 and investigate the implications of (7.2.4).

Example 7.2.1 (ctd.). Since for any $v \in V^+$, $P_\delta^N v(s_2) = v(s_2)$, (7.2.4) holds only if $v(s_2) = 0$. Therefore for v to satisfy both $v \leq Lv$ and (7.2.4) requires $v(s_2) = 0$ and $v(s_1) \leq 10$. Since $v^*(s_1) = 10$ and $v^*(s_2) = 0$, any v satisfying these conditions provides a lower bound for v^*.

As shown by this example, for finite S, (7.2.4) holds whenever $v(s) = 0$ for all s in a recurrent class of some stationary policy. For countable S, requiring $v(s) = 0$ on recurrent states does not ensure that (7.2.4) holds. Behavior on transient states must also be taken into account, as illustrated by the following simple example.

Example 7.2.2. Let $S = \{0, 1, 2, \ldots\}$, $A_s = \{a_s\}$, $p(s + 1|s, a_s) = 1$ for all $s \in S$, and $r(s, a_s) = (0.5)^{s+1}$ so that $v^*(0) = 1$. There is a single stationary policy δ^∞ and $P_\delta^N v(s) = v(s + N)$ so (7.2.4) requires $\lim_{n \to \infty} v(n) = 0$.

7.2.2 Identification of Optimal Policies

In this section we explore the relationship between the optimality equation and optimal policies in positive bounded models. The following result summarizes characterizations of optimal policies for these models. Part (a) follows from Theorems 7.1.6 and 7.2.2, while (b) follows from Theorem 7.1.7(b).

Theorem 7.2.5.

a. A policy $\pi^* \in \Pi^{HR}$ is optimal if and only if $\mathscr{L}v^{\pi^*} = v^{\pi^*}$.
b. If $d \in D^{MD}$ is conserving and

$$\lim_{N \to \infty} \sup P_d^N v^*(s) = 0 \qquad (7.2.5)$$

for all $s \in S$, then d^∞ is optimal.

The following example illustrates Theorem 7.2.5(b) and the need for condition (7.2.5) by showing that, in positive models, stationary policies derived from conserving decision rules need not be optimal.

Example 7.2.3. Let $S = \{s_1, s_2\}$, $A_{s_1} = \{a_{1,1}, a_{1,2}\}$, $A_{s_2} = \{a_{2,1}\}$, $p(s_1|s_1, a_{1,1}) = 1$, $r(s_1, a_{1,1}) = 0$, $p(s_2|s_1, a_{1,2}) = 1$, $r(s_1, a_{1,2}) = 1$, and $p(s_2|s_2, a_{2,1}) = 1$, $r(s_2, a_{2,1}) =$

0, and $p(j|s, a) = 0$ otherwise. Then the optimality equation is

$$v(s_1) = \max\{v(s_1), 1 + v(s_2)\} \qquad \text{and} \qquad v(s_2) = v(s_2).$$

Applying Theorem 7.2.3 shows that $v^*(s_1) = 1$ and $v^*(s_2) = 0$. Let δ be the decision rule which uses action $a_{1,1}$ in s_1 and γ be the decision rule which uses action $a_{1,2}$ in s_1. Then *both* decision rules are conserving but $v^{\gamma^\infty}(s_1) = 1 > 0 = v^{\delta^\infty}(s_1)$.

Since for all N

$$P_\delta^N = \begin{bmatrix} 1 & 0 \\ 0 & 1 \end{bmatrix} \qquad \text{and} \qquad P_\gamma^N = \begin{bmatrix} 0 & 1 \\ 0 & 1 \end{bmatrix},$$

$P_\delta^N v^*(s_1) = 1$, $P_\delta^N v^*(s_2) = 0$, $P_\gamma^N v^*(s_1) = 0$, and $P_\gamma^N v^*(s_2) = 0$. Therefore (7.2.5) holds for γ and not δ, so that γ^∞ is optimal as guaranteed by Theorem 7.2.5(b).

When the supremum in (7.1.5) is not attained, we seek ε-optimal policies. The following theorem shows how to identify them. A proof follows by combining several ideas which have been used in this chapter; we leave it as an exercise.

Theorem 7.2.6. Choose $\varepsilon > 0$ and let $\{\varepsilon_n\}$ be such that $\sum_{n=1}^\infty \varepsilon_n = \varepsilon$. Then, if $\pi = (d_1, d_2, \ldots) \in \Pi^{MD}$ satisfies

$$r_{d_n} + P_{d_n} v^* + \varepsilon_n e \geq \mathscr{L} v^*$$

and

$$\limsup_{N \to \infty} P_\pi^N v^* = 0, \tag{7.2.6}$$

then $v^\pi \geq v^* - \varepsilon e$.

*7.2.3 Existence of Optimal Policies

In discounted models, an optimal stationary policy exists under any condition which guarantees existence of a conserving decision rule. Unfortunately, this is not true for positive models, as demonstrated by an example of Strauch (1966), which shows that, in a countable-state, finite-action model, there need not exist an optimal policy. Examples from Blackwell (1967) and van der Wal (1981) show that there need not even exist randomized Markov ε-optimal policies.

In this section we provide conditions under which stationary optimal and ε-optimal policies exist, and introduce an alternative notion of ε-optimality which admits stationary optimal policies. We begin with examples. The first shows that there need not exist an optimal policy.

Example 7.2.4. Let $S = \{1, 2, \ldots\}$, $A_1 = \{a_1\}$, $p(1|1, a_1) = 1$, and $r(1, a_1) = 0$, and, for $s \geq 2$, let $A_s = \{a_{s,1}, a_{s,2}\}$, $p(1|s, a_{s,1}) = 1$, $p(s + 1|s, a_{s,2}) = 1$, and $r(s, a_{s,1}) = 1 - s^{-1}$ and $r(s, a_{s,2}) = 0$ (Fig. 7.2.3).

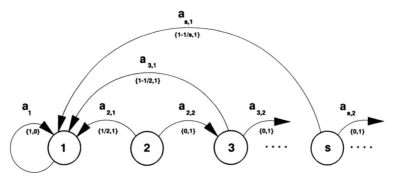

Figure 7.2.3   Symbolic representation of Example 7.2.4.

The optimality equation may be written as $v(1) = v(1)$ and

$$v(s) = \max\{1 - s^{-1} + v(1), v(s + 1)\}.$$

Any solution satisfies $v(1) = c$ and $v(s) = c + 1$ for $s > 1$, where c is an arbitrary positive constant. From Theorem 7.2.3, it follows that $v^*(1) = 0$ and $v^*(s) = 1$ for $s > 1$.

Observe that there exists *no* optimal policy but that the decision rule $\delta(s) = a_{s,2}$ is conserving for all $s > 1$. Even though it is conserving it does not satisfy (7.1.10), so that Theorem 7.1.7(b) does not apply. The stationary policy derived from this decision rule indefinitely delays absorption in state 1, so that the decision maker using this policy never receives a positive reward. Hence $v^{\delta^\infty}(s) = 0$ for all $s \in S$.

Note, however, that for any $\varepsilon > 0$, the stationary policy $(d_\varepsilon)^\infty$, where

$$d_\varepsilon(s) = \begin{cases} a_{s,1} & s \leq \varepsilon^{-1} \\ a_{s,2} & s > \varepsilon^{-1} \end{cases}$$

is ε-optimal.

The following example shows that there need not exist a deterministic stationary ε-optimal policy.

Example 7.2.5.   Let $S = \{0, 1, 2, \ldots\}$, $A_0 = \{a_0\}$, and $A_s = \{a_{s,1}, a_{s,2}\}$ for $s \geq 1$. When $s = 0$, $r(0, a_0) = 0$ and $p(0|0, a_0) = 1$. For $s \geq 1$, $r(s, a_{s,1}) = 0$, $p(s + 1|s, a_{s,1}) = p(0|s, a_{s,1}) = 0.5$, $r(s, a_{s,2}) = 2^s - 1$, and $p(0|s, a_{s,2}) = 1$ (Fig. 7.2.4). Since the optimality equation is given by $v(0) = v(0)$, and

$$v(s) = \max\{0.5v(0) + 0.5v(s + 1), 2^s - 1 + v(0)\},$$

$v^*(0) = 0$ and $v^*(s) = 2^s$ for $s \geq 1$. The stationary policy δ^∞, which uses action $a_{s,1}$

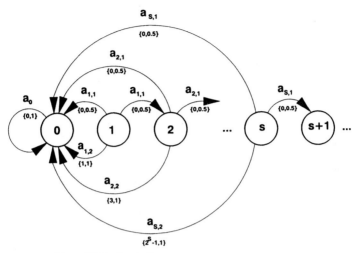

Figure 7.2.4 Symbolic representation of Example 7.2.5.

for all $s \geq 1$, has $v^{d^\infty}(s) = 0$ for $s \in S$, while any other stationary policy must choose $a_{s',2}$ for some $s' \in S$. Let γ be such a policy. Then $v^{\gamma^\infty}(s') = 2^{s'} - 1$, so it is not ε-optimal in s' for $0 < \varepsilon < 1$. Note, however, that for each $s \in S$, there exists a deterministic stationary ε-optimal policy but that it varies with the initial state.

We leave it to Problem 7.2 to show that, for this example, there exists a stationary randomized ε-optimal policy. Problems 7.3 and 7.4 provide van der Wal's modifications of this example in which there exist no randomized stationary or randomized Markov ε-optimal policies. Note that, in the latter example, for each $s \in S$ there exists a randomized Markov ε-optimal policy but that it varies with the initial state s.

Theorem 7.1.9 established existence of a deterministic stationary optimal policy when S and A_s for each $s \in S$ are finite. It appears that the only other condition which ensures existence of a stationary optimal policy in a positive bounded model is the existence of an optimal policy. We now establish that result.

Example 7.2.5 shows that, when S is countable, there exist choices for $\varepsilon > 0$ for which no stationary policy δ^∞ satisfies $v^{\delta^\infty} \geq v^* - \varepsilon e$. Ornstein (1969), using an intricate probabilistic argument, established the following result for a related class of models. We state it without proof. He shows also that the result may be false for uncountable S.

Theorem 7.2.7.   Suppose S is countable and, for each $s \in S$, A_s is a compact subset of a complete separable metric space and $r(s, a) \geq 0$ for all $s \in S$ and $a \in A_s$. Then for any $\varepsilon > 0$ there exists a $d \in D^{MD}$ such that

$$v^{d^\infty} \geq (1 - \varepsilon)v^*. \tag{7.2.7}$$

We refer to the optimality criteria implicit in (7.2.7) as *uniform near optimality*. It can be interpreted in two ways. A policy is uniformly nearly optimal, if either

a. it is $\varepsilon v^*(s)$-optimal at every $s \in S$, or
b. it attains a fixed percentage of the optimal value in every state.

Suppose $\varepsilon = 0.01$, then Theorem 7.2.7 implies that there exists a deterministic stationary policy that yields 99% of the optimal value in every state. An important consequence of this theorem follows.

Corollary 7.2.8. Suppose there exists an $M < \infty$ such that $v^*(s) < M$ for all $s \in S$. Then, for any $\varepsilon > 0$, there exists stationary policy d^∞ satisfying

$$v^{d^\infty} > v^* - \varepsilon e.$$

Proof. Apply Theorem 7.2.7 with $\varepsilon' = \varepsilon/M$. $\square$

The following lemmas, which we use to prove Theorem 7.2.11, have some independent interest. The first result generalizes the policy evaluation equation. Recall that the random variable Z_N represents the history at decision epoch N.

Lemma 7.2.9. Let $\pi^* = (d_1, d_2, \ldots) \in \Pi^{HR}$ be an optimal policy. Then, for all $s \in S$ and any N,

$$v^*(s) = E_s^{\pi^*}\left\{ \sum_{n=1}^{N} r(X_n, Y_n) \right\} + E_s^{\pi^*}\{v^*(X_{N+1})\} \qquad (7.2.8)$$

and

$$E^{\pi^*}\left\{ \sum_{n=N+1}^{\infty} r(X_n, Y_n) \Big| Z_{N+1} = (h_N, a, s') \right\} = v^*(s') \qquad (7.2.9)$$

for all $s' \in S$, for which $P_s^{\pi^*}\{X_{N+1} = s'\} > 0$.

Proof. By definition,

$$v^*(s) = E_s^{\pi^*}\left\{ \sum_{n=1}^{N} r(X_n, Y_n) \right\} + E_s^{\pi^*}\left\{ \sum_{n=N+1}^{\infty} r(X_n, Y_n) \right\}$$

$$\leq E_s^{\pi^*}\left\{ \sum_{n=1}^{N} r(X_n, Y_n) \right\} + E_s^{\pi^*}\{v^*(X_{N+1})\}.$$

We now establish the reverse inequality. Let π_N^* denote the policy $\{d_1, d_2, \ldots, d_N, \pi^*\}$, i.e., it uses π^* for the first N decision epochs and then starts over at decision epoch

$N + 1$ using the first decision rule in π^*. By hypothesis,

$$v^*(s) = v^{\pi^*}(s) \geq v^{\pi^*_N}(s) = E_s^{\pi^*_N}\left\{\sum_{n=1}^{N} r(X_n, Y_n)\right\} + E_s^{\pi^*_N}\{v^{\pi^*}(X_{N+1})\}$$

$$= E_s^{\pi^*}\left\{\sum_{n=1}^{N} r(X_n, Y_n)\right\} + E_s^{\pi^*}\{v^*(X_{N+1})\}$$

where the last equality follows since π^*_N and π^* agree through decision epochs $1, \ldots, N$ and the random variables in { }'s depend only on $d_1, \ldots, d_N$. Combining these inequalities yields (7.2.8).

To obtain (7.2.9) note that the above arguments imply

$$E_s^{\pi^*}\{v^*(X_{N+1})\} = E_s^{\pi^*}\left\{E^{\pi^*}\left\{\sum_{n=N+1}^{\infty} r(X_n, Y_n)\Big|Z_{N+1}\right\}\right\}. \qquad \square$$

Lemma 7.2.10. Suppose there exists an optimal policy in Π^{HR}. Then for each $s \in S$, there exists an optimal policy $\pi' = (\delta_1, \delta_2, \ldots) \in \Pi^{MD}$, for which δ_n is conserving.

Proof. Let $\pi^* = (d_1, d_2, \ldots)$ be an optimal policy, and let $\{h_n\}$ denote the realization of a history which has positive probability under π^*. By (7.2.9), for any N and any $s \in S$ which may be reached by π^* with positive probability,

$$v^*(s) = E^{\pi^*}\left\{\sum_{n=N}^{\infty} r(X_n, Y_n)\Big|Z_N = (h_{N-1}, a, s)\right\}$$

$$= \sum_{a \in A_s} q_{d_N(h_N)}(a)\left\{r(s, a) + \sum_{j \in S} p(j|s, a)\right.$$

$$\left. \times \left[E^{\pi^*}\left\{\sum_{n=N+1}^{\infty} r(X_n, Y_n)\Big|Z_{N+1} = (h_N, a, j)\right\}\right]\right\}.$$

Therefore, by (7.2.9),

$$v^*(s) = \sum_{a \in A_s} q_{d_N(h_N)}\left[r(s, a) + \sum_{j \in S} p(j|s, a)v^*(j)\right]$$

$$\leq \sup_{a \in A_s}\left\{r(s, a) + \sum_{j \in S} p(j|s, a)v^*(j)\right\} = v^*(s),$$

where the last equality follows from Theorem 7.2.5(a). Since the above equation shows that a randomized decision rule attains the supremum for each s, we can find a deterministic decision rule δ_N which attains the supremum. By definition this decision rule is conserving. $\square$

Theorem 7.2.11. Suppose S is countable, $r(s, a) \geq 0$ for all $a \in A_s$ and $s \in S$, and there exists an optimal policy. Then there exists a deterministic stationary optimal policy.

Proof. By Lemma 7.2.10, there exists an optimal policy in which each decision rule is conserving. Denote this policy by $(d_1, d_2, \ldots)$ and let D' be the subset of D^{MD} which contains decision rules $d_1, d_2, \ldots$ only, and Π' the set of policies constructed from decision rules in D' only. It follows from Lemma 7.2.10 that $v^* = \sup_{\pi \in \Pi'} v^{\pi}$. Therefore, for any ε, $1 > \varepsilon > 0$, by Theorem 7.2.7 there exists a $d \in D'$ such that $v^{d^{\infty}} \geq (1 - \varepsilon)v^*$. Hence

$$\lim_{N \to \infty} P_d^N v^* \leq \frac{1}{1 - \varepsilon} \lim_{N \to \infty} P_d^N \sum_{n=1}^{\infty} P_d^n r_d = \frac{1}{1 - \varepsilon} \lim_{N \to \infty} \sum_{n=N+1}^{\infty} P_d^n r_d = 0,$$

where the last equality follows since $v^{d^{\infty}}(s) < \infty$ by assumption. Therefore (7.2.5) holds and Theorem 7.2.5(b) implies the result. $\square$

7.2.4 Value Iteration

In this section we investigate convergence of value iteration for positive bounded models. Because of the unavailability of iteratively improving bounds on the optimal value, numerical implementation is impractical. For this reason our interest lies in using value iteration to establish the existence of structured optimal policies. *We assume countable S except where noted.*

We show that, by choosing a v^0 which satisfies $0 \leq v^0 \leq v^*$, value iteration converges to v^*. Note that, in contrast to discounted models, convergence is pointwise and that it need not converge for other choices of v^0.

Theorem 7.2.12. In a positive bounded model, let $v^0 = 0$ and set $v^{n+1} = \mathscr{L} v^n$. Then v^n converges pointwise and monotonically to v^*.

Proof. By Proposition 7.2.1(a), $\mathscr{L} 0 \geq 0$, so that, as a consequence of Lemma 7.1.5(a), $\{v^n\}$ increases monotonically. Further, because $v^0 = 0$, there exists a $\pi \in \Pi^{\mathrm{MD}}$, for which $v^n = v_n^{\pi}$ for all n. Therefore, as a consequence of Assumption 7.2.1, $v^n(s)$ is finite and $\lim_{n \to \infty} v^n(s) = v(s)$ exists for each $s \in S$. Since $v \geq v^n$, Lemma 7.1.5(a) implies that $\mathscr{L} v \geq \mathscr{L} v^n = v^{n+1}$ for all n. Therefore $\mathscr{L} v \geq v$.

For any $d \in D$, and all n,

$$L_d v^n \leq \mathscr{L} v^n = v^{n+1} \leq v \leq \mathscr{L} v. \tag{7.2.10}$$

As a consequence of Lemma 7.1.5(c), $\lim_{n \to \infty} L_d v^n = L_d v$. Combining this with (7.2.10) yields

$$\mathscr{L} v = \sup_{d \in D} L_d v \leq v \leq \mathscr{L} v,$$

which implies that v is a fixed point of $\mathscr{L}$.

It remains to show that $v = v^*$. Choosing $\varepsilon > 0$ and a sequence $\{\varepsilon_n\}$ which satisfies $\sum_{n=1}^{\infty} \varepsilon_n = \varepsilon$, we can inductively construct $\pi = (d_1, d_2, \ldots) \in \Pi^{MD}$, which satisfies

$$\sum_{k=0}^{n} P_\pi^k r_{d_{k+1}} + \sum_{k=1}^{n} \varepsilon_n e \geq \mathscr{L}^n 0 \tag{7.2.11}$$

for all n. Hence $v^\pi + \varepsilon e \geq v$, implying $v \leq v^*$. Since Theorem 7.2.3(a) established that v^* is the minimal solution of the optimality equation, $v = v^*$. □

The following corollary shows that value iteration converges for other choices of v^0.

Corollary 7.2.13. Suppose, for each $s \in S$, that $0 \leq v^0(s) \leq v^*(s)$, then value iteration converges.

Proof. By the monotonicity of $\mathscr{L}$, it follows that, for all n, $\mathscr{L}^n 0 \leq \mathscr{L}^n v^0 \leq \mathscr{L}^n v^* = v^*$. The result follows from Theorem 7.2.12 which establishes that $\mathscr{L}^n 0$ converges to v^*. □

7.2.5 Policy Iteration

In this section, we study the following policy iteration algorithm.

1. Set $n = 0$ and select a $d_0 \in D^{MD}$ with $r_{d_0} \geq 0$.
2. Obtain v^n by finding the minimal solution of

$$\left(I - P_{d_n}\right) v = r_{d_n}. \tag{7.2.12}$$

3. Choose

$$d_{n+1} \in \arg\max_{d \in D^{MD}} \left\{r_d + P_d v^n\right\}$$

setting $d_{n+1}(s) = d_n(s)$ whenever possible.
4. If $d_{n+1} = d_n$, stop and set $d^* = d_n$. Otherwise increment n by 1 and return to step 2.

This algorithm differs from that in the discounted case in several subtle ways. In step 1, we choose a d_0 with $r_{d_0} \geq 0$, so that $v^n \in V^+$ for all $n \geq 0$. As a result of Theorem 7.2.3(a), we can associate v^{d_n} with the *minimal* solution of (7.2.12) in step 2. At termination we set $d^* = d_n$ instead of to *any* conserving decision rule. In finite state models, we can find the minimal solution of (7.2.12) by setting v^n equal to 0 on the recurrent states of P_{d_n}.

The following result establishes the optimality of $(d^*)^\infty$.

Proposition 7.2.14.   Suppose

$$d^* \in \arg\max_{d \in D^{MD}} \{r_d + P_d v^{(d^*)^\infty}\}. \tag{7.2.13}$$

Then $(d^*)^\infty$ is optimal.

Proof.   As a consequence of (7.2.13), $v^{(d^*)^\infty} = L_{d^*} v^{(d^*)^\infty} = L v^{(d^*)^\infty}$. Since $v^{(d^*)^\infty} \leq v^*$, it must be the minimal solution of $Lv = v$, so, by Theorem 7.2.3(a), $v^{(d^*)^\infty} = v^*$.   □

Thus, whenever policy iteration terminates in step 4, it identifies an optimal stationary policy. Unfortunately this termination step might not be reached, as illustrated by the following countable-state, finite-action example, which shows that the sequence of values generated by the policy iteration need not be monotone and might even cycle.

Example 7.2.4 (ctd.).   Choose $d_0(s) = a_{s,2}$ for $s > 1$. Then $v^0(s) = 0$ for all $s \in S$. The next improvement step chooses $d_1(s) = a_{s,1}$ for $s > 1$, so that $v^1(s) = 1 - s^{-1}$ for $s > 1$, and $v^1(1) = 0$. At the subsequent improvement step, for $s > 1$,

$$\max\{1 - s^{-1}, 1 - (s+1)^{-1}\} = 1 - (s+1)^{-1},$$

so that $d_2(s) = a_{s,2}$ and $v^2(s) = 0$ for all s. Therefore policy iteration will cycle between these two stationary policies and values and not converge. Note also that neither of these policies are optimal.

We now show that for positive bounded models with S finite, if there is strict improvement in step 3 and the original decision rule is retained in the states in which no improvement is possible, the expected total reward of the stationary policy derived from the new decision rule exceeds that of the previous decision rule. Our proof is considerably more subtle than that of Proposition 6.4.1 in the discounted case, and utilizes the chain structure of finite-state Markov chains (see Appendix A). Such analyses are customarily reserved for average reward models. The proof relies on the observation that $r_d \leq 0$ on the recurrent states of P_d whenever Assumption 7.2.1 holds.

Proposition 7.2.15.   Suppose S is finite and, for d and $\delta \in D^{MD}$,

$$r_\delta(s) + P_\delta v^{d^\infty}(s) \geq v^{d^\infty}(s), \tag{7.2.14}$$

with $\delta(s) = d(s)$ for $s \in S' \equiv \{s \in S: L_\delta v^{d^\infty}(s) = v^{d^\infty}(s)\}$. Then $v^{\delta^\infty}(s) \geq v^{d^\infty}(s)$. Further, $v^{\delta^\infty}(s) > v^{d^\infty}(s)$ for all s at which strict inequality holds in (7.2.14).

Proof. Let R_δ and T_δ denote the sets of recurrent and transient states of P_δ. Since S is finite, $R_\delta \neq \varnothing$. Write P_δ in canonical form (Appendix A, A.7), where U corresponds to transitions within R_δ. Then (7.2.14) may be expressed in partitioned form as

$$\begin{bmatrix} y \\ z \end{bmatrix} + \begin{bmatrix} U & 0 \\ V & W \end{bmatrix} \begin{bmatrix} v_1^d \\ v_2^d \end{bmatrix} \geq \begin{bmatrix} v_1^d \\ v_2^d \end{bmatrix}. \tag{7.2.15}$$

Consider first the recurrent states of P_δ. We show that if s is recurrent under δ then it must have been recurrent under d. Proposition 7.2.1(e) implies that $r_\delta(s) \leq 0$ for all $s \in R_\delta$, so that $y \leq 0$. Multiplying out the first row of (7.2.15) shows that

$$y + U v_1^d \geq v_1^d \tag{7.2.16}$$

on R_δ. Denote the limiting distribution of U by μ, so that $\mu^T U = \mu^T$. Therefore

$$\mu^T y + \mu^T U v_1^d \geq \mu^T v_1^d$$

on R_δ, so that $\mu^T y \geq 0$ on R_δ. By the recurrence of δ on R_δ, $\mu(s) > 0$ for all $s \in R_\delta$, so that $y = 0$. Hence from (7.2.16) we can select a $u \geq 0$ for which

$$U v_1^d = v_1^d + u$$

on R_δ. Multiplying this equation by μ^T and applying the above argument establishes that $u = 0$. Therefore $L_\delta v_1^d = U v_1^d = v_1^d$, so $R_\delta \subset S'$ and it follows that $\delta = d$ on R_δ. Hence $v^{d^\infty}(s) = v^{\delta^\infty}(s) = 0$ for $s \in R_\delta$, or, equivalently, $v_1^d = 0$.

Expanding (7.2.15) on the transient states and taking into account that $v_1^d = 0$ yields $z + W v_2^d \geq v_2^d$ or $z \geq (I - W) v_2^d$. Choose $w \geq 0$ so that $z = (I - W) v_2^d + w$. By Proposition A.3(b), $(I - W)^{-1}$ exists, is non-negative, and satisfies $(I - W)^{-1} \geq I$. Consequently, on T_δ,

$$v^{\delta^\infty} = (I - W)^{-1} z = v^{d^\infty} + (I - W)^{-1} w \geq v^{d^\infty} + w.$$

Since $w(s) > 0$ for any s at which strict inequality holds in (7.2.15), the conclusion follows.   □

The following example shows that restriction that "$\delta(s) = d(s)$" for $s \in S'$ is necessary for the conclusion of Proposition 7.2.16 to be valid.

Example 7.2.3 (ctd.).   In this example, $r_\delta + P_\delta v^{\gamma^\infty} = v^{\gamma^\infty}$, but $v^{\delta^\infty}(s_1) < v^{\gamma^\infty}(s_1)$. Note that if we begin policy iteration with $d_0 = \gamma$ and use the rule "set $d_{n+1} = d_n$", we identify γ^∞ as optimal.

Theorem 7.2.16.   Suppose S and A_s for each $s \in S$ are finite, and let $\{v^n\}$ denote the sequence of iterates of the policy iteration algorithm. Then, for some finite N, $v^N = v^*$ and $(d^*)^\infty$ is optimal.

Proof. By Proposition 7.2.15, $v^{n+1} \geq v^n$, with strict inequality in at least one component until $d_{n+1} = d_n = d^*$. Because D^{MD} is finite, this must occur for some finite N. By Proposition 7.2.14, $(d^*)^\infty$ is optimal and $v^N = v^{(d_N)^\infty} = v^*$. □

Thus in finite-state and action positive bounded models, policy iteration obtains an optimal policy provided that we start with a policy with non-negative expected total reward, choose the minimal non-negative solution of the evaluation equation in the evaluation step, and follow the rule "set $d_{n+1} = d_n$ if possible" in the improvement step.

7.2.6 Modified Policy Iteration

In models with expected total-reward criterion, modified policy iteration is best viewed as a variant of value iteration in which, after carrying out a maximization to identify a v^n-improving decision rule, we iterate with the same decision rule several times before performing another maximization. The existence of a decision rule with non-negative rewards ensures the monotonicity and convergence of this scheme.

We allow S to be countable but assume (for simplicity) that A_s is finite for each $s \in S$. Choose a sequence of positive integers $\{m_n\}$ and define the sequence $\{v^n\}$ by

$$v^n = L_{d_n}^{m_n} v^{n-1},$$

where $d_n \in \arg\max_{d \in D} \{r_d + P_d v^{n-1}\}$. A proof of convergence of this algorithm follows that in the discounted case. For any $v \in V^+$, let $Bv \equiv \mathscr{L}v - v$.

Theorem 7.2.17. For any $v^0 \in V^+$ satisfying

i. $v^0 \leq v^*$, and

ii. $Bv^0 \geq 0$,

the sequence $\{v^n\}$ generated by modified policy iteration converges monotonically to v^*.

Proof. Let $\{u^n\}$ be the sequence of iterates of value iteration with $u^0 = v^0$. We show by induction that, for all n, $u^n \leq v^n \leq v^*$ and that $Bv^n \geq 0$. By assumption, the induction hypothesis holds for $n = 0$.

Assume it is valid for $m = 1, 2, \ldots, n$. Let δ be a v^n-improving decision rule. Since $Bv^n \geq 0$, $L_\delta v^n \geq v^n$, so that, by the monotonicity of L_δ, $v^n \leq L_\delta v^n \leq L_\delta^{m_n} v^n = v^{n+1}$. Therefore, by the monotonicity of $\mathscr{L}$, $u^{n+1} = \mathscr{L}u^n \leq \mathscr{L}v^n = L_\delta v^n \leq L_\delta^{m_n} v^n = v^{n+1}$ $\leq \mathscr{L}^{m_n} v^* = v^*$. To see that $Bv^{n+1} \geq 0$, apply Proposition 6.4.3 with $\lambda = 1$ to obtain

$$Bv^{n+1} \geq r_\delta + (P_\delta - I)v^{n+1} = r_\delta + (P_\delta - I)\left[v^n + \sum_{k=0}^{m_n} P_\delta^k Bv^n\right]$$

$$= Bv^n + (P_\delta - I)\left[\sum_{k=0}^{m_n} P_\delta^k Bv^n\right] = P_\delta^{m_n+1} Bv^n \geq 0.$$

Hence the induction hypothesis holds at $n + 1$. As a consequence of (i), we may apply

Corollary 7.2.13 to establish convergence of u^n to v^*, from which the convergence of v^n to v^* follows. □

In practice we may always choose $v^0 = 0$, since by assumption $\mathscr{L}0 \geq 0$. We leave it as an exercise to derive an analogous result for general action sets.

7.2.7 Linear Programming

This section discusses the use of linear programming in *finite-state and action* positive Markov decision problems. We assume familiarity with the properties of linear programs described in Appendix D.

Theorem 7.2.3(a) showed that v^* was the minimal $v \in V^+$ for which

$$v \geq r_d + P_d v \qquad (7.2.17)$$

for all $d \in D$. From this result we derive the following primal linear programming problem. In it, $\alpha(j), j \in S$, denotes arbitrary positive scalars which satisfy $\sum_{j \in S} \alpha(j) = 1$.

Primal Linear Program

$$\text{Minimize } \sum_{j \in S} \alpha(j) v(j)$$

subject to

$$v(s) - \sum_{j \in S} p(j|s, a) v(j) \geq r(s, a)$$

for $a \in A_s$ and $s \in S$, and $v(s) \geq 0$ for all $s \in S$.

The formulation is similar to that in the discounted case; the constraints of the primal problem are identical to those in the discounted case with $\lambda = 1$; however, we require the additional condition that $v \geq 0$. Consequently, in the following dual problem we replace equality constraints by inequalities.

Dual Linear Program

$$\text{Maximize } \sum_{s \in S} \sum_{a \in A_s} r(s, a) x(s, a)$$

subject to

$$\sum_{a \in A_j} x(j, a) - \sum_{s \in S} \sum_{a \in A_s} p(j|s, a) x(s, a) \leq \alpha(j) \qquad (7.2.18)$$

and $x(s, a) \geq 0$ for $a \in A_s$ and $s \in S$.

Because solving the primal LP to obtain v^* only provides an optimal policy when there is a unique conserving decision rule, we seek one through the dual. The following theorem shows that the dual has an optimal basic feasible solution and that, from it, we may obtain a stationary optimal policy. The simplex algorithm yields such a solution.

Theorem 7.2.18. Suppose Assumptions 7.2.1 and 7.2.2 hold, and S and A_s for each $s \in S$, are finite.

   a. Then there exists a finite optimal basic feasible solution x^* to the dual LP.
   b. For each $s \in S$, $x^*(s, a) > 0$ for at most one $a \in A_s$.
   c. If

$$d(s) = \begin{cases} a & \text{if } x^*(s, a) > 0 \quad \text{and} \quad s \in S^* \\ \text{arbitrary} & \text{if } s \in S/S^*, \end{cases} \tag{7.2.19}$$

where $S^* = \{s \in S: \sum_{a \in A_s} x(s, a_s) > 0\}$, then d^∞ is optimal.

Proof. The dual LP has the feasible solution $x(s, a) = 0$ for all $s \in S$ and $a \in A_s$. Because of the finiteness of S and A_s and Assumption 7.2.1, Theorem 7.1.9 establishes the finiteness of v^*. Assumption 7.2.2 implies that $v^* \geq 0$. Consequently Theorems D.1(a) and (b) establish existence of a finite basic feasible optimal solution to the dual LP.

For all $j \in S$, define the dual "slack" variable $y^*(j)$ by

$$y^*(j) = \alpha(j) - \sum_{a \in A_j} x^*(j, a) + \sum_{s \in S} \sum_{a \in A_s} p(j|s, a)x^*(s, a), \tag{7.2.20}$$

so that, for all $j \in S$,

$$y^*(j) + \sum_{a \in A_j} x^*(j, a) = \sum_{s \in S} \sum_{a \in A_s} p(j|s, a)x^*(s, a) + \alpha(j) \geq \alpha(j) > 0.$$

Note that the dual has $|S|$ constraints and that the vector with components $(x^*(s, a), y^*(s))$ is an extreme point of the set of (x, y) satisfying (7.2.20) (i.e., the augmented dual LP). Hence, for each $j \in S$, $x^*(j, a_j) > 0$ for at most one $a_j \in A_j$ and either $y^*(j) > 0$ or $x^*(j, a_j) > 0$, so that part (b) follows.

As a consequence of (b), (7.2.19) uniquely defines $d(j)$ for $j \in S^*$. Let $x_d^*(s) \equiv x^*(s, d(s))$. Rewriting (7.2.18) with $x = x_d^*$ gives

$$(x_d^*)^T \leq \alpha^T + (x_d^*)^T P_d.$$

Iterating this expression and applying it to r_d yields, for any $N \geq 0$.

$$(x_d^*)^T r_d \leq \alpha^T \sum_{n=1}^{N} P_d^{n-1} r_d + (x_d^*)^T P_d^N r_d.$$

Since by Assumption 7.2.1, v^{d^∞} is finite, the second term on the right in the above expression converges to 0 as $N \to \infty$, so that

$$(x_d^*)^T r_d \leq \alpha^T v^{d^\infty}. \tag{7.2.21}$$

Theorem D.1(c) implies that $\alpha^T v^* = (x_d^*)^T r_d$, so from (7.2.21) we conclude that d^∞ is optimal.   □

Note that the part (b) of the above theorem applies to *any* basic feasible solution. We now show that the stationary policy defined in the above theorem remains optimal for any choice of α. To do that, we show that there exists an optimal basic feasible solution with a particularly nice structure.

Proposition 7.2.19. Let x^* be an optimal solution to the dual for some α, and suppose d is derived from (7.2.19). Then

a. there exists an optimal basic feasible solution x^{**} for which

$$x^{**}(s, a) = \begin{cases} (\alpha_\tau)^T (I - P_d)_\tau^{-1}(s) & s \in T_d \text{ and } a = d(s) \\ 0 & \text{otherwise,} \end{cases} \tag{7.2.22}$$

where α_τ and $(I - P_d)_\tau$ denote the restriction of α and $(I - P_d)$ to the transient states of P_d, T_d;

b. d^∞ is optimal for all $\alpha' > 0$.

Proof. Let R_d and T_d denote the sets of recurrent and transient states of P_d. Let x^{**} satisfy (7.2.22). Then x^{**} is basic and feasible. Let v^{**} be the corresponding primal solution. Then since $\alpha_\tau > 0$, $x^{**}(s, d(s)) > 0$ for $s \in T_d$ so by Theorem D.1(d) (complementary slackness), v^{**} satisfies

$$v^{**}(s) = \begin{cases} (I - P_d)_\tau^{-1}(r_d)_\tau(s) & s \in T_d \\ 0 & s \in R_d \end{cases},$$

where $(r_d)_\tau$ denotes the restriction of r_d to transient states of P_d. Therefore $v^{**} = v^{d^\infty} = v^* \geq 0$. Since $\alpha^T v^{**} = (\alpha_\tau)^T (I - P_d)_\tau^{-1}(r_d)_\tau = x^{**} r_d$, Theorem D.1(c) establishes optimality of x^{**} so (a) follows.

We now show that for any $\alpha' > 0$, the basis corresponding to x^{**} remains feasible. Write $(x')^T (I - P_d) \leq (\alpha')^T$ in partitioned form:

$$\begin{bmatrix} u \\ z \end{bmatrix}^T \begin{bmatrix} U & W \\ 0 & Z \end{bmatrix} \leq \begin{bmatrix} \beta \\ \gamma \end{bmatrix}^T,$$

where the submatrix U corresponds to states in T_d, and Z to states in R_d. Results in

Appendix A concerning the class structure of Markov chains implies that the submatrices U, W, and Z possess the following properties.

1. U^{-1} exists and satisfies $U^{-1} \geq I$.
2. All entries of W are nonpositive.
3. There exists a $z > 0$ such that $z^T Z = 0$.

In (3), we can choose z to be any non-negative scalar multiple of the limiting distribution of the recurrent states of S.

Define u by $u^T = \beta^T U^{-1}$ and let z satisfy (3). Clearly $(x')^T = (u^T, z^T) \geq 0$. Further, $u^T U = \beta^T$ and, since $u > 0$, $u^T W + z^T Z < z^T Z = 0 < \gamma^T$. Therefore $(x')^T$ satisfies (7.2.18) and the non-negativity constraint.

Since $z = 0$ is feasible, the vector of slack variables corresponding to recurrent states of P_d can be set equal to γ. Hence by Theorem D.1(e) the basis which determined x^{**} remains optimal and (7.2.19) (with x' replacing x^*) uniquely defines d on its transient states and allows arbitrary choice of actions on recurrent states. Hence (b) follows. □

The proof of the above proposition sheds light on the structure of the set of feasible solutions to dual LPs in positive models. This set contains directions (corresponding to limiting distributions of recurrent subchains of transition matrices of Markovian decision rules), but, since Proposition 7.2.1(e) implies that rewards are bounded by 0 on recurrent classes, the dual objective function is bounded. Also, for any deterministic Markov decision rule the x defined through (7.2.22) is a basic feasible solution. Note that unlike in discounted models, several basic feasible solutions may correspond to the same decision rule. The following result provides an interpretation for dual variables corresponding to specific basic feasible solutions.

Corollary 7.2.20. Let x be a basic feasible solution to the dual LP, let d be the decision rule derived using (7.2.19), with x replacing x^{**} and let $N_{s,a}$ denote the number of times the system is in state s and chooses action a. Then there exists a basic feasible solution to the dual x' which satisfies

$$x'(s, a) = \sum_{j \in S} \alpha(j) \sum_{n=1}^{\infty} P^{d^\infty}(X_n = s, Y_n = a | X_1 = j) = E^{d^\infty}\{N_{s,a}\}$$

for each $s \in S$ which is transient under P_d.

The representation in Corollary 7.2.20 suggests that it is easy and natural to incorporate constraints on state action frequencies through the dual LP provided states are transient under all Markovian decision rules. In such cases optimal decision rules may be randomized.

Example 7.2.1 (ctd.). We solve this problem using linear programming. The dual LP is as follows:
Maximize

$$5x(s_1, a_{1,1}) + 8x(s_1, a_{1,2})$$

subject to

$$0.5x(s_1, a_{1,1}) + x(s_1, a_{1,2}) \leq 0.5,$$
$$- 0.5x(s_1, a_{1,1}) - x(s_1, a_{1,2}) \leq 0.5,$$

and $x(s_1, a_{1,1}) \geq 0$, $x(s_1, a_{1,2}) \geq 0$ and $x(s_2, a_{2,1}) \geq 0$. Note that $x(s_2, a_{2,1})$ appears only in the non-negativity constraint.

The solution of this problem is $x(s_1, a_{1,1}) = 1$, $x(s_1, a_{1,2}) = 0$, and $x(s_2, a_{2,1})$ arbitrary with objective function value 5. Consequently the optimal policy d^∞ has $d(s_1) = a_{1,1}$.

Starting in s_1, the expected length of time in s_1 using this policy is two periods, and since $\alpha(s_1) = 0.5$, $E^{d^\infty}\{N_{s,a}\} = 1$ as guaranteed by Corollary 7.2.20. We leave it as Problem 7.15 to investigate the effects of constraints such as $x(s_1, a_{1,1}) \leq 0.75$ on the LP solution and the optimal policy.

7.2.8 Optimal Stopping

Applications of the positive bounded model include

1. optimal stopping,
2. gambling, and
3. playing computer games.

In this section we focus on optimal stopping problems. We refer the reader to problems and references for other applications.

We analyze a time-homogeneous version of the model of Sec. 3.4. In it, the system moves between a set of states S' according to the transition probabilities $p(j|s)$ of a finite Markov chain. For each $s \in S'$, $A_s = \{C, Q\}$, where C denotes the continuation action and Q denotes the quitting or stopping action. Choosing C in state s causes the system to move to state $j \in S'$ with probability $p(j|s)$. Choosing Q moves the system to state Δ, at which it receives no subsequent rewards; $S \equiv S' \cup \Delta$ and $A_\Delta = \{C\}$.

Rewards may be

a. accumulated continuously until stopping, or
b. received only upon stopping.

A model with reward structure (a) may be converted to one with reward structure (b) by redefining the state space; however, we prefer a direct formulation.

An example of a model with reward structure (a) is one in which the reward is identically 1 until stopping. In it,

$$v^\pi(s) = E_s^\pi\left\{ \sum_{n=1}^{\infty} I_{(S')}(X_n) \right\} = E_s^\pi\{\tau\} - 1,$$

where τ denotes the random time at which the system enters the stopped state. Therefore, solving the optimal stopping problem maximizes the number of transitions until stopping. This reward structure may be suitable for computer games in which the objective might be viewed as playing as long as possible.

We begin this section by analyzing models in which rewards are received only upon termination. Therefore $r(s, Q) = g(s)$ for all $s \in S'$. Assumption 7.2.1 requires that $g(s) < \infty$ for all $s \in S'$, and that $g(s) \leq 0$ on recurrent states of S'. Further, $r(\Delta, C) = 0$. Our objective is to choose at which states to stop to maximize expected reward.

Applying Theorem 7.2.3(a) shows that v^* is the minimal solution in V^+ of

$$v(s) = \max\left\{ g(s) + v(\Delta), \sum_{j \in S'} p(j|s)v(j) \right\} \quad s \in S', \qquad (7.2.23)$$

and $v(\Delta) = v(\Delta)$. Consequently $v^*(\Delta) = 0$, and it follows that v^* is the minimal $v \geq 0$, satisfying

$$v(s) \geq g(s) \quad s \in S', \qquad (7.2.24)$$

$$v(s) \geq \sum_{j \in S'} p(j|s)v(j) \quad s \in S'. \qquad (7.2.25)$$

When v satisfies (7.2.24) we say that v *majorizes* g, and we call any v satisfying (7.2.25) on S' *excessive*. Therefore, in a positive bounded optimal stopping model, (7.2.24) and (7.2.25) imply that *the value v^* is the minimal excessive function which majorizes g.*

For finite S', Theorem 7.1.9 ensures that there exists a stationary optimal policy, so that $v^*(s)$ is uniformly bounded and equals 0 on recurrent states of S'. Consequently all policies satisfy (7.2.5), and Theorem 7.2.5(b) implies that a stationary policy derived from a conserving decision rule is optimal. We summarize this as follows.

Theorem 7.2.22

a. Suppose S' is countable, $g(s) < \infty$ for all $s \in S'$, and $g(s) \leq 0$ if s is a recurrent state. Then the value of the optimal stopping problem v^* is the minimal non-negative $v \geq g$ satisfying (7.2.25).

b. If, in addition, S' is finite, the stationary policy $(d^*)^\infty$ defined through

$$d^*(s) = \begin{cases} Q & s \in \{s' \in S' : v^*(s') = g(s')\} \\ C & \text{otherwise,} \end{cases}$$

is optimal.

The primal linear program of Sec. 7.2.7 provides an easy way of computing v^*. We illustrate this with the following example.

Example 7.2.6. We wish to determine an optimal stopping policy for the Markov chain in Fig. 7.2.5. In it, $S' = \{s_1, s_2, s_3, s_4, s_5\}$, and let $g(s_1) = 6$, $g(s_2) = 1$, $g(s_3) = 0$, $g(s_4) = 0$, and $g(s_5) = 1$. Note $g(s) = 0$ on recurrent states of S'.

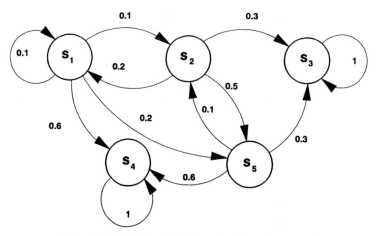

Figure 7.2.5 Markov chain for optimal stopping example.

We find v^* by solving the following linear program.
Minimize

$$v(s_1) + v(s_2) + v(s_3) + v(s_4) + v(s_5)$$

subject to

$$0.9v(s_1) - 0.1v(s_2) - 0.6v(s_4) - 0.2v(s_5) \geq 0,$$
$$-0.2v(s_1) + v(s_2) - 0.3v(s_3) - 0.5v(s_5) \geq 0,$$
$$-0.1v(s_2) - 0.3v(s_3) - 0.6v(s_4) + v(s_5) \geq 0,$$

and $v(s_1) \geq 6$, $v(s_2) \geq 1$, $v(s_5) \geq 1$, and $v(s) \geq 0$ for $s \in S'$.

Using the simplex algorithm, we find the optimal solution to be $v^*(s_1) = 6$, $v^*(s_2) = 1.7$, $v^*(s_3) = 0$, $v^*(s_4) = 0$, and $v^*(s_5) = 1$. Therefore the optimal stationary policy $(d^*)^\infty$ has $d^*(s) = Q$ for $s \neq s_2$ and $d^*(s_2) = C$.

When the Markov chain is a symmetric random walk with absorbing barriers, the optimal policy has a particularly elegant graphical representation. Suppose $S' = \{1, 2, \ldots, N\}$ and $p(j + 1|j) = p(j - 1|j) = 0.5$ for $j = 2, 3, \ldots, N - 1$ and $p(1|1) = p(N|N) = 1$. Then (7.2.25) becomes

$$v(j) \geq 0.5v(j - 1) + 0.5v(j + 1) \qquad 2 \leq j \leq N - 1. \qquad (7.2.26)$$

Since relationship (7.2.26) defines *concave* functions on the integers, it implies that v^* must be concave. Consequently we have the following result.

Corollary 7.2.23. Suppose the underlying Markov chain is a symmetric random walk on $S' = \{1, 2, \ldots, N\}$ with absorbing barriers 1 and N, that $g(s) < \infty$ for all

$s \in S$, $g(1) \le 0$, and $g(N) \le 0$. Then v^* is the smallest non-negative concave function which majorizes g.

We illustrate this result with the following example.

Example 7.2.7. Suppose the Markov chain is a random walk on $S' = \{1, 2, \ldots, 10\}$ with absorbing barriers 1 and 10, and $g = (0, 3, 5, 4, 7, 11, 8, 3, 6, 0)$. Then Corollary 7.2.23 implies that we can find v^* by the method suggested in Fig. 7.2.6.

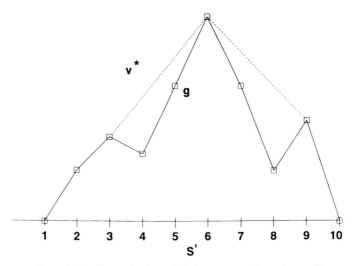

Figure 7.2.6 Determination of v^* for a symmetric random walk.

Note that $\Sigma \equiv \{s \in S': v^*(s) = g(s)\} = \{1, 2, 3, 6, 9, 10\}$, so the optimal policy stationary policy is to stop in states 1, 2, 3, 6, 9, and 10 and continue otherwise. Figure 7.2.6 suggests that we call Σ the *support set* of v^*. Hence the optimal policy is to stop when the system first enters the support set.

Models with Continuation Fees. We now consider optimal stopping problems with continuation fees. Assume that in each state there is a cost $c > 0$ for continuing for one more period. Therefore the optimality equations for this model become

$$v(s) = \max\left\{ g(s) + v(\Delta), -c + \sum_{j \in S'} p(j|s)v(j) \right\} \qquad s \in S'. \quad (7.2.27)$$

Assumption 7.2.1 requires that $g(s) < \infty$ for all $s \in S$. Since $-c > 0$, Assumption 7.2.2 requires that $g(s) \ge 0$. Observe that this is a positive bounded model although some $r(s, a) < 0$. Theorem 7.2.3(a) yields the following result.

Theorem 7.2.24. Suppose S' is countable and $0 \leq g(s) < \infty$ for all $s \in S'$. Then v^* is the minimal $v \in V^+$ satisfying $v(s) \geq g(s)$ for $s \in S'$ and

$$v(s) \geq -c + \sum_{j \in S'} p(j|s)v(j), \qquad s \in S'.$$

Since $v \geq g \geq 0$ on S', when S' is finite, we can find v^* directly by applying linear programming to the above system of equations. We now reanalyze Example 7.2.6 when it includes a continuation fee.

Example 7.2.6 (ctd.). Suppose that the continuation fee $c = 0.65$, then the optimal decision rule obtained in the no continuation fee model in Example 7.2.6 remains optimal, but, if $c > 0.70$, it is optimal to stop at *all* states.

Selling an Asset. We conclude this section by applying the above results to find an optimal policy for the asset selling problem of Sec. 3.4.2. We assume that offers arrive weekly, that the probability of receiving an offer of value j is $p(j)$, $j = 1, 2, \ldots, N$, and that the cost of retaining an asset for one week is c units. Assume that $c < N$; if not, we would accept the first offer. We define the *modified asset selling problem* to be one in which we may accept *any* offer up to and including the present offer. Our objective is to choose a stopping policy that maximizes the expected profit from the sale of the asset.

Decisions are made immediately after the arrival of an offer. Let states in S' represent the greatest offer to date; therefore $S' = \{1, 2, \ldots, N\}$. For $s < N$, $A_s = \{Q, C\}$, while $A_N = \{Q\}$ since N is the best offer possible. The optimality equation for this model is

$$v(s) = \max\left\{ s, -c + \sum_{j>s} p(j)v(j) + s \sum_{j \leq s} p(j) \right\}, \qquad 1 \leq s \leq N - 1,$$

and $v(N) = N$. Note that the last summation represents the probability that the subsequent offer falls below the best offer to date.

The following theorem solves this problem by showing that the optimal policy is to accept the first offer s for which $\sum_{j>s}(j - s)p(j) < c$. Let X denote a random variable with probability distribution $\{p(j); j = 1, 2, \ldots, N\}$. Since $\sum_{j>s}(j - s)p(j) = E\{(X - s)^+\}$, the optimal policy has the following interpretation. If the expected additional revenue from seeking the next offer exceeds the continuation cost, we wait another period; if not, we stop and accept the current offer.

Theorem 7.2.25. Let $s^* \equiv \min\{s: \sum_{j>s}(j - s)p(j) < c\}$. Then, if $E[X] > c$, $1 \leq s^* \leq N$ and the stationary policy $(d^*)^\infty$ derived from

$$d^*(s) \equiv \begin{cases} C & s < s^* \\ Q & s \geq s^* \end{cases}$$

is an optimal solution for the (unmodified) asset selling problem.

Proof. We use policy iteration to establish the optimality of this policy. We choose d_0 to be the decision rule which stops immediately in each state. Therefore $v^0(s) = s$ and $v^0(\Delta) = 0$. Now we evaluate

$$Lv^0(s) = \max\left\{s, -c + \sum_{j>s} jp(j) + s\sum_{j\leq s} p(j)\right\}$$

$$= \max\left\{s, -c + \sum_{j>s} (j - s)p(j) + s\right\}.$$

The quantity $\sum_{j>s}(j - s)p(j)$ is nonincreasing in s, equals $E[X]$ for $s = 0$, and equals 0 for $s = N$. Therefore if $E[X] > c$, $s^* \equiv \min\{s: \sum_{j>s}(j - s)p(j) < c\}$ exists and satisfies $1 \leq s^* \leq N$. Then

$$d_1(s) = \begin{cases} C & s < s^* \\ Q & s \geq s^*. \end{cases}$$

We need not solve for v^1 (which saves considerable tedious work), but note that $v^1(s) = s$ for $s \geq s^*$ and $v^1(s) > s$ for $s < s^*$. We now show that $d_2 = d_1$ from which we conclude that the $(d_1)^\infty$ is optimal.

For $s \geq s^*$,

$$Lv^1(s) = \max\left\{s, -c + \sum_{j>s} p(j)v^1(j) + s\sum_{j\leq s} p(j)\right\}.$$

Since $v^1(j) = j$ for $j \geq s^*$, adding and subtracting $s\sum_{j>s} p(j)$ shows that

$$Lv^1(s) = \max\left\{s, -c + \sum_{j>s} (j - s)p(j) + s\right\}.$$

By the definition of s^*, $\sum_{j>s}(j - s)p(j) - c < 0$, so the optimal action when $s \geq s^*$ is Q.

For $s < s^*$,

$$v^1(s) = \max\left\{s, -c + \sum_{j\geq s^*} jp(j) + \sum_{s<j<s^*} p(j)v^1(j) + s\sum_{j\leq s} p(j)\right\}.$$

Since $v^1(j) > j$ for $j < s^*$, the second expression in the max is bounded below by

$$-c + \sum_{j>s} (j - s)p(j) + s.$$

But, by the definition of s^*, this quantity exceeds s, so that optimal action is to continue. Therefore $d_2 = d_1$, and Proposition 7.2.14 establishes that the specified policy is optimal for the modified problem.

Since this policy is feasible for the unmodified problem, and the value of the modified problem exceeds that of the unmodified asset selling problem, it is also optimal for the unmodified problem.   □

We illustrate this result with the following simple example.

Example 7.2.8.   Let $S = \{1, 2, 3\}$, $p(1) = 0.2$, $p(2) = 0.4$, $p(3) = 0.4$ and $c = 1$. The expected offer size, $E[X] = 2.2$, so that $1 \leq s^* \leq 3$. Since $\Sigma_{j > 1}(j - 1)p(j) = 1.2$ and $\Sigma_{j > 2}(j - 2)p(j) = 0.4$, $s^* = 2$ and an optimal policy accepts the first offer of 2 or 3.

7.3   NEGATIVE MODELS

We call a MDP *negative* if $r(s, a) \leq 0$ for all $a \in A_s$ and $s \in S$ or, equivalently, if $r_d \leq 0$ for all $d \in D^{\mathrm{MD}}$. We do not restrict v^π to be finite for all policies and include in our analysis models in which $v^\pi(s) = -\infty$ for some $s \in S$ and $\pi \in \Pi^{\mathrm{HR}}$. If $v^\pi = -\infty$ for all $\pi \in \Pi^{\mathrm{HR}}$, all policies are equivalent under the expected total reward criterion and the average reward criteria would be preferable for distinguishing policies. Therefore we impose the following assumption.

Assumption 7.3.1.   There exists a $\pi \in \Pi^{\mathrm{HR}}$ for which $v^\pi(s) > -\infty$ for all $s \in S$.

Negative models usually represent problems in which the decision maker seeks to minimize expected total costs. We interpret $r(s, a)$ as a negative cost, so that maximization of the expected total reward corresponds to minimization of the expected total cost. In negative models, the decision maker seeks policies which drive the system to zero-cost states as quickly as possible while taking into account costs incurred along the way. Other applications include models in which the objective is to minimize the expected time required to reach a desirable state or to minimize the probability of reaching an undesirable state.

Analysis in this section parallels that in Sec. 7.2 as closely as possible. We assume *countable S*. Surprisingly, for negative models, optimal policies exist under weaker conditions than for positive models; however, they are more difficult to find because algorithms need not converge to v^*.

7.3.1   The Optimality Equation

This section explores the role of the optimality equation in negative models. We let $V^- = \{v: S \to R: v(s) \leq 0\}$. In contrast to V^+, we do not require elements of V^- to be bounded. The following easily proved proposition collects some useful properties of negative models, and parallels Proposition 7.2.1 for positive models.

Proposition 7.3.1.   In a negative MDP,

a. $\mathscr{L}0 \leq 0$;   (7.3.1)

b. $v^{d^\infty} \leq 0$ for all $d \in D^{\mathrm{MD}}$;

c. $v^* \le 0$;

d. $\mathscr{L}: V^- \to V^-$, and

e. for any $d \in D^{MD}$ if $v^{d^\infty}(s) > -\infty$ for all $s \in S$, then $r_d = 0$ on each recurrent state of P_d.

The theorem below provides lower bounds on v^* and allows us to characterize v^* as the maximal solution of the optimality equation.

Theorem 7.3.2. Suppose there exists a $v \in V^-$ for which $v \le \mathscr{L}v$. Then $v \le v^*$.

Proof. Specify $\varepsilon > 0$ and choose a sequence (ε_n) for which $\sum_{n=0}^\infty \varepsilon_n = \varepsilon$. Standard arguments such as those used in the proof of Theorem 7.2.4 imply that there exists a policy $\pi = (d_1, d_2, \dots) \in \Pi^{MD}$ with the property that, for any $N \ge 1$,

$$\sum_{n=1}^N P_\pi^{n-1} r_{d_n} + P_\pi^N v + \sum_{n=1}^N \varepsilon_n e \ge v.$$

Since $v \le 0$, $P_\pi^N v \le 0$ for all N. In light of Assumption 7.1.1, we can pass to the limit in the above expression and conclude that $v^\pi + \varepsilon e \ge v$. Therefore $v^* \ge v$. $\square$

Since Theorem 7.1.3 establishes that $v^* = \mathscr{L}v^*$, the above result implies that any other $v \in V^-$ which satisfies $v = \mathscr{L}v$ must be less than or equal to v^*. Thus we have the following important result.

Theorem 7.3.3.

a. The quantity v^* is the maximal solution of $v = \mathscr{L}v$ in V^-, and

b. for any $d \in D^{MR}$, v^{d^∞} is the maximal solution of $v = L_d v$ in V^-.

The following simple example illustrates the above concepts and shows that we need additional conditions to obtain an upper bound for v^*.

Example 7.3.1. Let $S = \{s_1, s_2\}$, $A_{s_1} = \{a_{1,1}, a_{1,2}\}$, $A_{s_2} = \{a_{2,1}\}$, $r(s_1, a_{1,1}) = 0$, $r(s_1, a_{1,2}) = -1$, $r(s_2, a_{2,1}) = 0$, $p(s_1|s_1, a_{1,1}) = 1$, $p(s_2|s_1, a_{1,2}) = 1$, $p(s_2|s_2, a_{2,1}) = 1$, and $p(j|s, a) = 0$ otherwise (Fig. 7.3.1). Let δ and γ denote the deterministic Markovian decision rules where $\delta(s_1) = a_{1,1}$, $\gamma(s_1) = a_{1,2}$ and $\delta(s_2) = \gamma(s_2) = a_{2,1}$. The optimality equation is given by

$$v(s_1) = \max\{v(s_1), -1 + v(s_2)\}$$

$$v(s_2) = v(s_2).$$

Since $v^{\gamma^\infty}(s_1) = -1$, δ^∞ is optimal with $v^{\delta^\infty}(s_1) = 0$, $v^{\delta^\infty}(s_2) = 0$.

Note that any $v \in V^-$ for which $v(s_1) \ge -1 + v(s_2)$ satisfies $\mathscr{L}v = v$, so that the optimality equation does not uniquely characterize v^*, but that $v^* = 0$ is the maximal

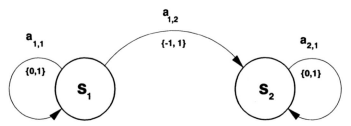

Figure 7.3.1 Symbolic representation of Example 7.3.1.

solution in V^-. The inequality $v \geq \mathscr{L}v$ does not provide upper bounds either. To see this, note that $v(s_1) = v(s_2) = -1$ satisfies $v \geq \mathscr{L}v$ but $v \leq v^*$.

The following result provides upper bounds for v^* and another characterization of the optimal value for negative models.

Proposition 7.3.4.

a. Suppose there exists a $v \in V^-$ for which $v \geq \mathscr{L}v$ and, for each $s \in S$

$$\limsup_{N\to\infty} E_s^\pi\{v(X_N)\} = \limsup_{N\to\infty} P_\pi^N v(s) = 0 \qquad (7.3.2)$$

for all $\pi \in \Pi^{MR}$. Then $v \geq v^*$.

b. Suppose $v \in V^-$ satisfies $\mathscr{L}v = v$ and (7.3.2) holds, then $v = v^*$.

Proof. To prove part (a), choose $\pi = (d_1, d_2, \ldots) \in \Pi^{MR}$. Lemma 7.1.2 together with $\mathscr{L}v \leq v$ implies that, for all N,

$$\sum_{n=1}^N P_\pi^{n-1} r_{d_n} + P_\pi^N v \leq v.$$

Noting (7.3.2) and Proposition 7.1.1 establishes the result.
Part (b) follows by combining part (a) with Theorem 7.3.2. □

Condition (7.3.2) requires that, for v to equal v^*, v must equal 0 on any states which are recurrent under any policy. Note in Example 7.3.1 that v^{γ^∞} satisfies the optimality equation but not (7.3.2), since s_1 is recurrent under δ^∞ and $v^{\gamma^\infty}(s_1) = -1$. Hence from Proposition 7.3.4 we conclude that γ^∞ cannot be optimal.

7.3.2 Identification and Existence of Optimal Policies

In negative models, we identify optimal policies in the same way as in discounted models; stationary policies derived from conserving decision rules are optimal. This is because $v^* \leq 0$, so that all policies are equalizing, that is, they satisfy (7.1.10). Applying Theorem 7.1.7 yields the following result.

Theorem 7.3.5. If $\delta \in d^{\text{MD}}$ satisfies

$$r_\delta + P_\delta v^* = v^*,$$

then δ^∞ is optimal.

Optimal policies exist in considerably greater generality than in the positive case; any condition which ensures the existence of conserving decision rules implies that existence of a stationary optimal policy. Example 7.2.4 showed that, in positive bounded models, there may not exist an optimal policy when S is countable and A_s is finite. Theorem 6.2.10 for discounted models provides conditions which ensure the existence of conserving decision rules and stationary optimal policies for negative models. We restate it below.

Theorem 7.3.6. Assume S is discrete (finite or countable) and either

a. A_s is finite for each $s \in S$,
b. A_s is compact, $r(s, a)$ is continuous in a for each $s \in S$, and $p(j|s, a)$ is continuous in a for each $j \in S$ and $s \in S$; or
c. A_s is compact, $r(s, a)$ is upper semicontinuous in a for each $s \in S$, and $p(j|s, a)$ is lower semicontinuous in a for each $j \in S$ and $s \in S$.

Then there exists an optimal deterministic stationary policy.

The following example with countable A_s shows that, even in finite-state models, there need not exist a stationary ε-optimal policy.

Example 7.3.2. Let $S = \{s_1, s_2, s_3\}$, $A_{s_1} = \{1, 2, \dots\}$, $A_{s_2} = \{a_{2,1}\}$ and $A_{s_3} = \{a_{3,1}\}$; $r(s_1, a) = 0$ for $a \in A_{s_1}$, $r(s_2, a_{2,1}) = -1$ and $r(s_3, a_{3,1}) = 0$; and $p(s_1|s_1, a) = p_a$, $p(s_2|s_1, a) = 1 - p_a$ for $a \in A_{s_1}$ where p_a are specified probabilities, $p(s_3|s_2, a_{2,1}) = 1$ and $p(s_3|s_3, a_{3,1}) = 1$ (Fig. 7.3.2).
For any policy π, $v^\pi(s_2) = -1$ and $v^\pi(s_3) = 0$. Pick $0 < \varepsilon < 1$ and set

$$p_a = (1 - \varepsilon)^{(\frac{1}{2})^a}.$$

Now let π' denote the policy which chooses the sequence of actions $\{1, 2, \dots\}$ in s_1. Then under π', the probability the system never leaves state s_1 is $\prod_{n=1}^\infty p_n = 1 - \varepsilon$.

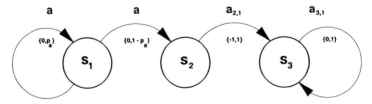

Figure 7.3.2   Symbolic representation of Example 7.3.2.

Therefore $v^{\pi'}(s_1) = (1 - \varepsilon) \cdot 0 + \varepsilon \cdot (-1) = -\varepsilon$. Since ε was arbitrary $v^*(s_1) = 0$ and π' which is in Π^{MD} is ε-optimal.

Now let δ^{∞} be any deterministic stationary policy and let α denote the action it chooses in s_1. Under this policy the system remains in state s_1 for N decision epochs with probability $(p_\alpha)^N$. Hence, with probability 1, the system eventually leaves state s_1 so that $v^{\delta^{\infty}}(s_1) = -1$. Therefore there exists no ε-optimal deterministic stationary policy for any $0 < \varepsilon < 1$. We leave it as an exercise to extend this observation to randomized stationary policies.

Since v^* satisfies the optimality equation, the proof of Theorem 7.3.2 establishes the following result.

Theorem 7.3.7. For any $\varepsilon > 0$, there exists a policy in Π^{MD} which is ε-optimal.

Finally, we have the following result, the proof of which is considerably easier than that for positive models.

Theorem 7.3.8. Suppose there exists an optimal policy; then there exists a deterministic stationary optimal policy.

Proof. Let $\pi^* \in \Pi^{HR}$ be an optimal policy. Write it as $\pi^* = (\delta, \pi')$ where $\delta \in D^{MR}$ denotes the decision rule it uses at the first decision epoch and $\pi' \in \Pi^{HR}$ denotes the policy it follows at subsequent decision epochs. Then, as a consequence of Lemmas 7.1.5(a) and 7.1.2,

$$v^{\pi^*} = L_\delta v^{\pi'} \le L_\delta v^{\pi^*} \le \sup_{d \in D^{MR}} \{r_d + P_d v^{\pi^*}\} = \sup_{d \in D^{MD}} \{r_d + P_d v^{\pi^*}\} = \mathscr{L} v^{\pi^*} = v^{\pi^*},$$

where the last equality follows from Theorem 7.1.3. Hence the supremum of $r_d + P_d v^{\pi^*}$ over D^{MR} is attained so by Lemma 7.1.2 there exists a $\delta' \in D^{MD}$ for which

$$r_{\delta'} + P_{\delta'} v^{\pi^*} = v^{\pi^*}$$

Theorem 7.3.5 establishes the optimality of $(\delta')^{\infty}$.   □

7.3.3 Value Iteration

As was the case for positive models, our interest in value iteration is primarily theoretical. Its main use is to determine structure of optimal policies. However, unlike the case of positive and discounted models, value iteration need not converge to v^*, as illustrated by the following example from Strauch (1966).

Example 7.3.3. Let $S = \{s_1, s_2, s_3, \ldots\}$, $A_{s_1} = \{a_{1,1}\}$, $A_{s_2} = \{a_{2,3}, a_{2,4}, a_{2,5}, \ldots\}$, $A_{s_3} = \{a: a = 2, 3, \ldots\}$, and $A_{s_j} = \{a_{j,1}\}$, $j \ge 4$. All transitions are deterministic except those from state s_3. There $p(s_1|s_3, a) = a^{-1}$ and $p(s_3|s_3, a) = 1 - a^{-1}$, for $a \in A_{s_3}$, otherwise $p(s_1|s_1, a_{1,1}) = 1$, $p(s_a|s_2, a) = 1$ for $a \in A_{s_2}$ and $p(s_{j-1}|s_j, a_{j,1}) = 1$. Rewards received in s_3 also depend on the state of the system at the subsequent decision epoch. Recall that $r(s, a, j)$ denotes the reward received in state s if action a

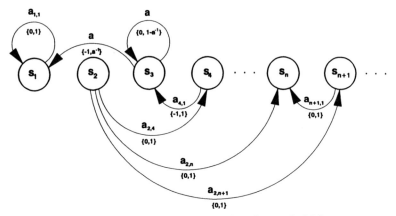

Figure 7.3.3 Symbolic representation of Example 7.3.3.

is chosen, and the next transition moves the system to state j. Let $r(s_3, a, s_1) = -1$ and $r(s_3, a, s_3) = 0$ for $a \in A_{s_3}$. All other rewards are 0 except $r(s_4, a_{4,1}, s_3) = -1$. Fig. 7.3.3 illustrates the transition and reward structure for this complicated example.

As in Example 7.3.2, we can choose a sequence of actions in s_3 which make the probability of a transition to s_1 arbitrarily small. This implies that $v^*(s_3) = 0$; otherwise, $v^*(s_1) = 0$ and $v^*(s_j) = -1$ for $j = 2$ and $j \geq 4$. Now apply value iteration beginning with $v^0 = 0$. For all n, $v^n = \mathscr{L}^n 0$ satisfies

$$v^n(s_j) = \begin{cases} 0 & j \leq 3 \\ -1 & j \geq 4 \end{cases}$$

since choosing a sufficiently large value for a in s_2 ensures that the system does not reach s_3 within n transitions. Therefore $0 = \lim_{n \to \infty} v^n(s_2) > v^*(s_2)$ and value iteration does not converge to v^*.

We provide some conditions which ensure convergence of value iteration in negative models. We begin with the following result which holds for *all* models with the expected total-reward criterion; however, we prove it only for negative models. We leave it as an exercise (Problem 7.1(a)) to prove it in general.

Proposition 7.3.9. Suppose $v^* \leq 0$; then

$$v^\infty \equiv \lim_{n \to \infty} \mathscr{L}^n 0 \geq v^* \tag{7.3.3}$$

Proof. By the monotonicity of $\mathscr{L}$, for any n, $\mathscr{L}^n 0 \geq \mathscr{L}^n v^* = v^*$. Therefore passing to the limit establishes (7.3.3). $\square$

Theorem 7.3.10. Suppose Assumption 7.3.1 holds and $r(s, a) \leq 0$ for all $s \in S$ and $a \in A_s$ and either

    a. S is finite, or

    b. S is countable and A_s is finite for each $s \in S$.

Then, if $v^0 = 0$, $v^{n+1} = \mathscr{L}v^n$ converges monotonically to v^*.

Proof. Since $v^1 = \mathscr{L}0 \leq 0$, $v^{n+1} = \mathscr{L}v^n \leq v^n$, so that $\{v^n\}$ is monotone and $\mathscr{L}v^\infty \leq v^\infty$ where v^∞ is defined in (7.3.3).

We now establish $\mathscr{L}v^\infty \geq v^\infty$. By Assumption 7.3.1, $v^*(s) > -\infty$ for all $s \in S$, so by Proposition 7.3.9, $v^\infty(s) > -\infty$ for all $s \in S$. Then, for $s \in S$

$$v^\infty(s) = \lim_{n \to \infty} \mathscr{L}v^n(s)$$

$$= \lim_{n \to \infty} \sup_{a \in A_s} \left\{ r(s, a) + \sum_{j \in S} p(j|s, a)v^\infty(j) + \sum_{j \in S} p(j|s, a)[v^n(j) - v^\infty(j)] \right\}$$

$$\leq \mathscr{L}v^\infty(s) + \lim_{n \to \infty} \sup_{a \in A_s} \left\{ \sum_{j \in S} p(j|s, a)[v^n(j) - v^\infty(j)] \right\}.$$

Hence, if

$$\Delta(s) \equiv \lim_{n \to \infty} \sup_{a \in A_s} \left\{ \sum_{j \in S} p(j|s, a)[v^n(j) - v^\infty(j)] \right\} = 0, \qquad (7.3.4)$$

the result follows.

We now show that (7.3.4) holds under (a) or (b). Under (a) v^n converges to v^∞ uniformly. This means that, for any $\varepsilon > 0$, there exists an N sufficiently large so that, for all $n \geq N$,

$$\Delta(s) \leq \max_{s \in S} \{v^n(s) - v^\infty(s)\} < \varepsilon.$$

Hence (7.3.4) holds.

Under (b), for each $s \in S'$, the finiteness of A_s ensures existence of an $a' \in A_s$ and a subsequence $\{n_k\}$ such that

$$\sup_{a \in A_s} \left\{ \sum_{j \in S} p(j|s, a)[v^{n_k}(j) - v^\infty(j)] \right\} = \sum_{j \in S} p(j|s, a')[v^{n_k}(j) - v^\infty(j)].$$

Therefore

$$\Delta(s) = \lim_{k \to \infty} \sum_{j \in S} p(j|s, a')[v^{n_k}(j) - v^\infty(j)],$$

so, by the Lebesgue Monotone Convergence Theorem, $\Delta(s) = 0$.

Therefore, under (a) or (b), $\mathscr{L}v^\infty = v^\infty$. By Proposition 7.3.9, $v^\infty \geq v^*$ and by Theorem 7.3.2, v^* is the maximal solution of $\mathscr{L}v = v$ in V^- so that $v^\infty = v^*$. $\square$

Note that if we relax Assumption 7.3.1 so that $v^*(s) = -\infty$ for some $s \in S$, then the iterates of value iteration $\{v^n(s)\}$, converge to $v^*(s)$ in finite-state models at all $s \in S$ for which $v^*(s) > -\infty$. The proof of the above theorem shows that value iteration converges whenever (7.3.4) holds. We restate this formally as follows.

Corollary 7.3.11.   Suppose in the negative model that $v^0 = 0$ and (7.3.4) holds. Then $v^{n+1} = \mathscr{L}v^n$ converges monotonically to v^*.

Value iteration also converges for starting values other than $v^0 = 0$.

Corollary 7.3.12.   Suppose (a) or (b) in Theorem 7.3.9 or (7.3.4) holds; then value iteration converges whenever $0 \geq v^0 \geq v^*$.

Proof.   By the monotonicity of $\mathscr{L}$, $\mathscr{L}^n 0 \geq \mathscr{L}^n v^0 \geq \mathscr{L}^n v^* = v^*$ for all n, which implies the result. $\square$

Problems 7.25 and 7.26 provide other conditions which ensure convergence of value iteration to v^*.

7.3.4 Policy Iteration

The policy iteration algorithm need not provide optimal policies in negative models because

a. there may exists a stationary policy δ^∞ for which $v^{\delta^\infty}(s) = -\infty$ for some $s \in S$, and
b. the expected total reward of a suboptimal policy may satisfy the optimality equation.

Even if we were able to circumvent (a), for example, by initiating the algorithm with a stationary policy which had a finite expected total reward, because of (b), we would have no assurance that at termination the resulting policy were optimal. We do not formally state an algorithm here. That of Sec. 7.2.5, in which v^n equals the maximal solution of $(I - P_{d_n})v = r_{d_n}$ in step 2, and we choose d_0 so that $v^{(d_0)^\infty}(s) > -\infty$ for all $s \in S$, would be the obvious candidate.

Let (v^n) be the sequence of iterates of such an algorithm. We use the following result to show that $v^n(s)$ is monotonically nondecreasing in n for each $s \in S$.

Proposition 7.3.13.   Let $v \in V^-$ and suppose $\mathscr{L}v \geq v$. Then, if

$$\delta \in \arg\max_{d \in D^{MD}} \{r_d + P_d v\} \tag{7.3.5}$$

$v^{\delta^\infty} \geq v$.

Proof. As a consequence of (7.3.5), and the assumption that $\mathscr{L}v \geq v$, $L_\delta v \geq v$, so that, for any N, $(L_\delta)^N v \geq v$. Expanding this and noting that $v \leq 0$ implies that

$$\sum_{n=0}^{N-1} P_\delta^n r_\delta \geq \sum_{n=0}^{N-1} P_\delta^n r_\delta + P_\delta^N v \geq v$$

Passing to the limit in the above expression establishes that $v^{\delta^\infty} \geq v$. Since, for any n, $\mathscr{L}v^n \geq L_{d_n} v^n = v^n$, Proposition 7.3.12 establishes that

$$v^{(d_{n+1})^\infty} = v^{n+1} \geq v^n.$$

Hence policy iteration generates a nondecreasing sequence of values.   □

We now show that, even in very simple examples, there exist policies with reward equal to $-\infty$, and that stopping the algorithm whenever

$$d^n \in \arg\max_{d \in D^{MD}} \{r_d + P_d v^n\} \tag{7.3.6}$$

as in step 4 of the algorithm of Sec. 7.2.5, may result in termination with a suboptimal policy.

Example 7.3.4. We consider the following modification of Example 7.3.1, in which there is an additional action $a_{2,2}$ in state s_2. The problem data follows; $S = \{s_1, s_2\}$, $A_{s_1} = \{a_{1,1}, a_{1,2}\}$, $A_{s_2} = \{a_{2,1}, a_{2,2}\}$, $r(s_1, a_{1,1}) = 0$, $r(s_1, a_{1,2}) = -1$, $r(s_2, a_{2,1}) = 0$, $r(s_2, a_{2,2}) = -3$, $p(s_1|s_1, a_{1,1}) = 1$, $p(s_2|s_1, a_{1,2}) = 1$, $p(s_2|s_2, a_{2,1}) = 1$, $p(s_1|s_2, a_{2,2}) = 1$ and $p(j|s, a) = 0$ otherwise (Fig. 7.3.4).

Consider first the policy η^∞ in which $\eta(s_1) = a_{1,2}$ and $\eta(s_2) = a_{2,2}$. Clearly $v^{\eta^\infty}(s) = -\infty$. Observe that the system of equations $(I - P_\eta)v = r_\eta$ is inconsistent, so that the evaluation step of the policy iteration algorithm cannot be implemented.

Now consider the policies δ^∞ and γ^∞ with $\delta(s_1) = a_{1,1}$, $\gamma(s_1) = a_{1,2}$ and $\delta(s_2) = \gamma(s_2) = a_{2,1}$. Apply policy iteration with $d^0 = \gamma$. The maximal solution in V^- of $(I - P_\gamma)v = r_\gamma$ is $v^0(s_1) = -1$, $v^0(s_2) = 0$. Since

$$r_\delta(s_1) + P_\delta v^0(s_1) = r_\gamma(s_1) + P_\gamma v^0(s_1) = -1$$

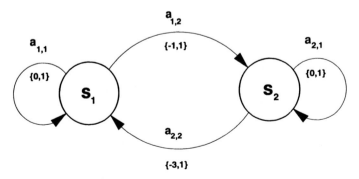

Figure 7.3.4 Symbolic representation of Example 7.3.4.

it follows that

$$d^0 = \gamma \in \underset{d \in D^{MD}}{\arg\max} \{r_d + P_d v^0\} = \{\gamma, \delta\}.$$

Hence the termination criterion (7.3.6) is satisfied but $v^0(s_1) = -1 < 0 = v^{\delta^\infty}(s_1)$, so that $(d_0)^\infty$ is not optimal.

Policy iteration fails in this example because both v^{γ^∞} and v^{δ^∞} satisfy $\mathscr{L}v = v$. Policy iteration finds a solution of the optimality equations but not the maximal one. Observe that (7.3.2) fails to hold with $v = v^0$, so that the conclusion of Proposition 7.3.4(b) that $v^0 = v^*$ need not hold.

The above example presents difficulties because the sets of recurrent and transient classes vary from policy to policy. Algorithms based on average and sensitive optimality criteria which account for chain structure are more appropriate. We regard the multichain policy iteration algorithm for finding a bias-optimal (0-discount optimal) policy (Chap. 10) as the appropriate form of the policy iteration algorithm for negative models. It takes into account policies with expected total reward equal to $-\infty$ by first seeking sets of policies with maximal average reward. Since by assumption there exists a policy with finite total expected reward, any average optimal policy will have average reward zero. The values of these policies all satisfy $\mathscr{L}v = v$. Then among such policies it finds a policy with greatest bias. Since any policy with finite total expected reward has average reward zero, its bias equals its total expected reward, so that maximizing bias corresponds to maximizing expected total reward.

If we were to apply the bias-optimal policy iteration algorithm to the above example with $d^0 = \gamma$, the algorithm would first identify the set $\{\gamma, \delta\}$ and then, within it, identify δ^∞ as bias optimal. We defer presentation of this algorithm to Sec. 10.4.

7.3.5 Modified Policy Iteration

In negative models, modified policy iteration behaves similarly to policy iteration; it need not converge to v^*. The following simple example shows that modified policy iteration may terminate at a suboptimal policy with a value which satisfies the optimality equation.

Recall that modified policy iteration may be represented by the recursion

$$v^n = L_{d_n}^{m_n} v^{n-1},$$

where $d_n \in D^{MD}$ is any v^{n-1} improving decision rule.

Example 7.3.5. Let $S = \{s_1, s_2, s_3\}$, $A_{s_1} = \{a_{1,1}, a_{1,2}\}$, $A_{s_2} = \{a_{2,1}\}$, $A_{s_3} = \{a_{3,1}\}$, $r(s_1, a_{1,1}) = 0$, $r(s_1, a_{1,2}) = 0$, $r(s_2, a_{2,1}) = -1$, $r(s_3, a_{3,1}) = 0$, $p(s_1|s_1, a_{1,1}) = 1$, $p(s_2|s_1, a_{1,2}) = 1$, $p(s_3|s_2, a_{2,1}) = 1$, $p(s_3|s_3, a_{3,1}) = 1$, and $p(j|s, a) = 0$ otherwise (Fig. 7.3.5).

Let δ_i denote the decision rule which uses action $a_{1,i}$ in s_1. Set $v^0 = 0$ and apply modified policy iteration with $m_1 = 2$. For any $v \in V^-$,

$$\mathscr{L}v(s_1) = \max\{v(s_1), v(s_2)\}, \qquad \mathscr{L}v(s_2) = -1 + v(s_3) \quad \text{and} \quad \mathscr{L}v(s_3) = v(s_3).$$

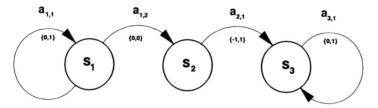

Figure 7.3.5 Symbolic representation of Example 7.3.5.

Since $L_{\delta_i}0(s_1) = 0 = \mathscr{L}0(s_1)$ for $i = 1$ and 2, we may set $d_1 = \delta_2$. We find that

$$v^1(s_1) = L_{\delta_2}^2 0(s_1) = -1, v^1(s_2) = L_{\delta_2}^2 0(s_2) = -1,$$

and

$$v^1(s_3) = L_{\delta_2}^2 0(s_3) = 0.$$

Repeating this calculation with $n = 2$ and $m_2 = 2$, we find that $v^2(s_1) = -1$, $v^2(s_2) = -1$, and $v^2(s_3) = 0$, so that $v^n(s_1)$ generated by modified policy iteration with $m_n = 2$ converges to -1 while $v^*(s_1) = 0$. Note that value iteration converges to v^*.

7.3.6 Linear Programming

The optimal expected return and optimal policies cannot be determined by direct application of linear programming in negative models. Theorem 7.3.3(a) characterizes the optimal value as the maximal v satisfying $v \leq \mathscr{L}v$. This translates to the following mathematical programming model:

Maximize

$$\sum_{j \in S} \alpha(j)v(j)$$

subject to

$$v(s) \leq \max_{a \in A_s}\left\{r(s, a) + \sum_{j \in S} p(j|s, a)v(j)\right\} \quad s \in S$$

and $v(s) \leq 0$ for $s \in S$.

Problem 7.22 provides an example of a model in which the above constraints are nonlinear and the feasible region nonconvex. Therefore complicated solution procedures are required to use mathematical programming to solve the negative model.

7.3.7 Optimal Parking

We use results above to find an optimal policy for the following colorful model we posed as Problem 3.9. Suppose you are driving down a long street seeking to find a parking spot as close as possible to a specified destination (Fig. 7.3.6).

Figure 7.3.6 The parking problem with destination J & D's Diner

Assume that

a. you may park on one side of the street and travel in one direction only,
b. you can only inspect one spot at a time,
c. the probability that a spot is vacant is $p > 0$, and
d. the occupancy status of a spot is independent of all other spots.

We formulate this as a negative model as follows. Let $S = \{[s, k] : s = \cdots -2,$ $-1, 0, 1, 2, \ldots, \ k = 0, 1\} \cup \{\Delta\}$, where s denotes the location of the parking spot; $s = 0$ represents the destination, and $s > 0$ $(s < 0)$ represents a parking spot s units in front of (past) the destination. Note that the driver proceeds from $s > 0$ towards $s < 0$. When $k = 0$, the spot is occupied, and when $k = 1$, the spot is vacant; Δ denotes the stopped state. The action sets are $A_{[s, 1]} = \{C, Q\}$, $A_{[s, 0]} = \{C\}$ and $A_\Delta = \{Q\}$; action C represents not parking (continuing), and action Q represents parking (quitting). Rewards equal 0 except when choosing action Q in $[s, 1]$, in which case $r([s, 1], Q) = -|s|$. Transition probabilities satisfy $p([s-1, 1] \mid [s, k], C) = p$, $p([s-1, 0] \mid [s, k], C) = 1 - p$, $p(\Delta \mid [s, 1], Q) = 1$, and $p(\Delta \mid \Delta, Q) = 1$. We represent these symbolically in Fig. 7.3.7.

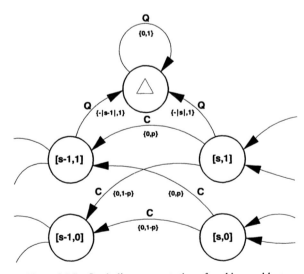

Figure 7.3.7 Symbolic representation of parking problem.

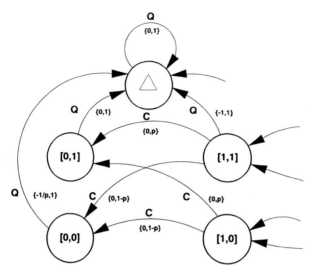

Figure 7.3.8 State reduction for the parking problem.

We simplify the problem by noting that an optimal policy must stop at the first vacant location at or beyond the destination, so that $v^*([0, 1]) = 0$, and

$$v^*([0,0]) = -\left[1p + 2p(1 - p) + 3p(1 - p)^2 + \cdots\right] = -1/p$$

Consequently we may reduce the problem to one with $S = \{[s, k]; \ s = 0, 1, 2, \ldots :$ $k = 0, 1\} \cup \{\Delta\}$ by setting $A_{[0,0]} = A_{[0,1]} = \{Q\}$, $r([0,0], Q) = -1/p$, $r([0,1], Q) = 0$, and $p(\Delta|[0,0], Q) = p(\Delta|[0,1], Q) = 1$. All other elements remain unaltered (see Fig. 7.3.8).

The optimality equation expressed in component notation becomes

$$v([0,0]) = -1/p + v(\Delta), \qquad v([0,1]) = v(\Delta),$$
$$v([s,0]) = pv([s - 1, 1]) + (1 - p)v([s - 1], 0)$$

and

$$v([s,1]) = \max\{-s + v(\Delta), pv([s - 1, 1]) + (1 - p)v([s - 1], 0)\}.$$

As a consequence of Theorem 7.3.3(a), v^* is the maximal solution of the optimality equation, so that $v^*(\Delta) = 0$. Changing signs we find that v^* satisfies

$$v^*([0,0]) = 1/p, \qquad v^*([0,1]) = 0,$$
$$v^*([s,0]) = pv^*([s - 1, 1]) + (1 - p)v^*([s - 1, 0]), \qquad s \geq 1, \quad (7.3.7)$$

and

$$v^*([s,1]) = \min\{s, v^*([s,0])\}, \qquad s \geq 1 \tag{7.3.8}$$

where we replace "max" by "min" and $-s$ by s in the last equation above.

We now determine the form of the optimal policy. Our analysis is similar to that for the secretary problem in Sec. 4.6.4.

Theorem 7.3.14. There exists an optimal stationary policy $(d^*)^\infty$ for the parking problem, in which

$$d^*([s,1]) = \begin{cases} Q & s \le M^* \\ C & s > M^*, \end{cases}$$

where

$$M^* = \max\{s : v^*([s,0]) \ge s\}. \tag{7.3.9}$$

Proof. By Theorem 7.3.5, if d^* is conserving, $(d^*)^\infty$ is optimal. Therefore an optimal policy may be determined by

$$d^*([s,1]) = \begin{cases} Q & s \le v^*([s,0],0) \\ C & s > v^*([s,0],0) \end{cases}.$$

We establish existence of M^* as follows. From (7.3.7) and (7.3.8), for $s \ge 1$

$$v^*([s,1]) \le v^*([s,0]) \le v^*([s-1,0]). \tag{7.3.10}$$

We show that if it is optimal to park in state s, then it is optimal to park in state $s - 1$. If $d^*([s,1]) = Q$, then, by (7.3.8), $s \le v^*([s,0])$ so by (7.3.10), $s - 1 \le s \le v^*([s,0]) \le v^*([s-1,0])$ so that $d^*([s-1,1]) = Q$. Next we show that if it is optimal to continue in s, then it is optimal to continue in $s + 1$. If $d^*([s,1]) = C$, $v^*([s,0]) < s$, so, by (7.3.10), $v^*([s+1,0]) \le v^*([s,0]) < s \le s + 1$. Hence d^* has the indicated form. Since $v^*([0,0]) > 0$, and $v^*([s,0])$ is nonincreasing given by M^* (7.3.9) is well defined.   □

Note that the optimal decision when $v^*([s,0]) = s$ is arbitrary; the policy described above parks as soon as possible. A more selective optimality criterion might distinguish these policies. Note also that the above result holds for any concave $r([s,1],Q)$ with maximum at $s = 0$.

We now investigate the relationship of M^* to p. Since by assumption $p > 0$, $M^* \ge 0$. From (7.3.7) and (7.3.8), it follows that

$$v^*([1,1]) = \min\{1, (1-p)/p\}.$$

Therefore if $(1-p)/p < 1$ or $p > \frac{1}{2}$, it is optimal to continue in location 1. Since in this case $v^*([0,0]) = 1/p > 0$, Theorem 7.3.14 shows that when $p > \frac{1}{2}$, $M^* = 0$, so when following the optimal policy we stop at the first vacant spot at or beyond our destination.

Through fairly involved manipulations, we can establish the following characterization of M^* as a function of p. We leave the proof as an exercise.

Proposition 7.3.15

a. There exists a decreasing sequence $\{q_n\}$, with $q_0 = 1$, such that $M^* = n$ for all p which satisfy $q_n \geq p > q_{n+1}$.

b. For $n \geq 1$,

$$q_n = 1 - \tfrac{1}{2}^{1/n}. \tag{7.3.11}$$

c. For any $p < 1$,

$$M^* = \text{int}\left[\frac{-\ln 2}{\ln(1 - p)}\right]$$

where $\text{int}[x]$ denotes the integer part of x.

Alternatively we may determine M^* for any fixed p by using recursion (7.3.7) to numerically determine M^* through (7.3.9). An advantage of such an approach is that it applies to any convex cost function and to any set of vacancy probabilities. We applied such an approach (using a spreadsheet) to obtain the representation of the optimal policy in Fig. 7.3.9.

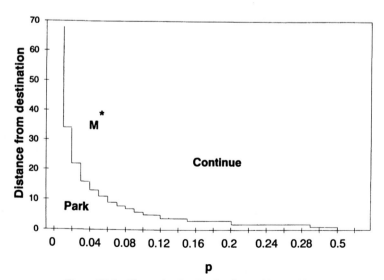

Figure 7.3.9 The optimal policy for the parking problem.

7.4 COMPARISON OF POSITIVE AND NEGATIVE MODELS

The following table summarizes results for positive bounded and negative models.

Comparison of Positive and Negative Models for Finite or Countable S

Result	Positive Bounded Model	Negative Model
Optimality equation	v^* minimal solution in V^+	v^* maximal solution in V^-
If d is conserving d^∞ is optimal	Only if $\limsup_{N \to \infty} P_d^N v^* = 0$	Always
Existence of optimal stationary policies	S and A_s finite or if optimal policy exists and $r \geq 0$	A_s finite or A_s compact, $r(s, \cdot)$ u.s.c. and $p(j\|s, \cdot)$ l.s.c., or if optimal policy exists
Existence of stationary ε-optimal policy	If v^* bounded	Not always, (but one in Π^{MD} exists)
Value iteration converges	$0 \leq v^0 \leq v^*$	$0 \geq v^0 \geq v^*$ and either A_s finite, or S finite, and $v^* > -\infty$
Policy iteration converges	Yes	Not always
Modified policy iteration converges	$0 \leq v^0 \leq v^*$ and $Bv^0 \geq 0$	Not always
Solution by linear programming	Yes	No

BIBLIOGRAPHIC REMARKS

The study of Markov decision process models with expected total-reward criterion originated with the book *How to Gamble if You Must* by Dubins and Savage (1976) which first appeared in 1965, and the papers of Strauch (1966) and Blackwell (1961 and 1967). The Dubins and Savage gambling model has a similar structure to an MDP, and its analysis raises many of the same challenges. To ensure that the expected total reward of each policy is well defined, Blackwell and Strauch distinguish positive and negative models. We have chosen to follow this approach in Chap. 7, but begin it by presenting some results which hold for any models in which the total-reward criterion is well defined. Our Assumption 7.1.1, which appears in Schal (1983) and Feinberg (1992), ensures this. Theorems 7.1.3, 7.1.7, and 7.1.9 are the key results. Our proof of Theorem 7.1.9 follows van der Wal (1981; Theorem 2.21). The expressions "conserving" and "equalizing" were coined by Dubins and Savage (1976).

Our development for positive models draws on results in Blackwell (1967 and 1970), Strauch (1966), Ornstein (1969), Hordijk (1974), Veinott (1969d), and van der Wal (1981 and 1984). Note that most of these authors, with the exception of Veinott, restrict attention to models in which $r(s, a) \geq 0$ for all state action pairs. We include several counterexamples which provide insight into many unexpected phenomena which occur in the total-reward models. Noteworthy are those of Blackwell (1967), Strauch (1966), Ornstein (1969), and van der Wal (1981). Ornstein (1969) established the existence of uniformly near-optimal policies for the Dubins-Savage gambling model with non-negative payoffs using probabilistic arguments. Blackwell (1970) extended this analysis by showing that existence of an optimal policy implies existence of a stationary optimal policy. Our proof of Theorem 7.2.11 is based on Hordijk (1974; Theorem 13.8), who extended Ornstein's results to MDP's with non-negative rewards, and Blackwell (1970, Result C). Lemma 7.2.9 presents the MDP analog of a result that pervades the Markov process literature (cf. Çinlar, 1975, p. 201).

Our analysis of value iteration follows Blackwell (1967), while that of policy iteration combines ideas in van Dawen (1986a) and Denardo and Fox (1968). Van der Wal (1981) considers the modified policy iteration algorithm for the positive model, while our analysis of linear programming draws on Kallenberg (1983). Note that Kallenberg considers only models for which $r(s, a) \geq 0$, but he does not restrict his analysis to bounded models. In the unbounded case, he develops an intricate procedure to solve models which include positive rewards on recurrent classes and consequently have unbounded dual LPs. We employ related methods in our proof of Proposition 7.2.15.

The section on optimal stopping for positive models draws on Dynkin and Yuskevich (1969) and Çinlar (1975). The asset selling problem originates with Karlin (1962). Ross (1983) provides examples of applications of positive model results to gambling problems.

Blackwell (1961), and Strauch (1966) in his Ph.D. dissertation under Blackwell's supervision, were the first to study the negative model. Section 7.3 relies heavily on Strauch's lengthy paper. Example 7.3.2 appeared in Blackwell (1961), and Example 7.3.3 as Example 6.1 in Strauch (1966). In addition to the above references, our analysis of value iteration draws on van der Wal (1981) and van Dawen (1985b). Whittle (1979, 1980a, and 1980b) and Hartley (1980) provide an alternative set of conditions which ensure convergence of value iteration. Van Dawen (1986a) provides a two-step policy iteration algorithm for negative models. His paper motivates our development in Sec. 10.4. Our recommendations concerning the use of policy iteration are in the spirit of Kallenberg (1983). Van der Wal (1981) analyzes the convergence of modified policy iteration in total-reward models; however, Example 7.3.5 originates here. I am unaware of the source of the parking problem in Sec. 7.3.6; Chow, Robbins, and Siegmund (1971) present it as an application of optimal stopping theory. Other contributions to the theory of negative models include Kreps and Porteus (1977), Schal (1975 and 1978), Demko and Hill (1981), and Maitra and Sudderth (1992).

Several authors, including Hordijk (1974), Schal (1975 and 1983), van Hee (1978), van der Wal (1981 and 1984), Feinberg and Sonin (1983 and 1984), van Dawen (1986a and 1986b), and Feinberg (1986, 1987, and 1992), analyze MDP's with total-reward criteria without distinguishing positive and negative models. The main issues addressed in these papers include providing conditions which ensure the existence of stationary uniformly near-optimal policies, and determining conditions when the supremum of the expected total reward over the set of stationary policies equals the value of the MDP. The papers of van der Wal and van Dawen extend Ornstein's approach, while Feinberg and Sonin use different methods.

PROBLEMS

7.1. a. For any model with expected total reward criterion show that $\liminf_{n \to \infty} \mathcal{L}^n 0 \geq v^*$.

   b. [van Hee, van der Wal, and Hordijk (1977)] Show for the following model that the above inequality is strict, so that v^* cannot be obtained using value iteration with $v^0 = 0$: $S = \{s_1, s_2, s_3\}$; $A_{s_1} = \{a_{1,1}, a_{1,2}\}$, $A_{s_2} = \{a_2\}$, $A_{s_3} = \{a_3\}$; $r(s_1, a_{1,1}) = 0$, $r(s_1, a_{1,2}) = 2$, $r(s_2, a_2) = -1$, and $r(s_3, a_3) = 0$; and $p(s_1|s_1, a_{1,1}) = 1$, $p(s_2|s_1, a_{1,2}) = 1$, $p(s_3|s_2, a_2) = 1$ and $p(s_3|s_3, a_3) = 1$.

   c. Show that the above model does not satisfy the hypotheses of either the positive or negative models.

   d. Show that this model can be solved using modified policy iteration with $m_n = 2$.

   e. Suppose instead that $r(s_1, a_{1,1}) = 1$ and $r(s_2, a_2) = -2$. Show that $v^{\pi} \leq 0$ for all π and that $\mathcal{L}^n 0$ converges to v^* for this model.

7.2. (van der Wal, 1981) Show for Example 7.2.5 that there exists a randomized stationary ε-optimal policy.

7.3. (van der Wal, 1981) Show that there exists no randomized stationary ε-optimal policy for the following modification of Example 7.2.5: $S = \{0, 1, 2, \cdots\}$, $A_0 = \{a_0\}$ and $A_s = \{a_{s,1}, a_{s,2}\}$ for $s \geq 1$. When $s = 0$, $r(0, a_0) = 0$ and $p(0|0, a_0) = 1$, For $s \geq 1$, $r(s, a_{s,1}) = 0$ and $p(s + 1|s, a_{s,1}) = \alpha_s$, $p(0|s, a_{s,1}) = 1 - \alpha_s$, where $\alpha_s = (1 + s^{-1})/(2 + 2(s + 1)^{-1})$; $r(s, a_{s,2}) = 2^s$ and $p(0|s, a_{s,2}) = 1$.

7.4. (van der Wal, 1981) Show that there exists no randomized Markov ε-optimal policy for the following example. $S = \{0\} \cup \{(\sigma, \tau)|\tau = 1, 2, \ldots, \sigma \leq \tau\}$, $A_0 = \{a_0\}$ and $A_{(\sigma, \sigma)} = \{a_{\sigma, 1}, a_{\sigma, 2}\}$ for $\sigma \geq 1$, and $A_{(\sigma, \tau)} = \{a_{(\sigma, \tau), 1}\}$ for $\tau \neq \sigma$ and $\tau \geq 1$. When $s = 0$, $r(0, a_0) = 0$; for $s \neq 0$, $r(s, a_{s,1}) = 0$; and for $s = (\sigma, \sigma)$, $r((\sigma, \sigma), a_{\sigma, 2}) = 2^{\sigma}$. Transition probabilities satisfy $p(0|0, a_0) = 1$, $p((\sigma + 1, \tau)|(\sigma, \tau), a_{(\sigma, \tau), 1}) = 1$ for $\sigma < \tau$ and $\tau = 1, 2, \ldots$, $p((\sigma + 1, \sigma + 1)|(\sigma, \sigma), a_{\sigma, 2}) = \alpha_{\sigma}$ and $p(0|(\sigma, \sigma), a_{\sigma, 2}) = 1 - \alpha^{\sigma}$, where α^{σ} is defined in the previous problem, and $p(0|(\sigma, \sigma), a_{s, 1}) = 1$.

7.5. Suppose S is finite, A_s is compact, and $r(s, a) \geq 0$ and $p(j|s, a)$ are continuous in a for each s. Show by example that there need not exist a stationary optimal policy.

7.6. Show in the positive model, if, for each $\pi \in \Pi^{MD}$, $v^{\pi}(s) < \infty$, that

$$\sup_{\pi \in \Pi^{HR}} v^{\pi}(s) < \infty.$$

7.7. Prove Lemma 7.1.8 for S countable.

7.8. Prove Theorem 7.2.6.

7.9. Show that there exists a $\pi \in \Pi^{MD}$ which satisfies (7.2.11).

7.10. Suppose, in the modified policy iteration algorithm, that we choose d_n so that

$$r_{d_n} + P_{d_n} v^n + \varepsilon_n e \geq \mathcal{L} v^n$$

for some $\varepsilon_n > 0$. Provide conditions on ε_n which ensure that such an algorithm converges for positive bounded models.

7.11. (Red and Black; Dubins and Savage, 1976). An individual plays the following game of chance. If he begins the game with s dollars, he may bet any amount $j \leq s$ dollars on each turn of the game. If he wins, which occurs with probability p, he receives his bet of j dollars plus an additional payoff of j dollars; while, if he loses, he receives 0. His objective is to maximize the probability that his fortune exceeds N. He may play the game repeatedly provided his fortune remains positive. When it reaches 0 he must stop playing.
  a. Formulate this as a positive bounded model.
  b. Find an optimal strategy for playing the game when $p \geq 0.5$.
  c. Find a policy which maximizes the expected length of the game.
  d. Suppose the only bet available is \$1 and $p = 0.5$. Use the approach of Example 7.2.7 to find an optimal strategy for playing this game.

7.12. a. Prove Corollary 7.2.20.
  b. Establish the relationship between feasible solutions to the dual LP and the quantity

$$\sum_{j \in S} \alpha(j) \sum_{n=1}^{\infty} P^{d^\infty}(X_n = s, Y_n = a | X_1 = j)$$

for $d \in D^{MR}$ in a positive bounded model.

7.13. Use policy iteration and value iteration to find v^* in Example 7.2.6.

7.14. (Positive models in which some rewards are negative) Consider a model with $S = \{s_1, s_2, s_3\}$; $A_{s_1} = \{a_1\}$, $A_{s_2} = \{a_{2,1}, a_{2,2}\}$, and $A_{s_3} = \{a_3\}$; $r(s_1, a_1) = 1$, $r(s_2, a_{2,1}) = -1$, $r(s_2, a_{2,2}) = 1$, and $r(s_3, a_3) = 0$; and $p(s_2 | s_1, a_1) = 1$, $p(s_1 | s_2, a_{2,1}) = 1$, $p(s_3 | s_2, a_{2,2}) = 1$, and $p(s_3 | s_3, a_3) = 1$ and all other transition probabilities equal to 0.
  Show that this model satisfies Assumptions 7.2.1 and 7.2.2, and find an optimal policy.

7.15. Solve the linear programming problem version of Example 7.2.1 with the additional constraint that $x(s_1, a_{1,1}) \leq 0.75$. Interpret the solution in terms of randomized policies and expected state action frequencies.

7.16. Solve the asset selling problem under the assumption that $p(j)$ is the probability mass function of a Poisson-distributed random variable with parameter 1 and $c = 0.5$.

7.17. Derive an analogous result to Theorem 7.2.25 for an asset selling problem in which successive offers are generated according to a Markov chain. That is, if the current offer is j, then the probability distribution for the offer in the subsequent period is $p(j|s)$.

7.18. (The rational burglar; Whittle, 1983) Each evening a burglar may decide to retire or to attempt another robbery. If he is successful, he receives a payoff according to a probability distribution $p(x)$ with mean μ. If he is caught, which occurs with probability q, he must forfeit his total fortune.

    a. Formulate this as an optimal stopping problem.

    b. Find an optimal policy for the burglar.

7.19. Show that there exists no ε-optimal randomized stationary policy in Example 7.3.2.

7.20. Verify the calculations in Example 7.3.5 for $n \geq 2$. Show that value iteration converges for this model.

7.21. (Eaton and Zadeh, 1962) Let η and δ be two decision rules. Define the new decision rule ω by

$$\omega(s) = \begin{cases} \delta(s) & \text{if } v^{\delta^{\infty}}(s) \geq v^{\eta^{\infty}}(s) \\ \eta(s) & \text{if } v^{\eta^{\infty}}(s) > v^{\delta^{\infty}}(s) \end{cases}.$$

    a. Show for negative models that $v^{\omega^{\infty}} \geq \max\{v^{\delta^{\infty}}, v^{\eta^{\infty}}\}$.

    b. Show by example that the result in (a) may not be valid in a positive bounded model.

7.22. (Mathematical programming formulation of the negative model.) Consider the following model with $S = \{s_1, s_2, s_3\}$, $A_{s_1} = \{a_{1,1}, a_{1,2}\}$, $A_{s_2} = \{a_{2,1}, a_{2,2}\}$, $A_{s_3} = \{a_{3,1}\}$, $r(s_1, a_{1,1}) = -1$, $r(s_1, a_{1,2}) = -3$, $r(s_2, a_{2,1}) = 0$, $r(s_2, a_{2,2}) = -1$, $r(s_3, a_{3,1}) = 0$, $p(s_2|s_1, a_{1,1}) = 1$, $p(s_3|s_1, a_{1,2}) = 1$, $p(s_1|s_2, a_{2,1}) = 1$, $p(s_3|s_2, a_{2,2}) = 1$, $p(s_3|s_3, a_{3,1}) = 1$ and $p(j|s, a) = 0$ otherwise.

    a. Give the optimality equation for this model.

    b. Find an optimal policy using value iteration.

    c. Formulate this as a mathematical programming model as in Section 7.3.6.

    d. Show that feasible region is not convex. It is easiest to show this by plotting the section of the feasible region for which $v(s_3) = 0$.

    e. Solve the mathematical program graphically.

7.23. (Rosenthal, White, and Young, 1978) Consider the following modification of Problem 6.48. There are $Q = 4$ work sites, with site 1 denoting the home office and 2, 3, and 4 denoting remote sites. The cost of relocating the equipment trailer is $d(k, j) = 300$ for $k \neq j$; the cost $c(k, j)$ of utilizing the equipment trailer is 100 if the repairman is at site $k > 1$ and trailer is at site $j \neq k$ with $j > 1$ and 200 if the repairman is at remote site $j > 1$ and the trailer is at the home office, site 1. If the repairman is at site 1, no work is carried out, so the cost of using the trailer if it is there equals 0. Assume that the probability of the repairman moves from site s to site j in one period, $p(j|s)$ is given by the matrix

$$P = \begin{bmatrix} 1.0 & 0.0 & 0.0 & 0.0 \\ 0.0 & 0.5 & 0.5 & 0.0 \\ 0.0 & 0.0 & 0.8 & 0.2 \\ 0.4 & 0.0 & 0.0 & 0.6 \end{bmatrix}$$

instead of that in Problem 6.48. Note that, in this formulation, state 1 is absorbing, so that once the repairman reaches the home office, no future decisions are made.

a. Formulate this as a negative Markov decision problem. Note that a finite-horizon formulation will not apply.

b. Find a relocation policy that minimizes the expected total cost in one cycle.

7.24. Show that, for any $n \geq 1$, $\mathscr{L}^n$ is a convex operator on V^-, that is, for any u and v in V^- and scalar α, $0 \leq \alpha \leq 1$,

$$\mathscr{L}^n(\alpha u + (1 - \alpha)v) \leq \alpha \mathscr{L}^n u + (1 - \alpha)\mathscr{L}^n v.$$

7.25. (Whittle, 1979; Hartley 1980) Let $v^\infty \equiv \lim_{n \to \infty} \mathscr{L}^n 0$. Proposition 7.3.8 shows for negative models that $v^\infty \geq v^*$. This problem provides conditions under which $v^* = v^\infty$.

Suppose in the negative model that there exists a real number $\alpha > 1$, such that for some $\pi \in \Pi^{MR}$, either

i. $v^\pi \geq \alpha \mathscr{L}^n 0$ for some $n \geq 1$, or

ii. $v^\pi \geq \alpha v^\infty$ and $v^\infty = \mathscr{L} v^\infty$.

Show by using the result in the previous problem that either of these conditions imply

1. $v^\infty = v^*$,

2. $\lim_{n \to \infty} \mathscr{L}^n u$ exists and equals v^* for any $u \in V^-$ for which $u \geq \alpha v^\infty$, and

3. v^* is the unique solution of $\mathscr{L} v = v$ in V^- which satisfies $v \geq \alpha v^\infty$.

7.26. Show using the result in Problem 7.24 that if there exist u and w in V^- for which $u \leq v^\pi$ for some $\pi \in \Pi^{MR}$ and $w \geq \mathscr{L}^n 0$ for some n and an $\alpha > 1$ for which $u \geq \alpha w$, then conclusions (1) to (3) of Problem 7.25 hold.

7.27. Verify inequality (7.3.10).

7.28. Prove Proposition 7.3.15.

7.29. Determine the relationship between M^* and p for a variant of the parking problem in which $r([s, 1], Q) = -s^2$.

7.30. Derive an optimal policy to the parking problem under the assumption that the vacancy probability depends on s through the relationship

$$p(s) = \frac{\alpha}{1 + e^{-\beta|s|}},$$

where $0 \leq \alpha \leq 1$ and $\beta \leq 0.5$.

7.31. Suppose S is countable and

$$w^*(s) = \sup_{\pi \in \Pi^{HR}} E^\pi \left\{ \sum_{n=1}^\infty r'(X_n^\pi, Y_n^\pi) \right\}$$

where $x' = \min(x, 0)$.

a. Show that value iteration and modified policy iteration converge to v^* whenever $w^* \leq v^0 \leq v^*$.

b. Determine the set of v^0 for which the value iteration scheme $\mathscr{L}^n v^0$ converges to v^* for the example in Problem 7.1b.

CHAPTER 8

Average Reward
and Related Criteria

When decisions are made frequently, so that the discount rate is very close to 1, or when performance criterion cannot easily be described in economic terms, the decision maker may prefer to compare policies on the basis of their average expected reward instead of their expected total discounted reward. Consequently, the average reward criterion occupies a cornerstone of queueing control theory especially when applied to controlling computer systems and communications networks. In such systems, the controller makes frequent decisions and usually assesses system performance on the basis of throughput rate or the average time a job or packet remains in the system. This optimality criterion may also be appropriate for inventory systems with frequent restocking decisions.

In this chapter, we focus on models with the expected average reward optimality criterion. We consider models with discrete-state spaces and focus primarily on finite-state models. Because the average reward criterion depends on the limiting behavior of the underlying stochastic processes, we distinguish models on the basis of this limiting behavior. Consequently we classify models on the basis of the chain structure of the class of stationary policies, and analyze these different model classes separately.

To illustrate key points and avoid many subtleties, this chapter analyzes finite- and countable-state models in which *all* stationary policies generate Markov chains with a single irreducible class. In this case a single optimality equation suffices to characterize optimal policies. We follow this with an analysis of finite-state models with more general chain structure.

In this chapter we assume

Assumption 8.0.1. *Stationary rewards and transition probabilities*; $r(s, a)$ and $p(j|s, a)$ do not depend on the stage;

Assumption 8.0.2. *Bounded rewards*; $|r(s, a)| \leq M < \infty$ for all $a \in A_s$ and $s \in S$ (except in Sec. 8.10),

Assumption 8.0.3. *Finite-state spaces* (except in Secs. 8.10 and 8.11).

Often we restate these assumptions for emphasis.

In contrast to discounted models, approaches for analyzing MDPs with average reward criterion vary with the class structure of Markov chains generated by stationary policies; we review this aspect of Markov chain theory in Appendix A, Secs. A.1-A.4. Sections A.5–A.6 of Appendix A serve as the basis for Sec. 8.2. Section 5.4 also provides some relevant introductory material.

8.1 OPTIMALITY CRITERIA

Referring to our analyses of models with discounted and total-reward criteria suggests that we define an average reward function $g^\pi(s)$ for each $\pi \in \Pi^{HR}$, seek a method for computing

$$g^*(s) = \sup_{\pi \in \Pi^{HR}} g^\pi(s) \qquad (8.1.1)$$

and finding a policy $\pi^* \in \Pi^{HR}$ for which

$$g^{\pi^*}(s) = g^*(s)$$

for all $s \in S$. Unfortunately this approach has some limitations as we show below.

8.1.1 The Average Reward of a Fixed Policy

Recall that for $\pi \in \Pi^{HR}$

$$v^\pi_{N+1}(s) = E^\pi_s \left\{ \sum_{t=1}^N r(X_t, Y_t) \right\} \qquad (8.1.2)$$

denotes the total reward up to decision epoch $N + 1$ or, equivalently, the total reward in an $N + 1$ period problem with terminal reward zero. Define the *average expected reward* of policy π by

$$g^\pi(s) = \lim_{N\to\infty} \frac{1}{N} v^\pi_{N+1}(s) = \lim_{N\to\infty} \frac{1}{N} \sum_{n=1}^N P^{n-1}_\pi r_{d_n}(s). \qquad (8.1.3)$$

As the following simple example illustrates, even in a finite-state example, this limit need not exist for some policies. Because of this, we require refined concepts of average optimality.

Example 8.1.1. Let $S = \{s_1, s_2\}$, $A_{s_1} = \{a_{1,1}, a_{1,2}\}$, $A_{s_2} = \{a_{2,1}, a_{2,2}\}$, $r(s_1, a_{1,1}) = 2$, $r(s_1, a_{1,2}) = 2$, $r(s_2, a_{2,1}) = -2$, $r(s_2, a_{2,2}) = -2$, $p(s_1|s_1, a_{1,1}) = 1$, $p(s_2|s_1, a_{1,2}) = 1$, $p(s_1|s_2, a_{2,1}) = 1$ and $p(s_2|s_2, a_{2,2}) = 1$ (Fig. 8.1.1).

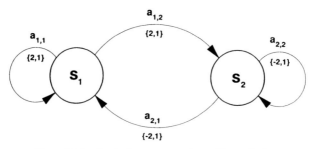

Figure 8.1.1   Symbolic representation of Example 8.1.1

Consider the history-dependent policy π which, on starting in s_1, remains in s_1 for one period, proceeds to s_2 and remains there for three periods, returns to s_1 and remains there for $3^2 = 9$ periods, proceeds to s_2 and remains there for $3^3 = 27$ periods, etc. Then direct computation shows that

$$\liminf_{N \to \infty} \frac{1}{N} v_{N+1}^\pi(s_1) = -1 \quad \text{and} \quad \limsup_{N \to \infty} \frac{1}{N} v_{N+1}^\pi(s_1) = 1,$$

so that the limit in (8.1.3) does not exist.

Note, however, for each of the four stationary policies in this model, the above lim sup and lim inf are identical and definition (8.1.3) is valid.

The example above suggests that the above approach is valid when we restrict attention to stationary policies. The result below confirms this. Note that we call a non-negative matrix *stochastic* if all of its row sums equal 1, i.e., P is stochastic if $Pe = e$.

Proposition 8.1.1.

a. Let S be countable. Let $d^\infty \in \Pi^{SR}$ and suppose that the limiting matrix of P_d, P_d^* is stochastic. Then the limit in (8.1.3) exists and

$$g^{d^\infty}(s) = \lim_{N \to \infty} \frac{1}{N} v_{N+1}^{d^\infty}(s) = P_d^* r_d(s). \tag{8.1.4}$$

b. If S is finite, (8.1.4) holds.

Proof.   Since

$$v_{N+1}^{d^\infty}(s) = \sum_{n=1}^{N} P_d^{n-1} r_d(s)$$

part (a) follows from (A.3) in Appendix A.4. Part (b) follows by noting that, in a finite-state Markov model, P_d^* is stochastic.   □

In Chapters 8 and 9, we will be concerned with finding policies which perform best in terms of limiting behavior of $N^{-1}v_{N+1}^{\pi}$. Since the limit in (8.1.3) need not exist, we introduce two related quantities. For $\pi \in \Pi^{HR}$, let $g_{+}^{\pi}(s)$ denote the *lim sup average reward* defined by

$$g_{+}^{\pi}(s) = \limsup_{N \to \infty} \frac{1}{N} v_{N+1}^{\pi}(s) \tag{8.1.5}$$

and let $g_{-}^{\pi}(s)$ denote the *liminf average reward* is defined by

$$g_{-}^{\pi}(s) = \liminf_{N \to \infty} \frac{1}{N} v_{N+1}^{\pi}(s). \tag{8.1.6}$$

Note that $g_{+}^{\pi}(s) \geq g_{-}^{\pi}(s)$ and that $g^{\pi}(s)$ exists if and only if $g_{+}^{\pi}(s) = g_{-}^{\pi}(s)$. We often refer to g^{π} as the *gain* of policy π. This expression has its origins in control engineering, in which it refers to the ratio of the output of a system to its input. We provide further justification for this designation in Sec. 8.2.

8.1.2 Average Optimality Criteria

We express three optimality criteria in terms of g_{+}^{π} and g_{-}^{π}. Note that they are *equivalent in finite-state models*. We say that a policy π^{*} is *average optimal* if

$$g_{-}^{\pi^{*}}(s) \geq g_{+}^{\pi}(s) \tag{8.1.7}$$

for all $s \in S$ and $\pi \in \Pi^{HR}$; is *lim sup average optimal* if

$$g_{+}^{\pi^{*}}(s) \geq g_{+}^{\pi}(s) \tag{8.1.8}$$

for all $s \in S$ and $\pi \in \Pi^{HR}$; and is *lim inf average optimal* if

$$g_{-}^{\pi^{*}}(s) \geq g_{-}^{\pi}(s) \tag{8.1.9}$$

for all $s \in S$ and $\pi \in \Pi^{HR}$.

We define the following additional quantities:

$$g_{+}^{*}(s) = \sup_{\pi \in \Pi^{HR}} g_{+}^{\pi}(s) \tag{8.1.10}$$

$$g_{-}^{*}(s) = \sup_{\pi \in \Pi^{HR}} g_{-}^{\pi}(s). \tag{8.1.11}$$

Lim inf average optimality corresponds to comparing policies in terms of worst case limiting performance, while lim sup average optimality corresponds to comparison in terms of best case performance. Average optimality is the strongest of these three criteria because it requires that

$$\liminf_{N \to \infty} \frac{1}{N} v_{N+1}^{\pi^{*}}(s) \geq \limsup_{N \to \infty} \frac{1}{N} v_{N+1}^{\pi}(s)$$

for all $\pi \in \Pi^{HR}$. This means that the lowest possible limiting average reward under policy π^* is as least as great as the best possible limiting average reward under any other policy. Clearly, if π^* is average optimal, it is lim sup and lim inf average optimal. Note that the lim sup and lim inf average optimality criteria are distinct; neither implies the other. As a consequence of Proposition 8.1.1, a *stationary* lim sup average optimal policy with stochastic limiting matrix is average optimal and hence lim inf average optimal.

Example 8.1.1 (ctd.). Observe that in this model the stationary policy $d^*(s_1) = a_{1,1}$, $d^*(s_2) = a_{2,2}$ is *average* optimal, with $g^*(s_1) = g^*(s_2) = 2$.

The following example illustrates the points in the preceding paragraph regarding the relationship of these optimality criteria.

Example 8.1.2. Let $S = \{-1, 0, 1, 2, \ldots\}$; $A_0 = \{a_{0,1}, a_{0,2}\}$ and $A_s = \{a_{s,1}\}$ for $s \neq 0$; and $r(0, a_{0,1}) = -1$, $r(0, a_{0,2}) = 0$, and $r(-1, a_{-1,1}) = 0$,

$$r(s, a_{s,1}) = 1, \qquad \sum_{k=0}^{n} 3^k \leq s \leq \sum_{k=0}^{n+1} 3^k - 1, \qquad \text{and } n = 0, 2, 4, \ldots$$

and

$$r(s, a_{s,1}) = -1, \qquad \sum_{k=0}^{n} 3^k \leq s \leq \sum_{k=0}^{n+1} 3^k - 1, \qquad \text{and } n = 1, 3, 5, \ldots.$$

$p(1|0, a_{0,1}) = 1$, $p(-1|0, a_{0,2}) = 1$, $p(-1|-1, a_{-1,1}) = 1$, and $p(s + 1|s, a_{s,1}) = 1$ for $s \geq 1$ (Fig. 8.1.2).

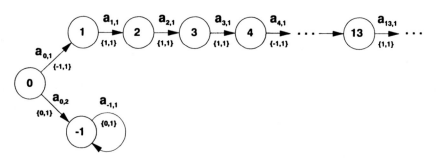

Figure 8.1.2 Symbolic representation of Example 8.1.2.

Action choice in state 0 determines the evolution of the system. Let $\delta(0) = a_{0,1}$ and $\gamma(0) = a_{0,2}$. Then

$$g_+^{\delta^\infty}(0) = \tfrac{1}{2}, \qquad g_-^{\delta^\infty}(0) = -\tfrac{1}{2},$$

and

$$g_+^{\gamma^\infty}(0) = g_-^{\gamma^\infty}(0) = g^{\gamma^\infty}(0) = 0.$$

Observe that

1. δ^∞ is lim sup average optimal,
2. γ^∞ is lim inf average optimal, and
3. no average optimal policy exists.

Note further that (8.1.4) does not hold for δ^∞ because P_δ^* is a matrix with all components equal to 0 and is not stochastic.

In this chapter, we will be concerned with establishing that average optimal policies exist and characterizing their limiting average reward. The following important result, which restates Theorem 5.5.3(b), allows us to restrict attention to Markov policies in the above definitions.

Theorem 8.1.2.   For each $\pi \in \Pi^{HR}$ and $s \in S$, there exists a $\pi' \in \Pi^{MR}$ for which

a. $g_+^{\pi'}(s) = g_+^{\pi}(s)$,
b. $g_-^{\pi'}(s) = g_-^{\pi}(s)$, and
c. $g^{\pi'}(s) = g^{\pi}(s)$ whenever $g_-^{\pi}(s) = g_+^{\pi}(s)$.

As a consequence of this result, we need only search over the set of Markov randomized policies when determining whether a particular policy is average, lim sup average, or lim inf average optimal. Consequently,

$$g^*(s) = \sup_{\pi \in \Pi^{MR}} g^{\pi}(s) \tag{8.1.12}$$

with analogous results for g_+^* and g_-^*.

8.2 MARKOV REWARD PROCESSES AND EVALUATION EQUATIONS

Let S denote a finite set. Let P denote the transition probability matrix of a Markov chain $\{X_t: t = 1, 2, \ldots\}$ and $r(s)$ a reward function. We refer to the bivariate stochastic process $\{(X_t, r(X_t)): t = 1, 2, \ldots\}$ as a *Markov reward process* (MRP). In Markov decision processes, each stationary policy d^∞ generates a MRP with transition matrix P_d and reward r_d. As a preliminary to deriving result for MDPs, we characterize the average reward and related quantities for a MRP and provide systems of equations which characterize them.

Our analysis in this section includes models with both aperiodic and periodic Markov chains. When chains are periodic, most limiting quantities do not exist. To account for this, we use Cesaro limits (Appendix A.4) instead. Recall that for a sequence (y_n) the Cesaro limit (denoted C-lim) is given by

$$C\text{-}\lim_{n\to\infty} = \lim_{n\to\infty} \frac{y_1 + y_2 + \cdots + y_n}{n}.$$

Note that the Cesaro limit equals the ordinary limit whenever it exists; however, it provides a reasonable limiting quantity whenever $\liminf_{n\to\infty} y_n$ and $\limsup_{n\to\infty} y_n$ are not equal. Including the Cesaro limit makes some of the formulas a bit more complex, but hopefully does not detract from understanding. When reading this chapter for the first time, you might wish to assume that all chains are aperiodic and all limits are ordinary limits. Feller (1950, p. 356) notes in his analysis of Markov chains that

"The modifications required for periodic chains are rather trite, but the formulations required become unpleasantly involved."

This point applies equally well to analysis of MDPs with average reward criterion, especially with respect to convergence of value iteration.

8.2.1 The Gain and Bias

Results in Section A.4 of Appendix A show that

$$\lim_{N\to\infty} \frac{1}{N} \sum_{t=1}^{N} P^{t-1} = P^*$$

exists when S is finite or countable. Therefore, whenever r is bounded and P^* is stochastic, the gain of the MRP satisfies

$$g(s) \equiv \lim_{N\to\infty} \frac{1}{N} E_s\left\{ \sum_{t=1}^{N} r(X_t) \right\} = \lim_{N\to\infty} \frac{1}{N} \sum_{t=1}^{N} P^{t-1} r(s) = P^* r(s) \quad (8.2.1)$$

for each $s \in S$. Example 5.1.2 showed that interchanging the limit and the expectation in (8.2.1) may not be justified in models with countable S and unbounded r. In that example, $P^* = 0$, so that $P^* r(s) = 0$ for $s \in S$ but

$$\lim_{N\to\infty} \frac{1}{N} E_s\left\{ \sum_{t=1}^{N} r(X_t) \right\}$$

does not exist.

Suppose P^* is stochastic. Since it has identical rows for states in the same closed irreducible recurrent class (Section A.4 of Appendix A), we have the following important result.

Proposition 8.2.1. Suppose P^* is stochastic. Then if j and k are in the same closed irreducible class, $g(j) = g(k)$. Further, if the chain is irreducible, or has a single recurrent class and possibly some transient states, $g(s)$ is a constant function.

When $g(s)$ is constant, we write ge to denote an $|S|$-vector with identical components g.

Assume now that S is finite and define the *bias*, h of the MRP by

$$h \equiv H_P r \tag{8.2.2}$$

where the fundamental matrix $H_P \equiv (I - P + P^*)^{-1}(I - P^*)$ is defined in (A.14) in Sec. A.5 of Appendix A. From (A.4), $PP^* = P^*$, so (8.2.1) implies that $Pg = PP^*r = P^*r = g$. Therefore, in an *aperiodic* Markov chain, it follows that

$$h = \sum_{t=0}^{\infty} (P^t - P^*)r = \sum_{t=0}^{\infty} P^t(r - g). \tag{8.2.3}$$

We may express this in component notation as

$$h(s) = E_s \left\{ \sum_{t=1}^{\infty} [r(X_t) - g(X_t)] \right\}. \tag{8.2.4}$$

From (8.2.1) and the definition of P^*, it follows that the gain represents the average reward per period for a system in steady state. Sometimes we refer to this quantity as the *stationary reward*. Therefore (8.2.4) allows interpretation of the bias as the expected total difference between the reward and the stationary reward. Alternatively, the first expression in (8.2.3) shows that the bias represents the difference between the total reward for a system that starts in state s and one in which the sth row of P^* determines the initial state. Since an aperiodic Markov chain approaches its steady-state exponentially fast, most of the difference in (8.2.4) will be earned during the first few transitions, so that we may regard the bias as a "transient" reward.

In *periodic* chains, (8.2.3) and (8.2.4) hold in the Cesaro limit sense, so that (8.2.4) may be interpreted as

$$h(s) = \lim_{N \to \infty} \frac{1}{N} \sum_{k=1}^{N} E_s \left\{ \sum_{t=1}^{k} [r(X_t) - g(X_t)] \right\}.$$

An alternative representation for h provides a different interpretation for the bias for *aperiodic* MRPs. Recall that v_{N+1} denotes the total expected reward over $N + 1$ periods in a system in which the terminal reward equals 0. That is,

$$v_{N+1} = \sum_{t=1}^{N} P^{t-1}r.$$

From (8.2.3),

$$h = \sum_{t=1}^{N} P^{t-1}r - Ng + \sum_{t=N+1}^{\infty} (P^{t-1} - P^*)r.$$

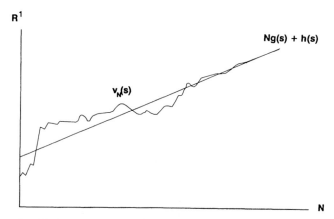

Figure 8.2.1   Graphical representation of (8.2.5). Vertical axis denotes the real line.

Theorem A.7(c) implies that the third term above converges to zero as $N \to \infty$, so that we may write

$$v_{N+1} = Ng + h + o(1) \tag{8.2.5}$$

where $o(1)$ denotes a vector with components which approach 0 pointwise as $N \to \infty$. Thus, as N becomes large, for each $s \in S$, $v_{N+1}(s)$ approaches a line with slope $g(s)$ and intercept $h(s)$ (Fig. 8.2.1).

For j and k in the same closed irreducible class, Proposition 8.1.1 implies that $g(j) = g(k)$, so that (8.2.5) implies that

$$h(j) - h(k) = \lim_{N \to \infty} \left[v_N(j) - v_N(k) \right].$$

Hence h equals the asymptotic relative difference in total reward that results from starting the process in state j instead of in state k. For this reason, we sometimes refer to h as the *relative value* vector.

For *periodic* chains, a similar line of reasoning shows that, for j and k in the same closed irreducible class,

$$h(j) - h(k) = \lim_{N \to \infty} \frac{1}{N} \sum_{n=1}^{N} \left[v_n(j) - v_n(k) \right]$$

so that $h(j) - h(k)$ gives the average relative difference in total reward. Another consequence of representation (8.2.5) is that $v_N - v_{N-1}$ eventually increases at rate g, providing further justification for referring to g as the gain of the process.

The result below follows immediately from (8.2.4).

Proposition 8.2.2. Suppose $g(s) = 0$ for all $s \in S$.

a. Then

$$h(s) = C - \lim_{N \to \infty} E_s \left\{ \sum_{t=1}^{N} r(X_t) \right\} = \lim_{N \to \infty} \frac{1}{N} \sum_{k=1}^{N} v_k(s)$$

and

b.

$$h(s) = \lim_{N \to \infty} E_s \left\{ \sum_{t=1}^{N} r(X_t) \right\} = \lim_{N \to \infty} v_{N+1}(s) \qquad (8.2.6)$$

whenever the limit exists.

We will use this result in Section 10.4 to further investigate behavior of policy iteration in models with total reward criterion. We showed in Chap. 7 that, in finite-state models, whenever $\lim_{N \to \infty} v_N^{d^\infty}(s)$ exists, $r(s) = 0$ at all states which are recurrent under P_d. Therefore in such models, *the bias equals the expected total reward*. Part (a) of this proposition refers to models which may cycle on recurrent states and accumulate positive and negative rewards which average out to 0. In that case the expected total-reward criterion may be inappropriate for comparing policies.

We demonstrate some of the above concepts by expanding on Example A.1 of Appendix A.

Example 8.2.1. Let $S = \{s_1, s_2\}$, and suppose a Markov chain has transition probability matrix

$$P = \begin{bmatrix} 0 & 1 \\ 1 & 0 \end{bmatrix}.$$

In Appendix A, we show that

$$P^* = \begin{bmatrix} \frac{1}{2} & \frac{1}{2} \\ \frac{1}{2} & \frac{1}{2} \end{bmatrix}, \qquad H_P = \begin{bmatrix} \frac{1}{4} & -\frac{1}{4} \\ -\frac{1}{4} & \frac{1}{4} \end{bmatrix}.$$

Suppose $r' = (1, -1)$. Then

$$g = \begin{bmatrix} 0 \\ 0 \end{bmatrix} \quad \text{and} \quad h = \begin{bmatrix} \frac{1}{2} \\ -\frac{1}{2} \end{bmatrix}.$$

We now interpret these quantities. Starting in s_1, the stream of rewards for this system is $(1, -1, 1, -1, \dots)$ so that the sequence of partial sums of rewards, $\{v_n(s_1)\}$ equals $(1, 0, 1, 0, \dots)$. Therefore the sequence of average rewards starting from s_1

equals $(1, 0, \frac{1}{3}, 0, \frac{1}{5}, 0, \dots)$. Hence the long-run average equals 0 as given by $g(s_1)$. The partial sum sequence $\{v_n(s_1)\}$ converges in the Cesaro sense, but not in the ordinary sense. The sequence of *average* partial sums equals $(1, \frac{1}{2}, \frac{2}{3}, \frac{1}{2}, \frac{3}{5}, \dots)$, so that its Cesaro limit equals $\frac{1}{2}$ as given by $h(s_1)$. Since the sequence of $v_n(s_2)$ equals $(-1, 0, -1, 0, \dots)$, the sequence $v_n(s_1) - v_n(s_2)$ equals $(2, 0, 2, 0, \dots)$, so that the average difference in total rewards between these two states equals 1 as given by $h(s_1) - h(s_2)$.

8.2.2 The Laurent Series Expansion

Laurent series expansions relate the gain and bias to the expected total discounted reward and provide a key tool for analyzing finite-state undiscounted models. Following Sec. A.6 of Appendix A, we parameterize the discounted reward in terms of the interest rate ρ instead of in terms of the discount rate λ. These quantities are related by $\lambda = (1 + \rho)^{-1}$ or $\rho = (1 - \lambda)\lambda^{-1}$; $0 \leq \lambda < 1$ implies $\rho > 0$. The quantity $1 + \rho$ represents the amount received at the start of the next period, when one unit is invested at the start of the present period and the interest rate equals ρ. Letting v_λ represent the total expected discounted reward of the MRP, it follows from (6.1.10) that

$$v_\lambda = (I - \lambda P)^{-1} r = (1 + \rho)(\rho I + [P - I])^{-1} r. \tag{8.2.7}$$

We refer to $(\rho I + [I - P])^{-1}$ as the *resolvent* of $I - P$ (at ρ). Multiplying the Laurent series expansion for the resolvent in Theorem A.8 by r, we obtain the following Laurent series expansion of v_λ.

Theorem 8.2.3. Assume finite S. Let ν denote the nonzero eigenvalue of $I - P$ with the smallest modulus. Then, for $0 < \rho < |\nu|$,

$$v_\lambda = (1 + \rho)\left[\rho^{-1} y_{-1} + \sum_{n=0}^{\infty} \rho^n y_n \right] \tag{8.2.8}$$

where $y_{-1} = P^* r = g$, $y_0 = H_P r = h$, and $y_n = (-1)^n H_P^{n+1} r$ for $n = 1, 2, \dots$.

Often, instead of writing $0 < \rho < |\nu|$, we say "for ρ sufficiently small." When analyzing MDPs with average reward criterion, it will suffice to use the following *truncated Laurent expansion* of v_λ.

Corollary 8.2.4. Let g and h represent the gain and bias of a MRP with finite S. Then if Assumption 8.0.2 holds

$$v_\lambda = (1 - \lambda)^{-1} g + h + f(\lambda), \tag{8.2.9}$$

where $f(\lambda)$ denotes a vector which converges to zero as $\lambda \uparrow 1$.

Proof. Expressing (8.2.8) on the λ scale and adding and subtracting h, we obtain

$$v_\lambda = \frac{1}{1-\lambda} g + h + \frac{1-\lambda}{\lambda} h + \frac{1}{\lambda} \sum_{n=0}^{\infty} (-1)^n \left[\frac{1-\lambda}{\lambda} \right]^n y_n.$$

Since the series in (8.2.8) converges for ρ sufficiently small, the last two terms above converge to 0 as $\lambda \uparrow 1$. □

The following corollary provides a powerful tool for extending structure and existence results from the discounted case to the average reward case. It follows immediately by multiplying both sides of (8.2.9) by $1 - \lambda$ and passing to the limit.

Corollary 8.2.5. Let g and v_λ represent the gain and expected discounted reward of a MRP. Then

$$g = \lim_{\lambda \uparrow 1} (1 - \lambda) v_\lambda. \tag{8.2.10}$$

The result in Corollary 8.2.5 has the following interesting probabilistic interpretation. Consider a *terminating* Markov reward process in which a random variable τ, with distribution independent of that of the Markov chain, determines the termination time and a reward $r(X_\tau)$ is received only at termination. Suppose τ has a geometric distribution parametrized as

$$P(\tau = n) = (1 - \lambda)\lambda^{n-1}, \qquad n = 1, 2, \ldots.$$

Then

$$E_s\{r(X_\tau)\} = \sum_{n=1}^{\infty} (1 - \lambda)\lambda^{n-1} E_s\{r(X_n)\} = (1 - \lambda) v_\lambda(s).$$

Since the geometric distribution corresponds to the time to first "failure" in a sequence of independent Bernoulli trials with failure probability $1 - \lambda$, as $\lambda \uparrow 1$ or $(1 - \lambda) \downarrow 0$, $P(\tau > M)$ approaches 1, for any M, so that

$$\lim_{\lambda \uparrow 1} (1 - \lambda) v_\lambda(s) = E_s\{r(X_\infty)\} = g(s).$$

This argument can be made formal to provide a probabilistic derivation of Corollary 8.2.5.

Example 8.2.1 (ctd.). The eigenvalues of $I - P$ are 0 and 2, so that, for $0 < \rho < 2$, the Laurent series expansion for v_λ satisfies

$$v_\lambda = (1 + \rho) \left(\frac{1}{\rho} \begin{bmatrix} 0 \\ 0 \end{bmatrix} + \begin{bmatrix} \frac{1}{2} \\ \frac{1}{2} \end{bmatrix} + \rho \begin{bmatrix} -\frac{1}{4} \\ \frac{1}{4} \end{bmatrix} + \rho^2 \begin{bmatrix} \frac{1}{8} \\ -\frac{1}{8} \end{bmatrix} + \cdots \right).$$

By direct calculation through inverting $(I - \lambda P)$ or summing the above series, we observe that

$$
v_\lambda = \begin{bmatrix} \dfrac{1+\rho}{2+\rho} \\ -\dfrac{1+\rho}{2+\rho} \end{bmatrix} = \begin{bmatrix} \dfrac{1}{1+\lambda} \\ -\dfrac{1}{1+\lambda} \end{bmatrix}.
$$

Note as $\rho \downarrow 0$ or $\lambda \uparrow 1$, that $(1 - \lambda)v_\lambda$ converges to $g = 0$, as indicated by Corollary 8.2.5 and that v_λ converges to h, as shown in Corollary 8.2.4.

8.2.3 Evaluation Equations

In all but the simplest examples, computation of g and h through direct evaluation of P^* and H_P may be inefficient. In this section we provide systems of equations which enable computation of these quantities as well as the higher order terms in the Laurent series expansion of v_λ. These equations serve as the basis for the optimality equations for models with average and sensitive optimality criteria.

We provide two approaches to deriving these equations; the first uses identities (A.4), (A.17) and (A.20) of Appendix A. The second approach uses the discounted reward evaluation equation and the Laurent series expansion for the discounted reward.

Theorem 8.2.6.   Let S be finite and let g and h denote the gain and bias of a MRP with transition matrix P and reward r.

a. Then

$$(I - P)g = 0 \tag{8.2.11}$$

and

$$g + (I - P)h = r \tag{8.2.12}$$

b. Suppose g and h satisfy (8.2.11) and (8.2.12), then $g = P^*r$ and $h = H_P r + u$ where $(I - P)u = 0$.

c. Suppose g and h satisfy (8.2.11), (8.2.12) and $P^*h = 0$, then $h = H_P r$.

Proof.   From (A.4) $(I - P)P^* = 0$, so multiplying r by $(I - P)P^*$ and noting that $g = P^*r$ establishes (8.2.11). From (A.17), $P^* + (I - P)H_P = I$ so applying both sides of this identity to r establishes (8.2.12).

To establish (b), apply P^* to (8.2.12) and add it to (8.2.11) to obtain

$$(I - P + P^*)g = P^*r.$$

Noting the non-singularity of $(I - P + P^*)$, (A.20) shows that

$$g = (I - P + P^*)^{-1}P^*r = Z_P P^*r = P^*r.$$

We have previously shown that $h = H_P r$ satisfies (8.2.12). Suppose h_1 also satisfies (8.2.12). Then $(I - P)(H_P r - h_1) = 0$, so h is unique up to an element of $\{u \in R^{|S|}: (I - P)u = 0\}$.

We now prove part (c). Suppose $P^*h = 0$, then adding this identity to (8.2.12) shows that

$$(I - P + P^*)h = r - g.$$

Therefore,

$$h = (I - P + P^*)^{-1}(I - P^*)r = H_P r. \qquad \square$$

The equations (8.2.11) and (8.2.12) uniquely characterize g and determine h up to an element of the null space of $I - P$. The condition $P^*h = 0$ doesn't provide much practical assistance in finding h because it requires prior determination of P^* (Appendix A.4). We discuss a more efficient approach for computing h below.

First, we show that we need not solve $(I - P)g = 0$ directly. If g satisfies $(I - P)g = 0$, then g is an element of the subspace of $R^{|S|}$ spanned by the eigenvectors of P corresponding to eigenvalue 1. Suppose there are m closed irreducible recurrent classes denoted by $C_1, C_2, \ldots, C_m$, then a basis for this subspace consists of vectors $u_1, u_2, \ldots, u_m$ for which $u_k(s) = 1$ if $s \in C_k$ and 0 otherwise. Thus, for s transient,

$$g(s) - \Sigma_{k \in T} p(k|s)g(k) = \Sigma_{k \in C_1 \cup \cdots \cup C_m} p(k|s)g(k). \qquad (8.2.13)$$

As a consequence of Proposition A.3, once we have found g on recurrent states, (8.2.13) uniquely determines g on transient states.

Therefore, we can find g using the following approach:

1. Classify the states of the Markov chain using the Fox-Landi algorithm from Section A.3,
2. Compute g on each recurrent class of P by solving (8.2.12) on each class separately with g constrained to be constant there, and
3. Compute g on transient states by solving (8.2.13).

When P is irreducible or unichain, g is constant and (8.2.11) becomes superfluous. In this case any solution of (8.2.11) is a scalar multiple of the unit vector so that we need not solve (8.2.11) and can find g uniquely by solving (8.2.12) alone. We state this as follows.

Corollary 8.2.7. Suppose P is unichain or irreducible. Then the average reward $P^*r = ge$ and it is uniquely determined by solving

$$ge + (I - P)h = r. \qquad (8.2.14)$$

Suppose g and h satisfy (8.2.14), then $g = P^*r$ and $h = H_P r + ke$ for arbitrary scalar k. Furthermore, if g and h satisfy (8.2.14) and $P^*h = 0$, then $h = H_P r$.

Another consequence of Theorem 8.2.6 is that (8.2.11) and (8.2.12) uniquely determine h up to an element of the null space of $I - P$. Since this space has dimension m, where m denotes the number of closed irreducible classes of P, this system of equations determines h up to m constants (one on each closed class and possibly all m on transient states). Therefore, any specification, such as $P^*h = 0$, which determines these m constants provides a unique representation for h, but it does not assure that $h = H_P r$. Consequently, for unichain P, we can find *relative values* $h(j) - h(k)$ by setting any component of h equal to zero and solving (8.2.14).

We illustrate the above concepts with the following multichain example.

Example 8.2.2. Suppose $S = \{s_1, \ldots, s_5\}$,

$$P = \begin{bmatrix} 1 & 0 & 0 & 0 & 0 \\ 0 & 0.4 & 0.6 & 0 & 0 \\ 0 & 0.7 & 0.3 & 0 & 0 \\ 0.5 & 0.2 & 0 & 0 & 0.3 \\ 0.2 & 0.3 & 0.3 & 0.2 & 0 \end{bmatrix} \quad \text{and} \quad r = \begin{bmatrix} 2 \\ 5 \\ 4 \\ 1 \\ 3 \end{bmatrix}.$$

Observe that P is in canonical form and that P has two closed irreducible recurrent classes, $C_1 = \{s_1\}, C_2 = \{s_2, s_3\}$ and transient states $T = \{s_4, s_5\}$. Solving $(I - P)g = 0$, does not determine $g(s_1)$, shows that $g(s_2) = g(s_3)$ and that

$$g(s_4) - 0.3g(s_5) = 0.5g(s_1) + 0.2g(s_2)$$

$$- 0.2g(s_4) + g(s_5) = 0.2g(s_1) + 0.3g(s_2) + 0.2g(s_3).$$

We may express (8.2.12) in component notation as

$$r(s) - g(s) + \sum_{j \in S} p(j|s)h(j) - h(s) = 0. \qquad (8.2.15)$$

Substituting P and r into this equation and noting the above relationships for g shows that $g(s_1) = r(s_1) = 2$,

$$5 - g(s_2) - 0.6h(s_2) + 0.6h(s_3) = 0,$$
$$4 - g(s_2) + 0.7h(s_2) - 0.7h(s_3) = 0, \qquad (8.2.16)$$

so that $g(s_2) = g(s_3) = 4.538$, and

$$g(s_4) - 0.3g(s_5) = 0.5 \times 2 + 0.2 \times 4.538 = 1.908,$$

$$- 0.2g(s_4) + g(s_5) = 0.2 \times 2 + 0.5 \times 4.538 = 2.669,$$

so that $g(s_4) = 2.882$ and $g(s_5) = 3.245$.

To determine h, (8.2.16) implies that

$$0.462 - 0.6h(s_2) + 0.6h(s_3) = 0,$$
$$- 0.538 + 0.7h(s_2) - 0.7h(s_3) = 0, \qquad (8.2.17)$$

so that one of these equations is redundant. On the transient states, (8.2.15) yields

$$-1.882 + 0.5h(s_1) + 0.2h(s_2) - h(s_4) + 0.3h(s_5) = 0$$
$$- 0.245 + 0.2h(s_1) + 0.3h(s_2) + 0.3h(s_3) + 0.2h(s_4) - h(s_5) = 0$$
(8.2.18)

As a consequence of Theorem 8.2.6, $P^*h = 0$ uniquely determines h. Using methods in Sec. A.4 shows that

$$P^* = \begin{bmatrix} 1 & 0 & 0 & 0 & 0 \\ 0 & 0.538 & 0.462 & 0 & 0 \\ 0 & 0.538 & 0.462 & 0 & 0 \\ 0.595 & 0.218 & 0.187 & 0 & 0 \\ 0.319 & 0.367 & 0.314 & 0 & 0 \end{bmatrix}$$

so that $h(s_1) = 0$, and

$$0.462 - 0.600h(s_2) + 0.600h(s_3) = 0,$$
$$0.538h(s_2) + 0.462h(s_3) = 0$$

Hence $h(s_2) = 0.354$ and $h(s_3) = -0.416$ and from (8.2.18) we obtain $h(s_4) = 2.011$ and $h(s_5) = 0.666$. Note also that $g = P^*r$.

We now provide a system of equations which uniquely characterize g, h, and the higher-order terms of the Laurent series expansions of v_λ. Note that (8.2.11) characterizes g as an element of the null space of $(I - P)$. By adding the additional equation (8.2.12), we uniquely determine g. This suggests that by adding another equation to the above system we can determine h uniquely. We formalize this observation.

Theorem 8.2.8. In a Markov reward process with transition matrix P and reward r, let y_n, $n = -1, 0, \ldots$ denote the coefficients of the Laurent series expansion of v_λ.

a. Then

$$(I - P)y_{-1} = 0, \tag{8.2.19}$$

$$y_{-1} + (I - P)y_0 = r, \tag{8.2.20}$$

and for $n = 1, 2, \ldots$

$$y_{n-1} + (I - P)y_n = 0. \tag{8.2.21}$$

b. Suppose for some $M \geq 0$, that $w_{-1}, w_0, \ldots, w_M$ satisfy (8.2.19), (8.2.20), and if $M \geq 1$, (8.2.21) for $n = 1, 2, \ldots, M$. Then $w_{-1} = y_{-1}$, $w_0 = y_0$, $\ldots$, $w_{M-1} = y_{M-1}$ and $w_M = y_M + u$ where $(I - P)u = 0$.

c. Suppose the hypotheses of part (b) hold and in addition $P^*w_M = 0$. Then $w_M = y_M$.

Proof. Since $(I - \lambda P)v_\lambda = r$ and $\lambda = (1 + \rho)^{-1}$ it follows from (8.2.8) that

$$(1 + \rho)\left(I - (1 + \rho)^{-1}P\right)\left[\rho^{-1}y_{-1} + \sum_{n=0}^{\infty}(-\rho)^n y_n\right] = r$$

Rearranging terms shows that

$$\rho^{-1}(I - P)y_{-1} + \sum_{n=0}^{\infty}(-\rho)^n[y_{n-1} + (I - P)y_n] = r$$

for $0 < \rho < |\nu|$ where ν was defined in Theorem 8.2.3. Equating terms with like powers of ρ shows that $y_{-1}, y_0, \ldots$ satisfy (8.2.19)–(8.2.21).

The proofs of parts (b) and (c) follow by inductively applying the argument in part (b) of Theorem 8.2.6. and noting from (A.18) and (A.19) that $Z_P(H_P)^n = (H_P)^{n+1}$. □

When we choose $M = 1$ in the above theorem we obtain the following system of equations for computing the bias h.

Corollary 8.2.9. Suppose u, v, and w satisfy

$$(I - P)u = 0, \quad u + (I - P)v = r \quad \text{and} \quad v + (I - P)w = 0 \quad (8.2.22)$$

then $u = g$, $v = h$, and $w = y_1 + z$ where $(I - P)z = 0$.

Example 8.2.2 (ctd.). We now compute the bias h using the approach of Corollary 8.2.9. Note that once we determine $h(s_1)$, $h(s_2)$, and $h(s_3)$, we may find $h(s_4)$ and $h(s_5)$ using (8.2.18). We express $h + (I - P)w = 0$ in component notation as

$$h(s) + w(s) - \sum_{j\in S}p(j|s)w(j) = 0 \qquad (8.2.23)$$

so that on substitution of the values for $p(j|s)$, we obtain

$$h(s_1) = 0$$
$$h(s_2) + 0.6w(s_2) - 0.6w(s_3) = 0$$
$$h(s_3) - 0.7w(s_2) + 0.7w(s_3) = 0$$

Hence $h(s_2)$ and $h(s_3)$ satisfy

$$0.7h(s_2) + 0.6h(s_3) = 0$$

Noting from (8.2.17) that

$$0.462 - 0.6h(s_2) + 0.6h(s_3) = 0$$

we obtain $h(s_2) = 0.355$ and $h(s_3) = -0.414$ as above. We leave it as an exercise to show that (8.2.21) determines the higher-order terms of the Laurent series expansion and that they agree with those computed directly.

8.3 CLASSIFICATION OF MARKOV DECISION PROCESSES

With the exception of a few results in Chap. 7, we have ignored the chain structure (Appendix A) of the transition matrices of Markov chains generated by stationary policies. In average reward models we can no longer take this liberty, since results and analyses depend on state accessibility patterns of chains corresponding to stationary policies. In this section we describe the different classes of models, discuss the relationship between them, describe a simple algorithm for classifying a model, and show how the model class effects the form of the optimal average reward.

8.3.1 Classification Schemes

We classify MDPs in two ways:

1. On the basis of the chain structure of the set of Markov chains induced by all stationary policies.
2. On the basis of patterns of states which are accessible from each other under *some* stationary policy.

We refer to any MDP as *general* and distinguish the following classes of models. We say that a MDP is

a. *Recurrent* or *ergodic* if the transition matrix corresponding to *every* deterministic stationary policy consists of a single recurrent class;

b. *Unichain* if the transition matrix corresponding to *every* deterministic stationary policy is unichain, that is, it consists of a single recurrent class plus a possibly empty set of transient states;

c. *Communicating* if, for every pair of states s and j in S, there exists a deterministic stationary policy d^∞ under which j is accessible from s, that is, $p_d^n(j|s) > 0$ for some $n \geq 1$;

d. *Weakly communicating* if there exists a *closed* set of states, with each state in that set accessible from every other state in that set under some deterministic stationary policy, plus a possibly empty set of states which is transient under every policy; and

e. *Multichain* if the transition matrix corresponding to *at least one* stationary policy contains two or more closed irreducible recurrent classes.

Many authors use the expressions *multichain* and *general* interchangably. We distinguish them and use the more restrictive notion of a multichain model above. In our terminology, unichain and recurrent models are special cases of general models but are distinct from multichain models. Weakly communicating models may be viewed as communicating models with "extra" transient states. Figure 8.3.1 represents the relationship between these classifications. Classifications appear in order of generality; the most general classification appears at the top. Connecting lines indicate that the lower class is included in the class above it.

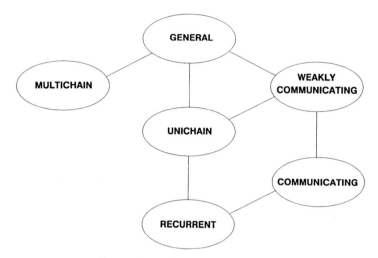

Figure 8.3.1 Hierarchy of MDP classes.

Unichain models generalize recurrent models while weakly communicating models generalize communicating models. Recurrent MDPs are communicating, and unichain MDPs are weakly communicating. Multichain MDPs may or may not be communicating, as the following inventory model illustrates. Refer to Sec. 3.2 for a formulation of the inventory model.

Example 8.3.1.   Demands $\{D_t\}$ are independent and identically distributed with $P\{D_t = 0\} = p$ and $P\{D_t = 1\} = 1 - p$, $0 < p < 1$; the warehouse has a capacity of three units, and unfilled demand is lost. Following Sec. 3.2, $S = \{0, 1, 2, 3\}$ and $A_s = \{0, \dots, 3 - s\}$. Under stationary policy d^∞ with $d(0) = 1$, $d(1) = 0$, $d(2) = 1$, and $d(3) = 0$, the Markov chain (Fig. 8.3.2) has two closed irreducible classes $\{0, 1\}$ and $\{2, 3\}$ so that the model is multichain. Alternatively, consider the stationary deterministic policy δ^∞ which orders three units when the stock level equals 0 and does not order otherwise; i.e., $\delta(0) = 3$, $\delta(1) = 0$, $\delta(2) = 0$, and $\delta(3) = 0$. Under this policy each state is accessible from each other state so that the model is communicat-

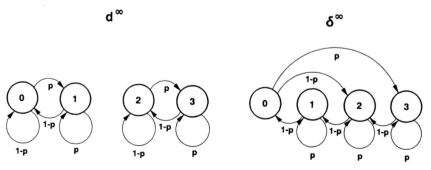

Figure 8.3.2   Markov chains corresponding to policies d^∞ and δ^∞ in Example 8.3.1.

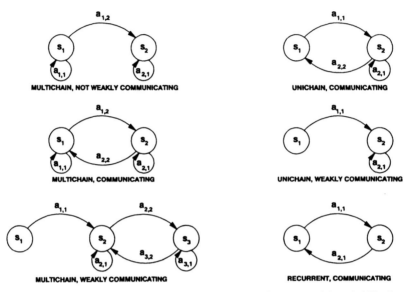

Figure 8.3.3   Examples of MDP classification. All transitions occur with probability 1.

ing (Fig. 8.3.2).

Figure 8.3.3 provides examples which illustrate each distinct model class.

8.3.2   Classifying a Markov Decision Process

In some applications, we might be able to determine whether all stationary policies are unichain or recurrent by inspection. If not, classification of a model as unichain or recurrent requires applying the Fox-Landi Algorithm of Appendix A, Sec. A.3 to the Markov chain corresponding to *each* stationary policy. Since there are $\Pi_{s \in S} |A_s| \leq N^{|S|}$ stationary policies where $N = \max_{s \in S} |A_s|$, in the worse case, classification using this approach would be an exponential task requiring $O(N^{|S|}|S|^2)$ comparisons. In practical applications in which the model structure is not obvious, classification by this approach would be prohibitive. In this case, we solve these MDPs using the more complex algorithms of Chap. 9 which do not require prior determination of chain structure.

Classifying models on the basis of communication properties is considerably easier and we now describe an approach. It relies on the following observation.

Proposition 8.3.1

a. An MDP is communicating if and only if there exists a randomized stationary policy which induces a recurrent Markov chain.

b. An MDP is weakly communicating if and only if there exists a randomized stationary policy which induces a Markov chain with a single closed irreducible class and a set of states which is transient under all stationary policies.

Noting this, we can classify a model as communicating, weakly communicating or general in the following way.

Model Classification Algorithm

1. Define the matrix Q as follows: set $q(j|s) = 1$ if $p(j|s, a) > 0$ for some $a \in A_s$; otherwise set $q(j|s) = 0$.
2. Apply the Fox-Landi Algorithm of Section A.4 to Q.
3. If the matrix Q consists of a single closed class and no transient states, classify the MDP as communicating; if it consists of a single closed class plus transient states, go to 4; otherwise classify the MDP as general.
4. Set $c(s) = 1$ if s is in the closed class; set $c(s) = 0$ otherwise. Repeat the following until $c(s)$ no longer changes for any $s \in S$: For each $s \in S$ for which $c(s) = 0$, set $c(s) = 1$ if $\sum_{j \in S} p(j|s, a)c(j) > 0$ for *all* $a \in A_s$.

   If $c(s) = 1$ for all $s \in S$, classify the MDP as weakly communicating; otherwise classify it as general.

Matrix Q has the same pattern of positive entries and zeros as the transition probability matrix of a randomized stationary policy which in each state assigns positive probability to all actions. The algorithm finds the closed classes of that policy. Proposition 8.3.1 justifies the classification in step 3. Note that, after forming Q, this algorithm requires $O(|S|^2)$ comparisons to determine if the MDP is communicating.

8.3.3   Model Classification and the Average Reward Criterion

We now show how the structure of the optimal average reward (gain) relates to model type. Proposition 8.2.1 implies that the optimal gain is constant in recurrent and unichain models. We show that, in communicating and weakly communicating models, whenever a stationary policy has nonconstant gain, we may construct a stationary policy with constant gain which dominates the nonconstant gain policy. This need not be the case when the model is multichain but not weakly communicating.

We summarize observations in the following theorem. We provide a rather formal proof below. It can best be understood by referring to an example such as the continuation of Example 8.3.1 below or Problem 8.7. The basic idea is that, in a weakly communicating model, any closed class can be reached from every other state.

Theorem 8.3.2.   Assume a weakly communicating model and let $d \in D^{\text{MD}}$.

a. Let C denote a closed, irreducible, and recurrent set of states in the Markov chain generated by the stationary policy d^∞. Then there exists a $\delta \in D^{\text{MD}}$ with $\delta(s) = d(s)$ for all $s \in C$, and for which the Markov chain generated by δ has C as its only closed, irreducible, recurrent class.

b. Suppose stationary policy d^∞ has $g^{d^\infty}(s) < g^{d^\infty}(s')$ for some s and s' in S. Then there exists a stationary policy δ^∞ for which $g^{\delta^\infty}(s) = g^{\delta^\infty}(s') \geq g^{d^\infty}(s')$.

c. Given any $d \in D^{\text{MD}}$, there exists a $\delta \in D^{\text{MD}}$ for which g^{δ^∞} is constant and $g^{\delta^\infty} \geq g^{d^\infty}$.

d. If there exists a stationary optimal policy, there exists a stationary optimal policy with constant gain.

Table 8.3.1   Relationship between model class and gain structure

Model Class	Optimal Gain	Gain of a Stationary Policy
Recurrent	Constant	Constant
Unichain	Constant	Constant
Communicating	Constant	Possibly nonconstant
Weakly Communicating	Constant	Possibly nonconstant
Multichain	Possibly nonconstant	Possibly nonconstant

Proof. Let T denote the set of states that is transient under every policy. From the definition of a weakly communicating model, there exists an $s \in S/(T \cup C)$ and an $a' \in A_s$ for which $\sum_{j \in C} p(j|s, a') > 0$. Set $\delta(s) = a'$. Augment C with s' and repeat this procedure until $\delta(s)$ is defined for all $s \in S/T$. By the definition of T, for each $s' \in T$, there exists an $a_{s'} \in A_{s'}$ for which $\sum_{j \in S/T} p(j|s', a_{s'}) > 0$. Set $\delta(s') = a_{s'}$ for each $s' \in T$. Then δ achieves the conclusions of part (a).

We now prove part (b). Let C be closed, irreducible, and recurrent under d^∞. Then, if $s' \in C$, it follows from (a) and Proposition 8.2.1 that there exists a $\delta \in D^{MD}$ for which g^{δ^∞} is constant and $g^{\delta^\infty}(s') = g^{d^\infty}(s')$ so the result follows with equality. If s' is transient under d^∞, then there exists an s'' which is recurrent under d^∞ with $g^{d^\infty}(s'') \geq g^{d^\infty}(s')$. The result now follows from (a) and Proposition 8.2.1.

Note that (c) follows easily from (b) and that (d) follows immediately from (c). □

In subsequent sections of Chaps. 8 and 9, we establish the optimality of stationary policies in finite-state models with average reward criterion. Therefore, in light of Theorem 8.3.2, we may summarize the relationship between model classification and optimal gain as in Table 8.3.1.

Note that we include the adjective "possibly" to account for unusual models in which the optimal gain is constant even though the optimal policy is multichain (Example 8.4.2 below with $r(s_2, a_{2,1})$ set equal to 3). We now return to Example 8.3.1 and illustrate the conclusions of Theorem 8.3.2.

Example 8.3.1 (ctd.). We add assumptions regarding costs and revenues. Let $K = 2$, $c(u) = 2u$, $h(u) = u$, and $f(u) = 8u$. Similar calculations to those in Sect. 3.2 show that $r(s, a)$ satisfies

$$r(s, a)$$

s \ a	0	1	2	3
0	0	-1	-4	-7
1	4	-2	-5	$\times$
2	4	-3	$\times$	$\times$
3	4	$\times$	$\times$	$\times$

where $\times$ denotes a nonfeasible action. Choose $p = 0.5$, so that $P(D_t = 0) = P(D_t = 1) = 0.5$. Thus

$$P_d = \begin{bmatrix} 0.5 & 0.5 & 0.0 & 0.0 \\ 0.5 & 0.5 & 0.0 & 0.0 \\ 0.0 & 0.0 & 0.5 & 0.5 \\ 0.0 & 0.0 & 0.5 & 0.5 \end{bmatrix} = P_d^*$$

and $(r_d)^T = (-1, 4, -3, 4)$. Since $g^{d^\infty} = P_d^* r_d$, $(g^{d^\infty})^T = (1.5, 1.5, 0.5, 0.5)$.

As suggested by the proof of Theorem 8.3.2, we can alter this policy by choosing a decision rule which leads the process from states 2 and 3 to the closed class $\{0, 1\}$. Let γ denote the decision rule which orders one unit in state 0 and zero units otherwise. For that decision rule,

$$P_\gamma = \begin{bmatrix} 0.5 & 0.5 & 0.0 & 0.0 \\ 0.5 & 0.5 & 0.0 & 0.0 \\ 0.0 & 0.5 & 0.5 & 0.0 \\ 0.0 & 0.0 & 0.5 & 0.5 \end{bmatrix}$$

so that

$$P_\gamma^* = \begin{bmatrix} 0.5 & 0.5 & 0.0 & 0.0 \\ 0.5 & 0.5 & 0.0 & 0.0 \\ 0.5 & 0.5 & 0.0 & 0.0 \\ 0.5 & 0.5 & 0.0 & 0.0 \end{bmatrix}.$$

Therefore $(g^{\gamma^\infty})^T = (1.5, 1.5, 1.5, 1.5)$ and $g^{\gamma^\infty} \geq g^{d^\infty}$. Observe that the Markov chain generated by γ^∞ is unichain. We leave it as an exercise to find the optimal policy. Our discussion above implies that it will have constant gain.

8.4 THE AVERAGE REWARD OPTIMALITY EQUATION—UNICHAIN MODELS

In this section we provide an optimality equation for unichain (and recurrent) Markov decision problems with average reward criterion. We analyze these models prior to multichain models, because for unichain models we may characterize optimal policies and their average rewards through a *single* optimality equation. The reason for this is that, in unichain models, *all* stationary policies have constant gain so that the MDP analog of (8.2.11),

$$\max_{d \in D} \{(P_d - I)g\} = 0,$$

holds for any constant g and provides no additional information regarding the structure of g. We will show that multichain models, be they communicating, weakly communicating, or noncommunicating, require the above equation in addition to the

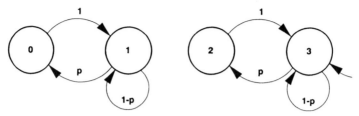

Figure 8.4.1 Transition diagram for a multichain queueing admission control policy.

equation

$$\max_{d \in D} \{ r_d - g + (P_d - I)h \} = 0 \tag{8.4.1}$$

to determine optimal policies.

We provide an example of a model with unichain stationary policies.

Example 8.4.1. Consider the queueing admission control model of Sec. 3.7.1. Assume that the "possible service distribution" $f(s)$ has countable support (i.e., it assigns positive probability to all non-negative integer values). For example, $f(s)$ might be a Poisson or geometric probability mass function. Under this assumption, the state 0, which corresponds to an empty queue, may be reached in one step with positive probability from any system state. Hence all policies are unichain.

On the other hand, if $f(s)$ has finite support, the model may include multichain policies so that the analyses of this chapter need not apply. As an example, suppose $f(0) = 1 - p$, $f(1) = p$, $0 < p < 1$, and jobs arrive deterministically at the rate of one per period. Then any stationary policy which accepts a job in state 0, no jobs in states 1 or 3, and a job in state 2 contains at least two recurrent classes (Fig. 8.4.1).

The development in this section parallels that in Sec. 6.2 as closely as possible. We focus primarily on finite-state and action models; however, results in Sec. 8.4.1 hold in greater generality. Our proof of the existence of a solution to (8.4.1) relies on the finiteness of the set of stationary policies; we consider models with compact action sets in Sec. 8.4.4.

8.4.1 The Optimality Equation

When the gain has equal components, we denote its value by the scalar g or the vector ge. The optimality equation in a unichain average reward MDP may be expressed in component notation as

$$0 = \max_{a \in A_s} \left\{ r(s, a) - g + \sum_{j \in S} p(j|s, a)h(j) - h(s) \right\} \tag{8.4.2}$$

or in matrix-vector and operator notation as

$$0 = \max_{d \in D} \{ r_d - ge + (P_d - I)h \} \equiv B(g, h). \tag{8.4.3}$$

We regard B as a mapping from $R^1 \times V$ to V, where, as before, V denotes the space of bounded real-valued functions on S. Note that the maximum in (8.4.3) is with respect to the componentwise partial order and $D \equiv D^{MD}$. When D consists of a single decision rule, this equation reduces to (8.2.12).

Before investigating properties of solutions of the optimality equation, we provide a *heuristic* derivation which we later formalize in Sec. 8.4.2. to demonstrate the existence of a solution. Begin with the optimality equation for the discounted reward

$$0 = \max_{d \in D} \{ r_d + (\lambda P_d - I) v_\lambda^* \}$$

and assume that v_λ^* has the partial Laurent series expansion

$$v_\lambda^* = (1 - \lambda)^{-1} g^* e + h + f(\lambda)$$

where, as before, $f(\lambda)$ denotes a vector which converges pointwise to 0 as $\lambda \uparrow 1$. By substituting this expression into the optimality equation, we obtain

$$0 = \max_{d \in D} \left\{ r_d + (\lambda P_d - I) \left[\frac{g^* e}{1 - \lambda} + h + f(\lambda) \right] \right\}$$

$$= \max_{d \in D} \{ r_d - g^* e + (\lambda P_d - I) h + f(\lambda) \}.$$

The second equation follows from the first by noting that, since g^* is constant, $P_d g^* e = g^* e$ for all $d \in D$. Taking the limit as $\lambda \uparrow 1$ suggests that (8.4.3) is the appropriate form for the optimality equation.

We now provide a different heuristic derivation based on the finite-horizon optimality equations. Eq. (8.2.5) suggests that

$$v_N^* = (N - 1) g^* e + h + o(1).$$

Rewrite the finite-horizon optimality equations (4.5.1) in this notation as

$$v_{N+1}^* = \max_{d \in D} \{ r_d + P_d v_N^* \}$$

and substitute in the above expression for v_N^* to obtain

$$N g^* e + h + o(1) = \max_{d \in D} \{ r_d + (N - 1) P_d g^* e + P_d h + o(1) \}.$$

Again, noting that $P_d g^* e = g^* e$ for all $d \in D$, rearranging terms and choosing N sufficiently large establishes (8.4.3).

We now provide bounds on g^* based on subsolutions and supersolutions of the optimality equation and show that, when the unichain optimality equation has a solution with bounded h, the average, lim inf average and lim sup average optimality criteria of Sec. 8.1.2 are equivalent. The following theorem is the average reward analog of Theorem 6.2.2 and one of the most important results for average reward models; part (c) gives the main result.

Theorem 8.4.1. Suppose S is countable.

a. If there exists a scalar g and an $h \in V$ which satisfy $B(g, h) \leq 0$, then

$$ge \geq g_+^* . \tag{8.4.4}$$

b. If there exists a scalar g and $h \in V$ which satisfy $B(g, h) \geq 0$, then

$$ge \leq \sup_{d \in D^{MD}} g_-^{d^\infty} \leq g_-^* . \tag{8.4.5}$$

c. If there exists a scalar g and an $h \in V$ for which $B(g, h) = 0$, then

$$ge = g^* = g_+^* = g_-^* . \tag{8.4.6}$$

Proof. Since $B(g, h) \leq 0$, Proposition 6.2.1 implies that

$$ge \geq r_d + (P_d - I)h \tag{8.4.7}$$

for all $d \in D^{MR}$. Let $\pi = (d_1, d_2, \dots) \in \Pi^{MR}$. Then, (8.4.7) implies that

$$ge \geq r_{d_1} + (P_{d_1} - I)h.$$

Applying (8.4.7) with $d = d_2$ and multiplying it by P_{d_1} yields

$$ge = gP_{d_1}e \geq P_{d_1}r_{d_2} + (P_{d_1}P_{d_2} - P_{d_1})h. \tag{8.4.8}$$

Repeating this argument (or using induction), for any $n \geq 2$ shows that

$$ge \geq P_{d_1}P_{d_2} \cdots P_{d_{n-1}}r_{d_n} + P_{d_1}P_{d_2} \cdots P_{d_{n-1}}(P_{d_n} - I)h.$$

Summing these expressions over n and noting (8.1.3) shows that, for all $\pi \in \Pi^{MR}$ and any N,

$$Nge \geq v_{N+1}^\pi + (P_N^\pi - I)h.$$

Since $h \in V$, $P_N^\pi h \in V$, so that $\lim_{N \to \infty} N^{-1}(P_N^\pi - I)h(s) = 0$ for each $s \in S$. Therefore,

$$ge \geq \limsup_{N \to \infty} \frac{1}{N}v_{N+1}^\pi = g_+^\pi$$

for all $\pi \in \Pi^{MR}$. Extension to $\pi \in \Pi^{HR}$ follows from Theorem 8.1.2, so (a) follows.

To establish part (b), note that $B(g, h) \geq 0$ implies there exists a $d^* \in D^{MD}$ for which

$$ge \leq r_{d^*} + (P_{d^*} - I)h.$$

Applying the above argument with P_{d^*} replacing P_{d_n} establishes that

$$ge \leq \liminf_{N \to \infty} \frac{1}{N} v_{N+1}^{(d^*)^\infty} = g_{-}^{(d^*)^\infty} \leq g_{-}^*,$$

from which part (b) follows.

Under (c), (a), and (b), hold so that $ge \leq g_{-}^* \leq g_{+}^* \leq ge$, from which (8.4.6) follows.   □

The constant gain assumption was used to establish the identity on the left-hand side of (8.4.8). Actually all we required for the proof to be valid was that $g \geq P_{d_N} g$. This holds with equality for constant g. Without such an assumption we need further conditions to assure that $g \geq P_{d_N} g$ for arbitrary g. We return to this point when we analyze general models in Chap. 9.

The above theorem holds without any assumptions about model classification, but it is most useful when we know *a priori* that the optimal gain does not depend on the initial state, as for example, in a unichain model. The bounds in parts (a) and (b) are always valid, but they are not of much practical significance when the quantities on the right-hand side of (8.4.4) and (8.4.5) vary with s. More importantly, the hypothesis of part (c) will be vacuous unless the optimal gain is constant. The following example illustrates these points.

Example 8.4.2. Let $S = \{s_1, s_2\}$; $A_{s_1} = \{a_{1,1}, a_{1,2}\}$ and $A_{s_2} = \{a_{2,1}\}$; and $r(s_1, a_{1,1}) = 1$, $r(s_1, a_{1,2}) = 3$, and $r(s_2, a_{2,1}) = 2$. All transitions are deterministic and as indicated in Fig. 8.4.1 (Fig. 8.4.2). Obviously the stationary policy which uses action $a_{1,2}$ in s_1 and $a_{2,1}$ in s_2 is optimal. Further the corresponding Markov chain is multichain with $g^*(s_1) = 3$, and $g^*(s_2) = 2$. For arbitrary g and h

$$B(g, h)(s_1) = \max\{1 - g + h(s_2) - h(s_1), 3 - g\}$$

$$B(g, h)(s_2) = 2 - g.$$

Note that $h(s_1) = h(s_2) = 0$, $g = 3$ satisfies $B(g, h) \leq 0$, so, by Theorem 8.4.1a, $ge \geq g^*$. If we choose $g = 2$, $B(g, h) \geq 0$, so that, by part (b) of the above theorem, $ge \leq g^*$. Thus we have established the bounds $2e \leq g^* \leq 3e$ but no tighter bounds

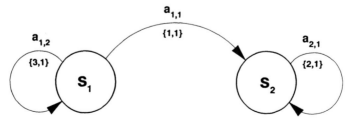

Figure 8.4.2 Symbolic representation of Example 8.4.2.

are available through the above theorem. Note further that there exists no scalar g for which $B(g, h) = 0$, so part (c) of Theorem 8.4.1 is vacuous.

If instead $r(s_1, a_{1,2}) = 2$, then $g^* = 2e$ even though the Markov chain generated by an optimal policy was multichain. In this case $g = 2$ and h as above satisfy $B(g, h) = 0$, so part (c) of Theorem 8.4.1 applies to this particular multichain model.

In the proof of Theorem 8.4.1 we used the assumption of bounded h to assure that $\lim_{N \to \infty} N^{-1} P_N^\pi h(s) = 0$ for all $s \in S$. For the proof to be valid we require only that this limit exists and equal 0. We express this formally as follows.

Corollary 8.4.2.   Suppose that

$$\lim_{N \to \infty} N^{-1} E_s^\pi \{ h(X_N) \} = 0 \tag{8.4.9}$$

for all $\pi \in \Pi^{HR}$ and $s \in S$. Then results (a) $-$ (c) in Theorem 8.4.1 hold.

Note that if the maximum over D is not attained, and "sup" replaces "max" in (8.4.4), (8.4.6), and (8.4.8), we can prove a similar result. We leave establishing the results of Theorem 8.4.1 under these weaker assumptions as an exercise. This theorem also extends to general S.

8.4.2   Existence of Solutions to the Optimality Equation

In this section we show that the unichain optimality equation (8.4.3) has a solution. Our approach formalizes the heuristic argument at the beginning of the preceding section. Other methods for demonstrating this result include policy iteration (Sec. 8.6), compactness of the set of stationary policies (Secs. 8.4.4 and 8.10), and approaches based on convex analysis and fixed point theory. Refer to the Bibliographic Remarks section of this chapter for references for the last two approaches.

Theorem 8.4.3.   Suppose S and A_s are finite Assumption 8.0.2 holds and the model is unichain.

a. Then there exists a $g \in R^1$ and an $h \in V$ for which

$$0 = \max_{d \in D} \{ r_d - ge + (P_d - I)h \}.$$

b. If (g', h') is any other solution of the average reward optimality equation, then $g = g'$.

Proof.   Choose a sequence of discount factors $\{\lambda_n\}$, $0 \le \lambda_n < 1$ with the property that $\lambda_n \uparrow 1$. By Theorem 6.2.10a, for each λ_n there exists a stationary discount optimal policy. Since D^{MD} is finite, we can choose a subsequence $\{\lambda'_n\}$ for which the same policy, δ^∞ is discount optimal for all λ'_n. Denote this subsequence by $\{\lambda_n\}$. Since δ^∞ is discount optimal for $\lambda = \lambda_n$, $v^*_{\lambda_n} = v^{\delta^\infty}_{\lambda_n}$. Therefore, for any $d \in D^{\text{MD}}$,

$$
\begin{aligned}
0 &= r_\delta + (\lambda_n P_\delta - I) v^{\delta^\infty}_{\lambda_n} = r_\delta + (\lambda_n P_\delta - I) v^*_{\lambda_n} \\
&= \max_{d' \in D} \left\{ r_{d'} + (\lambda_n P_{d'} - I) v^*_{\lambda_n} \right\} \\
&\geq r_d + (\lambda_n P_d - I) v^*_{\lambda_n} = r_d + (\lambda_n P_d - I) v^{\delta^\infty}_{\lambda_n}.
\end{aligned}
\tag{8.4.10}
$$

By Corollary 8.2.4,

$$
v^{\delta^\infty}_{\lambda_n} = (1 - \lambda_n)^{-1} g^{\delta^\infty} e + h^{\delta^\infty} + f(\lambda_n).
\tag{8.4.11}
$$

Noting that h^{δ^∞} is bounded implies

$$
\lambda_n P_d h^{\delta^\infty} = P_d h^{\delta^\infty} + (\lambda_n - 1) P_d h^{\delta^\infty} = P_d h^{\delta^\infty} + f'(\lambda_n)
\tag{8.4.12}
$$

for all $d \in D$, where $f'(\lambda)$ converges to 0 as $\lambda \uparrow 1$. Substituting (8.4.11) into the first expression in the sequence of equations (8.4.10), noting (8.4.12) and that $(\lambda_n P_d - I)e = (\lambda - 1)e$, implies

$$
\begin{aligned}
0 &= r_\delta + (\lambda_n P_\delta - I) \left\{ (1 - \lambda_n)^{-1} g^{\delta^\infty} e + h^{\delta^\infty} + f(\lambda_n) \right\} \\
&= r_\delta - g^{\delta^\infty} e + (P_\delta - I) h^{\delta^\infty} + f''(\lambda_n),
\end{aligned}
$$

where $f''(\lambda)$ converges to 0 as $\lambda \uparrow 1$. Performing similar operations to the last expression in the sequence of equations (8.4.10) establishes that

$$
\begin{aligned}
r_d + (\lambda_n P_d - I) & \left\{ (1 - \lambda_n)^{-1} g^{\delta^\infty} e + h^{\delta^\infty} + f(\lambda_n) \right\} \\
&= r_d - g^{\delta^\infty} e + (P_d - I) h^{\delta^\infty} + f_d(\lambda_n)
\end{aligned}
\tag{8.4.13}
$$

for all $d \in D$, where $f_d(\lambda)$ denotes a vector which converges to zero as $\lambda \uparrow 1$. Therefore

$$
r_\delta - g^{\delta^\infty} e + (P_\delta - I) h^{\delta^\infty} + f(\lambda_n) \geq r_d - g^{\delta^\infty} e + (P_d - I) h^{\delta^\infty} + f_d(\lambda_n).
$$

Taking the limit as $\lambda_n \uparrow 1$ shows that

$$
0 = r_\delta - g^{\delta^\infty} e + (P_\delta - I) h^{\delta^\infty} \geq r_d - g^{\delta^\infty} e + (P_d - I) h^{\delta^\infty}
$$

for each $d \in D$, from which result (a) follows.

To prove part (b), note that $g = g' = g^*$ follows from Theorem 8.4.1(c). □

Some comments regarding the above result and its proof follow. Note that, in addition to establishing that the unichain average reward optimality equation has a solution, the above proof identifies this solution as the gain and bias of the stationary policy δ^∞. Consequently, by Theorem 8.4.1(c), $g^{\delta^\infty} = g^*$ and δ^∞ is average optimal. Note also that this analysis applies to models in which "sup" replaces "max" in the optimality equations.

We have shown that $(g^{\delta^\infty}, h^{\delta^\infty})$ satisfies the optimality equation, however, the solution is not unique since $(g^{\delta^\infty}, h^{\delta^\infty} + ke)$ is also a solution for any scalar k.

Schweitzer and Federgruen (1978a) show that in unichain models, this gives a complete characterization of the set of solutions of the optimality equation.

Note that the argument used to establish Theorem 8.4.3 does not extend directly to countable-state models because the partial Laurent series expansions in Corollary 8.4.2 are not available without additional assumptions.

We conclude this section with an example which illustrates the optimality equation and properties of its solution.

Example 8.4.3. We analyze an infinite-horizon average reward version of the example in Fig. 3.1.1. There is a single decision in state s_2 and two decisions in s_1, $a_{1,1}$ and $a_{1,2}$. Let δ use action $a_{1,1}$ and γ use action $a_{1,2}$. Both δ^∞ and γ^∞ are unichain with recurrent state s_2 and transient state s_1. The optimality equations are

$$0 = \max\{5 - g - 0.5h(s_1) + 0.5h(s_2), \quad 10 - g - h(s_1) + h(s_2)\}, \quad (8.4.14)$$

$$0 = -1 - g.$$

Since the second equation above implies that $g^* = -1$, we know without any further analysis that δ^∞ and γ^∞ are average optimal. We proceed to solve the optimality equation. We apply Corollary 8.2.7 to find $(g^{\delta^\infty}, h^{\delta^\infty})$ and $(g^{\gamma^\infty}, h^{\gamma^\infty})$ by solving

$$ge + (I - P_d)h = r_d$$

subject to $P_\delta^* h = 0$ or $P_\gamma^* h = 0$. Noting that

$$P_\delta^* = P_\gamma^* = \begin{bmatrix} 0.0 & 1.0 \\ 0.0 & 1.0 \end{bmatrix}$$

implies that $h^{\delta^\infty}(s_2) = h^{\gamma^\infty}(s_2) = 0$. Consequently $g^{\delta^\infty} = g^{\gamma^\infty} = -1$, $h^{\delta^\infty}(s_1) = 12$ and $h^{\gamma^\infty}(s_1) = 11$. Observe that $(g^{\delta^\infty}, h^{\delta^\infty})$ satisfies (8.4.14), as does $(g^{\delta^\infty}, h^{\delta^\infty} + ke)$ for any constant k, but that $(g^{\gamma^\infty}, h^{\gamma^\infty})$ does not.

Hence, even though γ^∞ is average optimal, its gain and bias together do not satisfy the optimality equation. We elaborate on this point in Chap. 10.

8.4.3 Identification and Existence of Optimal Policies

In this section we show how to use the optimality equation to find a stationary optimal policy and discuss some results on the existence of optimal policies. We call a decision rule d_h *h-improving* if

$$d_h \in \arg\max_{d \in D} \{r_d + P_d h\}$$

or, equivalently,

$$r_{d_h} + P_{d_h} h = \max_{d \in D} \{r_d + P_d h\}.$$

This means that

$$r_{d_h} - ge + (P_{d_h} - I)h = \max_{d \in D}\{r_d - ge + (P_d - I)h\}. \qquad (8.4.15)$$

Theorem 8.4.4. Suppose there exists a scalar g^*, and an $h^* \in V$ for which $B(g^*, h^*) = 0$. Then, if d^* is h^*-improving, $(d^*)^\infty$ is average optimal.

Proof. By hypothesis

$$0 = r_{d^*} - g^*e + (P_{d^*} - I)h^*.$$

From Corollary 8.2.7 $g^{(d^*)^\infty} = g^*$ so that Theorem 8.4.1(c) establishes the optimality of $(d^*)^\infty$.   □

Example 8.4.3 shows that the converse of the above theorem need not hold. In it, γ^∞ is average optimal but not h^*-improving.

Theorem 8.4.3 established the existence of a solution to $B(g, h) = 0$ under the hypothesis of finite S and A_s. Combining this with the above result and Theorem 8.4.1 establishes the following existence results.

Theorem 8.4.5. Suppose S is finite and A_s is finite for each $s \in S$, $r(s, a)$ is bounded and the model is unichain. Then

   a. there exists a stationary average optimal policy,
   b. there exists a scalar g^* and an $h^* \in V$ for which $B(g^*, h^*) = 0$,
   c. any stationary policy derived from an h^*-improving decision rule is average optimal, and
   d. $g^*e = g_+^* = g_-^*$.

*8.4.4 Models with Compact Action Sets

In this section we establish the existence of solutions to the optimality equations, and the existence of optimal policies for unichain models with finite-state and compact action sets. The existence proof generalizes that of Theorem 8.4.3 and relies on continuity properties of the limiting matrix and deviation matrix in unichain models. We begin with assumptions and technical results.

Assumption 8.4.1. For each $s \in S$, $r(s, a)$ is a bounded, continuous function of a.

Assumption 8.4.2. For each $s \in S$ and $j \in S$, $p(j|s, a)$ is a continuous function of a.

The following result, which holds under weaker assumptions, relies on results in Appendix C which establish the continuity of matrix inverses and products.

Proposition 8.4.6. Let $\{P_n\}$ denote a sequence of unichain transition probability matrices and suppose

$$\lim_{n \to \infty} \|P_n - P\| = 0, \tag{8.4.16}$$

then

a. $\lim_{n \to \infty} \|P_n^* - P^*\| = 0$, and
b. $\lim_{n \to \infty} \|H_{P_n} - H_P\| = 0$.

Proof. For each n, P_n is unichain, so, by results in Appendix A, $P_n^* = q_n e^T$, where q_n is the unique solution of

$$q_n^T(I - P_n) = 0$$

subject to $q_n^T e = 1$. Let W_n denote the matrix which replaces the first column of $(I - P_n)$ by the column vector e, and W the same matrix formed from $I - P$. Thus q_n is the unique solution of

$$u^T W_n = z^T$$

where z denotes a column vector with 1 in the first component and 0 in the remaining components. Let q denote the unique solution of $q^T W = z^T$. Consequently, W_n^{-1} and W^{-1} exist, $q_n^T = z^T W_n^{-1}$, and $q^T = z^T W^{-1}$.

We show that q_n converges to q. By (8.4.16), $\lim_{n \to \infty} \|W_n - W\| = 0$, so, by Proposition C.5, $\lim_{n \to \infty} \|W_n^{-1} - W^{-1}\| = 0$. Therefore

$$\lim_{n \to \infty} q_n^T = \lim_{n \to \infty} z^T W_n^{-1} = z^T W^{-1} = q^T,$$

so (a) follows by noting that $P^* = qe^T$.

To establish (b), note that $H_{P_n} = (I - P_n + P_n^*)^{-1}(I - P_n^*)$. Since $(I - P - P^*)^{-1}$ exists, the result follows from part (a), Proposition C.5, and Lemma C.7. $\qquad\square$

The main result of this section follows. After taking some technical considerations into account, its proof follows Theorem 8.4.3.

Theorem 8.4.7. Suppose S is finite, A_s is compact, the model is unichain, and Assumptions 8.4.1 and 8.4.2 hold.

a. Then there exists a $g \in R^1$ and an $h \in V$ for which

$$0 = \max_{d \in D}\{r_d - ge + (P_d - I)h\}.$$

b. If (g', h') is any other solution of the average reward optimality equation, then $g = g'$.
c. There exists a $d^* \in D^{MD}$ for which $(d^*)^\infty$ is average optimal; further, if d is h-improving, then d^∞ is average optimal.

Proof. Choose a sequence of discount factors $\{\lambda_n\}$, $0 \le \lambda_n < 1$, with $\lambda_n \uparrow 1$. By Theorem 6.2.10 (a) for each λ_n, there exists a stationary discount optimal policy $(\delta_n)^\infty$. Since $D^{MD} = \times_{s \in S} A_s$ is compact, we can choose a subsequence (n_k) for which $\delta_{n_k}(s)$ converges to a limit $\delta(s) \in D^{MD}$ for each $s \in S$. To simplify subsequent notation, denote the subsequence by $\{\delta_k\}$.

By Assumption 8.4.2, $p_{\delta_k}(j|s)$ converges to $p_\delta(j|s)$ for each $(s, j) \in S \times S$, so, by the finiteness of S, (8.4.16) holds. Consequently, by Proposition 8.4.6, $P^*_{\delta_k}$ converges in norm to P^*_δ and H_{δ_k} converges in norm to H_δ. As a result of Assumption 8.4.1, r_{δ_k} converges in norm to r_δ, so, by Lemma C.6,

$$\lim_{k \to \infty} g^{(\delta_k)^\infty} e = \lim_{k \to \infty} P^*_{\delta_k} r_{\delta_k} = P^*_\delta r_\delta = g^{\delta^\infty} e \qquad (8.4.17)$$

and

$$\lim_{k \to \infty} h^{(\delta_k)^\infty} = \lim_{k \to \infty} H_{\delta_k} r_{\delta_k} = H_\delta r_\delta = h^{\delta^\infty}. \qquad (8.4.18)$$

As a consequence of the optimality of $(\delta_k)^\infty$, for any $d \in D$,

$$0 = r_{\delta_k} + (\lambda_k P_{\delta_k} - I) v^{(\delta_k)^\infty}_{\lambda_k} \ge r_d + (\lambda_k P_d - I) v^{(\delta_k)^\infty}_{\lambda_k}.$$

By Corollary 8.2.4,

$$v^{(\delta_k)^\infty}_{\lambda_k} = (1 - \lambda_k)^{-1} g^{(\delta_k)^\infty} e + h^{(\delta_k)^\infty} + f(\lambda_k)$$

so that

$$0 = r_{\delta_k} - g^{(\delta_k)^\infty} e + (P_{\delta_k} - I) h^{(\delta_k)^\infty} + f(\lambda_k)$$

$$\ge r_d - g^{(\delta_k)^\infty} e + (P_d - I) h^{(\delta_k)^\infty} + f_d(\lambda_k)$$

for each $d \in D$, where $f_d(\lambda)$ denotes a vector which converges to 0 as $\lambda \uparrow 1$. Hence taking the limit as $\lambda_k \uparrow 1$ shows that

$$0 = r_\delta - g^{\delta^\infty} e + (P_\delta - I) h^{\delta^\infty} \ge r_d - g^{\delta^\infty} e + (P_d - I) h^{\delta^\infty}$$

for all $d \in D$, so that $B(g^{\delta^\infty}, h^{\delta^\infty}) = 0$ from which result (a) follows.

Part (b) follows as in Theorem 8.4.3. Since $r(s, a) + \sum_{j \in S} p(j|s, a) h(j)$ is continuous in a, and A_s is compact, $\arg\max_{a \in A_s} \{r(s, a) + \sum_{j \in S} p(j|s, a) h(j)\}$ is nonempty for each $s \in S$, so that part (c) follows from Theorem 8.4.4. $\square$

Note that (8.4.17) and (8.4.18) establish the continuity of g^{d^∞} and h^{d^∞} in d, so that both of these quantities attain maxima on compact sets of decision rules.

8.5   VALUE ITERATION IN UNICHAIN MODELS

In this section we study value iteration for unichain average reward MDPs. We express the sequence of values generated by value iteration as

$$v^{n+1} = Lv^n, \tag{8.5.1}$$

where

$$Lv \equiv \max_{d \in D} \{r_d + P_d v\} \tag{8.5.2}$$

for $v \in V$. We assume throughout this section that the maximum is attained in (8.5.2). In discounted, positive, and negative models, we based value iteration on operators of this form, but analysis here requires more subtle arguments because we cannot appeal to contraction or monotonicity to establish convergence. In fact, in almost all practical applications, the sequence $\{v^n\}$ generated by (8.5.1) diverges or oscillates, so we must seek other ways for terminating algorithms and selecting optimal policies.

Our approach relies on material in Sec. 6.6, and we suggest reviewing it before proceeding. For a more general analysis of value iteration in average reward models, see Sec. 9.4. Throughout this section we assume finite S and A_s although results hold in greater generality.

8.5.1   The Value Iteration Algorithm

The following value iteration algorithm finds a stationary ε-optimal policy $(d^\varepsilon)^\infty$ and an approximation to its gain, when certain extra conditions are met.

Value Iteration Algorithm

1. Select $v^0 \in V$, specify $\varepsilon > 0$ and set $n = 0$.
2. For each $s \in S$, compute $v^{n+1}(s)$ by

$$v^{n+1}(s) = \max_{a \in A_s} \left\{ r(s, a) + \sum_{j \in S} p(j|s, a)v^n(j) \right\}. \tag{8.5.3}$$

3. If

$$sp(v^{n+1} - v^n) < \varepsilon, \tag{8.5.4}$$

   go to step 4. Otherwise increment n by 1 and return to step 2.
4. For each $s \in S$, choose

$$d_\varepsilon(s) \in \arg\max_{a \in A_s} \left\{ r(s, \dot{a}) + \sum_{j \in S} p(j|s, a)v^n(j) \right\} \tag{8.5.5}$$

   and stop.

The examples below illustrate the difficulties inherent in applying value iteration.

Example 8.5.1. Let $S = \{s_1, s_2\}$ and suppose there is a single decision rule d, with

$$r_d = \begin{bmatrix} 0 \\ 0 \end{bmatrix} \quad \text{and} \quad P_d = \begin{bmatrix} 0 & 1 \\ 1 & 0 \end{bmatrix}.$$

If

$$v^0 = \begin{bmatrix} a \\ b \end{bmatrix}, \quad \text{then} \quad v^n = P_d^n v^0 = \begin{cases} \begin{bmatrix} a \\ b \end{bmatrix} & n \text{ even} \\ \begin{bmatrix} b \\ a \end{bmatrix} & n \text{ odd,} \end{cases}$$

so unless $a = b$, $\lim_{n \to \infty} v^n$ does not exist and $\{v^n\}$ oscillates with period 2. Note that $\|v^{n+1} - v^n\| = |b - a|$ and $sp(v^{n+1} - v^n) = 2|b - a|$, so (8.5.4) is never satisfied in this example. Note also that each state is periodic with period 2.

Example 8.5.2. Consider the model in Fig. 3.1.1. In it, state s_2 is absorbing, $g^* = -1$, and $g^\pi = -1$ for any policy π. Choosing $v^0 = 0$, v^n equals

n	$v^n(s_1)$	$v^n(s_2)$	$sp(v^n - v^{n-1})$
0	0	0	NA
1	10	−1	11
2	9.5	−2	.5
3	8.75	−3	.25
4	7.875	−4	.125
5	6.9375	−5	.0625
⋮	⋮	⋮	⋮

so we see that, $v^n(s_1)$ approaches $12 - n$ and $v^n(s_2) = -n$.

Observe that if $\|v^{n+1} - v^n\| = 1$ for $n \geq 1$ but $sp(v^n - v^{n-1}) = 0.5^{n-1}$ for $n \geq 1$, so that stopping criterion (8.5.4) holds at $n = 5$. Note that v^n diverges linearly in n. More precisely,

$$v^n = ng^* + h,$$

where $h(s_1) = 12$ and $h(s_2) = 0$, as suggested in Example 8.4.3. Consequently, $g^* = v^{n+1}(s) - v^n(s)$ for $n \geq 1$ and $s \in S$.

The first example above shows that value iteration need not converge in models with periodic transition matrices, while the second example shows that the sequence may diverge but that $sp(v^{n+1} - v^n)$ may converge. In subsequent sections, we provide conditions which ensure the above algorithm terminates in a finite number of iterations, and that $(d_\varepsilon)^\infty$ is an ε-optimal policy.

8.5.2   Convergence of Value Iteration

In this section we provide conditions under which stopping criterion (8.5.4) holds for some finite n. Theorem 6.6.6 shows that

$$sp(v^{n+2} - v^{n+1}) \le \gamma sp(v^{n+1} - v^n), \tag{8.5.6}$$

where

$$\gamma = \max_{s \in S, a \in A_s, s' \in S, a' \in A_{s'}} \left[1 - \sum_{j \in S} \min\{p(j|s, a), p(j|s', a')\} \right]. \tag{8.5.7}$$

Consequently, if $\gamma < 1$, (8.5.6) ensures that in a finite number of iterations, stopping criterion (8.5.4) will be satisfied. The following example shows that γ may equal 1 in a unichain aperiodic model, but that value iteration may still converge. It motivates a more general condition which ensures convergence.

Example 8.5.3.   Let $S = \{s_1, s_2, s_3\}$; $A_{s_1} = \{a_{1,1}, a_{1,2}\}$, $A_{s_2} = \{a_{2,1}\}$, and $A_{s_3} = \{a_{3,1}\}$; $r(s_1, a_{1,1}) = 2$, $r(s_1, a_{1,2}) = 1$, $r(s_2, a_{2,1}) = 2$, and $r(s_3, a_{3,1}) = 3$; and $p(s_3|s_1, a_{1,1}) = 1$, $p(s_2|s_1, a_{1,2}) = 1$, $p(s_1|s_2, a_{2,1}) = 1$, and $p(s_1|s_3, a_{3,1}) = p(s_2|s_3, a_{3,1}) = p(s_3|s_3, a_{3,1}) = \frac{1}{3}$ (Fig. 8.5.1).

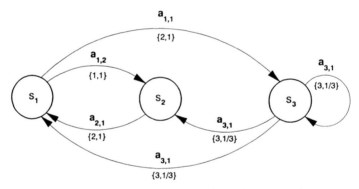

Figure 8.5.1   Symbolic representation of Example 8.5.3.

Let δ denote the decision rule which uses action $a_{1,1}$ in s_1, and η denote the decision rule which uses $a_{1,2}$ in s_1. Observe that

$$P_\delta = \begin{bmatrix} 0 & 0 & 1 \\ 1 & 0 & 0 \\ \frac{1}{3} & \frac{1}{3} & \frac{1}{3} \end{bmatrix}, \quad P_\eta = \begin{bmatrix} 0 & 1 & 0 \\ 1 & 0 & 0 \\ \frac{1}{3} & \frac{1}{3} & \frac{1}{3} \end{bmatrix},$$

so that $\gamma_\delta = \gamma_\eta = \gamma = 1$ where γ_d is defined in (6.6.5). Note also that, under η, states s_1 and s_2 are periodic with period 2 and s_3 is transient, while under δ all states are recurrent and aperiodic.

Applying value iteration with $(v^0)^T = (0, 0, 0)$ and $\varepsilon = 0.01$ yields

n	$v^n(s_1)$	$v^n(s_2)$	$v^n(s_3)$	$sp(v^{n+1} - v^n)$	$\max_{s \in S} \{v^{n+1}(s) - v^n(s)\}$	$\min_{s \in S} \{v^{n+1}(s) - v^n(s)\}$
0	0.000	0.000	0.000			
1	2.000	2.000	3.000	1.000	3.000	2.000
2	5.000	4.000	5.333	1.000	3.000	2.000
3	7.333	7.000	7.778	0.667	3.000	2.333
4	9.778	9.333	10.370	0.259	2.592	2.333
5	12.370	11.778	12.827	0.147	2.592	2.445
6	14.827	14.370	15.325	0.135	2.592	2.457
7	17.325	16.827	17.841	0.059	2.516	2.457
8	19.841	19.325	20.331	0.026	2.516	2.490
9	22.331	21.841	22.832	0.026	2.516	2.490
10	24.832	24.331	25.335	0.013	2.503	2.490
11	27.335	26.832	27.833	0.005	2.503	2.498
12	29.833	29.335	30.333	0.005	2.503	2.498

Observe that the value iteration algorithm stops in step 3 with $n = 11$ and identifies the 0.01-optimal policy δ^∞ through step 4. Note also that $sp(v^{n+1} - v^n)$ is monotone but that it does not decrease at each iteration.

The above example shows that value iteration may converge with respect to the span seminorm even though (8.5.6) holds with $\gamma = 1$. Our approach for showing that this algorithm stops in a finite number of iterations relies on the concept of a J-stage span contraction. We say that an operator $T: V \to V$ is a *J-stage span contraction* if there exists an α, $0 \leq \alpha < 1$ and a non-negative integer J for which

$$sp(T^J u - T^J v) \leq \alpha \, sp(u - v) \tag{8.5.8}$$

for all u and v in V.

Proposition 8.5.1. Suppose T is a J-stage span contraction. Then for any $v^0 \in V$, the sequence $v^n = T^n v^0$ satisfies

$$sp(v^{nJ+1} - v^{nJ}) \leq \alpha^n \, sp(v^1 - v^0) \tag{8.5.9}$$

for all non-negative integers n.

Proof. Choose $v = v^0$ and $u = Tv^0$ and iterate (8.5.8). □

The above proposition provides a simple property of J-stage span contractions. We leave it as an exercise to generalize Theorem 6.6.2 which establishes existence of a span fixed point and convergence of $T^n v^0$ to it. The following result generalizes Theorem 6.6.6.

Theorem 8.5.2. Suppose there exists an integer $J \geq 1$ such that, for every pair of deterministic Markov policies π_1 and π_2,

$$\eta(\pi_1, \pi_2) \equiv \min_{(s,u) \in S \times S} \sum_{j \in S} \min\{P_{\pi_1}^J(j|s), P_{\pi_2}^J(j|u)\} > 0. \qquad (8.5.10)$$

a. Then L defined in (8.5.2) is J-step contraction operator on V with contraction coefficient

$$\gamma' = 1 - \min_{\pi_1, \pi_2 \in \Pi^{MD}} \eta(\pi_1, \pi_2). \qquad (8.5.11)$$

b. For any $v^0 \in V$, let $v^n = L^n v^0$. Then, given $\varepsilon > 0$, there exists an N such that

$$sp(v^{nJ+1} - v^{nJ}) \leq \varepsilon$$

for all $n \geq N$.

Proof. The proof of Theorem 6.6.2 may be easily adapted to establish part (a). Part (b) follows by applying Proposition 8.5.1 with $\alpha = \gamma'$. ☐

Condition (8.5.10) means that starting in any pair of distinct states, policies π_1 and π_2 both reach at least one identical state with positive probability after J transitions. Since (8.5.10) must hold for any pair of policies, all stationary policies must be unichain and aperiodic under (8.5.10). This condition excludes periodic models such as that in Example 8.5.1 and multichain models.

Clearly it is not easy to verify (8.5.11) directly. The following theorem provides conditions which are easier to check and imply it. We leave it as an exercise to show that each of these conditions are distinct; that is, that any one of them does not imply any other.

Theorem 8.5.3. Suppose either

a. $0 \leq \gamma < 1$, where γ is given in (8.5.7),
b. there exists a state $s' \in S$ and an integer K such that, for any deterministic Markov policy π, $P_{\pi}^K(s'|s) > 0$ for all $s \in S$, or
c. all policies are unichain and $p(s|s, a) > 0$ for all $s \in S$ and $a \in A_s$.

Then (8.5.10) holds for all π_1 and π_2 in Π^{MD} and the conclusions of Theorem 8.5.2 follow.

Proof. If (a) holds, then (8.5.10) holds with $J = 1$. Since

$$\eta(\pi_1, \pi_2) \geq \min_{(s,u) \in S \times S} \{P_{\pi_1}^J(s'|s), P_{\pi_2}^J(s'|u)\},$$

(b) implies that (8.5.10) holds with $J = K$.

We now consider (c). Let $\pi_1 = (d_1, d_2, \ldots)$ and $\pi_2 = (f_1, f_2, \ldots)$ denote two deterministic Markovian policies, and choose $s_1 \neq s_2$ in S. Let

$$X_i(n) \equiv \{j \in S: P_{\pi_i}^n(j|s_i) > 0\}.$$

Let N denote the number of elements in S. We show by contradiction that $X_1(N) \cap X_2(N) \neq \emptyset$. Suppose $X_1(N) \cap X_2(N) = \emptyset$. Since $p(s|s, a) > 0$, $X_i(n) \subset X_i(n + 1)$ for all n, so that $X_1(n) \cap X_2(n) = \emptyset$ for all $1 \leq n \leq N$. Consequently, for some $m < N$, $X_1(m) = X_1(m + 1)$ and $X_2(m) = X_2(m + 1)$. This means that $X_1(m)$ is closed under d_m and $X_2(m)$ is closed under f_m. Construct a decision rule δ by

$$\delta(s) = \begin{cases} d_m(s) & s \in X_1(m) \\ f_m(s) & s \in X_2(m) \\ \text{arbitrary} & \text{otherwise.} \end{cases}$$

Then P_δ has at least two distinct closed classes contradicting hypothesis (c). Therefore $X_1(N) \cap X_2(N) \neq \emptyset$, so that, for some j', $P_{\pi_1}^N(j'|s_1) > 0$ and $P_{\pi_2}^N(j'|s_2) > 0$, which shows that (8.5.10) holds. Since π_1 and π_2 are arbitrary, the result follows. □

Thus under any of the conditions in Theorem 8.5.3, value iteration achieves stopping criterion (8.5.4) for any $\varepsilon > 0$. Section 8.5.5 provides a method for transforming any unichain model into one which satisfies condition (c).

We illustrate Theorem 8.5.3 by expanding on Example 8.5.3.

Example 8.5.3 (ctd.). Inspection of Example 8.5.3 reveals that, when starting the algorithm with $v^0 = 0$, action $a_{1,1}$ always achieves the maximum in s_1, so that, for $n > 0$,

$$r_\delta(s) + \sum_{j \in S} p_\delta(j|s)v^n(j) = \max_{a \in A_s} \left\{ r(s, a) + \sum_{j \in S} p(j|s, a)v^n(j) \right\}$$

for all $s \in S$. Consequently, in this example value iteration corresponds to iterating the fixed policy δ^∞. Note that P_δ satisfies neither hypotheses (a) nor (c) of Theorem 8.5.3. Since

$$P_\delta^2 = \begin{bmatrix} \frac{1}{3} & \frac{1}{3} & \frac{1}{3} \\ 0 & 0 & 1 \\ \frac{4}{9} & \frac{1}{9} & \frac{4}{9} \end{bmatrix}, \qquad P_\delta^3 = \begin{bmatrix} \frac{4}{9} & \frac{4}{9} & \frac{1}{9} \\ \frac{1}{3} & \frac{1}{3} & \frac{1}{3} \\ \frac{7}{27} & \frac{16}{27} & \frac{4}{27} \end{bmatrix},$$

(b) holds with $K = 2$ and $s' = s_3$. Note also that P_δ^3 satisfies (c).

Since P_η is periodic, (8.5.10), with $\pi_1 = \pi_2 = \eta^\infty$, does not hold for any J, so that this model does not satisfy (8.5.11).

The above example suggests the following result which we prove in greater generality in Sec. 9.4.

Theorem 8.5.4. Suppose that all stationary policies are unichain and that every optimal policy has an aperiodic transition matrix. Then, for all $v^0 \in V$ and any $\varepsilon > 0$, the sequence of $\{v^n\}$ generated by the value iteration algorithm satisfies (8.5.4) for some finite N.

8.5.3 Bounds on the Gain

The following proposition provides bounds on the optimal gain and ensures that, when value iteration stops, the policy derived through (8.5.5) is ε-optimal. Recall that δ is v-improving if $\delta \in \arg\max_{d \in D}\{r_d + P_d v\}$.

Theorem 8.5.5. Suppose the hypotheses of Theorem 8.4.5 hold, then for $v \in V$,

$$\min_{s \in S}[Lv(s) - v(s)] \leq g^{d^\infty} \leq g^* \leq \max_{s \in S}[Lv(s) - v(s)], \qquad (8.5.12)$$

where d is any v-improving decision rule.

Proof. For any v-improving $d \in D$,

$$g^{d^\infty} = P_d^* r_d = P_d^*[r_d + P_d v - v] = P_d^*[Lv - v] \geq \min_{s \in S}[Lv(s) - v(s)].$$

Since $g^{d^\infty} \leq g^*$, the two leftmost inequalities in (8.5.12) hold.

By Theorem 8.4.5, when S and A_s are finite, rewards are bounded, and the model is unichain, there exists a deterministic stationary optimal policy δ^∞, so that $g^{\delta^\infty} = g^*$. Therefore

$$g^* = g^{\delta^\infty} = P_\delta^* r_\delta = P_\delta^*[r_\delta + P_\delta v - v] \leq P_d^*[Lv - v] \leq \max_{s \in S}[Lv(s) - v(s)],$$

establishing the result.   □

We apply the above bounds to value iteration and obtain an approximation to g^*.

Theorem 8.5.6. Suppose (8.5.4) holds and d_ε satisfies (8.5.5).

a. Then $(d_\varepsilon)^\infty$ is an ε-optimal policy.
b. Define

$$g' = \tfrac{1}{2}\left[\max_{s \in S}(v^{n+1}(s) - v^n(s)) + \min_{s \in S}(v^{n+1}(s) - v^n(s))\right]. \qquad (8.5.13)$$

Then $|g' - g^*| < \varepsilon/2$ and $|g' - g^{(d_\varepsilon)^\infty}| < \varepsilon/2$.

Proof. Since $v^{n+1} = Lv^n$, applying Theorem 8.5.5 with $v = v^n$ shows that

$$\varepsilon > sp(v^{n+1} - v^n) \geq g^* - g^{(d_\varepsilon)^\infty},$$

which establishes part (a).

To prove part (b), note that, for scalars x, y, and z, if $x \leq y \leq z$ and $z - x < \varepsilon$, then

$$-\varepsilon/2 < \tfrac{1}{2}(x - z) \leq y - \tfrac{1}{2}(x + z) \leq \tfrac{1}{2}(z - x) < \varepsilon/2.$$

Therefore applying (8.5.12) with $v = v^n$ establishes the result.　　　□

Thus under any conditions which ensure that $sp(v^{n+1} - v^n) < \varepsilon$ for some finite n, value iteration identifies an *ε-optimal* policy and an approximation to its value. We summarize this result as follows.

Theorem 8.5.7.　Suppose (8.5.10) holds for all π_1 and π_2 in Π^{MD}, then, for any $\varepsilon > 0$, value iteration satisfies (8.5.4) in a finite number of iterations, identifies an optimal policy through (8.5.5), and obtains an approximation to g^* through (8.5.13). Further, for $n = 1, 2, \ldots$,

$$sp(v^{nJ+1} - v^{nJ}) \leq (\gamma')^n sp(v^1 - v^0). \tag{8.5.14}$$

Note that with the exception of the error bound, the results in Theorem 8.5.7 hold for any model in which g^* is constant, but require different methods of proof. We return to this point in Chap. 9.

Note also that (8.5.14) provides an estimate on the number of iterations required to obtain a specified degree of precision in estimating g^*, and establishes geometric convergence of value iteration, albeit along a subsequence. Note that, when $J = 1$, we have geometric convergence for the entire sequence.

We conclude this section by applying Theorem 8.5.6 to Example 8.5.3.

Example 8.5.3 (ctd.).　From Theorem 8.5.6 we conclude that δ^∞ is 0.01-optimal (in fact, it is optimal). Applying (8.5.13), shows that when $n = 11$, $g^{(d_\varepsilon)^\infty} \approx g^* \approx g' = 2.505$. Observe also, that subsequences of $sp(v^{n+1} - v^n)$ converge geometrically, but that for several values of n this quantity is the same at two successive iterates.

8.5.4　An Aperiodicity Transformation

Before applying value iteration, we need to know that the model satisfies conditions which ensure its convergence. This occurs under the hypothesis of Theorem 8.5.2, or any of the conditions in Theorem 8.5.3 which imply it. We show here that, through a simple transformation, all policies can be made aperiodic so that condition (c) in Theorem 8.5.3 holds. *In practice, if there is some reason to believe that a model contains policies with periodic transition probability matrices, apply this transformation.*

Define a transformed MDP with components indicated by " ~ " as follows. Choose τ satisfying $0 < \tau < 1$ and define $\tilde{S} = S$, $\tilde{A}_s = A_s$ for all $s \in S$,

$$\tilde{r}(s, a) = \tau r(s, a) \quad \text{for } a \in A_s \text{ and } s \in S,$$

and

$$\tilde{p}(j|s, a) = (1 - \tau)\delta(j|s) + \tau p(j|s,a) \quad \text{for } a \in A_s \text{ and } s \text{ and } j \text{ in } S$$

where $\delta(j|s)$ is 1 if $s = j$, and 0 otherwise. The effect of this data transformation is that, for any decision rule d,

$$\tilde{P}_d = (1 - \tau)I + \tau P_d \quad \text{and} \quad \tilde{r}_d = \tau r_d,$$

so that, under it, all transition probability matrices have strictly positive diagonal entries and are aperiodic. We can interpret the transformed model as being generated by a continuous-time system in which the decision maker observes the system state every τ units of time, or a discrete-time model in which, at each decision epoch, the system remains in the current state with probability τ regardless of the decision chosen. The following proposition relates the transformed and untransformed models and applies regardless of chain structure.

Proposition 8.5.8. For any decision rule d,

$$\tilde{P}_d^* = P_d^* \quad \text{and} \quad \tilde{g}^{d^\infty} = \tau g^{d^\infty}.$$

Proof. We use Theorem A.5(c) to establish the result. Clearly, P_d and $\tilde{P}_d$ have the same class structure. Since

$$P_d^* \tilde{P}_d = (1 - \tau)P_d^* + \tau P_d^* P_d = P_d^*$$

and, by a similar argument,

$$\tilde{P}_d P_d^* = P_d^*,$$

(A.4) holds, so Theorem A.5(c) implies that $\tilde{P}_d^* = P_d^*$.
The second identity follows by noting that

$$\tilde{g}^{d^\infty} = \tilde{P}_d^* \tilde{r}_d = \tau P_d^* r_d = \tau g^{d^\infty}. \qquad \square$$

The following corollary assures us that we obtain the same optimal policy in either model.

Corollary 8.5.9. The sets of average optimal stationary policies for the original and transformed problem are identical, and $\tilde{g}^* = \tau g^*$.

We illustrate this transformation through the following example.

Example 8.5.1 (ctd.). We form the transformed model

$$\tilde{r}_d = \begin{bmatrix} 0 \\ 0 \end{bmatrix} \quad \text{and} \quad \tilde{P}_d = \begin{bmatrix} 1 - \tau & \tau \\ \tau & 1 - \tau \end{bmatrix}.$$

The delta coefficient of $\overline{P}_d$ equals $|1 - 2\tau|$, so when $0 < \tau < 1$, Proposition 6.6.1 implies that

$$sp(v^{n-1} - v^n) \leq |1 - \tau| sp(v^n - v^{n-1})$$

and value iteration converges with respect to the span seminorm. Note that in models with more than two states, even after applying the aperiodicity transformation, γ defined by (8.5.7) may equal 1.

8.5.5 Relative Value Iteration

Even when a model satisfies (8.5.11) or has been transformed by the approach of the preceding section, convergence of value iteration can be quite slow because γ' is close to 1. Since v^n diverges linearly in n, this might cause numerical instability.

The following relative value iteration algorithm avoids this difficulty but does not enhance the rate of convergence with respect to the span seminorm.

Relative Value Iteration Algorithm

1. Select $u^0 \in V$, choose $s^* \in S$, specify $\varepsilon > 0$, set $w^0 = u^0 - u^0(s^*)e$, and set $n = 0$.
2. Set $u^{n+1} = Lw^n$ and $w^{n+1} = u^{n+1} - u^{n+1}(s^*)e$.
3. If $sp(u^{n+1} - u^n) < \varepsilon$, go to step 4. Otherwise increment n by 1 and return to step 2.
4. Choose $d_\varepsilon \in \arg\max_{d \in D}\{r_d + P_d u^n\}$.

This algorithm renormalizes u^n at each iteration by subtracting $u^n(s^*)$ from each component. This avoids the pointwise divergence of v^n of ordinary value iteration but does not effect the sequence of maximizing actions or the value of $sp\,(u^{n+1} - u^n)$. To see this, we apply relative value iteration to Example 8.5.2.

Example 8.5.2 (ctd.).   Choosing $s^* = s_2$ implies that $w^n(s_2) = 0$, $w^n(s_1) = 11$, $u^n(s_2) = -1$, and $u^n(s_1) = 10$ for all n. Consequently $sp(v^{n+1} - v^n) = sp(u^{n+1} - u^n) = sp(w^{n+1} - w^n) = 0$. Observe also that w^n converges to (in this case equals) an h which satisfies $B(g^*, h) = 0$.

The above example also shows that relative value iteration can be used to directly obtain an h which solves the optimality equation. Note, however, that by using value iteration we can obtain the same approximate solution by choosing $h(s) = v^n(s) - v^n(s^*)$ for n sufficiently large. Also we may vary s^* with n.

8.5.6 Action Elimination

In discounted models we combined bounds on v_λ^* with inequality (6.7.2) to obtain a procedure for identifying suboptimal actions. In average reward models we cannot directly use that approach; however, by using related methods, we can derive a useful one-step-ahead action elimination procedure. The following result is the average reward analog of Proposition 6.7.1.

Proposition 8.5.10. Suppose $B(g^*, h) = 0$ and

$$r(s, a') - g^* + \sum_{j \in S} p(j|s, a')h(j) - h(s) < 0. \tag{8.5.15}$$

Then any stationary policy which uses action a' in state s cannot be optimal.

To apply (8.5.15) to eliminate actions requires bounds for both g^* and h. Section 8.5.4 provides bounds on g^*, but no easily computable bounds are available for h, so we abandon this approach. Instead we provide a method for identifying actions which need not be evaluated when computing $v^{n+1}(s)$.

Define

$$L(s, a)v \equiv r(s, a) + \sum_{j \in S} p(j|s, a)v(j),$$

and recall that, for $v \in V$, $\Lambda(v) = \min_{s \in S} v(s)$ and $\Psi(v) = \max_{s \in S} v(s)$.

Proposition 8.5.11. Suppose v^n is known and, for $a' \in A_s$, there exist functions $u(s, a')$ and $l(s)$ for which

$$u(s, a') \geq L(s, a')v^n, \tag{8.5.16}$$

$$l(s) \leq Lv^n(s), \tag{8.5.17}$$

and

$$l(s) > u(s, a'). \tag{8.5.18}$$

Then

$$a' \notin \arg\max_{a \in A_s} \left\{ r(s, a) + \sum_{j \in S} p(j|s, a)v^n(j) \right\}$$

Proof. By hypothesis,

$$L(s, a')v^n \leq u(s, a') < l(s) \leq Lv^n(s),$$

which implies the result.   □

To use this result we require functions l and u, which satisfy (8.5.16)–(8.5.18) and may be evaluated with *little* effort *prior* to computing $L(s, a)v^n$ for any $a \in A_s$. The trick in developing an implementable procedure is for each state-action pair to store the value of $L(s, a)v^{n-k}$ the last time it was evaluated, and develop upper and lower bounds based on it. The following lemma provides such bounds.

Lemma 8.5.12. For $k = 1, 2, \ldots, n$,

$$L(s, a)v^n \leq L(s, a)v^{n-k} + \Psi(v^n - v^{n-k})$$

$$\leq L(s, a)v^{n-k} + \sum_{j=0}^{k-1} \Psi(v^{n-j} - v^{n-j-1}), \tag{8.5.19}$$

and

$$Lv^n(s) \geq v^n(s) + \Lambda(v^n - v^{n-k})$$

$$\geq v^n(s) + \sum_{j=0}^{k-1} \Lambda(v^{n-j} - v^{n-j-1}). \tag{8.5.20}$$

Proof. We derive (8.5.19); a derivation of (8.5.20) uses similar methods and is left as an exercise. We have

$$L(s, a)v^n(s) = r(s, a) + \sum_{j \in S} p(j|s, a)v^n(j) + \sum_{j \in S} p(j|s, a)v^{n-k}(j)$$

$$- \sum_{j \in S} p(j|s, a)v^{n-k}(j)$$

$$= r(s, a) + \sum_{j \in S} p(j|s, a)v^{n-k}(j) + \sum_{j \in S} p(j|s, a)\left[v^n(j) - v^{n-k}(j)\right]$$

$$\leq L(s, a)v^{n-k} + \sum_{j \in S} p(j|s,a)\Psi(v^n - v^{n-k}). \tag{8.5.21}$$

The first inequality in (8.5.19) follows immediately from (8.5.21) while the second inequality follows by writing

$$\left[v^n(j) - v^{n-k}(j)\right] = \left[v^n(j) - v^{n-1}(j)\right] + \cdots + \left[v^{n-k+1}(j) - v^{n-k}(j)\right]$$

prior to taking the maximum in (8.5.21). □

We combine the two results above to obtain inequalities that can be used for action elimination.

Theorem 8.5.13. Suppose, for $a' \in A_s$ and some $k = 1, 2, \ldots, n$, that either

$$v^n(s) - L(s, a')v^{n-k} - \sum_{j=0}^{k-1} sp(v^{n-j} - v^{n-j-1}) > 0 \tag{8.5.22}$$

or

$$v^n(s) - L(s, a')v^{n-k} - sp(v^n - v^{n-k}) > 0, \tag{8.5.23}$$

then

$$a' \notin \arg\max_{a \in A_s} \left\{ r(s, a) + \sum_{j \in S} p(j|s, a)v^n(j) \right\}. \tag{8.5.24}$$

Note that we can obtain sharper bounds if $\gamma < 1$, where γ is defined in (8.5.7).

A value iteration algorithm which includes the one-step action elimination proce-dure based on (8.5.22) follows. In it, $E_n(s)$ denotes the set of actions which need not be evaluated at iteration n because they satisfy (8.5.24).

Undiscounted Value Iteration with Action Elimination.

1. Select $v^0 \in V$, $\varepsilon > 0$ and set $n = 0$, $E_0(s) = \emptyset$ for $s \in S$ and $\Delta(s, a) = 0$ for all $a \in A_s$ and $s \in S$.
2. Set

$$v^{n+1}(s) = \max_{a \in A_s/E_n(s)} L(s, a)v^n. \qquad (8.5.25)$$

3. If

$$sp(v^{n+1} - v^n) < \varepsilon,$$

go to step 6. Otherwise continue.
4. (Update the elimination criterion) For $a \in A_s/E_n(s)$, set

$$\Delta(s, a) = L(s, a)v^n + \{v^{n+1}(s) - v^n(s)\} + sp(v^{n+1} - v^n), \qquad (8.5.26)$$

and, for $a \in E_n(s)$, set

$$\Delta(s, a) = \Delta(s, a) + sp(v^{n+1} - v^n).$$

5. (Action elimination) Replace $n + 1$ by n. For each $a \in A_s$ and $s \in S$, if

$$v^n(s) - \Delta(s, a') > 0,$$

include $a' \in E_n(s)$. Go to step 2.
6. For each $s \in S$, choose $d_\varepsilon(s)$ satisfying

$$d_\varepsilon(s) \in \underset{a \in A_s/E_n(s)}{\arg\max} \; L(s, a)v^n.$$

Some comments regarding this algorithm follow. The key feature of the algorithm is that, when computing $v^{n+1}(s)$ in (8.5.25), $L(s, a)$ is evaluated for non-eliminated actions only, namely those in $A_s/E_n(s)$. Since evaluating $L(s, a)$ requires $|S|$ multipli-cations and $|S|$ additions for each state action pair, this can result in large reductions in computational effort for problems with many actions. Updating the action elimina-tion criteria in Step 4 and checking whether an action should be eliminated requires little extra work because these quantities have already been evaluated in Step 2. Note that $\Delta(s, a)$ stores the negative of the expression in (8.5.22). We update it in (8.5.26) by subtracting $v^n(s)$ and adding $v^{n+1}(s)$.

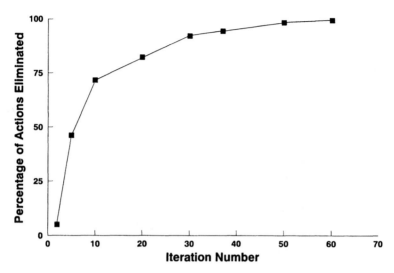

Figure 8.5.2 Percentage of actions eliminated using undiscounted value iteration algorithm with action elimination applied to Howard's automobile replacement model.

We have stated the algorithm in the above way for clarity. It could be implemented more efficiently by combining calculations in steps 2, 3, and 4. Note also that it applies equally well to *discounted* models and was referred to in Sec. 6.7.4.

Hastings (1976) reports results of applying the above algorithm to the Howard (1960) automobile replacement model. We summarize them in Fig. 8.5.2. Recall that this example has 41 states with 40 actions in each. Observe that, at iteration 30, the algorithm eliminated 92% of the actions, and at iteration 60, 99.8%. Note also that, at iteration 60, $sp(v^{n+1} - v^n) = 0.7$, which represents 0.5 percent of the optimal gain of 150.95. Figure 6.7.3 reports related results for the discounted model.

For ease of exposition, we based the algorithm on (8.5.22) instead of (8.5.23). To implement an elimination procedure based on (8.5.23) requires keeping track of the last time $L(s, a)v^m$ was evaluated for each action, and storing $v^n, v^{n-1}, \ldots, v^m$. Numerical experiments using these tighter bounds for discounted models (Puterman and Shin, 1982) suggest that this additional effort might be warranted.

8.6 POLICY ITERATION IN UNICHAIN MODELS

In this section we provide a policy iteration algorithm for unichain Markov decision processes. We assume a finite-state and action model and show that the iterates of the algorithm converge in a finite number of iterations to a solution of the optimality equation $B(g, h) = 0$, and identify an optimal stationary policy. This approach also provides a constructive proof of the existence of a solution to the optimality equation. Refer to the Bibliographic Remarks section for references to policy iteration in finite-state models with compact action sets.

8.6.1   The Algorithm

The Unichain Policy Iteration Algorithm

1. Set $n = 0$ and select an arbitrary decision rule $d_n \in D$.

2. (Policy evaluation) Obtain a scalar g_n and an $h_n \in V$ by solving

$$0 = r_{d_n} - ge + \left(P_{d_n} - I\right)h. \tag{8.6.1}$$

3. (Policy Improvement) Choose d_{n+1} to satisfy

$$d_{n+1} \in \arg\max_{d \in D}\{r_d + P_d h_n\}, \tag{8.6.2}$$

setting $d_{n+1} = d_n$ if possible.

4. If $d_{n+1} = d_n$, stop and set $d^* = d_n$. Otherwise increment n by 1 and return to step 2.

The above algorithm yields a sequence of decision rules $\{d_n\}$, scalars $\{g_n\}$, and vectors $\{h_n\}$. As a consequence of Corollary 8.2.7,

$$g_n = g^{(d_n)^\infty} \tag{8.6.3}$$

for each n and

$$h_n = H_{P_{d_n}} r_{d_n} + ke = h^{(d_n)^\infty} + ke, \tag{8.6.4}$$

where $h^{(d_n)^\infty}$ denotes the bias of the stationary policy $(d_n)^\infty$ and k is an arbitrary constant. Note that the choice of the additive constant has no effect on the maximizing decision rule in (8.6.2) since, for any vector h and any constant k,

$$r_d + P_d(h + ke) = r_d + P_d h + ke$$

for all $d \in D$.

Of course to implement policy iteration numerically, we require a specification which provides a unique h_n in step 2. In addition to (8.6.1), we may do the following:

a. Set

$$h_n(s_0) = 0 \tag{8.6.5}$$

for any fixed $s_0 \in S$.

b. Choose h_n to satisfy

$$P_{d_n}^* h_n = 0. \tag{8.6.6}$$

c. Choose h_n to satisfy

$$-h_n + \left(P_{d_n} - I\right)w = 0 \tag{8.6.7}$$

for some $w \in V$.

Alternatively we may use any available subroutine for solving the under-determined linear system

$$r_{d_n} + \left[-e \,\middle|\, P_{d_n} - I \right] \begin{bmatrix} g \\ h \end{bmatrix} = 0.$$

Clearly specification (a) is easiest to implement. Under it, the arbitrary constant equals $-h^{(d_n)^\infty}(s_0)$ so that, if h_n is determined through this specification,

$$h_n = h^{(d_n)^\infty} - h^{(d_n)^\infty}(s_0)e.$$

It is easy to see that we may implement the evaluation step under condition (8.6.5) by solving the linear system

$$r = (Q_{s_0})w, \tag{8.6.8}$$

where Q_{s_0} is the matrix $I - P_{d_n}$ with the column corresponding to state s_0 replaced by a column of 1's. The solution of (8.6.8) is unique, satisfies (8.6.5), has $g^{(d_n)^\infty}$ as its s_0th component and $h_n(s)$ for $s \neq s_0$ in its remaining components. Equation (8.6.8) may be solved by Gauss elimination or any other appropriate method.

When h_n is determined through (b), Corollary 8.2.7 implies that $h_n = h^{(d_n)^\infty}$, however, this approach requires computation of $P_{d_n}^*$.

Corollary 8.2.9 establishes that $h_n = h^{(d_n)^\infty}$ under specification (c). To find $h^{(d_n)^\infty}$ using approach (c) requires solution of a linear system with $2|S|$ rows and $2|S| + 1$ columns. Note that $(P_{d_n} - I)$ appears in both (8.6.1) and (8.6.7), so Gaussian iteration can be implemented efficiently on this combined system.

8.6.2 Convergence of Policy Iteration for Recurrent Models

We first consider models in which all policies have a single recurrent class and *no* transient states. In this case the sequence of gains of successive iterates increases monotonically until no further improvement is possible.

The following result applies to *any* model regardless of class structure. It shows that if improvement occurs in a recurrent state under stationary policy d_{n+1}, the gain for the improved policy is greater than the gain for the previous policy.

Proposition 8.6.1. Suppose $d_{n+1} \in \arg\max_{d \in D} \{r_d + P_d h_n\}$. Then

a. $g_{n+1}e = g_n e + P_{d_{n+1}}^* B(g_n, h_n)$. (8.6.9)

b. If $B(g_n, h_n)(s) > 0$ for a state s which is recurrent under d_{n+1}, then $g_{n+1} > g_n$.

c. If $B(g_{d_n}, h_{d_n})(s) = 0$ for all states s which are recurrent under d_{n+1}, then $g_{n+1} = g_n$.

Proof. Since

$$g_{n+1}e = P_{d_{n+1}}^* r_{d_{n+1}},$$

part (a) follows from adding and subtracting $g_n e$ to the right-hand side of this equation and using the results that $P_d^*(P_d - I) = 0$ and $P_d^* e = e$ for all $d \in D$ [cf. Appendix A, (A.4)] to obtain

$$g_{n+1}e = g_n e + P_{d_{n+1}}^* \left[r_{d_{n+1}} - g_n e + (P_{d_{n+1}} - I)h_n \right].$$

The result follows by noting that the quantity in parentheses on the right-hand side of the above expression equals $B(g_n, h_n)$.

Parts (b) and (c) are immediate consequences of (8.6.9) and the positivity of P_d^* on its recurrent states. $\square$

Equation (8.6.9) may be regarded as a "Newton method" representation for the gains at successive iterations of the policy iteration algorithm analogous to that in (6.4.11) for discounted models. An immediate consequence of parts (b) and (c) is the following convergence result.

Theorem 8.6.2. If all states are recurrent under every stationary policy, and the set of states and actions are finite, then policy iteration converges in a finite number of iterations to a solution (g^*, h) of the optimality equation $B(g, h) = 0$ and an average optimal stationary policy $(d^*)^\infty$.

Proof. By Proposition 8.6.1 the sequence of gains g_n is strictly increasing until a decision rule repeats. Since $g_n = g^{(d_n)^\infty}$ for some decision rule d_n and there are finitely many decision rules, the convergence criterion $d_{n+1} = d_n$ must be satisfied at some finite n. When this occurs, $B(g_n, h_n) = 0$, so, by Theorem 8.4.1(c), $g_n = g^*$ and, by Theorem 8.4.4, $(d^*)^\infty$ is average optimal. $\square$

When there are transient states associated with some (or all) stationary policies we require further analysis to establish convergence of policy iteration.

8.6.3 Convergence of Policy Iteration for Unichain Models

In unichain models in which some policies have transient states, we prove convergence of policy iteration by showing that iterates are monotone in the following lexicographic sense; either the gain increases strictly at successive iterations or, if there is no increase in gain, there is a strict increase in bias, until there is no further increase in either the gain or the bias. The analysis in this section relies on the structure of the limiting and fundamental matrices for unichain policies. To avoid unwieldly notation, let H_d denote the fundamental matrix of P_d [Appendix A, equation (A.14)].

Proposition 8.6.3. Suppose d_n is determined in the improvement step of the policy iteration algorithm, and h_n is any solution of (8.6.1). Then

$$h^{d_{n+1}^\infty} = h^{d_n^\infty} - P_{d_{n+1}}^* h^{d_n^\infty} + H_{d_{n+1}} B(g_n, h_n). \tag{8.6.10}$$

Proof. By definition,

$$h^{d_{n+1}^\infty} = H_{d_{n+1}} r_{d_{n+1}}$$

$$= H_{d_{n+1}}\left[r_{d_{n+1}} - g_n e + (P_{d_{n+1}} - I)h_n + g_n e - (P_{d_{n+1}} - I)h_n \right].$$

From (A.14), $H_d e = 0$ and, from (A.17), $H_d(P_d - I) = P_d^* - I$ for all $d \in D$. Since

$$B(g_n, h_n) = r_{d_{n+1}} - g_n e + (P_{d_{n+1}} - I)h_n,$$

it follows that

$$h^{d_{n+1}^\infty} = h_n - P_{d_{n+1}}^* h_n + H_{d_{n+1}} B(g_n, h_n).$$

To obtain (8.6.10), note that

$$(I - P_{d_{n+1}}^*)h_n = (I - P_{d_{n+1}}^*)h^{d_n^\infty},$$

since h_n and $h^{d_n^\infty}$ may differ by at most a constant times e.   □

The following technical lemma provides the structure of H_d for unichain models.

Lemma 8.6.4. Let $d \in D$. Suppose P_d is unichain and, in canonical form, R_d denotes recurrent states of P_d, T_d denotes its transient states, and H_d is expressed as

$$H_d = \begin{bmatrix} H_d^{RR} & H_d^{RT} \\ H_d^{TR} & H_d^{TT} \end{bmatrix}, \qquad (8.6.11)$$

where the partition corresponds to states in R_d and T_d. Then

  a. $H_d^{RT} = 0$;
  b. $H_d^{TT} = (I - P_d^{TT})^{-1}$, where P_d^{TT} denotes the restriction of P_d to its transient states, and
  c. if $u(s) = 0$ for $s \in R_d$ and $u(s) \geq 0$ for $s \in T_d$, then $H_d u \geq u \geq 0$.

Proof. By the unichain assumption, we may write

$$P_d = \begin{bmatrix} P_d^{RR} & 0 \\ P_d^{TR} & P_d^{TT} \end{bmatrix}$$

and

$$P_d^* = \begin{bmatrix} Q_d & 0 \\ Q_d & 0 \end{bmatrix}.$$

Since, from (A.14),

$$H_d = (I - P_d - P_d^*)^{-1}(I - P^*), \tag{8.6.12}$$

it follows from the above representations for P_d and P_d^* and a standard partitioned matrix inversion formula that

$$(I - P_d + P_d^*)^{-1} = \begin{bmatrix} W & 0 \\ X & Y \end{bmatrix}, \tag{8.6.13}$$

where W and X are particular derived matrices and $Y = (I - P_d^{TT})^{-1}$, which exists as a consequence of Proposition A.3. Since

$$I - P_d^* = \begin{bmatrix} I_R - Q_d & 0 \\ -Q_d & I_T \end{bmatrix},$$

where I_R and I_T denote appropriate restrictions of the the identity matrix, (a) and (b) follow from (8.6.12).

Proposition A.3b and part (b) of this lemma imply

$$H_d^{TT} = (I - P_d^{TT})^{-1} = \sum_{n=0}^{\infty} (P_d^{TT})^n \geq I,$$

so that part (c) follows by multiplication. $\square$

We now combine these technical results to establish the the monotonicity of the iterates of policy iteration.

Proposition 8.6.5. Suppose d_{n+1} is determined in step 3 of the policy iteration algorithm. If $B(g_n, h_n)(s) = 0$ for all s that are recurrent under d_{n+1}, and $B(g_n, h_n)(s_0) > 0$ for some s_0 which is transient under d_{n+1} then

$$h^{d_{n+1}^{\infty}}(s) > h^{d_n^{\infty}}(s),$$

for some s which is transient under d_{n+1}.

Proof. Since $B(g_n, h_n)(s) = 0$ for s recurrent under d_{n+1}, $d_n(s) = d_{n+1}(s)$ on recurrent states of d_{n+1}. Since $p_{d_{n+1}}(j|s) = 0$ if s is recurrent under d_{n+1} and j is transient under d_{n+1}, it follows that d_n and d_{n+1} have the same sets of recurrent states and $P_{d_{n+1}}^* = P_{d_n}^*$. Consequently,

$$P_{d_{n+1}}^* h^{d_n^{\infty}} = P_{d_n}^* h^{d_n^{\infty}} = 0,$$

where the second equality follows from the identity $P_d^* H_d = 0$ provided in (A.18).

As a consequence of the above equality and (8.6.10), it follows that

$$h^{d^{\infty}_{n+1}} = h^{d^{\infty}_n} + H_{d_{n+1}} B(g_n, h_n),$$
(8.6.14)

so the result follows Lemma 8.6.4(c).   □

The result in Proposition 8.6.5 means that if there is no change in the decision rule in states which are recurrent under the new policy, and a change in a state which is transient under the new policy, then the bias of the new policy will be greater than that of the previous policy in at least one component. Thus, at successive iterates, policy iteration determines a stationary policy with a larger gain and, if this is not possible, then a policy with a larger bias. If neither of these alternatives is possible, the algorithm terminates. This is the essence of the proof of the following result.

Theorem 8.6.6.   Suppose all stationary policies are unichain, and the set of states and actions are finite, then policy iteration converges in a finite number of iterations to a solution (g^*, h) of the optimality equation $B(g, h) = 0$ and an average optimal stationary policy $(d^*)^{\infty}$.

Proof.   By Propositions 8.6.1 and 8.6.5, at successive iterates of policy iteration either

a. $g^{(d_{n+1})^{\infty}} > g^{(d_n)^{\infty}}$,
b. $g^{(d_{n+1})^{\infty}} = g^{(d_n)^{\infty}}$ and $h^{(d_{n+1})^{\infty}}(s) > h^{(d_n)^{\infty}}(s)$ for some $s \in S$, or
c. $g^{(d_{n+1})^{\infty}} = g^{(d_n)^{\infty}}$ and $h^{(d_{n+1})^{\infty}} = h^{(d_n)^{\infty}}$.

Since there are only finitely many policies, (a) or (b), can only occur finitely many times and, consequently, (c) must be achieved in finitely many iterations. In this case, the algorithm terminates with a solution of the optimality equation and a stationary policy $(d^*)^{\infty}$ which is average optimal.   □

In addition to showing that policy iteration converges to a solution of the optimality equation, this theorem provides a *constructive proof of the existence of a solution to the optimality equation*. The above results also provide insight into the behavior of the iterates of policy iteration. If s is recurrent and j is transient for some $d \in D$, $p_d(j|s) = 0$. Consequently, once the optimality equation is satisfied on all states that are recurrent under a decision rule δ which attains the maximum in the improvement step, there will be no future changes in the gain. Hence any stationary policy which agrees with P_δ on its recurrent states is average optimal. We illustrate this together with other concepts through the following example.

Example 8.6.1.   Consider the model in Fig. 3.1.1. There are two actions in state s_1 and a single action in state s_2. Let δ use action $a_{1,1}$ in s_1, and γ use action $a_{1,2}$ in s_1. Both decision rules use action $a_{2,1}$ in s_2. Apply policy iteration with $d_0 = \gamma$. Then the evaluation equations become

$$0 = 10 - g - h(s_1) + h(s_2),$$
$$0 = -1 - g.$$

Setting $h(s_2) = 0$, we find $g = -1$, $h_0(s_1) = 11$, and $h_0(s_2) = 0$. The improvement equation becomes

$$\max\{5 + 0.5h_0(s_1) + 0.5h_0(s_2), 10 + h_0(s_2)\} = \max\{10.5, 10\} = 10.5$$

so that $d_1(s_1) = a_{1,1}$. The evaluation equations at the next pass through the algorithm become

$$0 = 5 - g - 0.5h(s_1) + 0.5h(s_2),$$
$$0 = -1 - g,$$

which have the solutions $g = -1$, $h_1(s_1) = 12$, and $h_1(s_2) = 0$. Repeating this argument establishes the optimality of δ^∞.

Note that s_2 is recurrent under d_1, s_1 is transient under d_1, $B(g_0, h_0)(s_2) = 0$, and $B(g_0, h_0)(s_1) > 0$, so that, as shown in Proposition 8.6.5, $h^{d_1^\infty}(s_1) = 12 > h^{d_0^\infty}(s_1) = 11$. Since the optimality equation held on the recurrent states of d_0 no further improvement is possible in the gain at subsequent iterations. This means that $(d_0)^\infty$ was also average optimal but its gain and bias do not solve the optimality equation at state s_1 which is given by

$$0 = \max\{5 - g - 0.5h(s_1) + 0.5h(s_2), 10 - g - h(s_1) + h(s_2)\}.$$

Thus policy iteration does more than find an average optimal policy, *it finds a solution of the optimality equation.*

The above example shows that after determining a policy with optimal average reward, policy iteration goes on to find one with optimal bias. Thus one might conjecture that policy iteration also finds a bias-optimal policy; that is, a policy with greatest bias among all policies with the same gain as δ. The following example shows that this supposition is false.

Example 8.6.2. Let $S = \{s_1, s_2\}$; $A_{s_1} = \{a_{1,1}, a_{1,2}\}$ and $A_{s_2} = \{a_{2,1}\}$; $r(s_1, a_{1,1}) = 4$, $r(s_1, a_{1,2}) = 0$, $r(s_2, a_{2,1}) = 8$; and $p(s_1|s_1, a_{1,1}) = 1$, $p(s_2|s_1, a_{1,2}) = 1$, and $p(s_1|s_2, a_{2,1}) = 1$. (Fig. 8.6.1)

Actions are chosen in state s_1 only. Let $\delta(s_1) = a_{1,2}$. Now apply policy iteration with $d_0 = \delta$. Solving the evaluation equations under the additional restriction $P_{d_0}^* h = 0$ yields $g^{(d_0)^\infty} = 4$, $h^{(d_0)^\infty}(s_1) = -2$ and $h^{(d_0)^\infty}(s_2) = 2$. At the subsequent improvement step we seek

$$\max\{4, -h^{(d_0)^\infty}(s_1) + h^{(d_0)^\infty}(s_2)\} = \max\{4, 2 + 2\} = 4,$$

so that $d_1 = d_0$ and we terminate the algorithm concluding that δ^∞ is average optimal.

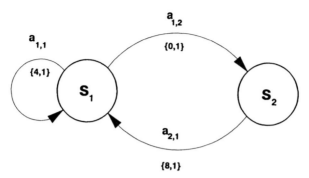

Figure 8.6.1 Symbolic representation of Example 8.6.2.

Now let γ denote the decision rule which uses action $a_{1,1}$ in state s_1. Then solving the evaluation equations shows that $g^{\gamma^{\infty}} = 4$, $h^{\gamma^{\infty}}(s_1) = 0$, and $h^{\gamma^{\infty}}(s_2) = 4$. Hence γ^{∞} is also average optimal but its bias $h^{\gamma^{\infty}}$ exceeds $h^{\delta^{\infty}}$ in both components. Therefore policy iteration may terminate with a policy that is not bias optimal.

The difficulty here is that both $h^{\delta^{\infty}}$ and $h^{\gamma^{\infty}}$ satisfy the optimality equation but each has a different set of recurrent states. Under δ^{∞}, s_1 and s_2 are recurrent while, under γ^{∞}, only s_1 is recurrent. Policy iteration has found the policy with the largest gain among all policies (in this case only δ^{∞}) with the same recurrent class as δ^{∞}.

The above example shows that policy iteration may terminate with a policy which is not bias optimal. It finds the policy with the largest bias among policies with the same set of recurrent states as the first optimal policy identified. To find bias optimal policies requires the solution of an additional optimality equation (Chapter 10, Corollary 10.2.10).

Demonstration of convergence in the nonfinite-state and/or action cases uses similar ideas to those stated above, but requires more complicated technical analysis to show convergence of an infinite sequence of gains and values.

8.7 MODIFIED POLICY ITERATION IN UNICHAIN MODELS

In average reward models, value iteration may converge very slowly, and policy iteration may be inefficient in models with many states because of the need to solve a $|S| \times |S|$ linear system of equations. As in discounted models, modified policy iteration provides a compromise between these two algorithms. It avoids the repeated evaluation of Lv^n when using value iteration step and avoids solving the linear system

$$0 = r_d - ge + (P_d - I)h$$

when using policy iteration. In discounted models, we analyzed modified policy iteration from the perspective of a variant of policy iteration, however, in average reward models, we view it as a variant of value iteration.

8.7.1   The Modified Policy Iteration Algorithm

The following modified policy iteration algorithm finds an ε-optimal policy. Let $\{m_n\}$ denote a sequence of non-negative integers.

The Modified Policy Iteration Algorithm (MPI)

1. Select $v^0 \in V$, specify $\varepsilon > 0$, and set $n = 0$.
2. (Policy Improvement) Choose d_{n+1} to satisfy

$$d_{n+1} \in \arg\max_{d \in D} \{r_d + P_d v^n\}, \qquad (8.7.1)$$

setting $d_{n+1} = d_n$ if possible.
3. (Partial Policy Evaluation)
   a. Set $k = 0$, and

$$u_n^0 = r_{d_{n+1}} + P_{d_{n+1}} v^n. \qquad (8.7.2)$$

   b. If

$$sp(u_n^0 - v^n) < \varepsilon, \qquad (8.7.3)$$

go to step 4. Otherwise go to (c).
   c. If $k = m_n$, go to (e). Otherwise, compute u_n^{k+1} by

$$u_n^{k+1} = r_{d_{n+1}} + P_{d_{n+1}} u_n^k \equiv L_{d_{n+1}} u_n^k. \qquad (8.7.4)$$

   d. Increment k by 1 and return to (c).
   e. Set $v^{n+1} = u_n^{m_n}$ and go to step 2.
4. Choose

$$d_\varepsilon \in \arg\max_{d \in D} \{r_d + P_d v^n\} \qquad (8.7.5)$$

and stop.

This algorithm combines features of both policy iteration and value iteration. Like value iteration, it is an iterative algorithm and begins with a value v^0. The stopping criterion used in step 3(b) is identical to that of value iteration; when it is satisfied, the designated policy $(d_\varepsilon)^\infty$ is ε-optimal. The computation of u_n^0 in step 3(a) requires no additional work because it has been evaluated in step 2 when determining the arg max in (8.7.2).

Like policy iteration, the algorithm contains an improvement step, step 2, and an evaluation step, step 3; however, the evaluation is not done exactly. Instead it is carried out iteratively in step 3(c), which is repeated m_n times at iteration n. As a consequence of the second equality in (8.7.4), step 3 may be represented by

$$v^{n+1} = (L_{d_{n+1}})^{m_n + 1} v^n. \tag{8.7.6}$$

The *order sequence* $\{m_n\}$ may be

a. fixed for all iterations ($m_n = m$);
b. chosen according to some prespecified pattern; or
c. selected adaptively, for example, by requiring $sp(u_n^{m_n+1} - u_n^{m_n}) < \varepsilon_n$, where ε_n is fixed or variable.

We show that the algorithm converges for any order sequence and discuss its rate of convergence in the next two sections.

8.7.2   Convergence of the Algorithm

We relate the algorithm to policy iteration and to value iteration. Clearly with $m_n = 0$ for all n, modified policy and value iteration are identical. Expression (8.7.4) shows that the evaluation step of the modified policy iteration algorithm corresponds to performing $m_{n-1} + 1$ value iteration steps with the *fixed* decision rule d_n. Under the assumption that, for all $d \in D$, $P_d \geq \alpha I$ for some $\alpha > 0$, the delta coefficient for $P_{d_n}^J$, which we denote by $\gamma_{d_n, J}$, and is defined in (6.6.5), satisfies $\gamma_{d_n, J} < 1$. Consequently

$$sp(u_n^{k+J+1} - u_n^{k+J}) \leq \gamma_{d_n, J} sp(u_n^{k+1} - u_n^k).$$

This implies that for any $\varepsilon' > 0$ and m sufficiently large, $sp(u_n^{m+1} - u_n^m) < \varepsilon'$, so that $sp(Lu_n^m - u_n^m)$ is close to zero. Therefore u_n^m is close to being a span fixed point of L_{d_n} so that

$$0 \approx r_{d_n} - ge + (P_{d_n} - I)u_n^m,$$

which is the policy evaluation equation of the unichain policy iteration algorithm. Therefore for m_n large, a modified policy iteration step and a policy iteration step yield almost identical values.

We now prove that modified policy iteration converges by showing that the stopping criterion in (8.7.3) is satisfied for some finite n. Our proof uses methods of Sec. 8.5; however, it is slightly more tedious because of the extra notation.

Theorem 8.7.1.   Suppose there exists an integer J for which γ' defined by (8.5.11) satisfies $\gamma' < 1$. Then, for any sequence $\{m_n\}$ and $\varepsilon > 0$, there exists an n for which (8.7.3) holds. Further the policy $(d_\varepsilon)^\infty$ is ε-optimal where d_ε is defined through (8.7.5).

Proof.   Let π denote the nonstationary policy corresponding to the sequence of decision rules generated by the modified policy iteration algorithm. It may be represented as $\pi = (d_1, \ldots, d_1, d_2, \ldots, d_2, d_3, \ldots, d_3, \ldots)$, where d_1 appears $m_0 + 1$ times, d_2 appears $m_1 + 1$ times, etc. Let w_j represent the sequence of iterates generated by applying the modified policy iteration algorithm, disregarding iteration number, that is, $w^j = u_0^j$ for $j \leq m_0 + 1$, and

$$w^{m_0 + \cdots + m_n + 1 + i} = u_n^i$$

otherwise. Therefore, by hypothesis,

$$sp(w^{nJ+1} - w^{nJ}) \leq (\gamma')^n sp(w^1 - w^0)$$

so that there exists an n for which $sp(w^{nJ+1} - w^{nJ}) < \varepsilon$. Since

$$sp(w^{k+1} - w^k) \leq sp(w^k - w^{k-1})$$

for all k, $sp(w^{nJ+j+1} - w^{nJ+j}) < \varepsilon$ for all $j \geq 1$. Therefore, for some n, (8.7.3) holds. The ε-optimality of $(d_\varepsilon)^\infty$ follows from Theorem 8.5.5.   □

From an applications perspective, we prefer the following immediate result.

Corollary 8.7.2.   Suppose any of the hypotheses of Theorem 8.5.3 are satisfied; then the conclusions of Theorem 8.7.1 follow.

In applications we may assure that condition (c) of Theorem 8.5.3 holds by applying the aperiodicity transformation of Sec. 8.5.5 prior to solving the problem. In addition, the action elimination procedure of Sec. 8.5.7 can be used to reduce the set of actions checked during the policy improvement step of the modified policy iteration algorithm.

8.7.3   Numerical Comparison of Algorithms

In this section we compare value iteration, policy iteration, and modified policy iteration, and investigate the effect of action elimination in the context of the following inventory model.

Example 8.7.1.   We solve a variant of the inventory model of Sec. 3.2 which arose from analysis of sales data from a cross section of Canadian Tire retail stores in the Vancouver metropolitan area (cf. Sec. 1.2). Preliminary analysis of the data suggests that *daily* demand for product #429410 (an ironing board) is time homogeneous and follows the probability distribution

Demand	0	1	2	3	4
Probability	573/630	53/630	2/630	1/630	1/630

Let the random variable Δ represent the demand on an arbitrary day, and set

$$p_k = P(\Delta = k).$$

Contrary to practice at Canadian Tire, we assume daily review of inventory levels and instantaneous delivery of orders. We arbitrarily assign the following parameter values: maximum capacity $M = 9$, revenue when u items are sold $f(u) = 30u$, ordering cost

$$O(u) = \begin{cases} 1 + 10u & u > 0 \\ 0 & u = 0, \end{cases}$$

and, holding cost, $h(u) = 0.01 + 0.01u$. Because of marketing considerations, we impose a penalty on a zero stock level of $z(0) = 300$ units. Under these assumptions, the expected reward function satisfies

$$r(s, a) = \sum_{j=0}^{s+a} f(j)P(\Delta = j) + [f(s + a)$$

$$-z(0)]P(\Delta \geq s + a) - O(a) - h(s + a).$$

Unfortunately, as in Example 8.3.1, when the inventory capacity M exceeds the maximal demand 4, the model is not unichain. To remedy this, we modify the probability distribution to achieve a unichain model yet remain faithful to the "observed" demand distribution. We perturb the above demand distribution by choosing $\gamma = 5\varepsilon + 630$, $\varepsilon = 0.000001$ and

$$p_0 = 573/\gamma, \quad p_1 = 53/\gamma, \quad p_2 = 2/\gamma, \quad p_3 = 1/\gamma, \quad p_4 = 1/\gamma,$$

$$p_5 = \varepsilon/\gamma, \quad p_6 = \varepsilon/\gamma, \quad p_7 = \varepsilon/\gamma, \quad p_7 = \varepsilon/\gamma, \quad p_9 = \varepsilon/\gamma.$$

Since decisions are made frequently and indefinitely, we choose to evaluate performance using the average reward criterion.

We solve the problem using the unichain policy iteration algorithm of Sec. 8.6, with the specification that $h_n(0) = 0$. We obtain an optimal gain $g^* = 1.9315$ in five iterations. Recalling that the state equals the inventory level when the order is placed, and the decision denotes the number of units to order, we obtain the optimal policy $(d^*)^\infty$ below.

s	$d^*(s)$	$h^*(s)$
0	8	0.0000
1	7	10.000
2	6	20.000
3	5	30.000
4	4	40.000
5	0	50.434
6	0	60.731
7	0	70.917
8	0	81.000
9	0	90.990

Table 8.7.1

Iteration	Value Iteration with action elim. span	% Elim.	Modified Policy Iteration $M_n = 5$ span	$M_n = 20$ span	Policy Iteration change in gain
1	23.02	57.8	23.02	23.02	
2	3.062	100.0	3.042	19.80	1.9317
3	3.054	93.3	3.610	0.5909	0.0896
4	3.050	84.4	2.481	0.1985	0.0003
5	3.046	82.2	0.9550	0.0459	0.0000
6	3.044	88.9	0.8581	0.0074	
7	3.042	77.8	0.7398	0.0008	
8	2.809	77.8	0.6133	0.0001	
9	2.033	82.2	0.4565	1.95e − 5	
10	2.032	75.6	0.3455	2.26e − 6	
15	1.835	64.4	0.0052		
20	1.014	62.2	0.0036		
25	0.9565	93.3	0.0002		
30	0.8781	84.4	1.31e − 5		
50	0.4414	68.9			
75	0.1087	84.4			
100	0.0163	97.8			
150	0.0001	100.0			
200	1.02e − 6	100.0			

As expected, the optimal policy has (σ, Σ) form. Whenever the inventory level falls below five units, the stock is immediately increased to eight units. The quantity $h^*(s)$ gives the relative value of initial stock position.

We also solve the problem using value iteration with the action elimination procedure of Sec. 8.5.7, and modified policy iteration. Table 8.7.1 compares the convergence of these algorithms. Reported results include $sp(v^{n+1} - v^n)$, denoted *span*, the percentage of actions which could be eliminated (total number of actions − number of states), denoted % Elim., and the change in gain (which approximates the span for policy iteration).

Note that to achieve accuracy to six decimal places, value iteration requires 200 iterations, modified policy iteration with $M_n = 5$, 35 iterations and with $M_n = 20$, 11 iterations. Policy iteration obtains an exact solution in five iterations, but at the cost of solving a linear system with $|S| - 1$ variables. Modified policy iteration and value iteration require computing $r_d + P_d v^n$ approximately 200 times for some d, but modified policy iteration avoids frequent maximization of this expression over a set of decision rules. When $M_n = 20$, only 11 such maximizations are carried out.

The results in Table 8.7.1 do not give a complete indication of performance, since each method has a different computational effort associated with each iteration. Table 8.7.2 provides the number of multiplications per iteration for each algorithm. In it we let $|A_s|$ denote the number of actions in state s, $|S|$ the number of states, and R the percentage of actions remaining after action elimination. Storage requirements of the algorithms are comparable, although action elimination requires storage of an $\sum_{s \in S} |A_s|$ array of comparison values.

Table 8.7.2. Comparison of Computational Effort Per Iteration When Solving Unichain MDPs Using Various Algorithms.

Algorithm	Multiplications/Iteration
Value Iteration (VI)	$\sum_{s \in S} |A_s| |S|$
VI with Action Elimination	$R \sum_{s \in S} |A_s| |S|$
Modified Policy Iteration	$(\sum_{s \in S} |A_s| + M_n + 1) |S|$
Policy Iteration	$\sum_{s \in S} |A_s| |S| + (\frac{1}{3}) |S|^3$

8.8 LINEAR PROGRAMMING IN UNICHAIN MODELS

This section provides a linear programming approach for solving finite-state and action unichain MDP's with average reward criterion. We assume finite S and A_s throughout.

As a consequence of Theorem 8.4.1(a), $g \geq g^*$ whenever there exist $(g, h) \in R^1 \times V$ satisfying

$$ge + (I - P_d)h \geq r_d \tag{8.8.1}$$

for all $d \in D$. Part (c) of that theorem implies that g^* is the smallest g for which there exists an $h \in V$ satisfying (8.8.1). This suggests the following linear programming model.

Primal Linear Program.

$$\text{Minimize } g$$

subject to

$$g + h(s) - \sum_{j \in S} p(j|s, a)h(j) \geq r(s, a), \qquad a \in A_s \quad \text{and} \quad s \in S,$$

with g and $h(s)$ unconstrained.

As in the discounted case, we analyze the model through the following dual linear program.

Dual Linear Program.

$$\text{Maximize } \sum_{s \in S} \sum_{a \in A_s} r(s, a)x(s, a) \tag{8.8.2}$$

subject to

$$\sum_{a \in A_j} x(j, a) - \sum_{s \in S} \sum_{a \in A_s} p(j|s, a)x(s, a) = 0, \quad j \in S \qquad (8.8.3)$$

$$\sum_{s \in S} \sum_{a \in A_s} x(s, a) = 1, \qquad (8.8.4)$$

and $x(s, a) \geq 0$ for $a \in A_s$, $s \in S$.

Observe that the dual differs from that in the discounted model in that it includes the additional constraint (8.8.4) and has $\lambda = 1$ and a zero right-hand side in (8.8.3). Note also that one of the equations in (8.8.3) is redundant.

8.8.1 Linear Programming for Recurrent Models

We first assume that all stationary policies have a single recurrent class and no transient states. We relate feasible solutions of the dual to stationary randomized policies in the average reward MDP. Our analysis relies heavily on Theorem A.2, which we restate in this context as follows.

Proposition 8.8.1. In a recurrent model, the system of equations

$$\sum_{s \in S} p_d(j|s)\pi(s) = \pi(j) \quad j \in S \qquad (8.8.5)$$

subject to

$$\sum_{j \in S} \pi(j) = 1 \qquad (8.8.6)$$

has a unique *positive* solution π_d for any $d \in D^{MR}$.

We refer to π_d as the *stationary distribution* of P_d. The following result establishes the relationship between feasible solutions to the dual LP and stationary policies. Recall that $q_{d(s)}(a)$ denotes the probability that randomized decision rule d chooses action a in state s.

Theorem 8.8.2. Suppose that the transition probability matrix of every stationary policy is irreducible.

a. For each $d \in D^{MR}$, $s \in S$ and $a \in A_s$ let

$$x_d(s, a) = q_{d(s)}(a)\pi_d(s) \qquad a \in A_s, \quad s \in S, \qquad (8.8.7)$$

where π_d is the unique solution of (8.8.5) subject to (8.8.6). Then $x_d(s, a)$ is a feasible solution to the dual problem.

b. Let x be any feasible solution to the dual linear program. Then for each $s \in S$, $\sum_{a \in A_s} x(s, a) > 0$. Define the randomized stationary policy d_x^∞ by

$$q_{d_x(s)}(a) = \frac{x(s, a)}{\sum_{a' \in A_s} x(s, a')}, \qquad a \in A_s, \quad s \in S. \qquad (8.8.8)$$

Then x_{d_x} is a feasible solution to the dual LP and $x_{d_x}(s, a) = x(s, a)$ for all $a \in A_s$ and $s \in S$.

Proof. Since $\sum_{a \in A_s} q_{d(s)}(a) = 1$, direct substitution into (8.8.3) and (8.8.4) establishes the feasibility of $x_d(s, a)$ defined by (8.8.7).

To establish (b), let

$$u(s) \equiv \sum_{a \in A_s} x(s, a)$$

and let $S' = \{s \in S : u(s) > 0\}$. We show that $S' = S$. For $s \in S'$, (8.8.8) implies that

$$x(s, a) = q_{d_x(s)}(a)u(s). \tag{8.8.9}$$

Substituting (8.8.9) into (8.8.3) and (8.8.4), and noting that $\sum_{a \in A_s} q_{d(s)}(a) = 1$ shows that, for $j \in S'$,

$$u(j) - \sum_{s \in S'} p_{d_x}(j|s)u(s) = 0$$

and

$$\sum_{s \in S'} u(s) = 1.$$

By Proposition 8.8.1, $u(s) = \pi_{d_x}(s) > 0$ for all $s \in S'$. Since $\sum_{s \in S'} \pi_{d_x}(s) = 1$, this implies that all recurrent states of P_{d_x} are contained in S'. Since all policies are recurrent, $S' = S$. From (8.8.9) and the demonstrated result that $u(s) = \pi_{d_x}(s)$, it follows that $x_{d_x}(s, a) = x(s,a)$ for all $a \in A_s$. □

This theorem establishes a one-to-one relationship between randomized stationary policies and feasible solutions to the dual problem through (8.8.7) and (8.8.8). As a result of (8.8.7), we interpret $x(s, a)$ as the stationary probability that the system is in state s and chooses action a under the randomized stationary policy generated by d_x. Consequently the objective function for the dual gives the expected stationary reward under this policy.

As in discounted models, *basic* feasible solutions to the dual correspond to deterministic stationary policies and have a nice representation.

Corollary 8.8.3

a. Let x be a basic feasible solution to the dual LP. Then $d_x \in D^{MD}$ and $d_x(s) = a$ if $x(s, a) > 0$.

b. Suppose that $d \in D^{MD}$, then $x_d = \pi_d$ is a basic feasible solution to the dual LP.

Proof. To prove part (a), note that a basic feasible solution has at most $|S|$ nonzero components. Since Theorem 8.8.2(b) establishes that $\sum_{a \in A_s} x(s, a) > 0$ for each $s \in S$, there must be one nonzero entry for each s. Part (b) follows immediately from part (a) of Theorem 8.8.2. □

Corollary 8.8.4. There exists a bounded optimal basic feasible solution x^* to the dual and the policy $(d_{x^*})^\infty$ in which $d_{x^*}(s) = a$ if $x^*(s, a) > 0$ is an optimal deterministic policy.

Proof. Since we assume $|r(s, a)| \leq M < \infty$, all feasible solutions of the dual are bounded. Therefore Theorem D.1(a) (Appendix D) implies that there exists an optimal basic feasible solution. Corollary 8.8.3 shows that it has the designated form.   □

Representation (8.8.7) shows that the initial-state distribution α which appeared in the LP formulation of the discounted model (Sec. 6.9) is not required in the average reward model. This is because the stationary distribution π_d does not depend on α.

In recurrent models, the relationship between policy iteration and the simplex algorithm is the same as under the discounted criterion. Policy iteration corresponds to the simplex method with block pivoting, and the simplex algorithm corresponds to a version of policy iteration in which the subsequent policy differs only in the state at which the quantity

$$\max_{a \in A_s} \left\{ r(s, a) + \sum_{j \in S} p(j|s, a)h(j) - h(s) \right\}$$

achieves its maximum value. Note that in practice, it may be difficult to identify optimal policies in large problems using linear programming because many non-zero $x^*(s, a)$ may be very close to zero.

We now illustrate these concepts by solving an example.

Example 8.8.1. Let $S = \{s_1, s_2\}$; $A_{s_1} = \{a_{1,1}\}$ and $A_{s_2} = \{a_{2,1}, a_{2,2}, a_{2,3}\}$; $r(s_1, a_{1,1}) = 4$, $r(s_2, a_{2,1}) = 0$, $r(s_2, a_{2,2}) = 1$, and $r(s_2, a_{2,3}) = 2$; and $p(s_1|s_1, a_{1,1}) = 0.4$, $p(s_2|s_1, a_{1,1}) = 0.6$, $p(s_1|s_2, a_{2,1}) = 1$, $p(s_1|s_2, a_{2,2}) = 0.8$, and $p(s_1|s_2, a_{2,3}) = 0.3$ (Fig. 8.8.1). The dual LP follows.

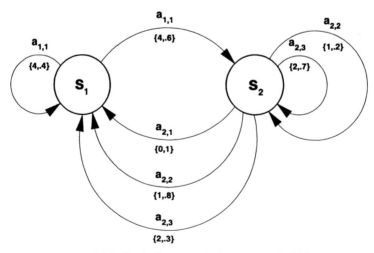

Figure 8.8.1   Symbolic representation of Example 8.8.1.

Maximize

$$4x(s_1, a_{1,1}) + 0x(s_2, a_{2,1}) + 1x(s_2, a_{2,2}) + 2x(s_2, a_{2,3})$$

subject to

$$x(s_1, a_{1,1}) - [0.4x(s_1, a_{1,1}) + 1x(s_2, a_{2,1})$$
$$+0.8x(s_2, a_{2,2}) + 0.3x(s_2, a_{2,3})] = 0,$$
$$\sum_{j=1}^{3} x(s_2, a_{2,j}) - [0.6x(s_1, a_{1,1}) + 0x(s_2, a_{2,1})$$
$$+0.2x(s_2, a_{2,2}) + 0.7x(s_2, a_{2,3})] = 0,$$
$$x(s_1, a_{1,1}) + x(s_2, a_{2,1}) + x(s_2, a_{2,2}) + x(s_2, a_{2,3}) = 1,$$
$$(8.8.10)$$

and $x(s, a) \geq 0$ for all $a \in A_s$ and $s \in S$. Note that the first or second constraint is redundant, so that the constraint set reduces to

$$0.6x(s_1, a_{1,1}) - 1x(s_2, a_{2,1}) - 0.8x(s_2, a_{2,2}) - 0.3x(s_2, a_{2,3}) = 0,$$

together with (8.8.10) and the non-negativity constraints.

Solving the LP yields $x^*(s_1, a_{1,1}) = 0.571$, $x^*(s_2, a_{2,1}) = 0$, $x^*(s_2, a_{2,2}) = 0.429$, and $x^*(s_3, a_{2,3}) = 0$ with objective function value of 2.713. Consequently, Corollary 8.8.4 establishes that the optimal policy is $(d_{x^*})^\infty$, where $d_{x^*}(s_1) = a_{1,1}$ and $d_{x^*}(s_2) = a_{2,2}$. Corollary 8.8.2(b) shows that $\pi_{d_x}(s_1) = 0.571$ and $\pi_{d_x}(s_2) = 0.429$. A direct calculation reveals that

$$g^{(d_x)^\infty} = 4 \times 0.571 + 1 \times 0.429 = 2.713.$$

We leave it as an exercise to interpret the primal variables which equal 2.143 and 2.713, respectively.

8.8.2   Linear Programming for Unichain Models

In this section we investigate the LP approach for unichain models in which some policies have a nonempty set of transient states. The following example shows that there need not be a one-to-one relationship between stationary policies and basic feasible solutions.

Example 8.8.2. We analyze the example in Fig. 3.1.1. The dual LP may be expressed as follows.
Maximize

$$5x(s_1, a_{1,1}) + 10x(s_1, a_{1,2}) - 1x(s_2, a_{2,1})$$

subject to

$$x(s_1, a_{1,1}) + x(s_1, a_{1,2}) - 0.5x(s_1, a_{1,1}) = 0,$$
$$x(s_2, a_{2,1}) - 0.5x(s_1, a_{1,1}) - x(s_1, a_{1,2}) - x(s_2, a_{2,1}) = 0,$$
$$x(s_1, a_{1,1}) + x(s_1, a_{1,2}) + x(s_2, a_{2,1}) = 1,$$

and $x(s, a) \geq 0$. Again note the redundant second constraint. The first constraint may be rewritten as

$$0.5x(s_1, a_{1,1}) + x(s_1, a_{1,2}) = 0.$$

It is easy to see that the only feasible solution and hence optimal solution is $x^*(s_1, a_{1,1}) = 0$, $x^*(s_1, a_{1,2}) = 0$, and $x^*(s_2, a_{2,1}) = 1$ with an objective function value of $g^* = -1$.

The approach in Corollary 8.8.4 does not provide an optimal policy because both $x^*(s_1, a_{1,1})$ and $x^*(s_1, a_{1,2})$ equal 0 so that no action is specified in state s_1. This is not a problem from the perspective of average reward, since it is irrelevant which action is selected in state s_1. Since s_1 is transient under both policies, action choice there does not effect the value of the average reward. We will see that, in models with transient states, the LP solution prescribes how to choose an optimal policy on its recurrent states and not on its transient states.

In this example, all policies have stationary distribution $\pi(s_1) = 0$ and $\pi(s_2) = 1$. Since this is the only feasible solution to the dual LP, *many* stationary policies correspond to the same feasible solution in contrast to the one-to-one relationship in recurrent models.

The following proposition, which characterizes the stationary distribution of a unichain policy, provides the basis for the main results of this section. It follows immediately from Theorem A.2 and Proposition A.3(b). Recall that R_d and T_d respectively, denote, the recurrent and transient states of P_d.

Proposition 8.8.5. In a unichain model, for any $d \in D^{MR}$, the system of equations (8.8.5) subject to (8.8.6) has the unique solution π_d in which $\pi_d(s) > 0$ for $s \in R_d$ and $\pi_d(s) = 0$ for $s \in T_d$.

The following results generalize Theorem 8.8.2 and Corollaries 8.8.3 and 8.8.4 to models with transient states. The proofs follow by minor modifications of arguments in the preceding section and are left as exercises.

Theorem 8.8.6. Suppose that the transition probability matrix of every stationary policy is unichain.

a. Then for $d \in D^{MR}$,

$$x_d(s, a) = \begin{cases} q_{d(s)}(a)\pi_d(s) & s \in R_d \\ 0 & s \in T_d \end{cases}$$

is a feasible solution to the dual LP.

b. Let $x(s, a)$ be a feasible solution to the dual,

$$S_x \equiv \left\{ s \in S : \sum_{a \in A_s} x(s, a) > 0 \right\} \tag{8.8.11}$$

and define d_x by (8.8.8) for $s \in S_x$ and arbitrary for $s \in S/S_x$. Then $S_x = R_{d_x}$ and $x_{d_x}(s, a) = x(s, a)$ for $a \in A_s$ and $s \in R_{d_x}$.

Corollary 8.8.7.

a. Let x be a basic feasible solution to the dual LP and suppose that d_x is defined as in Theorem 8.8.6(b). Then, for $s \in S_x$, $d_x(s)$ is deterministic and satisfies

$$d_x(s) = \begin{cases} a & \text{if } x(s, a) > 0 \quad \text{for } s \in S_x \\ \text{arbitrary} & \text{for } s \in S/S_x. \end{cases}$$

b. Suppose that $d(s)$ is deterministic; then $x_d = \pi_d$ is a basic feasible solution to the dual LP.

Corollary 8.8.8. There exists a bounded optimal basic feasible solution x^* to the dual and the policy $(d_{x^*})^\infty$ defined by

$$d_{x^*}(s) = \begin{cases} a & \text{if } x^*(s, a) > 0 \quad \text{for } s \in S_{x^*} \\ \text{arbitrary} & \text{for } s \in S/S_{x^*} \end{cases}$$

is an optimal policy.

In unichain models, the policy iteration algorithm finds a solution (g^*, h^*) of the optimality equation and the set of stationary policies

$$\left\{ \delta^\infty : \delta \in \arg\max_{d \in D} \{ r_\delta + P_\delta h^* \} \right\},$$

while solving the dual linear programming problem does not specify actions on transient states of an optimal policy. Therefore these two approaches may yield different optimal policies. From the perspective of average reward maximization, we may terminate policy iteration as soon as the optimality equation holds on the recurrent set of a maximizing policy and choose arbitrary actions on its complement. Such an approach would be analogous to the LP method. We illustrate this point with Example 8.8.2.

Example 8.8.2 (ctd.). We solve this model by policy iteration with $d_0(s_1) = a_{1,2}$. Solving the evaluation equations

$$\begin{aligned} 10 - g - h(s_1) + h(s_2) &= 0, \\ -1 - g &= 0, \end{aligned}$$

subject to $h(s_2) = 0$ yields $g_1 = -1$, $h_1(s_1) = 11$. This policy has recurrent set s_2 and, since no action choice is available there, the optimality equation holds on s_2. By the comment immediately preceding this example, this implies that from the perspective of the average reward maximization, we may select an arbitrary action on s_1 and still obtain an optimal policy. Using this approach results in the same degree of arbitrariness as using LP. On the other hand, a subsequent pass through the policy iteration algorithm specifies the decision rule with $d_1(s_1) = a_{1,1}$. Refer to the discussion at the end of Sec. 8.6.3 for more on this point.

8.9 STATE ACTION FREQUENCIES, CONSTRAINED MODELS, AND MODELS WITH VARIANCE CRITERIA

In this section we analyze constrained models and models in which the variance of returns impacts on the choice of the optimal policy. Our approach relies on the relationship between the sets of limiting average state action frequencies obtainable under policies of different types. We use the linear programming theory of the preceding section as the basis for analysis.

A large portion of this section focuses on limiting average state action frequencies. The reason for this apparent digression is as follows. In Sec. 8.8, Theorem 8.8.6 established that, in a unichain model, every feasible solution to the dual LP corresponds to some randomized stationary policy and that the dual variable $x(s, a)$ equals the limiting average probability that the system occupies state s and chooses action a under this policy. Therefore if we express constraints in terms of $x(s, a)$, and the constrained model does not necessarily have a *stationary* optimal policy, we cannot use linear programming (or any other finite algorithm) to find an optimal policy. In Sec. 8.9.1 we show, among other results, that in finite-state and action unichain models, given *any* policy, there exists a stationary randomized policy with the same limiting state action frequency. Hence we may use linear programming to find an optimal policy for a constrained model.

Constrained optimization occupies a cornerstone of applied mathematics, yet with the exception of constraints implicitly imposed through the action sets and a brief discussion in Sec. 6.9, we have not as yet considered Markov decision processes with constraints. Constraints arise naturally in all areas of MDP application, for example in the following ways.

- **a.** Queueing control—when maximizing throughput of one class of customers subject to constraints on the throughput for other classes of customers.
- **b.** Inventory control—when minimizing cost subject to a bound on the probability of being unable to fill arriving orders.
- **c.** Economic modeling—when maximizing utility subject to constraints on the probability of bankruptcy.
- **d.** Ecological modeling—when maximizing probability of survival of the current generation subject to constraints on the probability of survival of the subsequent generation.

From an applications perspective, the most important consequence of results in this section is that linear programming may be used to solve constrained average reward finite-state and action unichain models as well as models with variance criteria. *We assume a finite-state and action unichain model except where noted.*

*8.9.1 Limiting Average State Action Frequencies

In this section we define limiting average state-action frequencies and discuss the relationship between those attainable by different policy classes. Results rely on and extend material in Sec. 5.5.

For $\pi \in \Pi^{HR}$ and scalar α_i, $i \in S$ with $\alpha_i \geq 0$ and $\Sigma_{i \in S} \alpha_i = 1$, let

$$x_{\pi,\alpha}^T(s,a) \equiv \frac{1}{T} \sum_{n=1}^T \sum_{i \in S} \alpha_i P^\pi \{ X_n = s, Y_n = a \mid X_1 = i \} \qquad (8.9.1)$$

for $a \in A_s$ and $s \in S$ and $T = 1, 2, \ldots$. We refer to $x_{\pi,\alpha}^T$ as an *average state action frequency* and regard $x_{\pi,\alpha}^T$ as a finite vector with $\Sigma_{s \in S} |A_s|$ components $x_{\pi,\alpha}^T(s,a)$. We interpret α as an initial state distribution.

For each $\pi \in \Pi^{HR}$, let $\Xi^\pi(\alpha)$ denote the set of all limit points (subsequential limits) of $x_{\pi,\alpha}^T$. That is,

$$\Xi^\pi(\alpha) \equiv \left\{ x : x = \lim_{n \to \infty} x_{\pi,\alpha}^{t_n} \text{ for some sequence } \{t_n\} \text{ with } t_n \to \infty \right\}. \quad (8.9.2)$$

We define the limit in (8.9.2) in a componentwise sense. Since, for each T, $\Sigma_{s \in S} \Sigma_{a \in A_s} x_{\pi,\alpha}^T(s,a) = 1$, it follows that $\Sigma_{s \in S} \Sigma_{a \in A_s} x(s,a) = 1$ for all $x \in \Xi^\pi(\alpha)$. We refer to an $x \in \Xi^\pi(\alpha)$ as a *limiting average state action frequency*. In unichain models, the limiting average state action frequencies are independent of α, but in multichain models they may vary with α.

Note that, for the model in Fig. 3.1.1, for any α, $\Xi^\pi(\alpha)$ consists of the single element $x_{\pi,\alpha}(s_1, a_{1,1}) = x_{\pi,\alpha}(s_1, a_{1,2}) = 0$, and $x_{\pi,\alpha}(s_1, a_{2,1}) = 1$ for all policies π.

For $K = $ HR, HD, MR, MD, SR, and SD, and any initial distribution α, let

$$\Xi^K(\alpha) \equiv \left\{ x : x \in \Xi^\pi(\alpha) \text{ for some } \pi \in \Pi^K \right\} = \bigcup_{\pi \in \Pi^K} \Xi^\pi(\alpha).$$

These represent the set of limiting average state action frequencies obtainable by policies of a particular type. For example, $\Xi^{SR}(\alpha)$ denotes the set of limits obtainable from stationary randomized policies. Let $\Pi^1(\alpha)$ denote the set of policies for which $\lim_{T \to \infty} x_{\pi,\alpha}^T \equiv x_{\pi,\alpha}$ exists, that is, for each $\pi \in \Pi^1(\alpha)$, $\lim_{n \to \infty} x_{\pi'}^{t_n,\alpha}$ is identical for every subsequence. Let

$$\Xi^1(\alpha) \equiv \left\{ x_{\pi,\alpha} : \pi \in \Pi^1(\alpha) \right\}.$$

Whenever $\pi \in \Pi^1(\alpha)$, we can interpret quantities of the form

$$\sum_{s \in S} \sum_{a \in A_s} c(s,a) x_{\pi,\alpha}(s,a) \quad \text{or} \quad \sum_{s \in S} \sum_{a \in A_s} f(x_{\pi,\alpha}(s,a))$$

as long-run average rewards or costs associated with policy π and initial distribution α. This adds considerable generality to our models but limits algorithmic choice.

We now characterize the relationship between the sets Ξ^{SD}, Ξ^{SR}, Ξ^{MR}, Ξ^{HR}, and Ξ^1. We begin with a result relating Ξ^{SD}, Ξ^{SR}, and Ξ^1.

Proposition 8.9.1. Suppose S and A_s are finite.

a. Then for $d^\infty \in \Pi^{SR}$ or $d^\infty \in \Pi^{SD}$, $x_{d^\infty,\alpha}(s,a)$ exists and $\Xi^{SD}(\alpha) \subset \Xi^{SR}(\alpha) \subset \Xi^1(\alpha)$.

b. If $d \in D^{MR}$ has a unichain transition probability matrix, then

$$x_{d^{\infty},\alpha}(s,a) = q_{d(s)}(a)\pi_d(s), \tag{8.9.3}$$

where $\pi_d(s)$ denotes the stationary distribution of P_d.

c. In a unichain model, $\Xi^{SD}(\alpha)$, $\Xi^{SR}(\alpha)$, and $\Xi^1(\alpha)$, do not vary with α.

Proof. Since

$$x_{d^{\infty},\alpha}^T(s,a) = \frac{1}{T} \sum_{n=1}^{T} \sum_{j \in S} \alpha_j q_{d(s)}(a) P_d^{n-1}(s\,|\,j), \tag{8.9.4}$$

by results in Appendix A.4,

$$\lim_{T \to \infty} \frac{1}{T} \sum_{n=1}^{T} P_d^{n-1}(s\,|\,j) = p_d^*(s\,|\,j), \tag{8.9.5}$$

so (a) follows.

Since d is unichain, $p_d^*(s\,|\,j) = \pi_d(s)$ for all $j \in S$, (8.9.3) follows by substituting (8.9.5) into (8.9.4). Part (c) follows immediately from (b).   □

For a subset Ξ of Euclidean space, we let $(\Xi)^c$ denote the *closed convex hull* of Ξ. The means that $(\Xi)^c$ contains all convex combinations of elements in Ξ and all limit points of sequences in Ξ. Before proving the main results of this section, we state the following well-known result on separation properties of closed convex sets (cf. Luenberger 1969, p. 133).

Lemma 8.9.2.   Let Ξ be any closed convex subset of R^n, and suppose that $u \notin \Xi$. Then there exists a vector $r \in R^n$ such that

$$\sum_{i=1}^{n} r_i u_i > \sum_{i=1}^{n} r_i x_i$$

for all $x \in \Xi$.

The following theorem relates the sets of limiting state action frequencies. It holds regardless of chain structure, however, our proof requires existence of a deterministic stationary average optimal policy which we have only as yet established for unichain models. In fact, we obtain a stronger result under a unichain assumption (Corollary 8.9.5).

Theorem 8.9.3.   Let S and A_s be finite. Then, for each initial distribution α,

$$\left(\Xi^{SD}\right)^c(\alpha) = \left(\Xi^{SR}\right)^c(\alpha) = \Xi^1(\alpha) = \Xi^{MR}(\alpha) = \Xi^{HR}(\alpha).$$

Proof.   In the proof, we suppress α but assume it is fixed. Clearly $\Xi^{SD} \subset \Xi^{SR} \subset \Xi^{MR} \subset \Xi^{HR}$ and, from Proposition 8.9.1, $\Xi^{SD} \subset \Xi^{SR} \subset \Xi^1 \subset \Xi^{HR}$.

We show that $\Xi^{HR} \subset (\Xi^{SD})^c$. Suppose not, then there exists an $x \in \Xi^{HR}$ for which $x \in \Xi^{\pi'}$ for some $\pi' \in \Pi^{HR}$ and for which $x \notin (\Xi^{SD})^c$. Since $(\Xi^{SD})^c$ is closed and convex, by Lemma 8.9.2 there exists an $r(s, a)$ for which

$$\sum_{s \in S} \sum_{a \in A_s} r(s, a) x(s, a) > \sum_{s \in S} \sum_{a \in A_s} r(s, a) y(s, a) \qquad (8.9.6)$$

for all $y \in (\Xi^{SD})^c$. Consider an MDP with reward $r(s, a)$ satisfying (8.9.6). From Theorem 8.4.5 or Theorem 9.1.8, it follows that there exists an optimal stationary deterministic policy for this model. Consequently there exists a $d \in D^{MD}$ for which

$$g^{d^\infty} = \sum_{s \in S} \sum_{a \in A_s} r(s, a) x_{d^\infty}(s, a)$$

and

$$g^{d^\infty} \geq \sup_{\pi \in \Pi^{HR}} \left\{ \limsup_{T \to \infty} \frac{1}{T} \sum_{n=1}^{T} \sum_{j \in S} \sum_{s \in S} \sum_{a \in A_s} r(s, a) P^\pi \{ X_n = s, Y_n = a \mid X_1 = j \} \alpha_j \right\}.$$

Hence for π' defined above

$$g^{d^\infty} \geq \limsup_{T \to \infty} \frac{1}{T} \sum_{n=1}^{T} \sum_{j \in S} \sum_{s \in S} \sum_{a \in A_s} r(s, a) P^{\pi'} \{ X_n = s, Y_n = a \mid X_1 = j \} \alpha_j$$

$$\geq \sum_{s \in S} \sum_{a \in A_s} r(s, a) x(s, a),$$

contradicting (8.9.6). Thus $\Xi^{HR} \subset (\Xi^{SD})^c \subset (\Xi^{SR})^c$. Since $\Xi^{SR} \subset \Xi^{HR}$, we have that $(\Xi^{SR})^c \subset (\Xi^{HR})^c \subset (\Xi^{SD})^c$. Hence $(\Xi^{SD})^c = (\Xi^{SR})^c$.

We now show that $(\Xi^{SD})^c \subset \Xi^{MR}$. Let $x \in (\Xi^{SD})^c$ and represent Π^{SD} as $\Pi^{SD} = \{\delta_1^\infty, \delta_2^\infty, \ldots, \delta_n^\infty\}$. Then, for some $\{\beta_j\}$, $\beta_j \geq 0$ and $\sum_{j=1}^n \beta_j = 1$,

$$x(s, a) = \sum_{j=1}^{n} \beta_j x_{\delta_j^\infty}(s, a) \qquad (8.9.7)$$

for all $s \in S$ and $a \in A_s$. Let P' denote the probability distribution of the stochastic process $\{X_n, Y_n\}$ generated by an initial mixing which chooses the stochastic process corresponding to stationary policy δ_j^∞ with probability β_j. The proof of Theorem 5.5.1 can be easily modified to show that there exists a $\pi \in \Pi^{MR}$ for which

$$P^\pi \{ X_n = s, Y_n = a \mid X_1 = i \} = P' \{ X_n = s, Y_n = a \mid X_1 = i \}$$

$$= \sum_{j=1}^{n} \beta_j P^{\delta_j^\infty} \{ X_n = s, Y_n = a \mid X_1 = j \} \qquad (8.9.8)$$

for $n \geq 1$. Consequently $x_\pi = x$, so that $(\Xi^{SD})^c \subset \Xi^{MR} \subset \Xi^{HR}$. Therefore $(\Xi^{SD})^c = (\Xi^{SR})^c = \Xi^{MR} = \Xi^{HR}$.

We establish $(\Xi^{SD})^c \subset \Xi^1$. From (8.9.6), (8.9.7) and the result that $\Xi^{SD} \subset \Xi^1$,

$$
x(s, a) = \sum_{j=1}^{n} \beta_j \left\{ \lim_{T \to \infty} \frac{1}{T} \sum_{n=1}^{T} \sum_{i \in S} P^{\delta_j^{\infty}}\{X_n = s, Y_n = a \mid X_1 = i\} \alpha_i \right\}
$$

$$
= \lim_{T \to \infty} \frac{1}{T} \sum_{n=1}^{T} \sum_{i \in S} P^{\pi}\{X_n = s, Y_n = a \mid X_1 = i\} \alpha_i = x_{\pi}(s, a),
$$

so that $x \in \Xi^1$, from which the result follows.   □

Since $\Xi^{MR}(\alpha) = \Xi^{HR}(\alpha)$, this result implies that we may restrict attention to randomized Markov policies when maximizing a concave function or minimizing a convex function of the state action frequencies. Since $(\Xi^{SR})^c(\alpha) = \Xi^{MR}(\alpha)$, we may implement this optimization by searching over the closed convex hull of the state action frequencies of stationary policies. Unfortunately the maximum may occur at a point which *does not* correspond to a stationary policy, so that it is not directly obvious how to recover the maximizing policy. Results below show that, in *unichain* models, maxima occur at limiting state action frequencies of stationary randomized policies.

Another consequence of Theorem 8.9.3 is that in finite-state and action models, every (history-dependent randomized) policy has a unique limiting average state action frequency. Further, for these models, lim inf average optimality, lim sup average optimality, and average optimality are equivalent.

Let X denote the set of solutions of

$$
\sum_{a \in A_j} x(j, a) - \sum_{s \in S} \sum_{a \in A_s} p(j \mid s, a) x(s, a) = 0 \tag{8.9.9}
$$

for all $j \in S$,

$$
\sum_{s \in S} \sum_{a \in A_s} x(s, a) = 1, \tag{8.9.10}
$$

and $x(s, a) \geq 0$ for $s \in S$ and $a \in A_s$. That is,

$$
X \equiv \{x : x \geq 0 \text{ and satisfies (8.9.9) and (8.9.10)}\}.
$$

The linear programming results in Sec. 8.8.2 are based on the relationship between elements of X and randomized stationary policies in unichain models. We summarize them in the following theorem. Let $\text{ext}(X)$ denote the set of extreme points of X. Recall that $x \in \text{ext}(X)$ if x cannot be expressed as a convex combination of other elements in X.

Theorem 8.9.4.   In a finite-state and action unichain model, the following apply.

a. For any initial distribution α, $\Xi^{SR}(\alpha) = X$.
b. X is convex, bounded and closed.
c. $\text{ext}(X) = \Xi^{SD}(\alpha)$ for any α.

Proof. Theorem 8.8.6 and Proposition 8.9.1c establish that $X = \Xi^{SR}(\alpha)$ for all α. Clearly X is convex. Equality (8.9.10) and the non-negativity constraint imply that X is bounded and hence is closed.

Corollary 8.8.7(b) shows that $\Xi^{SD}(\alpha) \subset \text{ext}(X)$. We now establish the reverse equality. Suppose there exists an $x \in \Xi^{SD}(\alpha)$ for which $x \notin \text{ext}(X)$. Then, by Lemma 8.9.2, for all $y \in X$, (8.9.6) holds. Consider the dual LP with objective function $\sum_{s \in S}\sum_{a \in A_s} r(s, a)x'(s, a)$ and variables x'. It has an optimal solution $x^* \in \text{ext}(X)$ and, by Corollary 8.8.8, there exists an optimal deterministic stationary policy $(d^*)^\infty$ with $x_{(d^*)^\infty} = x^*$. This contradicts (8.9.6), from which the result follows.   □

For unichain models, the following result shows that we may restrict attention to limiting state action frequencies of stationary randomized policies when maximizing or minimizing a quantity which can be expressed as a function of limiting average state action frequencies. It is the main result of this section.

Corollary 8.9.5. In a unichain model, for any initial distribution α,

$$\Xi^{SR}(\alpha) = \Xi^1(\alpha) = \Xi^{MR}(\alpha) = \Xi^{HR}(\alpha) = X.$$

Proof. By Theorem 8.9.4, $\Xi^{SR}(\alpha)$ is closed and convex. Hence $(\Xi^{SR})^c(\alpha) = \Xi^{SR}(\alpha)$ and the result follows from Theorem 8.9.3.   □

The example below illustrates these concepts and shows that in a multichain model, $\Xi^{SR}(\alpha)$ may vary with α and may be a proper subset of $\Xi^{MR}(\alpha)$ and X.

Example 8.9.1. (Derman, 1963) Let $S = \{s_1, s_2\}$; $A_{s_1} = \{a_{1,1}, a_{1,2}\}$ and $A_{s_2} = \{a_{2,1}\}$; $r(s_1, a_{1,1}) = 1$, $r(s_1, a_{1,2}) = 0$, and $r(s_2, a_{2,1}) = 0$; and $p(s_1 \mid s_1, a_{1,1}) = 1$, $p(s_2 \mid s_1, a_{1,1}) = 0$, $p(s_1 \mid s_1, a_{1,2}) = 0$, $p(s_2 \mid s_1, a_{1,2}) = 1$, and $p(s_2 \mid s_2, a_{2,1}) = 1$. (Fig. 8.9.1) Represent a limiting average state action frequency by a vector $(x(s_1, a_{1,1})$, $x(s_1, a_{1,2}), x(s_2, a_{2,1}))^T$. Since under stationary policies either s_1 and s_2 are absorbing states, or s_1 is transient and s_2 absorbing, this is a multichain model.

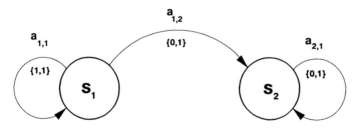

Figure 8.9.1 Symbolic representation of Example 8.9.1.

For α satisfying $\alpha_1 = 1$, $\alpha_2 = 0$, it follows that

$$\Xi^{SD}(\alpha) = \Xi^{SR}(\alpha) = \{(1,0,0)^T, (0,0,1)^T\},$$

and, when $\alpha_1' = 0$, $\alpha_2' = 1$,

$$\Xi^{SD}(\alpha') = \Xi^{SR}(\alpha') = \{(0,0,1)^T\}.$$

Observe that these sets vary with the initial distribution.

Now consider the randomized Markovian policy $\pi = (d_1, d_2, \dots)$ in which

$$q_{d_t(s_1)}(a_{1,1}) = e^{-(1/2)^t} = 1 - q_{d_t(s_1)}(a_{1,2})$$

for $t = 1, 2, \dots$. Under initial distribution $\alpha_1 = 1$, $\alpha_2 = 0$, the system remains in state s_1 and chooses action $a_{1,1}$ with probability

$$e^{-\sum_{t=1}^{\infty}(1/2)^t} = e^{-1} = x_{\pi, \alpha}(s_1, a_{1,1}).$$

Therefore $x_{\pi, \alpha}(s_1, a_{1,2}) = 0$ and $x_{\pi, \alpha}(s_2, a_{2,1}) = 1 - e^{-1}$, so that $x_{\pi, \alpha} \notin \Xi^{SR}(\alpha)$ and $\Xi^{SR}(\alpha)$ is a proper subset of $\Xi^{MR}(\alpha)$.

The following constraints define X:

$$x(s_1, a_{1,1}) + x(s_1, a_{1,2}) - x(s_1, a_{1,1}) = 0,$$
$$x(s_2, a_{2,1}) - x(s_1, a_{1,2}) - x(s_2, a_{2,1}) = 0,$$
$$x(s_1, a_{1,1}) + x(s_1, a_{1,2}) + x(s_2, a_{2,1}) = 1,$$

and $x(s, a) \geq 0$, so that

$$X = \{(p, 0, 1 - p) : 0 \leq p \leq 1\}$$

Hence it follows that $\Xi^{SR}(\alpha)$ is a proper subset of X and that $(\Xi^{SD})^c(\alpha) = (\Xi^{SR})^c(\alpha) = X$. We leave it as an exercise to verify that $\Xi^{MR}(\alpha) = X$.

8.9.2 Models with Constraints

In this section we consider unichain average reward models which include constraints of the form

$$c^{\pi}(s) \equiv \lim_{N \to \infty} \frac{1}{N} E_s^{\pi} \left\{ \sum_{t=1}^{N} c(X_t, Y_t) \right\} \leq \beta \qquad (8.9.11)$$

for $\pi \in \Pi^{HR}$. By arguments in the preceding section, for bounded $c(s, a)$ the limit above exists is identical for all initial state distributions and satisfies

$$c^{\pi} = \sum_{s \in S} \sum_{a \in A_s} c(s, a) x_{\pi}(s, a). \qquad (8.9.12)$$

Note we omit the dependence on s in (8.9.12). Given real-valued functions $c_i(s, a)$ and constants β_i, $i = 1, 2, \ldots, K$ defined the *constrained average reward Markov decision problem* to be that of choosing a $\pi \in \Pi^{HR}$ to

$$\text{Maximize } g^{\pi} \tag{8.9.13}$$

subject to

$$c_i^{\pi} \leq \beta_i, \qquad i = 1, 2, \ldots, K, \tag{8.9.14}$$

where c_i^{π} is defined by (8.9.11) with $c_i(s, a)$ replacing $c(s, a)$.

We solve this problem by relating it to the dual linear programming problem of Sec. 8.8 through (8.9.12). We consider the following *constrained dual linear program*:
Maximize

$$\sum_{s \in S} \sum_{a \in A_s} r(s, a) x(s, a) \tag{8.9.15}$$

subject to

$$\sum_{s \in S} \sum_{a \in A_s} c_i(s, a) x(s, a) \leq \beta_i, \qquad i = 1, 2, \ldots, K, \tag{8.9.16}$$

(8.9.9), (8.9.10), and $x(s, a) \geq 0$. The expression *constrained* LP may seem redundant but we use it to distinguish this model from that without (8.9.16).

Before analyzing these models, we provide some examples of constraints that may be represented by (8.9.16).

Example 8.9.2. Consider the inventory model with backlogging from Sec. 3.5. In it, the state represents the inventory on hand at the start of a month; negative states represent backlogged orders. Suppose that the manager seeks a policy which maximizes the long-run average profit, subject to the requirement that the average probability that the product is out of stock is at most β. This constraint may be expressed in form (8.9.16) as

$$\sum_{s \leq 0} \sum_{a \in A_s} x(s, a) \leq \beta.$$

Example 8.9.3. We consider a generalization of the queueing admission control model from Sec. 3.7.1 in the context of computer system load control. In a computer system, arriving jobs enter a potential service queue and are admitted for processing by a load controller. In systems with both batch and interactive jobs, interactive jobs will take precedence over batch jobs. Let s_1 denote the number of interactive jobs, and s_2 the number of batch jobs in the system. The system designer wishes to determine an operating policy for the controller which minimizes long-run average expected waiting time for interactive jobs subject to constraints on the average waiting time for batch jobs. Without providing details of the model dynamics and actions, we

express the constrained optimization problem as
  Minimize

$$\sum_{(s_1, s_2)\in S}\sum_{a\in A_{s_1, s_2}} s_1 x(s_1, s_2, a) \tag{8.9.17}$$

subject to

$$\sum_{(s_1, s_2)\in S}\sum_{a\in A_{s_1, s_2}} s_2 x(s_1, s_2, a) \le \beta, \tag{8.9.18}$$

where the quantities in (8.9.17) and (8.9.18) denote the respective long-run average queue lengths for the two job classes and are proportional to the average waiting times.

Example 8.9.4. Consider the model in Fig. 3.1.1. In it, state s_1 is transient under every policy so that, for all $\pi \in \Pi^{HR}$, Ξ^π contains the single vector $(0, 0, 1)$. Consequently no meaningful constraints may be expressed in terms of the limiting average state action frequencies.

The following theorem summarizes existence results for the constrained unichain average reward MDP.

Theorem 8.9.6.

a. Suppose there exists a solution x^* to the constrained linear programming problem. Then there exists an optimal policy $(d^*)^\infty \in \Pi^{SR}$ for the constrained Markov decision problem, where d^* satisfies

$$q_{d^*(s)}(a) = \begin{cases} x^*(s, a) \Big/ \displaystyle\sum_{a'\in A_s} x^*(s, a') & \text{if } \displaystyle\sum_{a'\in A_s} x^*(s, a') > 0 \\ \text{arbitrary} & \text{otherwise.} \end{cases} \tag{8.9.19}$$

b. Suppose there exists a $\pi^* \in \Pi^{HR}$ which is an optimal solution to the constrained MDP. Then there exists a $(d^*)^\infty \in \Pi^{SR}$ which is also optimal and $x_{(d^*)^\infty}$ is an optimal solution to the constrained linear program.

Proof. As in Sec. 8.9.1, X denotes the set of feasible solutions to the unichain average reward MDP without constraints (8.9.16). Since $x^* \in X$ and by Corollary 8.9.5, in a unichain model $X = \Xi^{SR}(\alpha)$ for any α, $x^* = x_{d^\infty}$ for some $d^\infty \in \Pi^{SR}$. Since x_{d^∞} satisfies (8.9.16), $c_i^{d^\infty} \le \beta_i$ for $i = 1, 2, \ldots, K$, so it follows that d^∞ is an optimal solution for the constrained MDP. Theorem 8.8.6(b) guarantees that $d = d^*$, so part (a) follows.

To establish part (b), we proceed as follows. By Corollary 8.9.5, $\Xi^{HR}(\alpha) = \Xi^{SR}(\alpha)$ for any α, so that there exists a $(d^*)^\infty \in \Pi^{SR}$ for which $x_{(d^*)^\infty} = x_{\pi^*}$. Consequently, by Theorem 8.8.6(b) and (8.9.12), $x_{(d^*)^\infty}$ is an optimal solution to the constrained LP. □

Note that the above theorem applies to models in which constraints (8.9.14) or (8.9.16) are satisfied. When they are not, both the constrained MDP and constrained LP are infeasible. In a recurrent model we have the following further result.

Corollary 8.9.7. Suppose in a recurrent model there exists a solution x^* to the constrained LP. Then the policy $(d^*)^\infty$ defined by

$$q_{d^*(s)}(a) = x^*(s, a) \Big/ \sum_{a' \in A_s} x^*(s, a')$$

for all $s \in S$ and $a \in A_s$, is optimal and chooses actions randomly in at most K states. Further, if $K = 1$, x^* is an optimal basic feasible solution and (8.9.16) holds with equality, then there exists an optimal stationary policy which chooses actions deterministically in all but one state, and in that state it randomizes between at most two actions.

Proof. From Theorem 8.8.2(b), $\sum_{a \in A_s} x^*(s, a) > 0$ for all $s \in S$, so that $q_{d^*(s)}$ is well defined. Since at most $S + K$ of (8.9.9), (8.9.10), and (8.9.16) hold with equality, $x^*(s, a) > 0$ in at most $|S| + K$ components since x^* is basic. Since $\sum_{a \in A_s} x^*(s, a) > 0$ for all $s \in S$, $x^*(s, a_s) > 0$ for at least one $a_s \in A_s$ for each $s \in S$ from which the results follows.   □

We illustrate these results by solving a recurrent constrained model.

Example 8.9.5. We consider a constrained version of the model in Example 8.8.1, in which the limiting occupancy probability for state s_2 is constrained to be no more than 0.4. This translates to the constraint

$$x(s_2, a_{2,1}) + x(s_2, a_{2,2}) + x(s_2, a_{2,3}) \le 0.4. \tag{8.9.20}$$

Solving the constrained model yields the optimal solution $x^*(s_1, a_{1,1}) = 0.6$, $x^*(s_2, a_{2,1}) = 0.2$, $x^*(s_2, a_{2,2}) = 0.2$, and $x^*(s_2, a_{2,3}) = 0$, so that by Corollary 8.9.7 the optimal stationary policy of $(d^*)^\infty$ uses action $a_{1,1}$ in s_1 and chooses action $a_{2,1}$ and $a_{2,2}$ with probability 0.5 each in s_2; that is,

$$q_{d^*(s_2)}(a_{2,1}) = q_{d^*(s_2)}(a_{2,2}) = 0.5.$$

Observe that the optimal policy randomizes between two actions since the single constraint is binding. Further, $g^{(d^*)^\infty} = 2.6$ and the dual variables equal 5 and 1 for (8.9.9) and (8.9.10) and 4 for constraint (8.9.20).

Recall that for the unconstrained model in Example 8.8.1, the optimal gain equalled 2.714, that the optimal policy chose action $a_{2,2}$ in s_2 with certainty, and that $x^*(s_2, a_{2,2}) = 0.429$ in violation of (8.9.20).

In economic applications, a decision maker may find randomization unappealing so that instead he may prefer to use a nonrandomized suboptimal policy or a policy which violates some constraints. In real time applications such as computer system load control or communications control, randomization may be incorporated in the implementation of the controller.

In Sec. 8.8 we derived the linear programming model as the dual of a primal LP generated from (8.8.1). We leave it as an exercise to derive and interpret the primal model which corresponds to the constrained linear program.

8.9.3 Variance Criteria

In many areas of application, a decision maker may wish to incorporate his attitude toward risk or variability when choosing a policy. One measure of risk is the variance of the rewards generated by a policy. Frequently one considers tradeoffs between return and risk when making decisions, and willingly foregoes an optimal level of return to reduce risk. Examples of this include a dynamic investment model in which the investor may accept a lower than optimal rate of return to achieve reduced variability in return, and a queueing control model, in which the controller might prefer a policy which results in greater but less variable expected waiting times. These mean-variance tradeoffs may be analyzed in the Markov decision process model using the following approach.

Assume a unichain model so that limiting state action frequencies are independent of the initial state distribution. For $\pi \in \Pi^{HR}$, define the *steady-state variance* by

$$\sigma_\pi^2 \equiv \sum_{s \in S} \sum_{a \in A_s} [r(s, a) - g^\pi]^2 x_\pi(s, a). \tag{8.9.21}$$

Since $\Xi^{HR} = \Xi^1$ (Theorem 8.9.3), $x_\pi(s, a)$ exists and is unique so that σ_π^2 is well-defined. From the definition of x_π, it follows that

$$\sigma_\pi^2 = \lim_{T \to \infty} \frac{1}{T} \sum_{t=1}^{T} E^\pi \{ (r(X_t, Y_t) - g^\pi)^2 \},$$

so σ_π^2 may be interpreted as the limiting average variance.

The following simple example illustrates this concept and motivates subsequent discussion.

Example 8.9.6. Let $S = \{s_1, s_2\}$; $A_{s_1} = \{a_{1,1}, a_{1,2}\}$ and $A_{s_2} = \{a_{2,1}\}$; $r(s_1, a_{1,1}) = 3$, $r(s_1, a_{1,2}) = 0$, and $r(s_2, a_{2,1}) = 8$; and $p(s_1 | s_1, a_{1,1}) = 0.9$, $p(s_2 | s_1, a_{1,1}) = 0.1$, $p(s_1 | s_1, a_{1,2}) = 0$, $p(s_2 | s_1, a_{1,2}) = 1$, and $p(s_1 | s_2, a_{2,1}) = 1$. (Fig. 8.9.2) Actions are chosen in s_1 only. Let $\delta(s_1) = a_{1,1}$ and $\gamma(s_1) = a_{1,2}$. The limiting average state action frequencies, gains, and steady-state variances for the two stationary policies are given by

π	$x_\pi(s_1, a_{1,1})$	$x_\pi(s_1, a_{1,2})$	$x_\pi(s_2, a_{2,1})$	g^π	σ_π^2
δ^∞	0.909	0	0.091	3.45	2.06
γ^∞	0	0.5	0.5	4	16

Observe that γ^∞ is gain optimal but has a considerably larger variance than δ^∞ so that a risk adverse decision maker may prefer δ^∞ to γ^∞.

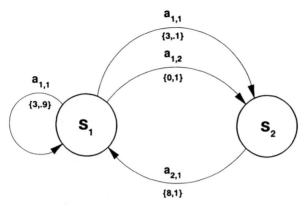

Figure 8.9.2 Symbolic representation of Example 8.9.6.

We can account for mean-variance tradeoffs in at least three ways.

a. We can select a policy which minimizes variance subject to a constraint that its average reward be at least some level.

b. We can select a policy which maximizes average reward subject to a constraint that its variance not exceed some level.

c. We can incorporate the variance as a penalty in the objective function and choose a policy to maximize the variance penalized reward.

In applications, a decision maker may prefer the first approach because the average reward is a tangible and directly interpretable quantity. We justify the third approach by noting that by varying the penalty we may obtain efficient or Pareto optimal policies. In this context we say that a policy is *efficient* or *Pareto optimal* if for a given level of average reward it has minimum variance, and for a given level of variance it has maximum average reward.

To apply the approach (a), we seek a policy $\pi \in \Pi^{HR}$ to

$$\text{Minimize } \sigma_\pi^2 \qquad (8.9.22)$$

subject to

$$g^\pi \geq \beta; \qquad (8.9.23)$$

To apply the second approach we seek a policy $\pi \in \Pi^{HR}$ to

$$\text{Maximize } g^\pi$$

subject to

$$\sigma_\pi^2 \leq \beta;$$

and to apply the third approach we seek a policy $\pi \in \Pi^{HR}$ to

$$\text{Maximize } \left[g^\pi - \gamma \sigma_\pi^2 \right] \tag{8.9.24}$$

for some $\gamma \geq 0$. We refer to the problem defined through (8.9.24) as a *variance penalized MDP*.

To implement any of these three approaches, express σ_π^2 as

$$\sigma_\pi^2 = \sum_{s \in S} \sum_{a \in A_s} r(s, a)^2 x_\pi(s, a) - \left(\sum_{s \in S} \sum_{a \in A_s} r(s, a) x_\pi(s, a) \right)^2.$$

The theorem below relates optimal policies for problem (8.9.22) subject to (8.9.23) to solutions of the following quadratic program (QP1).

Minimize

$$\sum_{s \in S} \sum_{a \in A_s} r(s, a)^2 x(s, a) - \left(\sum_{s \in S} \sum_{a \in A_s} r(s, a) x(s, a) \right)^2 \tag{8.9.25}$$

subject to

$$\sum_{s \in S} \sum_{a \in A_s} r(s, a) x(s, a) \geq \beta \tag{8.9.26}$$

and (8.9.9), (8.9.10), and $x(s, a) \geq 0$.

Note that if there is a feasible solution to QP1 there is an optimal solution since the objective function is a continuous function of x and the set of feasible solutions is a compact set. The proof of Theorem 8.9.6 also establishes the following theorem.

Theorem 8.9.8.

a. Suppose there exists an optimal solution x^* to QP1 and $(d^*)^\infty \in \Pi^{SR}$ is defined by (8.9.19). Then $(d^*)^\infty$ is an optimal solution to (8.9.22) subject to (8.9.23).

b. Suppose there exists a $\pi^* \in \Pi^{HR}$ which minimizes (8.9.22) subject to (8.9.23). Then there exists a $(d^*)^\infty \in \Pi^{SR}$ which minimizes (8.9.22) subject to (8.9.23) and $x_{(d^*)^\infty}$ is an optimal solution to QP1.

Corollary 8.9.9.

a. There exists an optimal policy for (8.9.22) subject to (8.9.23) which deterministically chooses actions in all but one state, and in that state it randomizes between at most two actions.

b. If (8.9.26) holds with strict equality, there exists an optimal deterministic stationary policy.

Proof. The proof of Corollary 8.9.7 establishes (a). Since the objective function is a concave function of x, the minimum occurs at an extreme point of the feasible region. When (8.9.26) holds with strict equality, (8.9.9), (8.9.10), and $x \geq 0$ define the feasible region, so by Theorem 8.9.4(c) there exists a $d^\infty \in \Pi^{SD}$ which achieves this minimum.

The variance penalized criterion may be maximized by solving QP2 below.

Maximize

$$\sum_{s \in S} \sum_{a \in A_s} \left[r(s, a) - \gamma r(s, a)^2 \right] x(s, a) + \gamma \left(\sum_{s \in S} \sum_{a \in A_s} r(s, a) x(s, a) \right)^2$$

(8.9.27)

subject to (8.9.9), (8.9.10), and $x(s, a) \geq 0$.

The following theorem relates solutions of QP2 to the variance penalized MDP. After noting that the function in (8.9.27) is convex, its proof follows by identical methods to results above.

Theorem 8.9.10.

a. Suppose there exists an optimal solution to QP2. Then there exists an optimal basic feasible solution x^* and a $(d^*)^\infty \in \Pi^{SD}$ satisfying (8.9.19), which is an optimal solution to the variance penalized MDP.

b. Suppose there exists a $\pi^* \in \Pi^{HR}$ which is optimal for the variance penalized MDP. Then there exists an optimal $(d^*)^\infty \in \Pi^{SD}$ for which $x_{(d^*)^\infty}$ is the optimal solution to QP2.

The following corollary establishes efficiency properties of optimal solution to the variance penalized MDP.

Corollary 8.9.11. Let π^* be an optimal solution of the variance penalized MDP with fixed $\gamma > 0$. Then there exists no $\pi \in \Pi^{HR}$ for which

a. $g^\pi \geq g^{\pi^*}$ and
b. $\sigma_\pi^2 \leq \sigma_{\pi^*}^2$

with strict equality in (a) or (b).

Proof. Suppose π satisfies (a) and (b) with strict equality in at least one. Then $g^\pi - \gamma \sigma_\pi^2 > g^{\pi^*} - \gamma \sigma_{\pi^*}^2$, which contradicts the optimality of π^*. □

Note that since QP2 identifies only stationary deterministic solutions to the variance penalized problem, it need not determine all efficient or Pareto optimal policies.

8.10   COUNTABLE-STATE MODELS

In this section, we analyze countable-state models with average reward criterion. These models are particularly relevant to queueing, inventory, and replacement systems, and have been the subject of a considerable amount of recent research. They require rather subtle analyses and rely on properties of the optimal value function for discounted models. The Laurent series expansion approach which we used to establish the existence of solutions to the optimality equation in finite-state models requires a considerably deeper theory than we present here. We direct the interested reader to references in the Bibliographic Remarks section of this chapter for more on that approach. Instead we use the following "differential discounted reward" approach. *We assume finite A_s, and with the exception of some examples, that $S = \{0, 1, 2, \ldots\}$.*

We begin with the following heuristic derivation of the optimality equations which is a bit different than that of Sec. 8.4. Since

$$v_\lambda^*(s) = \max_{a \in A_s} \left\{ r(s, a) + \lambda \sum_{j \in S} p(j \mid s, a) v_\lambda^*(j) \right\},$$

it follows that, for all s,

$$v_\lambda^*(s) - v_\lambda^*(0)$$
$$= \max_{a \in A_s} \left\{ r(s, a) - (1 - \lambda) v_\lambda^*(0) + \lambda \sum_{j \in S} p(j \mid s, a)[v_\lambda^*(j) - v_\lambda^*(0)] \right\}.$$

Letting λ approach 1, and defining g and h by

$$g \equiv \lim_{\lambda \uparrow 1} (1 - \lambda) v_\lambda^*(0) \qquad\qquad (8.10.1)$$

and

$$h(s) \equiv \lim_{\lambda \uparrow 1} [v_\lambda^*(s) - v_\lambda^*(0)] \qquad\qquad (8.10.2)$$

provided these limits exists, we obtain the average reward optimality equation

$$h(s) = \max_{a \in A_s} \left\{ r(s, a) - g + \sum_{j \in S} p(j \mid s, a) h(j) \right\}. \qquad\qquad (8.10.3)$$

The above argument glosses over many technicalities that will be discussed in this section. We will provide assumptions which ensure that the limits in (8.10.1) and (8.10.2) exist, that we can pass to the limit to obtain (8.10.3), and that g is finite. While our assumptions will be expressed in terms of

$$h_\lambda(s) \equiv v_\lambda^*(s) - v_\lambda^*(0), \qquad\qquad (8.10.4)$$

they will impose recurrence properties on Markov chains corresponding to stationary policies. Note that we have distinguished state 0. In applications, other reference states may be used.

We do not require that $r(s, a)$ be finite, but we will assume that for each s that it is bounded above (and not equal to $-\infty$ for any state-action pair). This will allow applications to queueing, inventory, and replacement models in which increased costs are associated with higher states, but not economic models with unbounded utility. We believe the approach herein may be modified to include such models.

Issues to be discussed in this section include the following.

a. Under what conditions do stationary average optimal policies exist?

b. Under what conditions does the optimality equation have a solution?

c. Under what conditions does the optimality equation have a *bounded* solution?

d. If the optimality equation has a bounded solution, what properties does it impose on the model?

Note that in countable-state models we distinguish the three average optimality criteria of Sec. 8.1.2.

8.10.1 Counterexamples

Even in very simple countable-state models, average optimal policies need not exist and, when they do, they need not be stationary. The following two examples illustrate these points.

Example 8.10.1. Let $S = \{1, 1', 2, 2', \ldots\}$, $A_s = \{a_{s,1}, a_{s,2}\}$ for $s = 1, 2, \ldots$, and $A_{s'} = \{a_{s',1}\}$ for $s' = 1', 2', \ldots$; $r(s, a_{s,1}) = r(s, a_{s,2}) = 0$ and $r(s', a_{s',1}) = 1 - 1/s'$; and $p(s + 1 \mid s, a_{s,1}) = 1$, $p(s' \mid s, a_{s,2}) = 1$, and $p(s' \mid s', a_{s',1}) = 1$, and all other transition probabilities equal zero (Fig. 8.10.1). Observe that $g_-^* \leq g_+^* \leq 1$ and that for each $d \in D^{MD}$, $g_+^{d^\infty} = g^{d^\infty} = g_-^{d^\infty}$ so that $g_-^* \geq \sup_{d \in D^{MD}} g^{d^\infty} = 1$. Therefore $g_-^* = g_+^* = g^* = 1$, but there exists no policy $\pi \in \Pi^{HR}$ for which $g^\pi = g^*$. Note, however, that for any $\varepsilon > 0$ there exists a deterministic stationary ε-optimal policy:

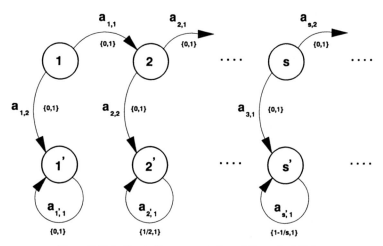

Figure 8.10.1 Symbolic representation of Example 8.10.1.

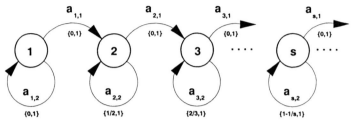

Figure 8.10.2   Symbolic representation of Example 8.10.2.

The example below shows that even when there exists a lim sup average optimal policy, there need not exist a deterministic stationary lim sup average optimal policy.

Example 8.10.2. Let $S = \{1, 2, 3, \ldots\}$ and $A_s = \{a_{s,1}, a_{s,2}\}$ for $s = 1, 2, \ldots$; $r(s, a_{s,1}) = 0$ and $r(s, a_{s,2}) = 1 - 1/s$; and $p(s + 1 \mid s, a_{s,1}) = 1$ and $p(s \mid s, a_{s,2}) = 1$ with all other transition probabilities equal to 0 (Fig. 8.10.2). As in the previous example $g^* = 1$, but in this case the history-dependent policy π^* which, for each $s \in S$, uses action $a_{s,2}$ s times in state s, and then uses action $a_{s,1}$, yields a reward stream (starting in state 1) of $(0, 0, \frac{1}{2}, \frac{1}{2}, 0, \frac{2}{3}, \frac{2}{3}, \frac{2}{3}, 0, \frac{3}{4}, \ldots)$, so that $g_-^{\pi^*} = 0$ and $g_+^{\pi^*} = 1$. Hence π^* is lim sup average optimal. Note that, under stationary deterministic policies, the only possible gains are 0 or $1 - 1/s$, so that there is no deterministic stationary lim sup average optimal policy.

Ross (1983) also provides an example in which there exists no stationary ε-optimal policy (Problem 8.35). In each of these multichain examples, the optimal gain results from indefinitely delaying absorption. We usually do not encounter this behavior in applications, and the conditions in the next section allow us to exclude it.

*8.10.2   Existence Results

This section provides conditions which guarantee existence of lim inf average optimal policies, and average optimal policies in countable-state models. In a departure from previous analyses, we establish the existence of a stationary optimal policy without first showing that the optimality equations has a solution. In fact, examples are available in which optimal stationary policies exist and the optimality equation does not have a solution.

Before proceeding, we state an important result for countable-state discounted models part of which we proved before and part of which follows from material in Sec. 7.3.

Theorem 8.10.1. Let S be countable and A_s finite for each $s \in S$, and suppose that either

   i. $-\infty < r(s, a) \le R < \infty$ for all $a \in A_s$ and $s \in S$, and $v_\lambda^*(s) > -\infty$ for all $s \in S$; or

   ii. Assumptions 6.10.1 and 6.10.2 hold.

Then, for $0 \le \lambda < 1$, v_λ^* satisfies the optimality equation

$$v(s) = \max_{a \in A_s} \left\{ r(s, a) + \lambda \sum_{j \in S} p(j \mid s, a) v(j) \right\}, \tag{8.10.5}$$

and, if for all $s \in S$,

$$d(s) \in \arg\max_{a \in A_s} \left\{ r(s, a) + \lambda \sum_{j \in S} p(j \mid s, a) v_\lambda^*(j) \right\},$$

then d^∞ is optimal. Further, if (i) holds, v_λ^* is the maximal solution of the optimality equation, and, if (ii) holds, then v_λ^* is the unique solution of (8.10.5) in V_w.

Proof. When (ii) holds, the result is a restatement of Theorem 6.10.4. Under (i), we convert this to a negative model using the following transformation suggested in Sec. 5.3. Let $S' = S \cup \{\Delta\}$, with $A_\Delta = \{a'\}$, $r'(\Delta, a') = 0$, and $r'(s, a) = r(s, a) - R$ for $s \in S$,

$$p'(j \mid s, a) = \begin{cases} \lambda p(j \mid s, a) & j \in S, \quad s \in S, \quad \text{and} \quad a \in A_s, \\ 1 - \lambda & j = \Delta, \quad s \in S, \quad \text{and} \quad a \in A_s, \end{cases}$$

and $p'(\Delta \mid \Delta, a') = 1$. As a result of the assumptions in (i), the transformed model satisfies the assumptions of the negative model in Sec. 7.3, so that the result follows from Theorems 7.3.3(a) and 7.3.5.   □

We analyze the model under the following assumptions.

Assumption 8.10.1.   For each $s \in S$, $-\infty < r(s, a) \le R < \infty$.

Assumption 8.10.2.   For each $s \in S$ and $0 \le \lambda < 1$, $v_\lambda^*(s) > -\infty$.

Assumption 8.10.3.   There exists a $K < \infty$ such that, for each $s \in S$, $h_\lambda(s) \le K$ for $0 \le \lambda < 1$.

Assumption 8.10.4.   There exists a non-negative function $M(s)$ such that

a. $M(s) < \infty$;
b. for each $s \in S$, $h_\lambda(s) \ge -M(s)$ for all λ, $0 \le \lambda < 1$; and
c. there exists an $a_0 \in A_0$ for which

$$\sum_{j \in S} p(j \mid 0, a_0) M(j) < \infty. \tag{8.10.6}$$

Assumption 8.10.4′. There exists a non-negative function $M(s)$ such that

a. $M(s) < \infty$;
b. for each $s \in S$, $h_\lambda(s) \geq -M(s)$ for all λ, $0 \leq \lambda < 1$; and
c. for each $s \in S$ and $a \in A_s$,

$$\sum_{j \in S} p(j \mid s, a)M(j) < \infty.$$

Under Assumptions 8.10.1–8.10.4, we show that there exists a lim inf average optimal stationary policy and, that if we replace Assumption 8.10.4 by Assumption 8.10.4′, that the optimality equation has a solution.

Clearly Assumption 8.10.1 holds in models with bounded rewards. In applications in inventory, queueing, and replacement, the cost of occupying states with larger values of s increases, so that the reward decreases with s and Assumption 8.10.1 applies. Assumption 8.10.2 is quite mild and holds in most realistic models. For example, if the assumptions in Sec. 6.10.1 hold, $|v_\lambda^*(s)| \leq Kw(s) < \infty$ for all $s \in S$ where K is defined in (6.10.8). Assumption 8.10.3 holds when $v_\lambda^*(s)$ does not increase with s, which is the case when rewards decrease (costs increase) as the state index increases. Instead of defining $h(s)$ in terms of the differences of $v_\lambda^*(s)$ and $v_\lambda^*(0)$ we may instead choose another convenient reference state, s_0, for which (8.10.6) holds with 0 replaced by s_0. Assumptions 8.10.4 and 8.10.4′ impose structure on the rewards and the Markov chains corresponding to stationary policies.

Theorem 8.10.7 below contains the main results of Sec. 8.10. Its proof requires some well-known technical results, all but one of which we state without proof. Proofs of Lemmas 8.10.2–8.10.5 may be found, for example, in Royden (1963). Note that, by a compact set, we mean a compact subset of a complete separable metric space. In this section we apply the following result with X_i finite, or a closed subset of the real line.

Lemma 8.10.2. (Tychonoff's Theorem). Let $\{X_i\}$ for $i = 0, 1, 2, \ldots$, denote a collection of compact sets. Then $X \equiv \times_{i=0}^\infty X_i$ is compact.

Lemma 8.10.3. Let $\{x_n\}$ denote a sequence of elements in a compact set X. Then $\{x_n\}$ contains a convergent subsequence.

Lemma 8.10.4. (Fatou's Lemma). Let $\{q(i)\}$ denote a probability distribution, $\{f_n(i)\}$ a sequence of non-negative real-valued functions on $\{0, 1, 2, \ldots\}$, and let $f(i) = \lim\inf_{n \to \infty} f_n(i)$. Then

$$\sum_{i=0}^\infty q(i)f(i) \leq \lim\inf_{n \to \infty} \sum_{i=0}^\infty q(i)f_n(i).$$

Lemma 8.10.5. (Lebesgue's Dominated Convergence Theorem). Let $\{q(i)\}$ denote a probability distribution, and let $\{f_n(i)\}$ denote a sequence of real-valued functions on $\{0, 1, 2, \ldots\}$ for which

$$\lim_{n \to \infty} f_n(i) = f(i)$$

for $i = 0, 1, 2, \ldots$, $|f_n(i)| \le g(i)$, and $\sum_{i=0}^{\infty} q(i)g(i) < \infty$. Then

$$\sum_{i=0}^{\infty} q(i)f(i) = \lim_{n \to \infty} \sum_{i=0}^{\infty} q(i)f_n(i).$$

We state the following result in a form suitable for latter application. Many variants are available; Powell and Shah (1972) contains a nice discussion of this and related results.

Lemma 8.10.6. Let $\{a_n\}$ denote a sequence of nonpositive numbers, and let $s_n = \sum_{j=1}^{n} a_j$. Then

$$\liminf_{n \to \infty} \frac{1}{n} s_n \le \liminf_{x \uparrow 1} (1 - x) \sum_{j=1}^{\infty} a_j x^{j-1}.$$

Proof. Suppose the radius of convergence of $\sum_{j=1}^{\infty} a_j x^{j-1}$ is greater than or equal to 1, in which case $\sum_{j=1}^{\infty} a_j x^{j-1}$ converges for $|x| < 1$. Let

$$f(x) = (1 - x) \sum_{j=1}^{\infty} a_j x^{j-1}.$$

Since $\sum_{j=1}^{\infty} x^{j-1}$ converges for $|x| < 1$,

$$(1 - x)^{-2} f(x) = (1 - x)^{-1} \sum_{j=1}^{\infty} a_j x^{j-1}$$

$$= \sum_{j=1}^{\infty} x^{j-1} \sum_{j=1}^{\infty} a_j x^{j-1}$$

$$= \sum_{j=1}^{\infty} s_j x^{j-1},$$

where the last equality follows by a standard formula for the product of two series (cf. Bartle, 1964, p. 384). Further $\sum_{j=1}^{\infty} s_j x^{j-1}$ converges for $|x| < 1$.

Therefore

$$f(x) = (1 - x)^2 \sum_{j=1}^{M-1} s_j x^{j-1} + (1 - x)^2 \sum_{j=M}^{\infty} s_j x^{j-1}$$

$$\ge (1 - x)^2 \sum_{j=1}^{M-1} s_j x^{j-1} + (1 - x)^2 \sum_{j=0}^{\infty} \left\{ \inf_{j \ge M} s_j / j \right\} j x^{j-1}$$

$$= (1 - x)^2 \sum_{j=1}^{M-1} s_j x^{j-1} + \inf_{j \ge M} s_j / j, \qquad (8.10.7)$$

where the inequality follows from the assumption $a_j \leq 0$ and from noting that $\sum_{j=1}^{\infty} jx^{j-1} = (1-x)^{-2}$ for $|x| < 1$. Since for $x \geq 0$

$$0 \geq (1-x)^2 \sum_{j=1}^{M-1} s_j x^{j-1} \geq (1-x)^2 (M-1)^2 \left\{ \min_{j \leq M-1} a_j \right\}$$

for each M, the first quantity in (8.10.7) can be made arbitrarily close to 0 by choosing x sufficiently close to 1. For M sufficiently large, $\inf_{j \geq M} s_j/j$ is arbitrarily close to $\liminf_{n \to \infty} s_n/n$, so it follows from (8.10.7) that

$$\liminf_{x \uparrow 1} f(x) \geq \liminf_{n \to \infty} s_n/n.$$

If the radius of convergence of $\sum_{j=1}^{\infty} a_j x^{j-1}$ is less than 1, $\{a_j\}$ diverges so that s_n/n diverges. Hence the result follows. $\square$

The following theorem contains our main result for countable-state models with average reward criterion. In it we refer to (8.10.9) as the *average reward optimality inequality*.

Theorem 8.10.7. Let $S = \{0, 1, 2, \ldots\}$, A_s be finite, and suppose Assumptions 8.10.1–8.10.4 hold. Then:

a. There exists a constant g satisfying

$$g = \lim_{\lambda \uparrow 1} (1 - \lambda) v_\lambda^*(s) \qquad (8.10.8)$$

for all $s \in S$, a vector h satisfying $-M(s) \leq h(s) \leq K$, and $h(0) = 0$ and a $d^* \in D$, for which

$$h(s) + g \leq r_{d^*}(s) + \sum_{j \in S} p_{d^*}(j \mid s) h(j)$$

$$\leq \max_{a \in A_s} \left\{ r(s, a) + \sum_{j \in S} p(j \mid s, a) h(j) \right\}. \qquad (8.10.9)$$

b. The policy $(d^*)^\infty$ is lim inf average optimal, and, if

$$d' \in \arg\max_{a \in A_s} \left\{ r(s, a) + \sum_{j \in S} p(j \mid s, a) h(j) \right\}, \qquad (8.10.10)$$

then $(d')^\infty$ is lim inf average optimal.

c. If Assumptions 8.10.1–8.10.3 and Assumption 8.10.4′ hold, then (8.10.9) holds with equality.

Proof. *In this proof we choose subsequences of subsequences. We denote all of them by the index n to simplify notation.*

Let $\{\lambda_n\}$ be a sequence of discount rates which increase to 1. Under Assumption 8.10.1, it follows from Theorem 8.10.1 that for each λ_n there exists a stationary optimal policy $(d_n)^\infty$ for the corresponding discounted model. Since each A_s is finite and hence compact, it follows from Lemma 8.10.2 that $A \equiv \times_{s \in S} A_s$ is compact. Since $d_n \in A$ for each n, there exists a subsequence $\{d_n\}$ for which $\lim_{n \to \infty} d_n(s)$ exists and equals $d^*(s) \in A_s$. Because each A_s is finite, for each $s \in S$ there exists an N_s with the property that, for all $n \geq N_s$, $d_n(s) = d^*(s)$.

By Assumption 8.10.1 and Theorem 8.10.1,

$$R \geq (1 - \lambda_n)v_{\lambda_n}^*(0) = r_{d_n}(0) + \lambda_n \sum_{j \in S} p_{d_n}(j \mid 0)v_{\lambda_n}^*(j) - \lambda_n v_{\lambda_n}^*(0)$$

$$= r_{d_n}(0) + \lambda_n \sum_{j \in S} p_{d_n}(j \mid 0)h_{\lambda_n}(j)$$

$$\geq r(0, a_0) - \sum_{j \in S} p(j \mid 0, a_0)M(j) \equiv H,$$

where the last inequality follows from Assumption 8.10.4 and the optimality of d_n^∞. Since Assumption 8.10.4(c) implies that $\sum_{j \in S} p(j \mid 0, a_0)M(j) < \infty$, $[H, R]$ is a compact subset of the real line and there exists a further subsequence of $\{\lambda_n\}$ for which

$$\lim_{n \to \infty} (1 - \lambda_n)v_{\lambda_n}^*(0) = g.$$

Since

$$(1 - \lambda_n)\big|v_{\lambda_n}^*(s) - v_{\lambda_n}^*(0)\big| = (1 - \lambda_n)\big|h_{\lambda_n}(s)\big| \leq (1 - \lambda_n)\max\{M(s), K\},$$

it follows that

$$\lim_{n \to \infty} (1 - \lambda_n)v_{\lambda_n}^*(s) = g.$$

Define $h(s)$ by

$$h(s) \equiv \limsup_{n \to \infty} h_{\lambda_n}(s). \tag{8.10.11}$$

For n sufficiently large, $d_n(s) = d^*(s)$, so that

$$h_{\lambda_n}(s) + (1 - \lambda_n)v_{\lambda_n}^*(0) = r_{d^*}(s) + \lambda_n \sum_{j \in S} p_{d^*}(j \mid s)h_{\lambda_n}(j). \tag{8.10.12}$$

Taking the lim sup on each side of (8.10.12), noting that $K - h_{\lambda_n}(s) \geq 0$, and applying Lemma 8.10.4 (Fatou's lemma) yields

$$h(s) + g = r_{d^*}(s) + \limsup_{n \to \infty} \lambda_n \sum_{j \in S} p_{d^*}(j \mid s)h_{\lambda_n}(j)$$

$$\leq r_{d^*}(s) + \sum_{j \in S} p_{d^*}(j \mid s)h(j) \leq \max_{a \in A_s}\left\{r(s, a) + \sum_{j \in S} p(j \mid s, a)h(j)\right\},$$

establishing (8.10.9).

Since

$$h(s) + g \leq r_{d*}(s) + \sum_{j \in S} p_{d*}(j \mid s)h(j) \leq R + K,$$

applying the same argument as in the proof of Theorem 8.7.1(b) establishes that $g \leq g_-^{(d*)^\infty}(s)$ for all $s \in S$. From Lemma 8.10.6, for all $\pi \in \Pi^{HR}$,

$$g_-^\pi(s) \leq \liminf_{\lambda \uparrow 1}(1 - \lambda)v_\lambda^\pi(s) \leq \liminf_{\lambda \uparrow 1}(1 - \lambda)v_\lambda^*(s) = g,$$

from which it follows that $g_-^{(d*)^\infty} = g$ and that $(d*)^\infty$ is lim inf average optimal. Hence we have established result (b). Since the original sequence of $\{\lambda_n\}$ was arbitrary, and every convergent subsequence has the same limit, (8.10.8) follows.

To establish part (c), note that from (8.10.12) it follows that, for all $a \in A_s$,

$$\lambda_n h_{\lambda_n}(s) + (1 - \lambda_n)v_{\lambda_n}^*(s) \geq r(s, a) + \lambda_n \sum_{j \in S} p(j \mid s, a)h_{\lambda_n}(j).$$

Since, for all $s \in S$, $h_\lambda(s) \in [-M(s), K]$, which is a compact subset of the real line, by Lemmas 8.10.2 and 8.10.3 we can choose a subsequence for which $\lim_{n \to \infty} h_{\lambda_n}(s)$ exists for all $s \in S$. Denote this limit by $h(s)$. Under Assumption 8.10.4', we may apply Lemma 8.10.5, so we may pass to the limit in the above expression to obtain

$$h(s) + g \geq r(s, a) + \lambda \sum_{j \in S} p(j \mid s, a)h(j),$$

from which it follows that

$$h(s) + g \geq \max_{a \in A_s} \left\{ r(s, a) + \sum_{j \in S} p(j \mid s, a)h(j) \right\}.$$

Combining this observation with (8.10.9) yields result (c).   □

When $h_\lambda(s)$ is uniformly bounded by a constant, Assumptions 8.10.2 and 8.10.4' hold, and we obtain the following corollary. It applies to finite-state models and to some countable-state models (see Theorem 8.10.10) with bounded rewards. The conclusion that there exists an average optimal policy (instead of a lim inf average optimal policy) follows from Theorem 8.4.1(c).

Corollary 8.10.8.   Suppose Assumption 8.10.1 holds and $|h_\lambda(s)| \leq K$ for $0 \leq \lambda < 1$ and all $s \in S$. Then there exists a constant g and a vector h with $h(0) = 0$ and $|h(s)| \leq K$ which satisfy the average reward optimality equation (8.10.3). Further, there exists an average optimal stationary policy and, if d' satisfies (8.10.10), then $(d')^\infty$ is average optimal.

Several authors have investigated conditions on the chain and reward structure of the Markov decision process which implies Assumptions 8.10.2–8.10.4; Cavazos-

Cadena and Sennott (1992) provide a nice overview of this work. The following condition appears to be the easiest to verify in applications. We state it without proof. In the problem section we provide an example which violates hypothesis (c) of the following theorem yet satisfies Assumptions 8.10.2–8.10.4.

Theorem 8.10.9. Suppose that

a. there exists a stationary policy d^∞ for which the derived Markov chain is positive recurrent,

b. $g^{d^\infty} > -\infty$, and

c. $\{s \in S : r(s, a) > g^{d^\infty}$ for some $a \in A_s\}$ is nonempty and finite.

Then Assumptions 8.10.2–8.10.4 hold.

The following theorem provides necessary and sufficient conditions for existence of a bounded solution to the optimality equation. In it, τ_G denotes the first passage time to the set G. Note that part (b) shows that existence of a bounded solution to the optimality equation implies a strong recurrence condition on the underlying Markov chains that is violated in many applications, especially in queueing control.

Theorem 8.10.10.

a. Suppose that rewards are bounded, that there exists a finite set G, and a finite positive M for which

$$E_s^{d^\infty}\{\tau_G\} < M \qquad (8.10.13)$$

for all $s \in S$ and $d \in D^{MD}$, and that, for each $d \in D^{MD}$, the Markov chain corresponding to d is unichain.
     Then $|h_\lambda(s)| \leq K$ for all $s \in S$ and $0 \leq \lambda < 1$, and there exists a bounded solution to the optimality equation (8.10.3).

b. Suppose that for every bounded $r(s, a)$, there exists a scalar g_r and a bounded h_r satisfying the optimality equation (8.10.3). Then, if for each $d \in D^{MD}$ the Markov chain corresponding to d^∞ has a unique positive recurrent class, there exists a finite set G and a finite constant M for which (8.10.13) holds for all $d \in D^{MD}$ and $s \in S$.

8.10.3   A Communications Model

In this section we apply results of the previous section to establish the existence of optimal policies in a single-server slotted communications system with average reward criterion.

Time is divided into discrete slots, and $k \geq 0$ packets (messages) arrive for transmission in any time slot with probability $p(k)$. The system behaves as a single-server queueing system with a deterministic service rate of one packet per slot. We assume that arriving packets may not be transmitted until the slot after which they arrive. The controller decides at the beginning of each slot whether or not to admit *all* arriving packets. Rejected packets are lost. Each accepted packet generates a reward of L units. The nondecreasing function $c(k)$ denotes the cost for holding k packets for one time slot.

We formulate this model as a Markov decision process as follows. Decision epochs correspond to the start of a time slot. The system state s denotes the number of packets available for transmission at the beginning of a slot (the queue length), so that $S = \{0, 1, \ldots\}$. For each $s \in S$, $A_s = \{0, 1\}$ with action 0 corresponding to rejecting, and action 1 corresponding to accepting, all arriving packets. The reward $r(s, a)$ satisfies $r(s, 0) = -c(k)$ and $r(s, 1) = L\mu - c(k)$, where μ denotes the mean of the arrival distribution. To simplify analysis, subtract $L\mu$ from $r(s, a)$ and redefine the reward as $r'(s, 0) = -L\mu - c(k)$ and $r'(s, 1) = -c(k)$. Note that this modification does not effect the optimal policy but alters its expected discounted reward by a constant. The transition probabilities satisfy $p(0 \mid 0, 0) = 1$ and $p(s - 1 \mid s, 0) = 1$ for $s \geq 1$; $p(k \mid 0, 1) = p(k)$ for $k \geq 0$; and $p(s + k - 1 \mid s, 1) = p(k)$ for $s \geq 1$. Therefore, with discount rate λ, v_λ^* satisfies

$$v_\lambda^*(0) = \max\left\{-L\mu - c(0) + \lambda v_\lambda^*(0), -c(0) + \lambda \sum_{k \geq 0} p(k) v_\lambda^*(k)\right\}$$

and   (8.10.14)

$$v_\lambda^*(s) = \max\left\{-L\mu - c(s) + \lambda v_\lambda^*(s - 1), -c(s) + \lambda \sum_{k \geq 0} p(k) v_\lambda^*(s + k - 1)\right\}$$

for $s \geq 1$.

We verify Assumptions 8.10.1–8.10.4. Since $r'(s, a) \leq 0$, Assumption 8.10.1 holds. Consequently $v_\lambda^*(s) \leq 0$. Let d^∞ denote the stationary policy that always rejects arriving packets. Assumption 8.10.2 holds because

$$v_\lambda^*(s) \geq v_\lambda^{d^\infty}(s) = -\sum_{j=0}^{s} \lambda^j c(s - j) - \frac{\lambda^{s+1}}{1 - \lambda} c(0) - \frac{\mu L}{1 - \lambda} > -\infty.$$

Because $c(k)$ is nondecreasing, $v_\lambda^*(s)$ is nonincreasing. To see this, note that Theorem 7.3.10 applies here with $0 \leq \lambda \leq 1$, so that the iterates of value iteration beginning with $v^0 = 0$ converge to v_λ^*. We leave it as an exercise to verify that if v^n is nonincreasing, then v^{n+1} defined by

$$v^{n+1}(0) = \max\left\{-L\mu - c(0) + \lambda v^n(0), -c(0) + \lambda \sum_{k \geq 0} p(k) v^n(k)\right\}$$

and   (8.10.15)

$$v^{n+1}(s) = \max\left\{-L\mu - c(s) + \lambda v^n(s - 1), -c(s) + \lambda \sum_{k \geq 0} p(k) v^n(s + k - 1)\right\}$$

is nonincreasing. Hence $v^n(s)$ is nonincreasing for all n, so it follows that $v_\lambda^*(s) = \lim_{n \to \infty} v^n(s)$ is nonincreasing (cf. Sec. 6.11). Thus, for all s, $v_\lambda^*(s) \leq v_\lambda^*(0)$. Since $h_\lambda(s) = v_\lambda^*(s) - v_\lambda^*(0) \leq 0$, it follows that Assumption 8.10.3 holds with $K = 0$.

From (8.10.14) and the nonpositivity of $v_\lambda^*(s)$,

$$v_\lambda^*(s) \geq - \sum_{j=1}^{s} \lambda^j [\mu L + c(s-j)] + \lambda^{s+1} v_\lambda^*(0)$$

$$\geq - \sum_{j=1}^{s} c(j) - (s-1)\mu L + v_\lambda^*(0),$$

so that, for $s > 0$,

$$h_\lambda(s) \geq - \sum_{j=0}^{s-1} c(j) - (s-1)\mu L \equiv -M(s)$$

for $0 \leq \lambda < 1$. Let $M(0) = 0$. To verify Assumption 8.10.4(c), note that $\sum_{j \in S} p(j \mid 0,0)M(j) = M(0) = 0$. Consequently parts (a) and (b) of Theorem 8.10.7 apply and there exists an optimal stationary policy, a scalar g, and an $h(s)$ with $h(0) = 0$ and $-M(s) \leq h(s) \leq 0$ which satisfy the optimality inequalities

$$h(0) + g \leq \max\left\{-L\mu - c(0) + h(0), -c(0) + \sum_{k \geq 0} p(k)h(k)\right\}$$

and   (8.10.16)

$$h(s) + g \leq \max\left\{-L\mu - c(s) + h(s-1), -c(s) + \sum_{k \geq 0} p(k)h(s+k-1)\right\}.$$

We now establish the existence of a solution to the optimality equation by providing conditions which imply Assumption 8.10.4'. With $a = 0$,

$$\sum_{j \in S} p(j \mid s,0)M(j) = M(j-1).$$

For $a = 1$, this assumption requires

$$\sum_{j \in S} p(j \mid s,1)M(j) = \sum_{k=0}^{\infty} p(k)\left[L\mu(s+k) + \sum_{j=1}^{s+k} c(j)\right] < \infty. \quad (8.10.17)$$

Suppose $c(k) = k$. Then

$$\sum_{j=1}^{s+k} c(j) = (k+s+1)(k+s)/2 = k^2/2 + (s+1/2)k + (s+1)s/2,$$

so that (8.10.17) holds whenever

$$\sum_{k=0}^{\infty} k^2 p(k) < \infty$$

(note that $\sum_{k=0}^{\infty} k^2 p(k) > \sum_{k=0}^{\infty} kp(k)$). Thus if the arrival distribution has a finite second moment and costs are linear in the state, there exists a solution to the optimality equation.

Since, under any policy $E_s^\pi\{\tau_{(0)}\} \geq s$, Theorem 8.10.10(b) implies that, even with bounded rewards, there need not exist a bounded solution to the optimality equation for this model.

8.10.4 A Replacement Model

Instead of showing that Assumptions 8.10.2–8.10.4 apply directly, we appeal to Theorem 8.10.9 to establish existence of an optimal policy for the replacement model of Sec. 6.10.4. Recall that, in the model, $S = \{0, 1, \dots\}$ represents the condition of the equipment at each decision epoch. State 0 corresponds to new equipment, and the greater the state the poorer the condition of the equipment. At each decision epoch the available actions are to replace the equipment ($a = 1$), or to continue to operate it as is ($a = 0$). Between decision epochs, the equipment deteriorates by i states with probability $p(i)$, so that the transition probabilities for this model satisfy

$$p(j \mid s, 0) = \begin{cases} 0 & j < s \\ p(j - s) & j \geq s, \end{cases}$$

and $p(j \mid s, 1) = p(j)$, $j \geq 0$. The reward is given by

$$r(s, a) = \begin{cases} R - c(s) & a = 0 \\ R - K - c(0) & a = 1, \end{cases}$$

where $R > 0$ denotes the fixed income per period, $K > 0$ the fixed replacement cost, and $c(s)$ the expected cost of operating equipment in state s for one period. We assume $c(s)$ is nondecreasing in s, that is, it costs more to operate it when in poorer condition.

We establish the existence of an optimal stationary policy by applying Theorem 8.10.9. Let d^∞ denote the stationary policy which replaces the equipment at every decision epoch so that $p_d(j \mid s) = p(j)$. Then, provided $p(j) > 0$ for all $j \in S$, $\tau_{(s)}$ follows a geometric distribution so that $E_s\{\tau_{(s)}\} = 1/p(s)$ and all states are positive recurrent under d^∞. Since this policy renews the system every period, $g^{d^\infty} = R - K - c(0)$. If $R - K - c(0) > R - c(s)$ for all s, the policy never to replace is optimal and we have achieved our objective. If not, the assumption that $c(s)$ is nondecreasing implies that $\{s \in S : r(s, a) > g^{d^\infty}$ for some $a \in A_s\} = \{s \in S : R - c(s) > R - K - c(0)\}$ is finite (see Fig. 6.10.1), so that this model satisfies condition (c) of Theorem 8.10.9, from which it follows that Assumptions 8.10.2–8.10.4 hold. Since $r(s, a) \leq R$, Assumption 8.10.1 is satisfied, so the existence of a lim inf average optimal stationary policy follows by Theorem 8.10.7. We leave it as an exercise to provide conditions under which there exists a solution to the optimality equation.

Note that Theorem 8.10.10(b) does *not* apply to this model because the underlying Markov chain is not positive recurrent under the policy "never replace." This policy will not be optimal if, for some s', $c(s) > K + c(0)$ for all $s \geq s'$. In this case we can replace the model by one which excludes the action "do not replace" for $s \geq s'$. The modified model will have the same optimal policy and gain as the original model and, if $\sum_{j \in s} jp(j) < \infty$, there exists a set G for which $E_s^\pi\{\tau_G\} \leq s'$ for all policies π. It follows from Theorem 8.10.10(a) that this model will have a bounded solution of the optimality equation whenever c is bounded. The distinguishing feature between this model and the one in the previous section is that, in this model, there is a policy which "returns the system to 0" in one transition, while in the communications model, every state between s and 0 must be occupied before reaching the empty state.

8.11 THE OPTIMALITY OF STRUCTURED POLICIES

This section provides a method for establishing the structure of optimal policies in models with average expected reward criterion. As has become customary in this chapter, we view the average reward model as a limit of a sequence of discounted models with a discount factor approaching 1. We extend structural results for discounted models in Sec. 6.11 by requiring that structure be preserved under limits as the discount rate approaches 1. We use this approach to provide conditions which ensure the optimality of monotone optimal policies and apply them to show that there exists a control limit policy which is optimal for the replacement model of Secs. 6.10.4 and 8.10.4.

8.11.1 General Theory

We follow the notation of Sec. 6.11. Recall that D^σ, Π^σ, and V^σ denote sets of structured decision rules, structured policies, and structured values, respectively. We state our result for the countable-state model of Sec. 8.10. That for finite-state models follows as a corollary.

Let $L_d v \equiv r_d + P_d v$. Let K and M be as defined in Assumptions 8.10.3 and 8.10.4, $V_K \equiv \{$real-valued $v: K \geq v(s) - v(0) \geq -M(s)\}$, and V_K^σ denote a structured subset of V_K.

Theorem 8.11.1. Let $S = \{0, 1, \ldots\}$, suppose Assumptions 8.10.1–8.10.4 hold, and $h \in V_K$ satisfies (8.10.9). Then, if

a. for any sequence $\{\lambda_n\}$, $0 \leq \lambda_n < 1$, for which $\lim_{n \to \infty} \lambda_n = 1$,

$$\lim_{n \to \infty} \left[v_{\lambda_n}^* - v_{\lambda_n}^*(0)e \right] \in V_K^\sigma; \tag{8.11.1}$$

and

b. $h \in V_K^\sigma$ implies that there exists a $d' \in D^\sigma \cap \arg\max_{d \in D} L_d h$.

Then $D^\sigma \cap \arg\max_{d \in D} \{r_d + P_d h\} \neq \varnothing$, and

$$d^* \in D^\sigma \cap \arg\max_{d \in D} \{r_d + P_d h\} \tag{8.11.2}$$

implies that $(d^*)^\infty$ is lim inf average optimal.

Proof. Let $\{\gamma_n\}$ denote a sequence of discount rates, for which $h(s)$ defined in (8.10.11) satisfies

$$h(s) = \lim_{n \to \infty} \left[v_{\gamma_n}^*(s) - v_{\gamma_n}^*(0) \right]$$

From hypothesis (a), $h(s) \in V_K^\sigma$, so, from (b), there exists a $d^* \in D^\sigma \cap$ arg max$_{d \in D} L_d h$. From Theorem 8.10.7(b), $(d^*)^\infty$ is lim inf average optimal, and the result follows. □

Corollary 8.11.2. Let S be finite, let all stationary policies have unichain transition matrices, suppose $|r(s, a)| \le \mu < \infty$ for $a \in A_s$ and $s \in S$, and let $h \in V$ satisfy the average reward optimality equation for some scalar g. Then, if

a. for any sequence $\{\lambda_n\}$, $0 \le \lambda_n < 1$, with $\lim_{n \to \infty} \lambda_n = 1$,

$$\lim_{n \to \infty} \left[v_{\lambda_n}^* - v_{\lambda_n}^*(s_0)e \right] \in V^\sigma$$

for some $s_0 \in S$; and
b. $h \in V^\sigma$ implies that there exists a $d' \in D^\sigma \cap$ arg max$_{d \in D} L_d h$.

Then $D^\sigma \cap$ arg max$_{d \in D} \{r_d + P_d h\} \ne \varnothing$, and

$$d^* \in D^\sigma \cap \arg \max_{d \in D} \{r_d + P_d h\} \tag{8.11.3}$$

implies that $(d^*)^\infty$ is average optimal.

Proof. Clearly Assumptions 8.10.1 and 8.10.2 hold. Assumptions 8.10.3 and 8.10.4 follow from Theorem 8.10.10(a). Therefore the result follows from Theorem 8.11.1 and Corollary 8.10.8. □

8.11.2 Optimal Monotone Policies

We can directly establish the optimality of a control limit policy for the replacement model (Secs. 6.10.4 and 8.10.4) by appealing to Theorem 8.11.1. Instead, we show in general that superadditivity and monotonicity conditions of Sec. 4.7 imply that a monotone policy is optimal. We assume that $A_s = A'$ for all $s \in S$. We present results for countable-state models, but the argument used to deduce Corollary 8.11.2 from Theorem 8.11.1 allows immediate application to finite-state models.

Theorem 8.11.3. Let $S = \{0, 1, \ldots\}$ and suppose Assumptions 8.10.1–8.10.4 hold, and that

a. $r(s, a)$ is nondecreasing in s for all $a \in A'$;
b. $q(k \mid s, a) \equiv \sum_{j=k}^{\infty} p(j \mid s, a)$ is nondecreasing in s for all $k \in S$ and $a \in A'$;
c. $r(s, a)$ is a superadditive (subadditive) function on $S \times A'$, and
d. $q(k \mid s, a)$ is a superadditive (subadditive) function on $S \times A'$ for all $k \in S$.

Then there exists a lim inf average optimal stationary policy $(d^*)^\infty$ with the property that $d^*(s)$ is nondecreasing (nonincreasing) in s. Further, when S is finite, $(d^*)^\infty$ is average optimal.

Proof. We establish this result by verifying the conditions of Theorem 8.11.1. Let $V_K^\sigma = \{v \in V_K : v$ is nondecreasing on $S\}$ and let $D^\sigma = \{d \in D : d$ is nondecreasing on $S\}$. It follows from the proof of Theorem 6.11.6 that, for every λ, $v_\lambda^*(s)$ is nondecreasing in s. Consequently for $0 \le \lambda < 1$, $v_\lambda^*(s) - v_\lambda^*(0)$ is nondecreasing in s, so that condition (a) of Theorem 8.11.1 is satisfied.

Theorem 4.7.4 establishes that for every $h \in V_K^\sigma$ there exists a $d' \in D^\sigma \cap$ arg max$_{d \in D} L_d v$, so that hypothesis (b) of Theorem 8.11.1 holds. Consequently Theorem 8.11.1 establishes the existence of a $d^* \in D^\sigma$ for which $(d^*)^\infty \in \Pi^\sigma$. □

A similar conclusion follows under the alternative conditions below.

Theorem 8.11.4. Let $S = \{0, 1, \dots\}$, suppose that Assumptions 8.10.1–8.10.4 hold, and further that

1. $r(s, a)$ is nonincreasing in s for all $a \in A'$,
2. $q(k \mid s, a)$ is nondecreasing in s for all $k \in S$ and $a \in A'$;
3. $r(s, a)$ is a superadditive function on $S \times A'$; and
4. $\sum_{j=0}^\infty p(j \mid s, a)u(j)$ is a superadditive function on $S \times A'$ for nonincreasing u.

Then there exists a lim inf average optimal stationary policy $(d^*)^\infty$ in which $d^*(s)$ is nondecreasing (nonincreasing) in s. Further, when S is finite, $(d^*)^\infty$ is average optimal.

The assumption that $A_s = A'$ for all s may be overly restrictive for some applications. As in Sec. 4.7, slight modifications of these theorems allow extensions to A_s which satisfy

a. $A_s \subset A'$ for all $s \in S$;
b. $A_s \subset A_{s'}$ for $s' \ge s$; and
c. for each s, $a \in A_s$ and $a' \le a$ implies $a' \in A_s$.

Example 8.11.1. We establish the optimality of control limit policies for the equipment replacement model of Sec. 6.10.4 by verifying the hypotheses of Theorem 8.11.4. The analysis in Sec. 8.10.4 establishes that Assumptions 8.10.1–8.10.4 hold. Section 4.7.5 verifies hypotheses 1–4 of Theorem 8.11.4 that

1. $r(s, a)$ is nonincreasing in s for $a = 0$ and 1,
2. $q(k \mid s, a)$ is nondecreasing,
3. $r(s, a)$ is superadditive, and
4. $\sum_{j=0}^\infty p(j \mid s, a)u(j)$ is superadditive for nonincreasing u.

Consequently, from Theorem 8.11.4, there exists a monotone lim inf average optimal stationary policy. Since there are only two actions, a control limit policy is lim inf average optimal. It can be found using the monotone policy iteration algorithm below.

We conclude this section with a policy iteration algorithm which finds a monotone optimal policy. We assume that the hypotheses of Theorem 8.11.3 or 8.11.4 hold, so that such a policy exists. Let $S = \{0, 1, \ldots, N\}$ with N finite and $A_s = A'$ for all $s \in S$. For countable S, we may regard this as solving an N-state approximation. Let D^σ denote the set of nondecreasing decision rules, and V_K^σ denote the subset of nondecreasing functions on S. We assume a unichain model.

The Unichain Monotone Policy Iteration Algorithm

1. Choose $d_0 \in D^\sigma$ and set $n = 0$.
2. Find a scalar g_n and an $h_n \in V_K^\sigma$ by solving

$$0 = r_{d_n} - ge + (P_{d_n} - I)h. \tag{8.11.4}$$

3. Set $s = 0$, and $A'_0 = A'$.
   a. Set

$$A_s^* = \arg\max_{a \in A'_s} \left\{ r(s, a) + \sum_{j \in S} p(j \mid s, a) h_n(j) \right\}.$$

   b. If $s = N$, go to step 3(d); otherwise set

$$A'_{s+1} = \{ a \in A'_s : a \geq \max[a' \in A_s^*] \}.$$

   c. Substitute $s + 1$ for s and return to step 3(a).
   d. Pick $d_{n+1} \in D^\sigma \cap \times_{s \in S} A_s^*$, setting $d_{n+1} = d_n$ if possible.
4. If $d_{n+1} = d_n$, stop and set d^* equal to d_n. Otherwise, replace n by $n + 1$ and return to step 2.

Theorem 8.6.6 together with Theorems 8.11.3 or 8.11.4 ensure that for finite A' this algorithm finds a monotone optimal policy. Note that since h_n determined by (8.11.3) is unique up to a constant, a specific choice of h_n will not effect its structure. The algorithm differs from the general policy iteration algorithm of Sec. 8.6 in that improvement is carried out over the derived sets A'_s which become smaller with increasing s. In the worse case, $A'_s = A'$ for all s, and the computational effort equals that of the general policy iteration algorithm. When an optimal decision rule is strictly increasing, the sets A'_s will decrease in size with increasing s and hence reduce the number of actions which need to be evaluated in the improvement step 3. If, at some $u \in S$, A_u^* contains a single element, say a^*, then no further maximization is necessary since that action will be optimal at all $s \geq u$. In such a case $d_{n+1}(s) = a^*$ for all $s \geq u$.

BIBLIOGRAPHIC REMARKS

Howard (1960) introduced the Markov decision process with average reward criterion, most likely independently of Gillette's (1957) study of infinite-horizon stochastic games with average reward criterion. In Howard's monograph, he recognized that distinct analyses were required for unichain and multichain MDPs, and provided policy iteration algorithms for solving each. Blackwell's important 1962 paper set the stage for future research in this area. In it, he used the truncated Laurent expansion to analyze the average reward model as a limit of discounted models, provided a refined notion of optimality which is now referred to as Blackwell optimality, and extended Howard's work on the policy iteration algorithm.

Flynn (1976) distinguishes and relates the different notions of average optimality which we discuss in Sec. 8.1.2. They previously appear in various forms in Derman (1964 and 1966), Ross (1968a), and Lippman (1968a). Example 8.1.2 which distinguishes these criteria is motivated by Example 3 in Dynkin and Yushkevich (1979, p. 182).

The average reward evaluation equations first appeared in Howard (1960). Blackwell (1962) and Veinott (1966a) provide other specifications for these equations which ensure that the quantity h in (8.2.12) equals the bias. The material on Laurent expansions in Sec. 8.2 draws on Veinott's (1969b) extension of Blackwell's results. Lamond and Puterman (1989) derive these equations using generalized inverse theory.

The classification schemes of Sec. 8.3 extend Howard's original model dichotomy. Bather (1973b) introduced the concept of a communicating MDP and, among other results, demonstrated the existence of an optimal stationary policy in finite-state communicating models. Hordijk (1974) used this concept to establish a similar result for countable-state models. Platzman (1977) introduced the concept of a weakly communicating model (he used the term *simply connected*) when analyzing value iteration for average reward models. Ross and Varadarajan (1991) provide an algorithm for decomposing a MDP into strongly communicating classes, and a set of states which are transient under all policies. Their decomposition is related to one proposed by Bather (1973c), Schweitzer and Federgruen (1978a), and Schweitzer (1984a).

The average reward optimality equation appears implicitly in Blackwell (1962); however, explicit statements appear for particular applications in Iglehart (1963b) and Taylor (1965), and in general in Derman (1966). We discuss these papers further below. Our existence proofs in Secs. 8.4.2 and 8.4.4 are based on concepts in Blackwell (1962). The continuity concepts in Sec. 8.4.4 originate with Hordijk (1974). Hordijk and Puterman (1987) use related ideas to establish the existence of a solution to the finite-state model with compact action sets.

We establish the existence of optimal stationary policies by first showing that there exists a solution to the optimality equation; other investigators use different approaches. Derman (1964) established this result directly by using a version of a result included here as Lemma 8.10.6; Martin-Lof (1967a) and Feinberg (1975) also studied this problem. Schweitzer (1983), Federgruen and Schweitzer (1984b),

Federgruen, Schweitzer, and Tijms (1983), and Schweitzer (1987) have established existence of solutions to the optimality equation in various settings using fixed-point theory. Borkar (1984 and 1989) uses convex analysis to establish existence of optimal policies. His monograph, Borkar (1991), and Sec. 5.3 of the Arapostathis *et al.* (1993) survey provide overviews of this work.

Our analysis of value iteration in Sec. 8.5 follows Federgruen, Schweitzer, and Tijms (1977) and van der Wal (1981). Good references for this topic include survey articles by Federgruen and Schweitzer (1978 and 1980) and books by Whittle (1983), Bertsekas (1987), and Hernandez-Lerma (1989). Important contributions to the theory of undiscounted value iteration include Bellman (1957), White (1963), Brown (1965), Schweitzer (1965), Lanery (1967) and Bather (1973a, b, c). Bellman (1957, p. 328–332) studied the asymptotic behavior of value iteration in a model with strictly positive transition matrices and no reward. White (1963) proposed the undiscounted relative value iteration algorithm and demonstrated its convergence under condition (b) of Theorem 8.5.3. Schweitzer (1965), Denardo (1973), and Federgruen, Schweitzer, and Tijms (1977) further refined this work.

Several authors including Brown (1965), Lanery (1967), Denardo (1973), and Bather (1973a, b, c) analyzed value iteration by investigating the asymptotic properties of $v^n - ng^*$. Schweitzer and Federgruen (1977) complete analysis of this problem by identifying the periodic structure of this sequence in addition to that of the set of maximizing policies. Brown (1965), Odoni (1969), and van der Wal (1981, pp. 123–124) show that, if $v^n - ng^*$ converges, then for n sufficiently large a v^n-improving policy is optimal. Examples in Brown (1965) and Bather (1973a) show that the sequence of v^n-improving decision rules can be very erratic: they can be strict subsets of the set of maximal gain decision rules for every n, they can oscillate periodically or even aperiodically within the set of maximal gain decision rules. When the $\lim_{n \to \infty} v^n - ng^*$ does not exist, Lanery (1967) provides an example where nonmaximal gain decision rules appear infinitely often in the sequence of v^n-improving decision rules, and Federgruen and Schweitzer (1980) refer to an unpublished example in which the sets of v^n-improving decision rules and the set of decision rules which are average optimal are disjoint. The aperiodicity transformation in Sec. 8.5.5 originates in Schweitzer (1971).

In unichain models, bounds on the optimal gain rate were given by Odoni (1969) and Hastings (1971). Federgruen, Schweitzer, and Tijms (1977) discuss approximations to h which allow using (8.5.15) in Proposition 8.5.10; however, evaluation requires many extra calculations which probably negate the benefits of such an action elimination algorithm. The action elimination algorithm in Sec. 8.5.7 was proposed by Hastings (1976), refined by Hubner (1977), applied to discounted models by Hastings and van Nunen (1977) and to modified policy iteration and policy iteration for discounted models by Puterman and Shin (1982). Lasserre (1994b) provides a permanent action elimination procedure for policy iteration in recurrent models. It is based on his variant of policy iteration (1994a), which changes only one action at each iteration.

The policy iteration algorithm was introduced by Howard (1960) for finite-state and action models, where he demonstrated finite convergence under the assumption that all policies are recurrent. Blackwell (1962) and Veinott (1966) provide refinements to the theory. In the countable-state case, Derman (1966) used policy iteration to constructively establish existence of a solution to the optimality equation under assumptions that all states are recurrent under each stationary policy and that the reward, gain, and bias are uniformly bounded on the set of stationary policies. Federgruen and Tijms (1978), Hordijk and Puterman (1987), and Dekker (1985) examined the convergence of policy iteration in unichain models with compact action spaces.

Van der Wal (1981, pp. 159–181) analyzed a modified policy iteration (he refers to it as value-oriented successive approximations) for average reward models under various chain structure assumptions. Our approach combines his results with those in Sec. 8.5. Ohno (1985) investigates some of its numerical properties.

De Ghellinck (1960) and Manne (1960) formulated the average reward model as a linear program and analyzed models in which the transition probability matrix for each stationary policy is ergodic. Other contributors to this theory include Denardo and Fox (1968), who consider both unichain and multichain problems, Denardo (1970), and Derman (1970). Kallenberg (1983) in his thesis provides a comprehensive and unified analysis of all aspects of MDP linear programming models. Some of those results also appear in Hordijk and Kallenberg (1979 and 1980). Sec. 8.8 follows Derman (1970) and Kallenberg (1983).

Constrained models have been the subject of considerable recent research activity. Derman was the first to analyze a constrained MDP model. His book (1970) provides a comprehensive presentation of his work on this topic. He introduced the state action frequency approach for analysis of these problems, and developed its relationship to linear programming. Kallenberg (1983) and Hordijk and Kallenberg (1984) develop further properties of sets of limiting state action frequencies, and extend the linear programming approach to include constrained multichain models. White (1974) and Beutler and Ross (1985) use Lagrange multipliers to analyze constrained models. The latter paper analyzes models with compact action sets for which the linear programming machinery is not available.

White (1988b) surveys models with mean, variance, and probabilistic optimality criteria and reviews the importance of and relationship between limits of state action frequencies in different classes of models. Altman and Schwartz (1991) extend the state action frequency approach to countable-state models. Section 8.9.4 on variance penalized models is based on Filar, Kallenberg, and Lee (1989). Altman and Schwartz (1991) and Bayal-Gursoy and Ross (1992) provide further results on variance penalized models.

The counterexamples in Sec. 8.10.1 provide insight into the countable-state average reward model. Whittle (1983, p. 118) notes

"The field of average cost optimization is a strange one. Counterexamples exist to almost all natural conjectures, yet these conjectures are the basis of a. proper intuition and are valid if reformulated right or if natural conditions are imposed."

Maitra (1964) provided Example 8.10.1. Examples 8.10.2 and that in Problem 8.35 appear in Ross (1983).

Taylor (1965) laid the groundwork for analysis of countable-state models with average reward criterion by using the differential discounted reward approach for analyzing replacement models. Most subsequent references build on his ideas. Ross (1968a, b) extended Taylor's approach to general countable-state and arbitrary-state models. He showed that the optimality equation has a solution when $h_\lambda(s) = v_\lambda^*(s) - v_\lambda^*(0)$ is uniformly bounded, and provided sufficient conditions for this bound to apply.

The text by Ross (1983) provides a nice summary of this work. Derman (1966) established the existence of a bounded solution to the optimality equation for a model with bounded rewards and strong recurrence conditions using an approach based on policy iteration.

Motivated by the discovery that the existence of a bounded solution to the optimality equation implies the existence of a stationary optimal policy, Derman and Veinott (1967) provided recurrence conditions on the underlying Markov chain related to those in Theorem 8.10.10(a), which implied the existence of a solution of the optimality equation. Subsequently Kushner (1971), Hordijk (1974), and Federgruen, Hordijk, and Tijms (1978) generalized these results in several directions. Thomas (1980) provides a nice survey of work in this area. Cavasoz-Cadena (1988 and 1989a) showed that existence of a bounded solution of the optimality equation in a class of MDP's implies uniformly bounded expected recurrence times to a finite set [Theorem 8.10.10(b)].

Lippman (1973 and 1975a) studied countable-stage average reward models with unbounded rewards. His approach was to combine results for discounted models as presented in Sec. 6.10 with strong recurrence conditions, so that the Taylor and Ross approach could be used. Federgruen, Hordijk, and Tijms (1979) and Federgruen, Schweitzer, and Tijms (1983) adopt a different approach for showing that a solution exists to the optimality equation, and that the derived stationary policy is optimal.

Our approach to countable-state models, especially in Secs. 8.10.2 and 8.10.3 follows Sennott (1989a), Ritt and Sennott (1992), and Cavasoz-Cadena (1991). Personal correspondence and discussion with Professor Sennott have greatly influenced this presentation. To quote Cavasoz-Cadena (1989b) regarding Sennott's (1989a) paper:

"To obtain her results, she follows the usual approach of examining the average case as a limit of discounted cases, but she does it more efficiently."

Cavasoz-Cadena (1989b), Schal (1992), Sennott (1993a and b), Ritt and Sennott (1992) and Cavasoz-Cadena and Sennott (1992) have made further refinements to this theory. The last paper relates many conditions which imply Assumptions 8.10.2–8.10.4, including those of Weber and Stidham (1987) which we have stated as Theorem 8.10.9. Cavasoz-Cadena (1991) provides a nice survey of many of these recent results. Thomas and Stengos (1985) investigate the use of finite-state approximations in countable-state average reward models along the lines of our analyses in Secs. 6.10.2 and 6.10.3 of discounted models.

A key tool in our analysis of Sec. 8.10.2 is Lemma 8.10.6. Derman (1962) and several other authors have used this, and related results, to establish the existence of average optimal policies. Apparently Gilette (1957) was the first to use such a result, in the context of stochastic games. This result, which relates Cesaro convergence to Abel convergence, is often attributed to Hardy and Littlewood and referred to as a Tauberian theorem. The standard reference for this result is Widder (1946); however, the proof herein follows Sennott (1986b) and Powell and Shah (1972). See also Sznadjer and Filar (1992) for more on this problem. Note that Niels Abel (1802–1829) who introduced many important summation concepts, some of which we apply here, was only 27 at his death.

Dekker (1985) and Dekker and Hordijk (1988, 1991, and 1992) use strong notions of recurrence to develop Laurent series expansions, and establish the existence of average optimal (as opposed to lim inf average optimal) stationary policies in countable-state models. Spieksma (1991) provides some queueing control models in which these recurrence conditions hold.

The idea of using the limit of discounted models to establish structural results for average reward models pervades the inventory, replacement, and queueing literature. Taylor (1965) and Derman (1970, pp. 121–125) provide illustrations of this approach. Tijms (1986, p. 221–245) provides examples of structured policy iteration algorithms.

There has been voluminous literature on average reward models, and no doubt several important references have been omitted from the above summary. The recent survey by Arapostathis, Borkar, Fernandez-Gaucherand, Ghosh, and Marcus (1993) provides a good overview. Denardo's (1973) article provides an accessible introduction to this subject matter, and the book by Dynkin and Yushkevich (1979) elegantly presents many further results.

We have chosen not to discuss average reward models with general-state spaces. Hernandez-Lerma (1989), Dynkin and Yushkevich (1979), Ritt and Sennott (1992), and Arapostathis *et al.* (1993) analyze the model at this level of generality and contain numerous references.

PROBLEMS

8.1. Complete the proofs of Theorem 8.2.8(b) and (c) by formally carrying out the suggested induction.

8.2. Prove identities (A.17)–(A.21) in Appendix A.

8.3. (Unsolved problem) Let u denote the expected total reward from a terminating Markov reward process in which the termination time ν follows a negative binomial distribution with success probability λ. Find an expression which relates this quantity to the expected discounted reward and terms in its Laurent series expansion.

8.4. Consider a model with $S = \{s_1, s_2\}$, $A_{s_1} = \{a_{1,1}, a_{1,2}\}$, $A_{s_2} = \{a_{2,1}, a_{2,2}, a_{2,3}\}$, $r(s_1, a_{1,1}) = 1$, $r(s_1, a_{1,2}) = 4$, $r(s_2, a_{2,1}) = 2$, $r(s_2, a_{2,2}) = 3$, $r(s_2, a_{2,3}) = 5$, $p(s_1 \mid s_1, a_{1,1}) = 1$, $p(s_1 \mid s_1, a_{1,2}) = 0.5$, $p(s_1 \mid s_2, a_{2,1}) = 1$, $p(s_1 \mid s_2, a_{2,2}) = 0$, and $p(s_1 \mid s_2, a_{2,3}) = 0.75$.

a. Determine the chain structure of each deterministic stationary policy.

b. Compute the gain and bias for each deterministic stationary policy by solving (8.2.11) and (8.2.12).

c. Show this is a communicating MDP and verify that its optimal gain is constant.

8.5. For Example 8.2.2, compute the y_1 term in the Laurent series expansion of v_λ using the following approaches,

a. the definition $y_1 = H_P^2 r$,

b. by solving (8.2.21) with $n = 1$ subject to the constraint $P^* y_1 = 0$, and

c. by solving (8.2.21) for both $n = 1$ and 2.

Compare the computational effort of each approach. Can you exploit the structure of P to simplify calculations further?

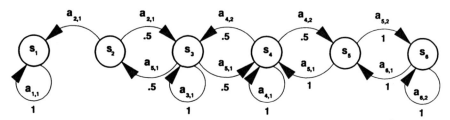

Figure 8.PR1 Symbolic representation of model for analysis in Problem 8.7. Numbers below arcs denote transition probabilities.

8.6. Show by example that a MDP can have a single closed communicating class yet be multichain.

8.7. (Ross and Varadarajan, 1991) Call a set of states C *strongly communicating* if
   i. it is recurrent under some stationary policy, and
   ii. it is not a proper subset of any set of states for which (i) holds.
   Consider the model in Figure 8.PR1.
   a. Use the algorithm of Sec. 8.3 to determine whether the model is weakly communicating.
   b. Find the strongly communicating classes for this model.
   c. Show that the model has two communicating classes but that only one is closed.

8.8. Suppose that, in a communicating MDP, we have a stationary policy d^∞ for which $g^{d^\infty}(s) = K$ for all s in some closed irreducible class C. Provide an algorithm to construct a stationary policy δ^∞ for which $g^{\delta^\infty}(s) = K$ for all $s \in S$.

8.9. This problem illustrates some of the concepts of Sec. 8.3.3. Consider the model in Figure 8.PR2

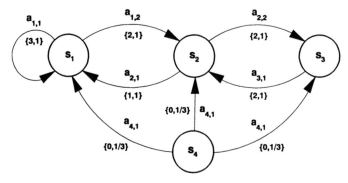

Figure 8.PR2 Symbolic representation of model for analysis in Problem 8.9.

a. Show that this model is weakly communicating.

b. Let $d(s_1) = a_{1,1}$, $d(s_2) = a_{2,2}$, $d(s_3) = a_{3,1}$, and $d(s_4) = a_{4,1}$. Compute $g^{d^{\infty}}$.

c. Using the approach in the proof of Lemma 8.3.2, find a $\delta \in D^{MD}$ for which $g^{\delta^{\infty}} \geq g^{d^{\infty}}$ with strict inequality in at least one component.

8.10. Prove an appropriate version of Theorem 8.4.1 when $B(g, h)$ is replaced by $\sup_{d \in D} \{r_d + ge + (P_d - I)h\}$. Note which portions of the proofs need modification.

8.11. Prove that, in a finite-state model, h_d is bounded whenever r_d is bounded. Obtain a bound on $\|h_d\|$.

8.12. Derive (8.5.20).

8.13. (Hubner, 1977) Derive the following alternatives to expressions (8.5.22) and (8.5.23) in Theorem 8.5.13:

$$v^n(s) - L(s, a)v^{n-k} - \sum_{j=0}^{k-1} \gamma^{j+1} sp(v^{n-j} - v^{n-j-1}) > 0 \quad (8.5.22')$$

and

$$v^n(s) - L(s, a)v^{n-k} - \gamma sp(v^n - v^{n-k}) > 0, \qquad (8.5.23')$$

where γ is defined in (8.5.7).

8.14. Show that solving (8.6.8) is equivalent to solving (8.6.1) subject to (8.6.5).

8.15. Explicitly derive the entries of $(I - P_d - P_d^*)^{-1}$ for a unichain d when expressed in partition form (8.6.13). Derive the entries of H_d when expressed as (8.6.11).

8.16. Consider the average reward version of the model in Problem 6.1.

a. Show that the model is recurrent.

b. Find a .01-optimal policy using value iteration.

c. Find an optimal policy using policy iteration.

d. Show that the model satisfies one of the hypotheses of Theorem 8.5.3 and conclude that value iteration converges. Determine its rate of convergence with respect to the span seminorm.

e. Find an optimal policy using modified policy iteration.

f. For what values of λ is the average optimal policy equal to the optimal policy for the discounted model?

g. Solve the problem using linear programming.

h. Solve a constrained version of the problem, in which the average cost of sending catalogs does not exceed $7.50 per period. Investigate the sensitivity of the optimal policy to this cost. Display your results graphically by plotting the optimal policy versus this cost.

8.17. a. Solve Problem 6.1 using the undiscounted value iteration algorithm with action elimination procedure of Sec. 8.5.7.

    b. Solve it using an action elimination algorithm based on (8.5.22), and compare the number of actions eliminated using each approach.

    c. Solve it using algorithms based on (8.5.22') and (8.5.23') from the preceding problem, and compare the number of actions eliminated to parts (a) and (b).

8.18. Carry out the analyses in the above two problems for an average reward version of Problem 6.48.

8.19. Consider an average reward version of the model in Problem 6.63. Verify that it is a unichain model and find an optimal policy for it.

8.20. Show through examples that each of the conditions in Theorem 8.5.3 is distinct (i.e., there exists a model which satisfies (a) but not (b), (a) but not (c), etc.).

8.21. Prove the following generalization of Theorem 6.6.1. Let $T : V \to V$ and suppose, for all u and v in V,

$$sp(Tu - Tv) \leq sp(u - v),$$

and that there exists an α, $0 \leq \alpha < 1$ and a non-negative integer J for which

$$sp(T^J u - T^J v) \leq \alpha^J sp(u - v)$$

for all u and v in V. Then each of the following results hold.

    a. There exists a $v^* \in V$ for which $sp(Tv^* - v^*) = 0$.

    b. For any $v^0 \in V$, the sequence $v^n = T^n v^0$ satisfies

$$\lim_{n \to \infty} sp(v^n - v^*) = 0$$

    c. For any $v^0 \in V$, and $k \geq 1$

$$sp(v^{nJ+k} - v^*) \leq \alpha^n sp(v^0 - v^*).$$

8.22. An alternative approach to value iteration proposed by Bather (1973b) and Hordijk and Tijms (1975) is to use the recursion

$$v^{n+1} = \max_{d \in D} \{ r_d + \lambda_n P_d v^n \}$$

where the discount rate $\lambda_n \uparrow 1$ in some sense. Show that such a scheme converges, and evaluate its performance for the data in Problem 6.1.

8.23. (Denardo, 1973) Let $S = \{s_1, s_2, s_3\}$; $A_{s_1} = \{a_{1,1}, a_{1,2}\}$, $A_{s_2} = \{a_{2,1}\}$, and $A_{s_3} = \{a_{3,1}\}$; $r(s_1, a_{1,1}) = r(s_1, a_{1,2}) = 0$, $r(s_2, a_{2,1}) = 3$, and $r(s_3, a_{3,1}) = 4$; and $p(s_1 \mid s_1, a_{1,1}) = p(s_2 \mid s_1, a_{1,1}) = \frac{1}{2}$, $p(s_1 \mid s_1, a_{1,2}) = \frac{2}{3}$, $p(s_3 \mid s_1, a_{1,1}) = \frac{1}{3}$, $p(s_1 \mid s_2, a_{2,1}) = 1$, and $p(s_1 \mid s_3, a_{3,1}) = 1$. Let δ choose action $a_{1,2}$ and γ choose action $a_{1,1}$ in s_1.

a. Show this model is unichain by determining the chain structure of both stationary policies.

b. Show that all states are recurrent under some randomized stationary policy.

c. Show that policy iteration may fail to find a bias-optimal policy; that is, a maximal gain policy which has greater bias than any other maximal gain policy.

8.24. Using the definitions and dual and primal linear programming models in Appendix D, show that the dual of the primal LP in Sec. 8.8 is as specified.

8.25. Prove Theorem 8.8.6 and Corollaries 8.8.7 and 8.8.8.

8.26. Consider a variant of the policy iteration algorithm which stops as soon as

$$d_n(s) \in \arg\max_{d \in D} \{r_d(s) + P_d h_n(s)\}$$

for all $s \in R_{d_n}$.

a. Show that this algorithm identifies an average optimal stationary policy.

b. Discuss how to implement the algorithm.

c. Apply this approach to solve the model in Problem 8.23.

8.27. Show that if all stationary deterministic policies are unichain, then all stationary randomized policies are unichain.

8.28. In a unichain model, show that $\Xi^{\mathrm{MD}}(\alpha) = \Xi^{\mathrm{MR}}(\alpha)$ for all α.

8.29. Verify all calculations in Example 8.8.1 and show that $X = \Xi^{\mathrm{MR}}$.

8.30. Determine $\Xi^{\mathrm{SD}}(\alpha)$, $\Xi^{\mathrm{SR}}(\alpha)$, and X for the model in Problem 6.1. Verify that $(\Xi^{\mathrm{SD}}(\alpha))^c = \Xi^{\mathrm{SR}}(\alpha) = X$ for this model, and that $\Xi^{\mathrm{SD}}(\alpha)$ and $\Xi^{\mathrm{SR}}(\alpha)$ do not vary with α.

8.31. **a.** Derive and interpret the primal linear program which corresponds to the dual problem in Sec. 8.8.2 when there is a single additional constraint, that is, $K = 1$.

b. (Unsolved Problem) Provide a policy iteration algorithm which solves the unichain constrained average reward model directly.

8.32. For the model in Example 8.8.1, do the following:

a. Solve QP1 of Sec. 8.9.3, and investigate the sensitivity of the optimal policy to β.

b. Solve QP2 of Sec. 8.9.3, and investigate the sensitivity of the optimal policy to γ. Does this approach determine all Pareto optimal solutions for this model?

8.33. Show that the objective function (8.9.25) for QP1 is a concave function of x and that, for QP2, the expression in (8.9.27) is a convex function of x.

8.34. Verify calculations in Example 8.10.2 and show that there exists a randomized stationary optimal policy.

8.35. (Ross, 1983, p. 91) Let $S = \{1, 1', 2, 2', \ldots\} \cup \{\Delta\}$, $A_s = \{a_{s,1}, a_{s,2}\}$, $A_{s'} = \{a_{s'}\}$, and $A_\Delta = \{a_\Delta\}$, with $p(s + 1 \mid s, a_{s,1}) = 1$, $p(s' \mid s, a_{s,2}) = p_s$, $p(\Delta \mid s, a_{s,2}) = 1 - p_s$, and $p(s' - 1 \mid s', a_{s'}) = 1$ for $s' \geq 2$, $p(1 \mid 1', a_{1'}) = 1$, and $p(\Delta \mid \Delta, a_\Delta) = 1$; and $r(s, a_{s,1}) = r(s, a_{s,2}) = 0$ for $s = 1, 2, \ldots$, $r(\Delta, a_\Delta) = 0$, and $r(s', a_{s'}) = 2$ for $s' = 1', 2', \ldots$. Assume further that $\Pi_{s=1}^{\infty} p_s = \frac{3}{4}$.

a. Draw a transition diagram for this model and determine its chain structure.

b. Show that, under any stationary policy, the average reward equals 0.

c. Consider the nonstationary policy π which initially chooses action $a_{s,2}$ in every state, chooses actions $a_{1,1}, a_{2,1}, \ldots, a_{n,1}$, upon its nth return to state 1, and then chooses action $a_{n,2}$. Show that $g^\pi = \frac{3}{4}$, and that there exists no stationary ε-optimal policy.

8.36. Provide conditions under which Assumptions 6.10.1 and 6.10.2 are satisfied in the communications model of Sec. 8.10.3.

8.37. Complete the analysis in Sec. 8.10.3 by showing that, if v^n is nonincreasing, v^{n+1} defined by (8.10.15) is nonincreasing.

8.38. Show that there exists an optimal stationary policy, and provide conditions under which the optimality equation holds for a variant of the packet communication model of Sec. 8.10.3 in which each rejected packet returns with probability p in the next time slot.

8.39. (Sennott, 1989a) Consider an infinite-horizon countable-state version of the service rate control model of Sec. 3.7.2, in which the service rate distribution is Bernoulli with probability pb, where b is chosen from a finite set B and $p^+ \equiv \max_{b \in B} pb < 1$ and $p^- \equiv \min_{b \in B} pb > 0$. This means that the probability of a service completion in a single time slot equals pb when action b is chosen. The cost of serving at rate pb for one period equals $c(b)$. There is no cost associated with changing the service rate. The revenue $R = 0$ and the holding cost $h(s)$ is linear in s. The objective is to determine a service rate policy which minimizes the long-run average expected cost.

a. Give the optimality equations for the discounted model.

b. Using Theorem 8.10.7, show that, if the second moment of the arrival distribution is finite and the mean arrival rate μ is less than p^+, there exists a lim inf average optimal stationary policy.

c. Show that (8.10.9) holds with *equality* if, in addition, $\mu < p^-$.

d. Provide conditions for the existence of a stationary lim inf average optimal policy and for the optimality equation to hold under the assumption that $h(s)$ is quadratic in s.

e. Determine the structure of an optimal policy under the linear holding cost assumption.

8.40. Show directly that Assumptions 8.10.1–8.10.4 hold for the replacement model of Sec. 8.10.4, and provide conditions under which Assumption 8.10.4′ holds.

8.41. Prove a version of Theorem 8.10.8(a) in which we assume that the set $G = \{0\}$.

8.42. Show that a control limit policy is optimal for the packet communication model of Sec. 8.10.3.

8.43. In the proof of Theorem 8.10.7, we applied Fatou's Lemma to show that

$$\limsup_{n \to \infty} \sum_{j \in S} p_{d^*}(j \mid s) h_{\lambda_n}(j) \le \sum_{j \in S} p_{d^*}(j \mid s) h(j),$$

where $h(s) = \limsup_{n \to \infty} h_{\lambda_n}(s)$ and $h_{\lambda_n}(s) \le K$. Verify this calculation using the statement of Fatou's Lemma in Lemma 8.10.4.

8.44. Show that an optimal policy exists in a variant of the packet communication model in which the action specifies the number of arriving packets to admit.

8.45. Show that a lim inf average optimal policy exists for a countable-state inventory model with no backlogging, fixed order cost K, holding cost function $h(s)$, and stationary demand distribution $\{p(k)\}$. Under which conditions does the optimality equation hold?

8.46. (Tim Lauck) Several bike paths connect The University of British Columbia to nearby residential areas. These bike paths deteriorate because of heavy wear and harsh weather. As an illustration of the methodology in Sec. 1.4, we provide the following model to determine optimal maintenance schedules.

Assume that 200-m segments of the roadway are classified into four states on the basis of wear at annual inspection. State 3 denotes heavily worn, state 2 denotes moderate wear, state 1 slight wear, and state 0 denotes no wear. On the basis of these observations, three maintenance actions can be chosen; action 0 denotes routine maintenance, action 1 corresponds to applying a thin overcoat of paving material, and action 2 corresponds to applying a thick overcoat. The effect of these actions are described in terms of the following transition probabilities. They give the probability that a segment of the bike path is in a particular state next year given the specified maintenance action is chosen in the current state this year:

Current State 0

	Next State			
Action	0	1	2	3
0	0.1	0.6	0.2	0.1
1	0.8	0.2	0	0
2	0.95	0.05	0	0

Current State 1

	Next State			
Action	0	1	2	3
0	0	0.1	0.6	0.3
1	0.7	0.2	0.1	0
2	0.85	0.1	0.05	0

Current State 2

	Next State			
Action	0	1	2	3
0	0	0	0.2	0.8
1	0.3	0.4	0.2	0.1
2	0.65	0.2	0.1	0.05

Current State 3

	Next State			
Action	0	1	2	3
0	0	0	0	1.0
1	0	0.6	0.2	0.2
2	0.5	0.5	0	0

Assume that action 0 costs 0, action 1 costs $1 per segment, and action 2 costs $2 per segment.

a. Verify that this model is unichain.

b. Show that if $p(0\,|\,0, 2) = 1$ instead of 0.95, that the model would be multichain and communicating.

c. Find a maintenance policy that minimizes long run average cost, subject to a constraint that no more than 10% of the segments can be in the heavily worn state.

d. Suppose that the maintenance department wishes to spend at most $.50 per segment per year. Find an optimal policy under this constraint and comment on its effect on the quality of the bike paths.

8.47. a. Find a policy to minimize long run average cost in the inventory model in Problem 4.33.

b. Instead of using a cost to penalize unsatisfied demand, solve a constrained version of the problem under the constraint that the expected proportion of demand not filled from stock on hand is at most .025.

CHAPTER 9

The Average Reward Criterion—Multichain and Communicating Models

The previous chapter introduced average models and emphasized the importance of taking into account chain structure when analyzing average reward models. In it, we concentrated on unichain models, that is, models in which all stationary policies have *at most one* recurrent class. The essential difference between unichain models and multichain models, which are the focus of this chapter, is that when Markov chains corresponding to stationary policies have more than one recurrent class, a single optimality equation may not be sufficient to characterize the optimal policy and its gain. Consequently, theory and algorithms are more complex than for unichain models.

As in previous chapters, we begin with an analysis of the optimality equation and its properties and then discuss policy iteration, linear programming, and value iteration. We conclude this chapter with a discussion of communicating and weakly communicating models which have wide applicability and some good computational properties.

Example 8.3.2 presented an inventory model which contains multichain decision rules, yet the optimal policy has unichain transition probabilities. One might suspect that this is always the case in realistic applications of Markov decision processes. The following example provides a model in which the optimal policy may indeed be multichain.

Example 9.0.1. Each month an individual must decide how to allocate his wealth between consumption and investment. There are a choice of several investment opportunities with differing payoffs and risks. Wealth at the start of the subsequent month equals current wealth less the amount of wealth spent on consumables, plus the payoff on investments, and additional wealth acquired through wages (or welfare). The individual's objective is to maximize average consumption subject to a lower bound on monthly consumption.

We model this as a Markov decision process as follows. Let the state represent the individual's wealth at the start of a month, and let the set of actions represent the quantity to consume and the available investment opportunities. Note that the set of actions becomes larger with increasing wealth.

Clearly there exists a set of low wealth states which are closed. In these states, the individual must allocate all wealth to consumables, so there is no opportunity to leave them. In intermediate and high wealth states, by choosing high levels of consumption and/or poor investments, the individual can reach the low wealth states and subsequently remain there forever. Instead, the individual can choose lower levels of consumption and more conservative investments, and remain in the set of intermediate and high wealth states. Under such a policy, these states would also be closed.

Hence the model is multichain and an optimal policy would have a multichain transition structure (Problem 9.6).

The theory of multichain models is not as complete as that for unichain models, and the analyses are often more subtle. Strong assumptions are required to analyze finite-state models with compact action sets and countable-state models. More general models require the theory of Markov chains for abstract spaces, which is well beyond the scope of this book. *Except where noted, we assume a model with a finite number of states and actions.* We suggest reviewing Secs. 8.2 and 8.3 before delving into this chapter.

9.1   AVERAGE REWARD OPTIMALITY EQUATIONS: MULTICHAIN MODELS

Theorem 8.2.6 in Sec. 8.2.3 showed that, for a fixed stationary policy d^∞, *two* linear equations uniquely characterize its gain and determine its bias up to a vector u satisfying $(I - P_d)u = 0$. A pair of optimality equations generalizes this result to MDP's. The following simple example shows that the unichain optimality equation need not have a solution in a multichain model.

Example 9.1.1.   Let $S = \{s_1, s_2, s_3\}$; $A_{s_1} = \{a_{1,1}, a_{1,2}\}$, $A_{s_2} = \{a_{2,1}, a_{2,2}\}$, and $A_{s_3} = \{a_{3,1}\}$; $r(s_1, a_{1,1}) = 3$, $r(s_1, a_{1,2}) = 1$, $r(s_2, a_{2,1}) = 0$, $r(s_2, a_{2,2}) = 1$, and $r(s_3, a_{3,1}) = 2$;   and $p(s_1|s_1, a_{1,1}) = 1$, $p(s_2|s_1, a_{1,2}) = 1$, $p(s_2|s_2, a_{2,1}) = 1$, $p(s_3|s_2, a_{2,2}) = 1$,   and $p(s_3|s_3, a_{3,1}) = 1$ (Fig. 9.1.1). Clearly this is a multichain

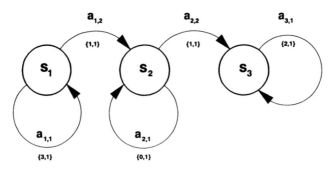

Figure 9.1.1   Symbolic representation of Example 9.1.1.

model. The unichain optimality equations (8.4.1) for this model may be written as

$$h(s_1) + g = \max\{3 + h(s_1), 1 + h(s_2)\},$$
$$h(s_2) + g = \max\{0 + h(s_2), 1 + h(s_3)\},$$
$$h(s_3) + g = 2 + h(s_3).$$

The third equation indicates that $g = 2$, which on substitution into the first equation shows that the system is inconsistent. Allowing nonconstant g is the first step to remedying this difficulty. However, without an additional system of equations, the system need not determine the optimal gain.

9.1.1 Multichain Optimality Equations

Multichain average reward MDP's require a pair of optimality equations to compute the optimal gain and determine optimal policies. The proof of Theorem 9.1.2 below shows why this is the case.

We refer to the system of equations

$$\max_{a \in A_s} \left\{ \sum_{j \in S} p(j|s, a)g(j) - g(s) \right\} = 0 \qquad (9.1.1)$$

and

$$\max_{a \in B_s} \left\{ r(s, a) - g(s) + \sum_{j \in S} p(j|s, a)h(j) - h(s) \right\} = 0, \qquad (9.1.2)$$

where $B_s = \{a' \in A_s: \sum_{j \in S} p(j|s, a')g(j) - g(s) = 0\}$, as the *multichain optimality equations*. We say that this system of equations is *nested* because the set of actions over which the maximum is sought in the second equation depends on the set of actions which attain the maximum in the first equation when g is substituted into it.

A solution to these nested equations is a pair of functions g and h which, for each $s \in S$, satisfy

$$\sum_{j \in S} p(j|s, a)g(j) - g(s) \le 0 \qquad (9.1.3)$$

for all $a \in A_s$ with equality holding in (9.1.3) for at least one $a' \in A_s$ and for each a' for which (9.1.3) holds with equality,

$$r(s, a') + \sum_{j \in S} p(j|s, a')h(j) - g(s) - h(s) \le 0 \qquad (9.1.4)$$

with equality holding in (9.1.4) for one such a'.

We express these equations in vector notation as

$$\max_{d \in D} \{(P_d - I)g\} = 0 \qquad (9.1.5)$$

and

$$\max_{d \in E} \{r_d - g + (P_d - I)h\} = 0, \qquad (9.1.6)$$

where $E = \{d \in D: d(s) \in B_s\}$.

We refer to (9.1.1) or (9.1.5) as the *first optimality equation* and (9.1.2) or (9.1.6) as the *second optimality equation*.

In unichain models and any other models in which all policies have constant gain, the first optimality equation is redundant, so that $B_s = A_s$ and $D = E$. This is because when all decision rules are unichain, g does not vary with s, so that $P_d g = g$ for all $d \in D$. In this case, the system of optimality equations reduces to the unichain optimality equation.

As the example below suggests, it is difficult to determine E explicitly, so instead we carry out some aspects of our analysis through the *modified optimality equations* in which A_s replaces B_s in (9.1.2) or, equivalently, D replaces E in (9.1.6). The modified optimality equations are (9.1.1) together with

$$\max_{a \in A_s} \left\{ r(s, a) - g(s) + \sum_{j \in S} p(j|s, a)h(j) - h(s) \right\} = 0, \qquad (9.1.7)$$

or (9.1.5) together with

$$\max_{d \in D} \{r_d - g + (P_d - I)h\} = 0. \qquad (9.1.8)$$

Note that, in the modified optimality equations, different actions may obtain the maximum in (9.1.1) and (9.1.7) for some s, in which case different decision rules may attain the maximum in (9.1.5) and (9.1.8). This suggests that solutions of the modified optimality equations might not directly determine optimal policies.

Note that the modified optimality equations hold if and only if

$$0 \geq \sum_{j \in S} p(j|s, a)g(j) - g(s) \qquad (9.1.9)$$

and

$$0 \geq r(s, a) + \sum_{j \in S} p(j|s, a)h(j) - h(s) - g(s) \qquad (9.1.10)$$

for all $s \in S$ and $a \in A_s$. On the other hand, the unmodified optimality equations require a *nonlinear* system of inequalities.

We use the optimality equations to develop a policy iteration algorithm, while we use the modified optimality equations to establish optimality properties of solutions and develop linear programming algorithms. *Note that in unichain models, there is no need to distinguish these two equation systems.*

We illustrate both sets of optimality equations for the model in Example 9.1.1.

Example 9.1.1. (ctd.). The *modified optimality equations* are

$$g(s_1) = \max\{g(s_1), g(s_2)\}$$
$$g(s_2) = \max\{g(s_2), g(s_3)\},$$
$$g(s_3) = g(s_3),$$
$$h(s_1) + g(s_1) = \max\{3 + h(s_1), 1 + h(s_2)\},$$
$$h(s_2) + g(s_2) = \max\{0 + h(s_2), 1 + h(s_3)\},$$
$$h(s_3) + g(s_3) = 2 + h(s_3).$$

The multichain optimality equations (9.1.3) and (9.1.4) are considerably more complex, because the equations involving h depend on which action attains the maximum in the first set of equations. They depend on the values of $g(s_1)$, $g(s_2)$, and $g(s_3)$. When $g(s_1) > g(s_2)$, equation

$$h(s_1) + g(s_1) = 3 + h(s_1)$$

replaces

$$h(s_1) + g(s_1) = \max\{3 + h(s_1), 1 + h(s_2)\},$$

and, if $g(s_1) = g(s_2)$, then

$$h(s_1) + g(s_1) = \max\{3 + h(s_1), 1 + h(s_2)\}.$$

Note that $g(s_2) > g(s_1)$ is inconsistent with the first optimality equation.

Observe that the first set of equations establishes relationships on g, so that its values are consistent with the transition structure of the underlying Markov chains.

9.1.2 Properties of Solutions of the Optimality Equations

In this section we establish that

i. The existence of a solution to the multichain optimality equations implies the existence of a solution to the modified optimality equations, and

ii. solutions of the modified optimality equations characterize the optimal average reward.

These results enable us to demonstrate the existence of an optimal policy after showing in the next section that the multichain optimality equations have a solution.

The following result establishes the relationship between the modified and unmodified optimality equations.

Proposition 9.1.1. Assume that S and A_s are finite, and that there exists bounded functions g^* and h^* which satisfy the optimality equations (9.1.1) and (9.1.2). Then there exists an $M > 0$ such that g^* and $h^* + Mg^*$ satisfy the modified optimality equations (9.1.1) and (9.1.7).

Proof. Suppose that (9.1.10) does not hold for some s and a. This means that

$$0 < r(s, a) + \sum_{j \in S} p(j|s, a)h^*(j) - h^*(s) - g^*(s) \equiv c.$$

Consequently there must have been a strict inequality at a in (9.1.9); that is, $a \in A_s/B_s$, and

$$0 > \sum_{j \in S} p(j|s, a)g^*(j) - g^*(s) \equiv d.$$

We now show that, by suitable choice of M, (9.1.7) holds with $h'(s) = h^*(s) + Mg^*(s)$. Since

$$r(s, a) + \sum_{j \in S} p(j|s, a)h'(j) - h'(s) - g^*(s)$$

$$= r(s, a) + \sum_{j \in S} p(j|s, a)[h^*(j) + Mg^*(j)] - [h^*(s) + Mg^*(s)] - g^*(s)$$

$$= c + Md$$

choosing $M \geq |c/d|$ ensures that

$$0 \geq r(s, a) + \sum_{j \in S} p(j|s, a)h'(j) - h'(s) - g^*(s). \tag{9.1.11}$$

Since there are only finitely many states and actions, we can choose an M for which (9.1.11) holds for all s and a.   □

Problem 9.1 provides conditions under which the proposition above holds for models with compact action sets. The following theorem enables us to establish average optimality properties of solutions of the modified optimality equations. It generalizes Theorem 8.4.1 for unichain models. Definitions of g_+^* and g_-^* appear in Sec. 8.1.

Comparison of the proof of Theorem 9.1.2 to that of Theorem 8.4.1 reveals the role of the first optimality equation. Note that Theorem 9.1.2 holds for countable S, but in the next subsection we only establish existence of the requisite g and h for finite S.

Theorem 9.1.2. Let S be countable.

a. Suppose there exists $(g, h) \in V \times V$ which satisfies

$$(P_d - I)g \leq 0 \tag{9.1.12}$$

and

$$r_d + (P_d - I)h - g \leq 0 \tag{9.1.13}$$

for all $d \in D^{MD}$. Then

$$g \geq g_+^*. \tag{9.1.14}$$

b. Suppose there exists $(g, h) \in V \times V$ and a $d' \in D^{MD}$ which satisfy

$$(P_{d'} - I)g \geq 0 \tag{9.1.15}$$

and

$$r_{d'} + (P_{d'} - I)h - g \geq 0. \tag{9.1.16}$$

Then

$$g \leq \sup_{d \in D} g_-^{d^{\infty}} \leq g_-^*. \tag{9.1.17}$$

c. Suppose there exists a $(g, h) \in V \times V$ which satisfies the modified optimality equations (9.1.5) and (9.1.8). Then

$$g = g_-^* = g_+^* = g^*. \tag{9.1.18}$$

Proof. From Proposition 6.2.1, (9.1.12) and (9.1.13) hold for all $d \in D^{MR}$. Let $\pi = (d_1, d_2, \dots) \in \Pi^{MR}$. Then, (9.1.13) implies that

$$g \geq r_{d_1} + (P_{d_1} - I)h.$$

Applying (9.1.13) with $d = d_2$ and multiplying this expression by P_{d_1} and applying inequality (9.1.12) yields

$$g \geq P_{d_1} g \geq P_{d_1} r_{d_2} + (P_{d_1} P_{d_2} - P_{d_1})h$$

(note the role of the first optimality equation here). Repeating this argument (or using induction) shows that, for $2 \leq n$,

$$g \geq P_{d_1} P_{d_2} \cdots P_{d_{n-1}} r_{d_n} + P_{d_1} P_{d_2} \cdots P_{d_{n-1}}(P_{d_n} - I)h.$$

Summing these expressions over n and noting (8.1.3) shows that, for all $\pi \in \Pi^{MR}$ and any $N \geq 1$,

$$Ng \geq v_{N+1}^{\pi} + (P_N^{\pi} - I)h.$$

Since $h \in V$, $P_N^{\pi} h \in V$, so that $\lim_{N \to \infty} N^{-1}(P_N^{\pi} - I)h(s) = 0$ for each $s \in S$. Therefore

$$g \geq \limsup_{N \to \infty} \frac{1}{N} v_{N+1}^{\pi}$$

for all $\pi \in \Pi^{MR}$. Extension to $\pi \in \Pi^{HR}$ follows from Theorem 5.5.3(c), so that (9.1.14) follows.

To prove (b), write (9.1.16) as

$$g \leq r_{d'} + (P_{d'} - I)h,$$

multiply this expression by $P_{d'}$, and apply (9.1.15) to establish that

$$g \leq P_{d'}g \leq P_{d'}r_{d'} + (P_{d'}^2 - P_{d'})h.$$

Repeating this argument [as in the proof of part (a)] and summing inequalities yields

$$ng \leq r_{d'} + P_{d'}r_{d'} + \cdots + P_{d'}^{n-1}r_{d'} + (P_{d'}^n - I)h.$$

Dividing both sides of this expression by n, and noting that $P_{d'}^n h \in V$, establishes that $g \leq g^{(d')^\infty}$, from which the result follows.

Since the hypotheses of (a) and (b) are satisfied when g and h satisfy the modified optimality equations, part (c) follows from combining (a) and (b). □

Part (c) of this theorem establishes two important consequences of the existence of a solution (g, h) to the modified optimality equations.

1. That g equals the optimal average expected reward.
2. That the optimal lim sup average reward equals the optimal lim inf average reward.

We now extend this result to solutions of the optimality equations by combining Theorem 9.1.2 with Proposition 9.1.1.

Theorem 9.1.3. Let S be countable and suppose $(g, h) \in V \times V$ satisfies the optimality equations (9.1.5) and (9.1.6). Then $g = g^* = g_+^* = g_-^*$.

9.1.3 Existence of Solutions to the Optimality Equations

Establishing the existence of solutions to the pair of optimality equations follows similar but more intricate arguments then for unichain models. The proof below accounts for the possibility that the gain need not be constant, and that different stationary policies may have different recurrent classes. As in the proof of Theorem 8.4.3, we relate results for the discounted models to that for the average reward model through the partial Laurent expansion of Corollary 8.2.4.

Theorem 9.1.4. Suppose S and A_s are finite. Then there exists a solution to the multichain optimality equations (9.1.1) and (9.1.2) [or (9.1.5) and (9.1.6)].

Proof. Let (λ_n) be a sequence of discount rates converging to 1 from below. By Theorem 6.2.10(a) there exists a stationary discount optimal policy for each. Under the finiteness assumption there exists only finitely many stationary policies, so there exists a subsequence $\{\lambda_{n'}\}$ for which the policy $(d^*)^\infty$ is discount optimal. From here on, denote the subsequence by $\{\lambda_n\}$.

Choose $d \in D$. As a consequence of Theorem 6.2.6,

$$0 = r_{d^*} + (\lambda_n P_{d^*} - I)v_{\lambda_n}^{(d^*)^\infty} \geq r_d + (\lambda_n P_d - I)v_{\lambda_n}^{(d^*)^\infty}$$

for $n = 0, 1, 2, \ldots$. Substituting the partial Laurent series expansion for $v_{\lambda_n}^{(d^*)^\infty}$ into the expression on the right-hand side above yields

$$0 \geq r_d + (\lambda_n P_d - I)\left[\frac{g^{(d^*)^\infty}}{1 - \lambda_n} + h^{(d^*)^\infty} + f(\lambda_n)\right],$$

where $f(\lambda_n)$ denotes an arbitrary vector which converges to 0 as $\lambda_n \to 1$. Expressing λ_n as $(\lambda_n - 1) + 1$, noting that $h^{(d^*)^\infty}$ is bounded, and rearranging terms yields

$$0 \geq \frac{(P_d - I)g^{(d^*)^\infty}}{1 - \lambda_n} + \left[r_d - g^{(d^*)^\infty} + (P_d - I)h^{(d^*)^\infty}\right] + f(\lambda_n). \quad (9.1.19)$$

For (9.1.19) to hold for all n requires that

$$(P_d - I)g^{(d^*)^\infty} \leq 0$$

and, if $(P_d - I)g^{(d^*)^\infty}(s) = 0$, that

$$r_d(s) - g^{(d^*)^\infty}(s) + (P_d - I)h^{(d^*)^\infty}(s) \leq 0.$$

Theorem 8.2.6 shows that $g^{(d^*)^\infty}$ and $h^{(d^*)^\infty}$ satisfy

$$0 = (P_{d^*} - I)g^{(d^*)^\infty}$$

and

$$0 = r^{d^*} - g^{(d^*)^\infty} + (P_{d^*} - I)h^{(d^*)^\infty}$$

Since d was arbitrary, $g^{(d^*)^\infty}$ and $h^{(d^*)^\infty}$ satisfy (9.1.1) and (9.1.2) [or (9.1.5) and (9.1.6)]. □

As a consequence of Proposition 9.1.1, we have the following results.

Corollary 9.1.5. Suppose S and A_s are finite, then there exists a solution to the modified optimality equations (9.1.1) and (9.1.7) [or (9.1.1) and (9.1.8)].

Theorem 9.1.6. Suppose S and A_s are finite. Then

a. $g^* = g_+^* = g_-^*$,
b. there exists an $h \in V$ for which (g^*, h) satisfy the optimality equations (9.1.5) and (9.1.6), and
c. there exists an $h' \in V$ for which (g^*, h') satisfy the modified optimality equations (9.1.5) and (9.1.8).

Proof. From Theorem 9.1.4, there exists a solution to the optimality equations in $V \times V$. From Theorem 9.1.3(c), it follows that $g = g^*$. Part (c) follows from Corollary 9.1.5.   □

9.1.4 Identification and Existence of Optimal Policies

In the previous section we established the existence of solutions to the optimality equations and modified optimality equations. We now show how to use such solutions to identify average optimal stationary policies.

Theorem 9.1.7.   Let S and A_s be finite.

a. Suppose $g \in V$ and $h \in V$ satisfy the multichain optimality equations, that

$$P_{d^*}g = g \qquad (9.1.20)$$

and

$$d^* \in \arg\max_{d \in E} \{r_d + P_d h\}. \qquad (9.1.21)$$

Then $(d^*)^\infty$ is average optimal.

b. Suppose g^* and $h^* \in V$ satisfy the modified optimality equations, that (9.1.20) holds, and

$$d^* \in \arg\max_{d \in D} \{r_d + P_d h^*\}. \qquad (9.1.22)$$

Then $(d^*)^\infty$ is average optimal.

Proof.   We prove part (a). Under the hypotheses of this theorem, g^* and h^* satisfy (8.2.11) and (8.2.12), so, from Theorem 8.2.6(b), it follows that $g = P_{d^*}^* r_{d^*} = g^{(d^*)^\infty}$. The result follows from Theorem 9.1.3. The proof of (b) is similar, and left as an exercise.   □

The difference between the two parts of this theorem is rather subtle. In part (a), we choose the arg max over the reduced decision set E. If we do not, the maximum of $r_d + P_d h^*$ over D may be greater than the maximum over E and an optimal d need not satisfy (9.1.21). In part (b), the arg max is sought over the entire decision set D.

Note that this result contains its unichain analog, Theorem 8.4.4, as a special case. This is because the gain is constant in unichain models, so that (9.1.20) holds for any d, so that $D = E$ and (9.1.21) becomes

$$d^* \in \arg\max_{d \in D} \{r_d + P_d h^*\}.$$

The following theorem establishes the existence of average optimal policies in finite models.

Theorem 9.1.8.   Suppose S is finite and A_s is finite for each $s \in S$. Then there exists a deterministic stationary optimal policy.

Proof. From Theorem 9.1.4, there exists a solution g^*, h^* to the optimality equations. Since D is finite, there exists a $d^* \in D$ which satisfies (9.1.20). Since $d^* \in E$, and E is finite, (9.1.21) holds and the result follows from Theorem 9.1.7.   □

We can summarize results regarding properties of the optimality equations as follows:

1. There exists a solution to the optimality equations in finite-state and -action models.
2. The optimality equation uniquely characterizes the optimal gain.
3. A stationary policy derived from a d^* satisfying (9.1.20) and (9.1.21) is optimal.

Note that analogous results hold for the modified optimality equations.
We now apply these results to Example 9.1.1.

Example 9.1.1 (ctd.).   Inspection shows that $g^*(s_1) = 3$, and $g^*(s_2) = g^*(s_3) = 2$. Consequently, the second set of optimality equations becomes

$$h(s_1) + g^*(s_1) = 3 + h(s_1)$$

$$h(s_2) + g^*(s_2) = \max\{0 + h(s_2), 1 + h(s_3)\},$$

$$h(s_3) + g^*(s_3) = 2 + h(s_3),$$

which has the solution $h(s_1) = a$, $h(s_2) = -1 + b$, and $h(s_3) = b$, where a and b are arbitrary constants. The optimal policy $(d^*)^\infty$ selects $a_{1,1}$ in s_1, $a_{2,2}$ in s_2, and $a_{3,1}$ in s_3. Following Theorem 8.2.6, its bias $h^{(d^*)^\infty}$ is obtained by setting $a = b = 0$ above. Observe that $h^{(d^*)^\infty}(s_2) = -1$ is the difference in the total reward obtained starting in s_2 instead of in s_3.

Note that as in unichain models $(d^*)^\infty$ may be average optimal but $g^{(d^*)^\infty}$ and $h^{(d^*)^\infty}$ need not satisfy the optimality equations. Example 8.4.3 provides an illustration of such a policy.

9.2   POLICY ITERATION FOR MULTICHAIN MODELS

In this section we provide, illustrate, and analyze a policy iteration algorithm for solving the multichain optimality equations. Because we seek a solution of a pair of equations, the algorithm is more complex than that for unichain models. Note that if we are unable to determine that all stationary policies are unichain prior to solution of the problem, we *must* use such a procedure to solve it. Another benefit of the analysis herein is that, in addition to demonstrating convergence of the algorithm, we constructively establish the existence of an optimal policy and a solution to the optimality equations.

9.2.1 The Algorithm

Policy iteration in the multichain case consists of an improvement and an evaluation step. In the improvement step, a decision rule is sought which provides a strict improvement through the first optimality equation and, if none is available, through the second optimality equation. When no improvement is possible through either equation, the algorithm stops. A formal statement of it follows.

The Multichain Policy Iteration Algorithm

1. Set $n = 0$ and select an arbitrary decision rule $d_0 \in D$.
2. (Policy evaluation) Obtain a $g_n \in V$ and an $h_n \in V$ which satisfy

$$(P_{d_n} - I)g = 0, \tag{9.2.1}$$

$$r_{d_n} - g + (P_{d_n} - I)h = 0 \tag{9.2.2}$$

subject to conditions 9.2.1, 9.2.2, or 9.2.3 below.
3. (Policy improvement)
   a. Choose $d_{n+1} \in D$ which satisfies

$$d_{n+1} \in \arg\max_{d \in D} \{P_d g_n\}, \tag{9.2.3}$$

   setting $d_{n+1} = d_n$ if possible. If $d_{n+1} = d_n$, go to (b); otherwise increment n by 1 and return to step 2.
   b. Choose $d_{n+1} \in D$ to satisfy

$$d_{n+1} \in \arg\max_{d \in D} \{r_d + P_d h_n\}, \tag{9.2.4}$$

   setting $d_{n+1} = d_n$ if possible.
4. If $d_{n+1} = d_n$, stop and set $d^* = d_n$. Otherwise, increment n by 1 and return to step 2.

The above algorithm yields a sequence of decision rules $\{d_n\}$ and corresponding gains $\{g_n\}$ with $g_n = g^{(d_n)^\infty}$. Equations (9.2.1) and (9.2.2) uniquely determine the gain, but h_n is unique up to a u satisfying

$$(P_{d_n} - I)u = 0.$$

Choosing h_n to fulfill any of the following conditions ensures convergence of the algorithm.

Condition 9.2.1. Augment (9.2.1) and (9.2.2) by the equation

$$-h + \left(P_{d_n} - I \right) w = 0 \tag{9.2.5}$$

and solve the system of three matrix equations.

Condition 9.2.2. Add the restriction that

$$P_{d_n}^* h = 0. \tag{9.2.6}$$

Condition 9.2.3. Determine the chain structure of P_{d_n} and denote its recurrent classes by $R_1, \ldots, R_{k_n}$. Solve (9.2.1) and (9.2.2) subject to the condition that $h_n(s_{j_i}) = 0$, where j_i denotes the minimal index such that $s_j \in R_i$ for $i = 1, \ldots, k_n$.

As a consequence of Corollary 8.2.9 and Theorem 8.2.6, Conditions 9.2.1 and 9.2.2 ensure that $h_n = h^{(d_n)^{\infty}}$, but solving (9.2.1), (9.2.2), and (9.2.5) requires less work than solving (9.2.1), (9.2.2), and (9.2.6). Since determining the chain structure of P_{d_n} by the Fox-Landi algorithm of Section A.3 of Appendix A requires at most $O(|S|^2)$ operations, the method implied by Condition 9.2.3 may be the most efficient. Note that when h_n satisfies (9.2.1), (9.2.2) and Condition 9.2.3, it need not equal $h^{(d_n)^{\infty}}$; however, its relative differences within recurrent classes are the same. Using the third approach, we specify states at which h_n equals 0. Other specifications may be used but arbitrary choice of h_n may cause cycling.

The improvement step of the algorithm consists of two phases. First, improvement is sought through the first optimality equation. If no *strict* improvement is possible, we seek an improved decision rule through the second optimality equation. When none is available, the algorithm terminates and, when an improvement occurs, the improved policy is evaluated at the subsequent iteration. The improvement step may be more transparent when expressed in component notation as follows.

a. For each $s \in S$, choose $d_{n+1}(s) \in D$ to satisfy

$$d_{n+1}(s) \in \arg\max_{a \in A_s} \left\{ \sum_{j \in S} p(j|s, a) g_n(j) \right\} \tag{9.2.7}$$

setting $d_{n+1}(s) = d_n(s)$ if possible. If $d_{n+1}(s) = d_n(s)$ for all $s \in S$, go to (b); otherwise increment n by 1 and return to step 2.

b. For each $s \in S$, choose $d_{n+1}(s) \in D$ to satisfy

$$d_{n+1}(s) \in \arg\max_{a \in A_s} \left\{ r(s, a) + \sum_{j \in S} p(j|s, a) h_n(j) \right\}, \tag{9.2.8}$$

setting $d_{n+1}(s) = d_n(s)$ if possible.

Note, that, for unichain models, the first equation in (9.2.1) and part (a) of the improvement step are redundant, so that the algorithm reduces to the unichain policy iteration algorithm.

9.2.2 An Example

Before proving the finite convergence of policy iteration, we illustrate this algorithm by solving Example 9.1.1. We use Condition 9.2.3 to determine a unique solution of the evaluation equations; however, in this example the first two conditions also give the same h_n values.

Set $n = 0$ and choose $d_0(s_1) = a_{1,2}$, $d_0(s_2) = a_{2,1}$ and $d_0(s_3) = a_{3,1}$. Observe that this policy generates a Markov chain with recurrent classes $\{s_2\}$ and $\{s_3\}$ and a transient class $\{s_1\}$. The general solution of (9.2.1) and (9.2.2) has $h_0(s_1) = a + 1$, $h_0(s_2) = a$, and $h_0(s_3) = b$, where a and b are arbitrary constants. Setting $h_0(s_2) = h_0(s_3) = 0$ yields

s	$g_0(s)$	$h_0(s)$
s_1	0	1
s_2	0	0
s_3	2	0

We now seek an improved policy through step 3(a). Equations (9.2.3) are

$$\arg\max\left\{ \sum_{j\in S} p(j|s_1, a_{1,1}) g_0(j), \sum_{j\in S} p(j|s_1, a_{1,2}) g_0(j) \right\}$$

$$= \arg\max\{0, 0\} = \{a_{1,1}, a_{1,2}\}$$

$$\arg\max\left\{ \sum_{j\in S} p(j|s_2, a_{2,1}) g_0(j), \sum_{j\in S} p(j|s_2, a_{2,2}) g_0(j) \right\}$$

$$= \arg\max\{0, 2\} = \{a_{2,2}\}$$

$$\arg\max\left\{ \sum_{j\in S} p(j|s_3, a_{3,1}) g_0(j) \right\} = \arg\max\{2\} = \{a_{3,1}\}.$$

Since $d_1(s_2) = a_{2,2} \neq d_0(s_2)$ we identify a new decision rule $d_1(s_1) = a_{1,2}$, $d_1(s_2) = a_{2,2}$, and $d_1(s_3) = a_{3,1}$. This policy generates a Markov chain with one recurrent class $\{s_3\}$ and a set of transient states $\{s_1, s_2\}$. Solving the evaluation equations with $h_1(s_3) = 0$ yields

s	$g_1(s)$	$h_1(s)$
s_1	2	-2
s_2	2	-1
s_3	2	0

We now seek an improved policy. Since g_1 is constant, step 3(a) does not identify an improved policy. We turn to step 3(b). Equation (9.2.8) become

$$\arg\max\left\{r(s_1,a_{1,1}) + \sum_{j\in S}p(j|s_1,a_{1,1})h_1(j), r(s_1,a_{1,2}) + \sum_{j\in S}p(j|s_1,a_{1,2})h_1(j)\right\}$$

$$= \arg\max\{3 + (-2), 1 + (-1)\} = \{a_{1,1}\}$$

and

$$\arg\max\left\{r(s_2,a_{2,1}) + \sum_{j\in S}p(j|s_2,a_{2,1})h_1(j), r(s_2,a_{2,2}) + \sum_{j\in S}p(j|s_2,a_{2,2})h_1(j)\right\}$$

$$= \arg\max\{0 + (-1), 1 + 0\} = \{a_{2,2}\}.$$

Since $d_2(s_1) = a_{1,1} \neq d_1(s_1)$, we identify a new decision rule $d_2(s_2) = a_{1,2}$, $d_1(s_2) = a_{2,2}$, and $d_1(s_3) = a_{3,1}$. This policy generates a Markov chain with two recurrent classes $\{s_1\}$ and $\{s_3\}$ and a set of transient states $\{s_2\}$. Solving the evaluation equations with $h_2(s_1) = h_2(s_3) = 0$ yields

s	$g_2(s)$	$h_2(s)$
s_1	3	0
s_2	2	-1
s_3	2	0

Since the subsequent pass through the algorithm identifies no further improvements, we identify $(d_2)^{\infty}$ as an optimal policy. Note that $g^{(d_2)^{\infty}} = g_2$ and $h^{(d_2)^{\infty}} = h_2$.

Observe that at each step of the algorithm g_n exceeds its previous value in at least one state. This need not always be the case. In other examples, an improved policy identified in step 3 may yield $g_{n+1} = g_n$ and $h_{n+1}(s) > h_n(s)$ for some s. Also note that, at the first pass through the algorithm, improvement occurred through step 3(a), and in the second pass through step 3(b).

9.2.3 Convergence of the Policy Iteration in Multichain Models

We take a rather circuitous route to establish convergence of multichain policy iteration. Our approach is to show that the algorithm generates a sequence of decision rules $\{d_n\}$ for which

$$\liminf_{\lambda\uparrow 1}\left[v_\lambda^{(d_{n+1})^{\infty}}(s) - v_\lambda^{(d_n)^{\infty}}(s)\right] > 0$$

for some s until no further improvements are possible. Since there are only finitely many deterministic stationary policies, this occurs in finitely many iterations.

We use the partial Laurent series expansion of $v_\lambda^{d^{\infty}}$ and the observation that, for scalars x and y, the quantity

$$\frac{x}{1-\lambda} + y$$

is positive for λ sufficiently close to 1 when $x > 0$ or $x = 0$ and $y > 0$.

Our approach in this section establishes convergence when $h_n = h^{(d_n)^\infty}$ which occurs if it is determined through Conditions 9.2.1 or 9.2.2. We establish convergence under Condition 9.2.3 in the next section.

We begin with some technical lemmas and results. For Markovian decision rules δ and d, let (δ, d^∞) denote the policy which uses δ at the first decision epoch and then uses d^∞. Recall that for $0 \le \lambda < 1$,

$$v_\lambda^{(\delta, d^\infty)} = r_\delta + \lambda P_\delta v_\lambda^{d^\infty}, \qquad (9.2.9)$$

and let

$$w_\lambda(\delta, d) \equiv v_\lambda^{(\delta, d^\infty)} - v_\lambda^{d^\infty} = r_\delta + (\lambda P_\delta - I) v_\lambda^{d^\infty}. \qquad (9.2.10)$$

The following lemma summarizes some key properties of $w_\lambda(\delta, d)$.

Lemma 9.2.1. Suppose $d \in D^{MD}$ and $\delta \in D^{MD}$. Then, for $0 \le \lambda < 1$

a. $(I - \lambda P_\delta)^{-1} w_\lambda(\delta, d) = v_\lambda^{\delta^\infty} - v_\lambda^{d^\infty}$.
b. If $\delta(s) = d(s)$, then $w_\lambda(\delta, d)(s) = 0$.

c.
$$w_\lambda(\delta, d) = \frac{(P_\delta - I) g^{d^\infty}}{1 - \lambda} + \left[r_\delta - g^{d^\infty} + (P_\delta - I) h^{d^\infty} \right] + f(\lambda) \quad (9.2.11)$$

where $f(\lambda) \to 0$ as $\lambda \uparrow 1$.

Proof. Part (a) follows from the second representation for $w_\lambda(\delta, d)$ in (9.2.10) and noting that $v_\lambda^{\delta^\infty} = (I - \lambda P_\delta)^{-1} r_\delta$. Part (b) follows by noting that

$$r_d(s) + \lambda P_d v_\lambda^{d^\infty}(s) = v_\lambda^{d^\infty}(s),$$

and part (c) follows from the same principles used to derive (9.1.19). $\qquad\square$

Proposition 9.2.2. Let $d \in D$ and suppose there exists a $\delta \in D$ for which $\delta(s) \ne d(s)$ implies that either

a.
$$P_\delta g^{d^\infty}(s) > P_d g^{d^\infty}(s) \qquad (9.2.12)$$

or

b. $P_\delta g^{d^\infty}(s) = P_d g^{d^\infty}(s)$ and

$$r_\delta(s) + P_\delta h^{d^\infty}(s) > r_d(s) + P_d h^{d^\infty}(s). \qquad (9.2.13)$$

Then

$$\liminf_{\lambda \uparrow 1} \left[v_\lambda^{\delta^\infty}(s) - v_\lambda^{d^\infty}(s) \right] > 0 \qquad (9.2.14)$$

for some $s \in S$.

Proof. Let $S_1 \subseteq S$ denote the set of states at which (a) holds. Then

$$P_\delta g^{d^\infty}(s) > P_d g^{d^\infty}(s) = g^{d^\infty}(s), \qquad s \in S$$

so that the first term in (9.2.11) is strictly positive for all $s \in S_1$ and equal to 0 for $s \in S/S_1$.

Let S_2 denote the set of states at which (b) holds. Then $S_2 \subseteq S/S_1$ and

$$r_\delta(s) + P_\delta h^{d^\infty}(s) > r_d(s) + P_d h^{d^\infty}(s) = h^{d^\infty}(s) + g^{d^\infty}(s),$$

so that the second expression in (9.2.11) is positive for all $s \in S_2$. By Lemma 9.2.1(b), $w_\lambda(s, d) = 0$ for $s \in S/(S_1 \cup S_2)$.

Consequently by choosing λ sufficiently close to 1, each component of $w_\lambda(\delta, d)$ can be made non-negative, with those in $S_1 \cup S_2$ being strictly positive. Therefore applying Lemma 9.2.1(a) and noting that $(I - \lambda P_\delta)^{-1} \geq I$ establishes the result.   □

Applying this result yields the following convergence theorem for multichain policy iteration.

Theorem 9.2.3. Suppose s and A_s for each $s \in S$ are finite, and for each $n, h_n = h^{(d_n)^\infty}$. Then the policy iteration algorithm terminates in a finite number of iterations, with a gain optimal stationary policy and a pair (g^*, h^*) which satisfy the optimality equations (9.1.5) and (9.1.6).

Proof. By Theorem 9.2.2, the algorithm generates a sequence of stationary policies $\{(d_n)^\infty\}$ with the property that

$$\liminf_{\lambda \uparrow 1} \left[v_\lambda^{(d_{n+1})^\infty}(s) - v_\lambda^{(d_n)^\infty}(s) \right] > 0 \qquad (9.2.15)$$

for some $s \in S$. Since there are only finitely many stationary policies, the algorithm must terminate in a finite number of iterations.

At termination, $d_{n+1} = d_n$, so that

$$0 = \left(P_{d_n} - I \right) g^{(d_n)^\infty} = \left(P_{d_{n+1}} - I \right) g^{(d_n)^\infty} = \max_{d \in D} \left\{ (P_d - I) g^{(d_n)^\infty} \right\}$$

and, for any $d_{n+1} \in E$,

$$h^{(d_n)^\infty} + g^{(d_n)^\infty} = r_{d_n} + P_{d_n} h^{(d_n)^\infty} = r_{d_{n+1}} + P_{d_{n+1}} h^{(d_n)^\infty} = \max_{d \in E} \left\{ r_d + P_d h^{(d_n)^\infty} \right\}.$$

Thus $g^* = g^{(d_n)^\infty}$, $h^* = h^{(d_n)^\infty}$ is a solution of the optimality equations and by Theorem 9.1.7(a), $(d_n)^\infty$ is average optimal.   □

The above proof is based on showing that (9.2.15) holds for successive policies, with it being strictly positive until the same decision rule repeats, at which point the algorithm stops. One might speculate that this implies that the gains of successive policies are strictly increasing. This is not the case because, for two stationary policies d^∞ and δ^∞

$$v_\lambda^{d^\infty} - v_\lambda^{\delta^\infty} = (1 - \lambda)^{-1}\left[g^{d^\infty} - g^{\delta^\infty}\right] + h^{d^\infty} - h^{\delta^\infty} + f(\lambda),$$

so that (9.2.15) does not exclude the possibility that the gains of two successive policies are identical and improvement occurs in the bias term.

Careful inspection of the above proof reveals that a variant of the policy iteration algorithm, in which the improvement step is altered so that in *each* state in which no improvement occurs in the first equation (9.2.7) we seek improvement in the second equation (9.2.8), yields a convergent sequence of policies. Taking such an approach in the example of Sec. 9.2.2 would enable us to choose the improved action $a_{1,2}$ in s_1 at the first pass through the algorithm.

9.2.4 Behavior of the Iterates of Policy Iteration

In this section we provide a different approach for establishing finite convergence of policy iteration to an optimal policy. It can be adapted to all three specifications of h_n in the evaluation step, and provides further insight into the properties of the iterates of the algorithm. We begin with the following lemma which follows from simple algebra and properties of the limiting matrix. When applying it we choose $d = d_n$ and $\delta = d_{n+1}$.

Lemma 9.2.4. Let $d \in D^{\text{MD}}$ and $\delta \in D^{\text{MD}}$ and define

$$u \equiv P_\delta g^{d^\infty} - g^{d^\infty},$$
$$v \equiv r_\delta + (P_\delta - I)\eta^{d^\infty} - g^{d^\infty},$$
$$\Delta g \equiv g^{\delta^\infty} - g^{d^\infty},$$
$$\Delta h \equiv \eta^{\delta^\infty} - \eta^{d^\infty},$$

where η^{d^∞} and η^{δ^∞} denote arbitrary solutions of the policy evaluation equations (8.2.11) and (8.2.12).

Then

$$\Delta g = u + P_\delta \Delta g, \tag{9.2.16}$$
$$\Delta h + \Delta g = v + P_\delta \Delta h, \tag{9.2.17}$$
$$0 = P_\delta^* u, \tag{9.2.18}$$
$$P_\delta^* \Delta g = P_\delta^* v. \tag{9.2.19}$$

Suppose we write P_δ in partitioned form (Appendix A) as

$$P_\delta = \begin{bmatrix} P_1 & 0 & 0 & \cdot & \cdot & 0 \\ 0 & P_2 & 0 & \cdot & \cdot & 0 \\ \cdot & & & \cdot & & \\ \cdot & & & & \cdot & \\ 0 & & & & P_m & 0 \\ Q_1 & Q_2 & \cdot & \cdot & Q_m & Q_{m+1} \end{bmatrix}, \tag{9.2.20}$$

where $P_1, P_2, \ldots, P_m$ correspond to transitions within closed recurrent classes, $Q_1, Q_2, \ldots, Q_m$ to transitions from transient to recurrent states, and Q_{m+1} to transitions between transient states. Then

$$P_\delta^* = \begin{bmatrix} P_1^* & 0 & 0 & \cdot & \cdot & 0 \\ 0 & P_2^* & 0 & \cdot & \cdot & 0 \\ \cdot & & & & & \\ \cdot & & & \cdot & & \\ 0 & & & & P_m^* & 0 \\ Q_1^* & Q_2^* & \cdot & \cdot & Q_m^* & 0 \end{bmatrix}. \tag{9.2.21}$$

Partition g^{δ^∞}, g^{d^∞}, η^{δ^∞}, η^{d^∞}, Δg, Δh, u, and v consistent with the above matrix partition, for example

$$u = \begin{bmatrix} u_1 \\ u_2 \\ \cdot \\ \cdot \\ u_{m+1} \end{bmatrix},$$

where u_i is a column vector of appropriate dimension.

Lemma 9.2.5

a. Let $1 \leq i \leq m$, and suppose $u_i = 0$ for $1 \leq i \leq m$. Then

$$\Delta g_i = P_i^* v_i.$$

b.
$$\Delta g_{m+1} = (I - Q_{m+1})^{-1} \left(u_{m+1} + \sum_{j=1}^{m} Q_j \Delta g_j \right) \tag{9.2.22}$$

c.
$$\Delta h_{m+1} = (I - Q_{m+1})^{-1} \left(v_{m+1} - \Delta g_{m+1} + \sum_{j=1}^{m} Q_j \Delta h_j \right). \tag{9.2.23}$$

Proof. To prove (a), note that if $u_i = 0$, then, by (9.2.16), $(I - P_i) \Delta g_i = 0$. Since P_i is the transition matrix of an irreducible and recurrent subchain, Δg_i must have equal components. Consequently

$$\Delta g_i = P_i^* \Delta g_i.$$

Multiply (9.2.17) by P_δ^*, partition it as in (9.2.21) and note that $P_i^* P_i = P_i^*$ yields $P_i^* \Delta g_i = P_i^* v_i$. Combining this with the previous equality establishes (a). Part (b) follows by a similar argument and is left as an exercise.

To obtain (c), write (9.2.16) and (9.2.17) in partitioned form, note from Proposition A.3 that $(I - Q_{m+1})^{-1}$ exists, and solve for Δg_{m+1} and Δh_{m+1}. □

The following theorem is the key result of this section. It establishes improvement properties of the policy iteration algorithm, and monotonicity in the sense that improvement occurs either through the gain or, if not, through h_n. We state the result when h_n is chosen to satisfy Condition 9.2.3, however, it can be easily modified to allow other specifications of the arbitrary constants.

Theorem 9.2.6.   Let $d \in D^{MD}$, and suppose that $\eta^{d^{\infty}}$ satisfies

$$r_d - g^{d^{\infty}} + (P_d - I)\eta^{d^{\infty}} = 0$$

subject to $\eta^{d^{\infty}}(s_{j_i}) = 0$ for $i = 1, \ldots, m$, where j_i denotes the minimal index for which $s_j \in R_i$.

a. If for some $\delta \in D^{MD}$

$$P_{\delta} g^{d^{\infty}}(s) > g^{d^{\infty}}(s) \tag{9.2.24}$$

and $\delta(s') = d(s')$ for all s' for which (9.2.24) holds with equality, then s is transient under δ^{∞} and $g^{\delta^{\infty}}(s) > g^{d^{\infty}}(s)$.

b. If for some $\delta \in D^{MD}$

$$P_{\delta} g^{d^{\infty}}(s) = g^{d^{\infty}}(s) \tag{9.2.25}$$

for all $s \in S$, and

$$r_{\delta}(s) - g^{d^{\infty}}(s) + (P_{\delta} - I)\eta^{d^{\infty}}(s) > 0, \tag{9.2.26}$$

where s is a recurrent state under δ^{∞} and in recurrent class R_i and $\delta(s') = d(s')$ for all s' for which (9.2.26) holds with equality, then

$$g^{\delta^{\infty}}(s) > g^{d^{\infty}}(s)$$

for all $s \in R_i$.

c. If for some $\delta \in D^{MD}$ (9.2.25) holds for all $s \in S$,

$$r_{\delta}(s) - g^{d^{\infty}}(s) + (P_{\delta} - I)\eta^{d^{\infty}}(s) = 0 \tag{9.2.27}$$

for all s which are recurrent under δ^{∞}, (9.2.26) holds for some s transient under δ^{∞}, and $\delta(s') = d(s')$ for all s' for which (9.2.27) holds, then

$$g^{\delta^{\infty}}(s) = g^{d^{\infty}}(s) \tag{9.2.28}$$

for all $s \in S$, and

$$\eta^{\delta^{\infty}}(s) > \eta^{d^{\infty}}(s). \tag{9.2.29}$$

Proof.

a. Under the specification for δ, $u \geq 0$. From (9.2.18), $P_\delta^* u = 0$. Since P_i^* has positive components for $i = 1, 2, \ldots, m$, this implies that $u_i = 0$, so that, if $u(s) > 0$, s is transient under δ^∞. Therefore $u_{m+1}(s) > 0$ for some s. Noting that $(I - Q_{m+1})^{-1} \geq I$ and applying (9.2.22) establishes this result.

b. Since $u = 0$, and $v_i(s) > 0$ for some s and i, $1 \leq i \leq m$, the result follows from Lemma 9.2.5(a).

c. Since $u = 0$ and $v_i = 0$, $i = 1, 2, \ldots, m$, Lemma 9.2.5(a) and (b) imply $\Delta g = 0$. Therefore, from (9.2.17), $(I - P_i)\Delta h_i = 0$, $i = 1, 2, \ldots, m$. Consequently Δh_i has equal components, and under the specification for the arbitrary constants in the hypothesis of the theorem, $\Delta h_i = 0$, $i = 1, 2, \ldots, m$. Since $v_{m+1}(s) > 0$ for some s, the result follows by applying (9.2.23). $\qquad\square$

Applying this theorem to the multichain policy iteration algorithm establishes the following:

a. The gains of successive policies are monotone nondecreasing.

b. If improvement occurs in state s' in step 3(a), of the algorithm, then s' is transient under d_{n+1}, and $g_{n+1}(s') > g_n(s')$. Further $g_{n+1}(s') > g_n(s')$ may hold for other states which are transient under d_{n+1}.

c. If no improvement occurs in step 3(a) of the algorithm, and it occurs in state s' in step 3(b) where s' is recurrent under d_{n+1}, then $g_{n+1}(s) > g_n(s)$ for all states in the recurrent class containing s' under d_{n+1}, and possibly at other states which are transient under d_{n+1}.

d. If no improvement occurs in step 3(a) of the algorithm, and it occurs in state s' in step 3(b) and s' is transient under d_{n+1}, then $h_{n+1}(s') > h_n(s')$.

The following important corollary follows immediately from the above discussion.

Corollary 9.2.7. Suppose the multichain policy iteration algorithm is implemented using Condition 9.2.3, to specify the arbitrary constants in the evaluation step. Then it terminates in a finite number of iterates and identifies an average optimal policy.

We illustrate these results through the example of Sect. 9.2.2.

Example 9.1.1 (ctd.). Observe that, at the first pass through the algorithm, improvement occurs in step 3(a) in state s_2 which is transient under $(d_1)^\infty$, with $g_1(s_2) > g_0(s_2)$ and $g_1(s_1) > g_0(s_1)$. At the second iteration, improvement occurs in step 3(b) in s_1 which is recurrent under $(d_2)^\infty$, with $g_2(s_1) > g_1(s_1)$. Note also that there is no pattern to the sequence of h_n's in this example.

The above analysis shows that improvements through step 3(a) cannot identify policies with additional recurrent classes so that in a model with compact action sets, an infinite number of improvements may occur in step 3(a). Consequently, the algorithm may not identify a recurrent class that is part of the optimal policy, and will converge to a suboptimal policy. Problem 9.9 illustrates this point.

9.3 LINEAR PROGRAMMING IN MULTICHAIN MODELS

This section discusses a linear programming approach to finding optimal policies in finite-state and action multichain models with average reward criterion. It extends results in Sec. 8.8 for unichain models.

We base the development on properties of solutions of the modified optimality equations established in Theorem 9.1.2. From Theorems 9.1.2(a) and 9.1.6(a) it follows that if g and h satisfy

$$g \geq P_d g, \tag{9.3.1}$$

$$h + g \geq r_d + P_d h \tag{9.3.2}$$

for all $d \in D^{MD}$, then $g \geq g^*$. Consequently the "minimal" g which satisfies (9.3.1) and for which there exists an h satisfying (9.3.2) equals the optimal average reward. Since, in multichain models, $g(s)$ may vary with s, we minimize a positive linear combination of components of g.

The above discussion suggests that the following primal linear programming problem characterizes g^*.

Primal linear program.

$$\text{Minimize } \sum_{j \in S} \alpha_j g(j)$$

subject to

$$g(s) \geq \sum_{j \in S} p(j|s, a) g(j) \qquad a \in A_s \text{ and } s \in S \tag{9.3.3}$$

and

$$g(s) \geq r(s, a) + \sum_{j \in S} p(j|s, a) h(j) - h(s) \qquad a \in A_s \text{ and } s \in S,$$

where $\alpha_j > 0$ and $\sum_{j \in S} \alpha_j = 1$.

We analyze the model through the following dual problem.

Dual linear program.

$$\text{Maximize } \sum_{s \in S} \sum_{a \in A_s} r(s, a) x(s, a) \tag{9.3.4}$$

subject to

$$\sum_{a \in A_j} x(j, a) - \sum_{s \in S} \sum_{a \in A_s} p(j|s, a) x(s, a) = 0, \qquad j \in S, \qquad (9.3.5)$$

$$\sum_{a \in A_j} x(j, a) + \sum_{a \in A_j} y(j, a) - \sum_{s \in S} \sum_{a \in A_s} p(j|s, a) y(s, a) = \alpha_j \qquad j \in S, \quad (9.3.6)$$

and $x(s, a) \geq 0$, $y(s, a) \geq 0$, $a \in A_s$, and $s \in S$.

This generalizes the unichain formulation in Sect. 8.8 by the inclusion of (9.3.3) in the primal problem and the corresponding additional set of variables $y(s, a)$ in the dual problem. Note that at least one of the equations in (9.3.5) is redundant and that (9.3.6) generalizes the constraint

$$\sum_{s \in S} \sum_{a \in A_s} x(s, a) = 1$$

in the unichain model. Note that by summing (9.3.6) over j, we obtain this equality.

Example 9.3.1. We provide the dual linear program for the model in Example 9.1.1. Choose $\alpha_1 = \alpha_2 = \alpha_3 = \frac{1}{3}$. Then the dual linear program becomes

Maximize $3x(s_1, a_{1,1}) + x(s_1, a_{1,2}) + 0x(s_2, a_{2,1}) + 1x(s_2, a_{2,2}) + 2x(s_3, a_{3,1})$

subject to

$$x(s_1, a_{1,2}) = 0,$$

$$x(s_2, a_{2,2}) - x(s_1, a_{1,2}) = 0,$$

$$x(s_2, a_{2,2}) = 0,$$

$$x(s_1, a_{1,1}) + x(s_1, a_{1,2}) + y(s_1, a_{1,2}) = \tfrac{1}{3},$$

$$x(s_2, a_{2,1}) + x(s_2, a_{2,2}) + y(s_2, a_{2,2}) - y(s_1, a_{1,2}) = \tfrac{1}{3},$$

$$x(s_3, a_{3,1}) - y(s_2, a_{2,2}) = \tfrac{1}{3},$$

and $x(s, a) \geq 0$ and $y(s, a) \geq 0$ for all a and s.

Observe that one of the first three equations is redundant and that $y(s_1, a_{1,1})$ and $y(s_2, a_{2,1})$ do not appear in any of the equations.

9.3.1 Dual Feasible Solutions and Randomized Decision Rules

Before establishing the relationship between optimal solutions and optimal policies, we discuss the relationship between feasible solutions and randomized decision rules. We use the following approach for generating decision rules from dual feasible solutions.

For any feasible solution (x, y) of the dual linear program, define a randomized stationary decision rule $d_{x,y}$ by

$$
q_{d_{x,y}(s)}(a) \equiv \begin{cases} x(s,a) \Big/ \sum\limits_{a \in A_s} x(s,a) & s \in S_x \\[2ex] y(s,a) \Big/ \sum\limits_{a \in A_s} y(s,a) & s \in S/S_x, \end{cases} \tag{9.3.7}
$$

where, as in Sec. 8.8.2, $S_x = \{s \in S: \sum_{a \in A_s} x(s,a) > 0\}$. Recall that $q_{d(s)}(a)$ denotes the probability of choosing action a under decision rule d when the system occupies state s.

Note that if, for each $s \in S_x$, $x(s, a) > 0$ for a single $a \in A_s$ and, for each s, $\in S/S_x$, $y(s, a) > 0$ for a single $a \in A_s$, then (9.3.7) generates the following deterministic decision rule:

$$
d_{x,y}(s) = \begin{cases} a & \text{if } x(s,a) > 0 \text{ and } s \in S_x \\ a' & \text{if } y(s,a') > 0 \text{ and } s \in S/S_x. \end{cases} \tag{9.3.8}
$$

We establish that $d_{x,y}$ is well defined through the following lemma.

Lemma 9.3.1. Suppose (x, y) is a feasible solution to the dual linear program. Then $\sum_{a \in A_s} y(s, a) > 0$, for $s \in S/S_x$.

Proof. Rewrite (9.3.6) as

$$
\sum_{a \in A_j} x(j, a) + \sum_{a \in A_j} y(j, a) = \alpha_j + \sum_{s \in S} \sum_{a \in A_s} p(j|s, a) y(s, a)
$$

and note that the right-hand side of this expression is strictly positive.   □

The following result relates the chain structure of the Markov chain generated by $(d_{x,y})^\infty$ to the sets S_x and S/S_x. Careful attention to summation subscripts will help you follow this involved proof.

Proposition 9.3.2. If (x, y) is a feasible solution to the dual linear program, then S_x is the set of recurrent states and S/S_x the set of transient states of the Markov chain generated by $(d_{x,y})^\infty$.

Proof. To simplify notation, write δ for $d_{x,y}$. Recall that R_δ and T_δ denote the sets of recurrent and transient states of P_δ. Letting $x(s) = \sum_{a \in A_s} x(s, a)$, it follows from (9.3.5) and (9.3.7) that

$$
x(j) = \sum_{s \in S} \sum_{a \in A_s} p(j|s, a) q_{\delta(s)}(a) x(s) = \sum_{s \in S} p_\delta(j|s) x(s).
$$

Any solution of this system is positive only on recurrent states of P_δ. Since, for $s \in S_x$, $x(s) > 0$, it follows that s is recurrent under P_δ. Hence $S_x \subseteq R_\delta$.

Noting that $x(j) = 0$ if $j \in S/S_x$, it follows from (9.3.5) that, for $j \in S/S_x$,

$$0 = x(j) = \sum_{s \in S_x} p_\delta(j|s)x(s).$$

Consequently, $p_\delta(j|s) = 0$ for $j \in S/S_x$ and $s \in S_x$, so that S_x is closed under P_δ. This implies that $T_\delta \subseteq S/S_x$.

Assume now that the inclusion is proper. Since S_x is closed, S/S_x must contain a closed recurrent set R. Hence $p_\delta(j|s) = 0$ if $j \in S/R$ and $s \in R$. Let $y(s) \equiv \sum_{a \in A_s} y(s, a)$. For $j \in S/S_x$, $q_{\delta(s)}(a)$ is determined from the second equality in (9.3.7), so that

$$0 = \sum_{j \in S/R} \sum_{s \in R} p_\delta(j|s)y(s) = \sum_{j \in S/R} \sum_{s \in R} \sum_{a \in A_s} p(j|s, a)y(s, a)$$

$$= \sum_{j \in S/R} \sum_{s \in S} \sum_{a \in A_s} p(j|s, a)y(s, a) - \sum_{j \in S/R} \sum_{s \in S/R} \sum_{a \in A_s} p(j|s, a)y(s, a)$$

$$= \sum_{j \in S/R} \left[-\alpha_j + x(j) - y(j)\right] - \sum_{j \in S/R} \sum_{s \in S} \sum_{a \in A_s} p(j|s, a)y(s, a)$$

$$+ \sum_{j \in S/R} \sum_{s \in R} \sum_{a \in A_s} p(j|s, a)y(s, a)$$

$$= - \sum_{j \in S/R} \alpha_j + \sum_{j \in S/R} x(j) + \sum_{j \in S/R} y(j) - \sum_{j \in S/R} y(j)$$

$$+ \sum_{j \in S/R} \sum_{s \in R} \sum_{a \in A_s} p(j|s, a)y(s, a),$$

where the equalities follow from appropriate substitution of (9.3.6) and simple algebra. Summing (9.3.6) over j shows that $\sum_{j \in S} x(j) = 1$ for any feasible x, and since $x(j) = 0$ for $j \in R$, $\sum_{j \in S/R} x(j) = 1$. Since $y(s, a) \geq 0$, the last expression above is non-negative, so it follows that

$$0 \geq 1 - \sum_{j \in S/R} \alpha_j,$$

contradicting the assumption that $\alpha_j > 0$. Consequently $R = \varnothing$, $S/S_x = T_\delta$ and $S_x = R_\delta$.   □

Problem 9.13 provides a model in which two distinct dual feasible solutions generate the same policy through (9.3.7) or (9.3.8). Because of this, we consider two feasible solutions (x_1, y_1) and (x_2, y_2) to be *equivalent* if they generate the same decision rule through (9.3.7), that is

$$q_{d_{x_1, y_1}(s)}(a) = q_{d_{x_2, y_2}(s)}(a) \tag{9.3.9}$$

for all $a \in A_s$ and $s \in S$. For decision rule d, let the set $C(d)$ denote the *equivalence class* of dual feasible solutions which generate d through (9.3.7). That is,

$$C(d) \equiv \{(x, y) : (x, y) \text{ are dual feasible and } q_{d_{x,y}} = q_d\}.$$

Now we reverse this process; starting with a decision rule, we show how to generate a dual feasible solution. Let $d \in D^{MR}$ and define $x_d(s, a)$ and $y_d(s, a)$ by

$$x_d(s, a) \equiv \sum_{j \in S} \alpha_j p_d^*(j|s) q_{d(s)}(a) \tag{9.3.10}$$

and

$$y_d(s, a) \equiv \sum_{j \in S} \left[\alpha_j h_d(s|j) + \gamma_j p_d^*(s|j) \right] q_{d(s)}(a), \tag{9.3.11}$$

where $h_d(j|s)$ is the (s, j)th element of H_d, the deviation matrix of d (Appendix A.5), and

$$\gamma_j \equiv \begin{cases} \max_{s \in R_i} \left\{ - \sum_{k \in S} \alpha_k h_d(s|k) \Big/ \sum_{k \in R_i} p_d^*(s|k) \right\} & j \in R_i \\ 0 & j \in T. \end{cases} \tag{9.3.12}$$

In (9.3.12), R_i denotes the ith recurrent class, $1 \le i \le m$, and T the transient states of the Markov chain generated by d^∞. Observe that γ is constant on each recurrent class and that, for each R_i, $y_d(s, a) = 0$ at the $s_i \in R_i$ where the maximum in (9.3.12) holds. The choice of γ_j is motivated by Proposition 9.1.1.

As in unichain models, we interpret $x_d(s, a)$ as the limiting probability under d^∞ that the system occupies state s and chooses action a when the initial state is chosen according to distribution $\{\alpha_j\}$. For transient states, $y_d(s, a)$ denotes the expected difference between the number of times the process occupies state s and chooses action a under P_d and P_d^*. Interpretation of $y_d(s, a)$ on recurrent states is not as obvious. Feasibility of (x_d, y_d) follows by direct substitution and properties (A.17)–(A.21) of the limiting and deviation matrices. We leave the details as an exercise.

Proposition 9.3.3. Let $d \in D^{MR}$. Then (x_d, y_d) is a feasible solution to the dual linear program.

It is natural to ask whether substituting (x_d, y_d) into (9.3.7) yields (d). The following theorem provides an affirmative answer to this query, and is the key result of this section. It establishes a one-to-one relationship between equivalence classes of feasible solutions and randomized Markovian decision rules.

Theorem 9.3.4.

a. Let $d \in D^{MR}$ and suppose x_d satisfies (9.3.10) and y_d satisfies (9.3.11). Then $(x_d, y_d) \in C(d)$.

b. Suppose (x, y) satisfy (9.3.5) and (9.3.6). Then $(x, y) \in C(d_{x,y})$.

Proof. Let $s \in S_{x_d}$. Then, from (9.3.10)

$$q_{d_{x_d,y_d}(s)}(a) = \frac{x_{d_{x_d,y_d}}(s,a)}{\sum_{a \in A_s} x_{d_{x_d,y_d}}(s,a)} = \frac{\sum_{j \in S} \alpha_j p_d^*(s|j) q_{d(s)}(a)}{\sum_{a \in A_s} \sum_{j \in S} \alpha_j p_d^*(s|j) q_{d(s)}(a)}$$

$$= \frac{q_{d(s)}(a)}{\sum_{a \in A_s} q_{d(s)}(a)} = q_{d(s)}(a). \tag{9.3.13}$$

For $s \in S/S_{x_d}$ repeat the above argument using (9.3.11) instead of (9.3.10), to conclude that $(x_d, y_d) \in C(d)$. This establishes (a). We leave the proof of (b) as an exercise.   □

9.3.2   Basic Feasible Solutions and Deterministic Decision Rules

Solving the dual linear program using the simplex algorithm, yields an optimal basic feasible solution. We now discuss the relationship between basic feasible solutions and deterministic Markovian decision rules.

Theorem 9.3.5.   Suppose $d \in D^{MD}$. Then (x_d, y_d) is a basic feasible solution of the dual linear program.

Proof. Suppose there exist basic feasible solutions (x_1, y_1) and (x_2, y_2) and λ with $0 < \lambda < 1$, for which $x_d = \lambda x_1 + (1 - \lambda) x_2$ and $y_d = \lambda y_1 + (1 - \lambda) y_2$. We show that $(x_1, y_1) = (x_2, y_2)$, from which it follows that (x_d, y_d) is a basic feasible solution.

Since $x_i \geq 0$ and $y_i \geq 0$ for $i = 1, 2$, $x_d(s, a) = 0$ implies $x_i(s, a) = 0$ for $i = 1, 2$, and $y_d(s, a) = 0$ implies $y_i(s, a) = 0$ for $i = 1, 2$. Therefore $x_i(s, a) = y_i(s, a) = 0$ whenever $a \neq d(s)$. As a consequence of this observation, (x_i, y_i) for $i = 1, 2$, and (x_d, y_d) satisfy the dual constraints which reduce to

$$x^T(I - P_d) = 0,$$

$$x^T + y^T(I - P_d) = \alpha^T,$$

where α denotes the vector with components α_j. As in the proof of Theorem (8.2.6), it follows from (A.4) and (A.17) in Appendix A that this system of equations uniquely determines x, so $x_1 = x_2$, and determines y up to one additive constant in each recurrent class of P_d. The defining equations (9.3.11) and (9.3.12) imply that $y_d(s', d(s')) = 0$ for some s' in each recurrent class of P_d. Therefore $y_1(s', d(s')) = y_2(s', d(s'))$ at that s'. Therefore it follows that $y_1(s, d(s)) = y_2(s, d(s)) = y_d(s, d(s))$ for all $s \in S$.   □

The following example shows that the converse of this theorem need not hold; that is, a basic feasible solution need not generate a deterministic decision rule through (9.3.7).

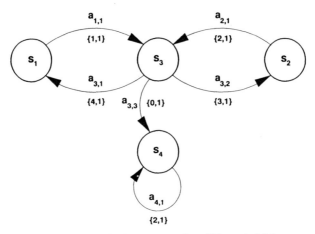

Figure 9.3.1 Symbolic representation of Example 9.3.2.

Example 9.3.2. Let $S = \{s_1, s_2, s_3, s_4\}$; $A_{s_1} = \{a_{1,1}\}$, $A_{s_2} = \{a_{2,1}\}$, $A_{s_3} = \{a_{3,1}, a_{3,2}, a_{3,3}\}$, and $A_{s_4} = \{a_{4,1}\}$; $r(s_1, a_{1,1}) = 1$, $r(s_2, a_{2,1}) = 2$, $r(s_3, a_{3,1}) = 4$, $r(s_3, a_{3,2}) = 3$, $r(s_3, a_{3,3}) = 0$, and $r(s_4, a_{4,1}) = 2$; and $p(s_3|s_1, a_{1,1}) = 1$, $p(s_3|s_2, a_{2,1}) = 1$, $p(s_1|s_3, a_{3,1}) = 1$, $p(s_2|s_3, a_{3,2}) = 1$, $p(s_4|s_3, a_{3,3}) = 1$, and $p(s_4|s_4, a_{4,1}) = 1$ (Fig. 9.3.1). Choosing $\alpha_1 = \alpha_2 = \alpha_3 = \alpha_4 = \frac{1}{4}$, we find that

(s, a)	$(s_1, a_{1,1})$	$(s_2, a_{2,1})$	$(s_3, a_{3,1})$	$(s_3, a_{3,2})$	$(s_3, a_{3,3})$	$(s_4, a_{4,1})$
$x(s, a)$	$\frac{1}{4}$	$\frac{1}{4}$	$\frac{1}{8}$	$\frac{1}{8}$	0	$\frac{1}{4}$
$y(s, a)$	0	0	0	0	0	0

is a basic feasible solution (in fact it is an optimal solution) to the dual linear program. Relationship (9.3.7) generates a randomized decision rule which chooses actions $a_{3,1}$ and $a_{3,2}$ with probability $\frac{1}{2}$ each in s_3.

9.3.3 Optimal Solutions and Policies

In this section, we examine the relationship between optimal solutions of the dual linear program and optimal policies for the average reward multichain Markov decision problem. We first establish the existence of an optimal solution to the dual linear program.

Theorem 9.3.6. There exists an optimal basic feasible solution (x^*, y^*) for the dual linear program, and an optimal basic feasible solution (g^*, h^*) for the primal linear program.

Proof. From Theorem 9.1.6(c), there exists a bounded h for which the pair (g^*, h) satisfies the primal constraints (9.3.3) and (9.3.4). As a consequence of Theorem 9.1.2(a), if (g', h') are also primal feasible, $\alpha^T g^* \le \alpha^T g'$, so it follows that the primal linear program has a bounded optimal solution. From Theorem D.1(a) and (c) (Appendix D), it follows that the dual has an optimal basic feasible solution.   □

The theorem below relates optimal solutions and optimal policies and is the main result of Sec. 9.3. Our proof of part (b) uses the following technical lemma.

Lemma 9.3.7.   Suppose (x, y) is an optimal solution to the dual LP, (g, h) an optimal solution to the primal LP, and

$$B_s \equiv \left\{ a \in A_s: q_{d_{x,y}(s)}(a) > 0 \right\}.$$

Then, for all $a \in B_s$,

$$g(s) - \sum_{j \in S} p(j|s, a)g(j) = 0, \qquad s \in S \qquad (9.3.14)$$

and

$$g(s) + h(s) - \sum_{j \in S} p(j|s, a)h(j) = r(s, a) \qquad s \in S_x. \qquad (9.3.15)$$

Proof.   For $s \in S_x$ and $a \in B_s$, $x(s, a) > 0$, and, for $s \in S/S_x$ and $a \in B_s$, $y(s, a) > 0$. Noting that the dual variables $y(s, a)$ correspond to the constraint

$$g(s) \geq \sum_{j \in S} p(j|s, a)g(j),$$

and $x(s, a)$ to the constraint

$$g(s) \geq r(s, a) + \sum_{j \in S} p(j|s, a)h(j) - h(s),$$

and applying the LP complementary slackness theorem Theorem D.1(d), it follows that (9.3.14) holds for $s \in S/S_x$ and (9.3.15) holds for $s \in S_x$. From (9.3.5),

$$0 = \sum_{j \in S} g(j) \left[\sum_{a \in A_j} x(j, a) - \sum_{s \in S} \sum_{a \in A_s} p(j|s, a)x(s, a) \right]$$

$$= \sum_{s \in S} \sum_{a \in A_s} x(s, a) \left[g(s) - \sum_{j \in S} g(j)p(j|s, a) \right].$$

Since g is primal feasible, the quantity in brackets in the last expression is non-negative, and, since $x(s, a) > 0$ for $s \in S_x$ and $a \in B_s$, (9.3.14) also holds for $s \in S_x$.   □

Theorem 9.3.8

  a. Suppose $(d^*)^\infty$ is an average optimal policy. Then (x_{d^*}, y_{d^*}) is an optimal solution to the dual linear program.
  b. Suppose (x^*, y^*) is an optimal solution to the dual linear program. Then $(d_{x^*, y^*})^\infty$ is a stationary average optimal policy.

Proof. From Proposition 9.3.3, (x_{d^*}, y_{d^*}) is dual feasible. Since the dual objective function satisfies

$$\sum_{s \in S} \sum_{a \in A_s} r(s, a) x_{d^*}(s, a) = \sum_{s \in S} \sum_{a \in A_s} r(s, a) \sum_{j \in S} \alpha_j p_{d^*}^*(s|j) q_{d^*(s)}(a)$$

$$= \sum_{s \in S} r_{d^*}(s) \sum_{j \in S} \alpha_j p_{d^*}^*(s|j) = \sum_{j \in S} \alpha_j g^{(d^*)^\infty}(j)$$

$$= \sum_{j \in S} \alpha_j g^*(j),$$

it follows from a LP duality theorem, Theorem D.1(b), that (x_{d^*}, y_{d^*}) is optimal.
   Suppose now that (x^*, y^*) is an optimal solution of the dual linear program. Let $\delta \equiv d_{x^*, y^*}$. We show that $g^{\delta^\infty} = g^*$. From Theorem 9.3.6, there exists an h for which (g^*, h) is an optimal solution of the primal linear program. From Proposition 9.3.2, $p_\delta^*(j|s) = 0$ if $j \in S/S_x$ so it follows from Proposition 8.1.1, (A.4), and (9.3.15) that, for all $s \in S$,

$$g^{\delta^\infty}(s) = \sum_{j \in S} p_\delta^*(j|s) r_\delta(j) = \sum_{j \in S_x} p_\delta^*(j|s) r_\delta(j) = \sum_{j \in S_x} p_\delta^*(j|s) \sum_{a \in A_j} q_{\delta(j)}(a) r(j, a)$$

$$= \sum_{j \in S_x} p_\delta^*(j|s) \left[g^*(j) + h(j) - \sum_{k \in S} p_\delta(k|j) h(k) \right]$$

$$= \sum_{j \in S_x} p_\delta^*(j|s) g^*(j). \tag{9.3.16}$$

From (9.3.14), for all $s \in S$,

$$g^*(s) = \sum_{j \in S} \sum_{a \in A_s} g^*(j) p(j|s, a) q_{\delta(s)}(a) = \sum_{j \in S} g^*(j) p_\delta(j|s).$$

Since $P_\delta^* = \lim_{N \to \infty} N^{-1} \sum_{i=0}^{N-1} P_\delta^i$ we obtain by iterating, summing and passing to the limit in the above expression that

$$g^*(s) = \sum_{j \in S} g^*(j) p_\delta^*(j|s),$$

which when combined with (9.3.16) establishes that $g^{\delta^\infty}(s) = g^*(s)$ for all $s \in S$. Hence δ^∞ is optimal.   □

   Note that part (a) of this theorem establishes that (x_{d^*}, y_{d^*}) is optimal for any vector α satisfying $\alpha_j > 0$ and $\sum_{j \in S} \alpha_j = 1$. We now apply the result in Theorem 9.3.8 to solve Example 9.3.1.

Example 9.3.1. (ctd.). We solve the dual linear program and obtain the optimal solution

(s, a)	$(s_1, a_{1,1})$	$(s_1, a_{1,2})$	$(s_2, a_{2,1})$	$(s_2, a_{2,2})$	$(s_3, a_{3,1})$
$x(s, a)$	$\frac{1}{3}$	0	0	0	$\frac{2}{3}$
$y(s, a)$	0	0	0	$\frac{1}{3}$	0

Through (9.3.7), we obtain the optimal policy $(d^*)^\infty$, where $d^*(s_1) = a_{1,1}$, $d^*(s_2) = a_{2,2}$, and $d^*(s_3) = a_{3,1}$. Observe that $S_x = \{s_1, s_3\}$ is the set of recurrent states of P_{d^*}.

The objective function value is 2.333 which equals $\sum_{j \in S} \alpha_j g^*(j)$. We find $g^* = (3, 2, 3)^T$ from the corresponding primal solution.

In the above example, the optimal solution of the dual linear program yields a deterministic stationary optimal policy. Example 9.3.2 showed that an optimal basic feasible solution may not generate a *deterministic* decision rule through (9.3.7). This means that, when solving a model using linear programming, we may not obtain a *deterministic* optimal stationary policy. Theorem 9.3.8(a) establishes that, for any optimal stationary policy d^∞, (x_d, y_d) is an optimal solution of the dual linear program, so, if an optimal solution does not generate a deterministic policy, we can find a different optimal solution which does. We illustrate this point in the context of Example 9.3.2.

Example 9.3.2. (ctd.). The solution in Example 9.3.2 is optimal, but it generates a randomized decision rule. The following basic solution is also optimal and generates a deterministic decision rule.

(s, a)	$(s_1, a_{1,1})$	$(s_2, a_{2,1})$	$(s_3, a_{3,1})$	$(s_3, a_{3,2})$	$(s_3, a_{3,3})$	$(s_4, a_{4,1})$
$x(s, a)$	$\frac{3}{8}$	0	$\frac{3}{8}$	0	0	$\frac{1}{4}$
$y(s, a)$	0	$\frac{1}{4}$	$\frac{1}{8}$	0	0	0

This solution corresponds to the deterministic decision rule $d^*(s_1) = a_{1,1}$, $d^*(s_2) = a_{2,1}$, $d^*(s_3) = a_{3,1}$ and $d^*(s_4) = a_{4,1}$.

When an optimal basic feasible solution does not yield a deterministic policy, we can obtain one by choosing d^* to satisfy

$$d^*(s) = \begin{cases} a_s & \text{if } x^*(s, a_s) > 0, \quad s \in S_{x^*} \\ a_s & \text{if } y^*(s, a_s) > 0, \quad s \in S/S_{x^*}. \end{cases}$$

If more than one $x^*(s, a) > 0$ for $s \in S_{x^*}$, or $y^*(s, a) > 0$ for $s \in S/S_{x^*}$, then multiple optimal stationary policies exist, and any policy chosen in the above manner is average optimal.

Note that it is also possible that a nonoptimal feasible solution yields an optimal policy through (9.3.7) (Problem 9.15).

As in recurrent average reward models and discounted models, policy iteration is equivalent to linear programming with block pivoting.

9.4   VALUE ITERATION

This section analyzes the value iteration algorithm for multichain models. We show that it converges under the assumption that all deterministic stationary optimal policies have aperiodic transition matrices. Unfortunately, this method cannot be used for numerical calculations because stopping criteria are not available unless the optimal gain is constant.

Recall that

$$Lv = \max_{d \in D} \{r_d + P_d v\} \tag{9.4.1}$$

and that the value iteration recursion can be expressed as

$$v^{n+1} = Lv^n. \tag{9.4.2}$$

9.4.1   Convergence of $v^n - ng^*$

In this section we analyze the limiting behavior of the sequence $\{e_n\}$ where

$$e_n \equiv v^n - ng^* - h^*,$$

and g^* and h^* satisfy the multichain optimality equations. When this limit exists it follows that

a. for n sufficiently large, $v^{n+1} - v^n \equiv g^n$ yields a good approximation to g^*,

b. for n sufficiently large, $v^n - ng^n$ yields a good approximation to h^*, and

c. for n sufficiently large, if $\delta \in \arg\max_{d \in D} \{r_d + P_d v^n\}$, then δ^∞ is close to being optimal.

Therefore the existence of this limit enables us in theory to obtain approximate solutions to the optimality equation and optimal policies. Before exploring the computational implications of this result further, we provide conditions which ensure convergence of $\{e_n\}$.

We begin by providing a bound on $v^n - ng^*$. When this bound holds, it follows that v^n/n converges to the optimal gain g^*.

Theorem 9.4.1.

a. Let $v^0 \in V$, and suppose $\{v^n\}$ satisfies (9.4.2). Then

$$\min_{s \in S} \{v^0(s) - h^*(s)\} \leq v^n(s) - ng^*(s) - h^*(s) \leq \max_{s \in S} \{v^0(s) - h^*(s)\}$$

for $n \geq 1$ and all $s \in S$, where $h^* \in V$ and $g^* \in V$ are solutions of the modified multichain optimality equations (9.1.5) and (9.1.8).

b. For all $v^0 \in V$,

$$\lim_{n \to \infty} n^{-1} v^n = g^*.$$

Proof. From Theorems 9.1.7b and 9.1.8, there exists a δ for which

$$g^* = P_\delta g^* \quad \text{and} \quad r_\delta = g^* + (I - P_\delta)h^*, \tag{9.4.3}$$

and δ^∞ is average optimal. Iterating (9.4.1) and applying the above relationships yields

$$v^n \geq \sum_{k=0}^{n-1} P_\delta^k r_\delta + P_\delta^n v^0 = \sum_{k=0}^{n-1} P_\delta^k [g^* + (I - P_\delta)h^*] + P_\delta^n v^0$$

$$= ng^* + h^* + P_\delta^n (v^0 - h^*) \geq ng^* + h^* + \min_{s \in S} [v^0(s) - h^*(s)]e,$$

which establishes the left-hand inequality.

As a consequence of Corollary 9.1.5, there exists a solution of the modified optimality equations so that

$$g^* \geq P_d g^* \quad \text{and} \quad h^* + g^* \geq r_d + P_d h^*$$

for all $d \in D$. Let $d_k \in \arg\max_{d \in D} \{r_d + P_d v^k\}$ for $k = 0, 1, \ldots$. Then for $\pi = (d_0, d_1, \ldots)$

$$v^n = \sum_{k=0}^{n-1} P_\pi^k r_{d_k} + P_\pi^n v^0 \leq \sum_{k=0}^{n-1} P_\pi^k [g^* + (I - P_{d_k})h^*] + P_\pi^n v^0$$

$$\leq ng^* + h^* + P_\pi^n (v^0 - h^*) \leq ng^* + h^* + \max_{s \in S} [v^0(s) - h^*(s)]e,$$

which establishes the upper inequality so that (a) follows. Part (b) follows by dividing the inequality in (a) by n, noting h^* is bounded and passing to the limit. □

We now use this result to establish existence of $\lim_{n \to \infty} [v^n - ng^* - h^*]$. We do this in two steps. The first result demonstrates convergence on recurrent states of optimal policies. The aperiodicity assumption is crucial.

Proposition 9.4.2. Suppose that there exists an optimal policy δ^∞ with an aperiodic transition probability matrix. Then, for any $s \in S$ which is recurrent under δ,

$$\lim_{n \to \infty} [v^n(s) - ng^*(s) - h^*(s)] \tag{9.4.4}$$

exists for all $v^0 \in V$.

Proof. From the definition of v^{n+1},

$$v^{n+1} \geq r_\delta + P_\delta v^n,$$

and, since δ satisfies (9.4.3),

$$v^{n+1} \geq h^* + g^* - P_\delta h^* + P_\delta v^n.$$

Rearranging terms, adding ng^* to both sides of the inequality, and noting the first equation in (9.4.3), yields $e_{n+1} \geq P_\delta e_n$. Hence, for any $m \geq n$,

$$e_m \geq P_\delta^{m-n} e_n. \tag{9.4.5}$$

From Theorem 9.4.1, $\{e_n\}$ is bounded. Let $x = \liminf_{n \to \infty} e_n$ and $y = \limsup_{n \to \infty} e_n$. Then there exist subsequences $\{e_{n_j}\}$ and $\{e_{n_k}\}$ which attain these respective limits. Since P_δ is aperiodic, $\lim_{n \to \infty} P_\delta^n = P_\delta^*$. Fixing n in (9.4.5) and letting m go to infinity along n_j, we obtain that $x \geq P_\delta^* e_n$, so it follows taking limits on the right-hand side along n_k that $x \geq P_\delta^* y$. Reversing the role of x and y establishes also that $y \geq P_\delta^* x$.

Multiplying these expressions by P_δ^* shows that $P_\delta^* x = P_\delta^* y$ or

$$P_\delta^*(y - x) = 0.$$

Since $y \geq x$ and $p_\delta^*(j|s) > 0$ if j is recurrent under δ, $x(j) = y(j)$ if j is recurrent under δ, and the result follows. □

We now extend this result to all $s \in S$. We use the following technical lemma, which states that eventually the value iteration operator chooses maximizing decision rules only from among those which satisfy $P_d g^* = g^*$.

Lemma 9.4.3. Let $\{v^n\}$ satisfy (9.4.2), and let $E = \{d \in D^{\text{MD}}: P_d g^* = g^*\}$. Then there exists an N such that, for $n \geq N$,

$$\max_{d \in D} \{r_d + P_d v^n\} = \max_{d \in E} \{r_d + P_d v^n\}.$$

Proof. Since there are only finitely many deterministic decision rules, there exists an N and a set D' such that, for $n \geq N$, $\arg\max_{d \in D}\{r_d + P_d v^n\} \in D'$. Pick a $d' \in D'$. Then there exists a subsequence $\{v_{n_k}\}$ such that

$$v^{n_k+1} = r_{d'} + P_{d'} v^{n_k}.$$

It follows that

$$(n_k + 1)^{-1} v^{n_k+1} = (n_k + 1)^{-1} r_{d'} + P_{d'} \left[(n_k + 1)^{-1} v^{n_k}\right],$$

so letting $k \to \infty$ and noting Theorem 9.4.1(b) implies that

$$g^* = P_{d'} g^*.$$

Hence $d' \in E$. □

The following is the main result on the convergence of value iteration in multi-chain models.

Theorem 9.4.4. Suppose that every average optimal stationary deterministic policy has an aperiodic transition matrix. Then the limit in (9.4.4) exists for all $s \in S$.

Proof. Let N be as defined in Lemma 9.4.3. Then, for $n \geq N$,

$$e_n = \max_{d \in D} \{r_d + P_d v^{n-1}\} - ng^* - h^* = \max_{d \in E} \{r_d + P_d v^{n-1}\} - ng^* - h^*$$

$$= \max_{d \in E} \{(r_d - g^* + P_d h^* - h^*) + P_d(v^{n-1} - h^* - (n-1)g^*)\}.$$

Letting $q_d \equiv r_d - g^* + P_d h^* - h^*$, the above is equivalent to

$$e_n = \max_{d \in E} \{q_d + P_d e_{n-1}\}. \tag{9.4.6}$$

Let $x = \liminf_{n \to \infty} e_n$ and, $y = \limsup_{n \to \infty} e_n$. Fix $s' \in S$. Then there exists a subsequence of $\{e_n\}$ such that $\lim_{k \to \infty} e_{n_k}(s') = y(s')$, and a further subsequence for which $\lim_{k \to \infty} e_{n_k-1}(s)$ exists for all $s \in S$. Set $w(s) = \lim_{k \to \infty} e_{n_k-1}(s)$. Given $\varepsilon > 0$, there exists a $K \geq N$ for which

$$e_{n_k}(s') \geq y(s') - \varepsilon \tag{9.4.7}$$

and

$$P_d e_{n_k-1}(s') \leq P_d w(s') + \varepsilon \tag{9.4.8}$$

for all $d \in E$ and $k \geq K$. Then from (9.4.6), (9.4.7), and (9.4.8), there exists a $\delta \in E$ for which

$$y(s') - \varepsilon \leq e_{n_k}(s') = q_\delta(s') + P_\delta e_{n_k-1}(s') \leq q_\delta(s') + P_\delta w(s') + \varepsilon$$
$$\leq q_\delta(s') + P_\delta y(s') + \varepsilon,$$

where the last inequality follows from the fact that $w \leq y$. Therefore

$$y(s') \leq \max_{d \in E} \{q_d(s') + P_d y(s')\} + 2\varepsilon.$$

Since ε and s' were arbitrary, it follows that, for all $s \in S$,

$$y(s) \leq \max_{d \in E} \{q_d(s) + P_d y(s)\}. \tag{9.4.9}$$

By a similar argument, we establish that

$$x(s) \geq \max_{d \in E} \{q_d(s) + P_d x(s)\}. \tag{9.4.10}$$

Now let $\delta \in E$ attain the maximum on the right-hand side of (9.4.9). Then it follows that

$$q_\delta + P_\delta x \le x \le y \le q_\delta + P_\delta y. \tag{9.4.11}$$

From this inequality,

$$0 \le y - x \le P_\delta(y - x).$$

Iterating this expression and applying (A.3) establishes that

$$0 \le y - x \le P_\delta^*(y - x). \tag{9.4.12}$$

Multiplying the right-hand inequality in (9.4.11) by P_δ^*, noting that $\delta \in E$, and recalling the definition of q_δ shows that

$$0 \le P_\delta^* q_\delta = P_\delta^* r_\delta - P_\delta^* g^* + P_\delta^* h^* - P_\delta^* h^* = P_\delta^* r_\delta - g^*,$$

implying the optimality of δ^∞. From Proposition 9.4.2, $\lim_{n \to \infty} e_n(s)$ exists for $s \in R_\delta$ so that $y = x$ on the recurrent states of P_δ. It follows from (9.4.12) and the fact that $p_\delta^*(j|s) = 0$ if j is transient that $y(s) = x(s)$ for all $s \in S$, which yields the result. □

The condition that all optimal policies are aperiodic cannot easily be verified without solving the problem, however, it certainly holds when all stationary policies have aperiodic transition matrices. Note that the aperiodicity transformation of Sec. 8.5.5 can be applied to achieve this.

9.4.2 Convergence of Value Iteration

We now investigate the implications of Theorem 9.4.4 from the perspective of the convergence of value iteration in multichain models, and summarize some earlier results.

Theorem 9.4.5. Suppose that every average optimal stationary deterministic policy has an aperiodic transition matrix. Then the following holds:

a. For any $v^0 \in V$,

$$\lim_{n \to \infty} [v^{n+1} - v^n] = g^*, \tag{9.4.13}$$

where g^* denotes the optimal gain.

b. For all n,

$$\max_{s' \in S} \{v^{n+1}(s') - v^n(s')\} \ge g^*(s) \ge g^{(d_n)^\infty}(s) \ge \min_{s' \in S} \{v^{n+1}(s') - v^n(s')\}$$

where $d_n \in \arg\max_{d \in D}\{r_d + P_d v^n\}$.

c. Let d_n be as defined in part (b). Then, for any $\varepsilon > 0$, there exists an N such that, for all $n \geq N$, $(d_n)^\infty$ is ε-optimal.

Proof. By Theorem 9.4.4, $\lim_{n \to \infty} \{v^n - ng^*\}$ exists under the hypotheses of the theorem. Thus

$$0 = \lim_{n \to \infty} \left[\left[v^{n+1} - (n+1)g^* \right] - \left[v^n - ng^* \right] \right] = \lim_{n \to \infty} \left[(v^{n+1} - v^n) - g^* \right],$$

so (a) follows.

Part (b), is a restatement of Theorem 8.5.5, and (c) follows immediately from (a) and (b). □

We apply value iteration to Example 9.1.1 and illustrate some features of this algorithm in multichain models.

Example 9.1.1 (ctd.). Apply value iteration beginning with $v^0 = 0$ to obtain the following iterates. Recall that $g^* = (3, 2, 2)^T$ in this model.

n	$v^n(s_1)$	$v^n(s_2)$	$v^n(s_3)$	$sp(v^n - v^{n-1})$
0	0	0	0	
1	3	1	2	2
2	6	3	4	1
3	9	5	6	1
4	12	7	8	1
.	.	.	.	.
.	.	.	.	.
.	.	.	.	.

Observe that, by iteration 2, $v^n - v^{n-1} = g^*$, but that $sp(v^n - v^{n-1}) = sp(g^*) \neq 0$. Therefore the span *does not* provide a valid stopping criterion when g^* has unequal components. We conjecture that if

$$\Delta_n \equiv sp(v^{n+1} - v^n) - sp(v^{n+2} - v^{n+1}) < \varepsilon,$$

then δ^∞ is ε-optimal for any $\delta \in \arg\max_{d \in D} \{r_d + P_d v^{n+1}\}$.

Observe also that after the first iteration, the same decision rule attains the maximum. This suggests that $v^n - ng^* = (0, -1, 0)^T$ approximates h^*.

We conclude this section with the following general result concerning the convergence of value iteration and the use of the span as a stopping criterion.

Corollary 9.4.6. Suppose that $g^*(s)$ does not vary with s, and that every average optimal stationary deterministic policy has an aperiodic transition matrix. Then, for any $v^0 \in V$,

$$\lim_{n \to \infty} sp(v^{n+1} - v^n) = 0 \qquad (9.4.14)$$

and $sp(v^{n+1} - v^n) \geq sp(v^{n+2} - v^{n+1})$.

As a consequence of Corollary 9.4.6 the span convergence criterion can be used to stop value iteration in models in which any of the following hold.

1. All deterministic stationary policies are recurrent and have aperiodic transition probability matrices.
2. All deterministic stationary policies are unichain and have aperiodic transition probability matrices.
3. All optimal deterministic stationary policies are unichain and have aperiodic transition probability matrices.

Note that, in Sec. 8.5, we established convergence of $sp(v^{n+1} - v^n)$ under the stronger assumption that all policies are unichain and $p(s|s, a) > 0$ for all $s \in S$ and $a \in A_s$. The results above provide a proof of Theorem 8.5.4.

9.5 COMMUNICATING MODELS

In many Markov decision process models, some policies are multichain, yet there exist policies under which each state is accessible from each other state. In our taxonomy of Markov decision processes in Sec. 8.3, we refer to models of this type as *communicating*. Example 8.3.1 provides an inventory model in which an undesirable policy partitions the state space into two closed classes. However, the model is communicating and, in it, "good" policies have a single closed class and possibly some transient states. Recall that we refer to a model as *weakly communicating* if the state space can be partitioned into a (possibly empty) set of states which are accessible from each other under some policy, and a set of states which are transient under all policies.

In communicating and weakly communicating models, optimal policies have state-independent gain, so analysis falls somewhere between that for unichain and multichain models. In this section we discuss this distinction and show how it effects policy iteration, linear programming, and value iteration. We suggest reviewing Sec. 8.3 prior to reading this section.

9.5.1 Policy Iteration

In a communicating model, Theorem 8.3.2 establishes the existence of an optimal policy with unichain transition structure. Therefore one might conjecture that we can solve such a problem using the unichain policy iteration algorithm. This would be the case if, whenever we initiated the multichain policy iteration algorithm of Section 9.2 with a unichain decision rule, the algorithm would always identify a unichain improvement. The following example shows that this need not happen.

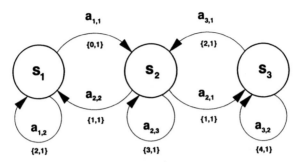

Figure 9.5.1 Symbolic representation of Example 9.5.1.

Example 9.5.1. Let $S = \{s_1, s_2, s_3\}$; $A_{s_1} = \{a_{1,1}, a_{1,2}\}$, $A_{s_2} = \{a_{2,1}, a_{2,2}, a_{2,3}\}$, and $A_{s_3} = \{a_{3,1}, a_{3,2}\}$; $r(s_1, a_{1,1}) = 0$, $r(s_1, a_{1,2}) = 2$, $r(s_2, a_{2,1}) = 1$, $r(s_2, a_{2,2}) = 1$, $r(s_2, a_{2,3}) = 3$, $r(s_3, a_{3,1}) = 2$, and $r(s_3, a_{3,2}) = 4$; and $p(s_2|s_1, a_{1,1}) = 1$, $p(s_1|s_1, a_{1,2}) = 1$, $p(s_3|s_2, a_{2,1}) = 1$, $p(s_1|s_2, a_{2,2}) = 1$, $p(s_2|s_2, a_{2,3}) = 1$, $p(s_2|s_3, a_{3,1}) = 1$, and $p(s_3|s_3, a_{3,2}) = 1$ (Fig. 9.5.1). Suppose we apply the multichain policy iteration algorithm with $d_0(s_1) = a_{1,2}$, $d_0(s_2) = a_{2,2}$, and $d_0(s_3) = a_{3,1}$. Observe that the policy corresponding to this decision rule generates a unichain process with recurrent class $\{s_1\}$ and transient states $\{s_2, s_3\}$. Then, solving the evaluation equations subject to Condition 9.2.1 yields $g_0 = (2, 2, 2)^T$ and $h_0 = (0, -1, -1)^T$. At the improvement step, 3(a) is unnecessary because g_0 is constant. Through step 3(b), we identify the improved decision rule $d_1(s_1) = a_{1,2}$, $d_1(s_2) = a_{2,3}$, and $d_{3,2}(s_3) = a_{3,2}$. This decision rule generates a multichain process with three recurrent classes $\{s_1\}$, $\{s_2\}$, $\{s_3\}$ and no transient states.

Note that, if we had instead attempted to solve this model with the unichain policy iteration algorithm, we would have proceeded as above but at the subsequent evaluation step we would be confronted with the inconsistent system of equations

$$h(s_1) + g = 2 + h(s_1),$$

$$h(s_2) + g = 3 + h(s_2),$$

$$h(s_3) + g = 4 + h(s_3)$$

Thus the unichain policy iteration algorithm need not provide a solution to a communicating model. We propose the following modification of the multichain algorithm, which exploits the communicating structure by finding a unichain improvement at each pass through the algorithm. We include a flag "UNICHAIN" which indicates whether or not the current policy is known to be unichain.

The Communicating Policy Iteration Algorithm.

1. Set $n = 0$ and select a $d_0 \in D$. If d_0 is unichain, set UNICHAIN = YES; otherwise set UNICHAIN = NO.
2. (Policy evaluation) If UNICHAIN = NO, go to 2a. Else go to 2b.

2a. Find vectors g_n and h_n by solving

$$(P_{d_n} - I)g_n = 0, \tag{9.5.1}$$

$$r_{d_n} - g_n + (P_{d_n} - I)h_n = 0, \tag{9.5.2}$$

subject to Condition 9.2.1, 9.2.2, or 9.2.3. Go to step 3.

2b. Find a scalar g_n and vector h_n by solving

$$r_{d_n} - g_n e + (P_{d_n} - I)h_n = 0, \tag{9.5.3}$$

subject to Conditions 9.2.1, 9.2.2, or 9.2.3. Go to step 3.

3. (Policy improvement) If g_n is constant, go to step 3b. Otherwise go to step 3a.

  3a. Let $S_0 = \{s \in S: g_n(s) = \max_{j \in S} g_n(j)\}$ and set $d_{n+1}(s) = d_n(s)$ for all $s \in S$. Set $T = S/S_0$ and $W = S_0$.

  i. If $T = \varnothing$, go to (iv).

  ii. Obtain an $s' \in T$ and $a \in A_s$ for which

$$\sum_{j \in W} p(j|s', a') > 0. \tag{9.5.4}$$

  iii. Set $T = T/\{s'\}$, $W = W \cup \{s'\}$, and $d_{n+1}(s') = a'$, and go to (i).

  iv. Set UNICHAIN = YES, increment n by 1, and go to step 2.

  3b. Choose

$$d_{n+1} \in \arg\max_{d \in D} \{r_d + P_d h_n\},$$

  setting $d_{n+1}(s) = d_n(s)$ if possible.
  If $d_{n+1} = d_n$ go to step 4. Otherwise set UNICHAIN = NO, increment n by 1, and go to step 2.

4. Set $d^* = d_n$.

A flow chart for this algorithm follows (Fig. 9.5.2).

We now discuss implementation of the algorithm by referring to the flow chart. Begin with an arbitrary decision rule. Choose it to be unichain if possible. Problem structure can guide this choice. If the decision rule is multichain, or its structure unknown, the UNICHAIN flag equals NO.

If UNICHAIN = YES, the evaluation and improvement steps proceed as in the unichain policy iteration algorithm, but since we cannot determine *a priori* that the improved decision rule generates a unichain process, we set UNICHAIN = NO and return to the evaluation step. When we enter the evaluation step with UNICHAIN = NO, we solve the multichain evaluation equations. If the gain is state independent, we carry out a unichain improvement through step 3(b). Otherwise we exploit the (weakly) communicating structure to identify a unichain improvement. Step 3(a) constructs a unichain decision rule with gain equal to $\max_{s \in S} g_n(s)$ by finding a path from each state to the maximal gain set S_0. It does this by augmenting this set one state at a time. We leave the precise rule for choosing this specification arbitrary, but

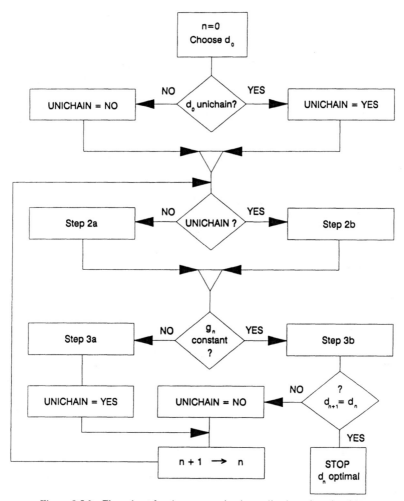

Figure 9.5.2 Flow chart for the communicating policy iteration algorithm.

any selection rule which achieves (9.5.4) suffices. We then set UNICHAIN = YES and repeat the process.

We illustrate this algorithm by solving Example 9.5.1.

Example 9.5.1 (ctd.). Since, in the above calculations, we began with a unichain decision rule, initially UNICHAIN = YES and the first pass was through the unichain loop. In step 3(b), we set UNICHAIN = NO and evaluate d_1 in step 2(a), finding $g_1 = (2, 3, 4)^T$. Since g_1 is nonconstant, we find an improvement through step 3(a) as follows.

Set $W = S_0 = \{s_3\}$, $T = \{s_1, s_2\}$, and $d_2(s_3) = a_{3,2}$. Since $p(s_3|s_2, a_{2,1}) > 0$, set $d_2(s_2) = a_{2,1}$, $W = \{s_2, s_3\}$, and $T = \{s_1\}$. Since $T \neq \varnothing$, we repeat this step and set

$d_2(s_1) = a_{1,1}$. We set UNICHAIN = YES, and the next pass through the algorithm reveals that $(d_2)^\infty$ is an optimal policy.

The following result establishes finite convergence of this algorithm in a unichain model.

Theorem 9.5.1. In a weakly communicating model, the communicating policy iteration algorithm terminates in a finite number of iterations with an optimal policy.

Proof. If UNICHAIN = YES, the proof of Theorem 8.6.6 establishes that either

a. $g_{n+1}(s) > g_n(s)$ for some $s \in S$, or
b. $g_{n+1}(s) = g_n(s)$ for all $s \in S$ and $h_{n+1}(s) > h_n(s)$ for some $s \in S$, or
c. $g_{n+1}(s) = g_n(s)$ and $h_{n+1}(s) = h_n(s)$.

If UNICHAIN = NO, $g_{n+1}(s) > g_n(s)$ for some $s \in S$. Consequently the algorithm cannot repeat decision rules unless (c) above holds. In this case, g_n and h_n satisfy the multichain optimality equations, so, by Theorem 9.1.7(a), $(d_n)^\infty = (d^*)^\infty$ is optimal. $\square$

Note that we can always use the multichain algorithm to solve a communicating model, but the above is more efficient. It avoids augmenting (9.5.2) with (9.5.1) when unnecessary, and it replaces improvement through (9.2.3) by a simple search procedure.

9.5.2 Linear Programming

Since g^* is constant in weakly communicating models, we would suspect some simplification in the linear programming approach. Since all $d \in D$ satisfy (9.3.1), this equation provides no information in weakly communicating models and we are left with the characterization of g^* as the minimal scalar g for which there exists an $h \in V$ satisfying

$$h + ge \geq r_d + P_d h$$

for all $d \in D$. This relationship characterized the unichain linear programming model of Sec. 8.8.2. We now investigate the implications of solving the dual unichain linear program (8.8.2)–(8.8.4) in a communicating model by applying it to Example 9.5.1.

Example 9.5.1 (ctd.). The dual (unichain) linear program for this model is the following.

Maximize $\quad x(s_1, a_{1,2}) + x(s_2, a_{2,1}) + x(s_2, a_{2,2})$

$$+ 3x(s_2, a_{2,3}) + 2x(s_3, a_{3,2}) + 4x(s_3, a_{3,2})$$

subject to

$$x(s_1, a_{1,1}) - x(s_2, a_{2,2}) = 0$$

$$-x(s_1, a_{1,1}) + x(s_2, a_{2,1}) + x(s_2, a_{2,2}) - x(s_3, a_{3,1}) = 0,$$

$$\sum_{s \in S} \sum_{a \in A_s} x(s, a) = 1,$$

and $x(s, a) \geq 0$, $s \in S$, and $a \in A_s$. The optimal solution has $x^*(s_3, a_{3,2}) = 1$ and $x^*(s, a) = 0$ for all other state action pairs. The objective function value equals 4.

Proceeding as if this were a unichain model, we apply Corollary 8.8.8 to choose an optimal policy (d^*), with $d^*(s) = a_{3,2}$, and arbitrary elsewhere. Clearly, this approach could generate a suboptimal policy because of the lack of specificity outside of $S_{x^*} = \{s \in S: \sum_{a \in A_s} x^*(s, a) > 0\}$.

In a unichain model, we can choose arbitrary actions in transient states because under any action the system eventually reaches the single recurrent class and achieves the maximal gain. In a communicating model, such an approach can result in suboptimal policies because it could keep the system outside of S_{x^*} indefinitely. Either of the following approaches alleviates this.

1. Solve the problem using the dual multichain linear program and the approach of Sec. 9.3.
2. Obtain a solution x^* of the unichain linear program. For $s \in S_{x^*}$, set $d^*(s) = a$ for some $a \in A_s$ satisfying $x^*(s, a) > 0$. On S/S_{x^*}, choose an action in each state which drives the system to S_{x^*} with positive probability. The search procedure in step 3(a) of the communicating policy iteration algorithm, with $S_0 = S_x$, would identify such actions.

In Example 9.5.1, the search procedure is easy to apply. Solving the dual as above yields $S_{x^*} = \{s_3\}$ and $d^*(s_3) = a_{3,2}$. Now apply step 3(a). Begin with $W = \{s_3\}$. Set $d^*(s_2) = a_{2,1}$ and $W = \{s_2, s_3\}$. Then set $d^*(s_1) = a_{1,1}$. This identifies the optimal policy.

The multichain linear programming approach, which requires a linear program with 14 variables and five constraints, yields the same optimal policy.

9.5.3 Value Iteration

The unichain value iteration algorithm of Sec. 8.5 applies directly to communicating and weakly communicating models. Because the optimal gain is constant, Corollary 9.4.6 implies that $sp(v^{n+1} - v^n)$ converges monotonically to 0 so that it provides a valid stopping criterion. As before, a sufficient condition for convergence is the aperiodicity of transition matrices of stationary deterministic optimal policies. This can always be achieved by applying the data transformation of 8.5.5 but, as Example 9.5.1 shows, it is not always necessary to do so.

Example 9.5.1. (ctd.). Choosing $v^0 = 0$, we obtain the following sequence of iterates.

n	$v^n(s_1)$	$v^n(s_2)$	$v^n(s_3)$	$sp(v^n - v^{n-1})$
0	0	0	0	
1	2	3	4	2
2	4	6	8	2
3	6	9	12	2
4	9	13	16	1
5	13	17	20	0

At $n = 5$, we obtain $v^n(s_i) - v^{n-1}(s_i) = g^* = 4$ for $i = 1, 2, 3$. Note also that $h^*(s_1) = v^5(s_1) - 5g^* = -7$ and $h^*(s_2) = v^5(s_2) - 5g^* = -3$.

BIBLIOGRAPHIC REMARKS

In his 1960 monograph, Howard recognized that multichain average reward models require more intricate analyses than unichain models. Among other results, he observed that a pair of optimality equations are required to characterize the optimal average reward. In an important and innovative paper, Blackwell (1962) provided a theoretical framework for analyzing multichain models. His observation that the average reward model may be viewed as a limit of expected discounted reward models, in which the discount rate approaches 1, stimulated extensive research on average reward models. He showed that the partial Laurent series expansion provided a link between these two models and incorporated some nonstandard Markov chain concepts (Kemeny and Snell, 1960) in his analyses. Denardo and Fox (1968) provided a linear programming formulation of the multichain model, and stated and proved a key result, which we state as Proposition 9.1.1. It relates solutions of the optimality equations and the modified optimality equations. The modified optimality equations were further studied by Yushkevich (1973), and play a key role in the development of the average reward models in Dynkin and Yushkevich (1979). Schweitzer and Federgruen (1978a) characterized the set of all solutions of the multichain optimality equations, and Federgruen and Schweitzer (1984b) established the existence of a solution to the optimality equations using fixed-point theory in finite-state models.

Schal (1992) provides an in-depth treatment of the countable-state multichain model with unbounded rewards and compact action sets. Under Liapunov-type growth conditions (Hordijk, 1974) on the expected mean recurrence time and the expected total reward until return to a given set, and continuity conditions on the rewards and transition probabilities, Schal establishes the existence of a solution to the optimality equation, the existence of average optimal policies, and a characterization of optimal policies. He views his work as an extension of Federgruen, Hordijk, and Tijms (1979) and Federgruen, Schweitzer, and Tijms (1983), in which similar hypotheses produce analogous results for the countable-state unichain models.

Howard (1960) presented the multichain policy iteration algorithm using Condition 9.2.3 to specify a solution to the evaluation equations. By analyzing the structure of the iterates of the algorithm, he proved that the average rewards obtained at successive passes through the algorithm were nondecreasing; however, he did not show that the algorithm terminates in finitely many steps. Veinott (1966a) completed this analysis by establishing that when the solution of the evaluation equations satisfies Condition 9.2.1 or, equivalently, 9.2.2, the algorithm cannot cycle, so that finite convergence is achieved. We present the Howard-Veinott approach in Sec. 9.2.4. Blackwell (1962) used the partial Laurent series expansion to demonstrate finite convergence under Condition 9.2.2. We base Sec. 9.2.3 on this analysis. Veinott (1969b and 1974) refined and extended Blackwell's approach; we discuss this work in Chap. 10. Denardo and Fox (1968) also show the convergence of the algorithm through detailed analysis of the chain structure. Federgruen and Spreen (1980) provide an efficient way to implement the algorithm which avoids determining the chain structure at the evaluation phase and efficiently solves the system of evaluation equations.

Dekker (1985 and 1987) investigated the convergence of multichain policy iteration in models with compact action sets. Among other results, he provides a clever example (Problem 9.9) which shows that the algorithm may converge to a suboptimal policy.

When the set of stationary policies is infinite, Schweitzer (1985b), in finite-state compact action models, and Zijm (1984a and 1984b), in countable-state models, establish existence of an average optimal policy and show that the boundedness of the bias vectors on the set of stationary average optimal policies implies the existence of solutions to the optimality equations. The proofs are based on initiating policy iteration at a gain optimal policy, and using the boundedness of the bias vectors to ensure convergence.

Denardo and Fox (1968) contains the first published formulation and analysis of the multichain linear programs which appear in Sec. 9.3. They attribute their formulation to an unpublished manuscript of Balinski (1961). Derman (1970) extends their results and shows that to find optimal policies requires solution of two linear programs and one search problem. These analyses are based on the primal linear program. Motivated by a statement on p. 84 of Derman's book:

"No satisfactory treatment of the dual problem for the multichain case has been published,"

Hordijk and Kallenberg (1979) and Kallenberg (1983) developed such a theory. We follow their approach in Sec. 9.3 and include several insightful examples from Kallenberg (1983) there and in the problem section. Hordijk and Kallenberg (1984) investigate the use of linear programming for studying multichain models with constraints, and Filar, Kallenberg, and Lee (1989) analyze variance penalized multichain models. Haviv (1993) provides an interesting example of a constrained multichain model in which the Principle of Optimality does not hold.

Many authors have studied value iteration in multichain average reward models, perhaps motivated by Howard's (1960) conjecture that

$$\lim_{n \to \infty} v^n - ng^* - h^*$$

exists, or by the challenge of accounting for the periodicity of $\{v^n\}$. Brown (1965) established the boundedness of $e_n \equiv v^n - ng^* - h^*$ (Theorem 9.4.1 herein), proved the important technical result we state as Lemma 9.4.3, and investigated the asymptotic periodicity of e_n. Lanery (1967) extended this analysis and showed among other results that

$$\lim_{n \to \infty} \left\{ v_{nd+r}(s) - (nd + r)g^*(s) - h^*(s) \right\}$$

exists on recurrent classes of the transition probability matrix of an optimal decision rule, where d is its period and $r = 0, 1, \ldots, d - 1$. Proposition 9.4.2 is a special case of his result. Schweitzer and Federgruen (1977 and 1979) provide an in-depth and complete analysis of this problem. They show that the above limit always exists in multichain models, establish a rate of convergence, and analyze the behavior of the sequence of the sets of decision rules which achieve the maximum at each iteration. Our presentation draws on Hordijk, Schweitzer, and Tijms (1975), who demonstrate convergence of e_n in countable-state models, and Schweitzer and Federguen (1977).

Brown (1965) initiated study of the asymptotic behavior of maximizing decisions by showing that they may oscillate periodically. Bather (1973) subsequently provided an example in which the sequence of maximizing decision rules oscillates aperiodically, and Lanery (1967) showed that this sequence may even contain nonoptimal decision rules. Surveys by Federgruen and Schweitzer (1978 and 1980) provide summaries of research on value iteration in average reward models.

From a computational perspective, the existence of a limit of e_n is of use only when the optimal gain does not depend on the state. Therefore, if the optimal policy has state-dependent gain, no stopping criteria are available for value iteration. Schweitzer (1984a) suggests a variant of the value iteration algorithm for a multichain model based on decomposing the model into a system of communicating models using a hierarchical algorithm from Bather (1973c). Federgruen and Schweitzer (1984) analyze multichain value iteration in the context of solving a generalization of the multichain optimality equations.

Denardo and Fox (1968, p. 472) observe that, in multichain models,

> "It is often the case that the optimal policy has a single ergodic subchain although the class of transient states can vary from policy to policy and nonoptimal policies may have multiple ergodic chains. Examples include (s, S) inventory problems."

Bather (1973a, b, c) identified a class of models which has this structure and analyzed them in considerable detail. He referred to these models as *communicating*. The concept of a weakly communicating model originates with Platzman (1977) (he calls them *simply connected*), who shows that these models can be solved using value iteration. Bather (1973c) provides an algorithm which decomposes a multichain model into a set of communicating models and a set of states that are transient under any policy; Ross and Varadarajan (1991) provide a related decomposition scheme. Kallenberg (1983) and Filar and Schultz (1988) discuss the linear programming approach to communicating models. Haviv and Puterman (1991) provide and discuss the communicating policy iteration algorithm of Sec. 9.5.1, and van der Wal (1981) establishes the convergence of modified policy iteration in a weakly communicating model under the assumption that $P_d \geq \alpha I$ for all $d \in D$ and some $\alpha > 0$.

PROBLEMS

9.1. Show that Proposition 9.1.1 holds in a finite-state model with compact action sets in which $p(j|s, a)$ is continuous in a for each s and j, and $r(s, a)$ is continuous in a for each s.

9.2. Prove Theorem 9.1.6b.

9.3. (A model with a compact action set) Let $S = \{s_1, s_2\}$; $A_{s_1} = \{u : 0 \le u \le 1\}$ and $A_{s_2} = \{a_{2,1}\}$; $r(s_1, a) = u$ and $r(s_2, a_{2,1}) = 0$; $p(s_1|s_1, u) = 1 - u^2$, $p(s_2|s_1, u) = u^2$, $p(s_1|s_1, a_{1,2}) = 0$, and $p(s_2|s_1, a_{1,2}) = 1$.

   a. Give the optimality equations for this model.

   b. Show, by direct evaluation of each stationary policy through $g^{d^\infty} = P_d^* r_d$ that the optimal gain $g^*(s) = 0$ for $s = s_1$ and s_2.

   c. Show that, upon substitution of g^* into the second optimality equation, it has no solution when $s = s_1$.

   d. Let d_k denote the decision rule which uses action k in state s_1. Show that $h^{(d_k)^\infty}(s_2) - h^{(d_k)^\infty}(s_1) = 1/k$, and conclude that $\lim_{k \to 0} h^{(d_k)^\infty} = \infty$.

   e. Show that $P_{d_k}^*$ has a discontinuity at $k = 0$.

9.4. Let $S = \{s_1, s_2\}$; $A_{s_1} = \{a_{1,1}, a_{1,2}\}$ and $A_{s_2} = \{a_{2,1}, a_{2,2}, a_{2,3}\}$; $r(s_1, a_{1,1}) = 1$, $r(s_1, a_{1,2}) = 4$, $r(s_2, a_{2,1}) = 2$, $r(s_2, a_{2,2}) = 3$, and $r(s_2, a_{2,3}) = 5$; and $p(s_1|s_1, a_{1,1}) = 1$, $p(s_1|s_1, a_{1,2}) = 0.5$, $p(s_1|s_2, a_{2,1}) = 1$, $p(s_2|s_2, a_{2,2}) = 1$, and $p(s_1|s_2, a_{2,3}) = 0.75$.

   a. For each stationary Markov policy, determine its chain structure and classify its states.

   b. Solve the problem using the multichain policy iteration algorithm, and using Conditions 9.1.1–9.1.3 for uniquely specifying h_n.

   c. Solve this problem using the linear programming formulation of the multi-chain MDP.

   d. Let δ denote the decision rule which chooses action $a_{1,2}$ in s_1, and $a_{2,3}$ in s_2. Compute x_δ and y_δ through (9.3.10) and (9.3.11), and verify that they are dual feasible.

   e. Show that is a communicating MDP and solve it using the communicating policy iteration algorithm.

   f. Solve the problem using value iteration starting with $v^0 = 0$. Use the bound in (8.5.4) to terminate the algorithm with a 0.1-optimal solution. Find g^* and h^* using the approach suggested at the start of Sec. 9.4.1 and compare them to values found in b.

9.5. Let $S = \{s_1, s_2, s_3\}$; $A_{s_1} = \{a_{1,1}\}$, $A_{s_2} = \{a_{2,1}\}$, and $A_{s_3} = \{a_{3,1}, a_{3,2}\}$; $r(s_1, a_{1,1}) = 1$, $r(s_2, a_{2,1}) = 2$, and $r(s_3, a_{3,1}) = r(s_3, a_{3,2}) = 0$; and $p(s_1|s_1, a_{1,1}) = 1$, $p(s_2|s_2, a_{2,1}) = 1$, $p(s_1|s_3, a_{3,1}) = 1$, and $p(s_2|s_3, a_{3,2}) = 1$. Show that the algorithm may cycle between optimal and suboptimal policies when implementing the policy iteration algorithm by choosing arbitrary solutions of (9.2.1) and (9.2.2) at each iteration.

9.6. Classify an individual's wealth as low (*L*), medium (*M*), or high (*H*). In state *L*, the individual has insufficient wealth to invest, so each period he consumes all earned income and as a consequence receives one unit of utility and remains in state *L*. In state *M*, three consumption options are available. Consuming at a high rate yields a utility of three units; at a moderate level, a utility of two units, and at a low level a utility of 1.5 units. If an individual in state *M* consumes at a low rate, with probability 0.95 he remains in state *M*, and with probability 0.05 his wealth rises to level *H* at the next period; if he consumes at a moderate rate, with probability 0.7 his wealth remains at a medium level, with probability 0.01 it rises to a high level, and with probability 0.29 it falls to a low level; and, finally, if he consumes at a high level, with probability 0.9 his wealth falls to a low level in the subsequent period and otherwise remains at a medium level. If an individual's wealth is high, he may consume at a moderate level or a high level. If he consumes at a moderate level, he receives two units of utility per period, and if he consumes at a high level, he receives three units of utility. By consuming at a moderate level, his wealth remains at a high level with probability 0.9, and falls to a moderate level with probability 0.1. Consuming at a high level, his wealth remains at a high level with probability 0.5, and falls to a moderate level with probability 0.5.

    a. Formulate the individual's consumption level choice problem as a Markov decision process.

    b. Verify that the model is neither unichain nor communicating.

    c. Find a policy that maximizes the individual's expected average utility using multichain policy iteration and linear programming.

    d. Suppose the individual discounts utility at rate λ, and uses the expected total discounted utility optimality criterion. Investigate the effect of the discount rate on the optimal consumption policy.

9.7. Prove Lemma 9.2.4 and Lemma 9.2.5b.

9.8. Prove Theorem 9.2.6 under the assumption that $\eta^{d^{\infty}}$ is chosen to satisfy either Condition 9.2.1 or 9.2.2.

9.9. (Dekker, 1987) Let $S = \{s_1, s_2, s_3, s_4\}$; $A_{s_1} = \{a_{1,1}\} \cup \{(x, y): x \geq 0, \ y \geq 0, \ x + y \leq 5, \ x \geq y^2\}$, $A_{s_2} = \{a_{2,1}\}$, $A_{s_3} = \{a_{3,1}\}$, and $A_{s_4} = \{a_{4,1}\}$; $r(s_1, a_{1,1}) = 6$, $r(s_1, (x, y)) = 0$, $r(s_2, a_{2,1}) = 1$, $r(s_3, a_{3,1}) = 2$, and $r(s_4, a_{4,1}) = 0$; $p(s_4|s_1, a_{1,1}) = 1$, $p(s_1|s_1, (x, y)) = 0.5 - x - y$; $p(s_2|s_1, (x, y)) = 0.25 + x$, $p(s_3|s_1, (x, y)) = 0.25 + y$, $p(s_2|s_2, a_{2,1}) = 1$, $p(s_3|s_3, a_{3,1}) = 1$, and $p(s_1|s_4, a_{4,1}) = 1$ (Fig. 9.PR1).

    a. Determine the optimal policy by inspection.

    b. Solve this problem by using the multichain policy iteration algorithm where the initial decision rule chooses action (x, y) in s_1. Show that it converges to a suboptimal policy.

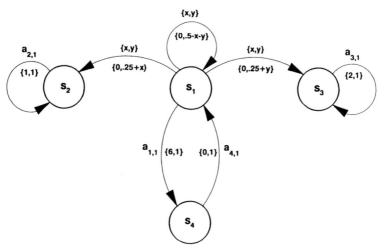

Figure 9.PR1 Symbolic representation of Problem 9.9.

9.10. Show that g^* is the minimal g for which there exists an h satisfying (9.3.1) and (9.3.2).

9.11. Prove Proposition 9.3.3 by substituting (x_d, y_d) into (9.3.5) and (9.3.6).

9.12. Suppose we add the constraint that $g(s_i) = g(s_j)$ for all i and j to the primal multichain linear programming. Show that the model reduces the unichain linear program, and that its dual agrees with that in Sec. 8.8.

9.13. (Kallenberg, 1983) Let $S = \{s_1, s_2, s_3, s_4\}$; $A_{s_1} = \{a_{1,1}, a_{1,2}\}$, $A_{s_2} = \{a_{2,1}, a_{2,2}\}$, $A_{s_3} = \{a_{3,1}\}$, and $A_{s_4} = \{a_{4,1}\}$; $r(s_1, a_{1,1}) = 4$, $r(s_1, a_{1,2}) = 2$, $r(s_2, a_{2,1}) = 1$, $r(s_2, a_{2,2}) = 2$, $r(s_3, a_{3,1}) = 1$, and $r(s_4, a_{4,1}) = 1$; and $p(s_2|s_1, a_{1,1}) = 1$, $p(s_4|s_1, a_{1,2}) = 1$, $p(s_2|s_2, a_{2,1}) = 1$, $p(s_3|s_2, a_{2,2}) = 1$, $p(s_1|s_3, a_{3,1}) = 1$, and $p(s_4|s_4, s_{4,1}) = 1$ (Fig. 9.PR2). Let $\alpha_j = \frac{1}{4}$, $1 \le j \le 4$.
a. Give the primal and dual linear programming formulations for this model.
b. Show that (x_1, y_1) and (x_2, y_2) below are dual feasible, and that each generates the same decision rule through (9.3.7).

(s, a)	$(s_1, a_{1,1})$	$(s_1, a_{1,2})$	$(s_2, a_{2,1})$	$(s_2, a_{2,2})$	$(s_3, a_{3,1})$	$(s_4, a_{4,1})$
$x_1(s, a)$	$\frac{1}{4}$	0	0	$\frac{1}{4}$	$\frac{1}{4}$	$\frac{1}{4}$
$y_1(s, a)$	0	0	0	0	0	0
$x_2(s, a)$	$\frac{1}{6}$	0	0	$\frac{1}{6}$	$\frac{1}{6}$	$\frac{1}{2}$
$y_2(s, a)$	0	$\frac{1}{4}$	0	$\frac{1}{12}$	$\frac{1}{6}$	0

c. Find an optimal policy using linear programming.

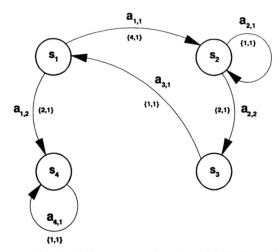

Figure 9.PR2   Symbolic representation of model in Problem 9.13.

9.14. Verify the calculations in Example 9.3.2.

9.15. (Kallenberg, 1983) Let $S = \{s_1, s_2, s_3\}$; $A_{s_1} = \{a_{1,1}, a_{1,2}\}$, $A_{s_2} = \{a_{2,1}, a_{2,2}\}$, and $A_{s_3} = \{a_{3,1}\}$; $r(s_1, a_{1,1}) = 1$, $r(s_1, a_{1,2}) = 0$, $r(s_2, a_{2,1}) = 0$, $r(s_2, a_{2,2}) = 0$, and $r(s_3, a_{3,1}) = 0$; and $p(s_2|s_1, a_{1,1}) = 1$, $p(s_3|s_1, a_{1,2}) = 1$, $p(s_1|s_2, a_{2,1}) = 1$, $p(s_2|s_2, a_{2,2}) = 1$, and $p(s_3|s_3, a_{3,1}) = 1$. Let $\alpha_j = \frac{1}{3}$, $1 \leq j \leq 3$.

  a. Give the primal and dual linear programs corresponding to this model.

  b. Find an optimal solution to the dual linear program, and obtain an optimal policy through (9.3.7).

  c. Show that (x, y) below generates an optimal policy through (9.3.7), and that it is a feasible but nonoptimal solution of the dual linear program.

(s, a)	$(s_1, a_{1,1})$	$(s_1, a_{1,2})$	$(s_2, a_{2,1})$	$(s_2, a_{2,2})$	$(s_3, a_{3,1})$
$x(s, a)$	$\frac{1}{6}$	0	0	$\frac{1}{6}$	$\frac{2}{3}$
$y(s, a)$	0	$\frac{1}{3}$	0	$\frac{1}{6}$	0

9.16. Suppose δ^∞ is a *deterministic* optimal policy generated through (9.3.7) from the dual linear programming solution (x_α, y_α) obtained with a specified vector α satisfying $\alpha_j > 0$ and $\sum_{j \in S} \alpha_j = 1$ on the right-hand side of (9.3.6). Show for any other vector α' that there exists an optimal solution $(x_{\alpha'}, y_{\alpha'})$ of the dual linear program which yields the same optimal policy.

Hint: Employ concepts from the proof of Theorem 9.3.5, apply Proposition 9.3.2, and note that

$$x_\alpha^T P_\delta^* = \alpha^T P_\delta^*,$$

that $x_\alpha(s) = 0$ on transient states of the Markov chain corresponding to P_δ, and that $(I - P_\delta)^{-1}$ exists on transient states of P_δ.

9.17. Establish inequalities (9.4.10) and (9.4.12) in the proof of Theorem 9.4.4.

9.18. Verify that the inventory model in Example 8.7.1 with the unmodified demand distribution is communicating, and solve it using the algorithm in Sec. 9.5. Compare the optimal policy and optimal gain to that in Sec. 8.7.1.

9.19. Find an optimal policy for an infinite-horizon version of the service rate control model in Problem 4.26 under the average reward criterion.

9.20. **a.** Complete the proof of part a of Theorem 9.3.4 by showing that (9.3.13) holds for $s \in S/S_{x_d}$ and $a \in A_s$.
 b. Prove Theorem 9.3.4b.

9.21. (Open Question) Find a stopping criterion for value iteration in multichain models.

9.22. Consider the model in Example 9.5.1 augmented with a state s_4 in which $A_{s_4} = \{a_{4,1}, a_{4,2}\}$, $r(s_4, a_{4,1}) = 5$, $r(s_4, a_{4,2}) = 1$, $p(s_1|s_4, a_{4,1}) = 1$, $p(s_3|s_4, a_{4,1}) = 1$.
 a. Show the model is weakly communicating.
 b. Solve it using the communicating policy iteration algorithm.
 c. Solve it using the two variants of linear programming discussed in Sec. 9.5.2.
 d. Solve it using value iteration.

9.23. (van der Wal, 1981) Develop a modified policy iteration algorithm for a communicating model and show that it converges if there exists an $\alpha > 0$ for which $P_d \geq \alpha I$ for all $d \in D^{MD}$.

CHAPTER 10

Sensitive Discount Optimality

In the previous two chapters we concentrated on characterizing and finding average optimal policies. The long-run average reward criterion focuses on the limiting or steady-state behavior of a system and ignores transient performance. The following simple example illustrates why this is a problem.

Example 10.0.1. Let $S = \{s_1, s_2\}$; $A_{s_1} = \{a_{1,1}, a_{1,2}\}$ and $A_{s_2} = \{a_{2,1}\}$; $r(s_1, a_{1,1}) = 1000$, $r(s_1, a_{1,2}) = 0$ and $r(s_2, a_{2,1}) = 0$; and $p(s_2|s_1, a_{1,1}) = 1$, $p(s_2|s_1, a_{1,2}) = 1$ and $p(s_2|s_2, a_{2,1}) = 1$ (Fig. 10.0.1).

Action choice in state s_1 determines two stationary policies. Let δ^{∞} choose $a_{1,1}$ and γ^{∞} choose $a_{1,2}$ there. Since $g^{\delta^{\infty}}(s) = g^{\gamma^{\infty}}(s) = 0$, both policies are average optimal, but of course, we would prefer δ^{∞} to γ^{∞} because it yields a reward of 1000 prior to absorption in s_2. The average reward criterion ignores this distinction.

We face a similar problem when solving inventory or queueing control problems. Such models usually have several average optimal policies because actions on transient states do not effect the average reward. For example, in an inventory control model, an optimal policy will keep the stock level below some target level. How we handle initial stock does not effect average reward so that many average optimal policies are available. Clearly, a decision maker might find some policies preferable to others.

We address this deficiency of the average reward criterion by proposing and investigating more selective optimality criteria. We introduced these concepts in Chap. 5, where we defined bias optimality, n-discount optimality, and Blackwell optimality. We refer to all of these as *sensitive discount optimality* criteria.

Our analysis uses the Laurent series expansion of the expected discounted reward of a *stationary* policy and deviates somewhat from that in earlier chapters with respect to the role of the optimality equations. In Chaps 6–9, we established optimality by showing that a solution of the optimality equations yields the optimal discounted reward, total reward, or average reward in the class of *all* (randomized history-dependent) policies. Here we do not take that approach because it is tedious to develop a Laurent series expansion for an arbitrary policy. Instead we proceed as

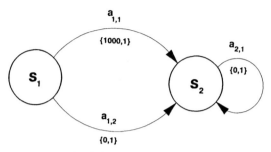

Figure 10.0.1   Symbolic representation of Example 10.0.1.

follows:

1. We show (nonconstructively) that there exists a *stationary* Blackwell optimal policy.

2. We show that a Blackwell optimal policy is n-discount optimal for any $n \geq -1$. This establishes the existence of a *stationary* n-discount optimal policy.

3. We characterize stationary n-discount optimal policies within the class of stationary policies.

4. We show that the optimality equations determine the coefficients of the Laurent series expansion of n-discount optimal policies and that they characterize stationary n-discount optimal policies.

5. We provide a finite policy iteration algorithm for computing a stationary n-discount optimal policy and a Blackwell optimal policy.

The results in this chapter unify a large segment of the research in finite-state and action Markov decision process models. The advanced reader might skip Chaps. 7–9 and proceed directly to this section. By taking this approach, you will acquire an elegant framework from which to analyze Markov decision processes; however, you might lose an appreciation of the significance of the results herein and an awareness of many special results and algorithms for models with expected total-reward and average reward criteria.

We suggest reviewing Secs. 5.4.3 and 8.2 prior to reading this chapter. *Throughout, we assume a model with a finite number of states and actions.*

10.1 EXISTENCE OF OPTIMAL POLICIES

This section defines and relates sensitive discount optimality criteria and carries out the first two steps of the above outline.

10.1.1 Definitions

We say that a policy $\pi^* \in \Pi^{HR}$ is *n-discount optimal* for some integer $n \geq 1$ if

$$\liminf_{\lambda \uparrow 1} (1 - \lambda)^{-n} [v_\lambda^{\pi^*} - v_\lambda^\pi] \geq 0 \tag{10.1.1}$$

for all $\pi \in \Pi^{HR}$.

From this definition it is easy to see the following.

Proposition 10.1.1. If a policy is *n*-discount optimal, then it is *m*-discount optimal for $m = -1, 0, \ldots, n$.

Special cases of this criterion are -1-*discount optimality* which holds if

$$\liminf_{\lambda \uparrow 1} (1 - \lambda) [v_\lambda^{\pi^*} - v_\lambda^\pi] \geq 0 \tag{10.1.2}$$

for all $\pi \in \Pi^{HR}$ and 0-*discount optimality* which holds if

$$\liminf_{\lambda \uparrow 1} [v_\lambda^{\pi^*} - v_\lambda^\pi] \geq 0 \tag{10.1.3}$$

for all $\pi \in \Pi^{HR}$.

We say that a policy $\pi^* \in \Pi^{HR}$ is *Blackwell optimal*, if there exists a λ^*, $0 \leq \lambda^* < 1$, for which

$$v_\lambda^{\pi^*} \geq v_\lambda^\pi$$

for all $\pi \in \Pi^{HR}$ and $\lambda^* \leq \lambda < 1$. From the above definitions it follows that Blackwell optimality implies 0-optimality. Since these two definitions appear similar, one might suspect that these two criteria are equivalent. The following example shows that this is not the case, and that the Blackwell optimality criterion is more selective. We return to this example at several points throughout this chapter.

Example 10.1.1. Let $S = \{s_1, s_2, s_3\}$; $A_{s_1} = \{a_{1,1}, a_{1,2}\}$, $A_{s_2} = \{a_{2,1}\}$, and $A_{s_3} = \{a_{3,1}\}$; $r(s_1, a_{1,1}) = 1$, $r(s_1, a_{1,2}) = 2$, $r(s_2, a_{2,1}) = 1$, and $r(s_3, a_{3,1}) = 0$; and $p(s_2|s_1, a_{1,1}) = 1$, $p(s_3|s_1, a_{1,2}) = 1$, $p(s_3|s_2, a_{2,1}) = 1$, and $p(s_3|s_3, a_{3,1}) = 1$ (Fig. 10.1.1).

Let δ^∞ choose $a_{1,2}$ in s_1, and γ^∞ choose $a_{1,1}$ in s_1. Since $v_\lambda^{\delta^\infty}(s_1) = 2$ and $v_\lambda^{\gamma^\infty}(s_1) = 1 + \lambda$, it follows that *both* policies are -1-discount optimal and 0-discount optimal but that *only* δ^∞ is Blackwell optimal with $\lambda^* = 0$.

Figure 10.1.2 illustrates this point. Both policies agree when $\lambda = 1$, but $v_\lambda^{\delta^\infty}(s_1) > v_\lambda^{\gamma^\infty}(s_1)$ for $0 \leq \lambda < 1$.

Since

$$v_\lambda^{\delta^\infty}(s_1) - v_\lambda^{\gamma^\infty}(s_1) = 1 - \lambda,$$

it follows that

$$(1 - \lambda)^{-n} [v_\lambda^{\delta^\infty}(s_1) - v_\lambda^{\gamma^\infty}(s_1)] = (1 - \lambda)^{-n+1}.$$

Therefore, δ^∞ is *n*-discount optimal for all $n \geq 1$ while γ^∞ is not.

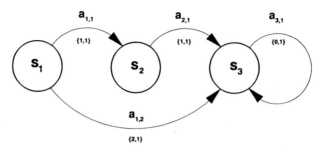

Figure 10.1.1   Symbolic representation of Example 10.1.1.

This example shows that the selectivity of the sensitive optimality criteria increase with n, and suggests that a Blackwell optimal policy is n-discount optimal for all n. In subsequent sections we show that these observations are always valid.

10.1.2   Blackwell Optimality

In this section we establish the existence of a stationary Blackwell optimal policy and investigate properties of such a policy. We extend the approach used in Sec. 8.4.2 to establish existence of a solution to the average reward optimality equation.

We begin with the following definition and technical results. We say that a real-valued function $f(x)$ is a *rational function* on an interval if $f(x) = p(x)/q(x)$, where $p(x)$ and $q(x)$ are polynomials of finite degree and $q(x) \neq 0$ for all x in that interval. The following easily proved lemma notes a key property of rational functions.

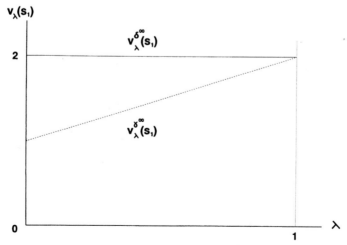

Figure 10.1.2   Graphical representation of $v_\lambda^{\delta^\infty}(s_1)$ and $v_\lambda^{\gamma^\infty}(s_1)$.

Lemma 10.1.2. Suppose $f(x)$ is a real-valued rational function on an interval I. Then if $f(x) = 0$ for some $x \in I$, either $f(x) = 0$ for all $x \in I$ or $f(x) = 0$ for at most finitely many $x \in I$.

Lemma 10.1.3. Let $d \in D^{MD}$. Then, for each $s \in S$, $v_\lambda^{d^\infty}(s)$ is a rational function of λ for $0 \le \lambda < 1$.

Proof. Since $v_\lambda^{d^\infty}$ is the unique solution of $(I - \lambda P_d)v = r_d$, by Cramer's rule it follows that $v_\lambda^{d^\infty}(s)$ is the ratio of two determinants with coefficients depending on λ. Consequently $v_\lambda^{d^\infty}(s)$ is the ratio of two polynomials of degree at most $|S|$. That the polynomial in the denominator does not vanish on $0 \le \lambda < 1$ follows from the invertibility of $(I - \lambda P_d)$. $\square$

Theorem 10.1.4. Suppose S and A_s for each $s \in S$ are finite. Then there exists a stationary deterministic Blackwell optimal policy.

Proof. Since D^{MD} is finite, there exists a sequence $\{\lambda_n\}$ converging to 1 for which there exists a $d^* \in D^{MD}$ with $(d^*)^\infty$ discount optimal for all λ_n. Then, for each $d \in D^{MD}$,

$$v_\lambda^{(d^*)^\infty}(s) - v_\lambda^{d^\infty}(s) \ge 0 \qquad (10.1.4)$$

for all $s \in S$ and $\lambda = \lambda_n$ for $n = 1, 2, \ldots$. Since by Lemma 10.1.3, each of these functions is a rational function of λ, the difference is a rational function of λ, so it follows from Lemma 10.1.1 that either it is identically 0 for all λ or equals 0 for at most finitely many λ's. Thus there exists a $\lambda_d < 1$ for which (10.1.4) holds for $\lambda_d \le \lambda < 1$. Since D^{MD} is finite, (10.1.4) holds for $\lambda^* \le \lambda < 1$ and all $d \in D^{MD}$ where $\lambda^* = \max_{d \in D} \lambda_d$. From Theorem 6.2.10 there exists a stationary discount optimal policy for any λ, so that

$$v_\lambda^{(d^*)^\infty}(s) - v_\lambda^\pi(s) \ge 0$$

for all $\pi \in \Pi^{HR}$ for $\lambda^* \le \lambda < 1$. Hence the result follows. $\square$

The following theorem gives some properties of stationary Blackwell optimal policies.

Theorem 10.1.5

a. There exists a stationary deterministic n-discount optimal policy for $n = -1, 0, 1, \ldots$.

b. Suppose $(d^*)^\infty$ is a Blackwell optimal policy. It is n-discount optimal for $n = -1, 0, 1, \ldots$.

Proof. We prove part (b) first. Since $(d^*)^\infty$ is Blackwell optimal,

$$(1 - \lambda)^{-n} \left[v_\lambda^{(d^*)^\infty} - v_\lambda^\pi \right] \ge 0 \qquad (10.1.5)$$

for $\lambda^* \le \lambda < 1$, and all $\pi \in \Pi^{HR}$ for $n = -1, 0, 1, \ldots$. Therefore

$$\liminf_{\lambda \uparrow 1} (1 - \lambda)^{-n} \left[v_\lambda^{(d^*)^\infty} - v_\lambda^\pi \right] \ge 0$$

establishing part (b).

Since there exists a deterministic stationary Blackwell optimal policy by Theorem 10.1.4, (a) follows from the observation in (b).   □

Part (b) of this result justifies referring to Blackwell optimal policies as ∞-discount optimal. Part (a) of this theorem implies that to find an n-discount optimal policy we need only search within the class of stationary deterministic policies. We exploit the consequences of this second observation in the next subsection.

10.1.2 Stationary n-discount Optimal Policies

In this section we use the Laurent series expansion to interpret the sensitive discount optimality criteria within the class of stationary policies.

Changing the scale of the discount factor simplifies subsequent notation and analysis. Define the interest rate ρ by

$$\rho = (1 - \lambda)/\lambda,$$

so that $\lambda = 1/(1 + \rho)$. Note that we should write $\lambda(\rho)$ instead of λ, but to simplify notation, we will suppress the argument ρ. When $0 \le \lambda < 1$, $\rho > 0$, and as $\lambda \uparrow 1$, $\rho \downarrow 0$. Since

$$\liminf_{\lambda \uparrow 1} (1 - \lambda)^{-n} \left[v_\lambda^{\pi^*} - v_\lambda^\pi \right] = \liminf_{\lambda \uparrow 1} \left[\frac{1 - \lambda}{\lambda} \right]^{-n} \left[v_\lambda^{\pi^*} - v_\lambda^\pi \right],$$

π^* is n-discount optimal if

$$\liminf_{\rho \downarrow 0} \rho^{-n} \left[v_\lambda^{\pi^*} - v_\lambda^\pi \right] \ge 0 \tag{10.1.6}$$

for all $\pi \in \Pi^{HR}$. Similarly π^* is Blackwell optimal if there exists a ρ^* for which $v_\lambda^{\pi^*} \ge v_\lambda^\pi$ for $0 < \rho \le \rho^*$.

For each $d \in D$, Theorem 8.2.3 yields the following Laurent series expansion for $v_\lambda^{d^\infty}$

$$v_\lambda^{d^\infty} = (1 + \rho) \left[\frac{g^{d^\infty}}{\rho} + h^{d^\infty} + \sum_{n=0}^\infty \rho^n y_n^{d^\infty} \right], \tag{10.1.7}$$

where $y_n^{d^\infty} = (-1)^n H_d^{n+1} r_d$ and $0 < \rho < \rho_d$ for some $\rho_d > 0$.

Let D_n^* denote the class of *stationary n-discount optimal policies* and D_∞^* denote the class of *stationary Blackwell optimal policies*. From Proposition 10.1.1, $D_n^* \subseteq D_m^*$ for $n \ge m$ and, as a result of Theorem 10.1.5, these sets are nonempty.

The following theorem uses the Laurent series representation above to characterize n-discount optimality within the class of stationary policies. It is a fundamental
result which ties together many Markov decision process concepts.

Theorem 10.1.6. Let $d^* \in D$.

a. Then $(d^*)^\infty \in D^*_{-1}$ if and only if $g^{(d^*)^\infty} \geq g^\pi_+$ for all $\pi \in \Pi^{HR}$.
b. Then $(d^*)^\infty \in D^*_0$ if and only if $h^{(d^*)^\infty} \geq h^{d^\infty}$ for all $d \in D^*_{-1}$.
c. Let $n \geq 1$. Then $(d^*)^\infty \in D^*_n$ if and only if $y^{(d^*)^\infty}_n \geq y^{d^\infty}_n$ for all $d \in D^*_{n-1}$.
d. Then $(d^*)^\infty \in D^*_n$ for all $n \geq -1$ if and only if $(d^*)^\infty$ is Blackwell optimal.

Proof. We prove part (a). Suppose $(d^*)^\infty \in D^*_{-1}$. From (10.1.7), it follows that, for
every $d \in D$,

$$
v^{(d^*)^\infty}_\lambda - v^{d^\infty}_\lambda = (1 + \rho) \left[\frac{g^{(d^*)^\infty} - g^{d^\infty}}{\rho} + h^{(d^*)^\infty} - h^{d^\infty} + \sum_{n=0}^{\infty} \rho^n \left(y^{(d^*)^\infty}_n - y^{d^\infty}_n \right) \right]
$$

$$(10.1.8)$$

Multiplying this expression by ρ shows that

$$
\rho \left[v^{(d^*)^\infty}_\lambda - v^{d^\infty}_\lambda \right] = g^{(d^*)^\infty} - g^{d^\infty} + o(\rho),
$$
$$(10.1.9)$$

where $o(\rho)$ is a vector with components which converge to 0 as ρ converges to 0.
Therefore the -1-discount optimality of $(d^*)^\infty$ implies that $g^{(d^*)^\infty} \geq g^{d^\infty}$ for all
$d \in D$. Extension of this relationship to the class of all policies follows from Theo-
rems 9.1.8 and 9.1.6(a).

To prove the reverse implication, note that the average optimality of $(d^*)^\infty$ and
(10.1.9) implies

$$
\liminf_{\rho \downarrow 0} \rho \left[v^{(d^*)^\infty}_\lambda - v^{d^\infty}_\lambda \right] \geq 0
$$

for all $d \in D$. Since Theorem 10.1.5(b) establishes existence of a -1-discount optimal
policy in Π^{SD}, it follows that $(d^*)^\infty$ is -1-discount optimal.

To establish the "only if" implication of (b), observe that, whenever $d^\infty \in D^*_{-1}$,
$g^{(d^*)^\infty} = g^{d^\infty}$ so that

$$
v^{(d^*)^\infty}_\lambda - v^{d^\infty}_\lambda = h^{(d^*)^\infty} - h^{d^\infty} + o(\rho).
$$
$$(10.1.10)$$

Consequently, if $(d^*)^\infty \in D^*_0$, $h^{(d^*)^\infty} \geq h^{d^\infty}$.

We now show that the "if" implication holds. Suppose $h^{(d^*)^\infty} \geq h^{d^\infty}$ for all
$d \in D^*_{-1}$. Then, from (10.1.10),

$$
\liminf_{\rho \downarrow 0} \left[v^{(d^*)^\infty}_\lambda - v^{d^\infty}_\lambda \right] \geq 0
$$

for all $d \in D^*_{-1}$. Theorem 10.1.5 implies the existence of a stationary 0-discount
optimal policy. Since $D^*_0 \subset D^*_{-1}$ it must be in D^*_{-1}, so that the result follows.

Part (c) follows by a similar argument and is left as an exercise.

We now prove part (d). Let $d \in D$ and suppose that, for some n, $d^\infty \notin D_n^*$. Let n' be the minimal n for which this holds. Since

$$\rho^{-n'}\left[v_\lambda^{(d^*)^\infty} - v_\lambda^{d^\infty}\right] = y_{n'}^{(d^*)^\infty} - y_{n'}^{d^\infty} + \sum_{k=n'+1}^{\infty} \rho^{k-n'}\left[y_k^{(d^*)^\infty} - y_k^{d^\infty}\right], \quad (10.1.11)$$

the n'-discount optimality of $(d^*)^\infty$ and the assumption that d^∞ is not n'-discount optimal implies that, for some $s \in S$, $x \equiv y_{n'}^{(d^*)^\infty}(s) - y_{n'}^{d^\infty}(s) > 0$. From (10.1.11), it follows that

$$v_\lambda^{(d^*)^\infty}(s) - v_\lambda^{d^\infty}(s) = \rho^{n'}x + \sum_{k=n'+1}^{\infty} \rho^k\left[y_k^{(d^*)^\infty}(s) - y_k^{d^\infty}(s)\right].$$

Since $x > 0$, we can find a ρ_d for which the above expression is positive for $0 < \rho \le \rho_d$.

Repeating the above argument, we obtain a ρ_d for each $d \in D$. Set $\rho^* = \min_{d \in D} \rho_d$. Since D is finite, $\rho^* > 0$. Therefore

$$v_\lambda^{(d^*)^\infty} - v_\lambda^{d^\infty} \ge 0$$

for all $d \in D^{\mathrm{MD}}$, and $0 \le \rho < \rho^*$. Combining this with Theorem 10.1.9 establishes the Blackwell optimality of $(d^*)^\infty$. The reverse implication is a restatement of Theorem 10.1.5(a).   □

Note that, because of results in Chaps. 8 and 9, the above result shows that stationary -1-discount optimal policies are optimal within the class of *all* policies, but that 0-discount optimality only implies that the bias is maximized within the class of *stationary* -1-discount optimal policies.

We conclude with the following important consequence of part (d) of the above theorem.

Corollary 10.1.7. Suppose that, for some $n \ge -1$, D_n^* contains a single policy. Then that policy is Blackwell optimal.

Proof. Let $D_n^* = \{d^\infty\}$. Since $D_\infty^* \ne \varnothing$ and $D_\infty^* \subset D_n^*$, $D_\infty^* = \{d^\infty\}$.   □

We explore the implications of this theorem in the following two examples.

Example 10.1.1 (ctd.). As noted in the previous analysis, $D_{-1}^* = D_0^* = \{\delta^\infty, \gamma^\infty\}$ and $D_n^* = \{\delta^\infty\}$ for $n \ge 1$. As a result of Corollary 10.1.7, δ^∞ is Blackwell optimal. We evaluate the first three terms in the Laurent series by applying Theorem 8.2.8. We find that $g^{\delta^\infty}(s) = g^{\gamma^\infty}(s) = 0$ for all $s \in S$, that $h^{\delta^\infty}(s_1) = h^{\gamma^\infty}(s_1) = 2$, $h^{\delta^\infty}(s_2) = h^{\gamma^\infty}(s_2) = 1$, and $h^{\delta^\infty}(s_3) = h^{\gamma^\infty}(s_3) = 0$, supporting the conclusions of Theorem 10.1.6(a) and (b).

Theorem 8.2.8 implies that $y_1^{\delta^\infty}$ is the unique solution of

$$-h^{\delta^\infty} + (P_\delta - I)y_1 = 0,$$

$$-y_1 + (P_\delta - I)y_2 = 0,$$

or, equivalently, of the first equation above subject to $P_\delta^* y_1 = 0$. Thus we find that $y_1^{\delta^\infty}(s_1) = -2$, $y_1^{\delta^\infty}(s_2) = -1$, and $y_1^{\delta^\infty}(s_3) = 0$. By a similar calculation, we find that $y_1^{\gamma^\infty}(s_1) = -3$, $y_1^{\gamma^\infty}(s_2) = -1$, and $y_1^{\gamma^\infty}(s_3) = 0$. Therefore $y_1^{\delta^\infty}(s_1) > y_1^{\gamma^\infty}(s_1)$, in agreement with the conclusion of Theorem 10.1.6(c) with $n = 1$.

Example 10.1.2. We now complete our analysis of the model in Fig. 3.1.1 from Sec. 3.1. In Sec. 8.4 (Example 8.4.3) we showed that $g^{\delta^\infty}(s) = g^{\gamma^\infty}(s)$ so that, from Theorem 10.1.6(a), $D_{-1}^* = \{\delta^\infty, \gamma^\infty\}$. Since $h^{\delta^\infty}(s_2) = h^{\gamma^\infty}(s_2)$, $h^{\delta^\infty}(s_1) = 12$, and $h^{\gamma^\infty}(s_1) = 11$, it follows from Theorem 10.1.6(b) that δ^∞ is 0-discount optimal. Consequently, from Corollary 10.1.7, δ^∞ is Blackwell optimal. Referring to the analysis of this model in Examples 6.1.1 and 6.2.1 shows that

$$\frac{(5 - 5.5\lambda)}{(1 - 0.5\lambda)(1 - \lambda)} = v_\lambda^{\delta^\infty}(s_1) > v_\lambda^{\gamma^\infty}(s_1) = \frac{10 - 11\lambda}{1 - \lambda}$$

for $10/11 < \lambda < 1$ with these quantities being equal at $\lambda = 10/11$ (Fig. 10.1.3).

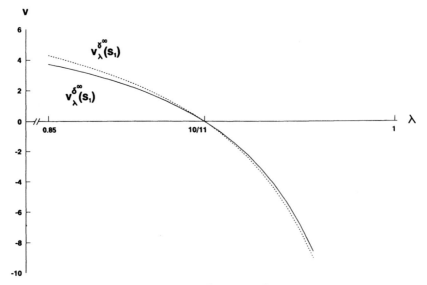

Figure 10.1.3 Graphical representation of $v_\lambda^{\delta^\infty}(s_1)$ and $v_\lambda^{\gamma^\infty}(s_1)$ as functions of λ in Example 10.1.2 showing that δ^∞ is Blackwell optimal.

10.2 OPTIMALITY EQUATIONS

In this section we provide a system of optimality equations which characterize the terms of the Laurent series of a deterministic stationary n-discount optimal policy, show that they have a solution and show how they can be used to find optimal policies. Since we have already established the existence of a deterministic stationary n-discount optimal policy for all n, and characterized its optimality properties with respect to the coefficients of the Laurent series expansion in Theorem 10.1.6, we need not use these equations to establish the existence of an optimal policy as in analyses with other optimality criteria.

10.2.1 Derivation of Sensitive Discount Optimality Equations

We derive the discount optimality equations by using the optimality properties of Blackwell optimal policies established in Theorem 10.1.6(d). To simplify notation, set $y_{-1}^{d^\infty} \equiv g^{d^\infty}$ and $y_0^{\delta^\infty} \equiv h^{\delta^\infty}$.

Suppose δ^∞ is Blackwell optimal. Then, by definition, δ^∞ is discount optimal for each ρ, $0 < \rho \le \rho^*$, so it satisfies

$$\max_{d \in D} \left\{ r_d + \left[(1 + \rho)^{-1} P_d - I \right] v_\lambda^{\delta^\infty} \right\} = 0.$$

Noting that $(1 + \rho)^{-1} P_d - I = (1 + \rho)^{-1}[(P_d - I) - \rho I]$, and that Theorem 8.2.3 implies that, for ρ near 0, $v_\lambda^{\delta^\infty}$ has the Laurent series expansion given in (10.1.7), we can substitute it into the above equation to obtain

$$\max_{d \in D} \left\{ r_d + (P_d - I - \rho I) \left[\sum_{n = -1}^{\infty} \rho^n y_n^{\delta^\infty} \right] \right\} = 0.$$

Rearranging terms yields

$$\max_{d \in D} \left\{ (P_d - I) \frac{y_{-1}^{\delta^\infty}}{\rho} + \left[r_d - y_{-1}^{\delta^\infty} + (P_d - I) y_0^{\delta^\infty} \right] \right.$$
$$\left. + \sum_{n = 1}^{\infty} \rho^n \left[-y_{n-1}^{\delta^\infty} + (P_d - I) y_n^{\delta^\infty} \right] \right\} = 0.$$

For the above equality to hold for all ρ near 0 requires that

$$(P_d - I) y_{-1}^{\delta^\infty} \le 0 \tag{10.2.1}$$

and, for those d for which $P_d y_{-1}^{\delta^\infty}(s') = y_{-1}^{\delta^\infty}(s')$ for some $s' \in S$,

$$r_d(s') - y^{\delta^\infty}(s') + (P_d - I) y^{\delta^\infty}(s') \le 0, \tag{10.2.2}$$

and, for those d for which both (10.2.1) and (10.2.2) hold with equality for some $s' \in S$,

$$-y_0^{\delta^{\infty}}(s') + (P_d - I)y_1^{\delta^{\infty}}(s') \leq 0, \tag{10.2.3}$$

and so forth.

This observation shows that the following system of inductively defined equations characterizes the coefficients of the Laurent series expansion of a Blackwell optimal policy:

$$0 = \max_{d \in D_{-2}} \{(P_d - I)y_{-1}\} \tag{10.2.4}$$

$$\max_{d \in D_{-1}(y_{-1})} \{r_d - y_{-1} + (P_d - I)y_0\} = 0, \tag{10.2.5}$$

$$\max_{d \in D_n(y_{-1}, \ldots, y_n)} \{-y_n + (P_d - I)y_{n+1}\} = 0 \tag{10.2.6}$$

where, for each $s \in S$, $D_{-2}(s) \equiv D$,

$$D_{-1}(y_{-1})(s) \equiv \underset{a \in D_{-2}(s)}{\arg\max} \left\{ \sum_{j \in S} p(j|s, a)y_{-1}(j) - y_{-1}(s) \right\}$$

$$D_0(y_{-1}, y_0)(s) \equiv \underset{a \in D_{-1}(y_{-1})(s)}{\arg\max} \left\{ r(s, a) - y_{-1}(s) + \sum_{j \in S} p(j|s, a)y_0(j) - y_0(s) \right\}$$

for $n \geq 1$,

$$D_n(y_{-1}, \ldots, y_n)(s) \equiv \underset{a \in D_{n-1}(y_{-1}, \ldots, y_{n-1})(s)}{\arg\max} \left\{ -y_{n-1}(s) + \sum_{j \in S} p(j|s, a)y_n(j) - y_n(s) \right\}$$

and

$$D_k(y_{-1}, \ldots, y_k) \equiv \times_{s \in S} D_k(y_{-1}, \ldots, y_k)(s) \tag{10.2.7}$$

for $k \geq -1$.

We refer to this system as the *sensitive discount optimality equations* and to the individual equations respectively as the (-1)th equation, 0th equation, 1th equation, etc. The sets of maximizing decision rules depend on the sequence of y_k's and consequently the system of equations is highly non-linear. Note that the "arg max" in the definition of $D_n(y_{-1}, \ldots, y_n)$ depends on $(y_{-1}, \ldots, y_{n-1})$ only through $D_{n-1}(y_{-1}, \ldots, y_{n-1})$.

Observe that the -1th and 0th equations (10.2.4) and (10.2.5) are the multichain average reward optimality equations. When defining them, we used the notation B_s instead of $D_{-1}(y_{-1})(s)$. From results in Chap. 9, if y_{-1} and y_0 satisfy these equations, and $d \in D_0(y_{-1}, y_0)$ then d^{∞} is average optimal. We generalize this observation to n-discount optimality below.

Combining our derivation above with Theorem 8.2.8 shows that the sequence of terms in the Laurent series expansion of a Blackwell optimal policy satisfy the entire system of sensitive discount optimality equations, and that solving the (-1)th through

nth equations (with $n \geq 0$) yields unique values for $y_{-1}, \ldots, y_{n-1}$. We state this observation formally below.

Theorem 10.2.1. Let δ^{∞} be a Blackwell optimal policy.

a. Then, for *all* $n \geq -1$, $(y_{-1}^{\delta^{\infty}}, y_0^{\delta^{\infty}}, \ldots, y_n^{\delta^{\infty}})$ satisfies the (-1)th, 0th, $\ldots$, nth optimality equations.

b. Suppose that $(y_{-1}, \ldots, y_n)$ satisfy the (-1)th, 0th, $\ldots$, nth optimality equations. Then $y_k = y_k^{\delta^{\infty}}$ for $k \leq n - 1$ and, if in addition either $P_\delta^* y_n = 0$ or

$$-y_{n-1} + (P_\delta - I)y_n = 0,$$

then $y_n = y_n^{\delta^{\infty}}$.

c. There exists a solution to the infinite system of sensitive discount optimality equations.

10.2.2 Lexicographic Ordering

In this section we introduce a partial ordering for real matrices and discuss its relationship to orderings of Laurent series expansions.

Let A be an arbitrary real $m \times n$ matrix with m finite and n *possibly infinite*. We say that A is *lexicographically non-negative* written $A \geq_{\mathscr{L}} 0$ if the first nonzero entry in each row of A is positive, and say that A is *lexicographically positive*, written $A >_{\mathscr{L}} 0$, if $A \geq_{\mathscr{L}} 0$ and $A \neq 0$. For two $m \times n$ matrices A and B, we write $A \geq_{\mathscr{L}} B$ and say that A is *lexicographically greater than B* if $A - B \geq_{\mathscr{L}} 0$, and write $A >_{\mathscr{L}} B$ and say that A is *strictly lexicographically greater than B* if $A - B >_{\mathscr{L}} 0$. We write $A <_{\mathscr{L}} 0$ if $-A >_{\mathscr{L}} 0$ and $A \leq_{\mathscr{L}} 0$ if $-A \geq_{\mathscr{L}} 0$.

For any $m \times n$ matrix A, let A_j denote the $m \times j$ submatrix of A which contains the first j columns of A. Observe that $A_n >_{\mathscr{L}} 0$ implies that $A_j \geq_{\mathscr{L}} 0$ for all $j \leq n$, and that $A >_{\mathscr{L}} 0$ if $A_j \geq_{\mathscr{L}} 0$ for all j and $A_k >_{\mathscr{L}} 0$ for some k.

The following example illustrates lexicographic ordering and its relationship to n-discount optimality. It motivates analyses to follow.

Example 10.2.1. Let x, y, and z be real numbers to be specified below, and set

$$A = \begin{bmatrix} x & y & -1 & 1 & 3 \\ 0 & 0 & 0 & 0 & z \end{bmatrix}.$$

Then $A >_{\mathscr{L}} 0$ if either $x > 0$ and $z \geq 0$ or $x = 0$, $y > 0$ and $z \geq 0$, and $A <_{\mathscr{L}} 0$ if (but not only if) $x = y = 0$ and $z \leq 0$.

We use this ordering to relate n-discount optimality to the lexicographic positivity in the following way. Let $a(j|s)$ denote the (s, j)th component of the matrix A and suppose that

$$A = \begin{bmatrix} 0 & 1 & -4 & 1 & 3 \\ 0 & 0 & 0 & 0 & 2 \end{bmatrix},$$

so that $A >_{\mathscr{L}} 0$. Label the columns of A as -1, 0, 1, 2, 3 and the rows as 1 and 2.

For any integer n,

$$\rho^{-n} \sum_{k=-1}^{3} \rho^k a(k|s) = \rho^{-n-1}a(-1|s) + \rho^{-n}a(0|s) + \cdots + \rho^{-n+3}a(3|s)$$

which, for $s = 1$, equals

$$\rho^{-n} - 4\rho^{-n+1} + \rho^{-n+2} + 3\rho^{-n+3} = \rho^{-n}(1 - 4\rho + \rho^2 + 3\rho^3).$$

Observe that, for $0 < \rho < \frac{1}{4}$, this quantity is positive for any integer n, that

$$\lim_{\rho \downarrow 0} \rho^{-n} \sum_{k=-1}^{3} \rho^k a(k|1) > 0$$

for $n \geq 0$, and

$$\lim_{\rho \downarrow 0} \rho^{-n} \sum_{k=-1}^{3} \rho^k a(k|1) = 0$$

for $n \leq -1$.

Suppose instead that

$$A = \begin{bmatrix} 0 & 0 & -4 & 1 & 3 \\ 0 & 0 & 0 & 0 & 2 \end{bmatrix},$$

so A is neither lexicographically non-negative, nor non-positive. Then, for $s = 1$,

$$\rho^{-n} \sum_{k=-1}^{3} \rho^k a(k|s) = -4\rho^{-n+1} + \rho^{-n+2} + 3\rho^{-n+3} = \rho^{-n+1}(-4 + \rho + 3\rho^2).$$

Therefore, for all $\rho > 0$ and integers n, this quantity is negative, but, for $n \leq 0$,

$$\lim_{\rho \downarrow 0} \rho^{-n} \sum_{k=-1}^{3} \rho^k a(k|1) = 0.$$

We now summarize some observations from this example. Let V_L denote the set of functions of ρ which have Laurent series expansions for all ρ in a neighborhood of 0; that is,

$$V_L \equiv \left\{ y_\lambda \in V: y_\lambda = (1 + \rho) \sum_{n=-1}^{\infty} \rho^n y_n \quad \text{for} \quad 0 < \rho \leq \rho_y \quad \text{and} \quad y_n \in V \right\}.$$

It follows from (10.1.7) that $v_\lambda^{d^\infty} \in V_L$ for all $d \in D$.

For $y_\lambda \in V_L$, let Y denote the $|S| \times \infty$ matrix with columns y_k for $k \geq -1$, so that Y_n denotes the finite submatrix

$$Y_n \equiv \begin{bmatrix} y_{-1}(s_1) & y_0(s_1) & \cdots & y_n(s_1) \\ y_{-1}(s_2) & y_0(s_2) & \cdots & y_n(s_2) \\ \vdots & \vdots & & \vdots \\ y_{-1}(s_m) & y_0(s_m) & \cdots & y_n(s_m) \end{bmatrix}.$$

The following easily proved lemma relates an ordering on V_L to lexicographic ordering.

Lemma 10.2.2. Let $y_\lambda \in V_L$.

a. Then

$$\liminf_{\rho \downarrow 0} \rho^{-n} y_\lambda \geq 0$$

if and only if $Y_n \geq_{\mathscr{L}} 0$.

b. There exists a $\rho' > 0$ such that $y_\lambda \geq 0$ for $0 < \rho \leq \rho'$ if and only if $Y \geq_{\mathscr{L}} 0$.

For $d \in D$, let Y^{d^∞} denote the $|S| \times \infty$ matrix with columns $y_k^{d^\infty}$, and let $Y_n^{d^\infty}$ denote its submatrix consisting of columns $-1, 0, \ldots, n$. Applying the above lemma yields the following restatement of Theorem 10.1.6.

Theorem 10.2.3

a. Let $n \geq -1$. Then d^∞ is n-discount optimal if and only if $Y_n^{d^\infty} \geq_{\mathscr{L}} Y_n^{\delta^\infty}$ for all $\delta \in D$.

b. The policy d^∞ is Blackwell optimal if and only if $Y^{d^\infty} \geq_{\mathscr{L}} Y^{\delta^\infty}$ for all $\delta \in D$.

10.2.3 Properties of Solutions of the Sensitive Optimality Equations

In Sec. 10.2.1, we established the existence of a solution to the system of sensitive discount optimality equations. We now explore optimality properties of solutions of subsets of these equations using lexicographic ordering concepts.

For $\delta \in D$, $k = -1, 0, 1, \ldots$, and $y_\lambda \in V_L$, define the operators $B_\delta^k : V_L \to V$ by

$$B_\delta^{-1} y_\lambda \equiv (P_\delta - I) y_{-1} \tag{10.2.8}$$

$$B_\delta^0 y_\lambda \equiv r_\delta - y_{-1} + (P_\delta - I) y_0, \tag{10.2.9}$$

and, for $n \geq 1$,

$$B_\delta^n y_\lambda \equiv -y_{n-1} + (P_\delta - I) y_n. \tag{10.2.10}$$

We restate Theorem 8.2.8 in this notation as follows.

Theorem 10.2.4. Let $y_\lambda \in V_L$ and suppose that $B_\delta^k y_\lambda = 0$ for $k = -1, \ldots, n$. Then $y_k = y_k^{\delta^\infty}$ for $k = -1, 0, \ldots, n-1$ and $y_n = y_n^{\delta^\infty} + u$, where $(P_\delta - I)u = 0$.

For $\delta \in D$ and $0 \leq \lambda < 1$, define the operator $B_\delta^\lambda : V_L \to V$ by

$$B_\delta^\lambda y_\lambda \equiv r_\delta + (\lambda P_\delta - I)y_\lambda. \tag{10.2.11}$$

Note that this operator was used extensively in Sec. 6.4 when analyzing the policy iteration algorithm in discounted models.

Following Sec. A.6 and Sec. C.2, define the *resolvent* of $P_\delta - I$ by

$$R_\delta^\rho \equiv (\rho I + [I - P_\delta])^{-1} = \lambda(I - \lambda P_\delta)^{-1}. \tag{10.2.12}$$

The following lemma expresses the limiting behavior of the resolvent as ρ approaches 0.

Lemma 10.2.5. For each $\delta \in D$,

$$\lim_{\rho \downarrow 0} \rho R_\delta^\rho = P_\delta^*. \tag{10.2.13}$$

Proof. From Theorem A.8, for $0 < \rho < \sigma(I - P_\delta)$,

$$\rho R_\delta^\rho = P_\delta^* + \sum_{n=0}^{\infty} \rho^{n+1} H_{P_\delta}.$$

Since each component of H_{P_δ} is bounded, and the series converges for small ρ, the limit in the above expression exists and satisfies (10.2.13). $\square$

The following lemma summarizes properties of B_δ^λ. It will be used to relate the difference in the coefficients of a Laurent series expansion of two policies to one-step improvements. It generalizes material in Sec. 9.2.3, especially Lemma 9.2.1.

Lemma 10.2.6. Suppose $\delta \in D$ and $y_\lambda \in V_L$.

a. Then there exists a ρ_y such that, for $0 < \rho \leq \rho_y$,

$$B_\delta^\lambda y_\lambda = (1 + \rho) \sum_{n=-1}^{\infty} \rho^n B_\delta^n y_\lambda. \tag{10.2.14}$$

b. For any λ, $0 \leq \lambda < 1$,

$$v_\lambda^{\delta^\infty} - y_\lambda = (I - \lambda P_\delta)^{-1} B_\delta^\lambda y_\lambda. \tag{10.2.15}$$

c. There exists a $\rho_y > 0$ such that, for $0 \leq \rho < \rho_y$,

$$v_\lambda^{\delta^\infty} - y_\lambda = (1 + \rho) R_\delta^\rho B_\delta^\lambda y_\lambda = R_\delta^\rho \sum_{n=-1}^{\infty} \rho^n B_\delta^n y_\lambda. \tag{10.2.16}$$

Proof. To establish (a), replace λ in (10.2.11) by $(1 + \rho)^{-1}$, expand y_λ in its Laurent series expansion in ρ, and combine terms in like powers of ρ. Result (b) follows by direct multiplication, and (c) by substituting (10.2.14) into (10.2.15) and noting the definition of the resolvent. $\square$

Let $\beta_\delta^n y_\lambda$ denote the $|S| \times (n + 2)$ matrix with components $B_\delta^j y_\lambda(s)$, $j = -1, 0, 1, \ldots, n$. That is, if $S = \{s_1, \ldots, s_m\}$, then

$$\beta_\delta^n y_\lambda \equiv \begin{bmatrix} B_\delta^{-1} y_\lambda(s_1) & B_\delta^0 y_\lambda(s_1) & \cdots & B_\delta^n y_\lambda(s_1) \\ B_\delta^{-1} y_\lambda(s_2) & B_\delta^0 y_\lambda(s_2) & \cdots & B_\delta^n y_\lambda(s_2) \\ \vdots & \vdots & & \vdots \\ B_\delta^{-1} y_\lambda(s_m) & B_\delta^0 y_\lambda(s_m) & \cdots & B_\delta^n y_\lambda(s_m) \end{bmatrix}.$$

We use concepts from the previous section to establish the following important result relating lexicographic non-negativity to n-discount optimality.

Proposition 10.2.7. Let $y_\lambda \in V_L$ and suppose for some $n \geq -1$ that

$$\beta_\delta^{n+1} y_\lambda \leq_{\mathscr{L}} 0,$$

then

$$\liminf_{\rho \downarrow 0} \rho^{-k} \left[v_\lambda^{\delta^\infty} - y_\lambda \right] \leq 0$$

for $k = -1, 0, \ldots, n$.

Proof. From (10.2.16),

$$\rho^{-n} \left[v_\lambda^{\delta^\infty} - y_\lambda \right] = \rho R_\delta^\rho \sum_{m = -1}^\infty \rho^{m-n-1} B_\delta^m y_\lambda. \tag{10.2.17}$$

Since

$$\sum_{m=-1}^\infty \rho^{m-n-1} B_\delta^m y_\lambda = \rho^{-n-2} B_\delta^{-1} y_\lambda + \rho^{-n-1} B_\delta^0 y_\lambda$$

$$+ \cdots + \rho^{-1} B_\delta^n y_\lambda + B_\delta^{n+1} y_\lambda + \rho B_\delta^{n+2} y_\lambda + \cdots \tag{10.2.18}$$

the hypothesis that $\beta_\delta^{n+1} y_\lambda \leq_{\mathscr{L}} 0$ implies that

$$\liminf_{\rho \downarrow 0} \sum_{m=-1}^\infty \rho^{m-n-1} B_\delta^m y_\lambda \leq 0. \tag{10.2.19}$$

For $u \leq 0$, $P_\delta^* u \leq 0$, so from Lemma 10.2.5, (10.2.19) and (10.2.17), the result follows. $\square$

Using the conclusion of this proposition, we obtain the main result of this section. The following theorem provides optimality properties of solutions of optimality equations $-1, \ldots, n$ and of maximizing decision rules.

Theorem 10.2.8. Let $n \geq -1$, $y_\lambda \in V_L$, and suppose that $(y_{-1}, y_0, \ldots, y_{n+1})$ satisfies the (-1)th, 0th, $\ldots, (n+1)$th optimality equations. Then the following hold.

a. For all $\delta \in D$,

$$\liminf_{\rho \downarrow 0} \rho^{-k}\left[v_\lambda^{\delta^\infty} - y_\lambda\right] \leq 0 \tag{10.2.20}$$

for $k \leq n$.
b. For all $\pi \in \Pi^{HR}$,

$$\liminf_{\rho \downarrow 0} \rho^{-k}\left[v_\lambda^\pi - y_\lambda\right] \leq 0$$

for $k \leq n$.
c. If $d \in D_{n+1}(y_{-1}, y_0, \ldots, y_{n+1})$, then d^∞ is n-discount optimal.

Proof. Since $(y_{-1}, y_0, \ldots, y_{n+1})$ satisfies the indicated optimality equations, it follows from (10.2.4)–(10.2.7) that, for all $\delta \in D$, $\beta_\delta^n y_\lambda \leq_\mathscr{L} 0$. Hence part (a) follows from Proposition 10.2.7.

Theorem 10.1.4(b) establishes the existence of a stationary n-discount optimal policy, so there exists a $d \in D$ for which

$$\liminf_{\rho \downarrow 0} \rho^{-k}\left[v_\lambda^\pi - v_\lambda^{d^\infty}\right] \leq 0$$

for all $\pi \in \Pi^{HR}$. Combining this observation with (10.2.20) yields part (b).

To establish part (c), note from Theorem 10.2.2, that for all

$$d \in D_{n+1}(y_{-1}, y_0, \ldots, y_{n+1}),$$

$y_k = y_k^{d^\infty}$ for $k = -1, 0, \ldots, n$. Hence, for any $\pi \in \Pi^{HR}$,

$$\liminf_{\rho \downarrow 0} \rho^{-k}\left[v_\lambda^\pi - y_\lambda\right] = \liminf_{\rho \downarrow 0} \rho^{-k}\left[v_\lambda^\pi - v_\lambda^{d^\infty}\right].$$

Noting this, it follows from part (b) that d^∞ is n-discount optimal.   □

One might conjecture that, whenever $(y_{-1}, y_0, \ldots, y_{n+1})$ satisfies the optimality equation, then $D_{n+1}(y_{-1}, y_0, \ldots, y_{n+1})$ contains all n-discount optimal policies. Example 10.1.2 shows that this conjecture is false even when $n = -1$.

Example 10.1.2 (ctd.). In this example, stationary policies δ^∞ and γ^∞ are -1-discount optimal. A solution of the (-1)th and 0th optimality equations is $y_{-1}^*(s) = -1$, $y_0^*(s_1) = 12$, $y_0^*(s_2) = 0$, and $D_0(y_{-1}^*, y_0^*) = \{\delta\}$. Thus γ^∞ is -1-discount optimal but $\gamma \notin D_0(y_{-1}^*, y_0^*)$.

As stated in the introduction to this chapter, these results generalize those in Chaps. 8 and 9, namely Theorems 9.1.2(a) (with S finite) and 9.1.7(a), which follow by choosing $n = -1$ in the above theorem. We restate them in the following corollary.

Corollary 10.2.9. Suppose that $y^*_{-1} \in V$ and $y^*_0 \in V$ satisfy

$$\max_{d \in D_{-2}} \{(P_d - I)y^*_{-1}\} = 0,$$

$$\max_{d \in D_-(y^*_{-1})} \{r_d - y^*_{-1} + (P_d - I)y^*_0\} = 0.$$

a. Then $y^*_{-1} \geq g^\pi_+$ for all $\pi \in \Pi^{HR}$.
b. If $d \in D_0(y^*_{-1}, y^*_0)$, d^∞ is average optimal.

This theorem also enables us to identify 0-discount or bias optimal policies by applying it with $n = 0$.

Corollary 10.2.10. Suppose that $y^*_{-1} \in V$, $y^*_0 \in V$ satisfies the hypothesis of Corollary 10.2.9, and, in addition, that there exists a $y^*_1 \in V$ which satisfies

$$\max_{d \in D_0(y^*_{-1}, y^*_0)} \{-y^*_0 + (P_d - I)y^*_1\} = 0.$$

a. Then $y^*_0 \geq h^{\delta^\infty}$ for any average optimal policy δ^∞.
b. If $y_\lambda \in V_L$ satisfies $y_k = y^*_k$ for $k = -1, 0, 1$, then, for all $\pi \in \Pi^{HR}$,

$$\liminf_{\rho \downarrow 0} [v^\pi_\lambda - y_\lambda] \leq 0.$$

c. If $d \in D_1(y^*_{-1}, y^*_0, y^*_1)$, then d^∞ is 0-discount optimal and maximizes the bias among all average optimal stationary policies.

We conclude this section with an example illustrating the above theory.

Example 10.2.2. We analyze the following model which appeared as Example 8.6.2. Recall that $S = \{s_1, s_2\}$; $A_{s_1} = \{a_{1,1}, a_{1,2}\}$ and $A_{s_2} = \{a_{2,1}\}$; $r(s_1, a_{1,1}) = 4$, $r(s_1, a_{1,2}) = 0$, and $r(s_2, a_{2,1}) = 8$; and $p(s_1|s_1, a_{1,1}) = 1$, $p(s_2|s_1, a_{1,2}) = 1$, and $p(s_1|s_2, a_{2,1}) = 1$. Stationary policies are determined by the action choice in state s_1; let δ choose action $a_{1,1}$ there, and γ choose action $a_{1,2}$.
From Theorem 10.2.4, with $n = 1$,

$$y^{\delta^\infty}_{-1} = \begin{bmatrix} 4 \\ 4 \end{bmatrix}, \quad y^{\delta^\infty}_0 = \begin{bmatrix} 0 \\ 4 \end{bmatrix}, \quad y^{\gamma^\infty}_{-1} = \begin{bmatrix} 4 \\ 4 \end{bmatrix}, \quad y^{\gamma^\infty}_0 = \begin{bmatrix} -2 \\ 2 \end{bmatrix}.$$

Hence, from Theorem 10.1.6, both δ^∞ and γ^∞ are -1-discount optimal, but only δ^∞ is 0-discount optimal.

We now examine the implications of Theorem 10.2.8. To simplify notation, let $u_k \equiv y_k^{\delta^\infty}$ and $w_k \equiv y_k^{\gamma^\infty}$ for $k \geq -1$. Observe that

$$D_{-1}(u_{-1})(s_1) = D_{-1}(w_{-1})(s_1) = \{a_{1,1}, a_{1,2}\},$$

that

$$D_0(u_{-1}, u_0)(s_1) = \underset{a \in D_{-1}(u_{-1})(s_1)}{\arg\max} \left\{ r(s_1, a) - u_{-1}(s_1) + \sum_{j \in S} p(j|s_1, a)u_0(j) - u_0(s_1) \right\}$$

$$= \arg\max\{4 - 4 + 0 - 0, 0 - 4 + 4 - 0\} = \{a_{1,1}, a_{1,2}\}.$$

Similarly, $D_0(w_{-1}, w_0)(s_1) = \{a_{1,1}, a_{1,2}\}$ so that Theorem 10.2.8 establishes the -1-discount optimality of both policies.

To identify a 0-discount optimal policy requires determining $D_1(u_{-1}, u_0, u_1)$ or $D_1(w_{-1}, w_0, w_1)$. Applying Theorem 10.2.4 with $n = 2$ shows that

$$u_1 = y_1^{\delta^\infty} = \begin{bmatrix} 0 \\ -4 \end{bmatrix}, \qquad w_1 = y_1^{\gamma^\infty} = \begin{bmatrix} 1 \\ -1 \end{bmatrix}.$$

Consequently,

$$D_1(u_{-1}, u_0, u_1)(s_1) = \underset{a \in D_0(u_{-1}, u_0)(s_1)}{\arg\max} \left\{ -u_0(s_1) + \sum_{j \in S} p(j|s_1, a)u_1(j) - u_1(s_1) \right\}$$

$$= \arg\max\{0 + 0 - 0, 0 - 4 - 0\} = \{a_{1,1}\},$$

and by a similar calculation we see that $D_1(w_{-1}, w_0, w_1)(s_1) = \{a_{1,1}\}$ establishing the 0-discount optimality of δ^∞.

Note that since δ^∞ is the unique 0-discount optimal policy. Corollary 10.1.7 shows that it is also Blackwell optimal. This observation provides further insight into this concept. Note that $y_{-1}^{\delta^\infty} = y_{-1}^{\gamma^\infty}$, that $y_0^{\delta^\infty} > y_0^{\gamma^\infty}$, but that $y_1^{\delta^\infty} < y_1^{\gamma^\infty}$. Thus a Blackwell optimal policy does not maximize *all* terms in the Laurent series of the expected discounted reward; it only maximizes the first term at which the policies differ.

Observe also, using notation of Sec. 10.2.2, that

$$Y^{\delta^\infty} = \begin{bmatrix} 4 & 0 & 0 & \cdots \\ 4 & 4 & -4 & \cdots \end{bmatrix}$$

and

$$Y^{\gamma^\infty} = \begin{bmatrix} 4 & -2 & 1 & \cdots \\ 4 & 2 & -1 & \cdots \end{bmatrix}.$$

Thus, as demonstrated in Theorem 10.2.3, $Y^{\delta^\infty} >_{\mathscr{L}} Y^{\gamma^\infty}$, $Y_{-1}^{\delta^\infty} = Y_{-1}^{\gamma^\infty}$ and $Y_k^{\delta^\infty} >_{\mathscr{L}} Y_k^{\gamma^\infty}$ for all $k \geq 0$. Note that Y^{δ^∞} is not greater than Y^{γ^∞} on a componentwise basis.

10.3   POLICY ITERATION

Results of the previous section show that we can find n-discount optimal policies by solving a system of $n + 1$ nested optimality equations. In this section, we provide a policy iteration algorithm which solves this system and finds an n-discount optimal policy. Further, we show how to use it to identify a Blackwell optimal (∞-discount optimal) policy in finitely many operations.

10.3.1   The Algorithm

The algorithm obtains the set of N-discount optimal policies in the following way. First, it identifies the set of -1-discount optimal policies. Then, within this set, it finds the subset of 0-discount optimal policies; then, within this set, it finds the set of 1-discount optimal policies. It continues in this way until it identifies the set of N-discount optimal policies. This algorithm generalizes the average reward multichain policy iteration algorithm of Sec. 9.3.3. For clarity, we state it in vector notation, although we implement it on a component by component basis. To simplify presentation, define

$$r_d^m = \begin{cases} r_d & m = 0 \\ 0 & m = -1, 1, 2, 3, \ldots \end{cases}$$

for $d \in D$ and $m \geq -1$.

The N-discount Optimality Policy Iteration Algorithm

1. Set $m = -1$, $D_{-1} = D$, $y_{-2}^* = 0$, $n = 0$, and select a $d_0 \in D$.
2. (Policy evaluation). Obtain y_m^n and y_{m+1}^n by solving

$$r_{d_n}^m - y_{m-1}^* + \left(P_{d_n} - I \right) y_m = 0, \tag{10.3.1}$$

$$r_{d_n}^{m+1} - y_m + (P_{d_n} - I) y_{m+1} = 0 \tag{10.3.2}$$

subject to either $P_{d_n}^* y_{m+1} = 0$ or

$$r_{d_n}^{m+2} - y_{m+1} + \left(P_{d_n} - I \right) y_{m+2} = 0. \tag{10.3.3}$$

3. (Policy improvement).
   a. (m-improvement) Choose

$$d_{n+1} \in \arg\max_{d \in D_m} \{ r_d^m + P_d y_m^n \}, \tag{10.3.4}$$

setting $d_{n+1}(s) = d_n(s)$ if possible. If $d_{n+1} = d_n$ go to (b); otherwise increment n by 1 and return to step 2.

b. $((m + 1)$-improvement) Choose

$$d_{n+1} \in \underset{d \in D_m}{\arg\max} \left\{ r_d^{m+1} + P_d y_{m+1}^n \right\} \tag{10.3.5}$$

setting $d_{n+1}(s) = d_n(s)$ if possible. If $d_{n+1} = d_n$, go to step 4; otherwise increment n by 1 and return to step 2.

4. Set

$$D_{m+1} = \underset{d \in D_m}{\arg\max} \left\{ r_d^{m+1} + P_d y_{m+1}^n \right\} \tag{10.3.6}$$

If D_{m+1} contains a single decision rule or $m = N$, stop. Otherwise, set $y_m^* = y_m^n$, increment m by 1, set $n = 0$, $d_0 = d_n$ and return to step 2.

Upon termination the algorithm yields the set

$$D_{N+1} = D_{N+1}(y_{-1}^*, y_0^*, \ldots, y_N^*, y_{N+1}^n).$$

Since $(y_{-1}^*, y_0^*, \ldots, y_N^*, y_{N+1}^n)$ satisfies the (-1)th through $(N + 1)$th optimality equations, it follows from Theorem 10.2.8(c) that any $d \in D_{N+1}$ is N-discount optimal. When $N = -1$, the algorithm duplicates the policy iteration algorithm of Sec. 9.2; however, we choose a solution of the evaluation equations (10.3.1) and (10.3.2), which ensures that y_{m+1}^n equals the $(m + 1)$th term in the Laurent series expansion of $v_\lambda^{(d_n)^\infty}$.

We refer to improvements obtained in step 3 as m or $m + 1$ improvements because they increase either the mth or $(m + 1)$th term in the Laurent series expansion of $B_\delta^\lambda v_\lambda^{(d_n)^\infty}$. Note that the specification in steps 3(a) and 3(b), that we set $d_{n+1}(s) = d_n(s)$ if possible, is crucial to our analysis below. Note also that in (10.3.4) and (10.3.5) we have deleted terms which do not effect the choice of the maximizing decision rule and that, for $m \neq 0$ (10.3.4) reduces to

$$d_{n+1} \in \underset{d \in D_m}{\arg\max} \left\{ P_d y_m^n \right\}$$

The flow chart in Fig. 10.3.1 summarizes the logic of this algorithm.

An alternative computational strategy would be at each pass through the improvement step to seek a decision rule δ which ensures that

$$B_\delta^{N+1} v_\lambda^{(d_n)^\infty} >_{\mathscr{L}} 0$$

In the worst case, this specification requires evaluation of the first $N + 3$ terms of the Laurent series expansion of $v_\lambda^{(d_n)^\infty}$ prior to seeking a -1 through $N + 1$ improvement. Since it is likely that improvements occur at low-order terms in the Laurent series expansion, this approach might be more efficient than that above. However, we are unaware of computational results which indicate the superiority of either of these methods.

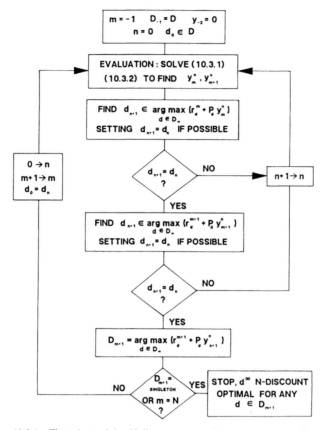

Figure 10.3.1 Flow chart of the N-discount optimality policy iteration algorithm.

10.3.2   An Example

Before demonstrating the convergence of the policy iteration algorithm, we illustrate it by finding a 0-discount optimal policy in Example 10.2.2.

Set $m = -1$, $D_{-1} = D$, $y^*_{-2} = 0$, and $n = 0$, and choose $d_0(s_1) = a_{1,2}$ and $d_0(s_2) = a_{2,1}$. We solve (10.3.1)–(10.3.3) to find

$$y^0_{-1} = \begin{bmatrix} 4 \\ 4 \end{bmatrix}, \qquad y^0_0 = \begin{bmatrix} -2 \\ 2 \end{bmatrix}.$$

We now seek an improvement through step 3. Observe that, since y^0_{-1} is constant, step 3(a) allows us to set $d_1 = d_0$. We then proceed to step 3(b). Since

$$\max_{d \in D_{-1}} \left\{ r_d(s_1) + P_d y^0_0(s_1) \right\} = \max\{4 - 2, 0 + 2\}$$

we can set $d_1 = d_0$. Therefore we go on to step 4. We set $D_0 = D_{-1}$ and, since neither stopping criterion in step 4 is satisfied, we set $y^*_{-1}(s_1) = y^*_{-1}(s_2) = 4$, $m = 0$, $d_0 = d_1$, and $n = 0$, and return to evaluation step 2.

The evaluation equations become

$$r_{d_0} - y^*_{-1} + (P_{d_0} - I) y^0_0 = 0, \qquad (10.3.7)$$

$$-y^0_0 + (P_{d_0} - I) y^0_1 = 0, \qquad (10.3.8)$$

which we solve under the additional constraint that

$$-y^0_1 + (P_{d_0} - I) y^0_2 = 0. \qquad (10.3.9)$$

Observe that we have already solved (10.3.7), so we need only solve (10.3.8) and (10.3.9) to determine y^0_1. Therefore we solve the system of equations

$$-2 + y^0_1(s_2) - y^0_1(s_1) = 0,$$

$$2 + y^0_1(s_1) - y^0_1(s_2) = 0,$$

$$-y^0_1(s_1) + y^0_2(s_2) - y^0_2(s_1) = 0,$$

$$-y^0_1(s_2) + y^0_2(s_1) - y^0_2(s_2) = 0.$$

to obtain $y^0_1(s_1) = 1$ and $y^0_1(s_2) = -1$.

We now seek a 0-improvement through step 3(a). We already performed this calculation at the preceding pass through the improvement step, so we proceed to step 3(b) and seek a 1-improvement. Since

$$\max_{d \in D_0} \left\{ P_d y^0_1(s_1) \right\} = \max\{1, -1\}$$

we identify the new maximizing action $a_{1,1}$ in state s_1. Set $d_1(s_1) = a_{1,1}$ and $d_1(s_2) = a_{2,1}$ and return to the evaluation step. Solve (10.3.1)–(10.3.3) to obtain

$$y^1_0 = \begin{bmatrix} 0 \\ 4 \end{bmatrix}, \qquad y^1_1 = \begin{bmatrix} 0 \\ -4 \end{bmatrix}.$$

Since

$$\max_{d \in D_0} \left\{ r_d(s_1) + P_d y^1_0(s_1) \right\} = \max\{4 - 0, 0 + 4\}$$

we can choose $d_2 = d_1$ in step 3(a), so we proceed to step 3(b) and evaluate

$$\max_{d \in D_0} \left\{ P_d y^1_1(s_1) \right\} = \max\{0, -4\}.$$

Because $d_2 = d_1$ again, we proceed to step 4. We set $D_1 = \{d_2\}$. Since it satisfies both stopping criteria, we stop and identify $(d_2)^\infty$ as 0-discount optimal policy. Since there is a unique 0-discount optimal policy, we conclude also that it is Blackwell optimal.

10.3.3 Convergence of the Algorithm

Our proof of the finite convergence of the algorithm generalizes that in Sec. 9.2.3. We show that, every time we identify a different decision rule in the improvement step,

$$\liminf_{\rho \downarrow 0} \rho^{-(N+1)} \left[v_\lambda^{(d_{n+1})^\infty} - v_\lambda^{(d_n)^\infty} \right] \geq 0, \tag{10.3.10}$$

with strict inequality in at least one component.

We begin with the following result which relates to Proposition 10.2.7. Note the subtle difference in the proof of these two results.

Proposition 10.3.1. Let $y_\lambda \in V_L$ and suppose for some n that

$$\beta_\delta^n y_\lambda >_\mathscr{L} 0 \tag{10.3.11}$$

and in addition that

$$\beta_\delta^\infty y_\lambda >_\mathscr{L} 0 \tag{10.3.12}$$

then

$$\liminf_{\rho \downarrow 0} \rho^{-n} \left[v_\lambda^{\delta^\infty} - y_\lambda \right] \geq 0, \tag{10.3.13}$$

with strict inequality in at least one component.

Proof. From (10.2.16), for $0 < \rho \leq \rho_y$,

$$\rho^{-n} \left[v_\lambda^{\delta^\infty} - y_\lambda \right] = R_\delta^\rho \sum_{k=-1}^{\infty} \rho^{k-n} B_\delta^k y_\lambda. \tag{10.3.14}$$

Note that (10.3.12) implies that, for ρ small enough,

$$\sum_{k=-1}^{\infty} \rho^k B_\delta^k y_\lambda \geq 0.$$

Since

$$R_\delta^\rho = \lambda (I - \lambda P_\delta)^{-1} = \lambda (I + \lambda P_\delta + \lambda P_\delta^2 + \cdots) \geq \lambda I = (1 + \rho)^{-1} I,$$

it follows from (10.3.14) that

$$\rho^{-n} \left[v_\lambda^{\delta^\infty} - y_\lambda \right] \geq (1 + \rho)^{-1} \sum_{k=-1}^{\infty} \rho^{k-n} B_\delta^k y_\lambda. \tag{10.3.15}$$

Because $\beta_\delta^n y_\lambda >_\mathscr{L} 0$, for each $s \in S$, there exists a $k \in \{-1, \ldots, n\}$ for which $B_\delta^j y_\lambda(s) = 0$ for $j \leq k$ and $B_\delta^{k+1} y_\lambda(s) > 0$. The conclusion (10.3.13) follows from this observation, (10.3.15), and the expansion

$$\sum_{k=-1}^{\infty} \rho^{k-n} B_\delta^k y_\lambda = \rho^{-n-1} B_\delta^{-1} y_\lambda + \rho^{-n} B_\delta^0 y_\lambda$$

$$+ \cdots + \rho^{-1} B_\delta^{n-1} y_\lambda + B_\delta^n y_\lambda + \rho B_\delta^{n+1} y_\lambda + \cdots. \qquad \square$$

We use this result to establish finite convergence of N-discount optimal policy iteration.

Theorem 10.3.2. Suppose S and A_s, for each $s \in S$, are finite. Then

a. the N-discount optimality policy iteration algorithm terminates in a finite number of iterations;

b. if $d \in D_{N+1}$, then d^∞ is N-discount optimal;

c. for $k \le N$, $y_k^* = y_k^{d^\infty}$, and

d. for $k \le N$, y_k^* satisfies the N-discount optimality equations.

Proof. Suppose at some pass through the algorithm that $d_{n+1} \neq d_n$. This implies that both

$$\beta_{d_{n+1}}^{N+1} v_\lambda^{(d_n)^\infty} >_{\mathcal{L}} 0$$

and

$$\beta_{d_{n+1}}^\infty v_\lambda^{(d_n)^\infty} >_{\mathcal{L}} 0$$

the second implication following because $d_{n+1}(s) = d_n(s)$ implies

$$B_{d_{n+1}}^k v_\lambda^{(d_n)^\infty}(s) = B_{d_n}^k v_\lambda^{(d_n)^\infty}(s) = 0$$

for $k \ge -1$. Therefore (10.3.10) holds with strict inequality for some $s \in S$ whenever $d_{n+1} \neq d_n$. Since there are only finitely many deterministic decision rules, the algorithm must terminate in finitely many iterations.

Upon termination $(y_{-1}^{(d_n)^\infty}, y_0^{(d_n)^\infty}, \ldots, y_{N+1}^{(d_n)^\infty})$ satisfies the $(-1$th) through $(N + 1)$th optimality equations, so from Theorem 10.2.8(b),

$$\liminf_{\rho \downarrow 0} \rho^{-N} \left[v_\lambda^\pi - v_\lambda^{(d_n)^\infty} \right] \le 0$$

for all $\pi \in \Pi^{HR}$. Part (b) follows by noting that, for any $d \in D_{N+1}$, $y_k^{d^\infty} = y_k^{(d_n)^\infty}$ for $k \le N$.

Part (c) follows from Theorem 10.2.8(c), and part (d) by construction. $\square$

The proof above shows that changing policies results in an improvement in the $(N + 1)$ discount optimality sense; however, upon termination, we end up with an

N-discount optimal policy. Recall that this is the same situation as in multichain models with average reward optimality criterion. In those models, not all improvements resulted in an increase in the gain; when it did not increase, the bias did. Note also that the proof above establishes the convergence of the alternate specification of the policy iteration algorithm proposed in Sec. 10.3.1.

10.3.4 Finding Blackwell Optimal Policies

Since Blackwell optimality corresponds to ∞-discount optimality, the analysis above suggests that finding a Blackwell optimal policy would require finding m-improvements for all $m \geq -1$. Examples 10.1.1 and 10.2.1 suggest that this is not necessary since, in Example 10.1.1, a 1-discount optimal policy is Blackwell optimal while, in Example 10.2.1, a 0-discount optimal policy is Blackwell optimal.

In this section we show that to find a Blackwell optimal policy requires finding at most a $(|S| - 2)$-discount optimal policy. This result is a consequence of the following two results from linear algebra theory. The first result is not well known and we provide its proof (Veinott, 1974). For the second refer to a linear algebra text, for example, Halmos (1958).

Lemma 10.3.3. Let H denote a real $m \times m$ matrix of rank p, and let U denote a q-dimensional subspace of R^m. Let $x \in R^m$, and suppose $H^k x \in U$ for $k = 1, \ldots, d$ for $d = \min(p, q)$; then $H^k x \in U$ for all $k \geq 1$.

Proof. First we show that $\{Hx, H^2 x, \ldots, H^n x\}$ is a linearly dependent set of vectors whenever $n \geq d + 1$. If $d = p$, the subspace of R^m spanned by the columns of H has dimension d, and, if $d = q$, there are at most q linearly independent vectors in U, so the assertion follows.

We show by induction on k that $H^{d+k} x$ is a linear combination of $\{Hx, H^2 x, \ldots, H^d x\}$ for all $k \geq 0$. Clearly it is true for $k = 0$. Assume it holds for $k = 1, 2, \ldots, n$. Then for some non-zero α_j's

$$H^{d+n} x = \sum_{j=1}^{d} \alpha_j H^j x.$$

Multiply this expression by H to obtain

$$H^{d+n+1} x = \sum_{j=1}^{d} \alpha_j H^{j+1} x$$

and note by the argument in the first paragraph that $H^{d+1} x \in U$. Hence the induction hypothesis is satisfied and the result follows. $\square$

Lemma 10.3.4. Suppose F and H are two real $m \times m$ matrices which satisfy

a. rank $(F + H) = m$, and
b. $FH = 0$.

Then rank(F) + rank$(H) = m$.

The following proposition contains the essential theoretical tool of this section. It shows that whenever the Laurent series expansions of the discounted reward of two stationary policies agree for at most $|S| - 2$ terms, then they agree for all terms.

Proposition 10.3.5. Let $\delta \in D$ and $d \in D$. Suppose

a. P_d has r recurrent classes, and
b. $y_k^{\delta^\infty} = y_k^{d^\infty}$ for $k = -1, 0, \ldots, |S| - r - 1$.

Then $y_k^{\delta^\infty} = y_k^{d^\infty}$ for all k.

Proof. From Appendix A, (A.18) and (A.19), and H_d and P_d^* satisfy the hypotheses of Lemma 10.3.4. Theorem A.5 shows that rank$(P_d^*) = r$, so it follows that rank$(H_d) = |S| - r$.

Recall that for any $d' \in D$, $y_k^{(d')^\infty} = H_{d'}^{k+1} r_{d'}$. By hypothesis (b), for $0 \leq k \leq |S| - r - 1$, $H_\delta^{k+1} r_\delta = H_d^{k+1} r_d$ so that

$$y_{k+1}^{\delta^\infty} - y_{k+1}^{d^\infty} = (-1)^{k+1}\left[H_\delta^{k+2} r_\delta - H_d^{k+2} r_d\right] = [H_\delta - H_d](-H_d)^{k+1} r_d. \tag{10.3.16}$$

As a consequence of the above equality, hypothesis (b) implies that, for $0 \leq k \leq |S| - r - 1$,

$$[H_\delta - H_d](-H_d)^{k+1} r_d = 0. \tag{10.3.17}$$

Apply Lemma 10.3.4 with $H = -H_d$, $x = r_d$ and $U = \{y \in R^{|S|} : [H_\delta - H_d]y = 0\}$ to conclude that (10.3.17) holds for all $k \geq 0$, which, from (10.3.16), establishes the result. □

The following theorem is the main result of this section: it completes the analysis of the sensitive optimality in finite-state and action models. It follows immediately from the above proposition and Theorem 10.1.5(c).

Theorem 10.3.6. Suppose $(d^*)^\infty$ is $(|S| - r - 1)$-discount optimal, and P_{d^*} has r recurrent classes. Then

a. $(d^*)^\infty$ is Blackwell optimal, and
b. $(d^*)^\infty$ is n-discount optimal for all n.

From a practical perspective, the following is the most important implication of this theorem.

Corollary 10.3.7. Suppose A_s for each $s \in S$ and S are finite. Then the $(|S| - r - 1)$-discount optimal policy iteration algorithm finds a stationary Blackwell optimal policy in finitely many iterations.

This corollary also establishes that existence of a Blackwell optimal policy (within the class of stationary policies) without resorting to Lemma 10.1.2 on rational functions.

We return to Examples 10.1.1 and 10.2.1 in the context of this theorem. Note that, in both of these examples, all stationary policies are unichain, so we choose $r = 1$ when applying Theorem 10.3.6. Since, in Example 10.1.1, $|S| = 3$, we conclude, as we observed, that any 1-discount optimal policy is Blackwell optimal and, in Example 10.2.1, any 0-discount (bias) optimal policy is Blackwell optimal. Of course in practice we solve problems with more states, so that in the worse case (c.f. Problem 10.6) we may need to find an $(|S| - 2)$-discount optimal policy to identify a Blackwell policy. However, in most applications, the policy iteration algorithm stops in step 4, when D_{m+1} is a singleton for a small value of m.

10.4 THE EXPECTED TOTAL-REWARD CRITERION REVISITED

In Chap. 7, we investigated the use of policy iteration algorithms for finding optimal policies in models with expected total reward criterion. We observed the following:

a. In positive models, proof of convergence required subtle analysis and, at termination, the set of maximizing decision rules may contain suboptimal policies (Example 7.2.3).

b. In negative models, stopping the algorithm when the same decision rule attains the maximum at two successive iterations may result in stopping with a suboptimal policy. (Example 7.3.4).

Also, in these models the evaluation step required obtaining either the maximal or minimal solution of

$$r_d + (P_d - I)v = 0.$$

To do this requires solution of a linear program, or a determination of the chain structure of P_d.

In this section we show that we can resolve these difficulties by solving these models using the 0-discount optimality policy iteration algorithm of the previous section. Proposition 8.2.2 contained the key observation: for any stationary policy with aperiodic transition probabilities and gain equal to 0, the bias equals the expected total reward. Consequently 0-discount optimality and expected total-reward optimality are equivalent for such models. In fact many of the difficulties encountered when analyzing models with expected total-reward criterion can be resolved by viewing them from this perspective.

Throughout this section we assume S and A_s to be finite.

10.4.1 Relationship of the Bias and Expected Total Reward

In Secs. 10.4.1 and 10.4.2, we consider models which satisfy

$$v_+^\pi(s) = E_s^\pi \left\{ \sum_{t=1}^\infty r^+(X_t, Y_t) \right\} < \infty \tag{10.4.1}$$

or

$$v_-^\pi(s) = E_s^\pi \left\{ \sum_{t=1}^\infty r^-(X_t, Y_t) \right\} < \infty \tag{10.4.2}$$

for all or some $\pi \in \Pi^{HR}$ where, as before, for a real number u, $u^+ = \max\{u, 0\}$ and $u^- = \max\{-u, 0\}$. Under either (10.4.1) or (10.4.2), $\lim_{N \to \infty} v_N^\pi$ exists, and (10.4.1) together with (10.4.2) assures that this limit is finite. The following easily proved proposition relates these conditions to the existence and value of the gain of a policy.

Proposition 10.4.1. Let S be finite and suppose (10.4.1) holds for all $\pi \in \Pi^{HR}$.

a. Then, for all $\pi \in \Pi^{HR}$, g^π exists and satisfies $g^\pi \leq 0$.
b. Suppose in addition that (10.4.2) holds for some $\pi' \in \Pi^{HR}$, then $g^{\pi'} = 0$ and the optimal gain satisfies $g^* = 0$.
c. Suppose in addition that (10.4.2) holds for all $\pi' \in \Pi^{HR}$; then $g^\pi = 0$ for all $\pi \in \Pi^{HR}$.

In this section, we distinguish models according to whether some or all policies satisfy both (10.4.1) and (10.4.2). The presence of stationary policies with negative gain when (10.4.2) does not hold necessitates an extra step in our analysis below. We refer to models in which (10.4.1) and (10.4.2) hold for all $\pi \in \Pi^{HR}$ as *zero gain models*.

Recall from (8.2.4) that, when $d \in D$ has an aperiodic transition probability matrix,

$$h^{d^\infty}(s) = E_s^{d^\infty} \left\{ \sum_{t=1}^\infty \left[r_d(X_t) - g^{d^\infty}(X_t) \right] \right\}. \tag{10.4.3}$$

When it is periodic,

$$h^{d^\infty}(s) = \lim_{N \to \infty} \frac{1}{N} \sum_{k=1}^N E_s^{d^\infty} \left\{ \sum_{t=1}^k \left[r_d(X_t) - g^{d^\infty}(X_t) \right] \right\}. \tag{10.4.4}$$

When both (10.4.1) and (10.4.2) hold, (10.4.3) together with Proposition 10.4.1 yields the following relationship between the bias and the expected total reward. It is the key observation for what follows.

Proposition 10.4.2. Let S be finite, let $d \in D^{MD}$, and suppose $v_+^{d^\infty}$ and $v_-^{d^\infty}$ are finite, then $v^{d^\infty} = h^{d^\infty}$.

10.4.2 Optimality Equations and Policy Iteration

Using the observation in Proposition 10.4.2, we apply Corollary 10.2.10 to obtain the following result, which characterizes the optimal expected total reward and identifies optimal stationary policies in models which satisfy (10.4.1) and (10.4.2).

Theorem 10.4.3. Suppose (10.4.1) and (10.4.2) hold for all $\pi \in \Pi^{HR}$, and S and A_s for each $s \in S$ are finite.

a. Then there exists a $v^* \in V$ and $w^* \in V$ which satisfy the system of equations

$$\max_{d \in D} \{ r_d + (P_d - I)v \} = 0$$

and

$$\max_{d \in F(v)} \{ -v + (P_d - I)w \} = 0,$$

where $F(v) \equiv \{ d \in D : r_d + (P_d - I)v = 0 \}$.

b. The quantity v^* equals the maximal expected total reward.

c. If

$$\delta \in \arg\max_{d \in F(v^*)} \{ P_d w^* \}$$

then δ^∞ is an optimal policy under the expected total-reward criterion.

The policy iteration algorithm below, which we adapt from that in Sec. 10.4.3, finds a stationary policy which maximizes the expected total reward in models which satisfy (10.4.1) and (10.4.2) for all $\pi \in \Pi^{HR}$. Its finite convergence is guaranteed by Theorem 10.3.2. It terminates with a set of decision rules D^* with the property that, if $d \in D^*$, d^∞ is optimal under the expected total-reward criterion.

Total-Reward Policy Iteration Algorithm for Zero Gain Models

1. Pick $d_0 \in D$, set $n = 0$.
2. (Policy evaluation). Obtain v^n and w^n by solving

$$r_{d_n} + (P_{d_n} - I)v = 0 \qquad (10.4.5)$$

and

$$-v + (P_{d_n} - I)w = 0 \qquad (10.4.6)$$

subject to either

$$P_{d_n}^* w = 0 \qquad (10.4.7)$$

or

$$-w + (P_{d_n} - I)u = 0. \qquad (10.4.8)$$

3. (Policy improvement).
   a. Choose

$$d_{n+1} \in \arg\max_{d \in D} \{ r_d + P_d v^n \}$$

setting $d_{n+1}(s) = d_n(s)$ if possible. If $d_{n+1} = d_n$, go to (b); otherwise increment n by 1 and go to step 2.

b. Choose

$$d_{n+1} \in \arg\max_{d \in D} \{P_d w^n\}$$

setting $d_{n+1}(s) = d_n(s)$ if possible. If $d_{n+1} = d_n$, go to step 4; otherwise increment n by 1 and go to step 2.

4. Set $v^* = v^n$ and

$$D^* = \arg\max_{d \in F(v^*)} \{P_d w^n\} \tag{10.4.9}$$

where $F(v^*) = \{d \in D: r_d + (P_d - I)v^* = 0\}$.

Some comments about step 4 may be illuminating. Note that, as a consequence of step 2,

$$r_{d_n} + (P_{d_n} - I)v^* = 0,$$

so that $d_n \in F(v^*)$. Note also that at the preceding pass through the improvement step 3(a), we evaluate $r_d + (P_d - I)v^n$ for all $d \in D$, so that we can retain those decision rules for which this quantity equals 0. This means that we determine $F(v^*)$ with no further effort. The reason for finding $F(v^*)$ explicitly at step 4 is that we have omitted the pass through the 0-discount optimal policy iteration algorithm which computes D_0. This set contains decision rules with gain and bias which satisfy the first two optimality equations. Therefore, without this extra step, we do not exclude suboptimal policies. Alternatively, if we are content with a single optimal policy, we can use $(d_n)^\infty$.

We illustrate this algorithm by solving Examples 7.2.3 and 7.3.1. They illustrate different features of the above algorithm.

Example 10.4.1.   (A positive model) We find an optimal policy in the model of Example 7.2.3 using the above algorithm. Recall that $S = \{s_1, s_2\}$, $A_{s_1} = \{a_{1,1}, a_{1,2}\}$, $A_{s_2} = \{a_{2,1}\}$, $r(s_1, a_{1,1}) = 0$, $r(s_1, a_{1,2}) = 1$, $r(s_2, a_{2,1}) = 0$, $p(s_1|s_1, a_{1,1}) = 1$, $p(s_2|s_1, a_{1,2}) = 1$, $p(s_2|s_2, a_{2,1}) = 1$, and $p(j|s, a) = 0$ otherwise. Observe that (10.4.1) and (10.4.2) hold for any policy π, so that $g^\pi = 0$ for all $\pi \in \Pi^{HR}$.

Begin policy iteration with $d_0(s_1) = a_{1,1}$, $d_0(s_2) = a_{2,1}$. Solving (10.4.5)–(10.4.7) we obtain $v^0(s_1) = v^0(s_2) = 0$ and $w^0(s_1) = w^0(s_2) = 0$. At improvement step 3(a), we identify the improved policy $d_1(s_1) = a_{1,2}$, $d_1(s_2) = a_{2,1}$ and return to step 2.

We evaluate d_1 to obtain $v^1(s_1) = 1$, $v^1(s_2) = 0$ and $w^1(s_1) = -1$, $w^1(s_2) = 0$. In the subsequent improvement step 3(a),

$$\max_{d \in D} \{r_d(s_1) + P_d v^1(s_1)\} = \max\{v^1(s_1), 1 + v^1(s_2)\} = \max\{1, 1\}$$

so that we can set $d_2 = d_1$. In 3(b),

$$\max_{d \in D} \left\{ P_d w^1(s_1) \right\} = \max\{w^1(s_1), w^1(s_2)\} = \max\{0, -1\}$$

so

$$\arg\max_{d \in D} \left\{ P_d w^1 \right\} = \{d_1\}.$$

We set $d_2 = d_1$ and proceed to step 4. We set $v^* = v^1$, note that $F(v^*) = \{d_1\}$ and conclude that $(d_1)^\infty$ maximizes the expected total reward and that v^1 and w^1 satisfy the optimality equations in Theorem 10.4.3.

Example 10.4.2. (A negative model) Let $S = \{s_1, s_2\}$, $A_{s_1} = \{a_{1,1}, a_{1,2}\}$, $A_{s_2} = \{a_{2,1}\}$, $r(s_1, a_{1,1}) = 0$, $r(s_1, a_{1,2}) = -1$, $r(s_2, a_{2,1}) = 0$, $p(s_1|s_1, a_{1,1}) = 1$, $p(s_2|s_1, a_{1,2}) = 1$, $p(s_2|s_2, a_{2,1}) = 1$, and $p(j|s, a) = 0$ otherwise. Observe that (10.4.1) and (10.4.2) hold for any policy π, so that $g^\pi = 0$ for all $\pi \in \Pi^{HR}$.

Initiate policy iteration with $d_0(s_1) = a_{1,2}$, $d_0(s_2) = a_{2,1}$. Solving (10.4.5)–(10.4.7), we obtain $v^0(s_1) = -1$, $v^0(s_2) = 0$ and $w^0(s_1) = 1$, $w^0(s_2) = 0$. At improvement step 3(a),

$$\max_{d \in D} \left\{ r_d(s_1) + P_d v^0(s_1) \right\} = \max\{v^0(s_1), -1 + v^0(s_2)\} = \max\{-1, -1\},$$

so we *do not* identify a new policy and we proceed to step 3(b). Since

$$\max_{d \in D} \left\{ P_d w^0(s_1) \right\} = \max\{w^0(s_1), w^0(s_2)\} = \max\{1, 0\},$$

we identify the improved policy with $d_1(s_1) = a_{1,1}$, $d_1(s_2) = a_{2,1}$, and return to step 2.

We evaluate d_1 to obtain $v^1(s_1) = 0$, $v^1(s_2) = 0$ and $w^1(s_1) = 0$, $w^1(s_2) = 0$. In the improvement step 3(a),

$$\max_{d \in D} \left\{ r_d(s_1) + P_d v^1(s_1) \right\} = \max\{v^1(s_1), -1 + v^1(s_2)\}$$

$$= \max\{0, -1\} \qquad (10.4.10)$$

so that we can set $d_2 = d_1$. In 3(b),

$$\max_{d \in D} \left\{ P_d w^1(s_1) \right\} = \max\{w^1(s_1), w^1(s_2)\} = \max\{0, 0\}$$

so

$$\arg\max_{d \in D} \left\{ P_d w^1 \right\} = D. \qquad (10.4.11)$$

Therefore we set $d_2 = d_1$ and proceed to step 4. Set $v^* = v^1$ and as a consequence of the calculation in (10.4.10), it follows that $F(v^*) = \{d_1\}$. Therefore

$\arg\max_{d \in F(v^*)}\{P_d w^1\} = \{d_1\}$ and $(d_1)^\infty$ maximizes the expected total reward. Note from (10.4.11) that $\arg\max_{d \in D}\{P_d w^1\}$ contains suboptimal policies, as noted above.

Observe that for both models, the algorithm identifies the unique optimal policy, however, in the negative model in Example 10.4.2, we require step 3(b) to obtain an improvement. In the positive model in Example 10.4.1, $F(v^*) = D$, so that without the specification for D^* in step 4, we would stop with a set of decision rules, some of which generate suboptimal policies. Thus this algorithm alleviates the two difficulties referred to at the beginning of this section.

We now relax the assumption that (10.4.2) holds for all $\pi \in \Pi^{HR}$. Instead we assume that it holds for at least one $d^\infty \in \Pi^{SD}$. This assumption allows models in which the expected total reward of some policies equal $-\infty$, so that the gain of these policies contains negative components. However, as a consequence of Proposition 10.4.1, the maximal gain for this model equals 0. Therefore we require the following two steps to find a policy which maximizes the total reward.

1. Apply the Multichain Average Reward Policy Iteration Algorithm of Sec. 9.2. to find a set of policies D' with zero gain.

2. Apply the Total-Reward Policy Iteration Algorithm for Zero Gain Models in which D' replaces D throughout.

Note that this is the 0-discount Optimal Policy Iteration Algorithm. The following modification of Example 7.3.1 provides a model in which (10.1.2) *does not* hold for some policies, so that we require the 0-Discount Optimal Policy Iteration Algorithm to solve it. Note also that it contains a periodic policy but, since its gain is negative, the above approach applies without modification.

Example 10.4.3. Let $S = \{s_1, s_2, s_3\}$, $A_{s_1} = \{a_{1,1}\}$, $A_{s_2} = \{a_{2,1}, a_{2,2}, a_{2,3}\}$, and $A_{s_3} = \{a_{3,1}\}$; $r(s_1, a_{1,1}) = -2$, $r(s_2, a_{2,1}) = 0$, $r(s_2, a_{2,2}) = 0$, $r(s_2, a_{2,3}) = -1$, and $r(s_3, a_{3,1}) = 0$; and $p(s_2|s_1, a_{1,1}) = 1$, $p(s_1|s_2, a_{2,1}) = 1$, $p(s_2|s_2, a_{2,2}) = 1$, $p(s_3|s_2, a_{2,3}) = 1$, and $p(s_3|s_3, a_{3,1}) = 1$ (Fig. 10.4.2). Observe that the deterministic stationary policy η^∞ which uses action $a_{2,1}$ in state s_2 has gain $g^{\eta^\infty}(s_1) = g^{\eta^\infty}(s_2) = -1$, $g^{\eta^\infty}(s_3) = 0$, and a periodic subchain. We leave it as an exercise to solve this problem with the 0-Discount Optimal Policy Iteration Algorithm above.

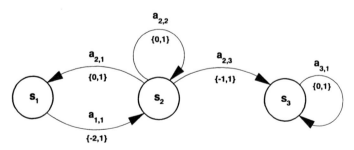

Figure 10.4.2 Symbolic representation of Example 10.4.3.

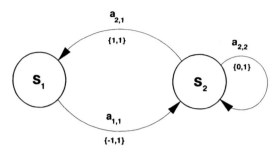

Figure 10.4.3   Symbolic representation of Example 10.4.4.

10.4.3   Finding Average Overtaking Optimal Policies

Conditions (10.4.1) and (10.4.2) exclude models with periodic optimal policies which have gain zero. In this section we discuss the implications of applying the above policy iteration algorithm to such models. The following example illustrates such a model.

Example 10.4.4.   Let $S = \{s_1, s_2\}$, $A_{s_1} = \{a_{1,1}\}$, and $A_{s_2} = \{a_{2,1}, a_{2,2}\}$; $r(s_1, a_{1,1}) = -1$, $r(s_2, a_{2,1}) = 1$, $r(s_2, a_{2,2}) = 0$; and $p(s_2|s_1, a_{1,1}) = 1$, $p(s_1|s_2, a_{2,1}) = 1$, and $p(s_2|s_2, a_{2,2}) = 1$ (Fig. 10.4.3).

Let δ^∞ denote the stationary policy which uses actions $a_{2,1}$ in state s_2, and let γ^∞ denote the stationary policy which uses action $a_{2,2}$ in s_2. Observe that (10.4.1) and (10.4.2) do not hold for δ^∞, so that $\lim_{N \to \infty} v_N^{\delta^\infty}$ does not exist. Since $\{v_N^{\delta^\infty}(s_1):$ $N \geq 1\} = \{-1, 0, -1, 0, -1, 0, \ldots\}$, $\limsup_{N \to \infty} v_N^{\delta^\infty}(s_1) = 0$ and $\liminf_{N \to \infty} v_N^{\delta^\infty}(s_1) = -1$. By a similar argument, $\limsup_{N \to \infty} v_N^{\delta^\infty}(s_2) = 1$ and $\liminf_{N \to \infty} v_N^{\delta^\infty}(s_2) = 0$. Note also that $v^{\gamma^\infty}(s_1) = -1$ and $v^{\gamma^\infty}(s_2) = 0$. Observe further that $g^{\delta^\infty}(s_1) = g^{\delta^\infty}(s_2) = g^{\gamma^\infty}(s_1) = g^{\gamma^\infty}(s_2) = 0$, so that the gain of δ^∞ equals 0 even though the limits in (10.4.1) and (10.4.2) do not exist.

Applying the zero gain policy iteration algorithm to this example identifies $D^* = \{\delta\}$.

A natural question to ask is: what do we achieve by solving this example with the zero gain or 0-discount optimal policy iteration algorithm? We answer this question by referring to (10.4.4). When the gain of a stationary policy equals zero, this expression reduces to

$$h^{d^\infty}(s) = \lim_{N \to \infty} \frac{1}{N} \sum_{k=1}^{N} E_s^{d^\infty} \left\{ \sum_{t=1}^{k} r_d(X_t) \right\} = \lim_{N \to \infty} \frac{1}{N} \sum_{k=1}^{N} v_k^{d^\infty}(s).$$

Thus, since the policy iteration algorithms find 0-discount optimal policies, as a consequence of Theorem 10.1.6(b) we have the following.

Theorem 10.4.4.   Suppose that the gain of every stationary policy equals 0 and $\delta \in D^*$, where D^* is determined in step 4 of the Total Reward Policy Iteration

Algorithm for Zero Gain Models. Then, for all $s \in S$,

$$\lim_{N \to \infty} \frac{1}{N} \sum_{k=1}^{N} \left[v_k^{\delta^\infty}(s) - v_k^{d^\infty}(s) \right] \geq 0 \tag{10.4.12}$$

for all $d \in D^{\text{MD}}$.

Recall that in Sec. 5.4 we referred to a policy π^* for which

$$\liminf_{N \to \infty} \frac{1}{N} \sum_{k=1}^{N} \left[v_k^{\pi^*}(s) - v_k^{\pi}(s) \right] \geq 0 \tag{10.4.13}$$

for all $\pi \in \Pi^{\text{HR}}$ as *average overtaking optimal*. Thus Theorem 10.4.4 means that the policy iteration algorithm finds an average overtaking optimal policy in the class of stationary policies. That this policy is optimal in the class of *all* policies requires a generalization of the argument used to prove Theorem 9.1.2. We leave this extension as an exercise. Hence, in Example 10.4.4, δ^∞ is average overtaking optimal.

BIBLIOGRAPHIC REMARKS

The material presented in this chapter has its roots in Blackwell's (1962) in-depth study of Howard's (1960) model and algorithms. Among other results, Blackwell introduced the criteria we now refer to as 0-discount optimality and Blackwell optimality (he referred to policies which achieve these criteria as *nearly optimal* and *optimal*, respectively), demonstrated the existence of a Blackwell optimal policy and convergence of multichain average reward policy iteration through use of a *partial* Laurent series expansion. That paper also raised the following challenging questions.

    a. What is the relationship between 0-discount optimal and Blackwell optimal policies?

    b. When are average optimal policies 0-discount optimal?

    c. How does one compute 0-discount optimal and Blackwell optimal policies?

Veinott (1966a), Veinott (1969b), and Miller and Veinott (1969) addressed these issues. In his 1966 paper, Veinott provided a policy iteration algorithm for finding a 0-discount optimal policy (note he refers to such a policy as 1-*optimal* in that paper). Miller and Veinott (1969) develop the *complete* Laurent series expansion, relate it to lexicographic ordering, and use this relationship to provide a finite policy iteration algorithm which finds a Blackwell optimal policy. Essentially they show that any $(|S| - 1)$-discount optimal policy is Blackwell optimal. In his comprehensive (and difficult) 1969 paper, Veinott among other results provides the link between nearly optimal and Blackwell optimal policies by introducing the concept of n-discount optimality and providing a policy iteration algorithm for finding a stationary n-discount optimal policy for any n. He further refines this work in Veinott (1974). This paper provides an accessible overview of the above work and a simplified presentation of the main results in Veinott (1969b). It also refines a result in Miller and Veinott (1969) by showing that any $(|S| - r - 1)$-discount optimal policy is Blackwell optimal. Lamond and Puterman (1989) establish this latter result using different methods.

Our presentation in this chapter borrows from all of these sources. Our approach differs somewhat in that we use the n-discount optimality equations in our analysis. They appear implicitly in Veinott (1969b) and explicitly in Dekker (1985).

Research on computational methods for models with sensitive optimality criteria has been quite limited. Denardo and Fox (1968), Denardo (1970a) and Kallenberg (1983) provide linear programming approaches for finding bias-optimal policies. Federgruen and Schweitzer (1984a) propose a value iteration approach for solving the nested system of optimality equations but do not investigate its implementation or numerical properties.

Sheu and Farn (1980) establish existence of a stationary 0-discount optimal policy in finite-state compact action models. Contributors to this theory in countable-state models include Hordijk and Sladky (1977), Wijngaard (1977), Mann (1985), Dekker (1985), Dekker and Hordijk (1988, 1991, and 1992), Cavasoz-Cadena and Lasserre (1988), Lasserre (1988), Hordijk and Spieksma (1989a) and Spieksma (1990 and 1991). Wijngaard (1977b) and Yushkevich (1993) explore these concepts in general state space models.

Section 10.4 is motivated by van Dawen's (1986a) analysis of the convergence of policy iteration in models with expected total-reward criteria. In that paper, he provides a rather complicated two-step policy iteration algorithm for finding an optimal policy and remarks that

"sometimes our methods of proof are more akin to those used for the average reward rather than the total reward criterion."

Motivated by this remark and our proof of Proposition 7.2.15, we observed that the 0-discount optimality policy iteration algorithm of 10.3 can be used to find an optimal policy in models with expected total-reward criterion. The validity of our extension of Theorem 10.4.4 to the class of all policies follows from Denrado and Miller (1968).

With the exception of Sec. 10.4.3, we do not discuss the related overtaking optimality criteria (Sec. 5.4.2). Veinott (1966a), Denardo and Miller (1968), Lippman (1968a), Sladky (1974), Hordijk and Sladky (1977), Denardo and Rothblum (1979), and van der Wal (1981) investigated the relationship between discount optimality, overtaking optimality, and average optimality. Sladky (1974) generalized average overtaking optimality and overtaking optimality to n-average optimality. Hordijk and Sladky (1977) show in countable-state models that a policy is n-discount optimal if and only if it is n-average optimal. An immediate consequence of this result is that average overtaking optimality and 0-discount optimality (bias optimality) are equivalent. The overtaking optimality criterion is more selective than average overtaking optimality. That is, if an overtaking optimal policy exists, then it is average overtaking optimal. An example from Brown (1965) shows that an overtaking optimal policy need not exist even in very simple settings. Denardo and Rothblum (1979) show that overtaking optimal policies exist whenever $D_0(y^*_{-1}, y^*_0)$ contains a single decision rule.

PROBLEMS

10.1. Prove Proposition 10.1.1.

10.2. Prove that if r_d is bounded and s is finite, all terms in the Laurent series expansion of $v_\lambda^{d^\infty}$ are bounded.

10.3. Find a 0-discount optimal policy and a Blackwell optimal policy in the model in Problem 6.1.

10.4. Prove Theorem 10.1.6(c).

10.5. Let $y_\lambda \in V_L$ and suppose that $(y_{-1}, y_0, \ldots, y_n)$ satisfy the -1th-nth discount optimality equations and δ^∞ is Blackwell optimal, then $B_\delta^\lambda y_\lambda \leq_{\mathscr{L}} 0$.

10.6. (Veinott, 1968) Let $S = \{1, 2, \ldots, N + 1\}$, $A_1 = \{a_{1,1}, a_{1,2}\}$, and $A_k = \{a_{k,1}\}$ for $k \geq 2$; $r(1, a_{1,1}) = \alpha_1$, $r(1, a_{1,2}) = 1$, and $r(k, a_{k,1}) = \alpha_k$ for $2 \leq k \leq N$ and $r(N + 1, a_{N+1,1}) = 0$; and $p(2|1, a_{1,1}) = 1$, $p(1|1, a_{1,2}) = p(2|1, a_{1,2}) = 0.5$, $p(k + 1|k, a_{1,k}) = 1$, and $p(N + 1|N + 1, a_{1,N+1}) = 1$, where α_k denote the coefficients in the power series expansion of $2\lambda(1 - \lambda)^{N-1}$.

Let δ denote the decision rule which uses action $a_{1,1}$ in state 1, and let γ denote the decision rule which uses action $a_{1,2}$ in state 1. Show that both δ^∞ and γ^∞ are m-discount optimal for $-1 \leq m \leq N - 2$, but that γ^∞ is m-discount optimal for $m \geq N - 1$.

10.7. (Blackwell, 1962). Let $S = \{s_1, s_2\}$; $A_{s_1} = \{a_{1,1}, a_{1,2}\}$; and $A_{s_2} = \{a_{2,1}\}$; $r(s_1, a_{1,1}) = 1$, $r(s_1, a_{1,2}) = 2$, and $r(s_2, a_{2,1}) = 0$; and $p(s_1|s_1, a_{1,1}) = 0.5$, $p(s_2|s_1, a_{1,1}) = 0.5$, $p(s_2|s_1, a_{1,2}) = 1$, and $p(s_2|s_2, a_{2,1}) = 1$. Let δ denote the decision rule which uses action $a_{1,1}$ in s_1, and let γ denote the decision rule which uses action $a_{1,2}$ in s_1.

a. Compute $v_\lambda^{\delta^\infty}$ and $v_\lambda^{\gamma^\infty}$.

b. Show, using definitions, that both δ^∞ and γ^∞ are (-1)- and 0-discount optimal, but that only γ^∞ is n-discount optimal for $n \geq 1$.

c. Plot $v_\lambda^{\delta^\infty}$ and $v_\lambda^{\gamma^\infty}$ as functions of λ, and show that γ^∞ is Blackwell optimal.

d. Verify that (10.4.1) and (10.4.2) hold for all policies and find the set of policies which maximize the expected total reward.

10.8. (Blackwell, 1962). Let $S = \{s_1, s_2\}$; $A_{s_1} = \{a_{1,1}, a_{1,2}\}$ and $A_{s_2} = \{a_{2,1}\}$; $r(s_1, a_{1,1}) = 3$, $r(s_1, a_{1,2}) = 6$, and $r(s_2, a_{2,1}) = -3$; and $p(s_1|s_1, a_{1,1}) = 0.5$, $p(s_2|s_1, a_{1,1}) = 0.5$, $p(s_2|s_1, a_{1,2}) = 1$, and $p(s_2|s_2, a_{2,1}) = p(s_2|s_2, a_{2,1}) = 0.5$. Let δ denote the decision rule which uses action $a_{1,1}$ in s_1, and let γ denote the decision rule which uses action $a_{1,2}$ in s_1.

a. Show that $v_\lambda^{\delta^\infty}$ and $v_\lambda^{\gamma^\infty}$ are rational functions of λ.

b. Write out the (-1)th, 0th, and 1th optimality equations for this model.

c. Find a 0-discount optimal policy using the 0-discount optimal policy iteration algorithm.

d. Verify that this model has zero optimal gain and that the policy identified in (c) is average overtaking optimal.

10.9. Find the matrices $Y_2^{d^\infty}$ for each of the stationary policies in Problem 8.4. Use them to determine a 1-discount optimal policy.

10.10. For all $n \geq -1$, find the sets of stationary n-discount optimal policies for the model in Example 5.4.2. Relate your observations to Theorem 10.3.6.

10.11. Find a policy which maximizes the expected total reward in Example 10.4.3, using the 0-discount optimal policy iteration algorithm.

10.12. (Denardo and Miller, 1968). Show that if a policy is 0-discount optimal, it is average overtaking optimal in the class of all policies. Hint: modify the proof of Theorems 8.4.1 and 9.1.2.

10.13. Find a 0-discount optimal and a Blackwell optimal policy for the model in Problem 8.23.

10.14. Find a 0-discount optimal and a Blackwell optimal policy for the model in Problem 9.18.

10.15. Find a 0-discount optimal and a Blackwell optimal policy for the model in Problem 4.26.

10.16. Find a 0-discount optimal and a Blackwell optimal policy for the model in Example 10.0.1.

10.17. Construct a finite queueing admission control model (Sec. 3.7) in which stationary policies using control units L and $L + 1$ are both average optimal. Are both of these control limit policies 0-discount optimal and Blackwell optimal?

10.18a. Show δ^∞ in Ex. 10.4.4 satisfies

$$\liminf_{N \to \infty} V_N^{\delta^\infty} \geq \limsup_{N \to \infty} V_N^{\gamma^\infty}$$

and that

$$\limsup_{N \to \infty} V_N^{\delta^\infty} \geq \limsup_{N \to \infty} V_N^{\gamma^\infty}$$

(with strict inequality in the second expression above).
b. What further results regarding the optimality of policies identified by the policy iteration algorithm of Section 10.4.2 does this suggest.

CHAPTER 11

Continuous-Time Models

In the models studied in previous chapters, the decision maker could choose actions only at a predetermined discrete set of time points; however, some applications, particularly in queueing control and equipment maintenance, are more naturally modeled by allowing action choice at random times in $[0, \infty]$. In this chapter we study such models. We classify them on the basis of the allowable decision epochs and intertransition time probability distributions.

The most general continuous-time models we consider in this book are *semi-Markov decision processes* (SMDP's). They generalize MDP's by

    a. allowing, or requiring, the decision maker to choose actions whenever the system state changes;

    b. modeling the system evolution in continuous time; and

    c. allowing the time spent in a particular state to follow an arbitrary probability distribution.

All of these aspects add generality to the model. Because of (c), we can use these methods to analyze $GI/M/c$ and $M/G/c$ queueing systems, and maintenance systems with nonexponential failure rates.

In semi-Markov decision processes, action choice determines the joint probability distribution of the subsequent state and the time between decision epochs. In its simplest form, the system evolves by remaining in a state for a random amount of time and then jumping to a different state. In greater generality, the system state may change several times between decision epochs, however, only the state at decision epochs is relevant to the decision maker. We refer to these models as semi-Markov because the distribution of the time to the next decision epoch and the state at that time depend on the past only through the state and action chosen at the current decision epoch and because the time between transitions may follow an arbitrary probability distribution.

We also study a controlled continuous-time Markov chain and refer to it as a *continuous-time Markov decision process* (CTMDP). This model may be viewed as a special case of a semi-Markov decision process in which intertransition times are exponentially distributed and actions are chosen at every transition.

In this chapter we restrict attention to models in which decision epochs may only occur after a distinguished set of transitions. When actions are allowed at any time, or

530

the horizon length is finite and fixed, continuous-time control theory methods are more suitable. Some authors refer to the model with restricted decision epochs as a *Markov renewal program* to distinguish it from a model with continuous-time control; however, in many applications, especially in CTMDP's, an optimal policy changes actions only when a transition occurs.

We assume some familiarity with semi-Markov process theory. Çinlar (1975) provides an excellent introduction at an appropriate level for this chapter. Here we restrict attention to infinite-horizon models with discounted and average reward optimality criteria and assume all models have *time-homogeneous* rewards and transition probabilities.

11.1 MODEL FORMULATION

In this section we formulate a semi-Markov decision process model. Our exposition follows that in Chap. 2 as closely as possible. Section 11.2 contains examples which illustrate this model and should be referred to while reading this section. That in Sec. 11.2.1 provides a particularly simple illustration.

11.1.1 Probabilistic Structure

Defining the system state requires some care. In queueing control models, such as those in Sec. 11.2, the system state may vary between decision epochs; however, we only allow action choice to depend on the system content at points of time when decisions may be implemented. From the perspective of this model, what transpires between decision epochs provides no relevant information to the decision maker. To this end we distinguish the *natural process* and the *semi-Markov decision process*. The natural process models the state evolution of the system as if it were observed continually throughout time, while the SMDP represents the evolution of the system state at decision epochs only (Fig. 11.1.1). The two processes agree at decision epochs. For example, in a queueing admission control model, the semi-Markov decision process describes the system state at arrival times only, while the natural process describes the system state at all time points. To determine rewards we need information about the queue size at all times. This is described by the natural process. To determine whether to admit a job, we need only know the number in the queue when a job enters the system. This is described by the SMDP.

Let S denote the finite- or countable-state space. Decision epochs occur at random points of time determined by the model description. For example, in a queueing system with admission control, decision epochs correspond to the time immediately following an arrival. If at some decision epoch the system occupies state $s \in S$, the decision maker must choose an action a from the set A_s. As a consequence of choosing action $a \in A_s$, the next decision epoch occurs at or before time t, and the system state at that decision epoch equals j with probability $Q(t, j|s, a)$. We use $Q(dt, j|s, a)$ to represent a time-differential.

In most applications $Q(t, j|s, a)$ is not provided directly. Instead the basic model quantities are $F(t|s, a)$ and either $P(j|s, a)$ or $p(j|t, s, a)$, where $F(t|s, a)$ denotes the probability that the next decision epoch occurs within t time units of the current

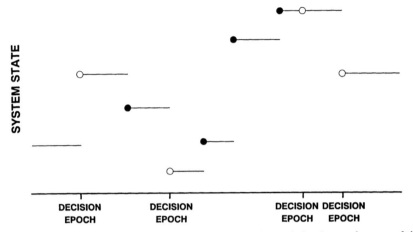

Figure 11.1.1 Typical sample path of the natural process. Open circles denote the state of the SMDP (at decision epochs). Circles indicate that sample paths of the natural process are right-continuous.

decision epoch, given that the decision maker chooses action a from A_s in state s at the current decision epoch. The quantity $p(j|t, s, a)$ denotes the probability that the natural process occupies state j, t time units after a decision epoch, given that action a was chosen in state s at the current decision epoch and that the next decision epoch has not occurred prior to time t.

Note that, in many applications, $F(t|s, a)$ may be independent of s or a. Although the semi-Markov decision process does not change state until the subsequent decision epoch, the natural process may change state several times between decision epochs. We use the transition probability $p(j|t, s, a)$ to compute cumulative rewards between decision epochs. If the state of the natural process does not change until the next decision epoch, $p(s|t, s, a) = 1$ for all t.

We also sometimes refer to the *embedded Markov decision process*. This process describes the evolution of the system at decision epochs only. Letting $P(j|s, a)$ denote the probability that the embedded Markov decision process occupies state j at the subsequent decision epoch when action a is chosen in state s at the current decision epoch, it follows that:

$$P(j|s, a) = Q(\infty, j|s, a) \tag{11.1.1}$$

and

$$Q(t, j|s, a) = P(j|s, a)F(t|s, a). \tag{11.1.2}$$

To avoid the possibility of an infinite number of decision epochs within finite time, we impose the following assumption:

Assumption 11.1.1   There exist $\varepsilon > 0$ and $\delta > 0$ such that

$$F(\delta|s, a) \leq 1 - \varepsilon \tag{11.1.3}$$

for all $s \in S$ and $a \in A_s$.

We distinguish two particular forms for $F(t|s, a)$. When

$$F(t|s, a) = 1 - e^{-\beta(s, a)t}, \tag{11.1.4}$$

we refer to this as a *continuous-time Markov decision process*. In such a model, the times between decision epochs are exponentially distributed with rate $\beta(s, a)$. When

$$F(t|s, a) = \begin{cases} 0 & t \leq t' \\ 1 & t > t' \end{cases}$$

for some fixed t' for all s and a, we obtain a discrete-time Markov decision process with decision epochs occurring every t' time units.

We now describe the evolution of the SMDP. At time t_0, the system occupies state s_0 and the decision maker chooses action a_0. As a consequence of this action choice, the system remains in s_0 for t_1 units of time at which point the system state changes to s_1 and the next decision epoch occurs. The decision maker chooses action a_2, and the same sequence of events occur. Figure 11.1.2 provides a representation of this process.

To avoid technicalities, choose $t_0 = 0$. We let $h_n \equiv (t_0, s_0, a_0, t_1, s_1, a_1, \ldots, t_n, s_n)$ denote the history of the semi-Markov decision process up to the nth decision epoch. Note that $h_0 = (t_0, s_0)$, and that

$$h_n = (h_{n-1}, a_{n-1}, t_n, s_n) \tag{11.1.5}$$

for $n \geq 1$. In contrast to discrete-time models, the history also contains the sojourn times $t_0, t_1, \ldots$. Equivalently, we can view discrete-time models as special cases of semi-Markov models, in which $t_n = t'$ for all n.

11.1.2 Rewards or Costs

The following reward structure appears to include most applications. When the decision maker chooses action a in state s, he receives a lump sum reward, or pays a

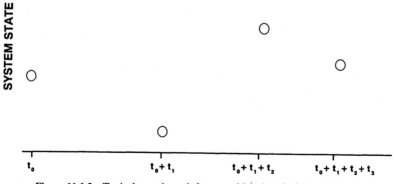

Figure 11.1.2   Typical sample path for a semi-Markov decision process.

lump sum cost $k(s, a)$. Further, he accrues a reward or incurs a cost at *rate* $c(j', s, a)$ as long as the *natural* process occupies state j', and action a was chosen in state s at the preceding decision epoch. In many applications $c(j', s, a)$ will be independent of j', s, or a. Since the model does not allow the action to change until the next decision epoch, this reward rate function applies throughout this period.

Under both the discounted and average reward criterion, we will transform the model to one with an expected reward $r(s, a)$ which depends only on the state of the semi-Markov decision process at a decision epoch and the action chosen.

11.1.3 Decision Rules and Policies

As in discrete-time models, we consider several classes of decision rules. They may be either deterministic or randomized and Markovian or history dependent. Note that history-dependent decision rules are defined in terms of the above expanded notion of history which includes the sojourn times. We adopt the same notation as in Sec. 2.1.4. We represent decision rules by d and classes of decision rules by D^K, where K denotes an abbreviation of the class as in Chap. 2; for example, D^{MR} denotes the set of randomized Markovian decision rules. We again denote the class of deterministic Markovian decision rules by D.

As in Sec. 2.1.5, policies are finite or infinite sequences of decision rules. These classes range in generality from randomized history dependent to deterministic stationary policies. We denote an individual policy by π and a class of policies by Π^K. To be consistent with earlier notation we write $\pi = (d_1, d_2, \dots)$ so that decision rule d_1 is used at $t_0 = 0$.

For each deterministic Markovian decision rule d, define $p_d(j|t, s) \equiv p(j|t, s, d(s))$, $F_d(t|s) \equiv F(t|s, d(s))$, $Q_d(t, j|s) \equiv Q(t, j|s, d(s))$, $k_d(s) = k(s, d(s))$, and $c_d(j', s) = c(j', s, d(s))$. The analogous quantities for randomized decision rules are defined as in Chap. 2.

*11.1.4 Induced Stochastic Processes

In this section we construct a probability model $\{\Omega, B(\Omega), P\}$ for the embedded process. The essential difference between the model here and that in Sec. 2.1.6 is that the description includes T which is a continuum.

Assume discrete S and A_s for each $s \in S$. Let $T = [0, \infty)$. The sample space Ω generalizes that in the discrete-time models by taking into account the sojourn times as well as the states and actions. Consequently, in an infinite-horizon model,

$$\Omega = T \times S \times A \times T \times S \times A \times \cdots = \{T \times S \times A\}^\infty,$$

and a typical sample path $\omega \in \Omega$ may be represented by

$$\omega = (t_0, s_0, a_0, t_1, s_1, a_1, \dots),$$

where $t_0 = 0$ and $s_n \in S$, $a_n \in A_{s_n}$ and $t_n \in T$ for all n.

In this model, the set of Borel sets of Ω is given by

$$B(\Omega) = B(\{T \times S \times A\}^{\infty}),$$

When S and A are discrete, $B(\Omega)$ is the smallest σ-algebra which contains all *finite* products of sets of the form (C, D, h) where C is a subinterval of T, D is a subset of S and h is a subset of A.

We define the coordinate random variables X_n, Y_n, and τ_n, which take values in S, A, and T, respectively, by

$$X_n(\omega) = s_n, \qquad Y_n(\omega) = a_n, \qquad \tau_n(\omega) = t_n.$$

We also require the *history process* Z_n which we define by $Z_0(\omega) = (t_0, s_0)$ and

$$Z_n(\omega) = (t_0, s_0, a_0, t_1, s_1, a_1, \ldots, t_n, s_n)$$

and the *elapsed time* process σ_n for $n \geq 0$ defined by

$$\sigma_n(\omega) = \sum_{j=0}^{n} t_j.$$

Note that

$$\sigma_n(\omega) = \sum_{j=0}^{n} \tau_n(\omega). \qquad (11.1.6)$$

For each $\pi \in \Pi^{HR}$ we construct a probability measure P^{π} on $\{\Omega, B(\Omega)\}$. Let $P_0(s)$ denote the initial state distribution. Then define conditional probabilities by

$$P^{\pi}\{X_0 = s\} = P_0(s) \qquad \text{and} \qquad P^{\pi}\{T_0 = 0\} = 1,$$

and, for $n \geq 0$,

$$P^{\pi}\{Y_n = a | Z_n = h_n\} = q_{d_{n+1}(h_n)}(a),$$

$$P^{\pi}\{X_{n+1} = j, \tau_{n+1} \in (t, t + dt) | Z_n = h_n, Y_n = a_n\} = Q(dt, j | s_n, a_n)$$

so that the probability of a sample path under π equals

$$P^{\pi}(t_0, s_0, a_0, t_1, s_1, a_1, t_2, s_2, \ldots)$$
$$= P_0(s_0) q_{d_1(h_0)}(a_0) Q(dt_1, s_1 | s_0, a_0) q_{d_2(h_1)}(a_1) Q(dt_2, s_2 | s_1, a_1) \ldots . \qquad (11.1.7)$$

For deterministic policies π, this simplifies to

$$P^{\pi}(t_0, s_0, a_0, t_1, s_1, a_1, t_2, s_2, \ldots)$$
$$= P_0(s_0) Q(dt_1, s_1 | s_0, a_0) Q(dt_2, s_2 | s_1, a_1) \ldots .$$

Conditional probabilities are constructed in a similar way to Sec. 2.1.6.

We let $E^{\pi}\{\cdot\}$ denote an expectation with respect to P^{π}. When $P_0(s) = 1$, so that $s_0 = s$, we write $E_s^{\pi}\{\cdot\}$ to denote this expectation.

For fixed Markovian π, the induced stochastic process $\{X_n; n \geq 0\}$ is a semi-Markov process. When $F(t|s, a)$ satisfies (11.1.4), it is a continuous-time Markov chain. With π history dependent, the process is considerably more complicated.

We conclude this section by stating, without proof, a generalization of Theorem 5.5.1 concerning the relationship between processes generated by history-dependent and Markovian policies. Let W_t denote the state of the natural process at time t.

Theorem 11.1.1. Let $\pi \in \Pi^{HR}$. Then, for each $s \in S$, there exists a $\pi' \in \Pi^{MR}$ for which

$$P^{\pi'}\{W_{\tau_n + t} = k, X_n = j, Y_n = a, \tau_n = u | X_1 = s\}$$
$$= P^{\pi}\{W_{\tau_n + t} = k, X_n = j, Y_n = a, \tau_n = u | X_1 = s\}$$

for $n = 0, 1, \ldots, 0 \leq t < u < \infty$, $j \in S$, $a \in A_j$, and $k \in S$.

11.2 APPLICATIONS

In this section we illustrate the semi-Markov decision process model through a simple two-state model and two continuous time queueing control models. Note that for the queueing models, the semi-Markov decision process formulations are more natural and more general than the discrete-time formulations in Sec. 3.7.

11.2.1 A Two-State Semi-Markov Decision Process

We consider a continuous-time variant of the model in Sec. 3.1. We assume the following timing of events. After choosing an action in a given state, the system remains there for an action-dependent random period of time. Then a transition occurs and the next action can be chosen.

As before, $S = \{s_1, s_2\}$, $A_{s_1} = \{a_{1,1}, a_{1,2}\}$, and $A_{s_2} = \{a_{2,1}\}$. Since transitions occur only at the end of a sojourn in a state, we specify transition probabilities for the embedded Markov decision process by $P(s_1|s_1, a_{1,1}) = P(s_2|s_1, a_{1,1}) = 0.5$, $P(s_2|s_1, a_{1,2}) = 1$, $P(s_1|s_2, a_{2,1}) = 0.1$, and $P(s_2|s_2, a_{2,1}) = 0.9$. We assume that the sojourn time in s_1 under $a_{1,1}$ is uniformly distributed on $[0, 2]$, denoted $U[0, 2]$, the sojourn time in s_1 under $a_{1,2}$ is $U[0, 4]$, and the sojourn time in s_2 under $a_{2,1}$ is $U[0, 3]$. Thus

$$F(t|s_1, a_{1,1}) = \begin{cases} t/2 & \text{if } 0 \leq t \leq 2 \\ 1 & \text{if } t > 2, \end{cases}$$

with similar definitions applying for $F(t|s_1, a_{1,2})$ and $F(t|s_2, a_{2,1})$

The lump sum rewards are given by $k(s_1, a_{1,1}) = 0$, $k(s_1, a_{1,2}) = -1$ and $k(s_2, a_{2,1}) = 0$ and the continuous reward rates are $c(s_1, s_1, a_{1,1}) = 5$, $c(s_1, s_1, a_{1,2}) = 10$, and $c(s_2, s_2, a_{2,1}) = -1$ (Fig. 11.2.1).

From (11.1.2) it follows that

$$Q(t, s_1, a_{1,1}) = \begin{cases} 0.25t & 0 \leq t \leq 2 \\ 0.5 & t > 2. \end{cases}$$

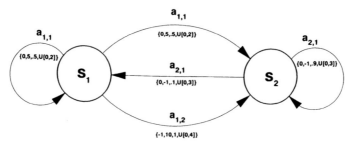

Figure 11.2.1 Symbolic representation of the model in Sec. 11.2.1. The expression in brackets represents the lump sum cost, the continuous cost rate, the transition probability, and the sojourn time distribution.

and

$$Q(t, s_2, a_{1,2}) = \begin{cases} 0.25t & 0 \le t \le 2 \\ 0.5 & t > 2. \end{cases}$$

11.2.2 Admission Control for a $G/M/1$ Queueing System

We now consider a continuous-time generalization of the model in Sec. 3.7.1. In a $G/M/1$ queueing system, interarrival times are independent and follow an arbitrary distribution, and service times at the single server are independent and exponentially distributed. A controller regulates the system load by accepting or rejecting arriving jobs or calls. If rejected, the arrival leaves the system (Fig. 3.7.1). This might serve as a simple model for a telephone switch or for the decision making process of a security guard at the door of our local ski store during its post-Christmas sale.

Let the state space for the natural process denote the number of jobs in the system (in service plus in the queue) at any time point. Either $S = \{0, 1, \ldots\}$ or $S = \{0, 1, \ldots, M\}$, depending on our assumptions regarding the system capacity M. We denote the interarrival time distribution function by $G(\cdot)$, its density or mass function by $g(\cdot)$, and, in light of Assumption 11.1.1, require that $G(0) < 1$. Further, we assume an exponential service rate with parameter μ independent of the number of jobs in the system. Each arriving job contributes R units of revenue and the system incurs a holding cost at rate $f(j)$ per unit time whenever there are j jobs in the system.

Decisions are required only when jobs enter the system. The embedded Markov decision process models the system state at these time points. When $S = \{0, 1, \ldots\}$, we set $A_s = \{0, 1\}$ for all $s \in S$. Action 0 denotes rejecting an arrival, while action 1 corresponds to accepting an arrival. When $S = \{0, 1, \ldots, M\}$, we may not admit jobs when the system is saturated so that $A_M = \{0\}$. Figure 11.2.2 represents a typical sample path of the natural process.

We specify $F(t|s, a)$ and $p(j|t, s, a)$ for this model. Since decisions are made only at arrival times,

$$F(t|s, a) = G(t)$$

independent of s and a. When the density exists, $F(dt|s, a) = g(t) \, dt$. In between arrivals, the natural state may change because of service completions. From elemen-

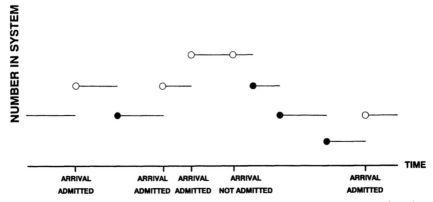

Figure 11.2.2 Sample path for the natural process in an admission control model. Open dots represent its state immediately following decision epochs.

tary probability theory, the number of service completions in t units of time follows a Poisson distribution with parameter μt. Consequently, the probabilistic evolution of the state of the natural process can be described by

$$p(j|t, s, 0) = \frac{e^{-\mu t}(\mu t)^{s-j}}{(s - j)!} \tag{11.2.1}$$

for $s \geq j > 0$. Letting

$$H(k|\mu, t) \equiv \frac{e^{-\mu t}(\mu t)^{k}}{k!},$$

it follows that $p(j|t, s, 1) = H(s + 1 - j|\mu, t)$ for $j \geq 1$, $p(0|t, s, 0) = \sum_{k \geq s} H(k|\mu, t)$, and $p(0|t, s, 1) = \sum_{k \geq s+1} H(k|\mu, t)$.

Finally,

$$k(s, a) = \begin{cases} 0 & a = 0 \\ R & a = 1, \end{cases}$$

and $c(j', s, a) = -f(j')$ where j' denotes the state of the natural process.

For this model

$$Q(t, j|s, 0) = \int_0^t \frac{e^{-\mu u}(\mu u)^{s-j}}{(s - j)!} G(du) \tag{11.2.2}$$

and from (11.1.1)

$$P(j|s, 0) = \int_0^\infty \frac{e^{-\mu u}(\mu u)^{s-j}}{(s - j)!} G(du)$$

Under reasonable assumptions on the cost rate $f(\cdot)$, we would expect that there exists an optimal policy of control limit form.

11.2.3   Service Rate Control in an $M/G/1$ Queueing System

We provide a continuous-time version of the model in Sec. 3.7.2. An $M/G/1$ queueing system has a single server, independent exponential interarrival times, and independent service times which follow an arbitrary distribution. In the controlled model, the controller regulates system load by varying the service rate; faster servers are more expensive.

We assume that interarrival times are exponential with parameter γ, and that service distributions $G_b(\cdot)$ with densities $g_b(\cdot)$ can be drawn from a *finite* set B. In most applications, b will represent a scale parameter; for example, we might specify G_b to be exponential with parameter b. Further, we assume that the controller may change the service rate only upon completion of a service, or on the arrival of a job to an empty system. Costs include a holding cost $f(j)$, a continuous service cost rate $d(b)$, and a fixed cost K for changing the service rate.

We denote the state of the natural process by $\langle s, b \rangle$, where s denotes the number of jobs in the system, and b denotes the index of the service distribution in use. The semi-Markov decision process describes these quantities at decision epochs. For either process $S = \{0, 1, \ldots\} \times B$ or $S = \{0, 1, \ldots, M\} \times B$ depending on whether or not the system has finite capacity. In each state $A_{\langle s, b \rangle} = B$; however, because of the choice of decision epochs, the embedded system cannot occupy state M when M is finite.

In this model, the sojourn time distribution explicitly depends on both the state of the system and the action. For $s \geq 1$, the next decision epoch occurs upon completion of a service, so that

$$F(t|\langle s, b \rangle, b') = G_{b'}(t),$$

while, if $s = 0$, the next opportunity to change the service rate occurs when a job arrives, so that

$$F(t|\langle 0, b \rangle, b') = 1 - e^{-\gamma t}.$$

To satisfy Assumption 11.1.1, we require that there exists an $\varepsilon > 0$ and a $\delta > 0$ for which $G_b(\delta) < 1 - \varepsilon$ for all $b \in B$. Since we assume finite B, this is equivalent to $G_b(0) < 1$ for all $b \in B$.

We now provide the transition probabilities for the natural process when $M = \infty$. When the queue is empty, the next transition occurs at an arrival so that

$$p(\langle 0, b' \rangle|t, \langle 0, b \rangle, b') = 1 \text{ and } P(\langle 1, b' \rangle|\langle 0, b \rangle, b')$$

In an analogous fashion to the admission control model, for $s > 0$,

$$p(\langle s + k, b' \rangle|t, \langle s, b \rangle, b') = \frac{e^{-\gamma t}(\gamma t)^k}{k!}.$$

The economic parameters of the model satisfy

$$k(\langle s, b \rangle, b') = \begin{cases} 0 & b' = b \\ -K & b' \neq b, \end{cases}$$

and the continuous reward rate

$$r(\langle j', b' \rangle, \langle s, b \rangle, b') = -f(j') - d(b').$$

11.3 DISCOUNTED MODELS

This section studies infinite-horizon discounted semi-Markov decision processes. In it, we generalize and apply results from Chap. 6 on discrete-time discounted Markov decision process models.

11.3.1 Model Formulation

We assume continuous-time discounting at rate $\alpha > 0$. This means that the present value of one unit received t time units in the future equals $e^{-\alpha t}$. By setting $e^{-\alpha} = \lambda$, where λ denotes the discrete-time discount rate, we see that $\lambda = 0.9$ corresponds to $\alpha = -\log(0.9) = 0.11$.

For policy $\pi \in \Pi^{HR}$, let $v_{\alpha}^{\pi}(s)$ denote the expected infinite-horizon discounted reward, given that the process occupies state s at the first decision epoch. Define it by

$$v_{\alpha}^{\pi}(s) \equiv E_s^{\pi} \left\{ \sum_{n=0}^{\infty} e^{-\alpha \sigma_n} \left[k(X_n, Y_n) + \int_{\sigma_n}^{\sigma_{n+1}} e^{-\alpha(t - \sigma_n)} c(W_t, X_n, Y_n) \, dt \right] \right\}. \quad (11.3.1)$$

In the above expression, $\sigma_0, \sigma_1, \ldots$ represents the times of successive decision epochs, and W_t denotes the state of the natural process at time t. The term $e^{-\alpha \sigma_n}$ transforms the expressions in [ ]'s to values at the first decision epoch. The first term in it corresponds to the lump sum portion of the reward, while the second term corresponds to the continuous portion of the reward which is received at rate $c(W_t, X_n, Y_n)$ between decision epochs n and $n + 1$.

Define the value of the discounted SMDP by

$$v_{\alpha}^*(s) \equiv \sup_{\pi \in \Pi^{HR}} v_{\alpha}^{\pi}(s)$$

for $s \in S$. We seek to characterize v_{α}^* through an optimality equation and find a policy π^* for which

$$v_{\alpha}^{\pi^*}(s) = v_{\alpha}^*(s)$$

for all $s \in S$.

We note that Theorem 11.1.1 implies that

$$v_{\alpha}^*(s) = \sup_{\pi \in \Pi^{MR}} v_{\alpha}^{\pi}(s). \quad (11.3.2)$$

11.3.2 Policy Evaluation

In this section we derive a generalization of the equation

$$v = r_d + \lambda P_d v,$$

which was used to evaluate the expected total reward of a stationary policy in a discrete-time discounted model.

For $s \in S$ and $a \in A_s$, let $r(s, a)$ denote the expected total discounted reward between two decision epochs, given that the system occupies state s at the first decision epoch and that the decision maker choose action a in state s. Because the rewards, transition probabilities, and sojourn times are time homogeneous,

$$r(s, a) = k(s, a) + E_s^a \left\{ \int_0^{\tau_1} e^{-\alpha t} c(W_t, s, a) \, dt \right\}, \tag{11.3.3}$$

where $E_s^a\{\cdot\}$ denotes the expectation with respect to the sojourn time distribution $F(t|s, a)$ and the probability distribution of the natural process under action a. For evaluation in applications the above expression for $r(s, a)$ becomes

$$r(s, a) = k(s, a) + \int_0^\infty \sum_{j \in S} \left[\int_0^u e^{-\alpha t} c(j, s, a) p(j|t, s, a) \, dt \right] F(du|s, a)$$

Note that $r(s, a)$ depends on the discount rate α.

For $d \in D$, let $r_d(s) \equiv r(s, d(s))$. In light of (11.3.3),

$$r_d(s) = k_d(s) + E_s^d \left\{ \int_0^{\tau_1} e^{-\alpha t} c_d(W_t, s) \right\}$$

$$= k_d(s) + \int_0^\infty \sum_{j \in S} \left[\int_0^u e^{-\alpha t} c_d(j, s) p_d(j|t, s) \, dt \right] F_d(du|s) \tag{11.3.4}$$

Let $\pi = (d_1, d_2, \ldots) \in \Pi^{HR}$. Then from (11.3.1) and (11.3.4), it follows that $v_\alpha^\pi(s)$ can be reexpressed as

$$v_\alpha^\pi(s) = E_s^\pi \left\{ \sum_{n=0}^\infty e^{-\alpha \sigma_n} r_{d_{n+1}}(X_n) \right\}. \tag{11.3.5}$$

Note that this expression has the same form as Eq. (6.1.1), which defined the expected discounted reward in a discrete-time model. The difference is that the decision-rule-dependent discount factor $e^{-\alpha \sigma_n}$ replaces the discount rate λ^{n-1}.

Letting $\pi' = (d_2, d_3, \ldots)$, it follows from (11.3.5) that

$$v_\alpha^\pi(s) = r_{d_1}(s) + E_s^\pi \{ e^{-\alpha \tau_1} v_\alpha^{\pi'}(X_1) \}$$

$$= r_{d_1}(s) + \sum_{j \in S} \int_0^\infty e^{-\alpha t} Q_{d_1}(dt, j|s) v_\alpha^{\pi'}(j). \tag{11.3.6}$$

For the stationary policy d^∞, $\pi' = d^\infty$, so that (11.3.6) becomes

$$v_\alpha^{d^\infty}(s) = r_d(s) + \sum_{j \in S} \int_0^\infty e^{-\alpha t} Q_d(dt, j|s) v_\alpha^{d^\infty}(j). \tag{11.3.7}$$

This can be expressed in matrix-vector notation as

$$v_\alpha^{d^\infty} = r_d + M_d v_\alpha^{d^\infty}, \tag{11.3.8}$$

where M_d denotes the $|S| \times |S|$ matrix with components

$$m_d(j|s) = \int_0^\infty e^{-\alpha t} Q_d(dt, j|s) \tag{11.3.9}$$

In greater generality, define $m(j|s, a)$ for $j \in S$, $s \in S$, and $a \in A_s$ by

$$m(j|s, a) \equiv \int_0^\infty e^{-\alpha t} Q(dt, j|s, a). \tag{11.3.10}$$

Then $m_d(j|s) = m(j|s, d(s))$.

Equation (11.3.7) together with (11.3.10) suggests that we can transform the discounted semi-Markov decision problem to a discrete-time model with primitive quantities S, A_s, $r(s, a)$, and $m(j|s, a)$. Denote the state- and action-dependent discount rate by $\lambda(s, a)$ and define it by

$$\lambda(s, a) \equiv \sum_{j \in S} m(j|s, a).$$

In contrast to the discrete-time model, $\lambda(s, a)$ varies with a, or, equivalently, the matrix M_d may have *unequal* row sums. It is an easy consequence of Assumption 11.1.1 to see that they are strictly bounded by 1. Note that the discounted Markov decision problem of Chap. 6 is a special case of this model in which

$$m(j|s, a) = \lambda p(j|s, a).$$

We now provide a method for evaluating the expected total discounted reward of a stationary policy. Recall that V denotes the set of bounded real-valued functions on S.

Theorem 11.3.1. Suppose, for all $s \in S$ and $a \in A_s$, that $|r(s, a)| \le M < \infty$ and there exists $\varepsilon > 0$ and $\delta > 0$ for which (11.1.3) holds. Then, for any $d \in D^{MR}$, $v_\alpha^{d^\infty}$ is the unique solution in V of

$$r_d + M_d v = v, \tag{11.3.11}$$

so that

$$v_\alpha^{d^\infty} = (I - M_d)^{-1} r_d. \tag{11.3.12}$$

Proof. From (11.1.3), it follows that (Problem 11.3a)

$$\sup_{s \in S} \sup_{a \in A_s} \lambda(s, a) \equiv \lambda^* < 1. \tag{11.3.13}$$

Hence, for each $d \in D^{MR}$, $\|M_d\| < 1$. This implies that $\sigma(M_d) < 1$, so it follows from Corollary C.4 that $(I - M_d)^{-1}$ exists. Hence $v' = (I - M_d)^{-1} r_d$ is the unique solution of (11.3.11). Corollary C.4 implies that

$$v' = \sum_{k=0}^{\infty} (M_d)^k r_d,$$

which yields (11.3.5) with π replaced by d^∞. Therefore $v' = v_\alpha^{d^\infty}$. $\qquad\square$

We illustrate these calculations by analyzing the model in Sec. 11.2.1.

Example 11.3.1. We assume a discount rate $\alpha = 0.1$ and compute $m(j|s, a)$ and $r(s, a)$. Since the natural process does not change state until the next decision epoch it follows that to evaluate $m(j|s, a)$ we need only compute $\int_0^\infty e^{-\alpha t} F(dt|s, a)$. For example,

$$\int_0^\infty e^{-\alpha t} F(dt|s_1, a_{1,1}) = \int_0^2 e^{-0.1t} \tfrac{1}{2} dt = \frac{1}{0.2}(1 - e^{-0.2}) = 0.906.$$

In a similar fashion we find that $\int_0^\infty e^{-\alpha t} F(dt|s_1, a_{1,2}) = 0.824$, and $\int_0^\infty e^{-\alpha t} F(dt|s_2, a_{2,1}) = 0.864$. Thus

$$m(s_1|s_1, a_{1,1}) = P(s_1|s_1, a_{1,1}) \int_0^\infty e^{-\alpha t} F(dt|s_1, a_{1,1}) = 0.5(0.906) = 0.453,$$

$m(s_2|s_1, a_{1,1}) = 0.453$, $m(s_1|s_1, a_{1,2}) = 0$, $m(s_2|s_1, a_{1,2}) = 0.824$, $m(s_1|s_2, a_{2,1}) = 0.086$, and $m(s_2|s_2, a_{2,1}) = 0.778$. Consequently $\lambda(s_1, a_{1,1}) = 0.906$, $\lambda(s_1, a_{1,2}) = 0.824$, and $\lambda(s_2, a_{2,1}) = 0.864$.

We now evaluate $r(s, a)$. Because in this model transitions occur at the end of the sojourn period, it follows from (11.3.3) that

$$r(s_1, a_{1,1}) = k(s_1, a_{1,1}) + c(s_1, s_1, a_{1,1}) \int_0^\infty \int_0^u e^{-\alpha t} dt \, F(du|s_1, a_{1,1})$$

$$= 0 + 5 \int_0^2 \int_0^u e^{-0.1t} dt \, \tfrac{1}{2} du = 4.683.$$

$r(s_1, a_{1,2}) = 16.580$ and $r(s_2, a_{2,1}) = -1.361$.

We let δ denote the deterministic decision rule which uses action $a_{1,1}$ in s_1, and γ that which uses $a_{1,2}$ in s_1. Then

$$M_\delta = \begin{bmatrix} 0.453 & 0.453 \\ 0.086 & 0.778 \end{bmatrix} \quad \text{and} \quad r_\delta = \begin{bmatrix} 4.683 \\ -1.361 \end{bmatrix}.$$

As noted above, M_δ has unequal row sums and $\|M_\delta\| = 0.906$. From Theorem 11.3.1,

$$v_\alpha^{\delta^\infty} = (I - M_\delta)^{-1} r_\delta = (5.141, -4.139)^T.$$

Similar calculations reveal that $v_\alpha^{\gamma^\infty} = (16.934, 0.429)^T$.

We now determine $m(j|s, a)$ and $r(s, a)$ for the admission control model in Sec. 11.2.2.

Example 11.3.2. We first provide general expressions for $m(j|s, a)$ and $r(s, a)$. In this model, $p(j|t, s, a)$ depends explicitly on t. From (11.2.1) and (11.3.10), it follows that, for $j > 0$,

$$m(j|s, 0) = \int_0^\infty e^{-\alpha t} \frac{e^{-\mu t}(\mu t)^{s-j}}{(s - j)!} G(dt) \qquad (11.3.14)$$

and

$$m(0|s, 0) = \sum_{k \geq s} \int_0^\infty e^{-\alpha t} \frac{e^{-\mu t}(\mu t)^k}{k!} G(dt). \qquad (11.3.15)$$

Since admitting a job immediately moves the system from state s to state $s + 1$, $m(j|s, 1) = m(j|s + 1, 0)$ for $j \geq 0$. When the interarrival time is exponential with parameter λ, we can explicitly evaluate $m(j|s, a)$. For example,

$$m(j|s, 0) = \int_0^\infty e^{-\alpha t} \frac{e^{-\mu t}(\mu t)^{s-j}}{(s - j)!} \lambda e^{-\lambda t} dt = \frac{\lambda \mu^{s-j}}{(\lambda + \mu + \alpha)^{s+1-j}}.$$

The transformed reward $r(s, a)$ may be derived in a similar way. Let N_t denote the number of service completions in t time units. Then, if s jobs were in the system at time 0 and none have been admitted,

$$W_t = [s - N_t]^+.$$

Thus, when $a = 0$, it follows from (11.3.3) that

$$r(s, 0) = -\int_0^\infty e^{-\alpha t} E_s^0\{f(W_t)\} G(dt) = -\int_0^\infty e^{-\alpha t} E\{f([s - N_t]^+)\} G(dt)$$

$$= -\int_0^\infty e^{-\alpha t} \sum_{k=0}^\infty f([s - k]^+) \frac{e^{-\mu t}(\mu t)^k}{k!} G(dt), \qquad (11.3.16)$$

where $E\{\cdot\}$ denotes the expectation with respect to the Poisson service distribution. Since admitting a service immediately increases the system state by 1 and yields a lump sum reward of R units,

$$r(s, 1) = R + r(s + 1, 0).$$

Observe that even in the case of linear holding costs and exponential interarrival times, $r(s, 0)$ is quite tedious to evaluate, especially for moderate to large values of s.

11.3.3 The Optimality Equation and Its Properties

By analogy to the discrete-time models, we would expect that the optimality equation for a discounted semi-Markov decision problem has the form

$$v = \max_{d \in D} \{r_d + M_d v\} \equiv Lv, \tag{11.3.17}$$

whenever the maximum in (11.3.17) is attained for each $v \in V$. In component form, we express this equation as

$$v(s) = \max_{a \in A_s} \left\{ r(s, a) + \sum_{j \in S} m(j|s, a) v(j) \right\}. \tag{11.3.18}$$

Under Assumption 11.1.1 and the assumption that $\|r_d\| \le M < \infty$ for all $d \in D$, L is a contraction operator on V, so consequently all results from Sec. 6.2 concerning existence of solutions of the optimality equations and optimal policies apply. We summarize these as follows.

Theorem 11.3.2. Suppose the hypotheses of Theorem 11.3.1 hold. Then the following is true.

a. The optimality equation (11.3.17) has the unique solution in V, v_α^*.

b. If the maximum in (11.3.17) is attained for each $v \in V$, then there exists a stationary deterministic optimal policy.

c. If

$$d^* \in \arg\max_{d \in D} \{r_d + M_d v_\alpha^*\}, \tag{11.3.19}$$

then $(d^*)^\infty$ is optimal.

d. There exists an optimal stationary deterministic policy whenever

   i. A_s is finite for each $s \in S$; or
   ii. A_s is compact, $r(s, a)$ is continuous in a for each $s \in S$, and, for each $j \in S$ and $s \in S$, $m(j|s, a)$ is continuous in a; or, more generally,
   iii. A_s is compact, $r(s, a)$ is upper semicontinuous in a for each $s \in S$, and, for each $j \in S$ and $s \in S$, $m(j|s, a)$ is lower semicontinuous in a.

We explore the consequences of this result to Examples 11.2.1 and 11.2.2.

Example 11.3.1 (ctd.). As a result of Theorem 11.3.2, the optimality equations

$$v(s_1) = \max\left\{ r(s_1, a_{1,1}) + \sum_{j \in S} m(j|s_1, a_{1,1}) v(j), r(s_1, a_{1,2}) + \sum_{j \in S} m(j|s_1, a_{1,2}) v(j) \right\},$$

$$v(s_2) = r(s_2, a_{2,1}) + \sum_{j \in S} m(j|s_2, a_{2,1}) v(j)$$

become

$$v(s_1) = \max\{4.683 + 0.453v(s_1) + 0.453v(s_2), 16.580 + 0.824v(s_2)\}$$
$$v(s_2) = -1.361 + 0.086v(s_1) + 0.778v(s_2).$$

From Theorem 11.3.2 we conclude that these equations have a unique solution and that there exists a stationary deterministic optimal policy which satisfies (11.3.19). Calculations in Sec. 11.3.2 show that γ^∞ is optimal and that $v_a^* = (16.934, 2.429)^T$.

Example 11.3.2 (ctd.). The optimality equations for this model become

$$v(s) = \max\left\{ r(s,0) + \sum_{j=0}^{s} m(j|s,0)v(j), R + r(s+1,0) + \sum_{j=0}^{s+1} m(j|s+1,0)v(j) \right\}.$$
$$(11.3.20)$$

When

$$\sup_{s \in S} |r(s,0)| < \infty, \qquad\qquad (11.3.21)$$

Theorem 11.3.2 implies that the optimality equation has the unique solution v_a^*, that there exists a stationary deterministic optimal policy, and that it can be found directly using (11.3.20). When (11.3.21) does not hold, we appeal to results in Sec. 11.3.5 below.

Note that the optimality equations have more intuitive appeal when expressed in random variable notation as

$$v(s) = \max\left\{ r(s,0) + \int_0^\infty e^{-\alpha t} E\{v([s - N_t]^+)\}G(dt),\right.$$
$$\left. R + r(s+1,0) + \int_0^\infty e^{-\alpha t} E\{v([s + 1 - N_t]^+)\}G(dt) \right\},$$

where N_t and $E\{\cdot\}$ are defined in the preceding section.

11.3.4 Algorithms

The value iteration, policy iteration, modified policy iteration, linear programming algorithms, and their variants in Chap. 6 apply directly to this model when we replace $\lambda p(j|s, a)$ by $m(j|s, a)$. Further, we can enhance performance of these algorithms by using action elimination procedures and stopping criteria based on results in Sec. 6.6. Note, however, that Proposition 6.6.1 regarding the span contraction properties of P_d does not apply directly because its proof requires that P_d have equal row sums.

An important consequence of the convergence of the value iteration is that we can use it to determine the structure of optimal policies as in Sec. 6.11. This will be especially important in countable-state models which cannot be solved numerically without truncation. We note that this approach provides a way of solving the queueing admission control problem in Sec. 11.2.2. Under the assumption that $f(\cdot)$ is convex and nondecreasing, Stidham (1978) established the optimality of a control limit policy. This means that there exists a constant $K \geq 0$ for which the stationary policy derived from the decision rule

$$\delta(s) = \begin{cases} 1 & s < K \\ 0 & s \geq K \end{cases}$$

is optimal. As a result of this, an optimal policy can be found by searching only within this class of policies; van Nunen and Puterman (1983) and Puterman and Thomas (1987) provide an efficient algorithm for doing this.

11.3.5 Unbounded Rewards

We assume $S = \{0, 1, 2, \dots\}$ and generalize results in Sec. 6.10 to include discounted semi-Markov decision problems. As above, we obtain them by replacing $\lambda p(j|s, a)$ with $m(j|s, a)$. We summarize these results in the following theorem.

Theorem 11.3.3. Suppose Assumption 11.1.1 holds, and there exists a positive real-valued function w on S for which $\inf_{s \in S} w(s) > 0$, and

i. there exists a constant $\mu < \infty$ for which

$$\sup_{a \in A_s} |r(s, a)| \leq \mu w(s)$$

for all $s \in S$;

ii. there exists a constant κ, $0 \leq \kappa < \infty$, for which

$$\sum_{j \in S} m(j|s, a)w(j) \leq \kappa w(s),$$

and

iii. there exists a β, $0 \leq \beta < 1$ and an integer J for which

$$\sum_{j \in S} \sum_{s_{J-1} \in S} \cdots \sum_{s_1 \in S} m_{d_J}(j|s_{J-1}) \cdots m_{d_1}(s_1|s)w(j) \leq \beta w(s)$$

for all $d_j \in D^{\mathrm{MD}}$.

Then

a. L defined in (11.3.17) is a J-stage contraction on V_w,

b. the optimality equation has a unique solution in V_w and it equals v_α^*,

c. if there exists a $d^* \in D^{\mathrm{MD}}$ satisfying

$$d^* \in \arg\max_{d \in D} \{r_d + M_d v_\alpha^*\},$$

then $(d^*)^\infty$ is optimal

d. for any $v^0 \in V_w$, $L^n v^0$ converges to v_α^*, and

e. for any $v^0 \in V_w$ satisfying $Lv^0 \geq v^0$, the sequence of iterates of policy iteration and modified policy iteration converge to v_α^*.

We leave it as an exercise to apply this result to the models in Secs. 11.2.2 and 11.2.3. Note that the structural results from Sec. 6.11 also apply to the discounted semi-Markov decision process model.

11.4 AVERAGE REWARD MODELS

In this section we analyze infinite-horizon semi-Markov decision processes with average reward criterion. As in the case of discrete-time problems, this theory is considerably more complex than that for discounted models. To simplify exposition we assume that, for every stationary policy, the embedded Markov chain has a *unichain* transition probability matrix. Under this assumption, the expected average reward of every stationary policy does not vary with the initial state. Most results apply to countable-state models; however, Sec. 11.4.5 deals with them explicitly.

11.4.1 Model Formulation

Let $\pi \in \Pi^{HR}$ and $s \in S$. For $t \geq 0$, let $v_t^\pi(s)$ denote the expected total reward generated by the process up to time t, given that the system occupies state s at time 0. Letting ν_u denote the number of decisions made up to time t, we have that

$$v_t^\pi(s) \equiv E_s^\pi \left\{ \int_0^t c(W_u, X_{\nu_u}, Y_{\nu_u}) \, du + \sum_{n=0}^{\nu_t - 1} k(X_n, Y_n) \right\}. \qquad (11.4.1)$$

Since, in this model we start counting time from the first decision epoch, $\nu_0 = 1$, and it remains at this value until the next decision epoch, which occurs at time τ_1. At the first decision epoch, the system generates a fixed reward $k(X_0, Y_0)$ and, until τ_1, it accumulates additional rewards at rate $c(W_u, X_{\nu_0}, Y_{\nu_0})$. After making the second decision, $\nu_t = 2$, and it retains that value until after the next decision epoch.

In this model we can define the average expected reward in two ways. We begin with the most natural definition. For each $\pi \in \Pi^{HR}$, define the average expected reward or gain by

$$g_C^\pi(s) \equiv \liminf_{t \to \infty} \frac{1}{t} v_t^\pi(s) \qquad s \in S. \qquad (11.4.2)$$

We include the subscript C (continuous) to distinguish it from an alternative criterion below. In Chap. 8 we distinguished the lim inf and lim sup of $t^{-1} v_t^\pi$ and denoted the analog of the above quantity by g_-^π; here we drop the subscript "$-$" to simplify notation.

We can also define the average expected reward by taking the limit inferior of the ratio of the expected total reward up to the nth decision epoch to the expected total time until the nth decision epoch as follows:

$$g^\pi(s) \equiv \liminf_{n \to \infty} \frac{E_s^\pi \left\{ \sum_{i=0}^n \left[k(X_i, Y_i) + \int_{\sigma_i}^{\sigma_{i+1}} c(W_t, X_i, Y_i) \, dt \right] \right\}}{E_s^\pi \left\{ \sum_{i=0}^n \tau_i \right\}}. \qquad (11.4.3)$$

We will analyze the semi-Markov decision process model in terms of this criterion, and then provide conditions under which the two criteria agree.

Our objective in this section is to characterize the optimal average expected reward

$$g^*(s) = \sup_{\pi \in \Pi^{HR}} g^\pi(s) \qquad s \in S$$

and identify a policy π^* with the property that

$$g^{\pi^*}(s) = g^*(s) \qquad s \in S.$$

As a result of Theorem 11.1.1,

$$g^*(s) = \sup_{\pi \in \Pi^{MR}} g^\pi(s) \qquad s \in S.$$

For $s \in S$ and $a \in A_s$, let $r(s, a)$ denote the expected total reward between two decision epochs, given that the system occupies state s at the first decision epoch and the decision maker chooses action a. In terms of the problem data,

$$r(s, a) = k(s, a) + E_s^a \left\{ \int_0^{\tau_1} c(W_t, s, a) \, dt \right\},$$

$$= k(s, a) + \int_0^\infty \sum_{j \in S} \left[\int_0^u c(j, s, a) p(j|t, s, a) \, dt \right] F(du|s, a) \quad (11.4.4)$$

where τ_1 denotes the random time until the first decision epoch. From the perspective of the average reward criterion, $r(s, a)$ contains all necessary information about the reward to analyze the model.

Modifying the definition of $m(j|s, a)$ in the previous section, let $m(j|s, a)$ denote the probability that the semi-Markov decision process occupies state j at the next decision epoch, given that the decision maker chooses action a in state s at the current decision epoch. It satisfies

$$m(j|s, a) = P(j|s, a). \quad (11.4.5)$$

where $P(j|s, a)$ denotes the transition probabilities for the embedded Markov decision process (11.1.1). Consequently, for each $a \in A_s$, $m(j|s, a)$ is a *transition probability function*. Because of this observation, many results from Chaps. 8 and 9 apply directly.

For each $s \in S$ and $a \in A_s$, define $y(s, a)$ by

$$y(s, a) \equiv E_s^a\{\tau_1\} = \int_0^\infty t \sum_{j \in S} Q(dt, j|s, a). \quad (11.4.6)$$

This quantity denotes the expected length of time until the next decision epoch, given that action a is chosen in state s at the current decision epoch. For $d \in D^{MR}$, let $r_d(s) = r(s, d(s))$, $m_d(j|s) = m(j|s, d(s))$, and $y_d(s) = y(s, d(s))$.

We leave it as an exercise to show that, under Assumption 11.1.1,

$$\inf_{s \in S, \, a \in A_s} y(s, a) > \varepsilon\delta > 0, \tag{11.4.7}$$

where ε and δ are defined in (11.1.1) and, as a result, for any finite $t > 0$,

$$E_s^a\{\nu_t\} < \infty.$$

These results mean that the expected time between decisions epochs is at least $\varepsilon\delta$, and that, in any finite interval, the expected number of decisions is finite.

In terms of the above notation, for $\pi = (d_1, d_2, \ldots)$ (11.4.3) becomes

$$g^\pi(s) = \liminf_{n \to \infty} \frac{E_s^\pi \left\{ \sum_{i=0}^n r_{d_{i+1}}(X_i) \right\}}{E_s^\pi \left\{ \sum_{i=0}^n \tau_i \right\}}. \tag{11.4.8}$$

To simplify the latter notation, define

$$R_n^\pi(s) \equiv \frac{E_s^\pi \left\{ \sum_{i=0}^n r_{d_{i+1}}(X_i) \right\}}{E_s^\pi \left\{ \sum_{i=0}^n \tau_i \right\}}. \tag{11.4.9}$$

11.4.2 Policy Evaluation

We begin with a heuristic derivation of the evaluation equation for stationary policies. In light of results for discrete-time models, suppose that there exists a function h^{d^∞} for which

$$v_t^{d^\infty} = h^{d^\infty} + g^{d^\infty}t + o(1). \tag{11.4.10}$$

Letting t denote an arbitrary decision epoch, and τ the length of time until the next decision epoch, it follows from (11.3.5)–(11.3.8), with $\alpha = 0$, that

$$E^{d^\infty}\{v_{t+\tau}^{d^\infty}\} = r_d + M_d v_t^{d^\infty}.$$

Substituting (11.4.10) into this expression yields

$$h^{d^\infty} + g^{d^\infty}\left[t + E^{d^\infty}\{\tau\}\right] + o(1) = r_d + M_d\left[h^{d^\infty} + g^{d^\infty}t + o(1)\right].$$

For t large enough it follows from the definition of y_d that

$$h^{d^\infty} + g^{d^\infty} y_d = r_d + M_d h^{d^\infty},$$

suggesting that an equation of the form

$$h = r_d - g y_d + M_d h$$

characterizes g^{d^∞} and h^{d^∞}. Note that this equation differs from that in the discrete-time model in that it contains the term y_d. This term accounts for the nonconstant times between decision epochs.

We now provide conditions under which the limit of $R_n^h(s)$ in (11.4.9) exists, and provide a useful representation for it. Part (a) does not require the unichain assumption.

Proposition 11.4.1. Suppose S is countable, and that Assumption 11.1.1 holds.

a. Then, for each $d \in D^{\text{MR}}$,

$$g^{d^\infty}(s) = \lim_{n \to \infty} \frac{E_s^{d^\infty}\left\{ \sum_{i=0}^{n} r_d(X_i) \right\}}{E_s^{d^\infty}\left\{ \sum_{i=0}^{n} \tau_i \right\}} = \frac{M_d^* r_d(s)}{M_d^* y_d(s)}, \qquad (11.4.11)$$

provided

$$M_d^* \equiv \lim_{n \to \infty} \frac{1}{n} \sum_{j=0}^{n-1} M_d^j \qquad (11.4.12)$$

is a transition probability matrix.

b. If M_d is unichain, $g^{d^\infty}(s)$ is a constant function of s.

Proof. Since M_d is a transition probability matrix, M_d^* exists (Appendix A). Under the assumption that

$$\sum_{j \in S} M_d^*(j|s) = 1,$$

it follows that

$$M_d^* r_d(s) = \lim_{n \to \infty} \frac{1}{n+1} E_s^{d^\infty}\left\{ \sum_{i=0}^{n} r_d(X_i) \right\}$$

and

$$M_d^* y_d(s) = \lim_{n \to \infty} \frac{1}{n+1} E_s^{d^\infty}\left\{ \sum_{i=0}^{n} \tau_i \right\}$$

Therefore (11.4.11) follows by taking ratios of these two quantities, noting that the limit of the ratios equals the ratio of the limits, and that, when taking the limit of the ratios, the factor $1/n + 1$ cancels from the numerator and denominator.

Part (b) follows since, under the unichain assumption, M_d^* has equal rows. □

The following theorem shows that solutions of the evaluation equations characterize g^{d^∞}.

Theorem 11.4.2. Assume that S is countable. Let $d \in D^{MR}$, and suppose that M_d is unichain, M_d^* is a probability matrix, and Assumption 11.1.1 holds.

If there exists a constant g and a function h satisfying

$$h = r_d - g y_d + M_d h, \tag{11.4.13}$$

then $g = g^{d^\infty}$.

Proof. Multiply both sides of (11.4.13) by M_d^* to obtain

$$M_d^* h = M_d^* r_d - g M_d^* y_d + M_d^* M_d h.$$

Noting that $M_d^* M_d = M_d^*$, that $y_d > \varepsilon\delta > 0$ under Assumption 11.1.1, and that rearranging terms yields

$$g = \frac{M_d^* r_d}{M_d^* y_d},$$

the result then follows from Proposition 11.4.1(a). □

We now establish the existence of a solution of these equations for *finite S* using a similar approach to that in Sec. 8.2. Section 11.4.5 provides conditions which ensure the existence of solutions to the evaluation equations for countable-state models with unbounded rewards.

Theorem 11.4.3. Let S be finite, $d \in D^{MR}$, and suppose that M_d is unichain. Then g^{d^∞} and $h^{d^\infty} \equiv H_d(r_d - g^{d^\infty} y_d)$ satisfy (11.4.13), where

$$H_d \equiv (I - M_d + M_d^*)^{-1}(I - M_d^*).$$

Proof. From Appendix A, (A.17), $M_d^* + (I - M_d)H_d = I$. Applying both sides of this expression to r_d and y_d yields

$$M_d^* r_d + (I - M_d)H_d r_d = r_d, \tag{11.4.14}$$

$$M_d^* y_d + (I - M_d)H_d y_d = y_d. \tag{11.4.15}$$

From Proposition 11.4.1, $M_d^* r_d = g^{d^\infty} M_d^* y_d$, so substituting this into (11.4.14) yields

$$g^{d^\infty} M_d^* y_d + (I - M_d) H_d r_d = r_d. \qquad (11.4.16)$$

Hence, writing (11.4.15) as

$$M_d^* y_d = y_d - (I - M_d) H_d y_d$$

and substituting this expression into Eq. (11.4.16) yields

$$g^{d^\infty} y_d + (I - M_d)\left[H_d\left(r_d - g^{d^\infty} y_d \right) \right] = r_d,$$

from which the result follows.   □

Note that the quantity $H_d(r_d - g^{d^\infty} y_d)$ generalizes the bias to semi-Markov decision problems. Problem 11.17 provides the multichain analog of this result for finite-state systems.

We now use the evaluation equations to find the average expected reward for stationary policies the model of Sec. 11.2.1.

Example 11.4.1. Direct calculations show that $y(s_1, a_{1,1}) = 1$, $y(s_1, a_{1,2}) = 2$, and $y(s_2, a_{2,1}) = 1.5$; and $r(s_1, a_{1,1}) = 5$, $r(s_1, a_{1,2}) = 19$, $r(s_2, a_{2,1}) = -1.5$, and, from (11.4.5), $m(j|s, a) = P(j|s, a)$ for all s and a.

Letting δ denote the decision rule which uses action $a_{1,1}$ in s_1, and γ denote the decision rule which uses action $a_{1,2}$ in s_1, we have

$$M_\delta = \begin{bmatrix} 0.5 & 0.5 \\ 0.1 & 0.9 \end{bmatrix}, \quad r_\delta = \begin{bmatrix} 5 \\ -1.5 \end{bmatrix}, \quad y_\delta = \begin{bmatrix} 1 \\ 1.5 \end{bmatrix},$$

so the evaluation equations become

$$h(s_1) = 5 - g + 0.5h(s_1) + 0.5h(s_2),$$
$$h(s_2) = -1.5 - 1.5g + 0.1h(s_1) + 0.9h(s_2).$$

Setting $h(s_2) = 0$, we find that $g^{\delta^\infty} = -0.29$ and $h(s_1) = 10.58$. Similarly, we find that $g^{\gamma^\infty} = 2.35$, and $h(s_1) = 14.3$. Results below establish that γ^∞ is average optimal.

We now evaluate $m(j|s, a)$, $y(s, a)$, and $r(s, a)$ for the admission control model of Sec. 11.2.2.

Example 11.4.2. In this model the state of the natural process changes between decision epochs, so calculation of $m(j|s, a)$ is not as direct as in the preceding example. From (11.2.2) and (11.4.5), it follows that $m(j|s, 0) = 0$ for $j > s$,

$$m(j|s, 0) = \int_0^\infty \frac{e^{-\mu t}(\mu t)^{s-j}}{(s-j)!} G(dt)$$

for $s \geq j > 0$, and

$$m(0|s,0) = \sum_{k \geq s} \int_0^\infty \frac{e^{-\mu t}(\mu t)^k}{k!} G(dt).$$

Since admitting a job moves the system immediately from state s to state $s + 1$, $m(j|s, 1) = m(j|s + 1, 0)$ for $j \geq 0$. When the interarrival time is exponential with parameter λ, we can explicitly evaluate $m(j|s, a)$. For example,

$$m(j|s,0) = \int_0^\infty \frac{e^{-\mu t}(\mu t)^{s-j}}{(s-j)!} \lambda e^{-\lambda t}\, dt = \frac{\lambda \mu^{s-j}}{(\lambda + \mu)^{s-j+1}}.$$

The transformed reward $r(s, a)$ may be derived in a similar way. Following a similar argument to that in Example 11.3.2,

$$r(s,0) = -\int_0^\infty \sum_{k=0}^\infty f([s-k]^+) \frac{e^{-\mu t}(\mu t)^k}{k!}\, dG(t).$$

Under the decision to admit an arrival,

$$r(s,1) = R + r(s+1,0).$$

Finally we note that, for all $s \in S$ and $a = 0$ or 1,

$$y(s,a) = \int_0^\infty \sum_{k=0}^\infty t \frac{e^{-\mu t}(\mu t)^k}{k!} G(dt) = \int_0^\infty t\, G(dt),$$

which is just the mean service time. Observe that $y(s, a)$ does not vary with s or a in this model.

11.4.3 Optimality Equations

By analogy to the discrete-time model, we would expect that the optimality equations for unichain models have the form

$$h = \max_{d \in D^{MD}} \{r_d - g y_d + M_d h\}, \tag{11.4.17}$$

whenever the maximum in (11.4.17) is attained. In component notation these equations may be expressed as

$$h(s) = \max_{a \in A_s} \left\{ r(s,a) - g y(s,a) + \sum_{j \in S} m(j|s,a) h(j) \right\}.$$

Note that the optimality equations differ from those for MDPs because of the inclusion of the expressions y_d or $y(s, a)$.

We now generalize Theorem 8.4.1 to semi-Markov decision process models. We include the proof because it shows how several of the above concepts interrelate.

Theorem 11.4.4. Let S be countable and suppose for each $d \in D^{\text{MD}}$ that M_d^* is a probability matrix.

a. If there exists a constant g and an $h \in V$ for which

$$h \geq \max_{d \in D^{\text{MD}}} \{r_d - g y_d + M_d h\}, \qquad (11.4.18)$$

then $g \geq g^*$.

b. If there exists a constant g and an $h \in V$ for which

$$h \leq \max_{d \in D^{\text{MD}}} \{r_d - g y_d + M_d h\}, \qquad (11.4.19)$$

then $g \leq g^*$.

c. If there exists a constant g and an $h \in V$ for which (11.4.17) holds, then $g = g^*$.

Proof. Let $\pi = (d_1, d_2, \dots) \in \Pi^{\text{MR}}$. Then, from Proposition 6.2.1 and (11.4.18),

$$h \geq r_{d_n} - g y_{d_n} + M_{d_n} h$$

for $n \geq 1$. Iterating this inequality yields

$$h \geq r_{d_1} + \sum_{k=1}^{N} M_{d_1} \cdots M_{d_k} r_{d_{k+1}} - g \left[y_{d_1} + \sum_{k=1}^{N} M_{d_1} \cdots M_{d_k} y_{d_{k+1}} \right]$$
$$+ M_{d_1} \cdots M_{d_{N+1}} h.$$

Rearranging terms and expressing the result in random variable notation yields

$$g(s) \geq \frac{E_s^{\pi} \left\{ \sum_{i=0}^{N} r_{d_{i+1}}(X_i) \right\} + E_s^{\pi} \{h(X_N) - h(s)\}}{E_s^{\pi} \left\{ \sum_{i=0}^{N} \tau_i \right\}}.$$

From (11.4.7), $E_s^{\pi} \{\sum_{i=0}^{N} \tau_i\} \geq (N+1)\varepsilon\delta$, so because $h \in V$, $h(s)$ is bounded and

$$\lim_{N \to \infty} \frac{E_s^{\pi} \{h(X_N) - h(s)\}}{E_s^{\pi} \left\{ \sum_{i=0}^{N} \tau_i \right\}} = 0.$$

Noting (11.4.9), $g(s) \geq \lim \sup_{N \to \infty} R_N^\pi(s) \geq g^\pi(s)$. Since π was arbitrary,

$$g(s) \geq \sup_{\pi \in \Pi^{MR}} g^\pi(s) = \sup_{\pi \in \Pi^{HR}} g^\pi(s).$$

We establish (b) as follows. Let $\delta \in \arg \max_{d \in D} \{r_d - g y_d + M_d h\}$. Then, applying the same argument as in the proof of Theorem 11.4.2, it follows that $g \leq g^{\delta^\infty} \leq g^*$. Part (c) follows by combining (a) and (b). $\square$

Inspection of the above proof reveals that we have actually established a stronger result. We have shown that, whenever the optimality equation has a bounded solution,

$$g(s) = \sup_{\pi \in \Pi^{HR}} \lim \sup_{N \to \infty} R_N^\pi(s) = \sup_{\pi \in \Pi^{HR}} \lim \inf_{N \to \infty} R_N^\pi(s).$$

Hence, as in Theorem 8.4.1, the existence of a solution to the optimality equation ensures the equivalence of lim inf average optimality, lim sup average optimality, and average optimality.

We now establish the existence of a solution to the optimality equation by transforming the model to a discrete-time model, and appealing to existence results from Chap. 8. The following transformation applies to both finite- and countable-state models. Denote all quantities in the transformed model with " $\sim$ ". Let $\tilde{S} = S$, $\tilde{A}_s = A_s$ for all $s \in S$,

$$\tilde{r}(s, a) \equiv r(s, a)/y(s, a) \tag{11.4.20}$$

and

$$\tilde{m}(j|s, a) \equiv \begin{cases} \eta m(j|s, a)/y(s, a) & j \neq s \\ 1 + \eta[m(s|s, a) - 1]/y(s, a) & j = s, \end{cases} \tag{11.4.21}$$

where η satisfies

$$0 < \eta < y(s, a)/(1 - m(s|s, a)) \tag{11.4.22}$$

for all $a \in A_s$ and $s \in S$ for which $m(s|s, a) < 1$. Further, let $\tilde{g}^\pi$ denote the average expected reward for policy π and $\tilde{g}^*$ denote the optimal gain for the transformed model.

The effect of this transformation is to convert rewards to a unit time basis, and then alter transition structure so that long-run average rewards of the discrete model and the semi-Markov decision process model agree. (See Sec. 11.5.1 for further elaboration on the interpretation of this transformation.) We relate the solutions of the optimality equations in these two models as follows.

Proposition 11.4.5. Let S be countable, and suppose that Assumption 11.1.1 holds.

a. Suppose $(\tilde{g}, \bar{h})$ satisfy the discrete-time optimality equations

$$h = \max_{d \in D^{MD}} \{\tilde{r}_d - ge + \tilde{M}_d h\}. \qquad (11.4.23)$$

Then $(\tilde{g}, \eta \bar{h})$ satisfy (11.4.17)
b. for each $d \in D^{MR}$, $\tilde{g}^{d^\infty} = g^{d^\infty}$, and
c. $\tilde{g}^* = g^*$.

Proof. From (11.4.23), it follows that, for all $d \in D^{MD}$,

$$\bar{h}(s) \geq \tilde{r}_d(s) - \tilde{g} + \sum_{j \in S} \tilde{m}_d(j|s)\bar{h}(j),$$

and that there exists a d^* for which equality holds. Hence, using the definitions of these quantities in (11.4.20) and (11.4.21),

$$\bar{h}(s) \geq \frac{r_d(s)}{y_d(s)} - \tilde{g} + \frac{\eta}{y_d(s)} \sum_{j \in S} m_d(j|s)\bar{h}(j) + \left[1 - \frac{\eta}{y_d(s)}\right]\bar{h}(s).$$

Multiplying through by $y_d(s)$ yields

$$\eta \bar{h}(s) \geq r_d(s) - \tilde{g} y_d(s) + \sum_{j \in S} m_d(j|s)\eta \bar{h}(j).$$

Since there exists a d^* for which equality holds in the above expression, $\tilde{g}$ and $\eta \bar{h}$ satisfy the optimality equation for the semi-Markov decision process.

Part (b) follows from part (a), Theorem 11.4.2, and Theorem 8.4.1 applied with $D^{MR} = \{d\}$; and part (c) follows from Theorems 11.4.4(c) and 8.4.1. $\qquad \square$

The following result summarizes properties of the optimality equations and existence of optimal policies for the *finite-state* semi-Markov decision process model under the average reward criterion. Theorems 8.4.3 and 8.4.7 guarantee the existence of solutions to the discrete-time optimality equations which, when combined with the above proposition, ensure the existence of solutions to the SMDP optimality equation. Parts (b) and (c) follow from Theorem 8.4.5. Part (d) follows from parts (a) to (c).

Theorem 11.4.6. Suppose that S is finite, Assumption 11.1.1 holds, M_d is unichain for each $d \in D^{MD}$, and the maximum in (11.4.17) is attained for each $g \in V$ and $h \in V$.

a. Then there exists a constant g and an $h \in V$ satisfying the optimality equation (11.4.17). Further, $g = g^*$.
b. There exists a stationary deterministic average optimal policy.

c. If g^* and h^* satisfy (11.4.17), and

$$d^* \in \arg\max_{d \in D} \{r_d - g^* y_d + M_d h^*\}, \tag{11.4.24}$$

then $(d^*)^\infty$ is average optimal.

d. There exists an optimal stationary deterministic average policy if

  i. A_s is finite for each $s \in S$, or

  ii. A_s is compact, $r(s, a)$ and $y(s, a)$ are continuous in a for each $s \in S$, and $m(j|s, a)$ is continuous in a for each $j \in S$ and $s \in S$.

We now discuss optimality with respect to g_C^π as defined in (11.4.2). Let $\tau_{(s)}$ denote the time until first return to state s, given that the system starts in state s. The following result from Ross (1970a, pp. 159–161; 1970b) relates the two criteria. We state it without proof.

Proposition 11.4.7. Suppose Assumption 11.1.1 holds, and $E_s^{d^\infty}\{\tau_{(s)}\} < \infty$. Then $g_C^{d^\infty} = g^{d^\infty}$.

The hypothesis is satisfied on recurrent sets of finite-state models, and consequently for unichain models under the following assumption.

Assumption 11.4.1. There exists a constant $K < \infty$ for which

$$\sup_{s \in S, \, a \in A_s} y(s, a) < K.$$

Corollary 11.4.8. Suppose that Assumption 11.4.1 and the hypotheses of Theorem 11.4.6 hold. Then, if d^* satisfies (11.4.24),

$$g^* = g_C^{(d^*)^\infty} = \sup_{\pi \in \Pi^{HR}} g_C^\pi.$$

11.4.4 Algorithms

We discuss the implementation of algorithms in finite-state and action models. The optimality equations can be solved directly by using the policy iteration algorithm of Sec. 8.6. The only modification is the inclusion of the quantity y_{d_n}, which multiplies g in (8.6.1) in the evaluation step, and the inclusion of the expression $-g_n y_d$ inside the "arg max" in the improvement step. Note that the finite convergence of this algorithm establishes Theorem 11.4.6 for finite-state and action models.

A linear programming formulation generalizes that in Sec. 8.8. The modifications required are that $y(s, a)$ multiplies g in the primal constraints, and $y(s, a)$ multiplies $x(s, a)$ inside the summation in constraint (8.8.4) in the dual.

To implement value iteration, we transform the model to an equivalent discrete-time model using the transformation in the preceding section and then apply the theory of Chap. 8. As in Sec. 8.5.5, the choice of η ensures that $\bar{m}(s|s, a) > 0$, so that all stationary policies have aperiodic chains and Theorem 8.5.4 establishes the convergence of value iteration. Consequently, we can use value iteration to solve the transformed model and find an optimal or ε-optimal policy for the SMDP.

11.4.5 Countable-State Models

In this section, we generalize results of Sec. 8.10 for countable-state Markov decision processes with unbounded rewards to semi-Markov decision processes. Proofs closely follows arguments in Sec. 8.10. We refer the reader to Sennott (1989b) and related papers for details. As in Sec. 8.10.2, we assume the following:

Assumption 11.4.2. There exists a constant R for which $-\infty < r(s, a) \leq R < \infty$ for all $s \in S$ and $a \in A_s$.

Assumption 11.4.3. For each $s \in S$, and $\alpha > 0$, $v_\alpha^*(s) > -\infty$.

Assumption 11.4.4. There exists a $K < \infty$ and an $\alpha_0 > 0$, such that, for each $s \in S$ and $\alpha_0 > \alpha > 0$,

$$h_\alpha(s) \equiv v_\alpha^*(s) - v_\alpha^*(0) \leq K.$$

Assumption 11.4.5. There exists a non-negative function $M(s)$ and an $\alpha_0 > 0$ such that

a. $M(s) < \infty$;
b. for each $s \in S$, $h_\alpha(s) \geq -M(s)$ for $\alpha_0 > \alpha > 0$; and
c. there exists an $a_0 \in A_0$ for which

$$\sum_{j \in S} m(j|0, a_0) M(j) < \infty.$$

Assumption 11.4.5'. There exists a non-negative function $M(s)$ such that

a. $M(s) < \infty$;
b. for each $s \in S$, $h_\alpha(s) \geq -M(s)$ for all $\alpha > 0$; and
c. for each $s \in S$ and $a \in A_s$,

$$\sum_{j \in S} m(j|s, a) M(j) < \infty.$$

The following theorem generalizes Theorem 8.10.7.

Theorem 11.4.8. Let $S = \{0, 1, 2, \ldots\}$, A_s be finite, and suppose that Assumptions 11.1.1 and 11.4.1–11.4.5 hold. Then the following are true.

a. There exists a constant g satisfying

$$g = \lim_{\alpha \downarrow 0} \alpha v_\alpha^*(s) \tag{11.4.25}$$

for all $s \in S$, a vector h satisfying $-M(s) \le h(s) \le K$ and $h(s) = 0$, and a $d^* \in D$, for which

$$h(s) \le r_{d^*}(s) - g y_{d^*}(s) + \sum_{j \in S} m_{d^*}(j|s) h(j)$$

$$\le \max_{a \in A_s} \left\{ r(s, a) - g y(s, a) + \sum_{j \in S} m(j|s, a) h(j) \right\}. \tag{11.4.26}$$

b. The policy $(d^*)^\infty$ is average optimal and, if

$$d' \in \arg\max_{a \in A_s} \left\{ r(s, a) - g y(s, a) + \sum_{j \in S} m(j|s, a) h(j) \right\}, \tag{11.4.27}$$

then $(d')^\infty$ is also average optimal.

c. If Assumptions 11.1.1, 11.4.1–11.4.4, and 11.4.5' hold, then (11.4.26) holds with equality.

When the conditions of this theorem are restricted to a model with a single action in each state, then it also establishes the existence of a solution to the evaluation equations (11.4.13) in countable-state models.

11.5 CONTINUOUS-TIME MARKOV DECISION PROCESSES

Continuous-time Markov decision processes (CTMDPs) may be regarded as a class of semi-Markov decision processes in which the time between decisions follows an exponential distribution. They generalize semi-Markov decision processes by allowing decisions to be made at any point of time. However, in infinite horizon models in which the reward depends only on the state and action chosen at a decision epoch and not on the duration of time in a state, it can be shown in the generality of the models considered here, that we may restrict attention to models in which decisions are only made at transition times. We adopt that viewpoint in this section.

We can analyze these models in three ways.

1. Discretize time and in view of the Markov property apply the discrete time methods of Chaps 1–10.
2. Use the semi-Markov decision process results of the preceding two sections directly.
3. Convert the model to an equivalent, more easily analyzed process through uniformization.

In this section we formulate the model, discuss the uniformization approach, and apply it to the admission control model. We assume finite or countable S.

11.5.1 Continuous-time Markov chains

Let S denote a countable set of states, and $X = \{X_t: t \geq 0\}$ a stochastic process with state space S. We say that $X = \{X_t: t \geq 0\}$ is a *continuous-time Markov chain* if, for any $u \geq 0$ and $j \in S$,

$$P\{X_{t+u} = j | X_v; t \geq v \geq 0\} = P\{X_{t+u} = j | X_t\}.$$

We say it is *time homogeneous* if, for all $u \geq 0$,

$$P\{X_{t+u} = j | X_u = s\} = P\{X_t = j | X_0 = s\} \equiv P^t(j|s).$$

Such a process evolves as follows. If, at some time t, $X_t = s$, the process remains in state s for a period of time determined by an exponential distribution with parameter $\beta(s)$, $0 \leq \beta(s) < \infty$, and then jumps to state j with probability $q(j|s)$. Let Q denote the matrix with components q.

We may summarize the probabilistic behavior of the process in terms of its *infinitesimal generator*. By an *infinitesimal generator* we mean a $|S| \times |S|$ matrix A with components

$$A(j|s) = \begin{cases} -[1 - q(s|s)]\beta(s) & j = s \\ q(j|s)\beta(s) & j \neq s. \end{cases}$$

Note that most natural formulations have $q(s|s) = 0$, which means that, at the end of a sojourn in state s, the system will jump to a different state. For what follows, it is important that we allow the system to occupy the same state before and after a jump.

The infinitesimal generator determines the probability distribution of the system state through the differential equations

$$\frac{d}{dt} P^t(j|s) = \sum_{k \in S} A(j|k) P^t(k|s)$$

or

$$\frac{d}{dt} P^t(j|s) = \sum_{k \in S} P^t(j|k) A(k|s),$$

which are often referred to as the Kolmogorov equations.

Consequently, processes with the same infinitesimal generator have identical finite-dimensional distributions, provided they have the same initial distribution. From our perspective, this suggests that if we modify the process to one that is easier to analyze and, at the same time, not alter its infinitesimal generator, then the modified process will have the same probabilistic structure.

When $\sup_{s \in S} \beta(s) < \infty$, we derive a modified process $\tilde{X}$ from X as follows. Choose a constant $c < \infty$ satisfying

$$\sup_{s \in S} [1 - q(s|s)]\beta(s) \leq c < \infty, \tag{11.5.1}$$

and define a process $\{\tilde{X}_t: t \geq 0\}$ with *state-independent* exponential sojourn times $\tilde{\beta}(j) = c$, and transition probabilities $\tilde{q}(j|s)$ given by

$$\tilde{q}(j|s) = \begin{cases} 1 - \dfrac{[1 - q(s|s)]\beta(s)}{c} & j = s \\ \dfrac{q(j|s)\beta(s)}{c} & j \neq s. \end{cases} \tag{11.5.2}$$

Since $\tilde{A}(j|s) = \tilde{q}(j|s)c$ for $j \neq s$, the infinitesimal generator $\tilde{A}$ of $\tilde{X}$ satisfies

$$\tilde{A} = A,$$

so that *the two processes are equal in distribution.* Note $\tilde{A} = c(I - Q)$.

We refer to $\tilde{X}$ as the *uniformization* of X because it has an identical (or uniform) sojourn time distribution in every state. The uniformization may be viewed as an equivalent process, in which the system state is observed at random times which are exponentially distributed with parameter c. Because of the Markov property, it begins anew at each observation point. As a result of the definition of c, we observe the system state more frequently in the uniformization than in the original system and, as a result, increase the probability that the system occupies the same state at different observation times. Alternatively, this transformation may be viewed as inducing extra or "fictitious" transitions from a state to itself.

We illustrate these points through the following simple example.

Example 11.5.1. Let $S = \{s_1, s_2\}$, and define a continuous-time Markov chain by $q(s_1|s_1) = 0$, $q(s_2|s_1) = 1$, $q(s_1|s_2) = 1$, $q(s_2|s_2) = 0$, $\beta(s_1) = 2$, and $\beta(s_2) = 0.8$ (Fig. 11.5.1). The sojourn time in s_1 follows an exponential distribution with parameter 2, denoted by $\exp(2)$, and that in s_2 follows an $\exp(0.8)$ distribution. Observe that there are no transitions from a state to itself in this model.

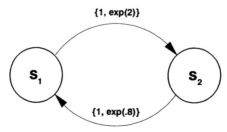

Figure 11.5.1 Symbolic representation of Example 11.5.1. The quantities in { }'s denote the transition probability and sojourn time distribution respectively.

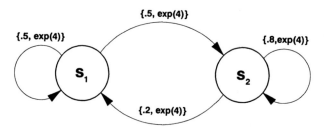

Figure 11.5.2. Symbolic representation of uniformization.

The infinitesimal generator for this process is given by

$$A = \begin{bmatrix} -2 & 2 \\ 0.8 & -0.8 \end{bmatrix}.$$

We now derive a uniformization $\tilde{X}$ for this process. Since $c = 4$ satisfies (11.5.1), it follows from (11.5.2) that

$$\tilde{Q} = \begin{bmatrix} 0.5 & 0.5 \\ 0.2 & 0.8 \end{bmatrix}.$$

We represent the uniformization in Fig. 11.5.2. We see by direct calculation that $\tilde{A} = A$, so that these two processes have the same probabilistic behavior.

In the uniformization, the mean time between system observations is 0.25 time units, so it is more likely that it has not changed state. This accounts for the added self-transitions in the uniformization.

11.5.2 The Discounted Model

In light of the comments at the start of Section 11.5.1, we may regard a continuous-time Markov decision process as a semi-Markov decision process in which decision epochs follow *each* state transition and the times between decision epochs are exponentially distributed. This means that the distribution of the time between decision epochs is given by

$$F(t|s, a) = 1 - e^{-\beta(s, a)t} \quad \text{for } t \geq 0.$$

Let $q(j|s, a)$ denote the probability that at the subsequent decision epoch the system occupies state j, if at the current decision epoch it occupies state s and the decision maker chooses action $a \in A_s$. Assume the same reward structure as in the previous sections. Since the system state does not change between decision epochs $r(s, a)$, the expected discounted reward between decision epochs satisfies

$$\begin{aligned}
r(s, a) &= k(s, a) + c(s, a) E_s^a \left\{ \int_0^{\tau_1} e^{-\alpha t} \, dt \right\} \\
&= k(s, a) + c(s, a) E_s^a \{ [1 - e^{-\alpha \tau_1}] / \alpha \} \\
&= k(s, a) + c(s, a) / [\alpha + \beta(s, a)].
\end{aligned} \tag{11.5.3}$$

For $d \in D^{MD}$, let $q_d(j|s) = q(j|s, d(s))$, $r_d(s) = r(s, d(s))$, $\beta_d(s) = \beta(s, d(s))$, and Q_d denote the matrix with components $q_d(j|s)$.

Let d^∞ denote a deterministic stationary policy. In a discounted continuous-time Markov decision process, (11.3.7) simplifies to

$$v_\alpha^{d^\infty}(s) = r_d(s) + \sum_{j \in S}\left[\int_0^\infty \beta_d(s)e^{-[\alpha + \beta_d(s)]t}\,dt\right]q_d(j|s)v_\alpha^{d^\infty}(j)$$

$$= r_d(s) + \frac{\beta_d(s)}{\beta_d(s) + \alpha}\sum_{j \in S}q_d(j|s)v_\alpha^{d^\infty}(j). \tag{11.5.4}$$

If $\beta(s, a) = c$ for all $s \in S$ and $a \in A_s$, $\beta_d(s) = c$ for all $s \in S$, so that (11.5.4) becomes

$$v_\alpha^{d^\infty}(s) = r_d(s) + \lambda \sum_{j \in S}q_d(j|s)v_\alpha^{d^\infty}(j), \tag{11.5.5}$$

with $\lambda \equiv c/(c + \alpha)$. Observe that this is identical to the *discrete-time* discounted evaluation equation (6.1.5), so it follows that the optimality equation has the form

$$v(s) = \max_{a \in A_s}\left\{r(s, a) + \lambda \sum_{j \in S}q(j|s, a)v(j)\right\}.$$

Hence all results of Chap. 6 apply directly to this model.

The uniformization of the preceding section may be applied to continuous-time Markov decision processes to obtain a model with constant transition rates, so that results and algorithms for discrete-time discounted models may be used directly. Since $r(s, a)$ depends explicitly on $\beta(s, a)$, we must modify it to ensure that the transformed system has the same expected total discounted rewards. With a continuous reward rate $c(s, a)$, we could handle this directly by redefining $r(s, a)$ in terms of the constant transition rate; however, because of the lump sum reward $k(s, a)$, it becomes more complicated.

We make the following assumption.

Assumption 11.5.1. There exists a constant $c < \infty$ for which

$$[1 - q(s|s, a)]\beta(s, a) \le c \tag{11.5.6}$$

for all $s \in S$ and $a \in A_s$.

Define a uniformization of the continuous-time Markov decision process with components denoted by " ~ ", as follows. Let $\tilde{S} = S$, $\tilde{A}_s = A_s$ for all $s \in \tilde{S}$, let

$$\tilde{r}(s, a) \equiv r(s, a)\frac{\alpha + \beta(s, a)}{\alpha + c} \tag{11.5.7}$$

and

$$\tilde{q}(j|s, a) = \begin{cases} 1 - \dfrac{[1 - q(s|s, a)]\beta(s, a)}{c}, & j = s \\[2ex] \dfrac{q(j|s, a)\beta(s, a)}{c}, & j \ne s. \end{cases} \tag{11.5.8}$$

Let $\bar{r}_d(s) \equiv \bar{r}(s, d(s))$, $\bar{q}_d(j|s) \equiv \bar{q}(j|s, d(s))$, and $\tilde{Q}_d$ denote the matrix with components $\tilde{q}_d(j|s)$. Further, let $\tilde{v}_\alpha^{d^\infty}$ denote the expected discounted reward for the transformed process. The following result relates the original and uniformized models.

Proposition 11.5.1. Suppose Assumption 11.5.1 holds, and that $\bar{r}$ and $\bar{q}$ satisfy (11.5.7) and (11.5.8). Then, for each $d \in D^{MR}$ and $s \in S$,

$$\tilde{v}_\alpha^{d^\infty}(s) = v_\alpha^{d^\infty}(s). \tag{11.5.9}$$

Proof. Fix $d \in D^{MR}$, let X denote the continuous-time Markov chain corresponding to d, and $\tilde{X}$ denote a uniformization. Because both processes have the same infinitesimal generator, all finite-dimensional distributions agree for each initial state distribution.

Let $\tilde{\nu}_s$ denote the number of transitions from state s to itself in the uniformization. Then, for $n \geq 0$,

$$P\{\tilde{\nu}_s = n | \tilde{X}_0 = s\} = \tilde{q}_d(s|s)^n [1 - \tilde{q}_d(s|s)].$$

In other words, $\tilde{\nu}_s$ follows a geometric distribution with parameter $\tilde{q}_d(s|s)$. Let $\tilde{\sigma}_n$ denote the transition times in the uniformization. The expected discounted reward received by $\tilde{X}$ conditional on starting in state s satisfies

$$E_s^d\left\{ \sum_{n=0}^{\tilde{\nu}_s} e^{-\alpha\tilde{\sigma}_n} \bar{r}_d(s) \right\} = \bar{r}_d(s) E_s^d\left\{ E\left(\sum_{n=0}^{\tilde{\nu}_s} e^{-\alpha\tilde{\sigma}_n} \Big| \tilde{\nu}_s \right) \right\} = \bar{r}_d(s) E_s^d\left\{ \sum_{n=0}^{\tilde{\nu}_s} \lambda^n \right\},$$

where $\lambda = c/(\alpha + c)$. Noting that the expectation in the last term is the generating function of a geometric random variable, it follows that

$$E_s^d\left\{ \sum_{n=0}^{\tilde{\nu}_s} \lambda^n \right\} = \frac{1}{1 - \lambda\tilde{q}_d(s|s)},$$

so that

$$E_s^d\left\{ \sum_{n=0}^{\tilde{\nu}_s} e^{-\alpha\tilde{\sigma}_n} \bar{r}_d(s) \right\} = \frac{\bar{r}_d(s)}{1 - \lambda\tilde{q}_d(s|s)}. \tag{11.5.10}$$

Similarly, for the untransformed process we have that

$$E_s^d\left\{ \sum_{n=0}^{\nu_s} e^{-\alpha\sigma_n} r_d(s) \right\} = \frac{r_d(s)}{1 - \lambda_s q_d(s|s)}, \tag{11.5.11}$$

where $\lambda_s = \beta_d(s)/(\beta_d(s) + \alpha)$.

Applying (11.5.7) and (11.5.8) establishes the equivalence of (11.5.10) and (11.5.11). Since the expected total discounted reward during sojourns in state s agree, and both processes have the same state occupancy distributions, (11.5.9) follows.   □

From (11.5.9) and Theorem 11.3.2, we obtain the following result on optimality properties of continuous-time Markov decision processes. Note that Assumption 11.5.1 implies Assumption 11.1.1.

Theorem 11.5.2. Suppose Assumption 11.5.1 holds, that, for all $s \in S$ and $a \in A_s$, $|r(s, a)| \leq M < \infty$, and $\lambda \equiv c/(\alpha + c)$. Then the following are true

a. Suppose

$$\tilde{L}v \equiv \max_{d \in D^{\mathrm{MD}}} \left\{ \tilde{r}_d + \frac{c}{\alpha + c} \tilde{Q}_d v \right\} \tag{11.5.12}$$

and the maximum in (11.5.12) exists for all $v \in V$. Then the optimality equation $v = \tilde{L}v$ has the unique solution $v_\alpha^* \in V$.

b. If the maximum in (11.5.12) is attained for each $v \in V$, there exists a stationary deterministic optimal policy.

c. If

$$d^* \in \arg\max_{d \in D} \left\{ \tilde{r}_d + \lambda \tilde{Q}_d v_\alpha^* \right\}, \tag{11.5.13}$$

then $(d^*)^\infty$ is optimal.

d. There exists an optimal stationary deterministic policy whenever
  i. A_s is finite for each $s \in S$; or
  ii. A_s is compact, $r(s, a)$ is continuous in a for each $s \in S$, $\beta(s, a)$ is continuous in a for each $s \in S$, and, for each $j \in S$ and $s \in S$, $q(j|s, a)$ is continuous in a.

e. For each $v^0 \in V$, $v^n = \tilde{L}^n v^0$ converges to v_α^*.

Note that all results of Secs. 6.10 and 6.11 also apply directly to this model. In particular, from Theorem 6.10.4 we obtain the following result for countable-state systems with unbounded rewards.

Theorem 11.5.3. Suppose $S = \{0, 1, \ldots\}$, Assumption 11.5.1 holds, and there exists a function $w(s)$ satisfying $\inf_{s \in S} w(s) > 0$ for which Assumptions 6.10.1 and 6.10.2 hold with $\lambda = c/(\alpha + c)$. Then the following are true.

a. If the maximum in (11.5.12) is attained for each $v \in V_w$, the optimality equation $v = \tilde{L}v$ has a unique solution in V_w which equals V_α^*.

b. If the maximum in (11.5.12) is attained for each $v \in V_w$, there exists a stationary deterministic optimal policy.

c. If d^* satisfies (11.5.13), then $(d^*)^\infty$ is optimal.

d. There exists an optimal stationary deterministic policy whenever
  i. A_s is finite for each $s \in S$; or
  ii. A_s is compact, $r(s, a)$ is continuous in a for each $s \in S$, $\beta(s, a)$ is continuous in a for each $s \in S$, and, for each $j \in S$ and $s \in S$, $q(j|s, a)$ is continuous in a.

e. For each $v^0 \in V_w$, $v^n = \tilde{L}^n v^0$ converges to v_α^*.

We summarize the analysis in this section through Fig. 11.5.3.

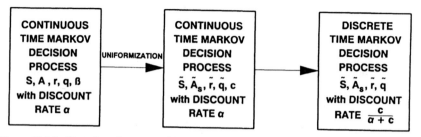

Figure 11.5.3 Transform from a continuous-time Markov decision process, to its uniformization to a discrete-time model.

Example 11.5.2. In this example, we illustrate the use of uniformization and transformation to a discrete time system by evaluating the expected total discounted reward in an augmented version of Example 11.5.1 which includes rewards. Since there is no action choice in this model, we drop the symbol a from the notation. As in Fig. 11.5.1, $\beta(s_1) = 2$, $\beta(s_2) = 0.8$, $q(s_2|s_1) = q(s_1|s_2) = 1$. We choose $c = 4$. We add the following rewards to the model: $k(s_1) = 3$, $k(s_2) = 5$, $c(s_1) = 2$ and $c(s_2) = 1$ and set $\alpha = 0.1$. Applying (11.5.3) yields

$$r(s_1) = 3 + 2/(0.1 + 2) = 3.952 \quad \text{and} \quad r(s_2) = 5 + 1/(0.1 + 0.8) = 6.111$$

and using transformation (11.5.7) yields

$$\bar{r}(s_1) = r(s_1)[(0.1 + 2)/(0.1 + 4)] = 2.024, \quad \text{and}$$
$$\bar{r}(s_2) = r(s_2)[(0.1 + 0.8)/(0.1 + 4)] = 1.3414.$$

As a result of Theorem 11.5.2, we can find v_α by solving

$$v_\alpha = \bar{r} + \frac{c}{c + \alpha}\bar{Q}v_\alpha$$

where $\bar{q}(s_1|s_1) = 0.5$, $\bar{q}(s_2|s_1) = 0.5$, $\bar{q}(s_1|s_2) = 0.2$ and $\bar{q}(s_2|s_2) = 0.8$. Solving this system of equations yields $v_\alpha(s_1) = 63.68$ and $v_\alpha(s_2) = 62.72$. We leave it as an exercise to verify that we obtain the same value for these quantities by applying the methods of Section 11.3.

11.5.3 The Average Reward Model

In this section we analyze a continuous-time Markov decision problem with average reward criterion under the assumption that the transition probability matrix for every stationary policy is unichain. The model formulation is identical to that in the previous section; however, the accumulated reward during a sojourn in state s after choosing action a is given by

$$r(s, a) = k(s, a) + c(s, a)E_s^a\{\tau_1\} = k(s, a) + c(s, a)/\beta(s, a). \quad (11.5.14)$$

To obtain the optimality equation, we formulate the average reward CTMDP as an average reward SMDP and then apply the uniformization transformation. We assume Q_d is unichain for all $d \in D$. These arguments generalize to the multichain case but will not be discussed here. In the notation of Section 11.4.4, $y(s, a) = 1/\beta(s, a)$ and $m(j|s, a) = q(j|s, a)$ so that the average reward SMDP optimality equation (11.4.17) may be written in component notation as

$$0 = \max_{a \in A_s} \{r(s, a) - g/\beta(s, a) + \sum_{j \in S} q(j|s, a)h(j) - h(s)\}.$$

Multiply the expression in { }'s by $\beta(s, a)/c$ where c satisfies (11.5.6) and apply similar calculations and arguments to those in the proof of Proposition 11.4.5 and rearrange terms to obtain the optimality equation

$$\tilde{h}(s) = \max_{a \in A_s} \{\tilde{r}(s, a) - \tilde{g} + \sum_{j \in S} \tilde{q}(j|s, a)\tilde{h}(j)\}$$

where $\tilde{g} = g/c$, $\tilde{h}(s) = h(s)$, $\tilde{q}(j|s, a)$ satisfies (11.5.8) and

$$\tilde{r}(s, a) = r(s, a)\beta(s, a)/c = k(s, a)\beta(s, a)/c + c(s, a)/c.$$

Since g denotes the optimal average reward per unit time and the expected time between transitions is $1/c$, we may interpret $\tilde{g}$ as the optimal average reward per transition. This optimality equation can be solved by the methods of Chapter 8.

Example 11.5.2. **(ctd.)** We now illustrate the above calculations in an average reward variant of Example 11.5.2. From (11.5.14) and (11.4.20) it follows that

$$\tilde{r}(s_1) = 3 \times 0.5 + 0.5 = 2 \quad \text{and} \quad \tilde{r}(s_2) = 5 \times 0.2 + 0.25 = 1.25.$$

As a consequence of Proposition 11.4.5(a), g can be found by solving

$$\tilde{h}(s_1) = 2 - \tilde{g} + 0.5\tilde{h}(s_1) + 0.5\tilde{h}(s_2)$$
$$\tilde{h}(s_2) = 1.25 - \tilde{g} + 0.2\tilde{h}(s_1) + 0.8\tilde{h}(s_2)$$

Solving the system of equations subject to $\tilde{h}(s_2) = 0$ yields $\tilde{h}(s_1) = 1.07$ and $\tilde{g} = 1.47$. Applying results from Sec. 11.4 directly yields the solution $g = 5.85$, $h(s_1) = 1.072$ and $h(s_2) = 0$. Consequently, $g = c\tilde{g}$ and $h = \tilde{h}$.

11.5.4 Queueing Admission Control

We now apply this theory to an $M/M/1$ version of the admission control model of Sec. 11.2.2, in which the interarrival times follow an $\exp(\gamma)$ distribution. To exploit the Markovian structure, we modify the state space to $S = \{0, 1, 2, \ldots\} \times \{0, 1\}$, and observe the system state at each arrival and departure. The system is in state $\langle s, 0 \rangle$ if there are s jobs in the system and no arrivals. We observe this state when a transition corresponds to a departure. In state $\langle s, 0 \rangle$, the only action is to continue, so we set $A_{\langle s, 0 \rangle} = \{0\}$. The state $\langle s, 1 \rangle$ occurs when there are s jobs in the system and a new job arrives. In state $\langle s, 1 \rangle$, the controller may admit or refuse service to the arrival, so that $A_{\langle s, 1 \rangle} = \{0, 1\}$ where action 0 corresponds to refusing service and 1 to admitting the arrival. Figure 11.5.4 illustrates the possible transitions, the transition probabilities, and the transition rates for this model.

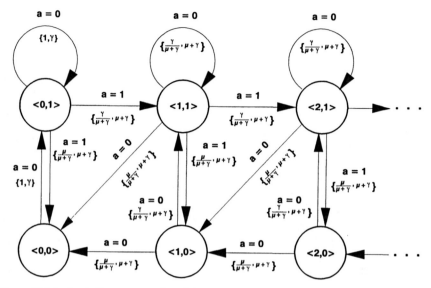

Figure 11.5.4 Symbolic representation of the state transition structure for the admission control model. The numbers in brackets represent the transition probability and the exponential transition rate.

Simple calculations reveal that

$$\beta(\langle s, b \rangle, a) = \begin{cases} \gamma & \text{if } s = 0, a = 0 \text{ and } b = 0, 1 \\ \gamma + \mu & \text{if } s = 0, a = 1 \text{ and } b = 1, \text{ or } s > 0. \end{cases}$$

To see this, observe that when the queue is empty, that is when $s = 0$, $b = 0$, and $a = 0$, or $s = 0$, $b = 1$, and $a = 0$, the next decision epoch occurs when the next job enters the system. This follows an $\exp(\gamma)$ distribution. In any other state, the next decision epoch occurs after either an arrival or service completion. If T_a denotes the time until the next arrival, and T_s denotes the time until the next service completion, then the time of the next decision epoch T_e satisfies $T_e = \min(T_a, T_s)$. Standard calculations show that T_e follows an $\exp(\gamma + \mu)$ distribution.

Transition probabilities satisfy

$$q(j|s, a) = \begin{cases} 1 & j = \langle 0, 1 \rangle, a = 0 \text{ and } s = \langle 0, 0 \rangle, \langle 0, 1 \rangle \\[2mm] \dfrac{\mu}{\mu + \gamma} & \begin{aligned} &j = \langle m - 1, 0 \rangle, s = \langle m, k \rangle, a = 0 \text{ for } m \geq 1 \text{ and } k = 0, 1, \text{ or} \\ &j = \langle m, 0 \rangle, s = \langle m, 1 \rangle, a = 1 \text{ for } m \geq 0 \end{aligned} \\[4mm] \dfrac{\gamma}{\mu + \gamma} & \begin{aligned} &j = \langle m + 1, 1 \rangle, s = \langle m, 1 \rangle, a = 1 \text{ for } m \geq 0, \text{ or} \\ &j = \langle m, 1 \rangle, s = \langle m, k \rangle, a = 0 \text{ for } m \geq 1 \text{ and } k = 0, 1, \end{aligned} \\[4mm] 0 & \text{otherwise.} \end{cases}$$

To see this, observe that, when the system is empty, the next decision epoch occurs when the system state becomes $\langle 0, 1 \rangle$. If at some point of time the system contains m jobs and no arrival is available for admission, or an arriving job has been refused service, then the state at the next decision epoch is either $\langle m, 1 \rangle$ or $\langle m - 1, 0 \rangle$, depending on whether there is an arrival or a service completion. If there are m jobs, and an arrival has been admitted, then the subsequent state is either $\langle m + 1, 1 \rangle$ or $\langle m, 0 \rangle$. The transition probabilities follow from noting that $P\{T_a \leq T_s\} = \gamma/(\mu + \gamma)$ and $P\{T_s \leq T_a\} = \mu/(\mu + \gamma)$.

With the exception of states $\langle 0, 0 \rangle$ and $\langle 0, 1 \rangle$, all transitions occur at rate $\mu + \gamma$. To uniformize the system, we choose $c = \mu + \gamma$ and alter the transition structure in these states. Applying (11.5.8) in $\langle 0, 0 \rangle$ yields the transformed probabilities:

$$\bar{q}(\langle 0, 1 \rangle | \langle 0, 0 \rangle, 0) = q(\langle 0, 1 \rangle | \langle 0, 0 \rangle, 0)\beta(\langle 0, 1 \rangle, 0)/c = \gamma/(\mu + \gamma),$$
$$\bar{q}(\langle 0, 0 \rangle | \langle 0, 0 \rangle, 0) = \mu/(\mu + \gamma).$$

Since, in the untransformed model, choosing $a = 0$ in state $\langle 0, 1 \rangle$ *instantaneously* moves the system to $\langle 0, 0 \rangle$,

$$\bar{q}(\langle 0, 0 \rangle | \langle 0, 1 \rangle, 0) = \mu/(\mu + \gamma),$$
$$\bar{q}(\langle 0, 1 \rangle | \langle 0, 1 \rangle, 0) = \gamma/(\mu + \gamma).$$

In the uniformized system, we observe the system more often when it is empty than in the untransformed system, so that this transformation increases the probability that it occupies $\langle 0, 0 \rangle$. We may also interpret this transformation as adding "fictitious" service completions in this state. In all other states,

$$\bar{q}(j|s, a) = q(j|s, a).$$

Figure 11.5.5 illustrates the transition structure for the uniformization of the admission control problem.

We now study the discounted model. Assuming $f(0) = 0$, and noting that $[\alpha + \beta(\langle s, b \rangle, a)]/[\alpha + c] = 1$ for $s \geq 1$ or $s = 0$, $a = b - 1$ it follows from (11.5.3) and (11.5.7) that the rewards in the uniformized system satisfy

$$\bar{r}(\langle 0, 0 \rangle, 0) = \bar{r}(\langle 0, 1 \rangle, 0) = 0,$$
$$\bar{r}(\langle s, 1 \rangle, 1) = R - f(s + 1)/(\mu + \gamma + \alpha) \quad s \geq 0,$$
$$\bar{r}(\langle s, 0 \rangle, 0) = \bar{r}(\langle s, 1 \rangle, 0) = -f(s)/(\mu + \gamma + \alpha) \quad s \geq 1.$$

Note that these agree with the rewards in the original system, since the uniformization only modifies the system when it is empty. Thus the optimality equations for this system become

$$v(\langle 0, 0 \rangle) = \frac{\mu + \gamma}{\mu + \gamma + \alpha}\left[\frac{\mu}{\mu + \gamma}v(\langle 0, 0 \rangle) + \frac{\gamma}{\mu + \gamma}v(\langle 0, 1 \rangle)\right]. \quad (11.5.15)$$

$$v(\langle s, 0 \rangle) = \frac{-f(s)}{\mu + \gamma + \alpha} + \frac{\mu + \gamma}{\mu + \gamma + \alpha}\left[\frac{\mu}{\mu + \gamma}v(\langle s - 1, 0 \rangle) + \frac{\gamma}{\mu + \gamma}v(\langle s, 1 \rangle)\right]$$

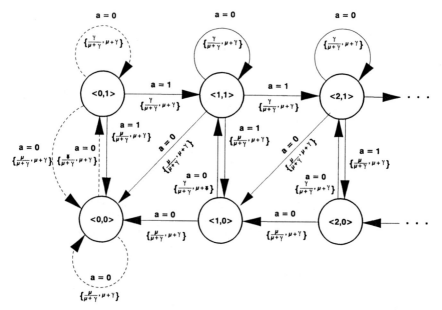

Figure 11.5.5 Uniformization of the admission control model with $c = \mu + \gamma$. Arcs denoted by - - - - have been altered from the original model.

for $s \geq 1$, and

$$
\begin{aligned}
v(\langle s,1\rangle) = \max\Bigg\{ & R - \frac{f(s+1)}{\mu + \gamma + \alpha} + \frac{\mu + \gamma}{\mu + \gamma + \alpha} \\
& \times \left[\frac{\mu}{\mu + \gamma} v(\langle s,0\rangle) + \frac{\gamma}{\mu + \gamma} v(\langle s+1,1\rangle) \right], - \frac{f(s)}{\mu + \gamma + \alpha} \\
& + \frac{\mu + \gamma}{\mu + \gamma + \alpha} \left[\frac{\mu}{\mu + \gamma} v(\langle s-1,0\rangle) + \frac{\gamma}{\mu + \gamma} v(\langle s,1\rangle) \right] \Bigg\}.
\end{aligned}
$$

$$(11.5.16)$$

When the hypotheses of Theorem 11.5.3 hold, it follows that this system of equations has a solution $v_\alpha^* \in V_w$ which can be found using value iteration.

Noting the expression for $v(\langle s,0\rangle)$, we see that the optimality equations simplify to (11.5.15) and

$$
v(\langle s,1\rangle) = \max\{R + v(\langle s+1\rangle,0), v(\langle s,0\rangle)\} \qquad (11.5.17)
$$

for $s \geq 0$. It follows from Theorem 11.5.3(c) that the stationary policy $(d^*)^\infty$ derived from the decision rule

$$d^*(s) = \begin{cases} 1 & v_\alpha^*(\langle s + 1, 0 \rangle) - v_\alpha^*(\langle s, 0 \rangle) > -R \\ 0 & v_\alpha^*(\langle s + 1, 0 \rangle) - v_\alpha^*(\langle s, 0 \rangle) \le -R \end{cases} \qquad (11.5.18)$$

is optimal. Hence if $v_\alpha^*(\langle s + 1, 0 \rangle) - v_\alpha^*(\langle s, 0 \rangle)$ is monotone nonincreasing, a control limit policy is optimal. We can establish this by induction as follows. Substitute $v^0 = 0$ into (11.5.16) to obtain

$$v^1(\langle s, 0 \rangle) = \frac{-f(s)}{\mu + \gamma + \alpha}.$$

From (11.5.18), a control limit policy is optimal if $f(s + 1) - f(s)$ is nondecreasing, which occurs if $f(s)$ is nondecreasing and convex. We leave it as an exercise to establish the optimality of a control limit policy under the assumption that $f(s)$ is convex nondecreasing, by using induction to show that $v^n(s)$ is convex for all n and then applying results from Sec. 6.11.

We now consider the admissions control model with expected average reward criterion. Following Sec. 11.5.3, we use $\tilde{q}$ as defined above and since the uniformization only affects states $(0, 0)$ and $(0, 1)$, it follows from (11.5.14) that $\tilde{r} = r$ and

$$\tilde{r}(\langle s, 1 \rangle, 1) = [R - f(s + 1)]/(\mu + \gamma), \quad s \ge 0$$

$$\tilde{r}(\langle s, 0 \rangle, 0) = \tilde{r}(\langle s, 1 \rangle, 0) = -f(s)]/(\mu + \gamma). \quad s \ge 1$$

The average reward optimality equations become

$$h(\langle 0,0 \rangle) + g = \frac{\mu}{\mu + \gamma} h(\langle 0,0 \rangle) + \frac{\gamma}{\mu + \gamma} h(\langle 0,1 \rangle),$$

$$h(\langle s,0 \rangle) + g = -f(s)(\mu + \gamma) + \frac{\mu}{\mu + \gamma} h(\langle s - 1,0 \rangle) + \frac{\gamma}{\mu + \gamma} h(\langle s,1 \rangle)$$

for $s \ge 1$, and

$$h(\langle s, 1 \rangle) + g = \max \left\{ [R - f(s + 1)]/(\gamma + \mu) + \frac{\mu}{\mu + \gamma} h(\langle s, 0 \rangle) \right.$$

$$+ \frac{\mu}{\mu + \gamma} h(\langle s + 1,1 \rangle), -f(s)/(\gamma + \mu)$$

$$\left. + \frac{\mu}{\mu + \gamma} h(\langle s - 1,0 \rangle) + \frac{\gamma}{\mu + \gamma} h(\langle s,1 \rangle) \right\}.$$

This system of equations has the same form as the optimality equations in a discrete-time model, so that the theory of Sec. 8.10 applies directly.

Observe that the Markovian formulation results in a model with a simpler reward and transition structure than that of the semi-Markov decision process formulation in Sec. 11.2.2. This has both theoretical and computational advantages. Since the transition matrices will be quite sparse, special algorithms can be developed to efficiently evaluate stationary policies or carry out value iteration. Also, in contrast to (11.3.16), the expected rewards between transitions can be easily evaluated. From a theoretical perspective, this simplified formulation may also simplify induction arguments necessary to establish the optimality of structured policies. For these reasons *one should apply uniformization when analyzing continuous-time Markov decision processes.*

BIBLIOGRAPHIC REMARKS

Semi-Markov decision processes, or Markov renewal programs, were introduced in Jewell (1963), Howard (1963), and deCani (1964). These models have been widely studied and applied, especially in queueing control. Fox (1966 and 1968), Denardo and Fox (1968), Ross (1970b), and Schweitzer (1971) were early and significant contributors to this theory. Ross (1970b) introduced Assumption 11.1.1, which appears to be fundamental. Lippman (1973 and 1975a) developed the weighted supremum norm approach for analyzing countable-state SMDP's with unbounded rewards. Note that we discussed this in detail in Sec. 6.10 in the context of discrete-time models. Porteus (1980a and 1983) investigated the efficiency of numerical methods for solving the evaluation equations in finite-state discounted models, paying particular attention to the effect of unequal row sums in the matrix M_d. The books by Ross (1970a), Heyman and Sobel (1984), and Tijms (1986) also discuss these models. The latter reference contains many interesting applications and insights. Çinlar (1975) provides an excellent introduction to the theory of uncontrolled Markov and semi-Markov processes. We base Sec. 11.5.1 on this reference.

As in the case of discrete-time models, the average reward model is more difficult to analyze. Numerous papers, most notably in the infinite-horizon average reward case, have been based on applying Schweitzer's data transformation (1971a) to convert SMDP's to MDP's. Federgruen and Tijms (1978) use this result to establish the existence of a solution to the countable-state optimality equations under bounded rewards and recurrence conditions, and to establish convergence of policy iteration and value iteration for this model. Federgruen, Hordijk, and Tijms (1979) analyze this model with unbounded rewards and a weaker recurrence condition. Federgruen, Schweitzer and Tijms (1983) further extend these results. Sennott (1989b) provides a more general set of conditions for this model which relate to our discussion in Sec. 8.10. Other contributors to the average reward theory include Sladky (1977), Lippman (1973 and 1975a), Yushkevich (1981), and Schal (1992). The last reference considers models with unbounded rewards and arbitrary chain structure.

Denardo (1971) investigated sensitive optimality criteria for finite-state SMDP's. Lamond (1986), rederives some of these results using generalized inverse methods. Dekker and Hordijk (1991) extend these analyses to countable-state models.

Chitgopekar (1969), Stone (1973), and Cantaluppi (1984a and b) study a generalization of a discounted semi-Markov decision process model in which reward and transition rates depend on the duration of time in a state, and decisions may be made at any time. Chitgopekar (1969) considers a model in which actions may change at a single point of time, while Stone (1973) allows the decision rules to be piecewise constant in the holding times. Cantaluppi (1984a) establishes the optimality of a policy which is stationary with respect to the state and holding times and shows that a stationary policy which is piecewise constant in the holding time is optimal when transition and reward rates are piecewise analytic in the holding time. Cantaluppi (1984b) provides a Gauss-Seidel algorithm for finding optimal policies in these models.

Howard (1960) introduced the continuous-time Markov decision process and provided policy iteration algorithms for solving finite state discounted and average reward models. Existence of optimal stationary policies within the class of measurable (with respect to the time index) Markovian policies were considered in various degrees of generality by Miller (1968a and b), Veinott (1969b), Kakumanu (1971 and 1972), Pliska (1975), and Doshi (1976). Rykov (1966) for finite state and action models, and Yushkevich and Feinberg (1979) for general models, establish the optimality of Markovian policies within the class of history-dependent policies. Lembersky (1974) studied the asymptotic behavior (with respect to the time horizon) of the optimal return and optimal policies for finite horizon, finite state, and action continuous-time Markov decision processes.

The use of uniformization for analyzing Markov processes dates back at least to Jensen (1953). Howard (1960), Veinott (1969b), and Schweitzer (1971a) implicitly use this approach to analyze continuous-time Markov decision processes. In a lengthy paper, Lippman (1975b) applies uniformization to characterize optimal policies in several exponential queueing control systems. Serfozo (1979) formalizes the use of this approach in the context of countable-state continuous-time models. We base Proposition 11.5.1 on this paper. Bertsekas (1987) and Walrand (1988) contain many interesting applications of the use of uniformization in queueing control models. Hordijk and van der Duyn Schouten (1983, 1984, and 1985) study a generalization of the above model which they refer to as a Markov decision drift process.

PROBLEMS

11.1. Compute $Q(j, t|s, a)$ for the model in Sec. 11.2.3 under the assumptions that $M < \infty$ and $M = \infty$.

11.2. Consider the following continuous-time machine operation and repair problem. A machine can be in one of two states; 0 denotes it is working, 1 that it is broken down. Two operating rates for the machine can be selected, either fast or slow. When the machine operates at the fast rate it yields greater revenue, but breaks down more frequently than when it operates at the slow rate. Two repair rates are available when the machine is broken down; one slow but cheap, and another fast but expensive.

Assume that failure time and repair time distributions are exponential and independent with failure rates and repair rates as follows.

State	Action	Exponential rate	Reward rate
0	slow	3 (breakdown)	5
0	fast	5 (breakdown)	8
1	slow	2 (repair)	−4
1	fast	7 (repair)	−12

    a. Formulate this as a semi-Markov decision process.

    b. Using the approach of Sec. 11.3, find a policy that maximizes the expected infinite-horizon discounted reward when the continuous-time discount rate is $\alpha = 0.1$.

    c. Using the approach of Sec. 11.4, find a policy that maximizes the long-run expected average reward.

    d. Give a uniformization of the model and solve it using the approach of Secs. 11.5.2 and 11.5.3.

11.3. Suppose that Assumption 11.1.1 holds.

    a. Show that, for all $s \in S$ and $a \in A_s$,

$$\lambda(s, a) \leq 1 - \varepsilon + \varepsilon e^{-\alpha\delta} < 1.$$

    b. Show that, for all $s \in S$ and $a \in A_s$,

$$E_s^a\{\tau_1\} = y(s, a) > \delta\varepsilon > 0,$$

    where τ_1 denotes the time until the first transition.

    c. Show that, for any finite t,

$$E_s^a\{\nu_t\} < \infty.$$

11.4. (Tijms, 1986, pp. 209–210). Let S be finite, and assume that the embedded chain corresponding to every decision rule d has a unichain transition probability matrix. Transform the semi-Markov decision process model according to (11.4.20)–(11.4.22), let $\tilde{\pi}_d$ denote the stationary distribution of the transformed model, and π_d denote the stationary distribution of the untransformed model.

    a. Show that the stationary distribution of the transformed model satisfies

$$\tilde{\pi}_d(s) = \frac{y_d(s)\pi_d(s)}{\sum\limits_{j \in S} y_d(j)\pi_d(j)}$$

    b. Use the result in (a) to show that $\tilde{g}^{d^\infty} = g^{d^\infty}$.

11.5. Establish the following result, which we used in the proof of Proposition 11.5.1. Suppose that ν has a geometric distribution with parameter q and that $0 < \lambda < 1$, then

$$E\left\{\sum_{k=0}^{\nu} \lambda^k\right\} = (1 - \lambda q)^{-1}.$$

11.6. Formulate the discounted semi-Markov decision process as a linear program, and solve the discounted version of Problem 11.2 using this approach.

11.7. (Lippman and Ross, 1971) The time between job offers received by a contractor follows an exponential distribution with rate γ. With each job is associated a reward r, and a random completion time T_r which has distribution function F_r. Jobs with reward r arrive with probability p_r. Assume that values for r are chosen from a finite set of alternatives, so that $\sum_{r=1}^{N} p_r = 1$. Assume further that the contractor can work on only one job at a time, that he cannot accept a new job until he completes his current job, and that he knows the reward and the distribution of the time to complete the job prior to deciding whether or not to accept it.

  a. Formulate this as a semi-Markov decision process.
  b. Give the optimality equation for this discounted model, show that an optimal policy exists, and determine its structure.
  c. Give the optimality equation for the model with average reward criterion, show that an optimal policy exists, and determine its structure.

11.8. Let X be a continuous-time Markov process, and $\bar{X}$ a uniformization. Show that the two processes have the same infinitesimal generator.

11.9. Consider the service rate control model of Sec. 11.2.3.

  a. Give the optimality equations for the discounted model, establish that an optimal policy exists, and determine its structure.
  b. Give the optimality equations for the average reward model, establish that an optimal policy exists, and determine its structure.

11.10. Consider a finite-state version of the service rate control model in which $S = \{0, 1, \ldots, 20\}$, $\gamma = 0.5$, $B = \{0, 1, 2\}$, G_b is deterministic with rates of 0.5, 1, and 2 jobs per unit time, $K = 3$, $f(s) = s$, and $d(b) = b$.

  a. Find an optimal policy for the discounted version of this model when $\alpha = 0.1$.
  b. Find an optimal policy for the average reward version of this model.
  c. Compare the efficiency of various algorithms for solving this problem.

11.11. Consider a modification of the model in the previous problem in which G_b follows an exponential distribution with rates as above.

  a. Use the uniformization transformation to obtain equivalent discounted and average reward models as in Sec. 11.5.
  b. Find optimal policies for each of these models.
  c. Compare the work needed to formulate and solve the uniformization to that needed to solve the problem directly using the semi-Markov decision process formulation.
  d. Solve a discrete-time version of the model under the assumption that decision epochs occur every 0.05 units of time. Compare the effort needed to solve the problem in this way to the uniformization and semi-Markov decision process approach.

11.12. (Open Problem) Derive an analog of Proposition 6.6.1, in which P_d is replaced by a matrix M_d which has unequal row sums as in Sec. 11.3.

11.13. Prove Theorem 11.1.1 under the assumption that S is countable.

11.14. Complete the inductive argument in Sec. 11.5.4 to establish that a control limit policy is optimal when f is convex nondecreasing in the discounted admission control model.

11.15. **a.** Formulate the admission control model under the assumption that a rejected job returns with fixed probability p, and that the return time follows an $\exp(\eta)$ distribution.

b. Find a uniformization of this model when the interarrival times follow an $\exp(\gamma)$ distribution, and compare it to that in Sec. 11.5.4.

11.16. **a.** Formulate a generalization of the admission control model of Sec. 11.2.2 in which there are c identical servers.

b. Show that there is an optimal control limit policy for a discounted version of this model in which arrival rates are exponential. Investigate its sensitivity to the number of servers.

c. Repeat (b) for a model with general arrival rates.

11.17. Consider an average reward finite-state semi-Markov decision process with no assumptions about the chain structure of the embedded processes.

a. Verify that, if g and h satisfy the evaluation equations

$$g = M_d g,$$

$$h = r_d - g y_d + M_d h,$$

then $g = g^{d^\infty}$.

b. Provide optimality equations for this model, and prove that they characterize the optimal average reward.

11.18. (Job routing) Upon arrival of a job, a controller routes it to one of two parallel queueing systems. Assume that interarrival times follow an $\exp(\gamma)$ distribution, and that the service times are $\exp(\mu_i)$ at queue i, $i = 1, 2$. Further assume that holding cost rates are linear with rate c_i at queue i.

a. Formulate this as a continuous-time Markov decision process.

b. Give the optimality equation for the discounted version of this problem, and show that it has a solution.

c. Determine the form of an optimal policy for the discounted problem, and investigate its sensitivity to c_1, c_2, μ_1, and μ_2.

d. Give the average reward optimality equation, and provide conditions under which it has a solution.

e. Repeat (c) for the average reward model.

11.19. Show that Assumption 11.5.1 implies Assumption 11.1.1.

11.20. Show that the expressions in (11.5.10) and (11.5.11) are equal.

11.21. In this problem we provide an alternative proof to Proposition 11.5.1 for finite S and bounded $r(s, a)$.

    a. Suppose

$$\bar{v}_\alpha^{d^\infty}(s) = \bar{r}_d(s) + \frac{c}{c + \alpha} \sum_{j \in S} \bar{q}_d(j|s) \bar{v}_\alpha^{d^\infty}(j)$$

    where $\bar{r}_d$ and $\bar{q}_d$ are defined through (11.5.7) and (11.5.8). Show that $\bar{v}_\alpha^{d^\infty}$ satisfies (11.5.4) by using the equivalences in (11.5.7) and (11.5.8).

    b. Show under Assumption 11.5.1 that (11.5.4) has a unique solution from which the result follows.

    c. Extend the above argument to S countable.

11.22. Consider a continuous-time Markov decision process with no lump sum rewards. Let $c(s, a)$ denote the continuous reward rate as in (11.5.3) and $A(j|s, a)$ denote the (s, j)th component of the infinitesimal generator when action a is chosen in state s. Let d denote a deterministic decision rule and $A_d(j|s) \equiv A(j|s, d(s))$. Let $\alpha > 0$ denote the continuous discount rate.

    a. (Howard, 1960) Show that $v_\alpha^{d^\infty}$ satisfies

$$\alpha v = r_d + A_d v$$

    b. (Howard, 1960) Show by setting $B_d = A_d + I$ that $v_\alpha^{d^\infty}$ satisfies

$$v = \bar{r}_d + \lambda B_d v$$

    where $\lambda = 1/(1 + \alpha)$ and $\bar{r}_d = \lambda r_d$ but that B_d need not be a probability matrix.

    c. Let $c \geq \max_{s \in S} \max_{a \in A_s} A(s|s, a)$ and $C_d = c^{-1} A_d + I$. Show that C_d is a probability matrix and that $v_\alpha^{d^\infty}$ satisfies

$$v = \bar{r}_d + \lambda C_d v$$

    where $\lambda = c/(c + \alpha)$ and $\bar{r}_d = (\alpha + c)^{-1} r_d$.

    d. Calculate $v_\alpha^{d^\infty}$ for the data in Example 11.5.2 under the modification $k(s_1) = k(s_2) = 0$ using the approaches in (a), (b) and (c).

11.23. Compute v_α in Example 11.5.2 using

    a. the approach suggested by (11.3.11) and

    b. the approach suggested by (11.5.4).

Afterword

Most of the models we have analyzed are infinite, but to achieve my objective of completing this book, I must make it finite. Because of this, it does not include material on several important topics. The most glaring omission is a discussion of *partially observed Markov decision processes* (POMDP's). These models differ from those discussed in the book in that the decision maker does not know the system state with certainty prior to making a decision. Applications of these models include equipment maintenance and replacement, cost control in accounting, quality control, fisheries management, design of teaching systems, medical decision making, sequential hypothesis testing in statistics, search for a hidden object, and computer file allocation.

In addition to states, actions, rewards, and transition probabilities, the POMDP model contains an observation or signal space Y, and a probability function $q(y|s, a, j)$ which represents the probability that the decision maker receives a signal $y \in Y$ at time $t + 1$, given that the unobserved system occupies state $s \in S$ at decision epoch t, the decision maker chooses action $a \in A$, and the unobserved system occupies state $j \in S$ at decision epoch $t + 1$. For example, the system state may represent the true defective rate of a production process, and the observation state may give the number of defects in a sample of n items from the output of the process, or the system state may measure the true health status of a patient, while the observation state contains the results of several diagnostic tests.

This model has been analyzed by transforming it to an equivalent continuous-state Markov decision process in which the system state is a probability distribution on the unobserved states in the POMDP, and the transition probabilities are derived through Bayes' rule (cf. Sec. 3.6.2). When the state space in the original POMDP contains two elements $\{s_1, s_2\}$, the state space in the transformed MDP is the unit interval, and the state equals the probability that the unobserved system occupies state s_1. In a model with N states, the state space is the N-dimensional unit simplex. Because of the continuity of the state space, algorithms are complicated and limited. Lovejoy (1991a) surveys computational aspects and notes

> "The significant applied potential for such processes remains largely unrealized, due to an historical lack of tractable solution methodologies."

Key references include papers by Astrom (1965), Aoki (1965), Dynkin (1965), Streibel (1965), Smallwood and Sondik (1973), and Sondik (1978), surveys by Monahan (1982) and Lovejoy (1991a), and books by Dynkin and Yushkevich (1979), Bertsekas (1987), Gheorghe (1990), and White (1993). Lane (1989) presents an interesting application of this model to decision making in a fishery and Hu, Lovejoy, and Shafer (1994) apply it to determine optimal dosing levels during drug therapy.

Other important topics not treated in this book and some references include combined parameter estimation and control models (Mandl, 1974; Kumar, 1985; and Hernandez-Lerma, 1989), discretization and aggregation methods (Bertsekas, 1975; Whitt, 1978, 1979a, and 1979b; Hinderer, 1978; and Mendelsohn, 1982), multicriteria models (Furukawa, 1980; White and Kim, 1980; Henig 1983; and Ghosh, 1990), sequential stochastic games (Shapley, 1953; and Heyman and Sobel, 1984), complexity issues (Papdimitriou and Tsitsiklis, 1987; and Tseng, 1990), and controlled diffusion processes (Mandl, 1968; Fleming and Rishel, 1975 and Kushner and Dupuis, 1992). Also, we have not provided an in-depth analysis of models with continuous or abstract state and action spaces (Bertsekas and Shreve, 1978; Dynkin and Yushkevich, 1979; and Hernandez-Lerma, 1989).

Significant areas of applications and references include bandit models (Gittins and Jones, 1974; Gittins, 1979; Whittle, 1980c; Ross, 1983; and Berry and Fristedt, 1985), linear quadratic control (Bertsekas, 1987), stochastic scheduling (Weiss, 1982; and Pinedo and Schrage, 1982), queueing control (Stidham, 1985; and Walrand, 1988), system reliability (Taylor, 1965; Derman, 1970; and Rust, 1987), and inventory control (Arrow, Karlin, and Scarf, 1958; Veinott, 1966c; and Porteus, 1990).

Notation

a	An action
A_s	Set of available actions in state s
A	$\bigcup_{s \in S} A_s$
$B(C)$	σ-algebra of Borel subsets of a set C
B	Operator $L - I$
$B_\lambda^\delta y$	The quantity $r_\delta + (\lambda P_\delta - I)y$ where y denotes an arbitrary vector in V
$B_\delta^n y$	Terms in the Laurent series expansion of $B_\lambda^\delta y$
c	A uniformization constant
$c(s, a)$	Cost in a constrained MDP (Chap. 8); expected total reward until the next decision epoch when action a is chosen in state s in an SMDP (Chap. 11)
$c(j, s, a)$	Reward rate when action a is chosen in state s and the natural process occupies state j in an SMDP
c^π	Expected long run average cost when using policy π in a constrained MDP
C_j	Closed ergodic class of a Markov chain
d	Decision rule
$d(s)$	Action chosen by decision rule d in state s
d_t	Decision rule at decision epoch t
d_v	A v-improving decision rule
d^∞	Stationary policy which uses decision rule d every period
D	Set of deterministic Markovian decision rules
$D_k(\cdot)$	Set of decision rules which obtain arg max in kth sensitive optimality equation
D^K	Set of decision rules where $K = MR, MD, HR, HD$
D_t^K	Set of decision rules for use at decision epoch t where $K = MR, MD, HR, HD$
D_v	Set of v-improving decision rules
D^σ	Set of structured decision rules
D_n^*	Set of v^n-improving decision rules (Chap. 6); set of n-discount optimal decision rules (Chap. 10)

e	Vector in which all components equal 1
$E_s^\pi\{\cdot\}$	Expected value with respect to policy π conditional on $X_1 = s$
$F(t\|s, a)$	Probability that the next decision epoch occurs within t time units given that action a is chosen in state s at the current decision epoch in an SMDP
$g^\pi(s)$	Gain or expected average reward of policy π conditional on $X_1 = s$
g_c^π	lim inf of average expected reward of policy π in an SMDP
g_+^π	lim sup of the average total expected reward under policy α
g_+^π	lim inf of the average total expected reward under policy π
g^*	Optimal average expected total reward or gain
g_+^*	Supremum of g_+^π over all policies
g_-^*	Supremum of g_-^π over all policies
h^π	Bias of policy π
H_P	Deviation matrix of the transition probability matrix P
H_d	Deviation matrix of P_d
H_t	Set of all histories up to time t
h_t	A history of system states and actions; $h_t = (s_1, a_1 \ldots, s_{t-1}, a_{t-1}, s_t)$
$h_x(s)$	Difference between $v_x^*(s)$ and $v_\lambda^*(0)$
I	Identity matrix or operator
I_k	$k \times k$ identity matrix
$I_{\{A\}}$	Indicator of set A
$k(s, a)$	Lump sum reward when action a is chosen in state s in an SMDP
L	Maximum return operator; $Lv = \max_{d \in D}\{r_d + \lambda P_d v\}$; $0 \le \lambda \le 1$
L_d	One period return operator corresponding to decision rule d
$m(j\|s, a)$	Discounted transition function in an SMDP
M_d	Matrix with components $m(j\|s, d(s))$ in an SMDP
$o(x)$	Vector or scalar function which converges to zero as x converges to zero
$O(\cdot)$	Order of a sequence (Chap. 6); ordering cost in an inventory model
$p(j\|s, a)$	Stationary version of $p_t(j\|s, a)$
$p(j\|t, s, a)$	Transition probability for natural process in an SMDP
$p_d(j\|s)$	$p(j\|s, d(s))$ for deterministic Markov decision rule d; a more general definition applied for randomized d
$p_t(j\|s, a)$	Probability system occupies state j at decision epoch $t + 1$ when action a is chosen in state s at decision epoch t
$p_d^*(j\|s)$	Component of the limiting matrix P_d^*
$P(j\|s, a)$	Transition probability of embedded MDP in an SMDP
P_d	Transition probability matrix with components $p_d(j\|s)$
P_π^J	J-step transition probability matrix under policy π
$P^\pi\{\cdot\}$	Probability distribution under policy π
$P_\Re^\pi\{\cdot\}$	Probability distribution on the set of rewards
$P_1\{\cdot\}$	Distribution of system state at the first decision epoch

P^*	Limiting matrix
P_d^*	Limiting matrix of P_d
$q(j\|s, a)$	Transition probability for the embedded chain in a CTMDP
$q_{d(s)}(\cdot)$	Probability distribution on the set of actions A_s when using randomized Markovian decision rule d in state s
Q_d	Component of a splitting of $I - \lambda P_d$ (Chap. 6); transition matrix in a CTMDP (Chap. 11)
$Q(t, j\|s, a)$	Joint probability that the state at the next decision epoch equals j and that the next decision epoch occurs at or before time t when action a is chosen in state s at the present decision epoch in an SMDP
$r(s, a)$	Stationary version of $r_t(s, a)$
$r(s, a, j)$	Stationary version of $r_t(s, a, j)$
$r_t(s, a)$	(Expected) present value of one period reward if system is in state s at decision epoch t and action a is chosen
$r_t(s, a, j)$	Present value of one period reward if system is in state s at decision epoch t, action a is chosen, and the state at decision epoch $t + 1$ is j
$r_N(s)$	Terminal reward in state s in a finite horizon problem
$r_d(s)$	$r(s, d(s))$ for Markov deterministic d; a more general definition applies for Markov randomized d
$r^+(s, a)$	$\max\{r(s, a), 0\}$
$r^-(s, a)$	$\max\{-r(s, a), 0\}$
R_+^1	The set $[0, \infty)$
R^n	n-dimensional Euclidean space
R_d	Component of a splitting of $I - \lambda P_d$ (Chap. 6); recurrent set of P_d (Chap. 7–10)
R^ρ	Resolvent of $P - I$
R_δ^ρ	Resolvent of $Pg - I$
s	A state
s_t	State of the system at time t or a generic state such as s_3
$s(Q)$	Spectrum of the operator Q
S	Set of states
t	A decision epoch
T	Set of decision epochs
T_d	Transient states under P_d
$u_t^\pi(s_t)$	Expected total reward under policy π from decision epoch t onward, if the system is in state s_t at decision epoch t
$u_t^*(s_t)$	Supremum over all policies of the expected total reward from decision epoch t onward, if system is in state s_t at decision epoch t
v	An element of a normed linear space of functions on S; if S is discrete it is a column vector
$v(s)$	Value of v at $s \in S$
v^*	Supremum over all policies of the expected total reward in an infinite horizon MDP

v^π	Expected total reward in an infinite horizon model under policy π
v_+^π	Expected total reward in an infinite horizon model when r^+ replaces r
v_-^π	Expected total reward in an infinite horizon model when r^- replaces r
v_α^*	Supremum of the expected total discounted reward in an infinite horizon SMDP
v_α^π	Expected total discounted reward under policy π in an infinite horizon SMDP
v_λ^*	Supremum of the expected total discounted reward in an infinite horizon MDP
v_λ^π	Expected total discounted reward under policy π in an infinite horizon MDP
v_N^π	Expected total reward under policy π in a finite horizon MDP
v_N^*	Supremum over all policies of expected total reward in a finite horizon MDP
$v_*^{N,u}$	Optimal expected total discounted reward in a countable state model truncated in state N
V	Space of bounded functions on S
V_B	Space of $v \in V$ for which $Bv \geq 0$
V_L	Space of $y_\lambda \in V$ that have Laurent series expansions for ρ near 1
V_M	Space of bounded measurable functions of S
V_w	Space of functions on S that have finite weighted supremum norm with respect to the bounding function w
V^σ	Set of structured value functions
V^+	Set of non-negative elements of V
V^-	Set of non-positive functions on S
w	A positive bounding function
W_t	Random variable representing the state of the natural process in a semi-Markov decision process model
$x(s,a)$	Variable in dual linear program
$x_{\pi,\alpha}^T(s,a)$	Average state action frequency over first T decision epochs under policy π and initial state distribution α
X_t	Random variable representing the state of system at time t
y_n^δ	nth term in Laurent series expansion of $v_\lambda^{\delta^\infty}$
$y(s,a)$	Variable in dual multichain linear program (Chap. 8); expected time until the next decision epoch in an SMDP (Chap. 11)
Y_t	Random variable representing action chosen at time t
Y^π	Matrix of coefficients of Laurent series expansion of v_λ^π
Y_n^π	Submatrix of Y^π consisting of columns -1 to n
Z_t	Random vector representing the history of a process up to time t
$\mathscr{L}$	Supremal return operator
$\mathscr{P}(A_s)$	Set of probability distributions on A_s
α	Continuous time interest rate
$\beta(s,a)$	Transition rate when action a is chosen in state s in a CTMDP

γ	Delta coefficient of composite transition matrix
γ_d	Delta coefficient of P_d
Δ	System state when a process is stopped
ε	A small quantity
λ	Discount rate in a discrete time model; $0 \leq \lambda < 1$
$\lambda(s, a)$	Effective discount rate when action a is chosen in state s in an SMDP
$\Lambda(v)$	Minimum component of a vector v
ν_t	Number of decisions up to time t in an SMDP
π	Policy $(d_1, d_2, \ldots, d_{N-1})$; $N \leq \infty$
π^*	Optimal policy
Π^K	A class of policies where $K = MR, MD, HR, HD, SR, SD$
Π^1	Set of policies for which limiting state action frequencies exist
Π^σ	Set of structured policies
$\sigma(C)$	Spectral radius of a matrix or operator C
$\sigma_s(C)$	Subdominant eigenvalue of a matrix or operator C
σ_n	Time of nth decision epoch
σ_π^2	Steady state variance of rewards under policy π
τ_G	First passage time to set G
τ_n	Time between decision epoch n and $n + 1$ in an SMDP
$\Psi(\cdot)$	Utility
ρ	Interest rate; $\rho = \lambda^{-1}(1 - \lambda)$ where λ denotes the discount rate
$\Upsilon(\cdot)$	Maximum component of a vector
$\Xi^K(\alpha)$	Set of limiting state action frequencies of policies in class K when initial state distribution is α
$\Xi^1(\alpha)$	Set of limiting state action frequencies for policies in Π^1
$\Xi^\pi(\alpha)$	Set of limiting state action frequencies under policy π
ω	Sample path of the state and action process
Ω	Set of all sample paths
$\Re$	Set of realizations of the reward process
$\binom{n}{r}$	A binomial coefficient
$\| \cdot \|$	A norm
$\| \cdot \|_w$	Weighted supremum norm
$\| \cdot \|$	Absolute value of a number or number of elements in a set
$[x]^+$	$\max(x, 0)$
$[x]^-$	$\min(x, 0)$
$*$	Superscript denoting an optimal policy, value or decision rule
$\times$	Cartesian product
$B^\#$	Generalized inverse of a matrix B
A/B	Subset of elements of A which are not contained in B
$\mathrm{sp}(\cdot)$	Span semi-norm
$\equiv$	Definition

u^T	Transpose of a vector u
arg max	Subset of elements at which the maximum of a function is obtained
□	Completion of a proof
$>_{\mathscr{L}}$	Lexicographic ordering
~	Expression denoting quantities in a transformed model
ext(X)	Set of extreme points of a set X in a linear space
X^c	Closed convex hull of a set X in a linear space
0	Scalar or vector zero

ABBREVIATIONS

CTMDP	Continuous-time Markov decision process
HD	History dependent deterministic
HR	History dependent randomized
LP	Linear program
l.s.c.	Lower semicontinuous
MDP	Markov decision process
MD	Markov deterministic
MR	Markov randomized
POMDP	Partially observed Markov decision process
SR	Stationary randomized
SMDP	Semi-Markov decision process
SD	Stationary deterministic
u.s.c.	Upper semicontinuous

APPENDIX A

Markov Chains

This appendix summarizes results from Markov chain theory that are particularly relevant for the analysis of MDPs, especially those with average reward optimality criterion. The material in Secs. A.1–A.2 and A.4 is quite standard; we refer the reader to Chung (1960), Kemeny and Snell (1960), Karlin (1969), and Çinlar (1975) for additional details. Section A.3 provides the Fox and Landi (1968) algorithm for classifying states of a Markov chain. Sections A.5 and A.6 use matrix decomposition theory to derive some fundamental results. References for those sections include Campbell and Meyer (1979), Berman and Plemmons (1979), Senata (1981), Kemeny (1981), Lamond (1986), and Lamond and Puterman (1989). In response to frequent questions of friends and students, the appendix concludes with a brief biography of A. A. Markov. Biographical sources include Ondar (1981) and bibliographic notes in Çinlar (1975).

A.1 BASIC DEFINITIONS

Let $\{X_n, n = 0, 1, 2, \ldots\}$ be a sequence of random variables which assume values in a discrete (finite or countable) state space S. We say that $\{X_n, n = 0, 1, 2, \ldots\}$ is a *Markov chain* if

$$P\{X_n = j_n | X_{n-1} = j_{n-1}, X_{n-2} = j_{n-2}, \ldots, X_0 = j_0\} = P\{X_n = j_n | X_{n-1} = j_{n-1}\}$$

for $n \geq 1$ and $j_k \in S$, $0 \leq k \leq n$. If $P\{X_n = j | X_{n-1} = s\}$ does not depend on n, we call the Markov chain *stationary* or *time homogeneous*. In this case we write $p(j|s) \equiv P\{X_n = j | X_{n-1} = s\}$ and refer to it as a *transition probability*. We call the matrix P with (s, j)th component $p(j|s)$, the *transition probability matrix* or *transition matrix*. In a stationary Markov chain, we denote the *m-step transition probability* by $p^m(j|s) = P\{X_{n+m} = j | X_n = s\}$. As a consequence of repeated application of the law of total probabilities, $p^m(j|s)$ is an element of the matrix P^m, the mth power of the matrix P.

For a real-valued function $g(\cdot)$ on S we denote the expected value of $g(X_n)$ by $E_s\{g(X_n)\} \equiv \sum_{j \in s} g(j) p^n(j|s)$ when $X_0 = s$. When g is the indicator of $A \subset S$, we let $P_s\{X_n \in A\} \equiv E_s\{I_{(A)}(X_n)\} = \sum_{j \in A} p^n(j|s)$.

A.2 CLASSIFICATION OF STATES

With each $s \in S$, associate the random variables ν_s and τ_s, which represent the number of visits and time of the first visit (first return if the chain starts in s), to state s. We classify states on the basis of $P_s\{\tau_s < \infty\}$ and $E_s\{\tau_s\}$ as follows:

	$P_s\{\tau_s < \infty\} < 1$	$P_s\{\tau_s < \infty\} = 1$
$E_s\{\tau_s\} < \infty$	not possible	positive recurrent
$E_s\{\tau_s\} = \infty$	transient	null recurrent

Often we do not distinguish positive and null recurrent states, and simply refer to them as recurrent. Sometimes we refer to recurrent states as *ergodic*. Note that s is recurrent if and only if

$$E_s\{\nu_s\} = \sum_{n=0}^{\infty} p^n(s|s) = \infty,$$

and transient if and only if $E_s\{\nu_s\} < \infty$.

We say that state $j \in S$ is *accessible* from state s ($s \to j$) if $p^n(j|s) > 0$ for some $n \geq 0$; otherwise we say that j is *inaccessible* from s. Note that $p^0(j|s) = 1$ if $s = j$ and 0 if $s \neq j$. State j *communicates* with state s if $s \to j$ and $j \to s$. Call a subset C of S a *closed set* if no state outside of C is accessible from any state in C. We call a closed set C *irreducible* if no proper subset of C is closed. Every recurrent state j is a member of some irreducible subset or class of S. Irreducible closed sets consisting of a single state are said to be *absorbing*. Closed irreducible sets may be regarded as distinct Markov chains.

We can partition the set of recurrent states (provided some are present) into disjoint closed irreducible sets C_k, $k = 1, 2, \ldots, m$ with m finite when S is finite and possibly infinite when S is countable. We can write S as $S = C_1 \cup C_2 \cup \cdots \cup C_m \cup T$, where T denotes the set of transient states that do not belong to any closed set.

After relabeling states if necessary, we can express any transition matrix P as

$$P = \begin{bmatrix} P_1 & 0 & 0 & \cdot & \cdot & 0 \\ 0 & P_2 & 0 & \cdot & \cdot & 0 \\ \cdot & & & \cdot & & \\ \cdot & & & & \cdot & \\ 0 & & & P_m & & 0 \\ Q_1 & Q_2 & \cdot & \cdot & Q_m & Q_{m+1} \end{bmatrix}, \tag{A.1}$$

where P_i corresponds to transitions between states in C_i, Q_i to transitions from states in T to states in C_i, and Q_{m+1} to transitions between states in T. Note that Q_i may be a matrix of zeros for some values of i. We refer to this representation as the *canonical form* of P. The algorithm in the next section provides an efficient method for classifying states and transforming a transition matrix to canonical form.

We use the expression *chain structure* to refer to the extent of the decomposition of the Markov chain into classes. We call a Markov chain *irreducible* if it consists of a single closed class. For *finite S* we use the expression *unichain* to refer to chains consisting of one closed irreducible set and a (possibly empty) set of transient states. Otherwise we say that a Markov chain is *multichain*.

Both recurrence and transience are class properties. This means that, in any closed irreducible class, all states are either transient, positive recurrent, or null recurrent. We note the following important results for finite-state Markov chains.

Theorem A.1. Suppose S is finite.

    a. Then any recurrent state is positive recurrent.

    b. There exists at least one positive recurrent class.

Consequently, in a finite irreducible chain, all states are positive recurrent. Note that countable-state Markov chains may have more than one closed irreducible set of transient states; for example, when $p(s + 2|s) = 1$ for $s = 0, 1, 2, \ldots$, the odd and even integers each form closed irreducible sets of transient states.

We refer to the greatest common divisor of all n for which $p^n(s|s) > 0$ as the *period* of s. Whenever $p(s|s) > 0$, s has period 1. We refer to such a state as *aperiodic*. Periodicity is a class property; all states in a closed irreducible class have the same period. If it exceeds 1, we call the class *periodic*; otherwise we refer to it as *aperiodic*. Similarly we call an irreducible chain periodic or aperiodic depending on the periodicity of its states. For example, the Markov chain with transition probability matrix

$$P = \begin{bmatrix} 0 & 1 \\ 1 & 0 \end{bmatrix} \qquad (A.2)$$

is irreducible, and each state has period 2.

A.3   CLASSIFYING THE STATES OF A FINITE MARKOV CHAIN

This section gives the Fox and Landi (1968) labeling algorithm for determining the closed irreducible classes and transient states of a finite Markov chain. We use this algorithm for computing the limiting matrix of a Markov chain (Section A.4) and for determining the class structure of a Markov decision process. We follow Fox and Landi; however, we state the algorithm more formally.

The algorithm cleverly exploits the following properties of a transition probability matrix.

a. State s is absorbing if and only if $p(s|s) > 0$ and $p(j|s) = 0$ or all $j \neq s$.

b. If state s is absorbing, and $p(s|k) > 0$, then state k is transient.

c. If state s is transient, and $p(s|k) > 0$, then state k is transient.

d. If state i communicates with j, and j communicates with k, then i communicates with k.

e. If i communicates with j, then i is transient if j is transient, and i is recurrent if j is recurrent.

Note that the above properties only depend on the pattern of 0 and positive entries, so that if only a decomposition is required, the matrix P may be replaced by an incidence matrix B with $b(j|i) = 1$ if $p(j|i) > 0$, and 0 otherwise.

Assume that $S = \{1, 2, \ldots, N\}$. In the algorithm, $L(i)$ denotes the label assigned to state i; recurrent states receive a label R, transient states T, and unlabeled states O. The set U denotes unlabeled states, and $S(i)$ denotes states which have been identified as communicating with i. We use the set W to indicate the indices of rows and columns in an aggregate matrix constructed in step 5. A verbal description of the algorithm follows it formal statement.

The Chain Decomposition Algorithm

1. *Initialization.* Set $S(i) = \{i\}$, $L(i) = O$ for $i = 1, 2, \ldots, N$, $U = S$, and $W = S$.

2. *Preliminary identification.*

　a. For each $i \in U$, if $p(i|i) > 0$ and $p(j|i) = 0$ for all $j \neq i$, set $L(i) = R$, and replace U by $U/\{i\}$.

　b. If $U = \varnothing$, go to step 6; otherwise, for each j for which $L(j) = R$, if $p(j|i) > 0$ for any $i \in U$, set $L(i) = T$ and replace U by $U/\{i\}$.

3. *Stopping.* If $U = \varnothing$, go to step 6; otherwise, set $r = 0$ and go to step 4.

4. *Path formation.*

　a. Select an $i \in U$, set $i_r = i$.

　b. Choose a state $j \neq i_r$ for which $p(j|i_r) > 0$; set $i_{r+1} = j$.

　　i. If $L(i_{r+1}) = T$, set $L(i) = T$ for $i \in S(i_0) \cup \cdots \cup S(i_{r+1})$, replace U by $U/(S(i_0) \cup \cdots \cup S(i_r))$, and go to step 3.

　　ii. If $L(i_{r+1}) \neq T$, and $i_{r+1} = i_k$ for some k, $0 \leq k \leq r$, go to step 5. Otherwise replace r by $r + 1$, and go to step 4(b).

5. *Path aggregation.*

　a. Replace $p(j|i_k)$ by $p(j|i_k) + p(j|i_{k+1}) + \cdots + p(j|i_r)$ for all $j \in W$.

　b. Replace $p(i_k|i)$ by $p(i_k|i) + p(i_{k+1}|i) + \cdots + p(i_r|i)$ for all $i \in W$.

　c. Replace $S(i_k)$ by $S(i_k) \cup \cdots \cup S(i_r)$.

　d. Replace W by $W/\{i_{k+1}, \ldots, i_r\}$. (i.e. Delete rows and columns $i_{k+1}, \ldots, i_r$ from the matrix).

e. If $p(i_k|i_k) > 0$ and $p(j|i_k) = 0$ for all $j \in W$, do the following.

   i. Set $L(i) = R$ for all $i \in S(i_k)$, and replace U by $U/S(i_k)$.

   ii. If $k > 0$, set $L(i) = T$ for all $i \in S(i_0) \cup \cdots \cup S(i_{k-1})$, and replace U by $U/(S(i_0) \cup \cdots \cup S(i_{k-1}))$.

   iii. For $j \in U$, if $p(i_h|j) > 0$ for some h, $0 \le h \le k$, set $L(i) = T$ for all $i \in S(j)$, and replace U by $U/S(j)$.

   iv. Go to Step 3.

f. Set $r = k$ and go to step 4.

6. *Classification.* For each $i \in W$ for which $L(i) = R$, $S(i)$ is a closed irreducible class. All other states are transient.

We may paraphrase steps 4 and 5 of the algorithm as follows. *Pick an unclassified state and begin a path starting at it. Extend the path until it either identifies a transient state or it cycles; that is, it repeats a previous state in the chain. In the former case, classify all states on the path as transient; otherwise, combine all states in the cycle and determine whether they are recurrent.* Note that steps 5(a)–5(d) are a rather formal way of saying "replace row and columns i_k by the sum of entries in rows and columns $i_k, i_{k+1}, \ldots, i_r$, and delete these rows and columns from the matrix." The above algorithm works with the full matrix but ignores the "deleted" columns.

If instead of using the transition probability matrix, we use the incidence matrix for computation, we replace steps 5(a) and 5(b) by a Boolean "or" operation; that is, if $u = 0$ and $v = 0$, then $u + v = 0$, otherwise $u + v = 1$. Using the incidence matrix allows for efficient storage of large matrices and faster calculation.

The above algorithm requires $O(|S|^2)$ comparisons. Since there are S^2 pairs of indices, this algorithm is extremely efficient.

A.4 THE LIMITING MATRIX

Results in this section apply to both finite- and countable-state Markov chains. Let $\{A_n: n \ge 0\}$ be a sequence of matrices. We write $\lim_{n \to \infty} A_n = A$ if $\lim_{n \to \infty} A_n(j|s) = A(j|s)$ for each (s, j) in $S \times S$. When analyzing periodic Markov chains, limits of this form do not exist and instead we consider the Cesaro limit which we define as follows. We say that A is the *Cesaro limit* (of order one) of $\{A_n: n \ge 0\}$ if

$$\lim_{N \to \infty} \frac{1}{N} \sum_{n=0}^{N-1} A_n = A,$$

and write

$$C - \lim_{N \to \infty} A_N = A$$

to distinguish this as a Cesaro limit. Sometimes, the ordinary limit is called a Cesaro limit of order zero.

Define the *limiting matrix* P^* by

$$P^* = C - \lim_{N \to \infty} P^N. \tag{A.3}$$

In component notation, where $p^*(j|s)$ denotes the $(j|s)$th element of P^*, this means that, for each s and j,

$$p^*(j|s) = \lim_{N \to \infty} \frac{1}{N} \sum_{n=1}^{N} p^{n-1}(j|s)$$

where, as above, p^{n-1} denotes a component of P^{n-1} and $p^0(j|s)$ is a component of a an $S \times S$ identity matrix. When P is aperiodic, $\lim_{N \to \infty} P^N$ exists and equals P^*.

We define P^* in terms of the Cesaro limit to account for the nonconvergence of powers of transition matrices of periodic chains. For example, with P given by (A.2), $P^{2n} = I$ and $P^{2n+1} = P$, so that $\lim_{N \to \infty} P^N$ does not exist but

$$C - \lim_{N \to \infty} P^N = P^* = \begin{bmatrix} 0.5 & 0.5 \\ 0.5 & 0.5 \end{bmatrix}.$$

Doob (1953) and Chung (1960, p. 33) prove that the limit in (A.3) exists. Doob's proof assumes finite S and uses matrix methods, while Chung's uses probabilistic argument. In the next section, we provide an alternative derivation for finite-state Markov chains using matrix decomposition theory.

The limiting matrix P^* satisfies the following equalities

$$PP^* = P^*P = P^*P^* = P^*. \tag{A.4}$$

Since $(P^*)^2 = P^*$, $(I - P^*)^2 = (I - P^*)$, and $P^*(I - P^*) = 0$, P^* and $(I - P^*)$ are orthogonal projection matrices. Note further that, for finite chains, P^* is a stochastic matrix (has row sums equal to 1), while in countable-state chains, some or all row sums might be less than 1, for example, when $p(s + 1|s) = 1$ for $s = 0, 1, \ldots$, and $p(j|s) = 0$ otherwise, then P^* is a zero matrix.

Since

$$E_s\{v_j\} = \sum_{n=1}^{\infty} p^{n-1}(j|s), \tag{A.5}$$

we may interpret $p^*(j|s)$ as the long-run fraction of time that the system occupies state j starting in state s. In aperiodic chains, $\lim_{N \to \infty} p^N(j|s)$ exists, in which case we may interpret $p^*(j|s)$ as the steady-state probability that the chain is in state j when it starts in state s.

We now show how to compute P^*. We begin with the following key result which holds for both finite and countable state chains. In it, q is a column vector and q^T denotes its transpose.

Theorem A.2. Suppose P is the transition matrix of a positive recurrent irreducible chain. Then the system of equations $q^T = q^T P$ subject to $\sum_{j \in s} q(j) = 1$, has a unique positive solution.

We call this solution the *stationary distribution* of P. Since $P^*P = P^*$, when P is recurrent and irreducible, P^* has identical rows and we can write

$$P^* = eq^T, \tag{A.6}$$

where q is the stationary distribution of P, e denotes a column vector of ones, and e^T its transpose.

The chain structure of the Markov chain determines the form of P^*. For P in canonical form (A.1),

$$P^* = \begin{bmatrix} P_1^* & 0 & 0 & \cdot & \cdot & 0 \\ 0 & P_2^* & 0 & \cdot & \cdot & 0 \\ & \cdot & & \cdot & & \cdot \\ & \cdot & & & \cdot & \cdot \\ 0 & & & & P_m^* & 0 \\ Q_1^* & Q_2^* & \cdot & \cdot & Q_m^* & 0 \end{bmatrix}, \tag{A.7}$$

where P_i^* is the limiting matrix of P_i. For each i, we find P_i^* by solving $q_i P_i = q_i$ subject to $q_i^T e_i = 1$, where e_i is a column vector of ones with $|C_i|$ components. Then, following (A.5), we set $P_i^* = q_i e_i^T$.

We determine $Q_1^*, \ldots, Q_m^*$ as follows. Assume P is in canonical form and write it as

$$P = \begin{bmatrix} U & 0 \\ V & W \end{bmatrix},$$

where U corresponds to transitions between recurrent states, W to transitions between transient states, and V to transitions from transient to recurrent states.

We use the following results [Çinlar (1975, p. 144–146) and Berman and Plemmons (1979, p. 223)] which follow from (A.5) and the definition of transience.

Proposition A.3. When S is finite

a. the spectral radius of W, $\sigma(W) < 1$, and
b. $(I - W)^{-1}$ exists and satisfies

$$(I - W)^{-1} = \sum_{n=0}^{\infty} W^n.$$

Theorem A.4. Let P be in canonical form. Then

$$Q_i^* = (I - Q_{m+1})^{-1} Q_i P_i^*.$$

Proof. Write P as

$$P = \begin{bmatrix} U & 0 \\ V & W \end{bmatrix},$$

and the corresponding limiting matrix as

$$P^* = \begin{bmatrix} U^* & 0 \\ V^* & 0 \end{bmatrix}.$$

Since $PP^* = P^*$,

$$VU^* + WV^* = V^*,$$

so that

$$V^* = (I - W)^{-1}VU^*,$$

where Proposition A.3 ensures the existence of $(I - W)^{-1}$. From (A.1) and (A.7), it follows that

$$Q_i^* = (I - Q_{m+1})^{-1}Q_i P_i^*. \qquad \square$$

Observe that for unichain P with closed irreducible recurrent class C_1 and transient states T, P^* has equal rows and

$$p^*(j|s) = \begin{cases} q(j) & j \in C_1 \\ 0 & j \in T, \end{cases}$$

where q satisfies $qP_1 = q$ subject to $\sum_{j \in C_1} q(j) = 1$.

In summary, to compute P^*, do the following:

1. Use the Fox-Landi algorithm to transform P to canonical form.
2. For each P_i, solve $q_i P_i = q_i$ subject to $q_i^T e_i = 1$, and set $P_i^* = q_i e_i^T$.
3. Set $Q_i^* = (I - Q_{m+1})^{-1}Q_i P_i^*$.

A.5 MATRIX DECOMPOSITION, THE DRAZIN INVERSE, AND THE DEVIATION MATRIX

The results in this section rely on decomposition of the transition matrix and are often referred to as "the modern theory of Markov chains." We follow Lamond and Puterman (1989), who draw on Campbell and Meyer (1979). Berman and Plemmons (1979) also cover some of this material and state

"...virtually everything that one would want to know about the chain can be determined by investigating a certain group inverse (the Drazin Inverse) and a limiting matrix...their introduction into the theory provides practical advantages over the more classical techniques and serves to unify the theory to a certain extent."

It appears that the results have only been established for finite-state Markov chains, therefore, *we assume finite S in this section*. An important research contribution would be to generalize these results to nonfinite S. The following theorem is fundamental. We denote the spectral radius of P by $\sigma(P)$ (see Sec. C.2 of Appendix C). Note that in the following results, our approach to matrix decomposition differs from that used to establish (A.1).

Theorem A.5.   Suppose S is finite and P has m recurrent classes.

a. Then 1 is an eigenvalue of P with (algebraic and geometric) multiplicity m and with m linearly independent eigenvectors.

b. There exists a nonsingular matrix W for which

$$P = W^{-1}\begin{bmatrix} Q & 0 \\ 0 & I \end{bmatrix}W, \tag{A.8}$$

where I is an $m \times m$ identity matrix and Q is an $(|S| - m) \times (|S| - m)$ matrix with the following properties.

1. 1 is not an eigenvalue of Q.
2. $\sigma(Q) \le 1$ and if all recurrent subchains of P are aperiodic, $\sigma(Q) < 1$.
3. $(I - Q)^{-1}$ exists.
4. $\sigma(I - Q) = \sigma(I - P)$.

c. There exists a unique matrix P^* which satisfies (A.4) and $P_i^* > 0$ for $i = 1, 2, \ldots, m$. It may be represented by

$$P^* = W^{-1}\begin{bmatrix} 0 & 0 \\ 0 & I \end{bmatrix}W. \tag{A.9}$$

Note in Theorem A.5(b) that even when $\sigma(Q) = 1$, the fact that 1 is not an eigenvalue of Q implies that 0 is not an eigenvalue of $I - Q$, so that $(I - Q)^{-1}$ exists. If $\sigma(Q) = 1$, $\lim_{n \to \infty} Q^n$ need not exist, so that

$$(I - Q)^{-1} = C - \lim_{N \to \infty} \sum_{n=0}^{N-1} Q^n.$$

We use Theorem A.5 to establish the following result.

Theorem A.6.   Suppose S is finite. Then the limit in (A.3) exists and equals P^*.

Proof.   From (A.8),

$$P^n = W^{-1}\begin{bmatrix} Q^n & 0 \\ 0 & I \end{bmatrix}W,$$

so that

$$\frac{1}{N} \sum_{n=0}^{N-1} P^n = W^{-1} \begin{bmatrix} \frac{1}{N} \sum_{n=0}^{N-1} Q^n & 0 \\ 0 & I \end{bmatrix} W.$$

The nonsingularity of $I - Q$ implies

$$\sum_{n=0}^{N-1} Q^n = (I - Q^N)(I - Q)^{-1}. \tag{A.10}$$

Because $\sigma(Q) \le 1$, Q^N is bounded, so that (A.10) implies $\sum_{n=0}^{N-1} Q^n$ is bounded and

$$\lim_{N \to \infty} \frac{1}{N} \sum_{n=0}^{N-1} Q^n = 0.$$

Therefore, from Theorem A.5(c), we conclude that

$$\lim_{N \to \infty} \frac{1}{N} \sum_{n=0}^{N-1} P^n = W^{-1} \begin{bmatrix} 0 & 0 \\ 0 & I \end{bmatrix} W = P^*. \qquad \square$$

In Markov decision process theory we frequently solve systems of the form

$$(I - P)v = r.$$

Since $(I - P)$ is singular, this system does not have a unique solution. We use the matrix representation in (A.8) to distinguish a solution which plays a key role in the theory of average reward Markov decision processes. From (A.8),

$$I - P = W^{-1} \begin{bmatrix} I - Q & 0 \\ 0 & 0 \end{bmatrix} W, \tag{A.11}$$

where 0 in the lower-right-hand corner denotes an $m \times m$ matrix of zeros. As a consequence of Theorem A.1(b), $m \ge 1$, so that $(I - P)^{-1}$ does not exist. Instead we seek a *generalized inverse* of $(I - P)$.

Given a matrix B with representation

$$B = W^{-1} \begin{bmatrix} C & 0 \\ 0 & 0 \end{bmatrix} W,$$

in which C is nonsingular, we define the matrix $B^\#$ by

$$B^\# = W^{-1} \begin{bmatrix} C^{-1} & 0 \\ 0 & 0 \end{bmatrix} W. \tag{A.12}$$

It is easy to see that $B^{\#}$ satisfies

$$B^{\#}BB^{\#} = B^{\#}, \qquad BB^{\#} = B^{\#}B, \qquad \text{and} \qquad BB^{\#}B = B. \qquad (A.13)$$

We call a matrix which satisfies (A.13) a *Drazin inverse* or *group inverse* of B. It is a particular generalized inverse of B. We now derive a representation for $(I - P)^{\#}$ and show how the matrix decomposition theory may be used to provide direct proofs of many key matrix identities. The following theorem defines the matrices Z_p and H_p.

Theorem A.7.

a. The matrix $(I - P + P^*)$ is nonsingular, with inverse denoted by Z_p.
b. The Drazin inverse of $(I - P)$, denoted by H_p, satisfies

$$(I - P)^{\#} = (I - P + P^*)^{-1}(I - P^*) \equiv H_p. \qquad (A.14)$$

c.

$$H_p = C - \lim_{N \to \infty} \sum_{n=0}^{N-1} (P^N - P^*), \qquad (A.15)$$

where the ordinary limit exists whenever P is aperiodic.

Proof. Combining (A.8) and (A.9) yields

$$I - P + P^* = W^{-1}\begin{bmatrix} I - Q & 0 \\ 0 & I \end{bmatrix}W.$$

Theorem A.5(b) implies that the matrix on the right-hand side is nonsingular, so that (a) follows.

To derive (b), note that the definition of the Drazin inverse implies

$$(I - P)^{\#} = W^{-1}\begin{bmatrix} (I - Q)^{-1} & 0 \\ 0 & 0 \end{bmatrix}W$$

$$= W^{-1}\begin{bmatrix} (I - Q)^{-1} & 0 \\ 0 & I \end{bmatrix}W - W^{-1}\begin{bmatrix} 0 & 0 \\ 0 & I \end{bmatrix}W$$

$$= (I - P + P^*)^{-1} - P^* = Z_p - P^*.$$

Expression (A.14) follows by noting that $P^* = Z_p P^*$.

The proof of (c) is slightly more involved, and we refer the reader to Lamond and Puterman (1989) for details. It relies on the identity

$$(P - P^*)^n = P^n - P^*, \qquad (A.16)$$

which holds for $n \geq 1$. $\qquad \square$

We refer to H_P as the *deviation matrix* and Z_P as the *fundamental matrix*. These play a key role in average reward MDP theory and computation. The following useful identities may be derived by similar methods to the proof of Theorem A.7(b) or directly from the power series expansion for $[I - (P - P^*)]^{-1}$:

$$(I - P)H_P = H_P(I - P) = I - P^*, \tag{A.17}$$

$$H_P P^* = P^* H_P = 0, \tag{A.18}$$

$$H_P = Z_P - P^*, \tag{A.19}$$

$$Z_P P^* = P^*, \tag{A.20}$$

$$P^* = I - (I - P)(I - P)^{\#}. \tag{A.21}$$

For aperiodic chains, Theorem A.7(c) implies that

$$H_P = \lim_{N \to \infty} \left[\sum_{k=0}^{N-1} P^k - NP^* \right],$$

so that we may interpret $H_P(j|s)$ as the difference between the expected number of visits to state j starting in s, and the expected number of visits to j starting in s for a chain with transition probability matrix P^*. Note that the Cesaro limit in (A.15) is equivalent to

$$H_P = \lim_{N \to \infty} \frac{1}{N} \sum_{k=0}^{N-1} \sum_{i=0}^{k} (P^k - P^*). \tag{A.22}$$

The following example illustrates some of these calculations and the need for the Cesaro limit when defining H_P.

Example A.1. Suppose

$$P = \begin{bmatrix} 0 & 1 \\ 1 & 0 \end{bmatrix}.$$

Decompose P as in (A.8) as

$$P = W^{-1} \begin{bmatrix} -1 & 0 \\ 0 & 1 \end{bmatrix} W,$$

where

$$W = \begin{bmatrix} \frac{1}{2} & -\frac{1}{2} \\ 1 & 1 \end{bmatrix} \quad \text{and} \quad W^{-1} = \begin{bmatrix} 1 & \frac{1}{2} \\ -1 & \frac{1}{2} \end{bmatrix}.$$

In the notation above $Q = -1$ and $m = 1$. Note that

$$P^* = W^{-1} \begin{bmatrix} 0 & 0 \\ 0 & 1 \end{bmatrix} W = \begin{bmatrix} \frac{1}{2} & \frac{1}{2} \\ \frac{1}{2} & \frac{1}{2} \end{bmatrix}$$

and

$$I - P = \begin{bmatrix} 1 & -1 \\ -1 & 1 \end{bmatrix} = W^{-1} \begin{bmatrix} 2 & 0 \\ 0 & 0 \end{bmatrix} W,$$

so that

$$H_P = (I - P)^{\#} = W^{-1} \begin{bmatrix} \frac{1}{2} & 0 \\ 0 & 0 \end{bmatrix} W = \begin{bmatrix} \frac{1}{4} & -\frac{1}{4} \\ -\frac{1}{4} & \frac{1}{4} \end{bmatrix}.$$

Since $P^{2n} - P^* = I - P^*$ and $P^{2n+1} = P - P^*$,

$$\lim_{N \to \infty} \sum_{n=0}^{N-1} (P^N - P^*)$$

does not exist in the ordinary sense and we require a Cesaro limit to define H_P as a limit of partial sums of matrix powers.

A.6 THE LAURENT SERIES EXPANSION OF THE RESOLVENT

The Laurent series expansion provides a powerful tool for analyzing undiscounted Markov decision processes. For $\rho > 0$, define the *resolvent* of $P - I$, (Sec. C.2) denoted R^ρ, by

$$R^\rho \equiv (\rho I + [I - P])^{-1}. \tag{A.23}$$

Letting $\lambda = (1 + \rho)^{-1}$, we see that

$$(I - \lambda P) = (1 + \rho)^{-1}(\rho I + [I - P]).$$

When $0 \leq \lambda < 1$, $\sigma(\lambda P) < 1$, so that $(I - \lambda P)^{-1}$ exists. Consequently the above identity implies that if $\rho > 0$, R^ρ exists. In our analysis of MDPs in Chap. 10, we will often appeal to the following easily derived identities:

$$(I - \lambda P)^{-1} = (1 + \rho)R^\rho$$

and

$$R^\rho = \lambda(I - \lambda P)^{-1}.$$

We use the matrix decomposition methods of the previous section to derive a series expansion for R_ρ.

Theorem A.8. For $0 < \rho < \sigma(I - P)$,

$$R^\rho = \rho^{-1}P^* + \sum_{n=0}^{\infty} (-\rho)^n H_P^{n+1} \tag{A.24}$$

Proof.   Let Q be defined through (A.8), and let $B = I - Q$. Then

$$\rho I + I - P = W^{-1}\begin{bmatrix} \rho I + B & 0 \\ 0 & \rho I \end{bmatrix} W,$$

so that

$$R^{\rho} = W^{-1}\begin{bmatrix} (\rho I + B)^{-1} & 0 \\ 0 & \rho^{-1}I \end{bmatrix} W$$

$$= \rho^{-1}W^{-1}\begin{bmatrix} 0 & 0 \\ 0 & I \end{bmatrix} W + W^{-1}\begin{bmatrix} (\rho I + B)^{-1} & 0 \\ 0 & 0 \end{bmatrix} W. \qquad (A.25)$$

From Theorem A.5(c), the first term in (A.25) equals $\rho^{-1}P^*$. Since

$$(\rho I + B)^{-1} = (I + \rho B^{-1})^{-1} B^{-1},$$

and whenever $\sigma(\rho B^{-1}) = [\rho/\sigma(I - Q)] < 1$ or $\rho < \sigma(I - Q) = \sigma(I - P)$,

$$(\rho I + B)^{-1} = \sum_{n=0}^{\infty} (-\rho)^n (B^{-1})^n,$$

the second term in (A.25) becomes

$$\sum_{n=0}^{\infty} (-\rho)^n W^{-1}\begin{bmatrix} (I - Q)^{-n} & 0 \\ 0 & 0 \end{bmatrix} W = \sum_{n=0}^{\infty} (-\rho)^n H_P^{n+1},$$

from which we conclude the result.   □

Example A.1 (ctd.).   Since $\sigma(I - Q) = 2$, it follows from Theorem A.7 that, for $\rho < 2$,

$$R_\rho = \rho^{-1}\begin{bmatrix} \frac{1}{2} & \frac{1}{2} \\ \frac{1}{2} & \frac{1}{2} \end{bmatrix} + \sum_{n=0}^{\infty} (-\rho)^n \begin{bmatrix} \frac{1}{4} & -\frac{1}{4} \\ -\frac{1}{4} & \frac{1}{4} \end{bmatrix}^{n+1}$$

A.7   A. A. MARKOV

The subject of this book derives its name from the Russian mathematician Andrei Andreivich Markov, who introduced the concept of a Markov chain in the early 20th century. His work in this area concerned what we now refer to as finite-state, irreducible, aperiodic Markov chains. In addition to considerable theoretical research on these models, he applied them to analyze Bernoulli and LaPlace's urn problem and the pattern of vowels and consonants in Russian novels. He did not develop Markov decision processes; this subject evolved much later.

A. A. Markov was born on July 14, 1856 in the province of Ryazan, and moved to St. Petersburg in the 1860's. In high school, he was a mediocre student but excelled in mathematics. He enrolled at the University of St. Petersburg in 1874 and was greatly influenced by the emminent mathematician P. L. Chebyshev. Markov received a gold medal for his paper "On the integration of differential equations with the aid of continued fractions," written prior to his graduation in 1878. Markov was awarded his doctoral degree in 1884 and went on to a distinguished career as Professor at The University of St. Petersburg. He became an Academician in the Russian Academy of Sciences in 1896 and died on July 20, 1922. His son, also named Andrei Andreivich Markov (1903–1979), was a emminent mathematician who made significant contributions to mathematical logic and the theory of algorithms.

In addition to his seminal work on dependent processes, which inspired Kolmogorv's development of the theory of stochastic processes, Markov studied important problems in probability and statistics. He formulated many fundamental concepts in probability theory while completing Chebyshev's proof of the central limit theorem. He was interested in laws of large numbers and developed them for sequences of dependent random variables. His justification of the method of least squares, now referred to as the Gauss-Markov Theorem, led to the modern statistical ideas of efficiency and unbiasedness. His writings in all areas emphasized his mathematical rigor.

Markov was also a tenacious social activist. He was often referred to (in rough translation) as "Andrei the irrepressible, who does not pull any punches." In 1902, he renounced all government honors when the Tsarist government expelled writer A. M. Gorky from the Russian Academy.

APPENDIX B

Semicontinuous Functions

In this book we use semicontinuous functions when providing sufficient conditions for a function to achieve its maximum on a compact set. They represent a broader class of functions than continuous functions but still possess many of the same properties. We state results in considerable generality, although in applications we usually consider real-valued functions on Euclidean space. Royden (1963) provides a nice mathematical background for material in this appendix. References in the Markov decision process literature include papers of Maitra (1968) and Himmelberg, Parthasarathy and van Vleck (1976), and the books of Hinderer (1970, pp. 31–35 and 113–117), Bertsekas and Shreve (1978), Dynkin and Yushkevich (1979), and Hernandez-Lerma (1989). Our presentation follows Maitra and Hinderer.

Let X be a complete separable metric space and f a real-valued function on X. We say that the real-valued function f is *upper semicontinuous* (u.s.c.) if, for any sequence $\{x_n\}$ of elements of X which converges to x^*,

$$\limsup_{n \to \infty} f(x_n) \le f(x^*). \tag{B.1}$$

The function f is said to be *lower semicontinuous* (l.s.c.) whenever $-f$ is u.s.c. or, equivalently, if $\liminf_{n \to \infty} f(x_n) \ge f(x^*)$. A continuous function is both l.s.c. and u.s.c. We summarize some basic properties of semicontinuous functions in the following proposition. A proof of part (e) appears in Maitra (1968).

Proposition B.1. Let X be a complete separable metric space.

a. If f and g are u.s.c. on X, then $f + g$ is u.s.c. on X.

b. If $f \ge 0$ and $g \ge 0$ are u.s.c. on X, then fg is u.s.c. on X.

c. If $f \ge 0$ is l.s.c. and $g \le 0$ is u.s.c. on X, then fg is u.s.c. on X.

d. If $\{f_n\}$ is a decreasing sequence of nonpositive u.s.c. functions on X, then $\lim_{n \to \infty} f_n$ is u.s.c. on X.

e. Suppose $\{f_n\}$ is a sequence of bounded u.s.c. functions on X, and f_n converges to f uniformly. Then f is u.s.c. on X.

The following theorem provides one of the key properties of u.s.c. functions: that it attains its maximum on compact sets.

Theorem B.2. Suppose C is a compact subset of X, and f is u.s.c. on X. Then there exists an x^* in C such that $f(x^*) \geq f(x)$ for all $x \in C$.

Proof. Let $y^* = \sup_{x \in C} f(x)$. Let $\{x_n\}$ be a sequence in C for which $\lim_{n \to \infty} f(x_n) = y^*$. Then, by the compactness of C, there exists a subsequence $\{x_{n_k}\}$ which has a limit x^*. By (B.1), $f(x^*) \geq \lim_{k \to \infty} f(x_{n_k}) = y^*$. Therefore $f(x^*) = y^*$. □

We frequently use the following result.

Proposition B.3. Let X be a countable set, Y a complete separable metric space, and $q(x, y)$ a bounded non-negative real-valued function that is l.s.c. in y for each $x \in X$. Let $f(x)$ be a bounded nonpositive real-valued function on X for which $\Sigma_{x \in X} f(x)$ is finite. Then

$$h(y) = \sum_{x \in X} f(x) q(x, y)$$

is u.s.c. on Y.

Proof. From Proposition B.1(b), for each $x \in X$, $f(x)q(x, y) \leq 0$ is u.s.c. Let $\{X_n\}$ be an increasing sequence of finite subsets of X such that $\cap_{n=1}^{\infty} X_n = X$. Then, by Proposition B.1(a),

$$h_n(y) \equiv \sum_{x \in X_n} f(x) q(x, y)$$

is u.s.c. for each n. Since $h_n(y) \leq 0$ is decreasing in n, by Proposition B.1(d), $h(y) = \lim_{n \to \infty} h_n(y)$ is u.s.c. □

A generalization of Proposition B.3 to nondiscrete sets follows. We say that a conditional probability $q(\cdot \mid y)$ on Borel sets of X (see Sec. 2.3.2) is *continuous* if $q(\cdot \mid y_n)$ converges weakly to $q(\cdot \mid y)$ whenever $\{y_n\}$ converges to y. This means that, for any bounded measurable function $f(\cdot)$ on X,

$$\lim_{n \to \infty} \int_X f(x) q(dx \mid y_n) = \int_X f(x) q(dx \mid y).$$

Proposition B.4. Let X and Y be complete separable metric spaces, $f(x)$ a bounded u.s.c. function on X, and $q(\cdot \mid y)$ a continuous conditional probability on the Borel sets of X. Then

$$h(y) \equiv \int_X f(x) w(dx \mid y)$$

is a u.s.c. function on Y.

We use the following result to show that upper semicontinuity is preserved under induction. It is the essence of applications of these concepts to Markov decision processes. A proof appears in Maitra (1968) and Hinderer (1970).

Theorem B.5. Let X be a complete separable metric space, Y a compact subset of a complete separable metric space, and $f(x, y)$ a real-valued upper semicontinuous function on $X \times Y$. Then

a.
$$g(x) \equiv \max_{y \in Y} f(x, y)$$

is upper semicontinuous on X, and

b. there exists a measurable h: $X \to Y$ for which

$$f(x, h(x)) = \max_{y \in Y} f(x, y).$$

We refer to part (b) of Theorem B.6 as a *selection theorem*. Himmelberg, Parthasarathy and vanVleck (1976) provide the following generalization.

Theorem B.6. Let X be a complete separable metric space, Y a compact subset of a complete separable metric space, and $f(x, y)$ a real-valued bounded measurable function on $X \times Y$ for which $f(x, y)$ is u.s.c. on Y for each $x \in X$.

a. Then

$$g(x) \equiv \max_{y \in Y} f(x, y)$$

is measurable on X.

b. Then there exists a measurable h: $X \to Y$ for which

$$f(x, h(x)) = \max_{y \in Y} f(x, y).$$

Several authors consider generalizations of the two previous results in which the set Y may vary with x. In that degree of generality, hypotheses are formulated in terms of the set $K \equiv \{(x, y): x \in X \text{ and } y \in Y_x\}$, where Y_x denotes a collection of sets index by x or a set-valued function on X. Such set-valued functions are sometimes referred to as *multifunctions*.

APPENDIX C

Normed Linear Spaces

Many results in this book, most notably the analysis of algorithms in Chap. 6, are conveniently carried out in the context of normed linear spaces. This appendix includes a brief review of relevant background material. It has been adapted from Yosida (1968) and Curtain and Pritchard (1977). Other relevant references include Liusternik and Sobolev (1961), Kato (1966), and Luenberger (1969).

C.1 LINEAR SPACES

Let V denote the set of bounded real-valued functions on S, that is, $v \in V$ if v: $S \to R$ and there exists a K_v such that $|v(s)| \leq K_v$ for all $s \in S$. Note that $\sup_{v \in V} K_v$ may equal $+\infty$. Let e denote the element of V which equals 1 in all components. For each $v \in V$, define the *norm* of v by

$$\|v\| = \sup_{s \in S} |v(s)|. \tag{C.1}$$

We refer to this norm as the *supremum norm* or *sup norm*. When S is finite, this supremum is attained and we often write "max" instead of "sup." Since V is closed under addition and scalar multiplication and is endowed with a norm, it is a *normed linear space*.

A sequence $\{v_n\} \subset V$ is said to be a *Cauchy sequence* if for every $\varepsilon > 0$ there exists an N such that, whenever $n > N$ and $m > N$, $\|v_n - v_m\| < \varepsilon$. We say that a normed linear space is *complete* if every Cauchy sequence contains a limit point in that space. The normed linear space V together with the norm $\|\cdot\|$ is a complete, normed linear space or *Banach Space*. When S is a subset of Euclidean space, or a Polish space (Sec. 2.3), V_m denotes the family of bounded measurable real-valued functions on S.

In Markov decision process theory, elements of V may represent rewards of decision rules, value functions of policies, or iterates generated by an algorithm. The norm on V provides a basis for measuring the proximity of such quantities. Let $\{v_n\}$ denote a sequence in V. We say that v_n *converges* to v whenever

$$\lim_{n \to \infty} \|v_n - v\| = 0.$$

This notion of convergence is often referred to as *strong convergence* or *uniform convergence*. Alternatively, $\{v_n\}$ *converges pointwise* to v if $v_n(s)$ converges to $v(s)$ for each $s \in S$. When S is finite, these convergence notions are equivalent.

To compare values of policies and make statements about the monotone convergence of algorithms, we assume further that V is *partially ordered* (Sec. 4.1.1), with partial order $\geq$ corresponding to componentwise ordering; that is, for $u \in V$ and $v \in V$, $u \geq v$ if $u(s) \geq v(s)$ for all $s \in S$. The space V together with a partial order is often referred to as a *partially ordered normed linear space*.

Let Q be a linear transformation on the normed linear space V. That is, for scalar α and β, and u and v in V, $Q(\alpha u + \beta v) = \alpha Qu + \beta Qv$. When S is discrete, Q is a matrix. A linear transformation Q on V is *bounded* if there exists a constant $K > 0$ such that, for all $v \in V$,

$$\|Qv\| \leq K\|v\|.$$

Define the norm of Q, denoted $\|Q\|$ by

$$\|Q\| = \sup\{\|Qv\|: \|v\| \leq 1, v \in V\}. \tag{C.2}$$

Consequently $\|Qv\| \leq \|Q\|\,\|v\|$.

When S is discrete, so that Q is a matrix with components $q(j|s)$, this definition implies that

$$\|Q\| = \sup_{s \in S} \sum_{j \in S} |q(j|s)|.$$

When Q is a probability matrix, $\|Q\| = 1$. In general, if P and Q are bounded linear transformations on V,

$$\|PQ\| \leq \|P\|\,\|Q\| \tag{C.3}$$

The *inverse* of Q (if it exists) is a bounded linear operator Q^{-1} satisfying $Q^{-1}Qv = QQ^{-1}v = v$ for all $v \in V$. Let $L(V)$ denote the set of bounded linear transformations on V. The following easily proven result provides a key property of $L(V)$.

Lemma C.1. Suppose V is Banach space, then $L(V)$ with norm defined by (C.2) is a Banach Space.

As a consequence of this lemma, every Cauchy sequence $\{Q_n\}$ of bounded linear transformations on V has a limit Q in V which satisfies

$$\lim_{n \to \infty} \|Q_n - Q\| = 0. \tag{C.4}$$

We say that a linear transformation $T \in L(V)$ is *positive* if $Tv \geq 0$ whenever $v \geq 0$. For Q and T in $L(V)$, write $T \geq Q$ if $T - Q \geq 0$.

C.2   EIGENVALUES AND EIGENVECTORS

A complex number ν is said to be an *eigenvalue* of the linear transformation $Q \in L(V)$ if there exists a $v \neq 0$ in V for which

$$(Q - \nu I)v = 0. \tag{C.5}$$

Any $v \neq 0$ satisfying (C.5) is said to be an *eigenvector* of Q. A complex number ν is said to be in the *resolvent set* of Q, denoted $\rho(Q)$, if $(\nu I - Q)^{-1}$ exists. In this case, the linear transformation $(\nu I - Q)^{-1}$ is called the *resolvent* of Q. For Q in $L(V)$, $\rho(Q)$ is nonempty. Define the *spectrum* of Q, denoted by $s(Q)$, to be the complement of $\rho(Q)$ in the set of complex numbers. When S is finite, $s(Q)$ equals the set of eigenvalues of Q, but it is a larger set in general settings.

Define the *spectral radius* of Q, denoted by $\sigma(Q)$, by

$$\sigma(Q) \equiv \lim_{n \to \infty} \|Q^n\|^{1/n}.$$

When $Q \in L(V)$ this limit exists. Clearly, $\sigma(Q) \geq 0$. It can be shown that $\sigma(Q) = \sup_{\nu \in s(Q)} |\nu|$. As a consequence of (C.3), $\|Q^n\|^{1/n} \leq \|Q\|$, so that $\sigma(Q) \leq \|Q\|$. Further, if P and Q commute,

$$\sigma(PQ) \leq \sigma(P)\sigma(Q).$$

Another useful property of the spectral radius is that, for $P, Q \in L(V)$,

$$\sigma(PQ) = \sigma(QP) \tag{C.6}$$

C.3   SERIES EXPANSIONS OF INVERSES

Theorem C.2 provides a sufficient condition for the existence of an inverse of a linear transformation Q and a Neumann series representation for it. The limit in (C.7) is in the sense of (C.4). The usual more restrictive conditions that ensure the existence of an inverse are expressed in Corollary C.3. We include a proof of Theorem C.2 because it contains arguments that are useful for comparing convergence rates of relative algorithms.

Theorem C.2.   Let Q be a bounded linear transformation on a Banach space V, and suppose $\sigma(I - Q) < 1$. Then Q^{-1} exists and satisfies

$$Q^{-1} = \lim_{N \to \infty} \sum_{n=0}^{N} (I - Q)^n. \tag{C.7}$$

Proof.   By hypothesis, there exists a $b < 1$ such that $\sigma(I - Q) < b < 1$. As a consequence of the definition of $\sigma(I - Q)$, given $\varepsilon > 0$, there exists an N^* such that, for $n \geq N^*$,

$$\left\| (I - Q)^n \right\|^{1/n} < b + \varepsilon < 1,$$

so that

$$\|(I - Q)^n\| < (b + \varepsilon)^n. \tag{C.8}$$

Now let $U_N = \sum_{n=0}^{N}(I - Q)^n$. Then, for $N > M \geq N^*$,

$$\|U_N - U_M\| = \left\| \sum_{n=M+1}^{N} (I - Q)^n \right\| \leq \sum_{n=M+1}^{N} \|(I - Q)^n\| \leq \sum_{n=M+1}^{N} (b + \varepsilon)^n.$$

Thus $\{U_N\}$ is a Cauchy sequence. From Lemma C.1, $L(V)$ is a Banach space, so there exists a U^* in $L(V)$ which satisfies

$$\lim_{N \to \infty} \| U_N - U^* \| = 0.$$

We now show that this limit equals Q^{-1}. Because

$$\|I - QU_N\| = \left\| I - \left[I - (I - Q)^{N+1} \right] \right\| \leq \left\| (I - Q)^{N+1} \right\|, \tag{C.9}$$

it follows from (C.8) that

$$\|I - QU^*\| = \lim_{N \to \infty} \|I - QU_N\| = 0.$$

Similarly, $\|I - U^*Q\| = 0$, so we conclude that $U^* = Q^{-1}$. $\qquad \square$

Since $\|I - Q\| > \sigma(I - Q)$, the following result follows immediately.

Corollary C.3. Suppose that $\|I - Q\| < 1$, then the conclusions of Theorem C.2 hold.

Usually we will be interested in the inverse of $I - Q$. We restate the following obvious consequence of Theorem C.2.

Corollary C.4. Let Q be a bounded linear transformation on a Banach space V, and suppose that $\sigma(Q) < 1$. Then $(I - Q)^{-1}$ exists and satisfies

$$(I - Q)^{-1} = \lim_{N \to \infty} \sum_{n=0}^{N} Q^n. \tag{C.10}$$

The following simple example illustrates these results.

Example C.1. Suppose

$$Q = \begin{bmatrix} 0.5 & 0.6 \\ 0.0 & 0.2 \end{bmatrix}.$$

Then $\|Q\| = 1.1$, but $\sigma(Q) = 0.5$. Consequently, by Corollary C.4, $(I - Q)^{-1}$ exists and satisfies (C.10).

C.4 CONTINUITY OF INVERSES AND PRODUCTS

In Chap. 8 we use the following results regarding continuity of matrix inverses and products. Note that they hold in greater generality then stated here. The proof of the first follows Ortega and Rheinboldt (1970, pp. 45–46).

Proposition C.5. Let $\{A_n\}$ be a sequence of matrices converging in norm to an invertible matrix A. Then for n sufficiently large A_n^{-1} exists and

$$\lim_{n \to \infty} \|A_n^{-1} - A^{-1}\| = 0.$$

Proof. First we establish the existence of A_n^{-1} and a bound on $\|A_n^{-1}\|$ in terms of $\|A^{-1}\| = \alpha$. Since $\|I - A^{-1}A_n\| \leq \|A^{-1}\| \|A - A_n\|$, given γ, $0 < \gamma < 1$, there exists an N such that $\|I - A^{-1}A_n\| < \gamma$ for $n \geq N$. Therefore, from Corollary C.3, $(A^{-1}A_n)^{-1}$ exists so that, for $n \geq N$, A_n^{-1} exists. Further, from (C.7),

$$\|A_n^{-1}\| = \left\|\left[I - \left(I - A^{-1}A_n\right)\right]^{-1}A^{-1}\right\| \leq \sum_{n=0}^{\infty} \alpha\gamma^n = \alpha/(1 - \gamma)$$

for $n \geq N$.

The result follows by noting that

$$\|A_n^{-1} - A^{-1}\| \leq \|A_n^{-1}\| \|A - A_n\| \|A^{-1}\|. \qquad \square$$

Lemma C.6. Let $\{A_n\}$ be a sequence of matrices which converges in norm to A with $\|A\| < \infty$, and $\{x_n\}$ a sequence of vectors which converges in norm to x with $\|x\| < \infty$. Then $\{A_n x_n\}$ converges in norm to Ax.

Proof. By the triangle inequality and properties of norms,

$$\|A_n x_n - Ax\| \leq \|A_n x_n - A_n x\| + \|A_n x - Ax\|$$

$$\leq \|A_n\| \|x_n - x\| + \|A_n - A\| \|x\|. \qquad (C.11)$$

Since $\|A_n\| \leq \|A\| + \|A - A_n\|$, given $\varepsilon > 0$, for n sufficiently large, $\|A_n\| < \|A\| + \varepsilon$. Therefore both expressions in (C.11) can be made arbitrarily small, establishing the result. $\square$

A similar proof establishes the following result.

Lemma C.7. Let $\{A_n\}$ and $\{B_n\}$ denote sequences of square matrices of the same dimension. Suppose that $\{A_n\}$ converges in norm to A with $\|A\| < \infty$, and that $\{B_n\}$ converges in norm to B with $\|B\| < \infty$. Then $\{A_n B_n\}$ converges in norm to AB.

APPENDIX D

Linear Programming

Relevant references for linear programming theory include Goldfarb and Todd (1989), Schrijver (1986), and Chvatal (1983). Kallenberg (1983) provides a nice summary of this material oriented to MDP application.

Let A be an $m \times n$ matrix, c a column vector in R^n, and b a column vector in R^m. The *primal linear programming (LP) problem* seeks an $x \in R^n$ which *minimizes* the objective function

$$c^T x$$

subject to constraints

$$Ax \geq b \quad \text{and} \quad x \geq 0. \tag{D.1}$$

Corresponding to the above problem, the *dual linear programming problem* seeks a $y \in R^m$ which *maximizes*

$$b^T y$$

subject to

$$A^T y \leq c \quad \text{and} \quad y \geq 0. \tag{D.2}$$

Call any x which satisfies (D.1) *primal feasible* and any y which satisfies (D.2) *dual feasible*. Call any primal feasible x^* for which $c^T x^* \leq c^T x$ for all primal feasible x *primal optimal*, and any dual feasible y^* for which $b^T y^* \geq b^T y$ for all dual feasible y *dual optimal*.

Define the *augmented* (or *standard*) *primal* LP problem as that of finding an $x \in R^n$ and $u \in R^m$ to minimize

$$c^T x$$

subject to

$$Ax - I_m u = b \quad \text{and} \quad x \geq 0, u \geq 0, \tag{D.3}$$

where I_k denotes a $k \times k$ identity matrix, and the *augmented dual* LP as that of finding a $y \in R^m$ and $v \in R^n$ to maximize

$$b^T y$$

subject to

$$A^T y + I_n v = c \quad \text{and} \quad y \geq 0, v \geq 0. \tag{D.4}$$

Refer to u and v as defined above as primal and dual *slack variables*.

If (x, u) satisfies (D.3) then x satisfies (D.1), and if x satisfies (D.1), then (x, u) satisfies (D.3) with $u = Ax - b$. Similarly, if (y, v) satisfies (D.4), then y satisfies (D.2), and if y satisfies (D.2), then (y, v) satisfies (D.4) with $v = c - A^T y$. We now relate optimal values for these two model forms. If (x^*, u^*) is optimal for the augmented primal, x^* is optimal for the primal, while, if x^* is optimal for the primal, then (x^*, u^*) is optimal for the augmented primal with $u = Ax^* - b$. Also, if (y^*, v^*) is optimal for the augmented dual, then y^* is optimal for the dual, and, if y^* is optimal for the dual, then (y^*, v^*) is optimal for the augmented dual with $v^* = c - A^T y^*$.

By a *basis*, we mean a maximal set of linearly independent columns of the augmented primal or dual constraint matrices $[A | -I_m]$ and $[A^T | I_n]$. Refer to the set of columns in the basis as *basic columns*. Let B (B^T) denote the submatrix of the augmented primal (dual) constraint matrix composed of the basic columns. By linear independence of its columns, B is invertible. We refer to the non-negative vector $w \in R^{m+n}$, in which all components not corresponding to the basic columns of $[A | -I_m]$ are set equal to 0 and the remaining components are set equal to $B^{-1}b$, as a *basic feasible solution* of the augmented primal LP. Similarly, a basic feasible solution of the dual LP is a non-negative $z \in R^{m+n}$, which equals 0 for nonbasic columns of $[A^T | I_n]$ and $(B^T)^{-1}c$ for basic columns.

We say that the vector $d \neq 0$ is a *direction* for the primal linear program if $d \geq 0$ and $Ad = 0$. When an *LP* contains directions, it may be unbounded. This is because if x is any feasible solution to the primal, then so is $x + kd$ for any positive scalar k.

With this terminology in place, we summarize the LP results we use in the book.

Theorem D.1.

    a. If either the primal or dual problem has a finite optimal solution, then it has an optimal solution which is a basic feasible solution.

    b. (Weak duality) If x is feasible for the primal, and y is feasible for the dual, then $c^T x \geq b^T y$.

    c. (Strong duality) If the primal problem has a bounded optimal solution x^*, then the dual has a bounded optimal solution y^*, and

$$c^T x^* = b^T y^*.$$

    d. (Complementary slackness) Necessary and sufficient conditions for (x^*, u^*) to be an optimal solution for the augmented primal, and (y^*, v^*) to be an optimal solution for the augmented dual, are that they are non-negative and satisfy

$$(y^*)^T u^* = (y^*)^T (Ax^* - b) = 0 \qquad\qquad (D.5)$$

and

$$(v^*)^T x^* = (c^T - (y^*)^T A)x^* = 0 \qquad\qquad (D.6)$$

Linear programming problems can be solved efficiently by the simplex method, its variants, and some recently proposed algorithms. Several excellent codes are available to implement these calculations.

Bibliography

S. C. Albright, Structural results for partially observable Markov decision processes, *Op. Res.* **27**, 1041–1053 (1978).

S. C. Albright and W. Winston, Markov models of advertising and pricing decisions, *Op. Res.* **27**, 668–681 (1979).

E. Altman and A. Shwartz, Markov decision problems and state-action frequencies, *SIAM J. Control Opt.* **29**, 786–809 (1991).

J. Anthonisse and H. Tijms, Exponential convergence of products of stochastic matrices, *J. Math. Anal. Appl.* **59**, 360–364 (1979).

M. Aoki, Optimal control of partially observable Markovian systems, *J. Franklin Inst.* **280**, 367–386 (1965).

A. Arapothasis, V. Borkar, E. Fernandez-Gaucherand, M. Ghosh, and S. Marcus, Discrete-time controlled Markov processes with average cost criterion: A survey, *SIAM J. Cont. Opt.* **31**, 282–344, (1993).

T. W. Archibald, K. I. M. McKinnon, and L. C. Thomas, Serial and parallel value iteration algorithms for discounted Markov decision processes, *Eur. J. Oper. Res.* **67**, 188–203, (1993).

K. J. Arrow, D. Blackwell, and M. A. Girshick, Bayes and minimax solutions of sequential decision problems, *Econometrica* **17**, 213–244 (1949).

K. J. Arrow, T. Harris, and J. Marschak, Optimal inventory policy, *Econometrica* **19**, 250–272 (1951).

K. J. Arrow, Historical background, in *Studies in the Mathematical Theory of Inventory and Production*, edited by K. J. Arrow, S. Karlin, and H. Scarf (Stanford University Press, Stanford, CA, 1958), pp. 3–15.

K. J. Arrow, S. Karlin, and H. Scarf (eds.), *Studies in the Mathematical Theory of Inventory and Production*, (Stanford University Press, Stanford, CA, 1958).

D. Assaf, Invariant problems in dynamic programming—average reward criterion, *Stoch. Proc. Appl.* **10**, 313–322 (1980).

D. Assaf, Extreme-point solutions in Markov decision processes, *J. Appl. Prob.* **20**, 835–842 (1983).

K. J. Astrom, Optimal control of Markov processes with incomplete state information, *J. Math. Anal. Appl.* **10**, 174–205 (1965).

M. Athans, The role and use of the stochastic linear-quadratic-Gaussian problem in control system design, *IEEE Trans. Autom. Control* **AC-16**, 529–552 (1971).

J. F. Baldwin and J. H. Sims-Williams, An on-line control scheme using a successive approximation in policy space approach, *J. Math. Anal. Appl.* **22**, 523–536 (1968).

M. Balinski, On solving discrete stochastic decision problems, Navy Supply System Research Study 2, Mathematica, Princeton, NJ, 1961.

R. G. Bartle, *The Elements of Real Analysis*, (Wiley, New York, 1964).

J. Bather, Optimal decision procedures for finite Markov chains I, *Adv. Appl. Prob.* **5**, 328–339 (1973a).

J. Bather, Optimal decision procedures for finite Markov chains II, *Adv. Appl. Prob.* **5**, 521–540 (1973b).

J. Bather, Optimal decision procedures for finite Markov chains III, *Adv. Appl. Prob.* **5**, 541–553 (1973c).

J. Bather, Dynamic programming, in *Analysis, Vol. 4, Handbook of Applicable Mathematics*, edited by W. Ledermann and S. Vajda (Wiley, New York, 1980).

M. Baykal-Gursoy and K. W. Ross, Variability sensitive Markov decision processes, *Math. Op. Res.* **17**, 558–571 (1992).

C. Bell, Characterization and computation of optimal policies for operating an $M/G/1$ queueing system with removable server: *Op. Res.* **16**, 208–218 (1971).

R. E. Bellman and D. Blackwell, On a particular non-zero sum game, RM-250, Rand Corp., Santa Monica, (1949).

R. E. Bellman and J. P. LaSalle, On non-zero sum games and stochastic processes, RM-212, Rand Corp., Santa Monica, (1949).

R. E. Bellman, The theory of dynamic programming, *Bull. Amer. Math. Soc.* **60**, 503–516, (1954).

R. Bellman, I. Glicksberg, and O. Gross, On the optimal inventory policy, *Man. Sci.* **2**, 83–104 (1955).

R. E. Bellman, A problem in the sequential design of experiments, *Sankhya* **16**, 221–229 (1956).

R. E. Bellman, *Dynamic Programming* (Princeton University Press, Princeton, NJ, 1957).

R. E. Bellman, *Adaptive Control Processes: A Guided Tour* (Princeton University Press, Princeton, NJ, 1961).

A. Berman and R. J. Plemmons, *Nonnegative Matrices in the Mathematical Sciences* (Academic, New York, 1979).

D. A. Berry and B. Fristedt, *Bandit Problems* (Chapman and Hall, London, 1985).

D. P. Bertsekas, Convergence of discretization procedures in dynamic programming, *IEEE Trans. Autom. Control* **20**, 415–419 (1975).

D. Bertsekas and S. Shreve, *Stochastic Optimal Control: The Discrete Time Case* (Academic, New York, 1978).

D. P. Bertsekas, *Dynamic Programming: Deterministic and Stochastic Models* (Prentice-Hall, Englewood Cliffs, NJ, 1987).

D. P. Bertsekas and D. A. Castanon, Adaptive aggregation for infinite horizon dynamic programming, *IEEE Trans. Autom. Control* **34**, 589–598 (1989).

D. P. Bertsekas and J. N. Tsitsiklis, An analysis of stochastic shortest path algorithms, *Math. Op. Res.* **16**, 580–595 (1991).

S. A. Bessler and A. F. Veinott, Jr., Optimal policy for a dynamic multi-echelon inventory model, *Nav. Res. Log. Quart.* **13**, 355–389 (1966).

F. J. Beutler and K. W. Ross, Optimal policies for controlled Markov chains with a constraint, *J. Math. Anal. Appl.* **112**, 236–252 (1985).

F. J. Beutler and K. W. Ross, Time average optimal constrained semi-Markov decision process, *Adv. Appl. Prob.* **18**, 341–359 (1986).

F. J. Beutler and K. W. Ross, Uniformization for semi-Markov decision processes under stationary policies, *J. Appl. Prob.* **24**, 399–420 (1987).

R. N. Bhattacharya and M. Majundar, Controlled semi-Markov model under long-run average rewards, *J. Stat. Planning Inference* **22**, 223–242 (1989).

R. Bielecki and J. A. Filar, Singularly perturbed Markov control problem: Limiting average cost, *Ann. Op. Res.* **28**, 153–168 (1991).

C. G. Bigelow, Bibliography on project planning and control by network analysis: 1959–1961, *Op. Res.* **10**, 728–731 (1962).

F. Black and M. Scholes, The pricing of options and corporate securities, *J. Pol. Econ.* **81**, 637–659 (1973).

T. R. Blackburn and D. R. Vaughan, Application of linear optimal control and filtering theory to the Saturn V launch vehicle, *IEEE Trans. Autom. Control* **AC-16**, 799–806 (1971).

D. Blackwell, On the functional equation of dynamic programming, *J. Math. Anal. Appl.* **2**, 273–276 (1961).

D. Blackwell, Discrete dynamic programming, *Ann. Math. Stat.* **33**, 719–726 (1962).

D. Blackwell, Memoryless strategies in finite-stage dynamic programming, *Ann. Math. Stat.* **35**, 863–865 (1964).

D. Blackwell, Discounted dynamic programming, *Ann. Math. Stat.* **36**, 226–235 (1965).

D. Blackwell, Positive dynamic programming, in *Proceedings of the 5th Berkeley Symposium on Mathematical Statistics and Probability*, Vol. 1 (University of California Press, Berkeley) pp. 415–418 (1967).

D. Blackwell, A Borel set not containing a graph, *Ann. Math. Stat.* **39**, 1345–1347 (1968).

D. Blackwell, On stationary policies, *J. R. Stat. Soc. Ser.* A, **133**, 33–38 (1970).

V. S. Borkar, Controlled Markov chains and stochastic networks, *SIAM J. Control Optim.*, **21**, 652–666 (1983).

V. S. Borkar, On minimum cost per unit time control of Markov chains, *SIAM J. Control Optim.*, **22**, 965–984 (1984).

V. S. Borkar, A convex analytic approach to Markov decision processes, *Prob. Theor. Relat. Fields*, **78**, 583–602 (1988).

V. S. Borkar, Control of Markov chains with long-run average cost criterion: The dynamic programming equations, *SIAM J. Control Optim.* **27**, 642–657 (1989).

V. S. Borkar, *Topics in Controlled Markov Chains*, Pitman Research Notes in Math. No. 240, Longman Scientific and Technical, Harlow, 1991.

L. Breiman, *Probability* (Addison-Wesley, Reading, MA, 1968).

B. W. Brown, On the iterative method of dynamic programming on a finite space discrete time Markov process, *Ann. Math. Stat.* **36**, 1279–1286 (1965).

S. L. Campbell and C. D. Meyer, Jr., *Generalized Inverses and Linear Transformations*, (Pitman, London, 1979).

L. Cantaluppi, Optimality of piecewise-constant policies in semi-Markov decision chains, *SIAM J. Control Opt.* **22**, 723–739 (1984a).

L. Cantaluppi, Computation of optimal policies in discounted semi-Markov decision processes, *OR Spektrum* **6**, 147–160 (1984b).

R. Cavazos-Cadena, Finite-state approximations for denumerable state discounted Markov decision processes, *Appl. Math. Opt.* **14**, 1–26 (1986).

R. Cavazos-Cadena, Necessary and sufficient conditions for a bounded solution to the optimality equation in average reward Markov decision chains, *Syst. Control Lett.* **10**, 71–78 (1988).

R. Cavazos-Cadena and J. Lasserre, Strong 1-optimal stationary policies in denumerable Markov decision processes, *Syst. Control Letters* **11**, 65–71 (1988).

R. Cavazos-Cadena, Necessary conditions for the optimality equations in average-reward Markov decision processes, *J. Appl. Math. Optim.* **19**, 97–112 (1989a).

R. Cavazos-Cadena, Weak conditions for the existence of optimal stationary policies in average Markov decisions chains with unbounded costs, *Kybernetika* **25**, 145–156 (1989b).

R. Cavazos-Cadena, Recent results on conditions for the existence of average optimal stationary policies, *Ann. Opns. Res.* **28**, 3–28 (1991a).

R. Cavazos-Cadena, Solution to the optimality equation in a class of Markov decision chains with the average cost criterion, *Kybernetika* **27**, 23–37 (1991b).

R. Cavazos-Cadena and L. I. Sennott, Comparing recent assumptions for the existence of optimal stationary policies, *Opns. Res. Lett.* **11**, 33–37 (1992).

A. Cayley, Mathematical questions with their solutions, No. 4528, *Educational Times*, **23**, 18 (1875).

S. T. Chanson, M. L. Puterman, and W. C. M. Wong, A Markov decision process model for computer system load control, *INFOR* **27**, 387–402 (1989).

H. Cheng, Computational methods for partially observed Markov decision processes, Ph.D. dissertation, The University of British Columbia, 1988.

R. Ya. Chitashvili, A controlled finite Markov chain with an arbitrary set of decisions, *Theory Prob. Appl.* **20**, 839–846 (1975).

S. S. Chitgopekar, Continuous time Markovian sequential control processes, *SIAM J. Control Optim.* **7**, 367–389 (1969).

K. J. Chung and M. J. Sobel, Risk sensitive discounted Markov decision processes, *SIAM J. Control Optim.* **25**, 49–62 (1986).

K. L. Chung, *Markov Chains with Stationary Transition Probabilities* (Springer-Verlag, New York, 1960).

Y. S. Chow, S. Moriguti, H. Robbins, and S. Samuels, Optimal selection based on relative rank, *Isr. J. Math.* **2**, 81–90 (1964).

Y. S. Chow, H. Robbins, and D. Siegmund, *Great Expectations: The Theory of Optimal Stopping* (Houghton-Mifflin, New York, 1971).

V. Chvatal, *Linear Programming* (Freeman, New York, 1983).

E. Çinlar, *Introduction to Stochastic Processes* (Prentice-Hall, Englewood Cliffs, NJ, 1975).

C. W. Clark, The lazy adaptable lions: a Markovian model of group foraging, *Anim. Behav.* **35**, 361–368 (1987).

C. W. Clark and D. A. Levy, Diel vertical migrations by juvenile sockeye salmon, *Am. Nat.* **131**, 271–290 (1988).

T. Colton, A model for selecting one of two medical treatments, *J. Am. Stat. Assoc.* **58**, 388–400 (1963).

J. C. Cox and M. Rubenstein, *Options Markets* (Prentice-Hall, Englewood Cliffs, NJ, 1985).

T. Crabill, D. Gross, and M. Magazine, A classified bibliography of research on optimal design and control of queues, *Op. Res.* **25**, 219–232 (1977).

R. Curtain and A. Pritchard, *Functional Analysis in Modern Applied Mathematics* (Academic, New York, 1977).

G. Dantzig, Discrete-variable extremum problems, *Op. Res.* **5**, 266 (1957).

J. S. De Cani, A dynamic programming algorithm for embedded Markov chains when the planning horizon is at infinity, *Man. Sci.* **10**, 716 (1964).

G. T. De Ghellinck, Les problèmes de décisions séquentielles, *Cah. Centre d'Etudes Rec. Oper.* **2**, 161–179 (1960).

R. Dekker, Denumerable Markov decision chains: Optimal policies for small interest rates, Ph.D. dissertation, The University of Leiden, (1985).

R. Dekker, Counterexamples for compact action Markov decision chains with average reward criteria, *Stoch. Models* **3**, 357–368 (1987).

R. Dekker and A. Hordijk, Average, sensitive and Blackwell optimal policies in denumerable Markov decision chains with unbounded rewards, *Math. Op. Res.* **13**, 395–421 (1988).

R. Dekker and A. Hordijk, Denumerable semi-Markov decision chains with small interest rates, *Ann. Opns. Res.* **28**, 185–212 (1991).

R. Dekker and A. Hordijk, Recurrence conditions for average and Blackwell optimality in denumerable state Markov decision chains, *Math. Op. Res.* **17**, 271–289 (1992).

R. Dembo and M. Haviv, Truncated policy iteration methods, *Op. Res. Lett.*, **3**, 243–246 (1984).

S. Demko and T. P. Hill, Decision processes with total cost criteria, *Ann. Prob.* **9**, 293–301 (1981).

S. Demko and T. P. Hill, On maximizing the average time at a goal, *Stoch. Proc. Appl.* **17**, 349–357 (1984).

E. V. Denardo, Contraction mappings in the theory underlying dynamic programming, *SIAM Rev.* **9**, 165–177 (1967).

E. V. Denardo and B. Fox, Multichain Markov renewal programs, *SIAM J. Appl. Math.* **16**, 468–487 (1968).

E. V. Denardo and B. L. Miller, An optimality condition for discrete dynamic programming with no discounting, *Ann. Math. Stat.* **39**, 1220–1227 (1968).

E. V. Denardo, Computing a bias-optimal policy in a discrete-time Markov decision problem, *Op. Res.* **18**, 272–289 (1970a).

E. V. Denardo, On linear programming in a Markov decision processes, *Man. Sci.* **16**, 281–288 (1970b).

E. V. Denardo, Markov renewal programs with small interest rates. *Ann. Math. Stat.* **42**, 477–496 (1971).

E. V. Denardo, A Markov decision problem, in *Mathematical Programming*, edited by T. C. Hu and S. M. Robinson (Academic, New York, 1973) pp. 33–68.

E. V. Denardo and U. G. Rothblum, Overtaking optimality for Markov decision chains, *Math. Op. Res.* **4**, 144–152 (1979).

E. V. Denardo and B. Fox, Shortest-route methods: 1—Reaching, pruning, and buckets, *Op. Res.*, **27**, 131–186 (1979a).

E. V. Denardo and B. Fox, Shortest-route methods: 2—Group knapsacks, expanded networks, and branch and bound, *Op. Res.* **27**, 548–566 (1979b).

E. V. Denardo, *Dynamic Programming: Models and Applications* (Prentice-Hall, Englewood Cliffs, NJ, 1982).

E. V. Denardo and U. G. Rothblum, Affine structure and invariant policies for dynamic programs, *Math. Op. Res.* **8**, 342–365 (1983).

F. D'Epenoux, Sur un problème de production et de stockage dans l'aléatoire, *Rev. Fr. Autom. Inf. Recherche Op.* **14**, 3 (1963); (Engl. trans.) *Man. Sci.* **10**, 98–108 (1963).

H. Deppe, On the existence of average optimal policies in semi-regenerative decision models, *Math. Op. Res.* **9**, 558–575 (1984).

C. Derman, On sequential decisions and Markov chains, *Man. Sci.* **9**, 16–24 (1962).

C. Derman, On sequential control processes, *Ann. Math. Stat.* **35**, 341–349 (1964).

C. Derman and R. E. Strauch, A note on memoryless rules for controlling sequential control processes, *Ann. Math. Stat.* **37**, 276–278 (1966).

C. Derman, Denumerable state Markovian decision processes—average cost criterion, *Ann. Math. Stat.* **37**, 1545–1554 (1966).

C. Derman and R. Strauch, A note on memoryless rules for controlling sequential decision processes, *Ann. Math. Stat.* **37**, 276–278 (1966).

C. Derman and A. F. Veinott, Jr., A solution to a countable system of equations arising in Markovian decision processes, *Ann. Math. Stat.* **38**, 582–584 (1967).

C. Derman, *Finite State Markovian Decision Processes* (Academic, New York, 1970).

E. W. Dijkstra, A note on two problems in connexion with graphs, *Numer. Math.* **1**, 269–271 (1959).

Y. M. J. Dirickx and M. R. Rao, Linear programming methods for computing gain-optimal policies in Markov decision models, *Cah. Centre d'Etudes Rec. Oper.* **21**, 133–142 (1979).

P. Dorato and A. H. Levis, Optimal linear regulators: The discrete-time case, *IEEE Trans. Autom. Control,* **AC-16**, 613–620 (1971).

B. T. Doshi, Continuous time control of Markov processes on an arbitrary state space: average return criterion, *Stoch. Proc. Appl.* **4**, 55–77 (1976).

R. M. Dressler and D. Tabak, Satellite tracking by combined optimal estimation and control techniques, *IEEE Trans. Autom. Control* **AC-16**, 833–840 (1971).

S. Dreyfus, An appraisal of some shortest-path algorithms, *Op. Res.* **17**, 395–412 (1969).

L. E. Dubins and L. J. Savage, *How to Gamble if You Must: Inequalities for Stochastic Processes* (McGraw-Hill, New York, 1965).

S. Durinovic, H. M. Lee, M. N. Katehakis, and J. A. Filar, Multiobjective Markov decisions process with average reward criterion, *Large Scale Syst.* **10**, 215–226 (1986).

A. Dvoretsky, J. Kiefer, and J. Wolfowitz, The inventory problem: I. Case of known distributions of demand, *Econometrica* **20**, 187 (1952).

A. Dvoretsky, J. Kiefer and J. Wolfowitz, On the optimal character of the (s, S) policy in inventory theory. *Econometrica*, **21**, 586–596, (1993).

E. B. Dynkin, Controlled random sequences, *Theor. Probability Appl.* **10**, 1–14 (1965).

E. B. Dynkin and A. A. Yushkevich, *Markov Processes: Theory and Problems* (Plenum, New York, 1969).

E. B. Dynkin and A. A. Yushkevich, *Controlled Markov Processes* (Springer-Verlag, New York, 1979).

J. E. Eagle, *A utility criterion for the Markov decision process*, Ph.D. dissertation, Stanford University, 1975.

J. E. Eagle, The optimal search for a moving target when the search path is constrained, *Opns. Res.* **32**, 1107–1115 (1984).

B. C. Eaves, Complementary pivot theory and Markov decision chains, in *Fixed Points: Algorithms and Applications*, edited by S. Karamardian (Academic, New York, 1977), pp. 59–85.

J. H. Eaton and L. A. Zadeh. Optimal pursuit strategies in discrete state probabalistic systems, *Trans. ASME, Series D,* 84, 23–29 (1962).

E. Erkut, Note: More on Morrison and Wheat's 'Pulling the Goalie Revisited,' *Interfaces* **17**, 121–123 (1987).

K. Fan, Subadditive functions on a distributive lattice and an extension of Szasz's inequality, *J. Math Anal. Appl.* **18**, 262–268 (1967).

A. Federgruen, P. J. Schweitzer, and H. C. Tijms, Contraction mappings underlying undiscounted Markov decision processes, *J. Math Anal. Appl.* **65**, 711–730 (1977).

A. Federgruen, A. Hordijk, and H. C. Tijms, A note on simultaneous recurrence conditions on a set of denumerable stochastic matrices, *J. Appl. Prob.* **15**, 842–847 (1978a).

A. Federgruen, A. Hordijk, and H. C. Tijms, Recurrence conditions in denumerable state Markov decision processes, in *Dynamic Programming and Its Applications*, edited by M. L. Puterman (Academic, New York, 1978b), pp. 3–22.

A. Federgruen and P. J. Schweitzer, Discounted and undiscounted value-iteration in Markov decision problems: A survey, in *Dynamic Programming and Its Applications*, edited by M. L. Puterman (Academic, New York, 1978), pp. 23–52.

A. Federgruen and H. C. Tijms, The optimality equation in average cost denumerable state semi-Markov decision problems, recurrency conditions and algorithms, *J. Appl. Prob.* **15**, 356–373 (1978).

A. Federgruen, A. Hordijk, and H. C. Tijms, Denumerable state semi-Markov decision processes with unbounded costs; average cost criterion, *Stoch. Proc. Appl.* **9**, 223–235 (1979).

A. Federgruen and P. J. Schweitzer, A survey of asymptotic value-iteration for undiscounted Markovian decision processes, in *Recent Developments in Markov Decision Processes*, edited by R. Hartley, L. C. Thomas, and D. J. White (Academic, New York, 1980), pp. 73–109.

A. Federgruen and D. Spreen, A new specification of the multichain policy iteration algorithm in undiscounted Markov renewal programs, *Man. Sci.* **26**, 1211–1217 (1980).

A. Federgruen, The rate of convergence for backwards products of a convergent sequence of finite Markov matrices, *Stoch. Proc. Appl.* **11**, 187–192 (1981).

A. Federgruen and P. J. Schweitzer, Nonstationary Markov decision problems with converging parameters, *J. Optim. Theor. Appl.* **34**, 207–241 (1981).

A. Federgruen, P. J. Schweitzer, and H. C. Tijms, Denumerable undiscounted semi-Markov decision processes with unbounded rewards, *Math. Op. Res.* **8**, 298–313 (1983).

A. Federgruen and J. P. Schweitzer, Successive approximation methods for solving nested functional equations in Markov decision problems, *Math. Op. Res.* **9**, 319–344 (1984a).

A. Federgruen and J. P. Schweitzer, A fixed point approach to undiscounted Markov renewal programs, *SIAM J. Alg. Disc. Meth.* **5**, 539–550 (1984b).

A. Federgruen and P. J. Schweitzer, Variational characterizations in Markov decision processes, *J. Math. Anal. Appl.* **117**, 326–357 (1986).

E. A. Feinberg, Stationary strategies in Borel dynamic programming, *Math. Op. Res.* **17**, 392–397 (1992).

E. A. Feinberg and A. Shwartz, Markov decision processes with weighted discounted criteria, *Math. Op. Res.* (to be published) (1994).

W. Feller, *An Introduction to Probability Theory and Its Applications*, Vol. *I*, 2nd ed. (Wiley, New York, 1957).

J. A. Filar and T. Schultz, Communicating MDPs: Equivalence and LP properties, *Op. Res. Lett.* **7**, 303–307 (1988).

J. A. Filar, L. C. M. Kallenberg, and H. M. Lee, Variance-penalized Markov decision processes, *Math. Op. Res.* **14**, 147–161 (1989).

C. H. Fine, A quality control model with learning effects, *Op. Res.* **36**, 437–444 (1968).

C. H. Fine and E. L. Porteus, Dynamic process improvement, *Op. Res.* **37**, 580–591 (1989).

L. Fisher and S. M. Ross, An example in denumerable decision processes, *Ann. Math. Stat.* **39**, 674–676 (1968).

W. H. Fleming and R. Rishel, *Deterministic and Stochastic Optimal Control* (Springer-Verlag, New York, 1975).

J. Flynn, Averaging versus discounting in dynamic programming: A counterexample, *Ann. Stat.* **2**, 411–413 (1974).

J. Flynn, Conditions for the equivalence of optimality criteria in dynamic programming, *Ann. Stat.* **4**, 936–953 (1976).

L. R. Ford and D. R. Fulkerson, *Flows in Networks* (Princeton University Press, Princeton, NJ, 1962).

B. L. Fox, Markov renewal programming by linear fractional programming, *SIAM J. Appl. Math.* **14**, 1418–1432 (1966).

B. L. Fox and D. M. Landi, An algorithm for identifying the ergodic subchains and transient states of a stochastic matrix, *Comm. ACM* **2**, 619–621 (1968).

B. L. Fox, (g, w)-Optima in Markov renewal programs, *Man. Sci.* **15**, 210–212 (1968).

B. L. Fox, Finite state approximations to denumerable state dynamic programs, *J. Math. Anal. Appl.* **34**, 665–670 (1971).

B. L. Fox, Discretizing dynamic programming, *J. Opt. Theor. Appl.* **11**, 228–234 (1973).

N. Furukawa, Markovian decision processes with compact action spaces, *Ann. Math. Stat.* **43**, 1612–1622 (1972).

N. Furukawa, Characterization of optimal policies in vector-valued Markovian decision processes, *Math. Op. Res.* **5**, 271 (1980).

D. Gale, On optimal development in a multi-sector economy, *Rev. Econ. Stud.* **34**, 1–18 (1967).

J. Gessford and S. Karlin, Optimal policies for hydroelectric operations, in *Studies in the Mathematical Theory of Inventory and Production*, edited by K. J. Arrow, S. Karlin, and H. Scarf (Stanford University Press, Stanford, 1958), pp. 179–200.

A. V. Gheorghe, *Decision Processes in Dynamic Probabilistic Systems*, (Kluwer Academic Publishers, Dordrecht, 1990).

M. K. Ghosh, Markov decision processes with multiple costs, *Op. Res. Lett.* **9**, **4**, 257–260 (1990).

K. D. Glazebrook, M. P. Bailey, and L. R. Whitaker, Cost rate heuristics for semi-Markov decision processes, *J. Appl. Prob.* **29**, 633–644 (1992).

J. Gilbert and F. Mosteller, Recognizing the maximum of a sequence, *J. Amer. Stat. Assoc.* **61**, 35–73 (1966).

R. G. Gillespie and T. Caraco, Risk-sensitive foraging strategies of two spider populations, *Ecology* **68**, 887–899 (1987).

D. Gillette, Stochastic games with zero stop probabilities, *Contributions to the Theory of Games, III, Annals of Mathematical Studies*, (Princeton University Press, Princeton), Vol. 39, pp. 71–187.

J. C. Gittins and D. M. Jones, A dynamic allocation index for the discounted multiarmed bandit problem, *Biometrika* **66**, 561–565 (1974).

J. Gittins and P. Nash, Scheduling, queues and dynamic, allocation indices, *Transactions of the Seventh Prague Conference on Information Theory, Statistical Decision Functions and Random Processes*, (D. Reidel, Dordrecht, 1977), pp. 191–202.

J. C. Gittins, Bandit processes and dynamic allocation indices, *J. R. Stat. Soc. Ser. B* **14**, 148–177 (1979).

K. Golabi, R. B. Kulkarni, and G. B. Way, A statewide pavement management system, *Interfaces* **12**, 5–21 (1982).

D. Goldfarb and M. J. Todd, Linear Programming, in *Handbook in Operations Research and Management Science, Vol. 1: Optimization*, edited by G. L. Nemhauser, A. H. G. Rinnooy Kan, and M. J. Todd (North-Holland, Amsterdam, 1989) pp. 73–170.

D. R. Grey, Non-negative matrices, dynamic programming and a harvesting problem, *J. Appl. Prob.* **21**, 685–694 (1984).

R. Grinold, Elimination of suboptimal actions in Markov decision problems, *Op. Res.* **21**, 848–851 (1973).

R. Grinold, Market value maximization and Markov dynamic programming, *Man. Sci.* **29**, 583–594 (1983).

P. R. Halmos, *Finite-Dimensional Vector Spaces*, 2nd ed. (Van Nostrand, Princeton, 1958).

J. M. Harrison, Discrete dynamic programming with unbounded rewards, *Ann. Math. Stat.* **43**, 636–644 (1972).

R. Hartley, L. C. Thomas, and D. J. White, (eds) *Recent Developments in Markov Decision Processes*, (Academic, New York, 1980).

R. Hartley, A simple proof of Whittle's bridging condition in dynamic programming, *J. Appl. Prob.* **17**, 1114–1116 (1980).

R. Hartley, A. C. Lavercombe, and L. C. Thomas, Computational comparison of policy iteration algorithms for discounted Markov decision processes, *Comput. Res.* **13**, 411–420 (1986).

N. A. J. Hastings, Some notes on dynamic programming and replacement, *Op. Res. Quart.* **19**, 453–464 (1968).

N. A. J. Hastings, Optimization of discounted Markov decision problems, *Op. Res. Quart.* **20**, 499–500 (1969).

N. A. J. Hastings, Bounds on the gain of a Markov decision process, *Op. Res.* **19**, 240–243 (1971).

N. A. J. Hastings and J. Mello, Tests for suboptimal actions in discounted Markov programming, *Man. Sci.* **19**, 1019–1022 (1973).

N. A. J. Hastings, A test for suboptimal actions in undiscounted Markov decision chains, *Man. Sci.* **23**, 87–91 (1976).

N. A. J. Hastings and J. A. E. E. van Nunen, The action elimination algorithm for Markov decision processes, in *Markov Decision Theory*, edited by H. C. Tijms and J. Wessels, Mathematical Centre Tract 93 (The Mathematical Centre, Amsterdam), pp. 161–170 (1977).

M. Haviv and L. V. D. Heyden, Perturbation bounds for the stationary probabilities of a finite Markov chain, *Adv. Appl. Prob.* **16**, 804–818 (1984).

M. Haviv, Block successive approximation for a discounted Markov decision model, *Stoch. Proc. Appl.* **19**, 151–160 (1985).

M. Haviv and M. L. Puterman, An improved algorithm for solving communicating average reward Markov decision processes, *Ann. Op. Res.* **28**, 229–242 (1991).

M. Haviv and M. L. Puterman, An unbiased estimator for the value of a discounted reward process, *Op. Res. Lett.* **11**, 267–272 (1992).

M. Haviv, On constrained Markov decision processes, Technical Report, Dept. of Econometrics, University of Sydney (1993).

M. I. Henig, Vector-valued dynamic programming, *SIAM J. Cont.* **21**, 490–499 (1983).

M. I. Henig, The shortest path problem with two objective functions, *Eur. J. Op. Res.* **25**, 281–291 (1985).

O. Hernandez-Lerma, *Adaptive Markov Control Processes* (Springer-Verlag, New York, 1989).

O. Hernandez-Lerma, J. C. Hennet, and J. B. Lasserre, Average cost Markov decision processes: Optimality conditions, *J. Math. Anal. Appl.* **158**, 396–406 (1991).

M. Hersh and S. P. Ladany, Optimal pole-valuting strategy, *Op. Res.* **37**, 172–175 (1989).

M. Herzberg and U. Yechiali, Criteria for selecting the relaxation factor of the value iteration algorithm for undiscounted Markov and semi-Markov decision processes, *Op. Res. Lett.* **10**, 193–202 (1991).

D. P. Heyman, Optimal operating policies for $M/G/1$ queueing systems, *Op. Res.* **16**, 362–382 (1968).

D. P. Heyman and M. J. Sobel, *Stochastic Models in Operations Research* (McGraw-Hill, New York, 1982), Vol. I.

D. P. Heyman and M. J. Sobel, *Stochastic Models in Operations Research* (McGraw-Hill, New York, 1984), Vol. II.

E. Hille and R. S. Phillips, *Functional Analysis and Semi-Groups* (American Mathematical Society, Providence, RI, 1957), Vol. 31.

C. J. Himmelberg, J. Parathasarathy, and F. S. van Vieck, Optimal plans for dynamic programming problems, *Math. Op. Res.* **1**, 390–394 (1976).

K. Hinderer, *Foundations of Non-Stationary Dynamic Programming with Discrete Time Parameter* (Springer-Verlag, New York, 1970).

K. Hinderer and G. Hubner, On approximate and exact solution for finite stage dynamic programs, in *Markov Decision Theory*, edited by H. Tijms and J. Wessels, (Mathematical Centre, Amsterdam, 1977) pp. 57–76.

M. Hlynka and J. Sheahan, The secretary problem for a random walk, *Stoch. Proc. Appl.* **28**, 317–325 (1988).

W. W. Hogan, Point-to-set maps in mathematical programming, *SIAM Rev.* **15**, 591–603 (1973).

W. J. Hopp, J. C. Bean, and R. L. Smith, A new optimality criterion for non-homogeneous Markov decision processes, *Op. Res.* **35**, 875–883 (1987).

W. J. Hopp, Sensitivity analysis in discrete dynamic programming, *J. Opt. Theor. Appl.* **56**, 257–269 (1988).

W. J. Hopp, Identifying forecast horizons in non-homogeneous Markov decision processes, *Op. Res.* **37**, 339–343 (1989).

A. Hordijk, *Dynamic Programming and Markov Potential Theory* (Mathematical Centre, Amsterdam, 1974).

A. Hordijk and H. Tijms, The method of successive approximations and Markovian decision problems, *Op. Res.* **22**, 519–521 (1974).

A. Hordijk and H. Tijms, A modified form of the iterative method of dynamic programming, *Ann. Stat.* **2**, 203–208 (1975).

A. Hordijk, P. J. Schweitzer, and H. Tijms, The asymptotic behaviour of the minimal total expected cost for the denumerable state Markov decision model, *J. Appl. Prob.* **12**, 298–305 (1975).

A. Hordijk and K. Sladky, Sensitive optimality criteria in countable state dynamic programming, *Math. Op. Res.* **2**, 1–14 (1977).

A. Hordijk and L. C. M. Kallenberg, Linear programming and Markov decision chains, *Man. Sci.* **25**, 352–362 (1979).

A. Hordijk and L. C. M. Kallenberg, On solving Markov decision problems by linear programming, in *Recent Developments in Markov Decision Processes*, edited by R. Hartley, L. C. Thomas, and D. J. White (Academic, New York, 1980), pp. 127–143.

A. Hordijk and F. A. van der Duyn Schouten, Average optimal policies in Markov decision drift processes with applications to a queueing and a replacement model, *Adv. Appl. Prob.* **15**, 274–303 (1983).

A. Hordijk and F. A. van der Duyn Schouten, Discretization and weak convergence in Markov decision drift processes, *Math. Op. Res.* **9**, 112–141 (1984).

A. Hordijk and L. C. M. Kallenberg, Constrained undiscounted stochastic dynamic programming, *Math. Op. Res.* **9**, 276–289 (1984).

A. Hordijk and F. A. van der Duyn Schouten, Markov decision drift processes: Conditions for optimality obtained by discretization, *Math. Op. Res.* **10**, 160–173 (1985).

A. Hordijk and M. L. Puterman, On the convergence of policy iteration in finite state undiscounted Markov decision processes: The unichain case, *Math. Op. Res.* **12**, 163–176 (1987).

A. Hordijk and F. Spieksma, Are limits of α-discount optimal policies Blackwell optimal? A counterexample. *Syst. Control Lett.* **13**, 31–41 (1989a).

A. Hordijk and F. Spieksma, Constrained admissions control to a queueing system *Adv. Appl. Prob.* **21**, 409–431 (1989b).

A. I. Houston and J. M. McNamara, Singing to attract a mate—a stochastic dynamic game, *J. Theor. Biol.* **129**, 171–180 (1986).

R. Howard, *Dynamic Programming and Markov Processes* (MIT Press, Cambridge, MA, 1960).

R. A. Howard, Semi-Markovian decision processes, *Proc. Intern. Stat. Inst.* (Ottawa, Canada, 1963).

R. A. Howard, *Dynamic Probabilistic Systems* (Wiley, New York, 1971).

R. A. Howard, Comments on origins and applications of Markov decision process, in *Dynamic Programming and its Applications*, edited by M. L. Puterman (Academic, New York, 1978).

R. A. Howard and J. E. Matheson, Risk sensitive Markov decision processes, *Man. Sci.* **18**, 356–369 (1972).

W. J. Hu, W. S. Lovejoy, and S. L. Shafer, Comparison of some suboptimal policies in medical drug therapy. *Operations Research* (to be published, 1994).

G. Hubner, Improved procedures for eliminating suboptimal actions in Markov programming by the use of contraction properties, in *Transactions of the Seventh Prague Conference on Information Theory, Statistical Decision Functions, Random Processes*, (D. Reidel, Dordrecht, 1977), pp. 257–263.

G. Hubner, Bounds and good policies in stationary finite-stage Markovian decision problems, *Adv. Appl. Prob.* **12**, 154–173 (1980).

G. Hubner, A unified approach to adaptive control of average reward Markov decision processes, *OR Spektrum* **10**, 161–166 (1988).

G. Hubner and M. Schal, Adaptive policy-iteration and policy-value-iteration for discounted Markovian decision processes, *Zeit. Op. Res.* **35**, 491–503 (1991).

A. Idzik, On Markov policies in continuous time discounted dynamic programming, in *Transactions of the Seventh Prague Conference on Information Theory, Statistical Decision Functions and Random Processes*, (D. Reidel, Dordrecht, 1977), pp. 265–275.

D. Iglehart, Optimality of (s, S) policies in the infinite horizon dynamic inventory problem, *Man. Sci.* **9**, 259–267 (1963).

R. Isaacs, *Games of Pursuit*, P-257, Rand Corp., Santa Monica, (1955).

R. Isaacs, *Differential Games* (Wiley, New York, 1965).

D. L. Isaacson and R. W. Madsen, *Markov Chains: Theory and Applications* (Wiley, New York, 1976).

S. C. Jaquette, Markov decision processes with a new optimality condition: discrete time, *Ann. Stat.* **1**, 496–505 (1973).

S. C. Jaquette, A utility criterion for Markov decision processes, *Man. Sci.* **23**, 43–49 (1976).

A. Jensen, Markov chains as an aid to the study of Markov processes, *Skand. Aktuariedtishr.* **36**, 87–91 (1953).

W. S. Jewell, Markov-renewal programming I: Formulation, finite return models; Markov-renewal programming II, infinite return models, example, *Op. Res.* **11**, 938–971 (1963).

E. Johnson, Optimality and computation of (s, S) policies in multi-item infinite horizon inventory problems, *Man. Sci.* **13**, 475–491 (1967).

E. Johnson, On (s, S) policies, *Man. Sci.* **15**, 80–101 (1968).

P. Kakumanu, Continuously discounted Markov decision model with countable state and action space, *Ann. Math. Stat.* **42**, 919–926 (1971).

P. Kakumanu, Non-discounted continuous time Markovian decision process with countable state space, *SIAM J. Control* **10**, 210–220 (1972).

P. Kakumanu, Continuous time Markovian decision processes; average criterion, *J. Math. Anal. Appl.* **52**, 173–188 (1975).

P. Kakumanu, Relation between continuous and discrete time Markov decision processes, *Nav. Res. Log. Quart.* **24**, 431–440 (1977).

R. Kalaba, On nonlinear differential equations, the maximum operation and monotone convergence, *J. Math. Mech.* **8**, 519–574 (1959).

D. Kalin, On the optimality of (σ, S) policies, *Math. Op. Res.* **5**, 293 (1980).

L. C. M. Kallenberg, *Linear Programming and Finite Markov Control Problems*, (Mathematical Centre, Amsterdam, 1983).

R. E. Kalman and R. W. Koepcke, Optimal synthesis of linear sampling control systems using generalized performance indexes, *Trans. ASME* **80**, 1820–1826 (1958).

L. V. Kantorovich, *Functional Analysis and Applied Mathematics*, Translated by C. D. Benster, NBS Report 1509, National Bureau of Standards, Los Angeles, 1952.

S. Karlin, The structure of dynamic programming models, *Naval Res. Log. Quart.* **2**, 285–294 (1955).

S. Karlin, Stochastic models and optimal policies for selling an asset, in *Studies in Applied Probability and Management Science*, edited by K. J. Arrow, S. Karlin, and H. Scarf (Stanford University Press, Palo Alto, CA) pp. 148–158 (1962).

S. Karlin, *A First Course in Stochastic Processes* (Academic, New York, 1969).

M. N. Katehakis and C. Derman, Optimal repair allocation in a series system, *Math. Op. Res.* **9**, 615–623 (1984).

M. N. Katehakis and A. F. Veinott, Jr., The multi-armed bandit problem: Decomposition and computation, *Math. Op. Res.* **12**, 262–268 (1985).

M. N. Katehakis and C. Derman, On the maintenance of systems composed of highly reliable components, *Man. Sci.* **35**, 551–560 (1989).

T. Kato, *Perturbation Theory for Linear Operations* (Springer-Verlag, New York, 1966).

R. W. Katz and A. H. Murphy, Qualtity/value relationships for imperfect weather forecasts in a prototype multistage decision-making model, *J. Forecasting* **9**, 75–86 (1990).

J. Kelley, Critical-path planning and scheduling: Mathematical basis, *Op. Res.* **9**, 296–300 (1961).

E. J. Kelly and P. L. Kennedy, A dynamic stochastic model of mate desertion, *Ecology*, **74**, 351–366 (1993).

J. G. Kemeny and J. L. Snell, *Finite Markov Chains* (Van Nostrand-Reinhold, New York, 1960).

J. G. Kemeny, Generalization of a fundamental matrix, *Lin. Alg. Appl.* **38**, 211–224 (1981).

J. Kiefer, Sequential minimax search for a maximum, *Proc. Amer. Math. Soc.* **4**, 502–506 (1953).

P. Kochel, A note on myopic solutions of Markovian decision processes and stochastic games, *Op. Res.* **30**, 1394–1398 (1985).

G. J. Koehler, A. B. Whinston, and G. P. Wright, The solution of Leontief substitution systems using matrix iterative methods, *Man. Sci.* 1295–1302 (1975).

G. J. Koehler, A case for relaxation methods in large scale linear programming, in *Large Scale Systems Theory and Applications*, edited by G. Guardabassi and A. Locatelli (IFAC, Pittsburgh, 1976), pp. 293–302.

G. J. Koehler, A complementarity approach for solving Leontief substitution systems and (generalized) Markov decision processes, *R.A.I.R.O. Recherche Op.* **13**, 75–80 (1979).

M. Kolonko, Strongly consistent estimation in a controlled Markov renewal model, *J. Appl. Prob.* **19**, 532–545 (1982).

M. Kolonko, Bounds for the regret loss in dynamic programming under adaptive control, *Z. Op. Res.* **27**, 17–37.

D. M. Kreps, Decision problems with expected utility criteria, I: Upper and lower convergent utility, *Math. Op. Res.* **2**, 45–53 (1977a).

D. M. Kreps, Decision problems with expected utility criteria, II: Stationarity, *Math. Op. Res.* **2**, 266–274 (1977b).

D. M. Kreps and E. Porteus, On the optimality of structured policies in countable stage decision processes. II: Positive and negative problems, *SIAM J. Appl. Math.* **32**, 457–466 (1977).

D. M. Kreps, Decision problems with expected utility criteria, III: Upper and lower transience, *SIAM J. Control Op.* **16**, 420–428 (1978).

D. M. Kreps and E. Porteus, Temporal resolution of uncertainty and dynamic choice theory, *Econometrica* **46**, 185–200 (1978).

D. M. Kreps and E. Porteus, Dynamic programming and dynamic choice theory, *Econometrica* **47**, 91–100 (1979).

N. V. Krylov, Construction of an optimal strategy for a finite controlled chain, *Theor. Prob.* **10**, 45–54 (1965).

P. R. Kumar, A survey of some results in stochastic adaptive control, *SIAM J. Control Opt.* **23**, 329–380 (1985).

P. R. Kumar and P. Varaiya, *Stochastic Systems: Estimation, Identification and Adaptive Control* (Prentice-Hall, Englewood Cliffs, NJ, 1986).

M. Kurano, The existence of a minimum pair of state and policy for Markov decision processes under the hypothesis of Doeblin, *SIAM J. Control Opt.* **27**, 296–307 (1989).

K. Kuratowski, *Topology* (Academic, New York, 1966).

B.-Z. Kurtaran and R. Sivan, Linear-quadratic-Gaussian control with one-step-delay sharing pattern, *IEEE Trans. Automat. Control* **19**, 571–574 (1974).

H. Kushner, Numerical methods for the solution of the degenerate nonlinear elliptic equations arising in optimal stochastic control theory, *IEEE Trans. Automat. Control* **13**, 344–353 (1968).

H. Kushner, *Introduction to Stochastic Control Theory* (Holt, Rinehart and Winston, New York, 1971).

H. Kushner and A. J. Kleinman, Accelerated procedures for the solution of discrete Markov control problems, *IEEE Trans. Automat. Control* **16**, 147–152 (1971).

H. Kushner, Numerical methods for stochastic control problems in continuous time, *SIAM J. Control Opt.* **28**, 999–1048 (1990).

H. J. Kushner and P. G. Dupuis, *Numerical methods for stochastic control problems in continuous time*, (Springer-Verlag, New York, 1992).

S. P. Ladany, Optimal starting height for pole-vaulting, *Op. Res.* **23**, 968–978 (1975).

B. F. Lamond, Matrix methods in queueing and dynamic programming, Ph.D. dissertation, The University of British Columbia, 1986.

B. F. Lamond and M. L. Puterman, Generalized inverses in discrete time Markov decision processes, *SIAM J. Matrix Anal. Appl.* **10**, 118–134 (1989).

B. F. Lamond, Optimal admission policies for a finite queue with bursty arrivals, *Ann. Op. Res.* **28**, 243–268 (1991).

B. F. Lamond and N. Drouin, MDPS—An interactive software package for the IBM personal computer, Special Document 92-108, Faculte des Sciences de l'Administration, Universite Laval (Quebec, Canada, 1992).

J. Lamperti, *Probability*, *A Survey of Mathematical Theory* (Benjamin, New York, 1966).

D. E. Lane, A partially observed model of decision making by fishermen, *Op. Res.* **37**, 240–254 (1989).

E. Lanery, Etude asymptotique des systèmes Markovien à commande, *Rev. Fr. Info. Res. Oper.* **1**, 3–56 (1967).

H.-J. Langen, Convergence of dynamic programming models, *Math. Op. Res.* **6**, 493–512 (1981).

J. B. Lasserre, Conditions for existence of average and Blackwell optimal stationary policies in denumerable Markov decision processes, *J. Math. Anal. Appl.* **136**, 479–490 (1988).

J. B. Lasserre, A new policy iteration scheme for MDPs using Schweitzer's format, *J. Appl. Prob.*, to be published (1994a).

J. B. Lasserre, Detecting optimal and non-optimal actions in average cost Markov decision processes, *J. Appl. Prob.* to be published (1994b).

L. Le Cam, in *More Mathematical People*, edited by D. J. Albens, G. L. Alexanderson, and C. Reed (Harcourt Brace Jovanovich, Boston, 1990).

A. Leizarowitz, Infinite horizon optimization for finite state Markov chain, *SIAM J. Control Op.* **25**, 1601–1681 (1987).

M. Lembersky, On maximal rewards and ε-optimal policies in continuous time Markov decision chains, *Ann. Stat.* **2**, 159–169 (1974).

S. A. Lippman, Criterion equivalence in discrete dynamic programming, *Op. Res.* **17**, 920–923 (1968a).

S. A. Lippman, On the set of optimal policies in discrete dynamic programming, *J. Math. Anal. Appl.* **24**, 440–445 (1968b).

S. A. Lippman and S. M. Ross, The streetwalker's dilemma: A job-shop model, *SIAM J. Appl. Math.* **20**, 336–342 (1971).

S. A. Lippman, Semi-Markov decision processes with unbounded rewards, *Man. Sci.* **19**, 717–731 (1973).

S. A. Lippman, Countable-state, continuous-time dynamic programming with structure, *Op. Res.* **24**, 447–490 (1974).

S. A. Lippman, On dynamic programming with unbounded rewards, *Man. Sci.* **21**, 1225–1233 (1975a).

S. A. Lippman, Applying a new device in the optimization of exponential queueing systems, *Op. Res.* **23**, 687–710 (1975b).

L. Liusternik and V. Sobolev, *Elements of Functional Analysis* (Ungar, New York, 1961).

C. E. Love, Source/Sink Inventory Control for Vehicle Retail Systems. *Comput. & Ops. Res.* **12**, 349–364 (1985).

W. S. Lovejoy, Some monotonicity results for partially observed Markov decision processes, *Op. Res.* **35**, 736–743 (1987).

W. S. Lovejoy, A survey of algorithmic methods for partially observed Markov decision processes, *Ann. Op. Res.* **28**, 47–66 (1991a).

W. S. Lovejoy, Computationally feasible bounds for partially observed Markov decision processes, *Op. Res.* **39**, 162–175 (1991b).

D. G. Luenberger, *Optimization by Vector Space Methods* (Wiley, New York, 1969).

J. MacQueen, A modified dynamic programming method for Markov decision problems, *J. Math. Anal. Appl.* **14**, 38–43 (1966).

A. Maitra, Dynamic programming for countable state systems, Ph.D. dissertation, University of California, Berkeley, 1964.

A. Maitra, Dynamic programming for countable state systems, *Sankhya* **27A**, 241–248 (1965).

A. Maitra, Discounted dynamic programming on compact metric spaces, *Sankhya* **30A**, 211–216 (1968).

A. Maitra, A note on positive dynamic programming, *Ann. Math. Stat.* **40**, 316–319 (1969).

A. Maitra and W. Sudderth, The optimal return operator in negative dynamic programming, *Math. Op. Res.* **17**, 921–931 (1992).

A. M. Makowski and A. Shwartz, Implementation issues for Markov decision processes, *IMA Stoch. Diff. Syst. Stoch. Control Theor. Appl.* **10**, 323–337 (1988).

P. Mandl, An iterative method for maximizing the characteristic root of positive matrices, *Rev. Roumaine Math. Pures Appl.* **XII**, 1317–1322 (1967).

P. Mandl, *Analytic Treatment of One-Dimensional Markov Processes* (Springer-Verlag, New York, 1968).

P. Mandl, Estimation and control in Markov chains, *Adv. Appl. Prob.* **6**, 40–60 (1974).

M. Mangel and C. W. Clark, Towards a unified foraging theory, *Ecology* **67**, 1127–1138 (1986).

M. Mangel, Oviposition site selection and clutch size in insects, *J. Math. Biol.* **25**, 1–22 (1987).

M. Mangel and C. W. Clark, *Dynamic Modeling in Behavioral Ecology* (Princeton University Press, Princeton, 1988).

A. Manne, Linear programming and sequential decisions, *Man. Sci.* **6**, 259–267 (1960).

E. Mann, Optimality equations and sensitive optimality in bounded Markov decision processes, *Optimization* **16**, 757–781 (1985).

E. Mann, The functional equations of undiscounted denumerable state Markov renewal programming, in *Semi-Markov Models*, edited by J. Janssen (Plenum, New York, 1986), pp. 79–96.

A. W. Marshall and I. Olkin, *Inequalities: Theory of Majorization and its Applications* (Academic, New York, 1979).

A. Martin-Lof, Existence of a stationary control for a Markov chain maximizing the average reward, *Op. Res.* **15**, 866–871 (1967a).

A. Martin-Lof, Optimal control of a continuous time Markov chain with periodic transition probabilities, *Op. Res.* **15**, 872–881 (1967b).

A. A. Markov and N. M. Nagorny, *The Theory of Algorithms* (Kluwer Academic, Dordrecht, 1988).

R. Mendelssohn, An iterative aggregation procedure for Markov decision processes, *Op. Res.* **30**, 62–73 (1982).

R. C. Merton, Theory of rational option pricing, *Bell J. Econ. Man. Sci.* **4**, 141–183 (1973).

B. L. Miller, Finite state continuous time Markov decision processes with a finite planning horizon, *SIAM J. Control Opt.* **6**, 266–280 (1968a).

B. L. Miller, Finite state continuous time Markov decision processes with an infinite planning horizon, *J. Math. Anal. Appl.* **22**, 552–569 (1968b).

B. L. Miller and A. F. Veinott, Jr., Discrete dynamic programming with a small interest rate, *Ann. Math. Stat.* **40**, 366–370 (1969).

B. L. Miller, Optimal consumption with a stochastic income, *Econometrica* **42**, 253–266 (1974).

H. Mine and S. Osaki, Some remarks on a Markovian decision process with an absorbing state, *J. Math. Anal. Appl.* **23**, 327–333 (1968).

G. J. Minty, A comment on the shortest route problem, *Op. Res.* **5**, 724 (1957).

G. E. Monahan, A survey of partially observable Markov decision processes: Theory, models, and algorithms, *Man. Sci.* **28**, 1–16 (1982).

D. G. Morrison, On the optimal time to pull the goalie: A Poisson model applied to a common strategy used in ice hockey, *TIMS Stud. Man. Sci.* **4**, 67–78 (1976).

D. G. Morrison, Misapplications reviews: Pulling the goalie revisited, *Interfaces* **16**, 28–34 (1986).

T. E. Morton, On the asymptotic convergence rate of cost differences for Markovian decision processes, *Op. Res.* **19**, 244–248 (1971).

T. E. Morton and W. E. Wecker, Discounting ergodicity and convergence for Markov decision processes, *Man. Sci.* **23**, 890–900 (1977).

T. Morton, The non-stationary infinite horizon inventory problem, *Man. Sci.* **24**, 1474–1482 (1978).

P. Naor, On the regulation of queue size by levying tolls, *Econometrica* **37**, 15–24 (1969).

J. M. Norman, Dynamic programming in tennis—When to use a fast serve, *J. Op. Res. Soc.* **36**, 75–77 (1985).

R. L. Nydick, Jr. and H. J. Weiss, More on Erkut's More on Morrison and Wheat's (Pulling the Goalie Revisited), *Interfaces* **19**, 45–48 (1989).

A. R. Odoni, On finding the maximal gain for Markov decision processes, *Op. Res.* **17**, 857–860 (1969).

K. Ohno, A unified approach to algorithms with a suboptimality test in discounted semi-Markov decision processes, *J. Op. Res. Soc. Jpn.* **24**, 296–324 (1981).

K. Ohno, Modified policy iteration algorithm with nonoptimality tests for undiscounted Markov decision processes, Working Paper, Dept. of Information System and Management Science, Konan University, Japan, 1985.

K. Ohno and K. Ichiki, Computing optimal policies for tandem queueing systems, *Op. Res.* **35**, 121–126 (1987).

K. O. Ondar, *The Correspondence Between A. A. Markov and A. A. Chuprov on the Theory of Probability and Mathematical Statistics* (Springer-Verlag, New York, 1981).

D. Ornstein, On the existence of stationary optimal strategies, *Proc. Am. Math. Soc.* **20**, 563–569 (1969).

J. M. Ortega and W. C. Rheinboldt, *Iterative Solutions of Nonlinear Equations in Several Variables* (Academic, New York, 1970).

S. Ozecki and S. R. Pliska, Optimal scheduling of inspections: A delayed Markov model with false positives and negatives, *Op. Res.* **39**, 261–273 (1991).

C. H. Papadimitriou and J. N. Tsitsiklis, The complexity of Markov decision processes, *Math. Op. Res.* **12**, 441–450 (1987).

M. Pinedo and L. Schrage, Stochastic shop scheduling: A survey, in *Deterministic and Stochastic Scheduling*, edited by M. Dempster, J. K. Lenstra, and A. Rinooy-Kan (Reidel, Dordrecht, 1982) pp. 181–196.

L. K. Platzman, Improved conditions for convergence in undiscounted Markov renewal programming, *Op. Res.* **25**, 529–533 (1977).

L. K. Platzman, Optimal finite horizon undiscounted control of finite probabilistic systems, *SIAM J. Control Optim.* **18**, 362–380 (1980).

L. K. Platzman, A feasible computational approach to infinite horizon partially observed Markov decision processes, Technical Note J-81-2, School of Industrial and Systems Engineering, Georgia Institute of Technology, Atlanta, GA, 1981.

S. R. Pliska, Controlled jump processes, *Stoch. Proc. Appl.* **3**, 259–282 (1975).

S. R. Pliska, Optimization of multitype branching processes, *Man. Sci.* **23**, 117–124 (1976).

S. R. Pliska, On the transient case for Markov decision processes with general state spaces, in *Dynamic Programming and Its Application*, edited by M. L. Puterman (Academic, New York, 1978), pp. 335–350.

S. R. Pliska, Accretive operators and Markovian decision processes, *Math. Op. Res.* **5**, 444–459 (1980).

M. Pollack and W. Wiebenson, Solutions of the shortest path problem—A review, *Op. Res.* **8**, 225–230 (1960).

M. Pollatschek and B. Avi-Itzhak, Algorithms for stochastic games with geometrical interpretation, *Man. Sci.* **15**, 399–413 (1969).

J. L. Popyack, R. L. Brown, and C. C. White, III, Discrete versions of an algorithm due to Variaya, *IEEE Trans. Autom. Control*, **24**, 503–504 (1979).

E. Porteus, Some bounds for discounted sequential decision processes, *Man. Sci.* **18**, 7–11 (1971).

E. Porteus, Bounds and transformations for discounted finite Markov decision chains, *Op. Res.* **23**, 761–784 (1975).

E. Porteus and J. Totten, Accelerated computation of the expected discounted return in a Markov chain, *Op. Res.* **26**, 350–358 (1978).

E. Porteus, Improved iterative computation of the expected discounted return in Markov and semi-Markov chains, *Z. Op. Res.* **24**, 155–170 (1980a).

E. Porteus, Overview of iterative methods for discounted finite Markov and semi-Markov decision chains, in *Recent Developments in Markov Decision Processes*, edited by R. Hartley, L. C. Thomas, and D. J. White (Academic, New York, 1980b), pp. 1–20.

E. Porteus, Computing the discounted return in Markov and semi-Markov chains, *Nav. Res. Log. Quart.* **28**, 567–578 (1981).

E. Porteus, Conditions for characterizing the structure of optimal strategies in infinite-horizon dynamic programs, *J. Opt. Theor. Appl.* **36**, 419–432 (1982).

E. Porteus, Survey of numerical methods for discounted finite Markov and semi-Markov chains, Notes from a talk at The Twelfth Conference on Stochastic Processes and Their Applications, Ithaca, NY, 1983.

E. Porteus, Stochastic inventory theory, in *Handbook of Operations Research*, edited by D. P. Heyman and M. J. Sobel (North-Holland, Amsterdam, 1990), Vol. 2, pp. 605–652.

R. E. Powell and S. M. Shah, *Summability Theory and Applications* (Van Nostrand Reinhold, London, 1972).

M. L. Puterman (ed.) *Dynamic Programming and Its Applications*, (Academic, New York, 1978).

M. L. Puterman and S. L. Brumelle, The analytic theory of policy iteration, in *Dynamic Programming and Its Application*, edited by M. L. Puterman (Academic, New York, 1978) pp. 91–114.

M. L. Puterman and M. C. Shin, Modified policy iteration algorithms for discounted Markov decision problems, *Man. Sci.* **24**, 1127–1137 (1978).

M. L. Puterman and S. L. Brumelle, On the convergence of policy iteration in stationary dynamic programming, *Math. Op. Res.* **4**, 60–69 (1979).

M. L. Puterman and M. C. Shin, Action elimination procedures for modified policy iteration algorithms, *Op. Res.* **30**, 301–318 (1982).

M. L. Puterman and L. C. Thomas, A note on computing optimal control limits for GI/M/1 queueing systems, *Man. Sci.* **33**, 939–943 (1987).

M. L. Puterman, Markov decision processes, in *Handbook of Operations Research*, edited by D. P. Heyman and M. J. Sobel (North-Holland, Amsterdam, 1990), Vol. 2, pp. 331–433.

W. T. Rasmussen, The candidate problem with unknown population size, *J. Appl. Prob.* **12**, 692–701 (1975).

D. Reetz, Solution of a Markovian decision problem by successive overrelaxation, *Z. Op. Res.* **17**, 29–32 (1973).

D. Reetz, A decision exclusive algorithm for a class of Markovian decision processes, *Z. Op. Res.* **20**, 125–131 (1976).

D. Reetz, Approximate solutions of a discounted Markovian decision process, *Dyn. Opt. Bonner Math. Schrift.* **98**, 77–92 (1977).

R. Righter, A resource allocation problem in a random environment, *Op. Res.* **37**, 329–338 (1989).

R. K. Ritt and L. I. Sennot, Optimal stationary policies in general state Markov decision chains with finite action set, *Math. Op. Res.* **17**, 901–909 (1992).

H. Robbins, Some aspects of the sequential design of experiments, *Bull. Amer. Math. Soc.* **58**, 527–536 (1952).

D. R. Robinson, Markov decision chains with unbounded costs and applications to the control of queues, *Adv. Appl. Prob.* **8**, 159–176 (1976).

D. R. Robinson, Optimality conditions for a Markov decision chain with unbounded cost, *J. Appl. Prob.* **17**, 996–1003 (1980).

G. Rode, Asymptotic properties of a finite state continuous time Markov decision process, *SIAM J. Control Opt.* **20**, 884–892 (1982).

I. V. Romanovsky, On the solvability of Bellman's equation for a Markovian decision process, *J. Math. Anal. Appl.* **42**, 485–498 (1973).

R. E. Rosenthal, J. A. White, and D. Young, Stochastic dynamic location analysis, *Man. Sci.* **24**, 645–653 (1978).

K. W. Ross, Randomized and past-dependent policies for Markov decision processes with multiple constraints, *Op. Res.* **37**, 474–477 (1989).

K. W. Ross and D. D. Yao, Optimal dynamic scheduling in Jackson networks, *IEEE Trans. Autom. Control* **34**, 47–53 (1989).

K. W. Ross and R. Varadarajan, Markov decision processes with sample path constraints: The communicating case, *Op. Res.* **37**, 780–790 (1989).

K. W. Ross and D. H. K. Tsang, Optimal circuit access policies in an ISDN environment: A Markov decision approach, *IEEE Trans. Commun.* **37**, 934–939 (1989a).

K. W. Ross and D. H. K. Tsang, The stochastic knapsack problem, *IEEE Trans. Commun.* **37**, 740–747 (1989b).

K. W. Ross and R. Varadarajan, Multichain Markov decision processes with a sample-path constraint: A decomposition approach, *Math. Op. Res.* **16**, 195–207 (1991).

S. M. Ross, Non-discounted denumerable Markovian decision models, *Ann. Math. Stat.* **39**, 412–423 (1968a).

S. M. Ross, Arbitrary state Markovian decision processes, *Ann. Math. Stat.* **39**, 2118–2122 (1968b).

S. M. Ross, *Applied Probability Models with Optimization Applications* (Holden Day, San Francisco, 1970a).

S. M. Ross, Average cost semi-Markov decision processes, *J. Appl. Prob.* **7**, 649–656 (1970b).

S. M. Ross, Quality control under Markovian deterioration, *Man. Sci.* **17**, 587–596 (1971).

S. M. Ross, *Introduction to Stochastic Dynamic Programming*, (Academic Press, New York, 1983).

U. G. Rothblum, Multiplicative Markov decision chains, Ph.D. dissertation, Stanford University (1975a).

U. G. Rothblum, Normalized Markov decision chains I: Sensitive discount optimality, *Op. Res.* **23**, 785–796 (1975b).

U. G. Rothblum and A. F. Veinott, Jr., Cumulative average optimality for normalized Markov decision chains, Working Paper, Dept. of Operations Research, Stanford University (1975).

U. G. Rothblum, Normalized Markov decision chains II: Optimality of nonstationary policies, *SIAM J. Control Opt.* **15**, 221–232 (1979a).

U. G. Rothblum, Iterated successive approximation for sequential decision processes, in *Stochastic Control and Optimization*, edited by J. W. B. van Overhagen and H. C. Tijms (Vrije University, Amsterdam, 1979b), pp. 30–32.

U. G. Rothblum and P. Whittle, Growth optimality for branching Markov decision chains, *Math. Op. Res.* **7**, 582–601 (1982).

U. G. Rothblum, Multiplicative Markov decision chains, *Math. Op. Res.* **9**, 6 (1984).

H. L. Royden, *Real Analysis* (MacMillan, New York, 1963).

J. Rust, Optimal replacement of GMC bus engines: An empirical model of Harold Zurcher, *Econometrica* **55**, 999–1033 (1987).

V. V. Rykov, Markov decision processes with finite state and decision spaces, *Theor. Prob. Appl.* **11**, 303–311 (1966).

K. Sawaki and A. Ichikawa, Optimal control for partially observable Markov decision processes over an infinite horizon, *J. Op. Res. Soc. Jpn*, **21**, 1–15 (1978).

K. Sawaki, Transformation of partially observable Markov decision process into piecewise linear ones, *J. Math. Anal. Appl.* **91**, 112–118 (1983).

H. E. Scarf, The optimality of (s, S) policies in the dynamic inventory problem, in *Studies in the Mathematical Theory of Inventory and Production*, edited by K. Arrow, S. Karlin, and P. Suppes (Stanford University Press, Stanford, 1960) pp. 196–202.

M. Schäl, On continuous dynamic programming with discrete time parameters, *Z. Wahrsch. Verw. Gebiete* **21**, 279–288 (1972).

M. Schäl, Conditions for optimality in dynamic programming and for the limit of n-stage optimal policies to be optimal, *Z. Wahrsch. Verw. Gebiete*, **32**, 179–196 (1975).

M. Schäl, On the optimality of (s, S) policies in dynamic inventory models with finite horizon, *SIAM J. Appl. Math.* **13**, 528–537 (1976).

M. Schäl, (ed.) *Dynamische Optimierung*, (Bonner Mathematische Schriften, Bonn, 1977).

M. Schäl, An operator theoretical treatment of negative dynamic programming, in *Dynamic Programming and its Applications*, edited by M. L. Puterman (Academic, New York, 1978) pp. 351–368.

M. Schäl, Estimation and control in discounted dynamic programming, *Stochastics*, **20**, 51–71 (1987).

M. Schäl, Stationary policies in dynamic programming models under compactness assumptions, *Math. Op. Res.* **8**, 366–372 (1983).

M. Schäl, On the second optimality equation for semi-Markov decision models, *Math. Op. Res.* **17**, 470–486 (1992).

A. Schrijver, *Linear and Integer Programming* (Wiley, New York, 1986).

P. J. Schweitzer, *Pertusrbation Theory and Markov Decision Chains*, Ph.D. dissertation, Massachusetts Institute of Technology (1965).

P. J. Schweitzer, Perturbation theory and finite Markov chains, *J. Appl. Prob.* **5**, 401–413 (1968).

P. J. Schweitzer, Perturbation theory and undiscounted Markov renewal programming, *Op. Res.* **17**, 716–727 (1969).

P. J. Schweitzer, Iterative solution of the functional equations of undiscounted Markov renewal programming, *J. Math. Anal. Appl.* **34**, 495–501 (1971a).

P. J. Schweitzer, Multiple policy improvements in undiscounted Markov renewal programming, *Op. Res.* **19**, 784–793 (1971b).

P. J. Schweitzer, Data transformations for Markov renewal programming, talk at National ORSA Meeting, Atlantic City, NJ (1972).

P. J. Schweitzer and A. Federgruen, The asymptotic behavior of value-iteration in Markov decision problems, *Math. Op. Res.* **2**, 360–381 (1977).

P. J. Schweitzer and A. Federgruen, The functional equations of undiscounted Markov renewal programming, *Math. Op. Res.* **3**, 308–321 (1978a).

P. J. Schweitzer and A. Federgruen, Foolproof convergence in multichain policy iteration, *J. Math. Anal. Appl.* **64**, 360–368 (1978b).

P. J. Schweitzer and A. Federgruen, Geometric convergence of value iteration in multichain Markov decision problems, *Adv. Appl. Prob.* **11**, 188–217 (1979).

P. J. Schweitzer, Solving MDP functional equations by lexicographic optimization, *RAIRO Recherche Op.* **16**, 91–98 (1982).

P. J. Schweitzer, On the solvability of Bellman's functional equations for Markov renewal programming, *J. Math. Anal. Appl.* **96**, 13–23 (1983).

P. J. Schweitzer, A value-iteration scheme for undiscounted multichain Markov renewal programs, *Z. Op. Res.* **28**, 143–152 (1984a).

P. J. Schweitzer, On the existence of relative values for undiscounted Markovian decision process with a scalar gain rate, *J. Math. Anal. Appl.* **104**, 67–78 (1984b).

P. J. Schweitzer, On the existence of relative values for undiscounted multichain Markov decision processes, *J. Math. Anal. Appl.* **102**, 449–455 (1984c).

P. J. Schweitzer, Iterative bounds on the relative value vector in undiscounted Markov renewal programming, *Z. Op. Res.* **29**, 269–284 (1985a).

P. J. Schweitzer, On undiscounted Markovian decision processes with compact action spaces, *RAIRO Recherche Op.* **19**, 71–86 (1985b).

P. J. Schweitzer, Solving Markovian decision processes by successive elimination of variables, Working paper no. QM8524, Graduate School of Management, University of Rochester, NY (1985c).

P. J. Schweitzer, The variational calculus and approximation in policy space for Markovian decision processes, *J. Math. Anal. Appl.* **111**, 14–25 (1985d).

P. J. Schweitzer, Generalized polynomial approximations in Markovian decision processes, *J. Math. Anal. Appl.* **110**, 568–582 (1985e).

P. J. Schweitzer, M. L. Puterman, and K. W. Kindle, Iterative aggregation-disaggregation procedures for discounted semi-Markov reward processes, *Op. Res.* **33**, 589–605 (1985).

P. J. Schweitzer and K. W. Kindle, Iterative aggregation for solving undiscounted semi-Markovian reward processes, *Commun. Stat. Stoch. Models* **2**, 1–41 (1986).

P. J. Schweitzer, A Brouwer fixed-point mapping approach to communicating Markov decision processes, *J. Math. Anal. Appl.* **123**, 117–130 (1987a).

P. J. Schweitzer, Bounds on the fixed point of a monotone contraction operator, *J. Math. Anal. Appl.* **123**, 376–388 (1987b).

E. Seneta, Finite approximations to infinite non-negative matrices, *Proc. Camb. Phil. Soc.* **63**, 983–992 (1967).

E. Seneta, *Non-negative Matrices and Markov Chains*, (Springer-Verlag, New York, 1981).

L. I. Sennott, A new condition for the existence of optimum stationary policies in average cost Markov decision processes, *Op. Res. Lett.* **5**, 17–23 (1986a).

L. I. Sennott, A new condition for the existence of optimum stationary policies in average cost Markov decision processes-unbounded costs case, in *Proceedings of the 25th, IEEE Conference on Decision and Control*, (Athens, Greece, 1986b) pp. 1719–1721.

L. I. Sennott, Average cost optimal stationary policies in infinite state Markov decision processes with unbounded costs, *Op. Res.* **37**, 626–633 (1989a).

L. I. Sennott, Average cost semi-Markov decision processes and the control of queueing systems, *Prob. Eng. Informat. Sci.* **3**, 247–272 (1989b).

L. I. Sennott, Value iteration in countable state average cost Markov decision processes with unbounded cost, *Ann. Op. Res.* **28**, 261–272 (1991).

L. I. Sennott, The average cost optimality equation and critical number policies, *Prob. Eng. Info. Sci.* **7**, 47–67, (1993a).

L. I. Sennott, Constrained average cost Markov decision chains, *Prob. Eng. Info. Sci.* **7**, 69–83, (1993b).

R. Serfozo, Monotone optimal policies for Markov decision processes, *Math. Prog. Study*, **6**, 202 (1976).

R. Serfozo, An equivalence between continuous and discrete time Markov decision processes, *Op. Res.* **27**, 616–620 (1979).

R. Serfozo, Optimal control of random walks, birth and death processes, and queues, *Adv. Appl. Prob.* **13**, 61–83 (1981).

J. Shapiro, Turnpike planning horizons for a Markovian decision model, *Man. Sci.* **14**, 292–300 (1968).

L. S. Shapley, Stochastic games, *Proc. Nat. Acad. Sci. U.S.A.* **39**, 1095–1100 (1953).

T. J. Sheskin, Successive approximations in value determination for a Markov decision process, *Op. Res.* **35**, 784–786 (1987).

S. E. Shreve and D. P. Bertsekas, Dynamic programming in Borel spaces, in *Dynamic Programming and its Applications*, edited by M. L. Puterman, (Academic, New York, 1978) pp. 115–130.

S. S. Sheu and K.-J. Farn, A sufficient condition for the existence of a stationary 1-optimal plan in compact action Markovian decision processes, in *Recent Developments in Markov Decision Processes*, edited by R. Hartley, L. C. Thomas, and D. J. White (Academic, New York, 1980a), pp. 111–126.

S. S. Sheu and K.-J. Farn, A sufficient condition for the existence of a stationary average optimal plan in compact action Markovian decision processes, *Proc. Nat. Sci. Council* **4**, 143–145 (1980b).

A. Shwartz and A. M. Makowski, Comparing policies in Markov decision processes: Mandl's lemma revisited, *Math. Op. Res.* **15**, 155–174 (1990).

J. Siededrsleben, Dynamically optimized replacement with a Markovian renewal process, *J. Appl. Prob.* **18**, 641–651 (1981).

K. Siegrist, Optimal occupation in the complete graph, *Prob. Eng. Inf. Sci.* **7**, 369–385 (1993).

H. A. Simon, Dynamic programming under uncertainty with a quadratic criterion function, *Econometrica* **24**, 74–81 (1956).

K. Sladky, On the set of optimal controls for Markov chains with rewards, *Kybernetika*, **10**, 350–367 (1974).

K. Sladky, On dynamic programming recursions for multiplicative Markov decision chains, *Math. Prog. Study*, **6**, 216–226 (1976).

K. Sladky, On the optimality conditions for semi-Markov decision processes, in *Transactions of the Seventh Prague Conference on Information Theory, Statistical Decision Functions and Random Processes*, (D. Reidel, Dordrecht, 1977), pp. 555–566.

R. Smallwood and E. Sondik, The optimal control of partially observed Markov processes over the finite horizon, *Op. Res.* **21**, 1071–1088 (1973).

M. Sniedovich, Some comments on preference order dynamic programming models, *J. Math. Anal. Appl.* **79**, 489–501 (1981).

M. J. Sobel, Optimal average-cost policy for a queue with start-up and shut-down costs, *Op. Res.* **17**, 145–162 (1969).

M. J. Sobel, The variance of discounted Markov decision processes, *J. Appl. Prob.* **19**, 794–802 (1982).

M. J. Sobel, Optimal operation of queues, in *Mathematical Methods in Queueing Theory*, edited by A. B. Clarke (Springer-Verlag, Berlin, 1984), pp. 233–261.

M. J. Sobel, Maximal mean/standard deviation ratio in an undiscounted MDP, *Op. Res. Lett.* **4**, 157–159 (1985).

M. J. Sobel, An overview of nonstandard criteria in Markov decision processes, in *Proceedings Airlie House NSF Conference, Airlie, Virginia*, (NSF, Washington, 1988).

E. J. Sondik, The optimal control of partially observable Markov processes, Ph.D. dissertation, Stanford University, 1971.

E. Sondik, The optimal control of partially observable Markov processes over the infinite horizon, *Op. Res.* **26**, 282–304 (1978).

F. Spieksma, Geometrically ergodic Markov chains and the optimal control of queues, Ph.D. dissertation, Leiden University, 1990.

F. Spieksma, The existence of sensitive optimal policies in two multi-dimensional queueing models, *Ann. Op. Res.* **28**, 273–296 (1991).

D. Stengos and L. C. Thomas, The blast furnaces problem, *European J. Operat. Res.* **4**, 330–336 (1980).

S. S. Stidham, Jr., Socially and individually optimal control of arrivals to a GI/M/1 queue, *Man. Sci.* **24**, 1598–1610 (1978).

S. S. Stidham, Jr., Optimal control of admission to a queueing system, *IEEE Trans. Autom. Control* **AC-30**, 705–713 (1989).

S. S. Stidham, Jr. and R. R. Weber, Monotonic and insensitive optimal policies for control of queues with unbounded costs, *Op. Res.* **87**, 611–625 (1989).

N. L. Stokey and R. E. Lucas, *Recursive Methods in Economic Dynamics* (Harvard University Press, Cambridge, MA, 1989).

L. D. Stone, Necessary and sufficient conditions for optimal control of semi-Markovian processes, *SIAM J. Control Opt.* **11**, 367–389 (1973).

R. Strauch, Negative dynamic programming, *Ann. Math. Stat.* **37**, 871–890 (1966).

C. Striebel, Sufficient statistics in the control of stochastic systems, *J. Math. Anal. Appl.* **12**, 576–592 (1965).

A. J. Swersey, A Markovian decision model for deciding how many fire companies to dispatch, *Man. Sci.* **28**, 352–365 (1982).

R. Sznajder and J. A. Filar, Some comments on a theorem of Hardy and Littlewood *J. Opt. Theor. Appl.*, **75**, 201–208 (1992).

H. M. Taylor, Markovian sequential replacement processes, *Ann. Math. Stat.* **36**, 1677–1694 (1965).

H. Theil, A note on certainty equivalence in dynamic planning, *Econometrica*, **25**, 346–349 (1957).

L. C. Thomas, Connectedness conditions for denumerable state Markov decision processes, *Recent Developments in Markov Decision Processes*, edited by R. Hartley, L. C. Thomas, and D. F. White (Academic, New York, 1980), pp. 181–204.

L. C. Thomas, Second order bounds for Markov decision processes, *J. Math. Anal. Appl.* **80**, 294–297 (1981).

L. C. Thomas, R. Hartley, and A. C. Lavercombe, Computational comparison of value iteration algorithms for discounted Markov decision processes, *Op. Res. Lett.* **2**, 72–76 (1983).

L. C. Thomas and D. Stengos, Finite state approximation algorithms for average cost denumerable state Markov decision processes, *OR Spek.* **7**, 27–37 (1985).

L. C. Thomas, Replacement of systems and components in renewal decision problems, *Op. Res.* **30**, 404–411 (1985).

W. R. Thompson, On the likelihood that one unknown probability exceeds another in view of the evidence of two samples, *Biometrika* **25**, 275–294 (1933).

W. R. Thompson, On the theory of apportionment, *Am. J. Math.* **57**, 450 (1935).

H. C. Tijms and J. Wessels (eds.) *Markov Decision Theory*, (The Mathematical Centre, Amsterdam, 1977).

H. C. Tijms and F. A. van der Duyn Schouten, A Markov decision algorithm for optimal inspection and revision in a maintenance system with partial information, *Eur. J. Op. Res.* **21**, 245–253 (1985).

H. C. Tijms, *Stochastic Modelling and Analysis, A Computational Approach* (Wiley, Chichester, England, 1986).

D. Topkis, Minimizing a submodular function on a lattice, *Op. Res.* **26**, 305–321 (1978).

P. Tseng, Solving H-horizon, stationary Markov decision problems in time proportional to $Log(H)$ *Op. Res. Lett.* **9**, 287–297 (1990).

R. van Dawen, A note on the characterization of optimal return functions and optimal strategies for gambling problems, *Ann. Stat.* **13**, 832–835 (1985a).

R. van Dawen, Gambling problems with finite state space, *Zeit. Ang. Math. Mech.* **65**, T299–T301 (1985b).

R. van Dawen, Negative dynamic programming, Preprint no. 758, University of Bonn (1985c).

R. van Dawen, Finite state dynamic programming with the total reward criterion, *Z. Op. Res.* **30**, A1–A4 (1986a).

R. van Dawen, Limit-optimal strategies in countable state decision problems, Preprint no. 779, University of Bonn (1986b).

R. van Dawen, Pointwise and uniformly good stationary strategies in dynamic programming models, *Math. Op. Res.* **11**, 521–535 (1986b).

J. S. Vandergraft, Newton's method for convex operations in partially ordered spaces, *SIAM J. Num. Anal.* **4**, 406–432 (1967).

J. van der Wal, A successive approximation algorithm for an undiscounted Markov decision process, *Computing* **17**, 157–162 (1976).

J. van der Wal and J. A. E. E. van Nunen, A note on the convergence of the value oriented successive approximations method, COSO Note R 77-05, Department of Mathematics, Eindhoven University of Technology (1977).

J. van der Wal, *Stochastic Dynamic Programming* (The Mathematical Centre, Amsterdam, 1981).

J. van der Wal, On stationary strategies in countable state total reward Markov decision processes, *Math. Op. Res.* **9**, 290–300 (1984).

J. van der Wal and P. J. Schweitzer, Iterative bounds on the equilibrium distribution of a finite Markov chain, *Prob. Eng. Inf. Sci.* **1**, 117–131 (1987).

K. M. van Hee, A. Hordijk, and J. van der Wal, Successive approximations for convergent dynamic programming, in *Markov Decision Theory*, edited by H. C. Tijms and J. Wessels (The Mathematical Centre, Amsterdam, 1976) pp. 183–211.

K. M. van Hee, Markov strategies in dynamic programming, *Math. Op. Res.* **3**, 37–41 (1978).

J. A. E. E. van Nunen, A set of successive approximation methods for discounted Markovian decision problems, *Z. Op. Res.* **20**, 203–208 (1976a).

J. A. E. E. van Nunen, *Contracting Markov Decision Processes*, (The Mathematical Centre, Amsterdam, 1976b).

J. A. E. E. van Nunen and J. Wessels, A note on dynamic programming with unbounded rewards, *Man. Sci.* **24**, 576–580 (1978).

J. van Nunen and M. L. Puterman, Computing optimal control limits for GI/M/s queueing systems with controlled arrivals, *Man. Sci.* **29**, 725–734 (1983).

C. C. van Weizsacker, Existence of optimal programs of accumulation for an infinite time horizon, *Rev. Econ. Stud.* **32**, 85–104 (1965).

R. Varaiya, J. C. Walrand, and C. Buyukkoc, Extensions of the multiarmed bandit problem: The discounted case, *IEEE Trans. Autom. Control* **AC-30**, 426–439 (1985).

P. Varaiya, Optimal and suboptimal stationary controls for Markov chains, *IEEE Trans. Autom. Control* **23**, 388–394 (1978).

R. Varga, *Matrix Iterative Analysis* (Prentice-Hall, Englewood Cliffs, NJ, 1962).

A. F. Veinott, Jr. and H. M. Wagner, Computing optimal (s, S) inventory policies, *Man. Sci.* **11**, 525–552 (1965).

A. F. Veinott, Jr., Optimal policies in dynamic, single product, nonstationary inventory model with several demand class, *Op. Res.* **13**, 761–778 (1965).

A. F. Veinott, Jr., On finding optimal policies in discrete dynamic programming with no discounting, *Ann. Math. Stat.* **37**, 1284–1294 (1966a).

A. F. Veinott, Jr., On the optimality of (*s*, *S*) inventory policies: New conditions and a new proof, *SIAM J. Appl. Math.* **14**, 1067–1083 (1966b).

A. F. Veinott, Jr., The status of mathematical inventory theory, *Man. Sci.* **12**, 745–777 (1966c).

A. F. Veinott, Jr., Extreme points of Leontief substitution systems, *Lin. Alg. Appl.* **1**, 181–194 (1968).

A. F. Veinott, Jr., Minimum concave-cost solution of Leontief substitution models of multi-facility inventory systems, *Op. Res.* **17**, 262–291 (1969a).

A. F. Veinott, Jr., On discrete dynamic programming with sensitive discount optimality criteria, *Ann. Math. Stat.* **40**, 1635–1660 (1969b).

A. F. Veinott, Jr., Optimally of myopic inventory policies for several substitute products, *Man. Sci.*, **15**, 284–304 (1969c).

A. F. Veinott, Jr., Course notes, Stanford University, unpublished, (1969d).

A. F. Veinott, Jr., Least *d*-majorized network flows with inventory and statistical applications, *Man. Sci.* **17**, 547–567 (1971).

A. F. Veinott, Jr., Markov decision chains, in *Studies in Optimization*, edited by G. B. Dantzig and B. C. Eaves (American Mathematical Association, Providence, RI, 1974) pp. 124–159.

I. Venenzia, Optimal insurance premium rates when the distribution of claims is unknown, *J. Appl. Prob.* **16**, 676–684 (1979).

L. M. M. Veugen, J. van der Wal, and J. Wessels, Aggregation and disaggregation in Markov decision models for inventory control, *Eur. J. Op. Res.* **20**, 248–254 (1985).

R. G. Vickson, Generalized value bonds and column reduction in finite Markov decision problems, *Op. Res.* **28**, 387–394 (1980).

A. Wald, *Sequential Analysis* (Wiley, New York, 1947).

A. Wald and J. Wolfowitz, Optimal character of the sequential probability ratio test, *Ann. Math. Stat.* **19**, 326–339 (1948).

K. H. Waldmann, On bounds for dynamic programs, *Math. Op. Res.* **10**, 220–232 (1985).

J. Walrand, *An Introduction to Queueing Networks*, (Prentice-Hall, Englewood Cliffs, NJ, 1988).

J. Walrand, Queueing networks, in *Handbook of Operations Research*, edited by D. P. Heyman and M. J. Sobel (North-Holland, Amsterdam, 1991), Vol. 2 pp. 519–604.

R. R. Weber and S. Stidham, Jr., Optimal control of services rates in networks of queues, *Adv. Appl. Prob.* **19**, 202–218 (1987).

G. Weiss, Multiserver stochastic scheduling, in *Deterministic and Stochastic Scheduling*, edited by M. Dempster, J. K. Lenstra, and A. Rinooy-Kan (Reidel, Dordrecht, 1982) pp. 157–179.

J. Wessels, Markov programming by successive approximations with respect to weighted supremum norms, *J. Math. Anal. Appl.* **58**, 326–335 (1977).

C. C. White, Procedures for the solution of a finite-horizon, partially observed, semi-Markov optimization problem, *Op. Res.* **24**, 348–358 (1976).

C. C. White, Monotone control laws for noisy, countable-state Markov chains, *Eur. J. Op. Res.* **5**, 124–132 (1980).

C. C. White, A Markov quality control process subject to partial observation, *Man. Sci.* **23**, 843–852 (1977).

C. C. White and K. Kim, Solution procedures for vector criterion Markov decision processes, *J. Large Scale Syst.* **1**, 129–140 (1980).

C. C. White, Sequential decision making under future preferences, *Op. Res.* **32**, 148–168 (1984).

C. C. White, L. C. Thomas, and W. T. Scherer, Reward revision for discounted Markov decision problems, *Op. Res.* **30**, 1299–1315 (1985).

C. C. White and W. T. Scherer, Solution procedures for partially observed Markov decision processes, *Op. Res.* **37**, 791–797 (1989).

C. C. White and D. J. White, Markov decision processes, *Eur. J. Op. Res.* **39**, 1–16 (1989).

D. J. White, Dynamic programming, Markov chains, and the method of successive approximations, *J. Math. Anal. Appl.* **6**, 373–376 (1963).

D. J. White, Dynamic programming and probabilistic constraints, *Op. Res.* **22**, 654–664 (1974).

D. J. White, Finite state approximations for denumerable state infinite horizon discounted Markov decision processes: The method of successive approximations, in *Recent Developments in Markov Decision Processes*, edited by R. Hartley, L. C. Thomas, and D. J. White (Academic, New York, 1980a) pp. 57–72.

D. J. White, Finite-state approximations for denumerable-state infinite horizon discounted Markov decision processes, *J. Math. Anal. Appl.* **74**, 292–295 (1980b).

D. J. White, Multi-objective infinite-horizon discounted Markov decision process, *J. Math. Anal. Appl.* **89**, 639–647 (1982a).

D. J. White, Finite state approximations for denumerable state infinite horizon discounted Markov decision process with unbounded rewards, *J. Math. Anal. Appl.* **86**, 292–306 (1982b).

D. J. White, Monotone value iteration for discounted finite Markov decision processes, *J. Math. Anal. Appl.* **109**, 311–324 (1985a).

D. J. White, Real applications of Markov decision processes, *Interfaces* **15**, 73–83 (1985b).

D. J. White, Further real applications of Markov decision processes, *Interfaces* **18**, 55–61 (1988a).

D. J. White, Mean, variance and probabilistic criteria in finite Markov decision processes: A review, *J. Opt. Theor. Appl.* **56**, 1–29 (1988b).

D. J. White, *Markov Decision Processes* (Wiley, Chichester, 1993).

W. Whitt, Approximations of dynamic programs, I, *Math. Op. Res.* **3**, 231–243 (1978).

W. Whitt, A priori bounds for approximations of Markov programs, *J. Math. Anal. Appl.* **71**, 297–302 (1979a).

W. Whitt, Approximations of dynamic programs, II, *Math. Op. Res.*, **4**, 179–185 (1979b).

P. Whittle, A simple condition for regularity in negative programming, *J. Appl. Prob.* **16**, 305–318 (1979).

P. Whittle, Stability and characterization condition in negative programming, *J. Appl. Prob.* **17**, 635–645 (1980a).

P. Whittle, Negative programming with unbounded costs: A simple condition for regularity, in *Recent Developments in Markov Decision Processes*, edited R. Hartley, L. C. Thomas, and D. J. White (Academic, New York, 1980b), pp. 23–34.

P. Whittle, Multi-armed bandits and the Gittins index, *J. R. Stat. Soc. Ser.* B, **42**, 143–149 (1980c).

P. Whittle, *Optimization Over Time, Dynamic Programming and Stochastic Control* (Wiley, New York, 1983), Vol. II.

P. Whittle and N. Komarova, Policy improvement and the Newton-Raphson algorithm, *Prob. Eng. Inf. Sci.* **2**, 249–255 (1988).

D. V. Widder, *The Laplace Transform*, (Princeton University Press, Princeton, 1946).

J. Wijngaard, Stationary Markov decision problems and perturbation theory for quasi-compact operators, *Math. Op. Res.* **2**, 91–102 (1977a).

J. Wijngaard, Sensitive optimality in stationary Markov decision chains on a general state space, in *Markov Decision Theory*, edited by H. C. Tijms and J. Wessels (Mathematics Centre, Amsterdam, pp. 85–94, 1977b).

B. K. Williams, MARKOV: A methodology for the solution of infinite time horizon Markov decision processes, *Appl. Stoch. Models Data Anal.* **4**, 253–271 (1988).

K. Yosida, *Functional Analysis*, (Springer-Verlag, New York, 1968).

D. Young, *Iterative Solutions of Large Linear Systems* (Academic, New York, 1971).

A. A. Yushkevich, On a class of strategies in general Markov decision models, *Theor. Prog. Appl.* **18**, 777–779 (1973).

A. A. Yushkevich and E. A. Feinberg, On homogeneous Markov models with continuous time and finite or countable state space, *Theor. Prob. Appl.* **24**, 156–161 (1979).

A. A. Yushkevich, On semi-Markov controlled models with an average reward criterion, *Theor. Prob. Appl.* **26**, 796–803 (1981).

A. A. Yushkevich, Blackwell optimal policies in a class of Markov decision processes with Borel state spaces, UNC-Charlotte, unpublished preliminary report (1993).

Y. Zheng, A simple proof of the optimality of (s, S) policies for infinite horizon inventory problems, *J. Appl. Prob.* **28**, 802–810 (1990).

Y. Zheng and A. Federgruen, Finding optimal (s, S) policies is about as simple as evaluating a single policy, *Op. Res.* **39**, 654–666 (1991).

H. Zijm, The optimality equations in multichain denumerable state Markov decision processes with the average cost criterion: The unbounded cost case, C.Q.M. Note 22, Centre for Quant. Methods, Philips B. V. Eindhoven (1984a).

H. Zijm, The optimality equations in multichain denumerable state Markov decision processes with the average cost criterion: The bounded cost case, *Stat. Dec.* **3**, 143–165 (1984b).

H. Zijm, Exponential convergence in undiscounted continuous-time Markov decision chains, *Math. Op. Res.* **12**, 700–717 (1987).

Index

643

WILEY SERIES IN PROBABILITY AND STATISTICS
ESTABLISHED BY WALTER A. SHEWHART AND SAMUEL S. WILKS

Editors: *David J. Balding, Noel A. C. Cressie, Nicholas I. Fisher,*
Iain M. Johnstone, J. B. Kadane, Geert Molenberghs. Louise M. Ryan,
David W. Scott, Adrian F. M. Smith, Jozef L. Teugels
Editors Emeriti: *Vic Barnett, J. Stuart Hunter, David G. Kendall*

The **Wiley Series in Probability and Statistics** is well established and authoritative. It covers many topics of current research interest in both pure and applied statistics and probability theory. Written by leading statisticians and institutions, the titles span both state-of-the-art developments in the field and classical methods.

Reflecting the wide range of current research in statistics, the series encompasses applied, methodological and theoretical statistics, ranging from applications and new techniques made possible by advances in computerized practice to rigorous treatment of theoretical approaches.

This series provides essential and invaluable reading for all statisticians, whether in academia, industry, government, or research.

*Now available in a lower priced paperback edition in the Wiley Classics Library.
†Now available in a lower priced paperback edition in the Wiley–Interscience Paperback Series.

† BELSLEY, KUH, and WELSCH · Regression Diagnostics: Identifying Influential Data and Sources of Collinearity

BENDAT and PIERSOL · Random Data: Analysis and Measurement Procedures, *Third Edition*

BERRY, CHALONER, and GEWEKE · Bayesian Analysis in Statistics and Econometrics: Essays in Honor of Arnold Zellner

BERNARDO and SMITH · Bayesian Theory

BHAT and MILLER · Elements of Applied Stochastic Processes, *Third Edition*

BHATTACHARYA and WAYMIRE · Stochastic Processes with Applications

† BIEMER, GROVES, LYBERG, MATHIOWETZ, and SUDMAN · Measurement Errors in Surveys

BILLINGSLEY · Convergence of Probability Measures, *Second Edition*

BILLINGSLEY · Probability and Measure, *Third Edition*

BIRKES and DODGE · Alternative Methods of Regression

BLISCHKE AND MURTHY (editors) · Case Studies in Reliability and Maintenance

BLISCHKE AND MURTHY · Reliability: Modeling, Prediction, and Optimization

BLOOMFIELD · Fourier Analysis of Time Series: An Introduction, *Second Edition*

BOLLEN · Structural Equations with Latent Variables

BOROVKOV · Ergodicity and Stability of Stochastic Processes

BOULEAU · Numerical Methods for Stochastic Processes

BOX · Bayesian Inference in Statistical Analysis

BOX · R. A. Fisher, the Life of a Scientist

BOX and DRAPER · Empirical Model-Building and Response Surfaces

* BOX and DRAPER · Evolutionary Operation: A Statistical Method for Process Improvement

BOX, HUNTER, and HUNTER · Statistics for Experimenters: An Introduction to Design, Data Analysis, and Model Building

BOX and LUCEÑO · Statistical Control by Monitoring and Feedback Adjustment

BRANDIMARTE · Numerical Methods in Finance: A MATLAB-Based Introduction

BROWN and HOLLANDER · Statistics: A Biomedical Introduction

BRUNNER, DOMHOF, and LANGER · Nonparametric Analysis of Longitudinal Data in Factorial Experiments

BUCKLEW · Large Deviation Techniques in Decision, Simulation, and Estimation

CAIROLI and DALANG · Sequential Stochastic Optimization

CASTILLO, HADI, BALAKRISHNAN, and SARABIA · Extreme Value and Related Models with Applications in Engineering and Science

CHAN · Time Series: Applications to Finance

CHATTERJEE and HADI · Sensitivity Analysis in Linear Regression

CHATTERJEE and PRICE · Regression Analysis by Example, *Third Edition*

CHERNICK · Bootstrap Methods: A Practitioner's Guide

CHERNICK and FRIIS · Introductory Biostatistics for the Health Sciences

CHILÈS and DELFINER · Geostatistics: Modeling Spatial Uncertainty

CHOW and LIU · Design and Analysis of Clinical Trials: Concepts and Methodologies, *Second Edition*

CLARKE and DISNEY · Probability and Random Processes: A First Course with Applications, *Second Edition*

* COCHRAN and COX · Experimental Designs, *Second Edition*

CONGDON · Applied Bayesian Modelling

CONGDON · Bayesian Statistical Modelling

CONOVER · Practical Nonparametric Statistics, *Third Edition*

COOK · Regression Graphics

COOK and WEISBERG · Applied Regression Including Computing and Graphics

COOK and WEISBERG · An Introduction to Regression Graphics

*Now available in a lower priced paperback edition in the Wiley Classics Library.

†Now available in a lower priced paperback edition in the Wiley–Interscience Paperback Series.

---

*Now available in a lower priced paperback edition in the Wiley Classics Library.

†Now available in a lower priced paperback edition in the Wiley–Interscience Paperback Series.

*Now available in a lower priced paperback edition in the Wiley Classics Library.

†Now available in a lower priced paperback edition in the Wiley–Interscience Paperback Series.

MAGNUS and NEUDECKER · Matrix Differential Calculus with Applications in Statistics and Econometrics, *Revised Edition*

MALLER and ZHOU · Survival Analysis with Long Term Survivors

MALLOWS · Design, Data, and Analysis by Some Friends of Cuthbert Daniel

MANN, SCHAFER, and SINGPURWALLA · Methods for Statistical Analysis of Reliability and Life Data

MANTON, WOODBURY, and TOLLEY · Statistical Applications Using Fuzzy Sets

MARCHETTE · Random Graphs for Statistical Pattern Recognition

MARDIA and JUPP · Directional Statistics

MASON, GUNST, and HESS · Statistical Design and Analysis of Experiments with Applications to Engineering and Science, *Second Edition*

McCULLOCH and SEARLE · Generalized, Linear, and Mixed Models

McFADDEN · Management of Data in Clinical Trials

* McLACHLAN · Discriminant Analysis and Statistical Pattern Recognition

McLACHLAN, DO, and AMBROISE · Analyzing Microarray Gene Expression Data

McLACHLAN and KRISHNAN · The EM Algorithm and Extensions

McLACHLAN and PEEL · Finite Mixture Models

McNEIL · Epidemiological Research Methods

MEEKER and ESCOBAR · Statistical Methods for Reliability Data

MEERSCHAERT and SCHEFFLER · Limit Distributions for Sums of Independent Random Vectors: Heavy Tails in Theory and Practice

MICKEY, DUNN, and CLARK · Applied Statistics: Analysis of Variance and Regression, *Third Edition*

* MILLER · Survival Analysis, *Second Edition*

MONTGOMERY, PECK, and VINING · Introduction to Linear Regression Analysis, *Third Edition*

MORGENTHALER and TUKEY · Configural Polysampling: A Route to Practical Robustness

MUIRHEAD · Aspects of Multivariate Statistical Theory

MULLER and STOYAN · Comparison Methods for Stochastic Models and Risks

MURRAY · X-STAT 2.0 Statistical Experimentation, Design Data Analysis, and Nonlinear Optimization

MURTHY, XIE, and JIANG · Weibull Models

MYERS and MONTGOMERY · Response Surface Methodology: Process and Product Optimization Using Designed Experiments, *Second Edition*

MYERS, MONTGOMERY, and VINING · Generalized Linear Models. With Applications in Engineering and the Sciences

† NELSON · Accelerated Testing, Statistical Models, Test Plans, and Data Analyses

† NELSON · Applied Life Data Analysis

NEWMAN · Biostatistical Methods in Epidemiology

OCHI · Applied Probability and Stochastic Processes in Engineering and Physical Sciences

OKABE, BOOTS, SUGIHARA, and CHIU · Spatial Tesselations: Concepts and Applications of Voronoi Diagrams, *Second Edition*

OLIVER and SMITH · Influence Diagrams, Belief Nets and Decision Analysis

PALTA · Quantitative Methods in Population Health: Extensions of Ordinary Regressions

PANKRATZ · Forecasting with Dynamic Regression Models

PANKRATZ · Forecasting with Univariate Box-Jenkins Models: Concepts and Cases

* PARZEN · Modern Probability Theory and Its Applications

PEÑA, TIAO, and TSAY · A Course in Time Series Analysis

PIANTADOSI · Clinical Trials: A Methodologic Perspective

PORT · Theoretical Probability for Applications

POURAHMADI · Foundations of Time Series Analysis and Prediction Theory

PRESS · Bayesian Statistics: Principles, Models, and Applications

---

*Now available in a lower priced paperback edition in the Wiley Classics Library.

†Now available in a lower priced paperback edition in the Wiley–Interscience Paperback Series.

*Now available in a lower priced paperback edition in the Wiley Classics Library.
†Now available in a lower priced paperback edition in the Wiley–Interscience Paperback Series.

STYAN · The Collected Papers of T. W. Anderson: 1943–1985

SUTTON, ABRAMS, JONES, SHELDON, and SONG · Methods for Meta-Analysis in Medical Research

TANAKA · Time Series Analysis: Nonstationary and Noninvertible Distribution Theory

THOMPSON · Empirical Model Building

THOMPSON · Sampling, *Second Edition*

THOMPSON · Simulation: A Modeler's Approach

THOMPSON and SEBER · Adaptive Sampling

THOMPSON, WILLIAMS, and FINDLAY · Models for Investors in Real World Markets

TIAO, BISGAARD, HILL, PEÑA, and STIGLER (editors) · Box on Quality and Discovery: with Design, Control, and Robustness

TIERNEY · LISP-STAT: An Object-Oriented Environment for Statistical Computing and Dynamic Graphics

TSAY · Analysis of Financial Time Series

UPTON and FINGLETON · Spatial Data Analysis by Example, Volume II: Categorical and Directional Data

VAN BELLE · Statistical Rules of Thumb

VAN BELLE, FISHER, HEAGERTY, and LUMLEY · Biostatistics: A Methodology for the Health Sciences, *Second Edition*

VESTRUP · The Theory of Measures and Integration

VIDAKOVIC · Statistical Modeling by Wavelets

VINOD and REAGLE · Preparing for the Worst: Incorporating Downside Risk in Stock Market Investments

WALLER and GOTWAY · Applied Spatial Statistics for Public Health Data

WEERAHANDI · Generalized Inference in Repeated Measures: Exact Methods in MANOVA and Mixed Models

WEISBERG · Applied Linear Regression, *Third Edition*

WELSH · Aspects of Statistical Inference

WESTFALL and YOUNG · Resampling-Based Multiple Testing: Examples and Methods for *p*-Value Adjustment

WHITTAKER · Graphical Models in Applied Multivariate Statistics

WINKER · Optimization Heuristics in Economics: Applications of Threshold Accepting

WONNACOTT and WONNACOTT · Econometrics, *Second Edition*

WOODING · Planning Pharmaceutical Clinical Trials: Basic Statistical Principles

WOODWORTH · Biostatistics: A Bayesian Introduction

WOOLSON and CLARKE · Statistical Methods for the Analysis of Biomedical Data, *Second Edition*

WU and HAMADA · Experiments: Planning, Analysis, and Parameter Design Optimization

YANG · The Construction Theory of Denumerable Markov Processes

* ZELLNER · An Introduction to Bayesian Inference in Econometrics

ZHOU, OBUCHOWSKI, and McCLISH · Statistical Methods in Diagnostic Medicine

*Now available in a lower priced paperback edition in the Wiley Classics Library.
†Now available in a lower priced paperback edition in the Wiley–Interscience Paperback Series.

Breinigsville, PA USA
19 January 2010
231032BV00003B/2/P

9 780471 727828